Israel's Exodus in Transdisciplinary Perspective

Quantitative Methods in the Humanities and Social Sciences

Quantitative Methods in the Humanities and Social Sciences is a book series designed to foster research-based conversation with all parts of the university campus from buildings of ivy-covered stone to technologically savvy walls of glass. Scholarship from international researchers and the esteemed editorial board represents the far-reaching applications of computational analysis, statistical models, computer-based programs, and other quantitative methods. Methods are integrated in a dialogue that is sensitive to the broader context of humanistic study and social science research. Scholars, including among others historians, archaeologists, classicists and linguists, promote this interdisciplinary approach. These texts teach new methodological approaches for contemporary research. Each volume exposes readers to a particular research method. Researchers and students then benefit from exposure to subtleties of the larger project or corpus of work in which the quantitative methods come to fruition.

For further volumes:
http://www.springer.com/series/11748

Thomas E. Levy • Thomas Schneider •
William H.C. Propp
Editors

Israel's Exodus in Transdisciplinary Perspective

Text, Archaeology, Culture, and Geoscience

Managing Editor: Brad C. Sparks

Springer

Editors
Thomas E. Levy
University of California
San Diego, California, USA

William H.C. Propp
University of California
San Diego, California, USA

Thomas Schneider
University of British Columbia
Vancouver, Canada

Managing Editor
Brad C. Sparks
Archaeological Research Group
Los Angeles-San Diego, CA, USA

Videos of the UCSD Exodus Conference of 2013, which pertain to this volume, can be found at http://exodus.calit2.net/

ISSN 2199-0956 ISSN 2199-0964 (electronic)
ISBN 978-3-319-04767-6 ISBN 978-3-319-04768-3 (eBook)
DOI 10.1007/978-3-319-04768-3
Springer Cham Heidelberg New York Dordrecht London

Library of Congress Control Number: 2014941943

Printed on acid-free paper

Springer is part of Springer Science+Business Media (www.springer.com)

Preface

The grand narrative about Israel's Exodus from Egypt, mainly preserved in the Book of Exodus in the Hebrew Bible, holds momentous significance for ancient Israel, Jewish identity, Christianity, Islam and Western thought, and deeply impacted the formation of academic disciplines studying the ancient Near East. In its role as a pervasive theme, it has inspired artistic and popular imagination. Finally, the Exodus lies at the heart of a controversy about the reliability and origins of the Biblical narrative, not the least in the context of research on the emergence of Israel after the end of the Late Bronze Age.

This volume is the most innovative gathering of thought assembled on the topic of Israel's Exodus from Egypt. In 9 sections, the volume presents papers first presented at Out of Egypt: Israel's Exodus Between Text and Memory, History and Imagination, a conference at the University of California, San Diego, May 31 to June 3, 2013. The transdisciplinary perspective this book takes combines an assessment of past research with current knowledge on the topic and new perspectives for future study. Research from Egyptologists, archaeologists, Biblical scholars, computer scientists, and geoscientists appears in active conversation throughout the various chapters of this book. The 44 contributions by leading scholars from the United States, Canada, Great Britain, Israel, Germany, Switzerland, and Italy, unite a diverse group of hermeneutic approaches. They pertain to the text and later reception of the Exodus narrative, including its Egyptian and Near Eastern parallels, function as cultural memory in the history of Israel, the interface of the Exodus question with the emergence of Israel and archaeological fieldwork, and exploration of the text's historicity. The historical geography and the environmental events described in the Exodus narrative and related texts receive thorough scientific analysis, reinforcing this volume's transdisciplinary character. An important section is devoted to cyberarchaeology, visualization techniques, and museological presentation of the Exodus.

Exodus research is evocative of "World Building," to use a phrase from Alex McDowell, production designer at the University of Southern California. World building is based on three pillars: "Storytelling is the most powerful system for the advancement of human capability due to its ability to allow the human imagination to precede the realization of thought; that all stories emerge logically and intuitively from the worlds that create them; and that new technologies powerfully enable us to sculpt the imagination into existence."[1] Through the practice of ancient world building described here, we hope to promote a new approach to transdisciplinary research.

Setting the Stage: Interdisciplinary Approaches to the Exodus Narrative

This volume commences with the conference's keynote lectures. *Jan Assmann* perceives Exodus as a myth about the origin of a people and a religion. This myth simultaneously symbolizes the revolutionary turn in human history from cosmotheism to monotheism, the transition from the idea of a god immanent in or equivalent to nature, to the concept of one transcendent god beyond nature. He also traces the incomplete realization of this concept in Western history, where the recurrence of cosmotheistic trends and the reactions to counter cosmotheism kept the myth and symbol of Exodus alive until modern times. *Ronald Hendel* situates the cultural memory of the Exodus in a dialectic between historical memory and ethnic self-fashioning. He locates the roots of the mnemohistory of the Exodus in the transition from the Late Bronze to the Iron Age. At that time, memories of the collapse of the Egyptian Empire in Canaan were transformed into a memory of liberation from Egyptian bondage that was narrativized as a myth of ethnic origins—expressed in its oldest form, the Song of the Sea, as Yahweh's victory over the chaos represented by pharaoh. The dishonorable storyline of the Exodus, with a people fleeing from slavery, is for *Manfred Bietak* an indication of elements that are historically credible. From this, he reviews the available evidence from Egypt on the settlement of "Proto-Israelites" during the later Ramesside period. He infers that such groups would have settled in Egypt simultaneously with the Proto-Israelites in Canaan. In his conception, the collective memory of the Proto-Israelites suffering in Canaan under Egyptian oppression and those suffering in Egypt merged in the genesis of Israel's myth of origin. The later belief in a stay of the Israelites at Tanis/Zoan was then spurred by the transfer of archaeological remains from Pi-Ramesse to Tanis and Bubastis. *Israel Finkelstein* reconstructs different historical stages of the narratives about Israel's wanderings in the southern desert. He focuses on the eighth century with Israelite activity along the Arabian trade route; the transformation of the desert narratives in Judah under Assyrian rule in the seventh century; their elaboration under the Egyptian 26th dynasty; and the loss of knowledge about the southern desert after 560 BCE. These narratives would have enriched a tradition of salvation from Egyptian rule that developed between the sixteenth and tenth centuries BCE. This tradition was transferred from the lowlands to the northern part of the central highlands, where it became a charter myth of the kingdom of

[1] You can read more about world building and the 5D Institute that Alex McDowell directs here: http://5dinstitute.org/about.

Israel. In contrast with these assessments that are skeptical of our ability to define a single Exodus event in the second millennium BCE and concentrate instead on the purpose and development of the narrative in the first millennium, *Lawrence Geraty* gives an overview of conservative and mainstream readings of the Exodus account. He attempts to situate an Exodus event in various historical contexts of the Egyptian Bronze Age, with a special focus on dates in the 18th and 19th dynasties, while also touching on other proposed Exodus dates between 2100 BCE and 650 BCE.

Science-Based Approaches to the Exodus

This section presents a scientific study of the eastern Nile delta and Sinai peninsula's ancient environment. This includes a thorough assessment of scenarios that could either underlie events detailed in the Exodus narrative (for instance, plagues and the parting of the sea) from a conservative point of view, or scenarios that could have inspired the imagery behind the Exodus plagues and miracles, imagery that also appears in apocalyptic and ritual texts from the ancient Near East. *Stephen O. Moshier and James K. Hoffmeier* reconstruct the changing ancient physical geography of the region due to dynamic interactions between the Nile river system, Mediterranean Sea, and the tectonics of the Red Sea rift system. By combining field geology, archaeology, digital topography, and satellite imagery with geographic information technology, they produce a new map that reveals different positions of the Mediterranean coastline, lagoons, and the existence of Pelusiac Nile distributaries, lakes, and wetlands. By creating an accurate geophysical map for Exodus research, this recreated geography helps to delineate the path of the ancient coastal road between Egypt and Palestine. *Mark Harris* gives a critical assessment of the interpretative strategies of the burgeoning number of popular, naturalistic readings of the Exodus text. In defiance of Biblical scholarship that emphasizes the complex genesis and character of the Exodus traditions, and without considering the ideological uses of a text, these strategies take the narrative at face value as reflecting apocalyptic natural catastrophes, particularly the eruption of Thera (Santorini) in the seventeenth century BCE. The remaining contributions of this section provide scientific data on a variety of natural disasters to determine whether they could have generated the phenomena described in the Exodus narrative. *Amos Salamon* and coauthors examine the main tsunamigenic sources in the Eastern Mediterranean that may explain the parting of the sea—how to make the sea dry and then inundate the land. Their simulations recreate the tsunami that followed the Thera eruption around 1600 BCE; the strong magnitude 8–8.5 earthquake of 365 CE in the Hellenic arc and the resulting tsunami that devastated Alexandria; and a voluminous Late Pleistocene submarine slump at the Nile cone that started with a significant drawback of the sea and was followed by a remarkable inundation. The Thera eruption is also crucial to the remaining papers. *Michael Dee* and coauthors discuss the relevance of radiocarbon dating to Exodus-related problems and present dates from three pertinent case studies; the dating of the Egyptian New Kingdom, the conquest of Canaanite cities as reported in the Book of Joshua, and the dating of the Thera eruption. *Malcolm Wiener's* article presents a reply to Dee's study, advocating for a much later eruption date (after 1530 BCE).

Cyber-Archaeology and Exodus

The four contributions to Section III explore the potential of visualizing the past, with the Exodus and recent archaeological research on Iron Age Jordan as a means to present original research for both scholarly and public dissemination in the format of the "museum of the future." The four contributions provide the background to an exhibition that accompanied the Exodus conference, prepared by a team of archaeologists, geoscientists, computer scientists, engineers, and digital media technologists under the direction of Thomas Levy. It was mounted in the Qualcomm Institute Theater at UC San Diego, a performance space reconfigured into a museum space that uses new visual and audio technologies. A large-format $8' \times 32'$ (2.3 m × 9.75 m) 64 million pixel screen and several other tiled display systems were used for the computer visualizations, and a new 50-megapixel 3D large-scale immersive display system called the WAVE had its premiere. New audio systems and content were developed by the Sonic Arts researchers to project archaeological and geological audio data. These cyber-archaeology papers, along with those in the science-based approaches to Exodus, provide the grist for ancient world building applied to the Exodus narrative.

The Exodus Narrative in Its Egyptian and Near Eastern Context

Bernard F. Batto proposes that the Exodus narrative was rewritten through redaction by priests in order to elevate the Exodus "event" to mythical status, in the exile or post-exilic times. In this conception, employing motifs of the Combat Myth prevalent throughout Mesopotamia and the Levant, Yahweh and pharaoh were profiled as the two antagonists in the battle between the creator and the chaos monster in the form of the primeval Sea. Pharaoh is identified with and defeated in the "Sea of End" (*yam sûp*), the Red Sea not Reed Sea, from which Yahweh emerges as the creator of Israel. *James Hoffmeier* reviews the long history of scholarly engagement with the Old Testament by Egyptologists since the nineteenth century, in particular their vivid and positive interest in the topic of the Exodus and the historicity of Israel's sojourn in Egypt. He calls for Egyptologists "to return to the debate to bring data from Egypt to bear on historical and geographical matters." *Susan Tower Hollis* studies the relationship between Egyptian stories and Biblical stories about Israel's sojourn in Egypt, most notably the comparison between the 19th dynasty "Tale of Two Brothers" from the Papyrus d'Orbiney (BM 10183) and the narrative of Joseph and Potiphar's wife in Genesis 39. *Scott B. Noegel* suggests the Egyptian sacred bark as the best non-Israelite parallel to the Ark of the Covenant that the late priestly tradition of Exodus 25 depicts as being manufactured at Mount Sinai. The Israelite conception of the Ark probably originated under Egyptian influence in the Late Bronze Age. *Gary Rendsburg* presents an assessment of Egyptian parallels to motifs occurring throughout Exodus 1–15, such as the hidden divine name, turning an inanimate object into a serpent or crocodile, the conversion of water to blood, the casting of darkness, the death of firstborn, the parting of waters, and death by drowning. Almost all these motifs are known only from Egypt without an echo in other ancient Near Eastern societies. Rendsburg imagines an educated Israelite writer and his well-informed Israelite audience were able to understand and enjoy the Egyptian cultural context of a composition that "both subverts Egyptian religious notions and simultaneously expresses Israel's national heritage in exquisite literary fashion." *Brad C. Sparks* concludes this section with a comprehensive compilation and discussion of scholarship since the nineteenth century that takes note of Exodus parallels in some 90 ancient Egyptian texts of different genres and time periods. Sparks argues that the historicity of the Exodus needs to be reevaluated in light of this Egyptian narrative material and intriguing associated iconography.

The Exodus Narrative as Text

This Section gives in-depth views of the starting point for Exodus discussions: the Exodus narrative as preserved in the book of Exodus. Eminent specialists on the exegesis of Exodus are represented in this section. *Christoph Berner* suggests that the complexity of the account of Exodus 1–15, a narrative that is not coherent in nature and whose details often stand in tension with each other (if not in flat contradiction), can best be understood as resulting from a process of continuous literary expansions (*Fortschreibungen*). He demonstrates through the references to the Israelites' forced labor, which he uses as a test case, that substantial parts of the narrative belong to an extensive post-priestly stage of textual redaction. Thus, the Exodus narrative reveals little about the historical circumstances of the Exodus, but all the more about how post-exilic scribes imagined it and participated in the literary development of the Biblical account. *Baruch Halpern* speaks about the Exodus as a fable that was inspired by possible events of Israel's past, although he contends that its historical genesis will be as irretrievable to us as its narrator. Halpern also emphasizes that the text's modern discussants must wield the tools necessary to confront the epistemological challenges that we face. The true question is: "What do we need to know in order to know what we want to know?" Faced with storytellers and their audiences who contributed historical detritus while adding artistic value to the story, the subject's sole value is to recover the story's magic: to understand Israel's modes of social thought over time and the culture that immortalized the Exodus. *Thomas Römer* addresses traditions about the origins of Yahweh. While the Exodus narrative was recorded in writing for the first time in Judah and seems to reflect the contemporaneous Assyrian oppression, the literary contours of the older Exodus tradition that came to Judah from Israel after 722 escape us. Hosea 12 shows Yahweh (the god of the Exodus) in opposition to the Jacob tradition, perhaps proof of the attempt to make the Exodus the foundation myth of Israel. The two accounts of Yahweh's revelation to Moses, as well as external evidence (Kuntillet Ajrud, evidence about the *Shasu Yahu*) preserve the memory that Yhwh was not an autochthonous deity but was introduced from the South. *Stephen C. Russell* proposes a new historical contextualization of Moses' appointment of officials to judge legal cases in Exod 18:13–26. Traditionally believed to reflect the social world of the monarchic period and to provide an etiology for the system of royal judges established by Jehoshaphat (2 Chron 19:5–10), this appointment more closely parallels the postexilic system of Ezra 7:12–26, where the Persian king Artaxerxes instructs Ezra to appoint judges who know the Mosaic law. Exod 18:13–26 is best understood as a postexilic expansion of Exodus 18; it constitutes a major bridge in the book of Exodus by summarizing the deliverance from Egypt and anticipating the revelation at Sinai. *Konrad Schmid* issues a call to acknowledge in our assessments the difference between the world of the narratives and the world of the narrators and to account for this difference in a methodologically controlled manner. His particular example is Exodus 1–15, a text debated in Hebrew Bible studies in terms of both its compositional and its historical evaluation. Despite all controversies, there is broad consensus about the Priestly texts in Exodus 1–15 (Exod 1:7, 13–14; 2:23*–25; 6:2–12; 7–11*; 14–15*). The article discusses several narrative peculiarities in the Priestly Exodus account that might be explained by seeing it as a Judean foundational myth of the early Persian period. The image of Egypt as unruly and needing to be tamed suggests a date prior to 525 BCE, when Cambyses conquered Egypt. Whereas Cyrus was a supporter of Jewish independence, the Pharaoh of Exodus is an "Anti-Cyrus."

The Exodus in Later Reception and Perception

This section deals with later Jewish, Christian, and Islamic traditions and interpretations of the Exodus. *Joel Allen* examines the "despoliation of Egypt" (Gen 15:14; Exod 3:21–22; 11:2–3 and 12:35–36), a motif that has evoked embarrassment for Jewish and Christian expositors. This essay examines key allegorical "text therapies" (e.g., the interpretation of the Egyptian plunder as *spiritual* treasures by which Jews and Christians sought to heal the text of its improprieties and warrant appropriation of the pagan classical heritage into the realm of the spirit). Philo, Origen and Augustine all sought to provide justification for those who wished to have the best of both the Biblical and classical worlds and to balance the faith of scripture with the reason of Greek philosophy. *René Bloch* examines the challenge that the Exodus story presented to Jewish-Hellenistic authors whose homeland was Egypt. How did they come to terms with the Biblical story of liberation from

Egyptian slavery and the longing for the promised land? His study looks at narrative differences in Philo's discussion of the Exodus. In particular, Philo read the story allegorically as a journey from the land of the body to the realms of the mind. This permitted him to control the meaning of the Exodus and stay in Egypt. *Caterina Moro* focuses on the extra-Biblical Exodus story of the Jewish-Hellenistic historian Artapanus. This story, unlike the Biblical Exodus account, provides an identity for Moses' Egyptian opponent, assigning him the name Chenephres. Moro examines the evidence on his possible historical equivalent, Khaneferra Sobekhotep IV of the 13th dynasty, and the fictional history that elevated him to the position of the pharaoh of the Exodus. *Babak Rahimi* proposes to understand the Quranic accounts of the Exodus as a salvational drama, in contrast to Classical Quran exegesis that has perceived the expulsion of the Israelites from Egypt as divine punishment imposed on them for their transgressions against God. Not only does the Quranic Exodus narrative contain many instances of God's blessing to the Israelites, the Exodus constitutes a metanarrative of spiritual liberation and represents a chronicle of God's presence. The Israelites' experience of a trial through adversity ultimately reveals divine grace and a promise for salvation. The article by *Pieter van der Horst* examines the pivotal role that the Exodus from Egypt played in Jewish-pagan polemics from the beginning of the Hellenistic period into the Imperial period. Pagan polemicists reversed the Biblical story of the Israelites' liberation from Egyptian bondage, portraying a negative image of Israelite origins and picturing them as misanthropes and atheists. Jewish-Hellenistic authors reacted to these attacks in writing through novels, dramas, and philosophical treatises.

The Exodus as Cultural Memory

It is in the first millennium BCE that the purported Exodus event became a written and cultural artifact, a canvas of continued ideological imagination. This section's contributions engage with the concept of cultural memory, whereby the Exodus story may be seen as a response to a society's religious and cultural needs during the monarchies of Israel and Judah and the exilic and post-exilic period. It may also be considered a vehicle of national identity. *Aren Maeir* discusses the variability of memories and their susceptibility to being shaped and changed. He discusses the Exodus tradition as a matrix of cultural memories, woven together and altered over a long period, thus defying any attempt to determine a single historical event that would correlate to the Exodus. It is not an ahistorical myth but rather reflects the many periods and contexts, in which the Exodus *mnemo-narratives* were formed—a multi-faceted "narrative complex" and space of memory shaped by the needs of Israel's identity. In a similar vein, *Victor Matthews'* contribution draws a distinction from efforts to determine the possible historicity of the Exodus event. Instead it concentrates on how and why collective memories are created, perpetuated, used and reused. While he inquires when and where the Exodus tradition originated, he concludes that current data make it impossible to provide a definitive answer and that what can be said has more to do with why the Exodus tradition was important to the Israelite community at various times. *William H.C. Propp* supports the view that the Exodus cannot be called "historical" and that the evidence is too diffuse to be adequately tested by the historical method; in consequence, we must resign ourselves to ignorance. By way of comparison, Propp adduces and contrasts another mythic tale of improbable, miraculous salvation to the Exodus story: the "Angel(s) of Mons" from World War I, where abundant information enables us to sift truth from fiction precisely and set both in historical context. *Donald B. Redford* puts the Biblical account of Israel's stay in and expulsion from Egypt in line with several other traditions regarding a "coming-out" from Egypt: Manetho's *Aegyptiaca*, the Bocchoris "leper"-tradition, Phoenician reminiscences, and asides in the Hebrew prophets. These traditions, extraneous to the book of Exodus, are the *Asiatic* folk-memory, "adjusted, distorted, inverted, with motivation reversed or imputed," of the Hyksos occupation and their expulsion after a reign of 108 years. Redford examines the contribution of these traditions to elucidating the original event and its perception through time. Finally, *William G. Dever* offers an archaeological critique of the "cultural memory" model of the Exodus-Conquest narrative. After reviewing what the Biblical writers actually knew about their origins, what they created as their memories, and what they may have forgotten about their past, he concludes that the narrative is a foundation myth whose basic elements—an immigration of "all Israel" from Egypt and conquest of the land—are invented. Instead of asking how these texts functioned socially, religiously, and culturally, he examines "what really happened" and points to Israel's emergence as an indigenous phenomenon within Canaan, with the Israelites being essentially Canaanites, displaced both geographically and ideologically.

The Exodus and the Emergence of Israel: New Perspectives from Biblical Studies and Archaeology

This section provides a link to questions about the historicity of the Exodus memory by moving the discussion from the Exodus text and its historical purposes in the first millennium BCE to the emergence of Israel in the later second millennium. *Emmanuel Anati* presents an overview of the archaeological survey of Mount Karkom (which he identifies as Mount Sinai) and surrounding valleys that resulted in the discovery of 1,300 new archaeological sites. The presence of altars, small sanctuaries and numerous other places of worship characterize Mount Karkom as a holy mountain of the Bronze Age at the foot of which large numbers of people seem to have lived temporarily. Drawing on several theoretical insights regarding the nature of social power and makeup of ancient states in his analysis of the Amarna letters, *Brendon Benz* presents an alternative hypothesis to recent scholarship which has hypothesized that Israel consisted of geographical, economic, and/

or political outsiders. Polities and populations of the Late Bronze Age Levant were more diverse than is generally recognized, with social power more widely distributed and often negotiated among a range of political players in an "egalitarian" manner. The various forms of political organization included those that consisted of populations defined by settled centers and those that were not. After highlighting points of continuity between the Late Bronze Age and the constituents of early Israel as they are depicted in some of the core passages of the Bible, Benz suggests that early Israel included a contingent of geographical, economic, and political insiders. *Avraham Faust* discusses the ethnogenesis of Israel with particular attention to the earliest Israel group, the one mentioned in the Merneptah stele, which according to Faust was composed mainly of Shasu pastoralists. Many different groups (including many Canaanites, and possibly a small "Exodus" group that had left Egypt) were amalgamated into what was to become Israel. The Exodus story was one of the traditions and practices that was useful in demarcating Israel from other groups and came thus to be adopted by "all Israel." *Daniel Fleming* looks at the phenomenon of herding over distance as a survival of earlier social strategies from Israel's background. In the Exodus story, Israel appears as settled in Egypt, with no indication of free movement with its livestock into and out of Egypt. For the Exodus writer, all pastoralism was local. However, elements of the deeper narrative structure follow the logic of long-range pastoralism (Moses and Israel's way into the wilderness; the move to herding bases in Canaan), with the Exodus from Egypt amounting to a change of operating base when an existing base was no longer available. Thus, the Exodus story may have been attached to and served social circles in which such migration could be celebrated. *Garrett Galvin* reviews certain aspects of the scholarship on the historicity of the Exodus event and the archaeological evidence from Egypt and Palestine, pointing to the powerful analogy of the establishment of the Philistines as a new nation in Palestine. He emphasizes that, despite a more cautionary approach to texts since the twentieth century Linguistic Turn in the philosophy of history, the Biblical Exodus narrative remains our main textual access to the question of a historical core. However, it is a myth of origin that has parallels in Greek, Roman and Germanic stories of ethnogenesis, a theological document operating within the conventions of historical narratives. *Christopher Hays* recognizes the numerous contacts that existed between ancient Israel and Egypt—memories of Egypt, which later became literary traditions in the Hebrew Bible, were handed down by some portion of the Iron Age proto-Israelite state, and textual and material data related to the earliest Israelite monarchy indicate an ongoing cultural relationship with Egypt. He adds an apparent similarity between the way the two nations conceptualized the extent of their kingdoms to the examples of Egyptian influence on early Israel. *Robert Mullins* reviews the three models that have been put forth to explain the appearance of Israel in the western highlands of the southern Levant: conquest, pastoral sedenterization, and social revolt. According to him, a fourth model, which takes into account the dissolution of the Egyptian empire at the end of the Late Bronze Age, provides a more satisfying explanation for what must have been a widespread, complex, and lengthy process. What we find in the Biblical text is a constructed history whereby later Israel enshrined and reshaped its past to create official memories of a culture and formulate a new vision for the future. *Nadav Na'aman* examines reasons for the contrast between the central place of the Exodus in Israelite memory and its questionable historical status. He suggests that the bondage, the suffering, and miraculous delivery from slavery actually took place in Canaan during Egypt's imperial reign over Palestine in the New Kingdom and the empire's demise in the twelfth century, and that the locus of these memories was later transferred from Canaan to Egypt. The bondage and liberation were experienced by the pastoral groups that later settled in the highlands of the Northern Kingdom, hence the central place of the Exodus tradition in the cultural memory of Israel's inhabitants. Since the process of settlement in the Judean highlands took place later and on a limited scale, the memory of the Exodus played only a minor role among Judah's inhabitants.

Conclusion

The concluding chapter of the volume by *Thomas Schneider* presents a glimpse of the complexity of Exodus research. Any attempt to trace and contextualize motifs of the narrative is obstructed by the complexity of the text's history. Exegetical certainties of the twentieth century have vanished in the crisis of pentateuchal research and given way to multiple scenarios of text composition and redaction, the interrelationship of major themes, and the provenance and historical context of phenomena mentioned in it. This general situation is exemplified by a study of Exodus 12 that at the same time aims to be a genuine contribution to Exodus research and a perspective at the end of the volume. The received text of Exodus 12 describes the last plague brought onto Egypt by Yahweh—the killing of Pharaoh's firstborn son and the firstlings of the country's livestock—by Yahweh or alternatively, his "destroyer" who strikes the Egyptians but spares the homes of the Israelites. Several aspects of the Passover protection ritual have not yet been explained in a satisfactory way. After giving an overview of the intricate exegetical situation, the study proposes a new approach to the text by drawing on parallels from Egyptian rituals which would have been appropriated by the text's authors for the Israelite cause. Pap. Cairo 58027, a ritual for the protection of Pharaoh at night, and rituals aimed at the "Plague of the Year" receive particular attention.

In addition to the exegetical approaches to Biblical studies presented here, the transdisciplinary methods illustrated in this volume demonstrate the great potential that scientific and quantitative methods have in answering long-standing questions in both the humanities and social sciences. Questions that have mystified generations of people who have been fascinated with the enigmatic Exodus of the Bible may now be examined using these approaches that are sure to bear fruit in the coming years. Indeed, transdisciplinary, or team science approaches will be in the forefront of cutting edge concepts that, with the integration of technology, will be brought together by researchers from the social sciences, humanities, natural sciences and engineering. The editors hope that the 43 contributions included in the volume will provide an inspiring point of departure for all future research on the question of Israel's Exodus.

San Diego, CA, USA — Thomas E. Levy
Vancouver, BC, Canada — Thomas Schneider
San Diego, CA, USA — William H.C. Propp

Acknowledgments

The editors would like to extend our warm appreciation to the many people whose help and support was essential in making both the Exodus conference and EX3 exhibition possible.

The Qualcomm Institute/California Institute for Telecommunications and Information Technology (Calit2) provided not just a physical place for our international conference and exhibition but the "transdisciplinary spirit" that has made the production of the Exodus Springer volume possible. Thanks to: Ramesh Rao, Director, Qualcomm Institute, Calit2; Tom DeFanti, Research Scientist—Scientific Visualization; Larry Smarr, Director, Calit2; Doug Ramsey, Director of Communications; Scott Blair, Graphic Designer; Jessica Block, GIS specialist; Dee Chieng, Fund Manager; Greg Dawe, Principal Development Engineer; Alice Carr, Project Support Coordinator; Tiffany Fox, Public Information Reprensentative; Eric Hamdan, Junior Development Engineer; Alexandra Hubenko Baker, Project Manager, NSF—IGERT TEECH program; Joseph Keefe, Research Project Manager; Nancy Lee, Business Office Manager; Albert Yu Min Lin, Assistant Research Scientist; Todd Margolis, Assistant Project Scientist; Yuki Marsden, Management Services Officer; Alex Matthews, Assistant Producer/Director; Christopher McFarland, Programmer/Analyst II; Johnny Nguyen, Executive Assistant; Amira Noeuv, Peter Otto, Acoustic Arts; Vanessa Pool, Project Support Coordinator; Andrew Scott Prudhomme, Programmer Analyst III; Tad Reynales, Manager, Technology Infrastructure; Karen Riggs-Saberton, Project Support Coordinator; Jürgen Schulze, Assistant Project Scientist; Zachary Seldess, Sonic Arts; Karen Stecher, Fund Manager; Trish Stone, Tour Director; and Philip P. Weber, Programmer Analyst III.

We are also grateful to Ken Garrett, National Geographic photographer for use of his photographs from Egypt; and Tim Sthal Photography for photographic exhibition mounting.

Special thanks to Debra England and Alina Levy and all the others whose help in organizing the Exodus conference was essential and greatly appreciated.

We would like to thank the Springer production staff for their hard work in producing this volume: Hannah Bracken, Brian Halm, Rekha Udaiyar, and Anitha Selvaraj.

Finally, we would also like to thank the UC San Diego Judaic Studies faculty for their support of the EX3 exhibition: Deborah Hertz, David

Goodblatt, Amelia Glaser, Lisa Lampert-Weissig, and program coordinator, Sally Hargate; thanks also to undergraduate student interns who served as docents for the EX3 exhibition that features in many of the papers in this volume: Aliya Hoff, Kat Huggins, James Darling and Anjali Phukan. Thanks also to Larry Geraty of La Sierra University and Willeke Wendrich of UCLA for their help and support especially in the early stages of conference planning.

Contents

List of Contributors

Joel S. Allen Dakota Wesleyan University, Mitchell, SD, USA
joallen@dwu.edu

Emmanuel Anati Centro Camuno di Studi Preistorici, Capo di Ponte, Italy; University of Lecce, Lecce, Italy
emmanuel.anati@gmail.com

Jan Assmann University of Konstanz, Constance, Germany; University of Heidelberg, Heidelberg, Germany
jan@assmanns.de

Bernard F. Batto DePauw University, Greencastle, IN, USA
bbatto@depauw.edu

Brendon C. Benz William Jewell College, Liberty, MO, USA
benzb@william.jewell.edu

Christoph Berner Theologische Fakultät, Universität Göttinge, Gottingen, Germany
email@christophberner.de

Manfred Bietak Institute of Egyptology, University of Vienna, Austria; Dept. of Egypt and the Levant, Institute for Oriental and European Archaeology, Austrian Academy of Sciences, Vienna, Austria
manfred.bietak@univie.ac.at

René Bloch Institute for Jewish Studies and Institute of Classical Studies, University of Bern, Switzerland
rene.bloch@theol.unibe.ch

Jessica Block Qualcomm Institute, University of California, San Diego (UCSD), La Jolla, CA, USA

Gregory L. Dawe Qualcomm Institute, UC San Diego, La Jolla, CA, USA

Michael W. Dee Research Laboratory for Archaeology, University of Oxford, Oxford, UK
michael.dee@rlaha.ox.ac.uk

Thomas A. DeFanti Qualcomm Institute, University of California, San Diego (UCSD), La Jolla, CA, USA

William G. Dever Near Eastern Studies, University of Arizona, Tucson, AZ, USA; Lycoming College Williamsport, PA, USA

Avraham Faust Institute of Archaeology, Bar-Ilan University, Cambridge, MA, USA; Near Eastern Languages and Civilizations Harvard University, Cambridge, MA, USA
avraham.faust@biu.ac.il

Israel Finkelstein Tel Aviv University, Tel Aviv, Israel
fink2@post.tau.ac.il

Daniel E. Fleming New York University, c/o Norwalk, CT, USA
daniel.fleming@nyu.edu

Garrett Galvin Franciscan School of Theology, Oceanside, CA, USA
garrettmgalvin@gmail.com

Lawrence T. Geraty LSU Foundation, La Sierra University (LSU), Riverside, CA, USA; American Schools of Oriental Research (ASOR) Boston, MA, USA
lgeraty@lasierra.edu

John Hall Division of Geological Hazards, Geological Survey of Israel, Jerusalem, Israel

Baruch Halpern University of Georgia, Athens, GA, USA
b.halpern@gmail.com; xapwn@uga.edu

Eric Hamdan Qualcomm Institute/California Institute of Telecommunications and Information Technology (Calit2), University of California, San Diego (UCSD), La Jolla, CA, USA

Mark Harris New College, School of Divinity, University of Edinburgh, Mound Place, Edinburgh, UK
Mark.Harris@ed.ac.uk

Christopher B. Hays Ancient Near Eastern Studies, Fuller Theological Seminary, Pasadena, CA, USA
hays@fuller.edu

Ronald Hendel University of California, Berkeley, CA, USA
hendel@berkeley.edu

Thomas F.G. Higham Oxford Radiocarbon Accelerator Unit, University of Oxford, Oxford, UK
thomas.higham@rlaha.ox.ac.uk

Aliya Hoff Qualcomm Institute, University of California, San Diego (UCSD), La Jolla, CA, USA

James K. Hoffmeier Divinity School, Trinity International University, Deerfield, IL, USA
jhoffmei@tiu.edu

Susan Tower Hollis SUNY Empire State College, State University, New York, New York, NY, USA
Susan.Hollis@esc.edu

Falko Kuester Qualcomm Institute, University of California, San Diego (UCSD), La Jolla, CA, USA

Thomas E. Levy Department of Anthropology, and Levantine & Cyber-Archaeology Laboratory; Qualcomm Institute/California Institute of Telecommunications and Information Technology (Calit2), University of California, San Diego (UCSD), La Jolla, San Diego, CA, USA
thomasevanlevy@gmail.com; tlevy@ucsd.edu

Aren M. Maeir Tell es-Safi/Gath Archaeological Project, Institute of Archaeology, Bar-Ilan University, Ramat Gan, Israel
arenmaeir@gmail.com

John Mangan Qualcomm Institute, University of California, San Diego (UCSD), La Jolla, CA, USA

Todd Margolis Qualcomm Institute, University of California, San Diego (UCSD), La Jolla, CA, USA

Victor H. Matthews College of Humanities and Public Affairs, Missouri State University, Springfield, MO, USA
VictorMatthews@MissouriState.edu

Floyd McCoy University of Hawaii, Windward, Kaneohe, CA, USA

Caterina Moro Sapienza Università di Roma, Rome, Italy
moroclarab@libero.it

Stephen O. Moshier Wheaton College, Wheaton, IL, USA
Stephen.Moshier@wheaton.edu

Robert A. Mullins Biblical Studies, Azusa Pacific University, Azusa, CA, USA
robertalanmullins@hotmail.com

Nadav Na'aman Tel Aviv University, Tel Aviv-Yafo, Israel
nnaaman@post.tau.ac.il

Scott B. Noegel Near Eastern Languages and Civilization (NELC), University of Washington, Seattle, WA, USA
snoegel@uw.edu

Peter Otto Qualcomm Institute/California Institute of Telecommunications and Information Technology (Calit2), University of California, San Diego (UCSD), La Jolla, CA, USA
potto@ucsd.edu

William H. C. Propp Department of History, University of California, San Diego, CA, USA
wpropp@ucsd.edu

Babak Rahimi University of Caifornia, San Diego (UCSD), La Jolla, CA, USA
brahimi@ucsd.edu

Christopher Bronk Ramsey Oxford Radiocarbon Accelerator Unit, University of Oxford, Oxford, UK
christopher.ramsey@rlaha.ox.ac.uk

Donald B. Redford Pennsylvania State University, University Park, PA, USA
mendes34@comcast.net

Gary A. Rendsburg Rutgers University, New Brunswick, NJ, USA
grends@rci.rutgers.edu

Thomas Römer Collège de France, Paris, France; University of Lausanne, Switzerland
thomas.romer@college-de-france.fr

Stephen C. Russell Department of History, John Jay College, CUNY, New York, NY, USA
stephen.christopher.russell@gmail.com

Amos Salamon Division of Geological Hazards, Geological Survey of Israel, Jerusalem, Israel
salamon@gsi.gov.il

Konrad Schmid University of Zurich, Zurich, Switzerland
konrad.schmid@access.uzh.ch

Thomas Schneider University of British Columbia, Vancouver, BC, Canada
thomas.schneider@ubc.ca

Jürgen P. Schulze Qualcomm Institute, UC San Diego, La Jolla, CA, USA

Zachary Seldess Qualcomm Institute/California Institute of Telecommunications and Information Technology (Calit2), University of California, San Diego (UCSD), La Jolla, CA, USA
zseldess@eng.ucsd.edu

Brad C. Sparks Archaeological Research Group, Los Angeles-San Diego, CA, USA
exodus@eng.ucsd.edu

David Srour Qualcomm Institute, University of California, San Diego (UCSD), La Jolla, CA, USA
dsrour@eng.ucsd.edu

Pieter W. van der Horst Department of Philosophy & Religious Studies, University of Utrecht, Utrecht, The Netherlands
pwvdh@xs4all.nl

Matthew L. Vincent Qualcomm Institute, University of California, San Diego (UCSD), La Jolla, San Diego, CA, USA; DigitalMed Universidad de Murcia, Spain

Steve Ward Institute of Geophysics and Planetary Physics, University of California (UCSD), Santa Cruz, CA, USA
wardsn@ucsc.edu

Philip Weber Qualcomm Institute, UC San Diego, La Jolla, CA, USA

Malcolm H. Wiener Institute for Aegean Prehistory (INSTAP), Philadelphia, PA, USA
mhwiener@villacandia.com

Part I

Setting the Stage: Interdisciplinary Approaches to the Exodus Narrative

1 Exodus and Memory

Jan Assmann

Abstract

The Biblical Book of Exodus is the narrative version of the great transformation from polytheism to Biblical monotheism in the Ancient World. The interest of the story, in which ancient Egypt plays such an important and sinister role, lies not in what really happened but how, by whom, when, in which form, and for what purpose it was told in the course of millennia. The story is about the revolutionary birth of both a people and a religion. It has a political and a religious aspect and both aspects are inseparably linked. It is a story of liberation (from Egypt) and to commitment (to "Law" and covenant)—from Egyptian slavery to Divine service. It involves a great amount of violence that is both of a political nature (Egyptian oppression of the Israelites, the "plagues" against the Egyptians) and of a religious one (the massacre after the cult of the Golden Calf)—the "founding violence" that typically accompanies the birth of something radically new.

The Biblical story of the Exodus of the children of Israel from Egypt is THE story, the story of stories, arguably the greatest, in any event the most consequential story ever told—though perhaps not literally experienced—in human history. It is a story that in its endless tellings and retellings, variations, and transformations changed and formed the human world in which we are living. Given this world-changing importance, it is only natural that scholarly attention has primarily focused on the question what really happened when the children of Israel went out of Egypt, i.e., what archaeological, epigraphic, and other evidence may tell us about its historical background. In this contribution, however, I will direct my attention in the opposite direction and not ask about what really happened, but who remembered the story, following a Latin scholastic hexameter teaching how to deal with historical sources:

J. Assmann (✉)
University of Konstanz, Konstanz, Germany

University of Heidelberg, Heidelberg, Germany
e-mail: jan@assmanns.de

T.E. Levy et al. (eds.), *Israel's Exodus in Transdisciplinary Perspective*,
Quantitative Methods in the Humanities and Social Sciences, DOI 10.1007/978-3-319-04768-3_1,

Quís, quid, ubí, quibus aúxiliís, cur, quómodo, quándo?
Who? what? where? by what means? why? how? and when?[1]

This is not to say that there is no historical background at all behind the story of the Exodus from Egypt and that it is futile to investigate all possible sources. On the contrary, it is quite probable that a great many historical experiences and memories lie behind and went into the Biblical story though certainly not this one gigantic and miraculous event of liberation, election, and revelation. It is only to say that the story acquired its world-changing momentum only in its reception history and whereas the historical events behind it may turn out to be rather trivial, its real importance is a question of memory rather than history (cf. Hendel 2001; see also Hendel, Chap. 5).

The theme of this story is not only (a) the origin of a people that still exists as the only ancient people that survived with its ethnic and religious identity through the fall of the Ancient World, but also, what is even more decisive, (b) the origin of monotheism that has become the prevailing religious orientation in most parts of the world. The story itself has become the model of many a story of liberation, emancipation, and salvation including its secular transformations such as Marxism and socialism (see Walzer 1985)—and even psychoanalysis if we think of the importance Sigmund Freud attached to Moses and Exodus.

In my book *Moses the Egyptian*, which I wrote in California 20 years ago, I tried to define the conceptual core of the Exodus narrative as the "Mosaic distinction" between true and false religion or true and false Gods (Assmann 1997; see also Assmann 2007, 2010). This theory has met with much criticism and I would not hold it any longer. The distinction as such, and as a defining feature of monotheism, still seems to me irrefutable, but I would no longer call it "mosaic."

It is true that the distinction between true and false in religion seems somehow implied in the prohibition of the worship of other gods and images, but it becomes a question of truth only later in antiquity with a certain concept of revelation. The Torah at Mt Sinai is not "revealed" but simply "given" and its power does not rest on its truth but on God's power and authority that has delivered Israel from Egyptian bondage. Here, Hobbes is right who stated that *auctoritas non veritas facit legem*. In its aspect of law, the torah is not about true and false, but about right and wrong.

The story of the Exodus draws several distinctions that have nothing to do with true and false. In itself it is divided in two parts: the liberation from Egyptian bondage (chs. 1–15) and the formation of the covenant on Mt Sinai (chs. 19–40). This looks like another distinction, that of liberation and binding, but the point of the narrative is that this is one and the same: liberation means binding and binding means liberation. The first part draws quite obviously the distinction between bondage and freedom, whereas the second part, the covenant, draws first the distinction between the chosen and the non-chosen, Israel and the peoples, and second, within the covenant, the distinction between friend and foe, those who love God and keep his commandments and those who don't. Both distinctions are very firmly drawn; Israel is separated from the nations to become a kingdom of priests and a holy nation (Exod 19:6), and to his friends God will show his mercy and loving-kindness up to the 1000th generation, but the sins of the second he will punish on the third and fourth generations, because God is a jealous God (Exod 20:5–6; Deut 5:9–10).

There is no question of truth and falsehood here. Decisive is the fact that the distinction between friend and foe and God's jealousy and wrath does only work within the covenant and must be carefully distinguished from the distinction between Israel and the other peoples. The other nations are neither foes nor their gods false and nonexistent. On the contrary, they

[1] The earliest exact quote is attributed to Cicero by Aquinas, *Sum. Theol.* II-I:Q7:3; cf. Cicero, *De inv.* 1.27 [1:41].

are very existent and Israel must forswear worshiping them. What God requests of his people is faithfulness, not truth and the metaphors and models for this unprecedented concept of covenant are sonship, matrimony, and political alliance. If there is any "Mosaic distinction," it is the distinction between matrimonial faithfulness and adultery, political loyalty and apostasy, filial love and rebellion, and, in this sense, between friend and foe, love and wrath.

There is, however, one very important case where the boundary between within and without is blurred and the distinction between friend and foe is extended to the outer, however restricted, sphere. This is the concept of "Holy War." In the case of Holy War, external peoples are promoted to the rank of enemies of God and objects of his wrath. This extension, however, is very restricted and concerns only the "seven peoples" who reside in the land that God has promised to give the Israelites (Deut 7:1–6). For these peoples, the tolerance towards the other nations is suspended and the Israelites are not only allowed but bound by sacred obligation to expel and exterminate them. The rules of Holy warfare are (1) the war is waged on divine command, (2) the enemy is consecrated to God, nothing must stay alive, no bounty must be taken, and everything is to be heaped up and burned on the market place.

This concept is specific neither to Israel nor to monotheism. In one form or the other it is common in the ancient world (Sa-Moon 1989; von der Way 1992; Lang 2011). It may be characterized as "occasional monolatry" in the sense of creating a specific and more intense, even monolatrous relationship between a warlord and a specific deity. In a certain situation which reminds one of Carl Schmitt's concept of "Ernstfall," (see Schmitt 1996) a people or a king puts all his hope on one singular deity and ensures his or her support by forswearing or consecrating the bounty (*herem* in Hebrew). It is very probable that the specific form of monotheism that originated in Israel with the early prophets at the end of the eighth century BC developed out of this custom of holy warfare and occasional monolatry.

The story of Exodus as we know it is a book of the Hebrew Bible, the second book of Moses or the Pentateuch, the Jewish Torah. But it is quite evident that before its integration into the Torah, it must have led a literary life of its own, and even before its *literary* life the story will certainly have circulated in *oral tradition* as a myth.[2]

Taking Exodus as a myth does not mean that we are dealing here with pure fiction without any historical core. Myths may very well be based on historical experiences. The decisive property of a myth is that it is a well known and widely shared foundational story irrespective of its historical or fictional base. Golgatha is a myth, but few people doubt that a historical person by the name of Jeshua ha-Nosri has in actual fact been executed by crucifixion. The same may apply to the Exodus from Egypt of a tribe by the name of Yisrael. But this is exactly the kind of question that I would like to put in brackets. My question, again, is not what really happened but who told the story, why, when, to whom, and how?

The first allusions to the myth occur with the early prophets, Hosea, Amos, and Micah[3]:

> "When Israel was a child, I loved him, and out of Egypt I called my son." (Hosea 11:1).
> "I brought you out of Egypt and led you 40 years through the desert." (Amos 2:10).
> "Did I not lead you out of Egypt and released you from serfdom and send before you Moses, Aaron and Miriam?" (Micah 6:4).

If we ask our mnemohistorical questions: who? when? why? the answers are obvious. These prophets were ardent mono-Yahwists, as I would like to call them. They were certainly not monotheists, because their core concept is loyalty, fidelity, faithfulness, and Hosea's core metaphor for this loyalty is matrimony respective to adultery. What is the point of faithfulness if

[2] For the textual history of Exodus cf., e.g., Schmid 1999.

[3] If those Biblical scholars are right who date the Song of the Sea (Exod 15) to a very early date (ninth century BC and earlier), because of its highly archaic language, this poem should count for the oldest allusion to the Exodus myth.

there are no other gods? What is the reproach of adultery if there are no other men with whom to betray the bridegroom or husband? Hosea's concept of loyalty presupposes a world full of other gods with whom Israel is all too prone to commit adultery. Another image is the sonship of Israel and it is in this context that the Exodus myth is alluded to: "When Israel was a child, I loved him, and out of Egypt I called my son." (Hos 11:1)[4]

What we may retain from this first allusion to the Exodus myth is the idea of a very close and intimate relationship between God and Israel, based on an act of salvation (from Egyptian bondage) and election (out of other nations and tribes). The prophets preach what one could call a "monotheism of faithfulness," acknowledging the existence of other gods but demanding exclusive and absolute loyalty to one specific god who proved himself the savior and liberator from Egyptian bondage. The prophets want to remind Israel, especially the Northern Kingdom, of this singular relationship.

When? This is the decisive question: in a time of utmost danger and affliction by the hand of the Assyrians. The prophets foresaw and witnessed the fall of the Northern Kingdom. If there ever was an "Ernstfall" in the Schmittian sense, it was now, at the end of the eighth century BC. It was the hour of decision and of "occasional monolatry": to put all one's hope on the One god able to help, to save, to liberate.

We must, however, not assume that this prophetic monotheism of loyalty and faithfulness became the general religion of Ancient Israel. What we are reading in the books of the early prophets is the voice of an opposition that met with strong rejection and even persecution (see Smith 1971). The Exodus-and-Moses narrative was the foundational story of this movement. When 100 years later the Assyrian Empire collapsed, there was a moment of hope. On the throne at Jerusalem sat Josiah, a king who was open to the new religious ideas. He was presented a book that turned up in the course of restoration work in the temple and that is commonly identified with the first version of Deuteronomy (2 Kg 22–23). This book gives the new idea of a monotheism of faithfulness the form of a political treaty which the authors adapted from Assyrian loyalty oaths and vassal treaties.[5] It is the same politicized concept of the ancient idea of faithfulness and love between Yahweh and Israel that we meet with in the Book of Exodus. The treaty or "covenant" is no longer a metaphor, such as matrimony or sonship, but the real thing. The whole system of political relations, between the gods and the king, the king and the subjects, the king and his vassals, is now transformed into the religious system of the monotheism of faithfulness and the one relation of god and people, i.e., the liberating god and the liberated people.

These political treaties are frequently based on a historical recapitulation that found the actual alliance in a friendly past (Baltzer 1964). The Exodus myth fulfills this function of historical frame. The various stipulations, commandments, and prohibitions that form the body of the treaty receive their meaning from the story of liberation from Egyptian bondage. Against the background of this story, the covenant appears as an instrument of freedom.

A political alliance between a god and a people is an absolutely new, unheard of, and unprecedented concept. As such, it requires a specific amount of historical motivation and explanation. This is the reason **why** the story is told. As stressed above, we are dealing here not with just "a" story, but with "THE" story, the foundation of the covenant that is the foundation of the people of Israel and of Jewish and Christian religion.

The revolutionary concept of the covenant between the people and God implies a triple

[4] Both images, by the way, come from the Egyptian and Babylonian imagery of sacred kingship. In Egypt, Pharaoh is held to be the son of god and in Babylonia, the king is wedded to the divine world by a *hieros gamos*.

[5] See Otto 1999; Steymans 1995. King Manasseh must have been among the vassals who swore loyalty to Esarhaddon, see Steymans 2006; Otto 2007: 119.

process of theologization: (1) the theologization of the *political* concept of alliance, (2) the transformation of a *secular* law code into *ius divinum*—the torah—and of (3) *human* history into sacred history.[6] Therefore, the torah has these three aspects and functions: of a law code, a treaty, and sacred historiography.

In Deuteronomy, the Exodus narrative functions only as a frame. The main theme of the book is the law of the covenant. It is, however, highly probable that at the same time, in the second half of the seventh century BC, there existed also an early literary version of the Exodus narrative. At the beginning of the sixth century, the catastrophe of the Northern Kingdom repeated itself with respect to the Southern Kingdom of Judah. Jerusalem was conquered, the temple destroyed, and the elite deported into exile. This time, however, they were able to take with them a body of literature, a codification of their sacred traditions which allowed them to survive 50 and more years of Babylonian exile. It is during this time that the new religion, the monotheism of faithfulness, became the dominating belief and practice.[7]

When the exiles returned to Jerusalem, they brought with them their new code of religious beliefs and practices, which was also a criterion of identity and belonging. Being a Jew and belonging to Israel as the people of God was now defined by observance of the Law. The books of Ezra and Nehemiah report the conflicts that arose with the population that remained in the land and had in ignorance of this new code of Jewishness adopted Canaanite customs, formed mixed marriages, and begot Jewish-Canaanite children. The Exodus narrative with its strong and exclusivist ideas of liberation, election, covenant, friend and foe now acquired a very decisive meaning, provoking a more liberal counter-narrative.

In the same way as the book of Ruth may be read as a counter-narrative to Ezra's and Nehemiah's policy of forced divorce, the narrative of the patriarchs appears as a counter-narrative to the Exodus-Moses narrative (see esp., Schmid 1999). In this myth, God also forms a covenant, but not with a people but with an individual, Abraham, to whom is promised to become the ancestor of a people. This covenant, however, is not spelled out in a body of legislation. The only criterion of belonging is circumcision as the "sign of the covenant," and, of course, Abrahamic descent. A common element of both myths of origin is allochthony, foreign provenance. Abraham is called to Canaan from Mesopotamia, Israel from Egypt. In both narratives, the motif of foreign provenance is an expression of the defining difference between the people of the covenant—in the patriarch myth a family, in the Exodus myth a nation—and the surrounding peoples. The relation of the immigrants to their host country, however, is completely different.

In the Exodus myth, the land is to be conquered and its inhabitants to be slain or expelled. In the patriarch myth, the land is to be bought, the relations between the Abraham family and the inhabitants are friendly. Whereas in the Exodus myth the Canaanite deities are abhorred, Abraham and Melchizedek, the priest-king of (Jeru)Salem, find out that their god is one and the same. The god of the patriarch myth is the universal creator of heaven and earth; the god of the Exodus myth is the very particular liberator "who brought thee out of Egypt, the house of bondage." The patriarch myth is determined by a spirit of liberalism, humanism, and pacifism; the Exodus myth shows a spirit of revolutionary radicalism

[6] Ronald Hendel points out to me that even in the Code of Hammurapi the (secular) law receives a divine foundation since the king is shown before Shamash, the god of the sun and of justice to whom he is responsible. However, Hammurapi, not Shamash, acts as legislator here, whereas in the torah the laws are given by Yahweh, not by Moses. Hammurapi is bound to formulate his laws in conformity with the divine idea of justice, whereas Moses is bound to promulgate the divine laws in conformity with Yahweh's dictation.

[7] Up to this point, the monotheism of faithfulness as propagated by the early prophets was just a—much contested—minority position within a generally syncretistic Israel worshipping other gods (Ba'alîm and Asherôth) besides Yahweh. Only among the exile community did it achieve a position of dominance.

implying a lot of violence: first the violent treatment of the Israelites by the Egyptians; then the violent treatment of the Egyptians by God in the ten plagues; after the liberation the sometimes violent opposition that Moses and Aaron meet with the "murmuring" people; and after the formation of the covenant the extremely violent reactions of the "jealous god" against lawbreakers, defectors, and disobedience.

It is easy to imagine parties behind these stories, a more radical one promoting the Exodus narrative and a more liberal one promoting the patriarch narrative.[8]

At a later stage, both narratives are brought together, making the patriarch myth the prehistory of the Exodus myth, combining both by means of the Joseph novella. This combined narrative must have led a literary life of its own before being integrated in a history of huge scope, starting after the model of the Babylonian and Egyptian king-lists with cosmogony and integrating the Babylonian story of the flood.

In the Book of Exodus, the compact myth is unfolded in a sequence of core scenes: (1) the suffering of the children of Israel in Egypt, the house of serfdom, (2) the birth, upbringing, flight, and vocation of Moses as savior, (3) the negotiations of Moses and Aaron with Pharaoh and the ten plagues by means of which God forces Pharaoh to yield, (4) the exodus proper, from the night of passover to the miracle at the sea of rushes, (5) the revelation of the Law at Mt Sinai with the crisis of the Golden Calf, and (6) instructions for the tabernacle.

We are dealing here with a careful composition, with a beginning, middle, and end. The tabernacle is a perfect ending of the story that could have ended there. This motif concludes the emergence of a new religion by describing its institution. It fulfills the promise of God to dwell among his people. This is far more decisive than what follows. In Leviticus and Numbers, the story continues with (7): the 40 years of wandering in the wilderness, more legislation, and more crises. The severest crises are the episode of the spies leading to God's verdict to ban the present generation from entering the Promised Land, and the scene at Shittim, the last station before entering the Promised Land, where the Israelites accept an invitation by the Moabites to join in a feast of their god Baal Peor, and 24,000 are slain by a plague in consequence of the transgression. Deuteronomy is a summarizing recapitulation on the eve of crossing the Jordan.

The last scene, (8) the conquest, is told in the book of Joshua which is cut off from the Torah proper and relegated to the second order, the prophets. The Torah ends with the death of Moses. This is highly significant. The story that begins with the suffering of the children of Israel in the hands of the Egyptians ends, not with the conquest of Canaan, but with the death of Moses, turning the sacred narrative into a biography of Moses.

Narrative structure is determined by the correspondence of beginning and end in terms of lack—lack liquidated. The lack is clearly represented by the suffering of Israel in Egypt. It is liquidated by the lifework of Moses who has turned a mass of slaves into the people of God and has instituted a covenant in form of a law, a cult, and a temple. This status the Israelites have achieved even before entering the Promised Land, and it is, therefore, independent of their dwelling there. The point of the narrative is not conquest—from destitution to possession—but liberation: from serfdom to freedom. The Bible is careful in drawing the distinction between savior and conqueror and in assigning the conqueror to the second rank.

The lasting achievement of Moses is the covenant that God has formed through his mediation with the people. This goal has been achieved on Mt Sinai, in the no-man's-land between Egypt and Palestine, especially with the construction of the tabernacle that ensures God's presence among his people, notably a portable sanctuary. The covenant has only to be *remembered* in the Promised Land in order to enjoy the freedom that the liberation from Egyptian serfdom has bestowed on the people.

[8] See also Bernhard Lang, *Buch der Kriege*, 10–13; 45–47. The late date of the patriarch stories vis à vis the Exodus story follows from the scarcity of references to Abraham outside the book of Genesis.

To be and to remain free means to stay within the covenant and its stipulations. To abandon the covenant means to fall into the hands of other slaveholders and symbolically to return to Egypt.

Perhaps the most remarkable and strangest section of the Exodus narrative is on the ten plagues. In the economy of the narrative, the scene fulfills two functions: it compensates the Israelites for their suffering by punishing their tormentors, the Egyptians, and it makes clear beyond any doubt that the Israelites have not been expelled but delivered from Egypt. Nevertheless, one major plague would have fulfilled this function. Why ten of them? Their sequence, too, does not show a clear climactic logic.

1. Turning the water of the Nile into blood
2. Frogs
3. Lice
4. Insects
5. Pestilence striking livestock
6. Boils hitting man and beast
7. Hail smiting man, beast, and plants
8. Locusts
9. Darkness

The tenth plague, the killing of the firstborn, is set apart by a totally different form of narration. I shall come back to that.

The multiplication of the motif of the plague by the factor 10 has a clear mnemonic function. Like the ten commandments, it is based on the human hands with their ten fingers.[9] However, the plagues are not grouped into two pentads (such as the 10 commandments in Jewish counting) but in three triads plus the tenth plague that stands apart. They are grouped in triads by the formula "in the morning" and other markers. The plagues are *signs* to be remembered like the ten commandments. It is not one punishing and liberating event. It is a message to be forever retained and taken to heart.

The theme of memory is central in the Book of Exodus. As a historical narrative, it is in itself an act of memory. It remembers an event of the past that according to Biblical chronology took place in the fifteenth century BC, thus in the Late Bronze Age. As we have seen with the early prophets Amos and Hosea, this memory was already alive in the late eighth century BC, in the time of Homer, who also looked back to the late Bronze Age in telling the story of the Trojan War. The eighth and seventh centuries were generally a time of looking back across the break that the end of the Bronze Age and the first centuries of the Iron Age had brought about in the Mediterranean and Near Eastern World. In Egypt, we are dealing with a period of a very pronounced archaism. Texts were copied and architectural, sculptural, and pictorial models were carefully followed that date back to the second and third millennia BC. The Neo-Assyrian empire even turned into a digging society trying by means of systematic excavations to reach to the traces of the Sargonid Empire, the twenty-third century BC that was held to be a Golden Age and a model of cultural and political perfection (Maul 2001; Jonker 1995).

This was a time of general reorientation where the past began to matter in various conspicuous forms as a "normative past" that must by all means be remembered and followed as a source of political, legal, religious, and artistic models and norms. For Israel, the Exodus fulfilled precisely this function of a normative past—in such a degree of normativity, however, that has no parallel in Egypt, Mesopotamia, and Greece. For Israel looked not only back like its neighbors, it looked also forward. The story of Exodus is a story of promise. The element of promise distinguishes the covenant from other treaties and law-codes.

Normally, a law is coupled with a sanction. The commandments and prohibitions of the covenant, however, are additionally associated with a promise. Keeping the covenant will be rewarded by the possession of, and blissful life in, the Promised Land, meaning reproduction, fertility, victory over enemies, peace, and prosperity. The treaty at Mt Sinai looks back to the Exodus from Egypt and forward to an unlimited future in the Promised Land—on the condition of staying faithfully by the covenant and its

[9] The Mishnaic collection of proverbs *Pirqê Avôt* has in its 5th section a collection of decades, three of which occur in the Exodus narrative: the ten plagues, ten commandments, and ten cases of "murmuring" of the people during their wandering in the wilderness.

613 statutes, commandments, and prohibitions. All depends on this one condition: that the covenant will not be neglected or even broken.

In order to secure the keeping of the covenant, a mnemotechnique has to be devised. This corresponds to traditional usage. Treaties have to be laid down in writing on durable material, e.g., on a silver tablet to be deposited in the temple but also—and this is decisive—to be read aloud at regular intervals before the two parties. The Assyrian king Esarhaddon devised yet another ritual of commemoration. He summoned his subjects and vassals to the capital in order to swear an oath of loyalty to his designated successor Assurbanipal. Foreseeing, however, that the change of frame, when the subjects and vassals will have returned to their various homes, will cause forgetting, Esarhaddon devised a mnemonic ritual:

> Water from a sarsaru jar, she [i.e., Ishtar] gave them to drink,
> A goblet she half filled with water from the sarsaru jar and gave it them saying:
> You speak in your heart: Ishtar, a narrow one [i.e., watchful—or locally restricted?] is she.
> But then you will go away to your towns and your districts,
> You will eat bread and forget these oaths.
> But as soon as you drink from this water,
> You will remind yourself and you will keep this swearing-in which I have enacted on behalf of king Esarhaddon.[10]

The Book of Exodus contains instructions for a similar though much more elaborate ritual of commemoration. This is contained in Chapter 12 and 13 including the report of the tenth plague, the killing of the firstborn in Egypt.

> And the LORD spake unto Moses and Aaron in the land of Egypt, saying, This month *shall be* unto you the beginning of months: it *shall be* the first month of the year to you. Speak ye unto all the congregation of Israel, saying, In the tenth *day* of this month they shall take to them every man a lamb, according to the house of *their* fathers, a lamb for an house: and ye shall keep it up until the fourteenth day of the same month: and the whole assembly of the congregation of Israel shall kill it in the evening. And they shall take of the blood, and strike *it* on the two side posts and on the upper door post of the houses, wherein they shall eat it. And they shall eat the flesh in that night, roast with fire, and unleavened bread; *and* with bitter *herbs* they shall eat it. And thus shall ye eat it; *with* your loins girded, your shoes on your feet, and your staff in your hand; and ye shall eat it in haste: it *is* the LORD'S passover. Seven days shall ye eat unleavened bread; even the first day ye shall put away leaven out of your houses: for whosoever eateth leavened bread from the first day until the seventh day, that soul shall be cut off from Israel. And ye shall observe *the feast of* unleavened bread; for in this selfsame day have I brought your armies out of the land of Egypt: therefore shall ye observe this day in your generations by an ordinance for ever. (Exod 12:1–18, KJV, verse numbers omitted)

In the same way that the Sarsaru ritual is a ritual of drinking water that reminds the drinkers of the oath they have sworn; the Passover is a ritual of eating unleavened bread that reminds the eater of their hasty departure from Egypt when they had no time to add yeast to their dough. For the same commemorative reason, the ritual has to be performed in the family and not in the temple or synagogue, because the Israelites spent this night in their homes when the killing angel of the Lord haunted the houses of the Egyptians.

The mnemotechnique that Moses devised in order to constantly remind the people of the covenant, its various obligations, and of the story that frames and explains it, surpasses by far anything comparable in the ancient world (see Assmann 2011: 193–205). Like Esarhaddon, Moses foresees that the people will forget their obligations once they will live in the Promised Land, eat bread, and get saturated.[11]

[10] Quoted and translated after Otto 1999: 82.

[11] Deuteronomy is especially rich in passages that bespeak the anxiety of forgetting through the change of place, e.g.: "Take heed to thyself that thou forget not the Lord thy God, so as not to keep his commands, and his judgments, and ordinances, which I command thee this day: lest when thou hast eaten and art full, and hast built goodly houses, and dwelt in them; . . . thou shouldest be exalted in heart, and forget the Lord thy God, who brought thee out of the land of Egypt, out of the house of bondage." (Deut 8:11–14)

The Book of Exodus—as well as the myth behind it—is, therefore, not only a *feat* of memory, remembering an event however decisive of the distant past. But it is also and above all the *foundation* of a memory, i.e., part and object of a mnemotechnique that frames and supports the covenant.

> And thou shalt shew thy son in that day, saying, *This is done* because of that *which* the LORD did unto me when I came forth out of Egypt. [9] And it shall be for a sign unto thee upon thine hand, and for a memorial between thine eyes, that the LORD'S law may be in thy mouth: for with a strong hand hath the LORD brought thee out of Egypt.[10] Thou shalt therefore keep this ordinance in his season from year to year. (Exod 13:8–10, cf. 16)

The Exodus is THE decisive memory never to fall into oblivion, and the Book of Exodus is the codification of that memory. "Remember the Exodus" means "remember the covenant" and vice versa. To remember the Exodus and the covenant means always to remember the promise, to look into the future.

In later (medieval) times, in the diaspora, this ritual prescription has been fleshed out in great detail in form of the Seder haggadah.[12] In the Jewish tradition, the memory of the Exodus lives on in two forms: firstly as part of synagogal recitation in which weekly portions (*parashot*) from the entire Torah are read in the course of the year, and secondly in the form of an annual celebration taking place not on the synagogue but at home, with the *pater familias* (and not the Rabbi) acting as master of ceremonies. Moses is scarcely mentioned in the Seder haggadah. This makes the biggest difference between the Book of Exodus where Moses is the protagonist and the myth of Exodus as reenacted in the Seder ceremony.

The Jewish Seder, on the first night of Pesach, is the festive and liturgical realization of the commandment "Thou shalt teach your son and your son's son", viz., that we have been slaves in Egypt and that the Lord redeemed us from bondage with a strong hand and an outstretched arm. It is a teach-in to remember the connection between history and covenant, law and liberation. The story must be told and the questions be asked in the "we" and "us"-key. Why do we perform these rites and obey these laws? Because we have been slaves in Egypt.

In the same way as this "we" includes every Jew in addition to those who once emigrated from Egypt 3500 years ago, the concepts of Egypt and Pharaoh extend to every form of oppression and violence where and whenever they occur. A Jew is someone who was liberated from Egypt and who is free insofar as he/she commits himself/herself to the covenant and its prescriptions. In liturgical memory, history is turned into myth, into a set of archetypal patterns with regard to which the present is made transparent so that they shine through and render the present readable. In the *New York Times* one could read some years ago:

> For thousands of years, Jews have affirmed that by participating in the Passover Seder, we not only remember the Exodus, but actually relive it, bringing its transformative power into our own lives.[13]

This is an excellent definition of liturgical memory. "In every generation," the Pesach Haggadah prescribes, "a man should look upon himself as if he came forth from Egypt" (Shire et al. 1998: 36). The [illegible] identity through identification. It is about the transformation of history into memory, to make a certain past "our" past and to let everyone participate in or even identify with this past as "his/her" past. One could even go so far as to speak of a transformation of semantic memory, i.e., something we have learned, into episodic memory, something we have lived, albeit in the form of a ritual play, of an "as if."

The function of the Seder is to provide a frame for remembering the Exodus, not only by liturgical recitation of the written texts of the Haggadah, but also and above all by improvised "conversational remembering" (Middleton 1997). Frames, as Erving Goffman has shown, organize

[12] I am using the Hebrew-German edition *Die Pessach Haggada* (Shire et al. 1998). Translations mine.

[13] Ad of the journal *Tikkun* in *New York Times* of March 22, 2002.

our everyday life (Goffman 1974). Thus, they relieve us from reflection and enable spontaneous action. With the Seder, we move on to the level of non-everyday behavior. This shift from an everyday frame to a festive and an exceptional one is explicitly marked and foregrounded in the Haggadah, the script for the feast. The arrangements have to be so exceptional that they strike the minds of the uninitiated, and the youngest child has to ask the question that will trigger the chain of explanations and commemorations: "Why is this night so different from all other nights?" (Shire et al. 1998: 14). This question addresses precisely the point of framing; it is the question of somebody who lacks the cue: "What is going on here?" The Seder starts with a festive enactment of a frameshift.

Difference is a key word in the Seder ceremony. God is praised for having made a difference: between this night and all other nights, "between the sacred and the profane, between light and darkness, between Shabbat and the other 6 days of the weeks, and between Jews and Gentiles" (Shire et al. 1998: 12)—and between serfdom (*ʿavodah*) and freedom (*ḥerût*), which is the basic theme of the story to be remembered.[14]

All these differences are to be made understandable and palpable through the one difference which is sensually staged and brought to the forefront by the striking exceptionality and unfamiliarity of the arrangements and actions, of "what is going on." The children, the uninitiated, are provoked to ask, and the answers given serve the function of an initiation, of conveying and acquiring a new identity. This connection between question, answer, and identity is made clear by the "Midrash of the four sons" (Shire et al. 1998: 18). At several places in the Torah, there occurs the prescription of what to answer when one's son asks about the meaning of the Law or one particular law. These passages are collected in this Midrash and attributed to four types of sons: the wise one, the wicked one, the simpleton, and the one who does not know how to ask.

> The wise one—what does he say? "What are the testimonies, and the statutes and the laws that the LORD our God commanded you?" (Deuteronomy 6,20). So you tell him about the laws of Pesach, that one may not eat anything whatsoever after the Pesach sacrifice.
> The wicked one—what does he say? "What is this service to you?" (Exodus 12,26). "To you", and not to him. And since he excluded himself from the people at large, he denies the foundation of our faith. So you blunt his teeth and tell him, "It is because of this that the LORD acted for me when I came forth out of Egypt" (Exodus 13:8). "For me", and not for him; had he been there, he would not have been redeemed.
> The simple son—what does he say? "What is this?" (Exodus 13:14). "Tell him, 'with a strong hand God took us out from Egypt, from the house of slavery'" (ibid.).
> As for the one who does not know how to ask, you must begin for him, as it is written "and thou shalt tell thy son in that day, saying: It is because of this that the LORD acted for me when I came forth out of Egypt" (Exodus 13:8).

The Midrash of the four sons is a mini-drama about memory, history, and identity. The identity question is expressed by the play with the personal pronouns: I and me, us and our, you and he. The entire ceremony is about telling the story. This is history as it is remembered and told, not as it might have happened. The Seder provides a frame for telling and explaining the story. The important questions to ask are pretty much the same as those codified in the Latin scholastic hexameter quoted above:

> Quís, quid, ubí, quibus aúxiliís, cur, quómodo, quándo?
> Who? what? where? by what means? why? how? and when?

Who tells the story? The father and the adult participants who play the role of the emigrants from Egypt. To whom? To the children who have to learn to identify with the group of the liberated slaves and to say "we" and "us" with respect to the ancient story. "Why?" Because it is this story that tells us who we are. When? On the occasion of the annual return of the time when this event is believed to have happened, the spring time of the

[14] *Herût* "freedom" is not a Biblical term. The Bible uses the word *avodah* "service" both for the Egyptian serfdom and for the service of God. It opposes the liberating service of God and the oppressive service of Pharaoh.

offering of the first fruits. By which means, in which form? In the form of a "symposium" (the Haggadah prescribes or recommends to eat and drink in "reclining posture", i.e., in Greek and Roman style: Shire et al. 1998: 12) and in a combination of liturgical and conversational remembering.

Even the recital of the ten plagues forms part of the Seder liturgy, spilling some drops of wine with every mention of a plague (Shire et al. 1998: 27):

1. blood (*dam*)
2. frogs (*tz*^{e}*fardea*ʿ)
3. lice (*kinnîm*)
4. insects (ʿ*arov*)
5. pestilence (*dævær*)
6. boils (*sh*^{e}*hîn*)
7. hail (*barad*)
8. locusts (*arbæh*)
9. darkness (*ḥôshek*)
10. killing the firstborn (*makkat b*^{e}*kôrot*)

Trauma and triumph go together in liturgical memory. The triumph culminates in the crossing of the Red Sea where the persecuting Egyptians are drowned. This is the decisive act of liberation. The keyword is *b*^{e}*-yad ḥazaqah* "with strong arm." Again and again this formula recurs in the liturgy. Its theological meaning is to represent the liberation as God's and not Moses'—work, as a sign of God's power (Hoffmeier 1997).

Liturgical memory—in the same way as cultural memory—provides a society with a connective structure working both in the social and temporal dimensions. In the social dimension, it works as a social cement binding human beings to fellow human beings and creates a common space of experience, expectation, and action that provides trust, confidence, and orientation. In the temporal dimension, cultural connectivity works as a principle of continuity linking past, present, and future, in that it creates meaning, memory, and expectation by integrating the images and stories of the past into an ever-progressing present. This aspect is the basis of myths and historical narratives such as the Exodus from Egypt.

Both aspects, the normative/social and the narrative/temporal one—the aspect of instruction and the aspect of narration—consolidate belonging or identity, enable an individual to say "we."

In the Seder feast, however, the past is not only remembered but performed. The celebration does not scrupulously follow a fixed model, a ritual prescription, but it re-presents or "presentifies," in the sense of making present, by a form of actual reliving. The recitation of the Haggadah is complemented by all kinds of improvised contributions about "our" sufferings in Egypt and the delights of liberation.

The themes of promise and future are also very prominent in the Seder liturgy that closes with the proclamation le-shanah ha-ba'ah bi-yerushalayim "next year in Jerusalem!" (Shire et al. 1998: 52)—the expression of hope founded on memory. Only he who remembers is able to look with confidence into the future.

This is the utopian aspect of the Exodus narrative. Like so many utopian texts, Exodus starts with a departure, with leaving home, setting out for an unknown goal in order to finally and in most cases unexpectedly arrive at an island where ideal conditions prevail. In Bacon's *Nova Atlantis* which is typical of the genre in this respect, the newcomers have to undergo a moral transformation in order to be accepted into the new community and its ideal constitution and institutions. If we apply this pattern to the Exodus, the parallels but also the differences become obvious. The departure is not for the absolutely unknown, there is a clearly indicated goal, first Mt Sinai and then Canaan. Nevertheless, there is a departure, there is an ideal constitution—to be received at Mt Sinai—and there is the land of milk and honey, a clear model of Cockaigne, the Schlaraffenland. The Book of Exodus, to be sure, is not meant as a utopia, such as, e.g., Plato's *nomoi*. The constitution as spelled out in the *saefaer* ha-berît is to be real, and not ideal, is to be lived and not just aspired to. The Promised Land is not some fictional island of bliss but a very real geographic unit. Still, there is a utopian element in the book and the myth of Exodus that is responsible for its extraordinary radiance and its being so much alive inside and outside of Judaism.

The puritans in the early seventeenth century, the time when Francis Bacon wrote *Nova Atlantis*, crossed the Atlantic Ocean and set out for America as a New Promised Land, identifying with the children of Israel going out of Egypt. This was an act of memory as much as it was a revolutionary step forward into something new, a new society, a new constitution, a new attempt at becoming the people of God and performing the covenant as laid down in the Bible. The same may be said of the Puritan revolution, the civil wars, and Oliver Cromwell's protectorate from 1642 to 1659.

Exodus—as a myth, a book, and a symbol—refers to that revolutionary turn in the history of a large part of mankind we are used to describe as the turn from polytheism to monotheism. The Exodus from Egypt is the narrative articulation of this act of emancipation, disembedding and distancing of a much larger scope. It is the move from what I have proposed to call "cosmotheism" (Assmann 1993), where the divine is conceived of and worshipped as immanent in nature—leading ultimately to the idea that nature or cosmos is God, the visible manifestation of a hidden deity—to a religion that draws a categorical distinction between God and the world, defining god as transcendent in the sense of strict extra-mundaneity.

The Exodus of the children of Israel from Egypt stands for the emancipation of humanity from its embeddedness in the world, its political, natural, and cultural powers, and for the emancipation of the divine from mundane immanence. Cosmotheism seems to me to be a far more adequate term than polytheism. Most "pagan" religions may be characterized as "cosmogonic monotheism": they recognize *one* God as origin of the world including heaven and earth, gods and men, and emphasize the oneness of god and the unity of the world. In antiquity, this basic religious conviction led to the idea of a supreme being that is both "hypercosmic" and "cosmic," transcendent and immanent, transcendent in its oneness and immanent in its differentiated multiplicity. The world that turned monotheistic with the Christianization of the Roman Empire had already come to emphasize the unity of god. The turn or exodus was not from polytheism but from cosmotheism to monotheism.

Seen in this light, we realize that this exodus has never fully been completed. There have always been relapses, counter-movements in the direction, not of poly- but of cosmotheism. The most powerful of these cosmotheistic trends is neo-platonism in its various branches such as hermeticism and all kinds of mystic and esoteric traditions including Kabbalah. The persisting presence of cosmotheism in Western tradition made it necessary to renew the power and pathos of Exodus in several waves of iconoclasm, emancipation, and even emigration, starting with the Reformation, especially in its extreme form of Calvinism and Puritanism, and Enlightenment, especially in its pronounced anti-clericalism (*écrasez l'infame*). It was this indefeatable, at times latent, at times manifest, continuity of cosmotheism that kept the idea, the myth, the book, and the symbol of Exodus alive.

References

Assmann, Jan. 1993. *Monotheismus und Kosmotheismus. Altägyptische Formen eines 'Denkens des Einen und ihre europäische Rezeptionsgeschichte*, Sitzungsberichte der Heidelberger Akademie der Wissenschaften, Jahrgang 1993, Bericht 2. Heidelberg: Universitätsverlag Winter GmbH.

———. 1997. *Moses the Egyptian: The Memory of Egypt in Western Monotheism*. Cambridge, MA: Harvard Univ. Press.

———. 2007. *Of God and Gods. Egypt, Israel and the Rise of Monotheism*. Madison, WI: Wisconsin Univ. Press.

———. 2010. *The Price of Monotheism*. Stanford, CA: Stanford Univ. Press.

———. 2011. *Cultural Memory and Early Civilization*, New York, NY: Cambridge University Press.

Baltzer, Klaus. 1964. *Das Bundesformular*, Wissenschaftliche Monographien zum Alten und Neuen Testament (WMANT), vol. 4, 2nd ed. Neukirchen: Neukirchener Verlag.

Goffman, Erving. 1974. *Frame Analysis. An Essay on the Organization of Experience*. New York, NY: Harper and Row.

Hendel, Ronald. 2001. The Exodus in Biblical Memory. *Journal of Biblical Literature* 120(4): 601–622.

Hoffmeier, James K. 1997. *Israel in Egypt. The Evidence for the Authenticity of the Exodus Tradition*. New York, NY: Oxford Univ. Press.

Jonker, Gerdien. 1995. *The Topography of Remembrance. The Dead, Tradition and Collective Memory in Mesopotamia*. Leyden: Brill Academic Publishers.

Lang, Bernhard. 2011. *Buch der Kriege—Buch des Himmels. Kleine Schriften zur Exegese und Theologie*. Louvain: Peeters.

Maul, Stefan M. 2001. Altertum in Mesopotamien. Beiträge zu den Sektionsthemen und Diskussionen. In *Die Gegenwart des Altertums*, ed. D. Kuhn and H. Stahl, 117–124. Heidelberg: Formen und Funktionen des Altertumsbezugs in den Hochkulturen der Alten Welt.

Middleton, David. 1997. Conversational Remembering and Uncertainty: Interdependencies of Experience as Individual and Collective Concerns in Team Work. *Journal of language and Social Psychology* 16(4): 389–410.

Otto, Eckart. 1999. *Das Deuteronomium. Politische Theologie und Rechtsreform in Juda und Assyrien*. Berlin: de Gruyter.

———. 2007. *Das Gesetz des Mose*. Darmstadt: WBG.

Sa-Moon, Kang. 1989. *Divine War in the Old Testament and in the Ancient Near East*. Berlin: De Gruyter.

Schmid, Konrad. 1999. *Erzväter und Exodus. Untersuchungen zur doppelten Begründung der Herkunft Israels innerhalb der Geschichtsbücher des Alten Testaments*, Wissenschaftliche Monographien zum Alten und Neuen Testament (WMANT), vol. 81. Neukirchen: Neukirchener Verlag.

Schmitt, Carl. 1996. *The Concept of the Political*. Trans. George Schwab. Chicago Univ. Press (orig. 1932).

Shire, Rabbi Michael, et al. (eds.). 1998. *Die Pessach Haggada*. Berlin: Hentrich & Hentrich.

Smith, Morton. 1971. *Palestinian Parties and Politics That Shaped the Old Testament*. New York, NY: Columbia University Press.

Steymans, Hans Ulrich. 1995. *Deuteronomium 28 und die adê zur Thronfolgeregelung Asarhaddons. Segen und Fluch im Alten Orient und in Israel*. Fribourg, CH: Fribourg Univ. Press.

———. 2006. Die literarische und historische Bedeutung der Thronfolgeregelung Asarhaddons. In *Die deuteronomistischen Geschichtswerke*, ed. J. Gertz et al., 331–349. Berlin: de Gruyter.

von der Way, Thomas 1992. *Göttergericht und "heiliger Krieg" im Alten Ägypten*, SAGA, vol. 4. Heidelberg: Heidelberger Orientverlag.

Walzer, Michael. 1985. *Exodus and Revolution*. New York, NY: Basic Books.

2 On the Historicity of the Exodus: What Egyptology Today Can Contribute to Assessing the Biblical Account of the Sojourn in Egypt

Manfred Bietak

Abstract

The storyline of the Exodus, of a people fleeing from a humiliating slavery, suggests elements that are historically credible. Normally, it is tales of glory and victory that are preserved in narratives from one generation to the next. The salvation from this servitude and misery created a bond among this people of Israel. This chapter attempts to use knowledge of Egyptian historical geography and new archaeological data from Ancient Egypt to identify some of the layers of the biblical Exodus tradition.

This study reviews the available evidence from Egypt on the settlement of "Proto-Israelites" during the later Ramesside period. Such groups are proposed to have settled in Egypt simultaneously with the Proto-Israelites in Canaan. The collective memory of the Proto-Israelites suffering in Canaan under Egyptian oppression and those suffering in Egypt merged in the genesis of Israel's story of origin from the transformation of oral tradition into written text. The later belief in a stay of the Israelites at Tanis/Zoan was inspired by the transfer of archaeological remains from Pi-Ramesse to Tanis and Bubastis.

Nadav Na'aman recently put forth a new theory regarding the bondage of the Children of Israel. Following Redford (1997: 59–62), he sees a problem in the emergence of a fundamental collective memory such as the Exodus story in an environment of fragmented groups—not clustered in a single location in Egypt, but rather disbanded and dispersed all over the country according to their employment as enforced workmen at different pharaonic projects (Na'aman 2011: 59, 61). In this case, bondage could not have been lived through and pictured as a single collective experience of the sort described in the Bible. He believes that such a communal experience could only

For editing this manuscript I am indebted to: Thomas Schneider; for English editing to Adrian Melman; for reading the manuscript and for valuable comments I would like to thank Friedrich Reiterer (Univ. Salzburg), Orly Goldwasser (Hebrew University Jerusalem), Claus Jurman (OREA), Malcolm H. Wiener, and Irit Ziffer (Eretz Israel Museum Tel Aviv). Special thanks go to Idan Dershowitz (Hebrew University of Jerusalem), who was most helpful with the Bible texts and citations.

M. Bietak (✉)
Institute of Egyptology, University of Vienna, Austria

Dept. of Egypt and the Levant, Institute for Oriental and European Archaeology, Austrian Academy of Sciences, Vienna, Austria
e-mail: manfred.bietak@univie.ac.at

T.E. Levy et al. (eds.), *Israel's Exodus in Transdisciplinary Perspective*, Quantitative Methods in the Humanities and Social Sciences, DOI 10.1007/978-3-319-04768-3_2,

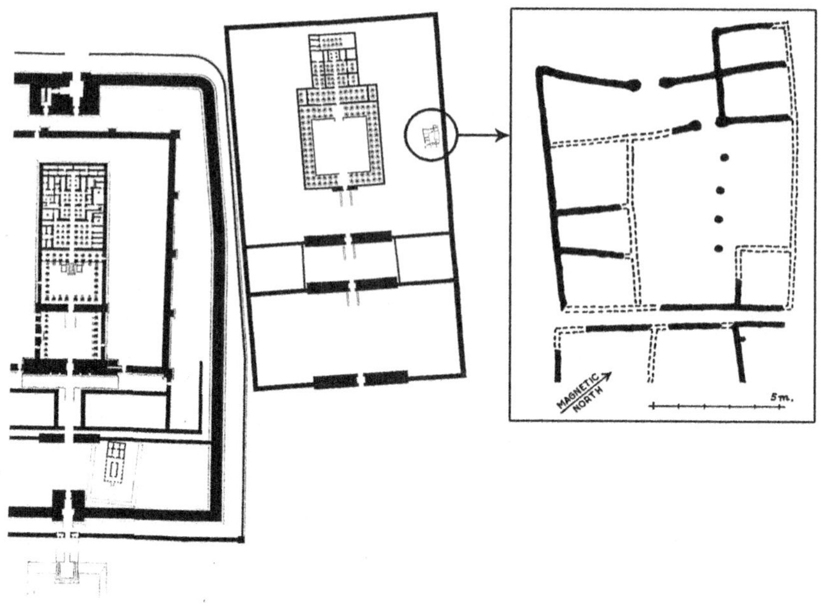

Fig. 2.1 A Four-Room House in Western Thebes (after Hölscher 1939: Fig. 59)

have been shared by people in Canaan, and in greater numbers, during the Egyptian occupation and its oppressive administration, which also involved deportations of groups of people to Egypt. According to Na'aman, the oppression of Egyptian rule in Canaan found its way into the Hebrews' collective memory in an altered form, with Canaan and Egypt interchanged, and it is in this "reversed" route that it made it into the Biblical text.[1] Na'aman also expressed the opinion that the narrative was remodeled according to the realities of the late eighth and seventh centuries in Canaan, integrating the experience with the Assyrian oppression and deportations.

As explained in the following, the experience of severe domination in Canaan may have been integrated and fused into a single collective memory together with a real history of ordeals in Egypt itself.

Archaeological evidence of deportation of a group of people of the same cultural background to Egypt can be found in the discovery of the so-called Four-Room House in Western Thebes (Fig. 2.1) within the precinct of the temple of kings Aya and Horemheb (Bietak 1991, 2000: 179–182; 2003a, b, 2010: 53–54). At the northern edge of the temenos the University of Chicago had discovered vestiges of a house with the prototypical plan (Hölscher 1939: 68–72, Fig. 59). A part of a second one, with inverted room arrangement, seems to have been uncovered southeast of the first one. The construction was not of stone but of wattle and daub. The room arrangement was on three sides around a central courtyard or center room. Even the series of postholes on one side of the central unit is definitely a typical feature of this type of Four-Room House considered by many of our colleagues as an ethnic marker for the presence of Israelites.[2] Its plan leaves no doubt about the origin of this type of house in Canaan and finds no parallel or similarity among Egyptian contemporary houses.

According to the Chicago excavators who were unaware of what they had found, the structure can most likely be dated to the twelfth or the eleventh century BC, the time of Ramses IV

[1] Redford (1992: 257–258) points out, however, that no mention of Egypt's presence or supremacy over Canaan is mentioned in the Bible.

[2] Shiloh 1970, 1973; Fritz 1977; Braemer 1982:192–105; Stager 1985b; Finkelstein 1988: 236–259; Netzer 1992; Holladay 1997; Bunimovitz and Faust 2002, 2003.

(c. 1164–1156 BC)[3] whose mortuary temple was built just north of the precinct of Aya and Horemheb (Hölscher 1939: 115–117, Weinstein 1994: 279–281; Ullmann 2002: 529–530). The Chicago team suggested that the makeshift construction of the Four-Room House was for workmen who had the task of pulling down the temple of Aya and Horemheb (Anthes in Hölscher 1939: 116–7) for quarrying building material for a new temple project of Ramses IV or, at the latest, for works on the Khonsu temple in Karnak during the early 21st Dynasty (eleventh century BC). Given the immediate proximity of the building project of Ramses IV and the fact that a part of the temenos wall of Aya and Horemheb had to be removed for this new construction, the likelihood is high that the demolition of the Aya and Horemheb temple is to be dated to the twelfth century BC.

The time of construction of the putative workmen's quarters fits in perfectly with the chronology of the Four-Room Houses in Canaan where they appear with the Iron Age I in the twelfth century, with a precursor at the western foot of the Judaean mountains at Tel Batash (Mazar 1997: 59–66) by the end of the thirteenth century. At the end of the Late Bronze Age transition to Iron Age I, the Four-Room House also appears with a distinct example in Transjordan at Tall al-'Umayri (Tell el-'Umayri), south of Amman, and is followed by another example dating back to Iron Age I (Herr 1999, 2000, 2001, 2009; Clark 2007; Herr and Clark 2001).[4] Other Iron Age I examples of Four-Room Houses were found at al-Lahun in the same area of the central Jordanian plateau (Swinnen 2009).

The only atypical feature of our model in Thebes seems, *prima facie*, to be the entrance through the back room, opening up locally to the northwest. The first speculation is that this may be a concession to the usual opening up of Egyptian houses towards the north to keep the building cool with the prevailing northerly winds under the very hot climatic conditions in Upper Egypt. Such variations in the position of doors are also observable in other Four-Room Houses in Canaan (Fritz 1980: Fig. 1; Fritz and Kempinski 1976: Fig. 2; 1983/III: plans 3,5,6,9). In Canaan, however, the whole scheme of the house is reversed, with the courtyard facing north. But exactly the same orientation and alignment as our house in Thebes, with the same kind of opening through the broad room north of the house, can be found in one of the oldest prototypes at Tel Batash (stratum VII), at the end of the Late Bronze Age/transition to Iron Age I, going back to a patrician house of the Late Bronze Age (stratum VIII) (Mazar 1997: 59–66).

If we try to think about our workmen's place of origin, a persuasive feasible guess would be that they had been Shosu (Shasu, *Š3św*) bedouins[5] from the desert of Seïr,[6] against whom Ramses III had waged a campaign and taken many prisoners of war just shortly before, according to Papyrus Harris (I.76:9–11) (Erichsen 1933: 93; Giveon 1971: 134–137; 267–71; Grandet 1993: 202–207; Grandet 1994/vol. 1, 337, vol. 2: 243–244, n. 921). Ramses III was in the habit of allocating them as slaves to temples all over the country. As the desert of Seïr is associated with Yahweh in biblical sources (Deut. 33:2; Ju. 5:4–5; c.f. Habakuk 3:3)[7] and possibly in Egyptian texts from the fourteenth, perhaps even as early as the

[3] Chronology in this article after Schneider 2008a, 2010.

[4] The presence of the Four-Room House in Tall al-'Umayri and al-Lahun as the only outliers during Iron Age I mooted the idea that the settlers there were Proto-Israelites from the tribe of Reuben which, according to the Bible (Num. 32, Josh. 13), was allotted land east of the Jordan (Herr 1999, Herr and Clark 2001).

[5] On the Shosu bedouins in general with all relevant documents see Giveon 1971; see also Helck 1968; 1971: 335; Weippert 1974; Görg 1976; 1979; 1997: 45–51.

[6] According to the understanding of the Bible in Transjordan east of the Arabah (Gen. 32:3–4; Gen. 36:8; Deut. 33:2).

[7] Herrmann 1967: 213–4; Axelsson 1987: 57–60; Görg 1989: 183–7; Lemaire 2003: 19–25. For a different opinion: De Moor 1997: 110–207 who ties up Yahweh worship to the migrations of Proto-Israelites to Bashan and from there to the south, to Edom.

fifteenth, century BC onwards,[8] it seems plausible to consider the Seïr region to be the possible origin of Yahweh worship, perhaps even of the core group of Proto-Israelites in Egypt.[9]

Can we identify the builders of our Four-Room House at Western Thebes as Proto-Israelites? The Four-Room House is considered prototypical for early Israel during the whole period of Iron Age I and II (see fn. 4). It appears first in what is considered the early settlements of the Proto-Israelites in the central hill country of Canaan at the beginning of Iron Age I, perhaps by the end of the Late Bronze Age. It was only later taken over in Philistia and also later seems to appear in Transjordan during Iron Age II.[10] The exceptions are the already mentioned examples at Tall al-'Umayri and al-Lahun in the central Jordanian plateau, south of Amman (see above). There is still no evidence available of the early presence of the Four-Room House in Iron Age I in southern Transjordan, but Tall al-'Umayri and al-Lahun have taught us to expect surprises.

Identification of the builders of the Four-Room House in Western Thebes as Proto-Israelites seems perfectly possible as the architectural prototype, and the likeliest origin of its builders points in that direction. But the conclusion is not necessarily cogent as we cannot prove that they came from Seïr. We are also uncertain whether the early exponents of the Yahweh cult are identifiable as Proto-Israelites,[11] and we are also not yet convinced that this architectural type was used only by the Proto-Israelites during Iron Age I. We can say, however, that the builders of this house in Thebes must have been people who were culturally closely related to the Proto-Israelites, even though the ethno-genesis of this group had not been completed by that time (Faust 2006).

The spread of the Iron Age culture associated with early Israel in the central hill country during the twelfth century BC and the example of a Four-Room House in Western Thebes from the same period can be taken as circumstantial evidence that the conditions for a sojourn in Egypt should be envisioned for the twelfth century BC.[12] The fact that this house type appears in Western Thebes shows that such groups of Near Easterners did stay together, even after distribution as slaves over the country.

However, prisoners of war or temple slaves in Western Thebes are not part of the message transmitted in the Books of Genesis and Exodus. The Theban scenario seems to be a kind of random example symptomatic of the ordinary course of Egyptian events. More in keeping with the biblical tradition is the oft-discussed event described in Papyrus Anastasi VI (4.11–6.5) in which a group of Shosu Bedouins from Edom is granted permission by the Egyptian authorities to pass the border stronghold in the region of Tjeku[13] in what is today the Wadi Tumilat and to proceed with

[8] On the identification of *Yhw3* in the topographical lists of the temples of Amenhotep III at Soleb and of Ramses II at Amara with the tetragrammaton there is a long bibliography, see Giveon 1971, 24–28, doc. 6a; Görg 1989; Leclant 1991; Redford 1992: 272–273, n. 71. See for a different chronologically matching Egyptian source: Schneider 2008b. Astour (1979) has pointed out, however, that what is construed as Seïr in the topographical lists is written differently (*S-'-r-r* instead of *S-'-i-r*) and because of other toponyms known from campaigns of Thutmose III, such as Labana *(R-b-na)*, he prefers locating the Shosu lands of the Soleb and Amara lists, including *Yhw3*, in Lebanon/Syria.

[9] According to Biblical tradition early Yahweh worship and the Proto-Israelites do not necessarily tie up with each other. Moses meets Yahweh when among the Midianites (Ex. 2:1–22).

[10] I would like to thank for this information Peter Fischer (Univ. Gothenburg excavation at Tall Abu al-Kharaz in the Jordan valley) and Monique Denise Vincent (Univ. of Chicago).

[11] See fn. 9.

[12] See also for other reasons Rendsburg 1992; Knauf 1994: 103–106. Knauf sees in the figure of the Asiatic Chancellor Bay at the end of the 19th Dynasty a parallel with the biblical Joseph and the xenophobic policy of king Sethnakhte at the beginning of the 20th Dynasty. This is the kind of ambience that makes the best fit for relocating the period of the Exodus into the 20th Dynasty, as stated by Knauf. See also Görg 1997: 63–67; De Moor 1997: 227–240, who sees in Chancellor Bay the figure of Moses and, in Sethnakhte, the Pharaoh of the Exodus.

[13] The hieroglyphic classifiers for Tjeku in this text are the throw stick and the hill-country/foreign land. Tjeku must, at that time, have been neither a town nor a fortress (so Helck 1965: 37–38) which would require other classifiers. The text first depicts the letting through of "a tribe of Shosu from Edom the border fortress of Merenptah *ḥtp-ḥr-M3'.t* which is in Tjeku, as far as the lakes of Per-Atum (Pithom)" with the same description "of Merenptah *ḥtp-ḥr-M3'.t* which is in Tjeku" which

their livestock to the Lakes[14] of Pithom to keep their flocks alive (Gardiner 1937: 76–77; Caminos 1954: 293–295; Giveon 1971: 131–134).[15] This papyrus dates back to the end of the thirteenth century BC, more precisely to the fifth year of Sethos (Seti) II (ca. 1209 BC). These people should not be taken as Edomites, but as Shosu from Edom. As Yahweh is said to have risen in Seïr in Edom (Deut. 33:2; Ju. 5:4; see also Habakuk 3:3)[16] these pastoralists appear at the right time and from the right corner to be considered the gene pool of Midianites or Proto-Israelites.[17] Nevertheless the incident described in this papyrus can only be seen as a symptomatic event, recorded and preserved by accident as a template record for apprentice scribes.

However, one has to wonder in this context why the toponym the "Lakes of Pithom" is referred to by Egyptian scribes with the Semitic loanword *b-r-k-w.t* in syllabic writing (Wörterbuch 1:466; Hoch 1994: 106–7, no. 131) and not with its Egyptian name.[18] This is indeed remarkable. We gain the suspicion that this word had become something of a toponym in a region inhabited long enough by a Semitic-speaking population to supplant the original Egyptian name with an idiomatic expression of their own. In support of this proposition, the point needs to be made that, in the Onomasticon Amenope (Amenemopet), word no. 33 *b-r-k-t* is followed by no. 34 *ḫnty(w)* "borderland," "frontier"—which would be a meaningful association (Gardiner 1947/I: 8*, no. 34). Furthermore another Semitic expression *s-g-r* (Hoch 1994: 270–1, no. 385) is used for an enclosure or a fortification of a fortress in the same region (p. Anastasi V.19, 7). Even Tjeku, the name of the region of Wadi Tumilat, is regarded by many as an Egyptian rendering of the biblical Sukkot.[19]

---

has again the classifier of the throw stick and the foreign land.

[14] During an investigation of the paleogeography of the Eastern Delta (Bietak 1975: 88–90) it was discovered in topographic contour maps that the western part of Wadi Tumilat, just west of Tell el-Retaba was in antiquity the basin of a large natural overflow lake of 18 km length and 1.8 km width. It must have been a stretch of water that was suitable for fishing. Only now does it become understandable why this lake was eponymous and had given its name to this region as "Eastern Harpoon Nome" (for the nomes, see Helck 1974: 172–4) from the time of the Old Kingdom onwards. This lake was fed by a side branch of the easternmost Nile branch. In late Ramesside times it seems, according to information from Papyrus Anastasi VI: 4:11–6:5, that the lake had ramified into several smaller lakes (plural *b-r-k-ȝ*-w.t), most likely owing to a reduced water intake or sedimentation. The contours of the Wadi basin indeed show a separation of the lake into subunits by accumulation of ridges of sediments. The description of those waters is written with a Semitic loanword in syllabic writing: *b-r-k-ȝ*-w.t. The modern translation influenced by Hebrew בְּרֵכוֹת "pools" or "water reservoirs" must be assumed wrong, not only because of the paleoenvironment explained above. In the onomastic Papyrus Golenischeff 1.9–10 the *b-r-k-ȝ*-w.t are associated with and directly follow the expression *nwy* "wave" (Wörterbuch, lemma no. 56570, DZA 22.903.660). What is more, the Demotic Fayum Papyrus (Papyruszettel 7 or 7a) mentions a toponym *b-r-g-t* with the compound classifiers of a channel signifying a stretch of water and the town. It has been pinpointed as a habitat of the crocodile and is most probably the precursor of the Birket Qarun in the Fayum (DZA 22.903.700). A pool is significant neither for making distinct waves nor as a habitat for crocodiles. I would venture to translate this loanword, as in the Arabic "birka," as a lake, and thus a larger body of water, as paleogeographic studies have shown for the western portion of Wadi Tumilat. I am grateful to Orly Goldwasser from the Hebrew University in Jerusalem for helping me with the research into this loanword.

[15] As long as the Exodus was dated to the time of the 19th Dynasty (reigns of Ramses II and Merenptah) the event of Papyrus Anastasi VI was considered an illustration long after the supposed event of how the Children of Israel may have previously entered Egypt. Dating the likely time frame of the sojourn to a later juncture, these Shosu from Edom ought to have moved far more into the focus of biblical scholars.

[16] See fn. 9.

[17] For the time period when Hebrew is considered a derivative of Canaanite language, see the eminent linguist Anson F. Rainey (2007) who concludes a Transjordanian origin of early Hebrew.

[18] There is no evidence that this expression by that time had, as a loanword, entered the general Egyptian lexicon around the period of the cited pap. Anastasi VI (see Hoch 1994: 106–7, no. 131). The only other example of the 19th Dynasty is indeed questionable, the word being incomplete (Kitchen 1979=KRI II: 689,3).

[19] Identification of Tjeku with Sukkot has been generally approved, but has also met with strong reservations from A.H. Gardiner (1922: 213, n.2) because Tjeku is not written in syllabic script.

All of this could be viewed as some evidence that this borderland had been settled by a Semitic-speaking population. Earlier inhabitants of this region, supposed to have spoken a Canaanite dialect, were exponents of the Syro-Palestinian Middle Bronze Age culture during the Hyksos Period whose sites were surveyed by the Canadian Wadi Tumilat expedition (Holladay 1982; Redmount 2000).[20] However, texts with Semitic toponyms do not appear until the Ramesside Period and may have derived from later settlers.[21] One wonders if under such evidence Wadi Tumilat could not be deemed, with all due reservation, a paradigm for the biblical scenario of the land of Goshen (Gen. 45:10; 46:28–29, 34; 47:1, 4, 6, 27; 50:8; Exod. 8:18; 9:26). Indeed, Sarah I. Groll cautiously postulated the identification of a possible Egyptian rendering of the biblical toponym Goshen in Papyrus Anastasi IV (1b:1–2) referring to a stormy lake by the name of *gsm* in association with waves[22] (Groll 1998: 190).[23] This name *gsm* seems to have been used as a toponym and is considered to have been most probably a Semitic loanword (Hoch 1994: 354, no. 522). It is in this connection that it is most important to note that, in the Septuagint translation of the Pentateuch into Greek, this name seems to be retained as Γεσεμ (Gesem).[24] The lake must have been a wide stretch of water. This would fit in perfectly with the western part of Wadi Tumilat with its large overflow lake (Fig. 2.2). The Septuagint of Genesis 46:28–29 changes Goshen to Heroonpolis and appears to locate the land of Goshen at Wadi Tumilat. Heroonpolis is most likely identifiable with Tell el-Maskhuta (Redford 1982; Kettenhofen 1989).[25]

More tangible is the well-known reference to Israel as a people (not as toponym) on a stela of the fifth year of Merenptah (ca. 1219 BC) shortly before the twelfth century (Fecht 1983; Hornung

[20] The loanwords in this region are unlikely to originate from the Hyksos Period. We have no evidence from the "Lakes of Pithom" in this region before the late 19th Dynasty. Besides Tjeku (Gauthier 1925–1931, VI: 83) which may be an old name, but preserved only from Thutmose IV onwards, it seems that at least in this region the Semitic names are a relatively recent 19th Dynasty acquisition.

[21] There are also Semitic toponyms in the northern borderland such as *Ṯ3rw*, the frontier fortress, written in syllabic writing instead of the Middle Kingdom "Way of Horus" fortress. On the fortress name with the Semitic designation Migdol (*mktr*) see Gardiner 1919, 1947: 24; Seguin 2007: 77–95.

[22] We have already encountered waves in connection with the *b-r-k-3-w.t*—possibly a synonym of the *gsm* lake—in the Golenisheff Glossary (see fn. 14).

[23] Redford (1987: 140) tries to explain the toponym Goshen as originating from *Gasmu*, the name of the Arab dynasty of the Kingdom of Qedar in northwestern Arabia. They controlled the eastern accesses to Egypt from the seventh to the sixth centuries BC. This is bound to be completely anachronistic. So Redford sees this as evidence of a late composition of the Pentateuch. The name Gesem as a King of the Arabs also appears in Judea, namely, in Nehemiah 2:19; 6:1–2; 6:6. I cannot see, however, the point of applying such purely phonetic similarities if there are other explanations that are in keeping with the pool of Egyptian topographical names.

[24] I am indebted to Idan Dershowitz in making this link.

[25] Doubt has been cast on this identification by Bleiberg (1983). He argues that the width of the Isthmus of Suez is situated by Strabo XVII: 1.21 and 26 and other sources between Pelusium and the "recess of the Arabian Gulf near Heroopolis," which implies that Heroo(n)polis would be situated at the northern edge of the Gulf of Suez. This seems to be a mistake as the *Itinerarium Antonini Augusti* lists as stops on the way (a day's march away from each other) of the road along the Red Sea canal *Heliopolis–Hero–Serapiu–Clysma* (Suez), which would put Hero(polis) far away from the Gulf of Suez into the reaches of Wadi Tumilat. The itinerary of the pilgrimage of Egeria (Aetheria) in the late fourth century is rather confusing to use as she does a mix of contemporary and biblical toponyms, but shows again that Heroonpolis is situated far away from the Gulf of Suez. Heroonpolis, called by her Hero, is listed on her way back from the Sinai to Egypt in the following way: after *Clysma* (Suez) follows *Migdol–Baal Zephon–Etham–Succot–Pithom–Hero–Rameses–Arabia* (Gardiner 1918: 263, Vretska 1958; Wilkinson 1999). As Heroonpolis was still a lively and well-known settlement at that time and is described by Egeria as being in the territory of Egypt after they had left the desert and the territory of the Sarazenes, its position in the Wadi Tumilat should be clear, even if some of the stations seem to have been mixed up in her diary. The explanation of the confusion about the position of Heroonpolis is that it was the capital of the nome Heroopolites which extended along the Red Sea canal as far as the Arabian Gulf (Gulf of Suez). The only larger site known from late Saïte to Christian Period in the Wadi is Tell el-Maskhuta.

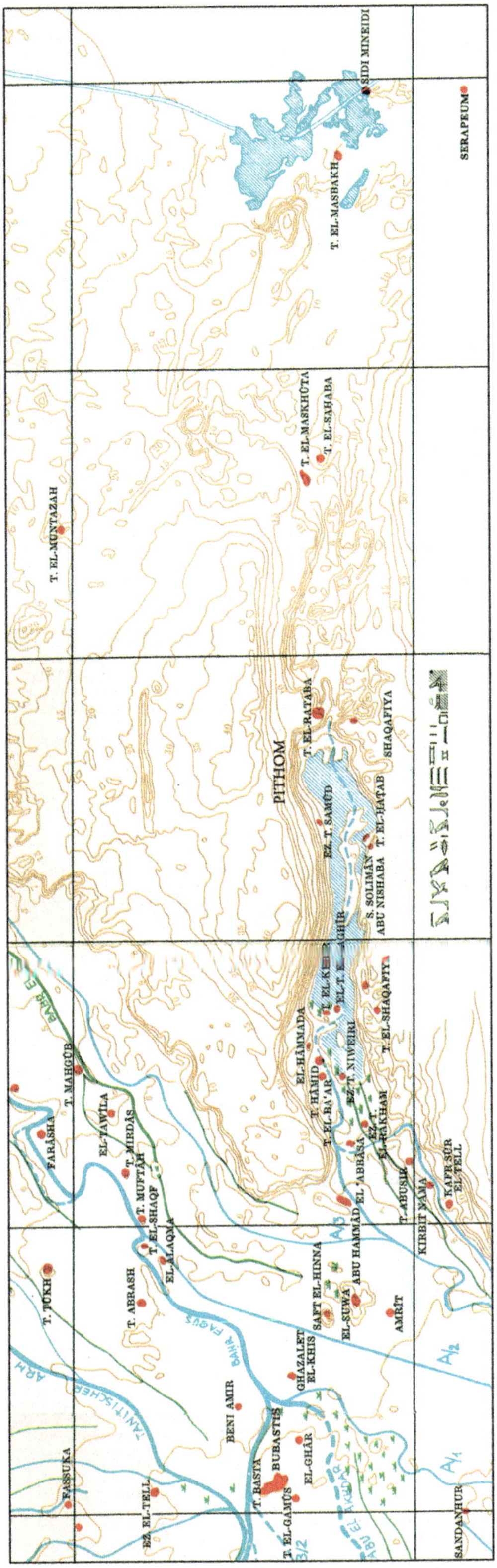

Fig. 2.2 Paleogeography of Wadi Tumilat (after Bietak 1975: plan 4)

1983; Von der Way 1992; Kitchen 2003: 33–37; 2004).[26] The reference to Israel is modeled on the toponyms of Canaan,[27] Ashkelon, Gezer, and Yeno'am,[28] the latter located near the Lake of Galilee or in the Jordan Valley (Ahituv 1984: 198–200; Na'aman 1977; Görg 1997: 45, n. 20; Wimmer 2002). It has therefore been very reasonably posited that Israel, which had not been written with the classifier[29] of a settlement as had the toponyms on this stela but with the classifier of "people" and therefore not yet tied to settlements, was at that time likely to have still been beyond Jordan (Engel 1979a; Rainey 2001). While Israel Finkelstein (1988, 1991) suggests that the settlers making an appearance in the central hill country in Iron Age I had already been there during the Late Bronze Age as pastoralists,[30] other researchers opt traditionally for the region of Transjordan as the origins of this Iron Age culture and where a transition from a strong representation of the Late Bronze Age to the Iron Age culture seems to be more obvious than in Cisjordan (van der Steen 1996). Repeated patterns of immigration of Bedouins from the desert to Cisjordan changing from nomadism to sedentary life and agriculture have also been observable until recent times (van der Steen 1995, 1999).

As regards the claim by several colleagues that the biblical narrative of the Exodus is a late saga, composed in the Saïtic or even the Persian Period and with some memories echoed of the kingship of Tanis, it would be interesting to investigate how far the topography, as reflected in Genesis and Exodus, would fit in with the geographical conditions of the Eastern Delta during this period or if older memories which reflect topographic conditions of the Ramesside Period are detectable. Donald Redford concluded that the latter is not the case and that the geographic information supplied by Genesis and Exodus would relocate the scene to the Eastern Delta of the Saïte and Persian Periods (Redford 1963: 408–418; 1970: 203–6; also van Seters 1992: 332; 2001: 260; Thompson 1987: 192–4).[31] I cannot agree with him on this.

Major changes occurred to the water system of the Eastern Delta in the late 20th Dynasty. Along the easternmost Nile branch, the classical Bubastic branch, the "Waters of Re'" of the Ancient Egyptians, there is not a single site north of Bubastis that is datable to the Third Intermediate Period. There is a long hiatus between ca. 1000 and 600 BC. Occupation did not start until the late Saïte Period again (Bietak 1975: 77–87, 107–9).[32] It is patent that late in the 20th Dynasty the Pelusiac branch had silted up and been blocked in its lower reaches (Bietak 1975: 215–216). This culminated in a relocation of the residence from Pi-Ramesse to Tanis. From the 21st Dynasty onwards (after ca. 1086 BC) Tanis was the new residence and harbor town. Pi-Ramesse was abandoned and served as a quarry, particularly during the 22nd Dynasty (ca. 962–736 BC) when not only Tanis but also Bubastis and other towns of the Libyan Dynasty had their monumental buildings fitted with spoliae (reused blocks) from Pi-Ramesse (Habachi 1954: 559; 2001: 84–95, 115–116; van Seters 1966: 126–137; Uphill 1969: 28–29; Bietak 1975: 31–32, 213–220). These are likely to have been shipped during the time of the flood along the remnants of the easternmost Nile branch and along a canal leading

[26] For appraisals of the Israel Stela and of war reliefs attributed to the Merenptah campaign see Engel 1979a; Stager 1985; Yurco 1986; Hasel 1994; Rainey 2001; Gilmour and Kitchen 2012: 8–11.

[27] Canaan seems not to mean the town Canaan near Gaza but the whole Egyptian-held province, south of Upe.

[28] This toponym is also mentioned in the Beth Shean Stela (no. 12) of year 1 of Seti I in connection with a punitive campaign of the king mentioning the toponyms Beth Shean, Rehov, Hamma, and Peher and should be sought in the Jordan Valley, west, south, or east of the Lake of Galilee. See Wimmer 2002: 6 n. 19.

[29] For the role of classifiers in the Egyptian language, see Goldwasser 2002.

[30] In his support, it should be mentioned that the Four-Room House seems to appear at Tel Batash at the same time as in Transjordan, if not even earlier (Mazar 1997: 59–66).

[31] For criticism of the argument of a late redaction of Genesis and Exodus, see Kitchen 1973 and Hoffmeier 1996: 77–106 (especially 97–8).

[32] The stratigraphy of Tell el-Dab'a clearly shows that the site had been abandoned early in the Third Intermediate Period and had stayed unoccupied until the late Saïte Period. See the recent excavations of M. Lehmann at Tell el-Dab'a (Lehmann 2011, 2012/3).

from Qantir to Tanis (Bietak 1975: 81, 214). This of course had various repercussions on the importance of towns that were being commemorated; in this case, Tanis and Bubastis were the new major centers during the Third Intermediate Period.

The shipment of the spoliae was also attended by cult images and temple inscriptions of Pi-Ramesse being taken to Bubastis and Tanis during the 22nd Dynasty. After a lengthy interval[33] at both sites in the 30th Dynasty and Ptolemaic times, secondary cults developed (Yoyotte 1971–1972; Bietak 1975: 212, 219–220). This is inferred from inscriptions on statues of the third century BC at Tanis mentioning priests of Amun of Ramses of Pi-Ramesse[34] and in Bubastis from epigraphic evidence of the fourth century BC of Pre‘ of the Waters of Re‘ of Ramses and of Ptah-Tatenen of Ramses.[35] This means that in early Ptolemaic times Tanis was most likely to have been perceived as the site of Pi-Ramesse, which had perished over 700 years previously. At Bubastis, people thought during the 30th Dynasty that the ruins of this famous former capital had been situated near that place, probably further east at Pi-Soptu (Per-Sopdu) or Tell el-Kebir at the western end of Wadi Tumilat.

As I postulated in 1975 in *Tell el-Dab‘a II*,[36] Jewish exiles in Egypt trying to reconstruct a memory of their forefathers followed contemporary Egyptian opinions in identifying Zoan (Gardiner 1918: 264) with Pi-Ramesse and the Fields of Zoan[37] (Gardiner 1918: 200, 246–8, 264) with the Sea of Reeds[38] (Psalm 78:12, 43) (Bietak 1975: 217–221; 2010: 186). The Septuagint clearly identified the land of Goshen and the route of the Exodus with the Wadi Tumilat and the predecessor of Lake Timsah, called by the Egyptians *km-wr*, the "Great Black" (Gauthier 1925–1931/V: 202), as the Sea of Reeds. Another interpretation of the Exodus route in the Septuagint (Exod. 1:11)—originally most likely to be a gloss that found its way into the main text—seems to lead from Heliopolis, thought to be Pithom (Pi-Atum), to the Gulf of Suez and the Red Sea (Gardiner 1918: 265, n.2).

It is possible that Pelusium was also on the receiving end of blocks quarried at Pi-Ramesse and had its secondary cults of the gods of Ramses as suggested by the fact this site had been identified by the editors of versions of Targum Pseudo-Jonathan as the town of Ramses (Diez Macho 1968: 314, 315, 481, 628; 1970: 4, 5, 76, 77, 407, 440). All of this goes to show that later editors clearly had Pi-Ramesse in mind when referring to the biblical town of Raamses. Donald B. Redford argues—using reasoning predicated on vocalization—that the missing Pi- or Per- in the biblical text suggests that it is not Pi-Ramesse that was being referenced, but one of the many other toponyms constructed with the name of Ramses (Redford 1963: 409–413; for rejection on other grounds, see Lemche 1994).[39]

[33] The long silence on Pi-Ramesse can be explained by the new political landscape in Egypt which also brought about a change in religious politics. Ramses II deliberately created at Pi-Ramesse new ideologies by moving away from Thebes and by creating his own religious concept binding the gods strictly to himself by the epithet God X of Ramses (of Pi-Ramesse). The main temple of Pi-Ramesse had not been dedicated to Amun alone, but to Amun-Re‘-Harakhti-Atum. After Ramses II, the politico-religious hierarchy returned to the Theban concept of theology and dispensed with Ramses II's.

[34] Gardiner 1918: 199–200, 250–1; Borchardt 1930: 32–34, 41–43; Porter and Moss 1934: 26; more recently Zivie-Coche 2004: 79–182.

[35] Naville 1991: pl. 48 B.

[36] Bietak 1975: 220; 1981: 279; 1987: 164–6; supported by Weimar 1985: 265–8; picked up on by Wente 1992: 617, see Hoffmeier 1996: 118.

[37] Corresponds to Egyptian *sḫ.t-Ḏ‘*, which became later *sḫ.t-Ḏ‘n.t*.

[38] It is also possible that the "Field of Zoan" is a synecdochic reference to Egypt per se, but the first reference in Psalm 78 clearly addresses the miracle of the "crossing of the sea" and so has to be seen in association with a location of the Ramses town at Tanis. We meet the same association again in early Ptolemaic times with two statues of the namesake Teos at Tanis, both of whom were, besides, bearers of other titles such as "priests of Amun of Ramses of Pi-Ramesse" (see fn. 34)—of the famous residence of the Ramessides which had perished over 700 years previously. There is no evidence that the location of the site had ever been known at that time.

[39] Van Seters (2001: 266) suggests that the name Raamses/Ramesse in the Bible originates from one of the many cult places of the gods of Ramses of Pi-Ramesse which he claims flourished in the 1st millennium BC in the Delta. However, we have no evidence of such cult places during the 1st millennium, except secondary cults of the gods of Ramses (of Pi-Ramesse) at Bubastis and Tanis during the

However, as the most eminent Egyptian philologist Alan H. Gardiner has shown (Gardiner 1918: 137–8, 180, 265; also Helck 1965: 41–43; Groll 1998: 189–190), the Pi- or the Per- can be omitted if followed by the expression of *pꜣ-dmi* "the town" or preceded by *nꜣy* "those of" (Gardiner 1918: 188) in which event the "Ramesse" left would be practically identical with the biblical toponym of Raamses/Rameses.[40] Several biblical passages putting the living area of the children of Israel near the palace and center of administration show that the famous residence Pi-Ramesse was meant.[41] In Genesis 45:10, Joseph grants his family the land of Goshen: "You shall settle in the land of Goshen, and you shall be near me." Similarly, in Exodus 2:1–11, the birth legend of Moses (Rendsburg 2006: 201–19) puts the Hebrews' location near the palace, as Moses' mother placed him close to where Pharaoh's daughter would be bathing (Gardiner 1918: 265). The fact that Moses and Aaron were quickly summoned by Pharaoh and were able in no time at all to appear at court the same evening after another plague had blighted Egypt (Exod. 8:20) shows that the royal residence in this biblical narrative was envisioned as lying close to where Moses and Aaron were living.

This biblical tradition appears to identify the Hebrews' place of residence with Pi-Ramesse situated at Qantir and Tell el-Dab'a.[42]

Pithom was a toponym known from the Ramesside Period onwards.[43] It ties up with Per-Atum (the house of the primeval sun god Atum) and is most likely identifiable with the fortress and town at Tell el-Retaba in the Wadi Tumilat, including impressive Ramesside remains (Petrie 1906: 31–34; Rzepka et al. 2009; Rzepka et al. 2011). The location of Pithom at Wadi Tumilat is, apart from pap. Anastasi VI, inferable from Herodotus II:158 (there as Patumos). Its identification with Tell el-Maskhuta (Redford 1982; Holladay 1982, 1999: 201) can be at best considered a relocation of the temple of Pi-Atum from Tell el-Retaba to Tell el-Maskhuta. From the Saïte Period onwards the former site was abandoned and the latter site was settled and assumed an important status in conjunction with the construction of the Red Sea canal under Necho and later during the Persian Period under Darius. Tell el-Maskhuta is favored only by the Pithom Stela from the age of Ptolemy II (Porter and Moss 1934: 54; Thiers 2007), and even this monument according to Gardiner is not cogently identified with Tell el-Maskhuta (Gardiner 1918: 268).

A debatable point is the statue of a priest from Tell el-Maskhuta mentioning a "temple of Per-Atum, the great, in Tjeku" (Naville 1903: pl. 5). It is dated to the 26th Dynasty (Borchardt 1911–1930/II: 113) and could fit narrowly into the period around the beginning of the late Saïte foundation at Tell el-Maskhuta. Be that as it may, a block of Ramses II mentioning Tjeku could have been ferried to Tell el-Maskhuta from its original site later identifiable as Pithom. As Tell el-Maskhuta had been occupied not before late Saïte and Persian times,[44] the only place at Wadi Tumilat we know had experienced an occupation from the Ramesside and the Third Intermediate Periods, and had been endowed with a temple, is Tell el-Retaba. According to a relief and inscription from this site depicting Ramses II sacrificing enemies in front of god Atum, "Lord of Tj(k)w" (and on the opposite side to the god Seth), it seems certain that the temple was dedicated to the god Atum (Petrie 1906: pl. 29–30). Such reliefs are normally carved into the face of the temple pylon and show the dedication of the shrine. This would mean that this temple at Tell el-Retaba would have been a *Pr-Itm* (Per-Atum) and would coincide with Pithom during the period in question. A group statue of Ramses II with Atum was also found at the spot (Petrie 1906: pl. 32).

fourth and third centuries BC and which were not associated with toponyms apart from "of Pi-Ramesse" at Tanis. So this argument makes no sense.

[40] See also the transition of the toponym Pi-Saptu (house/domain of Soped/Sopdu) to modern Saft el-Henneh.

[41] I am indebted for the biblical references to Idan Dershowitz, Hebrew University of Jerusalem.

[42] Habachi 1954, 2001; van Seters 1966; Bietak 1975.

[43] Naville 1903; Petrie 1906; Gardiner 1918: 268–70; Uphill 1968; 1969; Cazelles and Leclant 1972; Redford 1963, 1982; Bleiberg 1983; Hoffmeier 1996: 119–121.

[44] Except for a Middle Bronze Age occupation.

The Priestly source pinpoints the Exodus' starting point as Raamses (Exod. 12:37, Num. 33:3, 5). In this case, the only way of reaching the highway to Canaan, the "Way of Horus" or the "Way of the Land of the Philistines," would have been by moving north-north-eastwards between the easternmost Nile branch and the Bahr el Baqar drainage system. The latter is the largest natural drainage system in the Delta, which poured into expansive overflow lakes with year-to-year, seasonal swamps in antiquity (Figs. 2.3 and 2.4). This route would have led to the Isthmus of Qantara which, when crossed, would lead out into a desert track that had been well guarded during the Ramesside Period by fortresses (see the recent reconstruction of the physical landscape by Hoffmeier 2006; Hoffmeier and Moshier 2006; Moshier and Hoffmeier, Chap. 8). The only way of avoiding border control was to cross the Ballah Lakes, south of the isthmus.

The Ballah Lakes could, during the active period of the Pelusiac, have formed a single and wider stretch of water than during the period just before the construction of the ancient Suez canal. They are the likeliest contender for the Yam Suph יַם־סוּף (Bietak 1975: 135–9, 217–8, n. 598, 619), the Sea of Reeds of the Bible,[45] also mentioned in the alternative as the "Sea" יָם (Psalm 78:13, 53; 114:3, 5). This Yam Suph is in all likelihood identifiable with the Egyptian toponym *p3-ṯwf*, "the Papyrus thicket," known from the Ramesside Period in Papyrus Sallier (I:4,9), pap. Anastasi III (2:11–12), pap. Anastasi IV (15:6) (written without the article *p3*), and pap. Anastasi VIII (3,4) and from the Onomasticon Amenope at the end of the 20th Dynasty (Gardiner 1947/II: no. 418).[46] It is therefore another Ramesside toponym commemorated in the Exodus narrative. In pap. Anastasi III (2:11–12), *p3-ṯwf* is again mentioned in an eulogy on Pi-Ramesse together with the Shi-Hor—a toponym shared by Egypt and the Bible—these two stretches of water as the purveyors of papyrus and rushes for the Ramses town.

In keeping with the narrative of the Bible, and to avoid the Way of the Land of the Philistines, i.e., the Horus Road, fugitives were compelled to turn back and move on towards the Sea of Reeds. This could be achieved after bypassing the overflow lake of the Bahr el-Baqar system (Bietak 1975: 90–91). It is only this reconstruction that fits the scheme. According to the Exodus compilers' ideas, the Sea of Reeds was located east of Raamses. Exodus 10:19 describes, in conjunction with the end of the 8th plague, that strong westerly winds carried the locusts away into the Sea of Reeds.[47]

Moving along the Exodus itinerary from Raamses to Sukkot, we are up against a problem

---

[45] In the bible Yam Suph may signify an expanse of water around the eastern border region of Egypt (Exod. 10:19, 13:18; 15:4), but it may also designate the Red Sea in the Gulf of Suez (Exod. 10:19; 13:18; Num. 33:10–11) or in the Gulf of Aqaba (cf. Exod. 23:31; Num. 14:25; 21:4; Deut.1:40; 2:1).

[46] On the waters called *(p3)-ṯwf* and its connection to Yam Suph there is an abundant literature: See particularly Wörterbuch V:359; Gardiner 1918: 186, 198–9; Gardiner 1947/II: 201–2, no. 418; Gauthier 1925–1931/VI: 72; Lambdin 1953; Ward 1974; Vervenne 1995; Hoffmeier 1996: 210–5; Dayan 1998; Groll 1998: 190. Vervenne has accumulated a near-complete reference list of opinions. A case can be made for claiming that this term must have originally been a distinct concept as its definite article defines a specific place (for application of definite articles in Late Egyptian in connections with personal names (PNs), see Goldwasser 2010). The *p3-ṯwf* is also written with a town classifier, which shows that it had been a closely defined toponym (Gauthier 1925–31/VI: 72). In the Onomasticon Amenope (Golenischeff Glossary) *p3-ṯwf* is listed among toponyms between the "Fields of Dja'" and Tjaru—which means that it is at the eastern edge of the Delta and is toponymic in nature. Dayan (1998) points out that the Ptolemaic Papyrus Spiegelberg mentions *p3-ṯwf* twice in connection with Asiatics from *p3-ṯwf*, which should be taken as an indication of a borderland between Egypt and Asia. It is considered as an Egyptian loanword in Hebrew (Lambdin 1953). Ward concludes that the word has old Semitic roots (Ward 1974).

[47] This scenario would fit in with Pi-Ramesse at Qantir (Tell el-Dab'a) and the Ballah Lakes as the Sea of Reeds. It would also be consistent with Tanis and Lake Manzala (Menzaleh) and also fits in with the approach of locating Raamses near the Western end of Wadi Tumilat and Lake Timsah. However, it would be a harder proposition to identify the Yam Suph with the Red Sea.

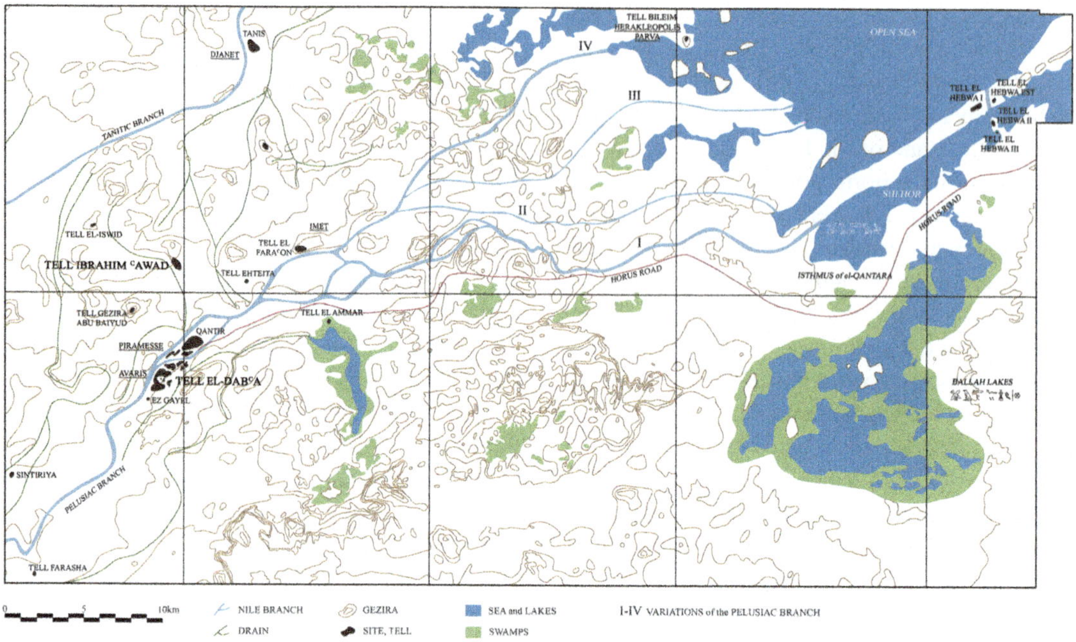

Fig. 2.3 The historical landscape of the Eastern Delta (revised from Bietak 1975: 10)

if the latter was identifiable with the Egyptian toponym of Tjeku[48] and which must be associated with the district of Wadi Tumilat (Gauthier 1925–1931/VI: 83). If this identification is correct, then the Exodus track from Raamses to Sukkot is not a feasible proposition because there would have been no way of avoiding crossing the sprawling swamps of Bahr el-Baqar. Furthermore, Wadi Tumilat is not on the way from Pi-Ramesse to the Sinai.

There are two possible explanations:

1. Sukkot which, in Hebrew, means temporary shelters, is not identifiable with Tjeku.[49]
2. Tjeku is to be associated with Sukkot and originates from an old Semitic toponym which had already merged previously into the Egyptian vocabulary, explaining the fact that it is not rendered in syllabic writing. This might perchance imply a conflation of two traditions. One is associated with the town Raamses, and the other is an interpretation of an Exodus route that found its way into the text, starting from the second store city of Pithom, namely, in Sukkot, moving to the East and then crossing the precursor of Lake Timsah as the Sea of Reeds. In this case the deflection from the "Way to the Land of the Philistines" דֶּרֶךְ אֶרֶץ פְּלִשְׁתִּים is a less reasonable idea, but it seems possible that two different versions of routes may have been conflated in the Exodus writers' minds to wit one starting point from Raamses and the other starting point from Sukkot. Such ambivalence is also reflected in the two different traditions over the land allotted to the Children of Israel: Raamses[50] (Gen. 47:11, Exod. 12:37, Num. 33:3,5) and Goshen (Gen. 45:10; 46:28, 29, 34; 47:6).

In conclusion, the toponyms of Raamses, Pithom, and Yam Suph correspond to the Ramesside toponyms of Pi-Ramesse, Pi-Atum, and (Pa-)Tjuf, respectively. This correlation is not only in name alone; to some extent, fragments of their geographical information were apparently

[48] See fn. 19.

[49] See fn. 19.

[50] In the Septuagint, also in Gen. 47:11, Ραμεσση (Ramesse) appears instead of Goshen.

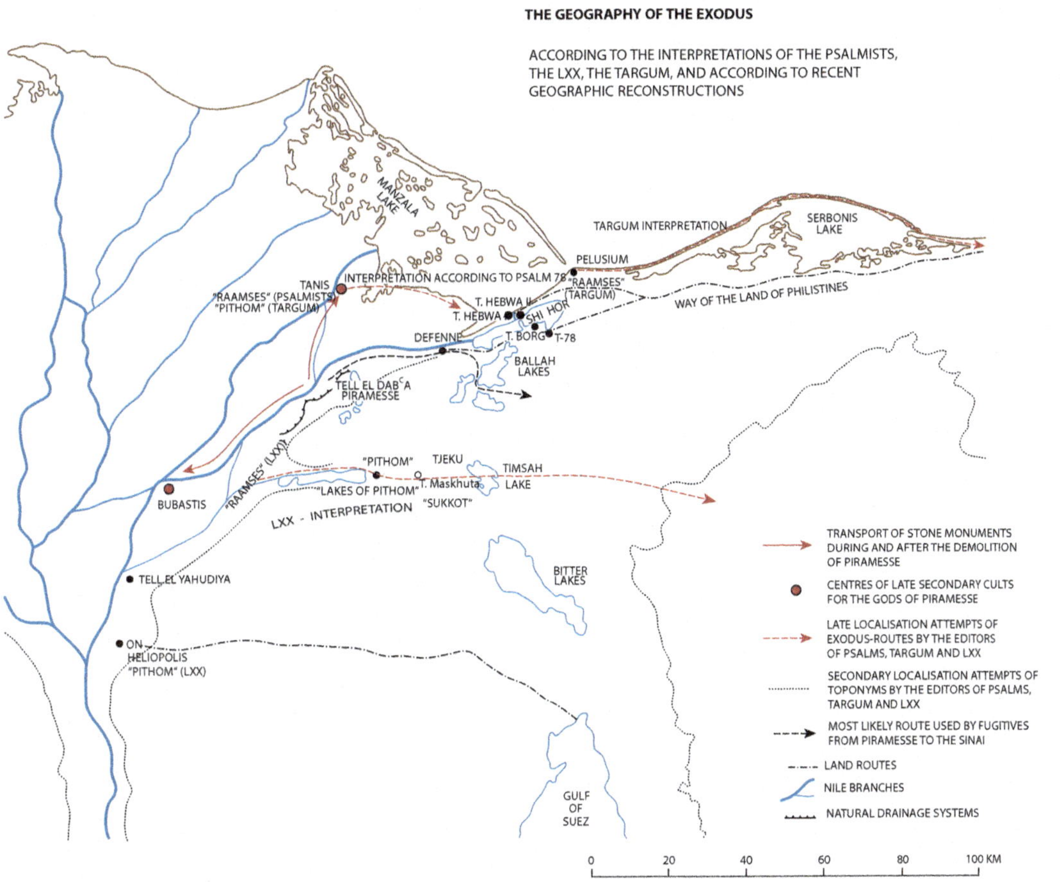

Fig. 2.4 Geographic considerations on the first part of the Exodus Route according to the interpretations of Psalmists, LXX, Targum, and according to recent geographic feasibility studies (revised from Bietak 1975: Fig. 45)

transmitted to the compilers.[51] Ramesside documents also share other important toponyms with the Bible, such as Shi-Hor as an important stretch of water at the eastern edge of the Delta near (Pa-)Tjuf (Gardiner 1947/II: 201–2, no. 418).[52] The region of Seïr in Transjordan is also mentioned in Ramesside and in biblical contexts (Giveon 1971: doc. 16a, 25, 38).

Most of the abovementioned toponyms also appear after the Ramesside Period and can be used to explain a much later composition. Groll (1998: 189) has, however, pointed out that it is the combination of the toponyms Pi-Ramesse, Pi-Atum, Tjeku, and Pa-Tjuf that occurs in Ramesside texts alone and not later. And it is important to stress that it is this very medley of toponyms that also appears in the Pentateuch. Moreover, Pi-Ramesse[53] is absent from texts after the 20th Dynasty and resurfaces only after a lengthy absence, not until the third century BC

[51] If Tjeku corresponds to Sukkot that would be another biblical toponym that appears in Ramesside era texts.

[52] I will leave out from this article any discussion about the identification of the biblical toponyms of Etham, Ba'al Zephon, and Pi-ha-Hiroth as the opinions are too controversial and the basis for an answer is too flimsy.

[53] Redford (1963: 409; 1992: 314–5) claims that Pi-Ramesse is also mentioned in post-Ramesside texts such as in the Onomasticon Amenope of the 21st Dynasty. This was corrected by Helck (1965: 47) who points out that the Onomasticon Amenope imitates a model of the 19th Dynasty. Gardiner (1947/I: 24) dates it to the late 20th Dynasty. The rest of the references go back no earlier than the fourth and third centuries BC (cf. fn. 39, above).

when it appears on statues of dignitaries from Tanis who were priests of secondary cults of gods of Ramses from Pi-Ramesse.[54] Those secondary cults were brought about by the abovementioned transporting to Tanis of stone inscriptions and statuary from Pi-Ramesse at Qantir. But those inscriptions do not seem to have attracted interest until much later when, during the 30th Dynasty, Egypt's glorious past became a part of the Late Period ideology when there was an attempt to return to sources from their past. This shows that the presence of the toponym Raamses in the Books of Genesis and Exodus must have been adopted from a tradition older than the Third Intermediate and Saïte Periods. The said changes in the physical and political landscape of the Eastern Delta, including new major centers and toponyms, were incorporated only later into the Bible (e.g., Psalm 78:12, 43), while the start of the itinerary in Exodus (13:17–18; 14:2) reflects the topographical conditions of the Ramesside Period.

That is why the likeliest alternative is to date the memory connected with a possible sojourn of Israel in Egypt to the late Ramesside Period, i.e., the 20th Dynasty.[55] The sojourn of a group of Proto-Israelites in Egypt may have happened at the same time or even later than the beginning of the settlement in Canaan. Such a late date would enable the toponym of the "Way of the Land of Philistines" in the Exodus tale (Exod. 13:17) to be explained as not an anachronism, since by that time the Philistines and other Sea Peoples had been able to seize a fair portion of coastal Canaan in the fifth year of Ramesses III (ca. 1190 BC) and soon afterwards to form Philistia (Stadelmann 1968; Bietak 1994; Stager 1998). So the transfer of the name "Way of Horus" to the "Way of the Land of the Philistines" from the mid-20th Dynasty would be consonant with the historical events.

To conclude:

1. We have evidence for the presence of population groups of the southern Levant in Egypt in the late Ramesside Period that were culturally and ethnically close to what represents early Israel in the Iron Age. We have to bear in mind that, in all likelihood, the ethnogenesis of Israel was concurrent and had not yet by that time been finalized. According to Egyptian texts, a number of Near Easterners came into the country as prisoners of war. Others immigrated peacefully as cattle and livestock keepers.
2. Scanty archaeological remains such as one or two Four-Room Houses in Western Thebes during the 20th Dynasty (twelfth century BC) show that people of the same or very similar Iron Age culture and who started to spread at the same time in the Judaean hill country and in the Negev were in Egypt during the 20th Dynasty. Most probably the latter were temple slaves, originating from a raid of Ramses III in Transjordan.
3. There are areas at the eastern borders of Egypt that had Semitic toponyms used even by Egyptian scribes. This situation can only be reasonably explained by assuming that Semitic-speaking people inhabited such areas during the New Kingdom. One has to mention particularly in this respect Wadi Tumilat, which would fulfill in every respect the model of the land of Goshen (or the land of Ramses) in the Bible.
4. If we assume a sojourn of early Israelites in Egypt, the likeliest period is the late New Kingdom (late thirteenth and twelfth century BC). This chapter supports the opinion that, in the late thirteenth century, people identified in the Merenptah stela as Israel were still in Transjordan and had not yet settled in Canaan. It is however likely that some groups of their kinsmen had already entered Canaan. It is also likely that the settlement in Canaan and at the boundary areas of Egypt happened simultaneously and not consecutively.
5. At least some ideas about the topographical conditions of the Eastern Delta reflected in Genesis and the Exodus story go back to the Ramesside Period.

[54] Cf. fn. 34.

[55] See fn. 12.

6. The town Raamses in the Bible is likely to represent a memory of Pi-Ramesse, the Eastern Delta capital of the 19th and 20th Dynasty identified with Tell el-Dab'a–Qantir.
7. Pithom in the Bible can be identified with the only known settlement of the Ramesside Period in the Wadi Tumilat, at Tell el-Retaba, not with Tell el-Maskhuta which according to the archaeological record did not yet exist.
8. The collective memory of the Egyptian oppression endured by the segment of early Israel in Canaan seems to have merged with the experience of suffering by Proto-Israelites in Egypt. The excessive brick making fits in only with large Egyptian construction sites tied to the Egyptian toponyms Pithom and Raamses. In Canaan, the bulk of Egyptian constructions made of brick was concentrated on relatively small fortresses while in Canaanite towns stone as building material was indispensable but not mentioned in the relevant biblical texts. The late arrival of the Iron Age culture identified with early Israel at the end of the thirteenth and during the twelfth centuries BC slipped inexorably into full political decline in the southern Levant.
9. The quarrying and dissemination of inscribed stone blocks and of statues from Pi-Ramesse to Tanis, Bubastis, and other places spawned secondary cults of the gods of Ramses in the fourth century in Bubastis and of the gods of Ramses of Pi-Ramesse in the third century BC at Tanis. This development goes hand in hand with the thought that Pi-Ramesse—whose location was forgotten in between—was at Tanis and/or near Bubastis in the time of the 30th Dynasty. Such opinions were advocated by the diaspora in exile and developed into the northern and southern theories of the Exodus routes at that time.

Excursus on the Hyksos

Flavius Josephus (*Contra Apionem* I:26–31) identified the early Israelites and the Exodus with the Hyksos and their expulsion. Such an identification—which we have to reject on chronological grounds—may stem from the memories of a Western Semitic population living in the eastern Delta for quite a length of time, from the late 12th Dynasty (ca. 1830 BC) until the Ramesside Period.[56] It is not a new idea, but the 400 or the 430 years attested as the time of sojourn of the Israelites in Egypt (Gen. 15:13 and Exod. 12:40)[57] are surprisingly similar to the 400 years, month 4, and day 4 on the famous Stela of 400 Years[58]—a commemoration of the rule of the god Seth of Avaris (Sethe 1930; Montet 1931; Stadelmann 1965; Goedicke 1966, 1981; von Beckerath 1993; Kitchen 1996: 116–117; Kitchen 1999: 168–172; Bietak 2004).

The 400 Year Stela originates from the temple of Seth in Avaris, but was found dislocated at Tanis where the god Seth is not depicted as an Egyptian god but—with his horns, high crown with a long pommel, and western Asiatic kilt with tassels—is clearly defined as a Canaanite god who was identifiable as the Canaanite storm god Ba'al Zephon (Bietak 1990).[59] The Egyptian storm god Seth became an *interpretatio aegyptiaca* of the Canaanite storm god (Bietak 1990; Redford 1992: 117). Most of the abovementioned researchers see in the Stela of 400 Years an event commemorating a temple era in the time of Horemheb (ca. 1300 BC) which would be about 400 years after the first evidence surfaced of this cult in Avaris under king Nehesy (ca. 1700 BC).[60] Because the stela depicts Seti I

[56] For the continuation of the western Asiatic population from the Hyksos Period until Ramesside times at Avaris-Piramesse, see Bietak 2010: 164–171.

[57] The duration of the Israelites in Egypt is otherwise defined as four generations (Gen. 15:16; Exod. 6:16–20), the first and the last generation of which did not spend most of their life-span in Egypt—which would, realistically speaking, mean a stay of three generations, ca. 90 years.

[58] Engel 1979b: 191–193; Schneider 2008a, b: n. 119.

[59] For a thorough account of the statuary of Seth, particularly sculpture originating in the Delta, see Souroussian 2006.

[60] A temple precinct with Canaanite types of temples was constructed precisely in the period of the 14th Dynasty (Phase F at Tell el-Dab'a) according to the Austrian excavations at Tell el-Dab'a. This precinct was replaced by a larger compound in the New Kingdom (Bietak 2009: 213–220).

and mentions both him and his father Ramses I[61] as viziers, this event must have happened during the reign of Horemheb. It was not until well into the reign of Ramses II that this event had been commemorated on a stela, probably supplanting an older stela from the reign of Horemheb.[62]

Given the fact that a locally cut cylinder seal depicting Ba'al Zephon, found at Tell el-Dab'a (Porada 1984), proves that the cult of this storm god was at Avaris as early as the time of the late Middle Kingdom, the Stela of 400 Years can be deemed just one proof of the continuity of Canaanite cults from the Pre-Hyksos and the Hyksos Periods, through the time of the major Egyptian harbor stronghold Peru-nefer during the 18th Dynasty, and up to the Ramesside Period (Stadelmann 1967: 148; Helck 1971: 446–473; Bietak 2010: 165–171).[63] Amenhotep II was a special patron, not only of the naval stronghold of Peru-nefer but also of Canaanite cults in Memphis and in the Eastern Delta (Stadelmann 1967: 32–47, 99–110, 147–150; Collombert and Coulon 2000: 217; Hoffmeier and Kitchen 2007). He compared himself with Ba'al (Schneider 2003: 161).

This longue durée of a western Semitic population living in Egypt may have been the reason for Josephus identifying them with the Proto-Israelites. Predicated on a text ascribed to Hekataios (Hecataeus) of Abdera (Diodorus of Sicily XL:3)[64] it is even possible to argue, but not particularly cogently, that still earlier compilers who had been used as sources for Manetho had made such a connection. However, the population under Hyksos rule was an urban society allied to trade and seafaring and, for a certain period, ruled Egypt (ca. 1640–1530 BC). They experienced the glory of controlling the Delta and a part of the Nile valley for over 100 years. However, this is in no way in keeping with the tradition of the Israelites and their experience of oppression in Egypt. That is why an association of the Hyksos and their people with the Proto-Israelites should be dismissed, but their presence in Egypt and the commemoration of the cult of Ba'al Zephon/alias Seth of 400 years may have merged and been fused into the memory of the time span of the long sojourn of the early Israelites in Egypt.

[61] Because the name of the mother of the Vizier Seti on the stela is different from the name of the mother of king Seti I, Stadelmann (1965) argues that the two viziers depicted on the stela have to represent forefathers of Ramses I and Seti I. Vizier Seti on the stela is supposed to be the father of king Ramses I and vizier Paramesse the grandfather of Ramses I. The strength of the hypothesis is in the name difference of the mother of Seti I, and its weakness is that these ancestors are not known as viziers.

[62] Bietak 2009: 213–220.

[63] For Canaanite cults at Pi-Ramesse, see Daressy 1928–1929: 326; Stadelmann 1967: 148–150; Uphill 1984: 200–2; 212; 233–4; 245–6, 2523; pap. Anastasi III: 2.8 mentions a waterway at Pi-Ramesse called "the Waters of Ba'al."

[64] De Vaux 1971: 353, n. 35.

References

Ahituv, Sh. 1984. *Canaanite Toponyms in Ancient Egyptian Documents*. Jerusalem: The Magnes Press, the Hebrew University.

Astour, Michael C. 1979. Yahwe in Egyptian Topographical Lists. In *Festschrift Elmar Edel*, Ägypten und Altes Testament, vol. 1, 17–34. Bamberg: O. Harrassowitz.

Axelsson, L.E. 1987. *The Lord Rose up from Seïr: Studies in the History and Traditions of the Negev and Southern Judah*, Coniectanea Biblica Old Testament Series, vol. 25. Stockholm: Almquist & Wiksell.

Bietak, Manfred. 1975. *Tell el-Dab'a II. Der Fundort im Rahmen einer archäologisch-geographischen Untersuchung über das ägyptische Ostdelta*, UZK, vol. II. Vienna: ÖAW.

———. 1981. *Avaris and Piramesse, Archaeological Exploration in the Eastern Nile Delta*, Ninth Mortimer Wheeler Archaeological Lecture. The British Academy. Oxford: Oxford Univ. Press. (Repr. from *Proceedings of the British Academy London* 56, 1979, 225–290). - 2nd enlarged ed. Oxford 1986.

———. 1987. Comments on the Exodus. In *Egypt, Israel, Sinai. Archaeological and Historical Relationships in the Biblical Period*, ed. A.F. Rainey, 163–171. Tel Aviv: Tel Aviv University.

———. 1990. Zur Herkunft des Seth von Avaris. *Egypt and the Levant* 1: 9–16.

———. 1991. An Iron Age Four Room House in Ramesside Egypt. *Eretz Israel (A. Biran Volume)* 23: 10–12.

———. 1994. The Sea Peoples and the End of the Egyptian Administration in Canaan. In *Biblical Archaeology Today II, Proceedings of the Second International Congress on Biblical Archaeology, Jerusalem, June–July 1990*, ed. A. Biran and J. Aviram, 299–306. Jerusalem: Magnes Press.

———. 2000. Der Aufenthalt "Israels" in Ägypten und der Zeitpunkt der "Landnahme" aus heutiger archäologischer Sicht. *Egypt and the Levant* 10: 179–195.

———. 2003a. Israel in Ägypten - Ein architektonischer Nachweis? In *Von der Bauforschung zur Denkmalpflege*, Fs für A. Machatschek, ed. M. Kubelik and M. Schwarz, 1–6. Vienna: Phoibos.

———. 2003b. Israelites found in Egypt? *Biblical Archaeology Review* 29(5): 40–42, 82–83.

———. 2004. Kat. Nr. 391, Abguß der 400-Jahr-Stele. In *Pharaonen und Fremde, Dynastien im Dunkel*, Catalogue of the 194th Special Exhibition of the Historisches Museum of the city of Vienna, in the City Hall of Vienna, Volkshalle, 8. September - 23. October 1994, 279–281. Vienna: Historisches Museum der Stadt Wien.

———. 2009. Near Eastern Sanctuaries in the Eastern Nile Delta. In *Baal*, Hors-Série, vol. VI, 209–228. Beirut: Ministry of Culture of Lebanon.

———. 2010. From Where Came the Hyksos and Where Did They Go. In *The Second Intermediate Period (Thirteenth—Seventeenth Dynasties): Current Research, Future Prospects*, OLA, vol. 192, ed. M. Marée, 139–181. Leuven: Peeters.

Bleiberg, Edward L. 1983. The Location of Pithom and Succoth. In *Egyptological Miscellanies – A Tribute to Professor Ronald J. Williams*, The Ancient World, vol. VI, ed. J.K. Hoffmeier and E.S. Meltzer, 1–4, 21–27. Chicago: Ares.

Borchardt, Ludwig. 1911–1930. *Statuen und Statuetten von Königen und Privatleuten im Museum von Kairo, Nr. 1-1294, Teil 1-3: Text und Tafeln zu Nr.1-380, 381-653, 654-950*. CG 55, 77, 88. Berlin: Reichsdruckerei.

Braemer, Frank. 1982. *L'architecture domestique du Levant à l'age du fer*. Éditions Recherche sur les civilisations, Cah. No, 8. Paris

Bunimovitz, Shlomo, and Avraham Faust. 2002. Ideology in Stone: Understanding the Four-Room House. *Biblical Archaeology Review* 28(4): 32–41. 59–60.

———. 2003. Building Identity. The Four Room House and the Israelite Mind. In *Symbiosis, Symbolism, and the Power of the Past. Canaan, Ancient Israel, and Their Neighbors from the Late Bronze Age through Roman Palaestina*, ed. W.G. Dever and S. Gitin, 411–423. Eisenbrauns: Winona Lake, Ind.

Caminos, Ricardo C. 1954. *Late Egyptian Miscellanies*, Brown Egyptological Studies, vol. I. Oxford: Oxford Univ. Press.

Cazelles, Henri, and Jean Leclant. 1972. Pithom. In *Supplément au Dictionnaire de la Bible*, 42. Nr.: 8. Paris: Letouzey & Ané.

Clark, Douglas R. 2007. Cultural Interaction Through the Windows of the Four Room House. In *Studies in the History and Archaeology of Jordan* vol. 9, 103–112.

Collombert, Philippe, and Laurent Coulon. 2000. Les dieux contre la mer, le début du "papyrus d'Astarte" (pBN 202). *Bulletin de l'Institut Français d'Arché ologie Orientale* 100: 193–242.

Daressy, Georges. 1928–1929. Les branches du Nil dans la XVIIIe dynastie. *Bulletin de la Société Gé ographique d'Égypte* 16: 225–254, 293–329.

Dayan, Galit. 1998. The Term "*p3-twf*" in the Spiegelberg Papyrus. In *Jerusalem Studies in Egyptology*, ed. I. Shirun-Grumach, 133–135. Wiesbaden: Harrassowitz.

De Moor, Johannes C. 1997. *The Rise of Yahwism: The Roots of Israelite Monotheism*, 2nd ed. Leuven: Louvain Univ. Press.

de Vaux, Roland. 1971. *Histoire Ancienne d'Israel*, Études Bibliques. Paris: Lecoffre.

Diez Macho, Alejandro. 1968. *Neophyti 1. I. Targum Palestinense, Ms de la Biblioteca Vaticana*,T. I, Génesis. Madrid-Barcelona : Consejo Superior de Investigaciones Científicas.

———. 1970. *Neophyti 1.II. Targum Palestinense, Ms de la Biblioteca Vaticana*, T. II, Exodo. Madrid-Barcelona : Consejo Superior de Investigaciones Científicas.

Engel, Helmut. 1979a. Die Siegesstele des Merenptah. Kritischer Überblick über die verschiedenen Versionen historischer Auswertung des Schlußabschnittes. *Biblica* 60: 373–394.

———. 1979b. *Die Vorfahren Israels in Ägypten: Forschungsgeschichtlicher Überblick über die Darstellungen seit Richard Lepsius (1849)*, Frankfurter Theologische Studien, vol. 27. Frankfurt am Main: Knecht.

Erichsen, Wolja. 1933. *Papyrus Harris I, Hieroglyphische Transkription*, Bibliotheca Aegyptiaca, vol. V. Brussels: Fondation égyptologique reine Élisabeth.

Faust, Avraham. 2006. *Israel's Ethnogenesis: Settlement, Interaction, Expansion and Resistance*. London: Equinox.

Fecht, Gerhard, 1983 Die Israelstele, Gestalt und Aussage. In *Fontes atque pontes: Eine Festgabe für Helmut Brunner*, ÄAT, vol. 5, ed. M. Görg, 106–138. Wiesbaden: Harrassowitz.

Finkelstein, Israel. 1988. *The Archaeology of the Israelite Settlement*. Jerusalem: Israel Exploration Society.

———. 1991. The Emergence of Early Israel: Anthropology, Environment and Archaeology. *Journal of the American Oriental Society* 110: 677–686.

Fritz, Volkmar. 1977. Bestimmung und Herkunft des Pfeilerhauses in Israel. *Zeitschrift des Deutschen Palästina-Vereins* 93: 30–45.

———. 1980. Die kulturhistorische Bedeutung der früheisenzeitlichen Siedlung auf der Ḫirbet el-*Mšāš* und das Problem der Landnahme. *Zeitschrift des Deutschen Palästina-Vereins* 96: 121–135.

Fritz, Volkmar, and Aharon Kempinski. 1976. Vorbericht über die Ausgrabungen auf der Ḫirbet el-Mšāš (Tel Māśōś), 3. Kampagne 1975. *Zeitschrift des Deutschen Palästina-Vereins* 93: 83–104.

———. 1983. *Ergebnisse der Ausgrabungen auf der Ḫirbet el-Mšāš (Tel Māśōś) 1972–1975*, Abhandlungen des Deutschen Palästina Vereins 1-3, vol. I–III. Wiesbaden: O. Harrassowitz.

Gardiner, Alan H. 1918. The Delta Residence of the Ramessides. *Journal of Egyptian Archaeology* 5: 127–138, 179–200, 242–271.

———. 1919. The Ancient Military Road between Egypt and Palestine. *Journal of Egyptian Archaeology* 6: 99–116.

———. 1922. The Geography of the Exodus. In *Recueil d'études égyptologiques dédiées à la mémoire de Jean-François Champollion*, Bibliothèque de l'École des Hautes Études. 4. Section: Sciences Historiques et Philologiques, vol. 234, 203–215. Paris: Champion.

———. 1937. *Late Egyptian Miscellanies*, Bibliotheca Aegyptiaca, vol. 7. Brussels: Fondation égyptologique Reine Élisabeth.

———. 1947. *Ancient Egyptian Onomastica*, vol. I–II. Oxford: Oxford Univ. Press.

Gauthier, Henri. 1925–1931. *Dictionnaire des noms gé ographiques contenues dans les textes hié-roglyphiques*, vol. I–VII. Cairo: Société royale de géographie d'Égypte, Reimpression Osnabrück 1975: O. Zeller.

Gilmour, Garth, and Kenneth A. Kitchen. 2012. Pharaoh Sety II and Egyptian Political Relations with Canaan at the End of the Late Bronze Age. *Israel Exploration Journal* 62(1): 1–21.

Giveon, Raphael. 1971. *Les bédouins Shosou des documents égyptiens*, Documenta et Monumenta. Orientis Antiqui, vol. XVIII. Leiden: Brill.

Goedicke, Hans. 1966. Some Remarks on the 400-year Stela. In *CdÉ* vol. 41, 23–39.

———. 1981. The 400-Year-Stela Reconsidered. *Bulletin of the Egyptological Seminar* 3 (Columbia University NY): 25–42.

Goldwasser, Orly. 2002. *Lovers, Prophets and Giraffes—Wor[l]d Classification in Ancient Egypt*, Göttinger Orientforschungen IV. Reihe Ägypten, vol. 38. Wiesbaden: O. Harrassowitz.

———. 2010. The Aten is the 'Energy of Light': New Evidence from the Script. *Journal of the American Research Center in Egypt* 46: 159–165.

Görg, Manfred. 1976. Zur Geschichte der *Š3św*. *Orientalia* 45: 426–428.

———. 1979. Tuthmosis III. und die *Š3św*-Region. *Journal of Near Eastern Studies* 38: 199–202.

———. 1989. Jahwe—ein Toponym? In *Beiträge zur Zeitgeschichte der Anfänge Israels. Dokumente—Materialien—Notizen*, ÄAT, vol. 2, 180–187. Wiesbaden: O. Harrassowitz.

———. 1997. *Die Beziehungen zwischen dem Alten Israel und Ägypten. Von den Anfängen bis zum Exil*, Erträge der Forschung, vol. 290. Darmstadt: Wissenschaftliche Buchgesellschaft.

Grandet, Pierre. 1993. *Ramsès III, Histoire d'un règne*. Paris: Pygmalion.

———. 1994. *Le papyrus Harris I*. BdE CIX/i-ii, Cairo.

Groll Sarah, Israelit. 1998. The Egyptian Background of the Exodus and the Crossing of the Reed Sea: A New Reading of Papyrus Anastasi VIII. In *Jerusalem Studies in Egyptology*, ÄAT, vol. 40, ed. I. Shirun-Grumach, 173–192. Wiesbaden: O. Harrassowitz.

Habachi, Labib. 1954. Khataʿna-Qantir: Importance. *Annales du service des antiquités de l'Égypte* 52: 443–559.

———. 2001. *Tell el-Dab'a I: Tell el-Dab'a and Qantir: The Site and its Connection with Avaris and Piramesse*, UZK, vol. II, ed. E.-M. Engel, P. Jánosi, and C. Mlinar, Vienna: ÖAW.

Hasel, Michael G. 1994. Israel in the Merenptah Stele. *Bulletin of the American Schools of Oriental Research* 296: 45–61.

Helck, Wolfgang. 1965. *Ṯ3rw* und die Ramsesstadt. *Vetus Testamentum* 15: 35–48.

———. 1968. Die Bedrohung Palästinas durch einwandernde Gruppen am Ende der 18. und am Anfang der 19. Dynastie. *Vetus Testamentum* 18: 472–480.

———. 1971. *Die Beziehungen Ägyptens zu Vorderasien im 3. und 2. Jahrtausend v. Chr. ÄA 5*, 2nd ed. Wiesbaden: O. Harrassowitz.

———. 1974. *Die altägyptischen Gaue*, Beihefte zum Tübinger Atlas des Vorderen Orients, B.5. Wiesbaden: L. Reichert.

———. 2001. The history of the Collared Pithos at Tell el-'Umayri, Jordan. In *Studies in the Archaeology of Israel and Neighboring Lands in Memory of Douglas Esse*, SAOC, vol. 59, ed. S.M. Wolff, 237–250. Chicago: Chicago University Press.

Herr, Larry G. 1999. Tall al-'Umayri and the Reubenite Hypothesis. *Eretz Israel* 26, F.M. Cross volume: 64*–67*.

———. 2000. The Settlement and Fortification of Tell Al-'Umayri in Jordan during the LB I/ Iron I Transition. In *The Archaeology of Jordan and Beyond, Essays in Memory of James A. Sauer*, ed. J.A. Sauer, L.E. Stager, J.A. Greene, and M.D. Coogan, 167–178. Harvard Semitic Museum Publications. Winona Lake, IN: Eisenbrauns.

———. 2009. The House of the Father at Iron I Tall al-'Umayri, Jordan. In *Exploring the Longue Durée: Essays in Honor of Lawrence* E. Stager, ed. J.D. Schloen, 191–198. Winona Lake, IN: Eisenbrauns.

Herr, Larry G., and Douglas R. Clark. 2001. Excavating the Tribe of Reuben. In *BAR*, vol. 27(2), 36–47, 64, 66.

Herrmann, Siegfried. 1967. Der Name Jhw3 in den Inschriften von Soleb. In *Fourth World Congress of Jewish Studies, Papers*, vol. I, 213–216. Jerusalem: World Union of Jewish Studies.

Hoch, James E. 1994. *Semitic Words in Egyptian Texts of the New Kingdom and Third Intermediate Period*. Princeton, NJ: Princeton University Press.

Hoffmeier, James K. 1996. *Israel in Egypt: The Evidence for the Authenticity of the Exodus Tradition*. New York, NY: Oxford University Press.

———. 2006. "The Walls of the Ruler" in Egyptian Literature and the Archaeological Record: Investigating Egypt's Eastern Frontier in the Bronze Age. *Bulletin of the American Schools of Oriental Research* 343: 1–20.

Hoffmeier, James K., and Kenneth A. Kitchen. 2007. Reshep and Astarte in North Sinai: A Recently

Discovered Stela from Tell el-Borg. *Egypt and the Levant* 17: 127–136.

Hoffmeier, James K., and Stephen O. Moshier. 2006. New Paleo-Envoironmental Evidence from North Sinai to Complement Manfred Bietak's Map of the Eastern Delta and Some Historical Implications. In *Timelines: Studies in Honour of Manfred Bietak*, OLA, vol. 143/ II, 167–176. Leuven: Peeters. OLA 143/II.

Holladay, John S. 1982. *Tell el-Maskhuta: Preliminary Report on the Wadi Tumilat Project 1978–1979*, Cities of the Delta, vol. 3. Malibu, CA: Undena.

———. 1997. Four Room House. In *The Oxford Encyclopedia of Archaeology in the Near East*, ed. E.M. Meyers, 337–342. Oxford: Oxford University Press.

———. 1999. Tell el-Maskhuta. In *Encyclopedia of the Archaeology of Ancient Egypt*, ed. K.A. Bard, 786–789. London-New York, NY.

Hölscher, U. 1939. *The Excavation of Medinet Habu, vol. 2: The Temples of the Eighteenth Dynasty*, OIP, vol. 41. Chicago: University of Chicago Press.

Hornung, Erik. 1983. Die Israelstele des Merenptah. In *Fontes atque pontes: Eine Festgabe für Helmut Brunner*, ÄAT, vol. 5, ed. M. Görg, 224–233. Wiesbaden: Harrassowitz.

Kettenhofen, E. 1989. Einige Beobachtungen Zu Heroonpolis. *Orientalia Lovaniensia Periodica* 20: 75–97.

Kitchen, Kenneth A. 1973. Review of D.B. Redford, *A Study of the Biblical Story of Joseph*. Vetus Testamentum Supplements, vol. 20. Leiden: Brill.

———. 1979. *Ramesside Inscriptions, Historical and Biographical II:16*. Oxford: Blackwell.

———. 1996. *Ramesside Inscriptions, Translated and Annotated, Translations, vol. II. Ramses II, Royal Inscriptions*. Oxford: Blackwells.

———. 1999. *Ramesside Inscriptions, Translated and Annotated, Notes and Comments II. Rameses, Royal Inscriptions*. Oxford: Blackwells.

———. 2003. *On the Reliability of the Old Testament*. Grand Rapids, MI: Wm.B. Eerdmans.

———. 2004. The Victories of Merenptah, and the Nature of Their Record. *Journal for the Study of the Old Testament* 28(3): 259–272.

Knauff, Ernst A. 1994. *Die Umwelt des Alten Testamentes*. Neuer Stuttgarter Kommentar AT Bd. 29. Stuttgart.

Lambdin, Thomas O. 1953. Egyptian Loan Words in the Old Testament. *Journal of the American Oriental Society* 73(3): 145–155.

Leclant, Jean. 1991. Le "Tétragramme" à l'époque d'Aménophis III. In *Near Eastern Studies dedicated to H.I.H. Prince Takahito Mikasa on the Occasion of His Seventy-Fifth Birthday*, 215–219. Wiesbaden: O. Harrassowitz.

Lehmann, Manuela. 2011. Vorbericht über die Grabungstätigkeiten der Herbstkampagne 2009 im Areal A/I von Tell el-Dab'a. *Egypt and the Levant* 21: 47–65.

———. 2012/3. Vorbericht über die Grabungstätigkeiten der Frühjahrskampagne 2011 im Areal A/I von Tell el-Dab'a. *Egypt and the Levant* 22/3: 75–86.

Lemaire, André. 2003. *La naissance de monotheism*, point de vue d'un historien. Montrouge (Hauts-de-Seine): Bayard.

Lemche, Niels P. 1994. Is it still Possible to Write a History of Ancient Israel? *Scandinavian Journal of the Old Testament* 8: 165–190.

Mazar, Amihai. 1997. *Timnah (Tel Batash) I: Stratigraphy and Architecture*, Qedem, vol. 37. Jerusalem: Hebrew University of Jerusalem.

Montet, Pierre. 1931. La stèle de l'an 400 retrouvée. *Kêmi* Vol. 4 191–215.

Na'aman, Nadav. 1977. Yeno'am. In: *Tel Aviv* vol. 4, 168–177.

———. 2011. The Exodus Story: Between Historical Memory and Historical Composition. *Journal of Ancient Near Eastern Religions* 11: 39–69.

Naville, Edouard. 1991. *Bubastis (1887–1889)*, 8th Memoir of the Egypt Exploration Fund. London: Kegan Paul, Trench, Trübner & Co.

———. 1903. *The Store-City of Pithom*, 1st Memoir of the Egypt Exploration Fund, 4th ed. London: Kegan Paul, Trench, Trübner & Co.

Netzer, Ehud. 1992. Domestic Architecture in the Iron Age. In *The Architecture of Ancient Israel, From the Prehistoric to the Persian Periods*, ed. A. Kempinski and R. Reich, 193–201. Jerusalem: IES.

Petrie, William M. Flinders. 1906. *Hyksos and Israelite Cities*, with a chapter by J.G. Duncan. London: School of Archaeology UCL

Porada, Edith. 1984. The Cylinder Seal from Tell el-Dab'a. *American Journal of Archaeology* 88: 485–488.

Porter, B., and R.L.B. Moss. 1934. *Topographical Bibliography of Hieroglyphic Inscriptions, Reliefs and Paintings, vol. IV, Lower and Middle Egypt*. Oxford: Oxford Univ. Press.

Rainey, Anson F. 2001. Israel in Merenptah's Inscriptions and Relief. *Israel Exploration Journal* 51(1): 57–75.

———. 2007. Whence Came the Israelites and Their Language? *Israel Explotation Journal* 57(1): 41–64.

Redford, Donald B. 1963. Exodus I.11. *Vetus Testamentum* 13: 401–418.

———. 1970. *A Study of the Biblical Story of Joseph*, Vetus Testamentum Supplements, vol. 20. Leiden: Brill.

———. 1982. Pithom. Cols. 1054–1058. In *Lexikon der Ägyptologie*, vol. IV, ed. W. Helck and W. Westendorf. Wiesbaden: O. Harrassowitz.

———. 1987. An Egyptological Perspective on the Exodus. In *Egypt, Israel, Sinai. Archaeological and Historical Relationships in the Biblical Period*, ed. A.F. Rainey, 137–161. Tel Aviv: Tel Aviv University Press.

———. 1992. *Egypt, Canaan, and Israel*. Princeton, NJ: Princeton Univ. Press.

———. 1997. Observations on the Sojourn of the Bene-Israel. In *Exodus: The Egyptian Evidence*, ed. E.S. Frerichs and L.H. Lesko, 57–66. Winona Lake, IN: Eisenbrauns.

Redmount, Carol A. 2000. *On an Egyptian/Asiatic Frontier: An Archaeological History of the Wadi Tumilat*, vols. I–IV. PhD. Dissertation, Univ. of Chicago Ill. 1989. UMI Diss Service. Ann Arbor, MI: UMI

Rendsburg, Gary A. 1992. The Date of the Exodus and the Conquest/Settlement: The Case for the 1100s. *Vetus Testamentum* 47(4): 510–527.

———. 2006. Moses as Equal to Pharaoh. In *Text, Artifact and Image: Revealing Ancient Israelite Religion*, ed. G. Beckman and T.J. Lewis, 201–219. Providence, RI: Brown University.

Rzepka, S., A. Wodzinska, J. Hudec, and T. Herbich. 2009. Tell el-Retaba 2007–2008. *Egypt and the Levant* 19: 241–280.

Rzepka, S., A. Wodzinska, C. Mallelson, J. Hudec, L. Jarmuzek, W. Malkowski, and M. Bogacki. 2011. New Kingdom and Third Intermediate Period in Tell el-Retaba. *Egypt and the Levant* 21: 129–184.

Schneider, Thomas. 2003. Foreign Egypt: Egyptology and the 'Concept of Cultural Appropriation. *Egypt and the Levant* 13: 155–161.

———. 2008a. Das Ende Der kurzen Chronologie: Eine kritische Bilanz der Debatte zur absoluten Datierung des Mittleren Reiches und der Zweiten Zwischenzeit. *Egypt and the Levant* 18: 275–313.

———. 2008b. The First Documented Occurrence of the God Yahweh? (Book of the Dead. Princeton 'Roll 5'). *Journal of Ancient Near Eastern Religion* 7: 113–120.

———. 2010. Contributions to the Chronology of the New Kingdom and the Third Intermediate Period. *Egypt and the Levant* 20: 373–409.

Seguin, Joffrey. 2007. *Le Migdol du Proche-Orient à l'Égypte*, Les institutions dans l'Égypte ancienne, vol. 3. Paris: Sorbonne.

Sethe, Kurt. 1930. Der Denkstein mit dem Datum des Jahres 400 der Ära von Tanis. *ZÄS* 65: 85–89.

Shiloh, Y. 1970. The Four Room House: Its Situation and Function in the Israelite City. *Israel Exploration Journal* 20: 180–190.

———. 1973. The Four-Room House—The Israelite Type-House? *Eretz Israel* 11: 277–285.

Souroussian, Hourig. 2006. Seth fils de Nout et Seth d'Avaris dans la statuaire royale ramesside. In *Timelines, Studies in Honour of Manfred Bietak*, OLA 149/I, vol. I, ed. E. Czerny et al., 331–354. Leuven: Peeters.

Stadelmann, Rainer. 1965. Die 400-Jahr Stele. *CdE* 40, no. 78: 46–60.

———. 1967. *Syrisch-palästinensische Gottheiten in Ägypten*, Probleme der Ägyptologie, vol. 5. Leiden: Brill.

———. 1968. Die Abwehr der Seevölker unter Ramses III. *Saeculum* 19(2): 156–171.

Stager, Lawrence E. 1985a. Merenptah, Israel and Sea Peoples. New Light on an Old Relief. *Eretz Israel* 18: 46*–64*.

———. 1985b. The Archaeology of the Family in Ancient Israel. *Bulletin of the American Schools of Oriental Research* 260: 1–35.

———. 1998. The Impact of the Sea Peoples in Canaan (1185–1050 BCE). In *The Archaeology of the Society in the Holy Land*, ed. T. Levy, 332–448. London: Continuum International Publishing Group.

Swinnen, Ingrid M. 2009. The Iron Age I Settlement and its residential Houses at El-Lahun in Moab, Jordan. *Bulletin of the American Schools of Oriental Research* 354: 29–53.

Thiers, Christophe. 2007. *Ptolémée Philadelphe et les prêtres d'Atoum de Tjékou. Nouvelle édition commentée de la "stèle de Pithom" (CGC 22183)*. Orientalia Monspeliensia XVII. Montpellier.

Thompson, Thomas L. 1987. *The Origin Tradition of Ancient Israel: The Literary Formation of Genesis and Exodus 1–23*. JSOT Supplement no. 55. Sheffield: Sheffield Press.

Ullmann, Martina. 2002. *König für die Ewigkeit—Häuser der Millionen von Jahren: Eine Untersuchung zu Königskult und Tempeltypologie in Ägypten*, ÄAT, vol. 51. Wiesbaden: O. Harrassowitz.

Uphill, Eric P. 1968. Pithom and Raamses: Their Location and Significance. *Journal of Near Eastern Studies* 27: 291–316.

———. 1969. Pithom and Raamses: Their Location and Significance. *Journal of Near Eastern Studies* 28: 15–39.

———. 1984. *The Temples of Per Ramesses*. Warminster: Aris & Phillips.

Van der Steen, Eveline J. 1995. Aspects of Nomadism and Settlement in the Central Jordan Valley. *Palestine Exploration Quarterly* 127: 141–158.

———. 1996. The Central East Jordan Valley in the Late Bronze and Early Iron Ages. *Bulletin of the American Schools of Oriental Research* 302: 51–74.

———. 1999. Survival and Adaptation: Life East of the Jordan in the Transition from the Late Bronze Age to the Early Iron Age. *Palestine Exploration Quarterly* 131: 176–192.

Van Seters, John. 1966. *The Hyksos: A New Investigation*. New Haven, CT: Yale University Press.

———. 1992. *Prologue to History: The Yahwist As Historian in Genesis*. Louisville, KY: John Knox Press.

———. 2001. The Geography of the Exodus. In *The Land that I Will Show You. Essays on History and Archaeology of the Ancient Near East in Honour of J. Maxwell Mille*, Journal for the Study of the Old Testament Supplement Series, vol. 343, ed. J.A. Dearman and M. P. Graham, 255–276. Sheffield: Sheffield Academic Press.

Vervenne, Marc. 1995. The Lexeme סוּף (sûph) and the phrase יַם סוּף (yam sûph): A Brief Reflection on the Etymology and Semantics of a Key Word in the Hebrew Exodus Tradition. In *Immigration and Emigration within the Ancient Near East. Festschrift E. Lipinsky*, OLA, vol. 65, ed. K. van Lerberghe and A. Schoors, 403–429. Leuven: Peeters.

Von Beckerath, Jürgen. 1993. Nochmals die Vierhundert-Stele. *Orientalia* 62: 400–403.

Von der Way, Thomas. 1992. *Göttergericht und "Heiliger" Krieg im Alten Ägypten. Die Inschriften des Merenptah zum Libyerkrieg des Jahres 5*, SAGA, vol. 4. Heidelberg: Heidelberger Orientverlag.

Vretska, K. 1958. *Die Pilgerfahrt der Aetheria (Peregrinatio Aetheriae)*. Klosterneuburg: Bernina.

Ward, William A. 1974. The Semitic biconsonantal root *sp* and the common origin of Egyptian *čwf* and Hebrew *sûp*: "marsh(-plant)". *Vetus Testamentum* 24(3): 339–349.

Weimar, P. 1985. *Die Meerwundererzählung: Eine redaktionskritische Analyse von Ex 13,17–14,31*, ÄAT, vol. 9. Wiesbaden: O. Harrassowitz.

Weinstein, James M. 1994. *Foundation Deposits in Ancient Egypt*. Ph.D. diss., University of Pennsylvania 1973: UMI Ann Arbor, MI.

Weippert, Manfred. 1974. Semitische Nomaden des zweiten Jahrtausends: Über die *Š3św* der ägyptischen Quellen. *Biblica* 55(265–280): 427–433.

Wente, Edward. 1992. Rameses. In *Anchor Bible Dictionary*, vol. V, ed. David Noel Freedman, 617. New York, NY: Doubleday.

Wilkinson, John (ed.). 1999. *Egeria's Travels*, 3rd ed. Warminster: Aris & Phillips.

Wimmer, Stefan. 2002. A New Stela of Ramesses II in Jordan in the Context of Egyptian Royal Stelae in the Levant. Paper presented at the Third International Conference on the Archaeology of the Ancient Near East (3ICAANE), Paris, 18 April 2002, <http://www.stefan-jakob-wimmer.de/3ICAANE_wimmer.pdf>

Yoyotte, Jean. 1971–1972. Pi-Ramsès et Tanis. In *Annuaire de l'École Pratique des Hautes Études, Section des Sciences Religieuses*, vol. 79, 167–173.

Yurco, Frank. 1986. Merenptah's Canaanite Campaign. *Journal of the American Research Center in Egypt* 23: 189–215.

Zivie-Coche, Christiane. 2004. *Statues et autobiographies de dignitaries: Tanis à l'époque ptolémaïque*, Travaux récents sur le Tell Sân el-Hagar, vol. 3. Paris: Cybele.

3 The Wilderness Narrative and Itineraries and the Evolution of the Exodus Tradition

Israel Finkelstein

Abstract

This chapter examines the Exodus and wandering tradition from the perspective of the archaeology of several pivotal sites in the desert. It poses the question, "What, how, and when did the biblical authors know about the southern desert?" The answer helps to reconstruct the history of the Exodus-wandering tradition from its vague beginning as salvation-from-Egypt memories in sixteenth to tenth century BCE Canaan, through the involvement of the Northern Kingdom along the desert trade routes in the first half of the eighth century, and the presence of Judahites in the south during the "Assyrian Century," to the Priestly scribes in post-exilic times.

Introduction

Scholars that have attempted to deal with the historical reality behind the Exodus and desert wandering narrative (for the two being connected see, e.g., Dozeman 2000: 64) are for the most part divided into two camps. Members of one camp adhere to the traditional research notion that the biblical material portrays the situation in the Late Bronze Age, in the thirteenth century BCE—the time calculated according to the logic of biblical chronology (e.g., Kitchen 1998; Halpern 1993; Hoffmeier 1997, 2005). These scholars face two major problems:

First, it is clear today that there was no significant scribal activity in Ancient Israel until close to 800 BCE (Finkelstein and Sass 2013), and hence, they need to assume an oral transmission of the story with all its details over a period of four centuries with no infiltration of realities from the time passed.

Second, there is no single piece of evidence to support a Late Bronze Age origin of the tradition that cannot be understood against the background of other, later periods (e.g., Na'aman 2011: 56–60).

Members of the second camp propose that the text describes realities that fit the time of compilation of the text—in late-monarchic to post-exilic days (Redford 1992: 408–422; Van Seters 2001; Finkelstein and Silberman 2001: 48–71; Liverani 2005: 277–282). The main difficulty that these researchers face is in explaining the strong tradition of both the Exodus and the

I. Finkelstein (✉)
Tel Aviv University, Tel Aviv, Israel
e-mail: fink2@post.tau.ac.il

T.E. Levy et al. (eds.), *Israel's Exodus in Transdisciplinary Perspective*,
Quantitative Methods in the Humanities and Social Sciences, DOI 10.1007/978-3-319-04768-3_3,

desert experience in the writings of the northern prophets of the eighth century BCE.[1]

The following points need to be taken into consideration when dealing with the biblical texts of the Exodus-wandering tradition:

1. This tradition had an important status in Northern Israel as early as the eighth century BCE (e.g., Hoffman 1983; 1989; Van der Toorn 1996: 287–315; Dozeman 2000);
2. It contains an "inner" literary history (e.g., Dozeman 1989; Römer 2003; Carr 2012 for Moses);
3. Originally it was independent from – and earlier than – the patriarchal stories;
4. The two blocks—patriarchs and Exodus – were connected at a relatively late date;
5. In its present form the narrative represents Priestly (or even late Priestly and/or post-Priestly) compilations (for Points 3-5 see, e.g., Römer 1990; Gertz 2000: 380–388; Kratz 2005: 248–308; Schmid 2010; 2012; various articles in Dozeman and Schmid 2006; Römer and Schmid 2007).

Here, I wish to look at the subject from the standpoint of archaeology, but to take a different track from the usual "hunt" for Late Bronze finds at sites and regions mentioned in the text. I wish to deal with the wilderness wandering material through the lens of toponyms that appear in the narrative and the itinerary lists in Exodus, Numbers, and Deuteronomy. These itineraries have been studied intensively on issues such as structure, sources, redactions, and geography (recently Roskop 2011; Dozeman 2011; and more below).

Yet the question of historical reality behind them has seldom been addressed. In other words, though the itinerary lists belong to late redactions of the Pentateuch, for reasons which will be explained below, they are probably based on earlier sources. I will try to identify these earlier materials in an attempt to reconstruct the history of the Exodus-wandering tradition.

In order to do this, I ask the following question: What could biblical authors of different schools and times have known about the southern desert? This question becomes especially acute when one looks at the settled lands: late-monarchic, exilic, and post-exilic authors were well-acquainted with the geography of Judah-Yehud and adjacent areas such as the plateau of Benjamin and Moab; but even for regions not far off such as the northern Gilead (Finkelstein, Koch and Lipschits 2012) and northern Samaria their knowledge was limited and fragmentary.

If this was the case for regions close to the world of the authors, what does this say about the desolate and remote desert areas, located hundreds of kilometers away? What could have been the source of knowledge about them?

The Itineraries

Sites in the desert visited during the period of the wandering are mentioned in several verses in the Pentateuch, in the narrative in Exodus-Numbers (Exod 12:37; 13:20; 14:2; 15:22–23; 16:1; 17:1; 19:2; Num 10:12; 11:35; 12:14; 20:1, 22; 21:10–12), in the summary list of Numbers 33:1–49 and in several verses in Deuteronomy (1:1–2; 1:46–2:1; 2:14; 10:6–7). This material was the subject of intensive research, regarding themes such as division into geographical sections (Walsh 1977); meaning of the formulae used in the text (Coats 1972); relations between the narrative and the summary list (Noth 1940; Cross 1973: 301–321; Kallai 1998; Davies 1983; North 2001; Lee 2003; Roskop 2011: 223–232); the genre of itineraries in the Bible and in ancient Near Eastern texts (Coats 1972; Davies 1974; Roskop 2011); identification of specific places, and the route in the desert (e.g., Davies 1979, 1990; North 2001); and questions of sources and redactions (Noth 1940; 1968: 242–246; Coats 1972; Davies 1983; 1995: 341–343; Römer 2007; Lee 2008). Regarding the latter, scholars assumed that the itineraries are based on earlier materials (Noth 1968: 243; 1972: 224–227; Fritz 1970: 116–117; Davies 1983; 1995: 342).

[1] For a somewhat different version, emphasizing an earlier reality in the Iron Age, in the days of Jeroboam I, see Van der Toorn 1996: 287–315; Albertz 2001.

One way to reveal these early materials and their background is to look at the archaeology of places mentioned in the narrative and lists that can be securely identified. As far as I can judge, east of the Nile Delta and south of Moab there are only four such places: Kadesh-Barnea (=Tell el-Qudeirat in northeastern Sinai), Ezion-Geber (=Tell el-Kheleifeh between Aqaba and Eilat), Punon (=Khirbet Faynan in the eastern Arabah south of the Dead Sea), and the land of Edom. Two additional sites in the south have the potential of shedding light on the question posed here, regarding the acquaintance of biblical authors with the southern desert: En Hazeva in the western Arabah south of the Dead Sea and Kuntillet ʻAjrud in northeastern Sinai.

The Sites

Kadesh-Barnea

This site is a small mound in the oasis of Ein el-Qudeirat. It was almost fully excavated (final report in Cohen and Bernick-Greenberg 2007). The publication and other recent treatments of the finds opened the way for reappraisal (Finkelstein 2010a); the points relevant to this chapter are summarized below.

Sub-stratum 4c represents the earliest occupation at the site. Painted Qurayyah pottery and other Iron I finds (Singer-Avitz 2008) must have originated from this settlement; they probably date to the twelfth century BCE. Radiocarbon results from seed samples that provide dates in the tenth century BCE (Gilboa et al. 2009) should be affiliated with this settlement. Sub-strata 4b and 4a represent a settlement that covers much of the sequence of the Iron IIA (the late 10th to early eighth century BCE). The different phases of Stratum 4 feature the remains of a small settlement that was probably inhabited by local desert people. The Sub-stratum 4b settlement belongs to the phenomenon of the Negev Highlands sites of the Iron IIA—a wave of sedentarization of desert people connected to the prosperity of the copper industry in the Arabah (Finkelstein 1995: 103–126; Martin and Finkelstein 2013). The archaeological data are insufficient for deciding whether the site was also inhabited in the first half of the eighth century BCE.

Strata 3–2 are the most important for this discussion. They feature the remains of a well-preserved Iron IIB-C rectangular fortress, surrounded by a solid wall built as a foundation for a system of casemates. This fortress was erected in the second half of the eighth century (i.e., contemporary with the Assyrian take-over of the region), and continued to function until ca. 600 BCE. Cohen and Bernick-Greenberg (2007: 13) identified the fortress as a Judahite administration center built along the Dharb el-Ghazza trade route that led from the Red Sea to the Mediterranean coast (Fig. 3.1). Na'aman (1991: 48–49; 2001: 268) argued that the construction of the fortress was initiated by the Assyrians and that it was staffed by people from the Assyrian vassal kingdoms. I see the site as a Judahite fortress that was commissioned by—and functioned in the service of—the Assyrian administration (below). Taking into account the overall geopolitical situation in the Levant in the second half of the seventh century, it is reasonable to assume that Egypt of the 26th dynasty took control in the south, including Kadesh-Barnea, after the Assyrian pull-out from the region. The destruction of Stratum 2 should be affiliated with the Babylonian assault in 604/603 BCE, or with the fall of Judah in the early sixth century.

With the destruction of the fortress, Kadesh-Barnea lost its importance. Scanty remains, apparently associated with pottery typical of the end of the Iron Age, were retrieved in one area over the destruction layer of Stratum 2. Meager Persian period remains were found above the Stratum 2 destruction in several sectors of the site; they include a Yehud seal impression that belongs to the Vanderhooft and Lipschits (2007) Group 14, dated to the fouth–third centuries BCE.

Ezion-Geber

Ezion-Geber is identified with the Iron Age site of Tell el-Kheleifeh on the northern tip of the Gulf of Aqaba. No other Iron Age site is known

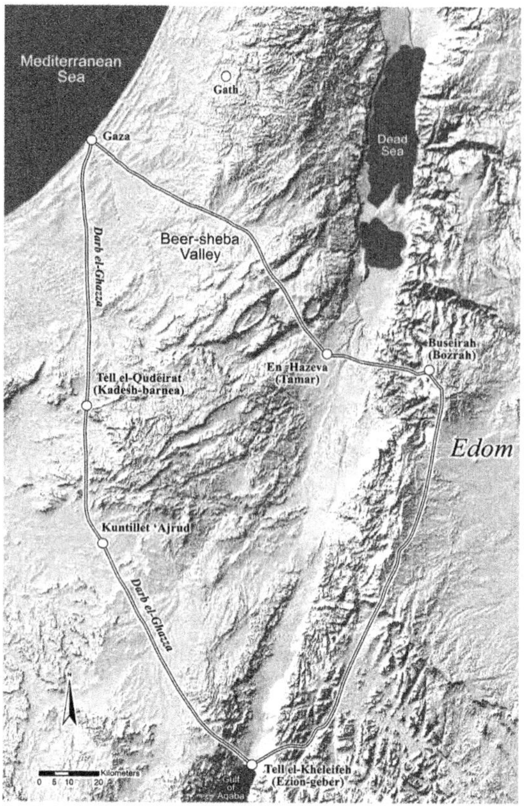

Fig. 3.1 Main sites and ancient roads in the south mentioned in this chapter

in this region. Nelson Glueck uncovered much of the site between 1938 and 1940 (summary in Glueck 1965). He separated the remains into five periods of activity and dated them from the tenth to the fifth centuries BCE, with every Judahite monarch mentioned in relation to activities in the region granted a stratum. Glueck interpreted the remains of the first period as evidence for a large copper smelting industry from the days of King Solomon.

A few Qurayyah Ware sherds (Pratico 1993: 49–50) may attest to some activity in the twelfth century BCE, possibly related to mining at Timna (Bienkowski 2001a: 261; Singer-Avitz 2008: 78). Glueck's association of the site with Solomon and the copper industry proved to be an illusion. A thorough study of the finds by Pratico (1993) revealed no evidence of smelting activity. No less important, no tenth century finds have been uncovered at Tell el-Kheleifeh; the first significant settlement there was established in the eighth century BCE.

Architecturally the site features two main structures: a ca. 45 × 45 m casemate fortress and a later, larger fortress, ca. 75 × 75 m in size, with a solid wall, which Glueck and Pratico describe as a later "fortified settlement." The former should probably be dated to the first half of the eighth century and possibly affiliated with the account in 2 Kings 14:22 that King Uzziah "built Elath and restored it to Judah." This could have taken place under Israelite hegemony (Finkelstein forthcoming a). Based on the Iron IIB-C pottery retrieved at the site and the similarity to En Hazeva, the latter should be interpreted as an Assyrian fortress.

The Qosanal seal impressions and Ostracon 6043 (Divito 1993) show that in ca. 600 BCE the site was inhabited by Edomites. Attic Greek sherds and a few Aramaic ostraca testify to some activity in the Persian period. The nature of this activity has not been fully clarified. In any event, after the Assyrian withdrawal from the Levant in the late seventh century, or the take-over of Edom by Nabonidus in the mid-sixth century at the latest, the site must have declined in importance.

Punon

This site is mentioned in the Hebrew Bible only in the summary list of the desert "stations" (Num 33:42–43). It should be identified with the multi-period mound of Khirbet Faynan, located in the eastern Arabah, ca. 50 km south of the Dead Sea (Roman-Byzantine Phaeno). This is the largest site in the region, covering an area of 15 ha. Surveys of the site collected Nabatean, Roman, and Byzantine pottery (Barker et al. 1997: 21). The area north of the mound revealed a large number of Iron Age sherds, some of which appear to date "before the seventh century BC" (Mattingly et al. 2007: 278–279). Late Iron Age sherds were found to the southwest of the tell (Barker et al. 1998: 20–21). Hauptmann (2007: 97) described an Iron Age slag heap immediately to the east of the mound. Three ^{14}C

determinations for samples collected there provided dates that fit the Iron IIA. The first season of excavations at the site revealed a head of an anthropomorphic figurine, somewhat similar to the Iron IIC figurines found at Horvat Qitmit (Levy et al. 2012).

A strong Iron IIA copper industry was recorded at other sites in the vicinity, chiefly among them Khirbet en-Nahas (Levy et al. 2014). The date of the fortress that dominates the surface of this site—Iron IIA or Iron IIB-C—is debated (Levy et al. 2014 for the former; Finkelstein and Piasetzky 2006 for the latter date).

Edom

Geographically, the biblical text refers to the territory of Edom as extending also west of the Arabah; the towns of the biblical Negeb (the Beer-sheba Valley) are denoted as located "toward the boundary of Edom" (בגבול אדום—Josh 15:21). Archaeology indicates that in the Iron IIC the material culture of Edom extended to sites bordering on Judah, such as Horvat Qitmit and En Hazeva. Yet the biblical text and archaeology put the heartland of Edom in the south Jordanian highlands—the only region south of the Dead Sea that is amenable to significant agricultural activity.[2]

Excavations and surveys conducted on the Edomite plateau revealed no evidence of permanent activity in the Late Bronze Age (Bienkowski 2001a: 257, 265). At least a few Iron I sites are known in Edom. Even if my initial evaluation (Finkelstein 1992) was somehow exaggerated (Bienkowski 2001a), it is clear that Iron I sherds were revealed at Buseirah and Tawilan (Qurayyah Ware—Bienkowski 2001a: 262). Thus far, not a single Iron IIA site has been detected on the Edomite plateau. A wave of settlement there commenced in the late eighth century BCE, probably with the transfer of the main Arabian trade route by the Assyrians from the Dharb el-Ghazza to the plateau east of the Arabah (below). Settlement activity then intensified and reached a peak in the late seventh and early sixth centuries BCE (Bienkowski 2001a). Edom was taken over by Nabonidus in 553 BCE (Bartlett 1989: 157–161; Bienkowski 2001b). This was followed by reduction in human activity in the Persian period (MacDonald et al. 2004: 58: Bienkowski 2001b, reporting Persian period material only in Buseirah and Tawilan).

En Hazeva

This site, probably the location of biblical Tamar (Aharoni 1963), was thoroughly excavated (for the most updated and detailed description of the remains see Cohen and Yisrael 1995). The excavators described the Iron Age finds as representing three fortresses. Scanty remains (Stratum 6) uncovered under the gate of the main fortress (below) were interpreted as belonging to an early fortress from the tenth century BCE.[3]

The main layer (Stratum 5) features the remains of a large Iron Age II casemate fortress with square towers in its corners and a four-entry gate, measuring 100 × 100 m and protected by an earthen glacis and a moat. The excavators described two phases, the earlier featuring a 50 × 50 m casemate structure, which was later incorporated into the northeastern corner of the large fortress. They affiliated the fortress with the kingdom of Judah and dated it to the ninth–eighth centuries BCE. Stratum 4 features the scanty remains of a smaller fortress found over the remains of Stratum 5; it was dated to

[2] Levy describes the area of Wadi Faynan as lowlands Edom (e.g., Levy et al. 2014). As far as I can judge, activity in this region, though close to the plateau geographically, should be associated with territories further to the northwest rather than with Edom (Finkelstein 2005).

[3] Current excavations at the site have revealed evidence for pre-Assyrian-period activity, possibly somewhat similar to the early layer at Tell el-Qudeirat (Tali Erickson-Gini, personal communication).

the late seventh and early sixth century and affiliated with King Josiah of Judah. A small shrine with a rich assemblage of cult vessels was unearthed immediately outside of the north wall of the Stratum 5 fortress. The finds were interpreted as belonging to an Edomite shrine and dated to the time of Stratum 4. No Persian period finds have been reported thus far.

The lack of a final report on the results of the excavations hinders any attempt to reconstruct the history of En Hazeva. Still, the following observations can be suggested: So far no evidence has been presented for the existence of an Iron IIA layer at the site. The vestiges ascribed to Stratum 6 do not seem to represent a fortress (for a similar situation at Kadesh-Barnea see Finkelstein 2010a). The remains of Stratum 5 should be seen as the substructure of a fortress (Ussishkin 2010), which dates to the late eighth and/or seventh century BCE, and which should be affiliated with the Assyrian control of the Arabian trade route via the Edomite plateau to the coast (Na'aman 2001: 267–268). The fortress features certain similarities in layout and method of construction to the one unearthed by Glueck at Tell el-Kheleifeh (above). It was probably manned by locals—Edomites and possibly also Judahites. The nature of the remains of Stratum 4 (whether a fortress at all) and their date have not been fully clarified. The shrine with the locally produced cult vessels indeed dates to the end-phase of the Iron Age; it was devoted to the Edomite deity Qos (Beck 1996; Ben-Arieh 2011).

Kuntillet 'Ajrud

Kuntillet 'Ajrud, possibly the most important site for this discussion, is located on one of the branches of the Dharb el-Ghazza, ca. 50 km south of Ein el-Qudeirat (Kadesh-Barnea). The site dates to the first half of the eighth century BCE. This is clear from the pottery assemblage (Ayalon 1995; Freud 2008; contra Singer-Avitz 2006), the inscriptions (Lemaire 1984), and the evaluation of radiocarbon results (Finkelstein and Piasetzky 2008, with bibliography of previous works). New, yet unpublished short-lived radiocarbon determinations support this date (Boaretto in a lecture at Tel Aviv University, January 2013; for different interpretations of the site's function see Meshel 2012: 68; Ahituv et al. 2012; Na'aman 2012 with references to previous discussions). The finds point to a strong connection with the Northern Kingdom (overview in Mastin 2011; for the pottery see Ayalon 1995; Gunneweg, Perlman and Meshel 1985; for the inscriptions Lemaire 1984; Mastin 2004–2007; 2009; Ahituv et al. 2012: 95, 126–129; Na'aman 2012); a certain connection to Judah is demonstrated in the pottery assemblage (Ayalon 1995; Gunneweg et al. 1985).[4]

Regarding the inscriptions (Ahituv et al. 2012; Na'aman 2012 and bibliography of previous works), the most important for the theme of this chapter are the references to YHWH of Samaria, which appears once, in Inscription 3.1 (see also Inscription 3.8); YHWH of Teman or YHWH of the Teman (Inscriptions 3.6, 3.9, once in each, twice in Inscription 4.1.1 [three times according to Na'aman 2012: 10]); and possibly to a king of Israel in Inscriptions 3.1, 3.6, 3.9 and in an inscription that was omitted from the final publication (Na'aman 2012: 4–5, 8–9). To these I should add Na'aman's proposal (2012: 12–14) that plaster Inscription 4.3 refers to the Exodus story.

Regarding the drawings, the most significant for this chapter is the possible appearance of the king of Israel sitting on a throne on the plaster at the entrance wall (Beck 2000: 180–181; Na'aman 2012: 2–3). Ornan (in press) has recently interpreted more of the drawings as representing royal scenes.

[4] For conflicting views regarding the language of the inscriptions written in Phoenician script, and the identity of the writers—whether Hebrew written by Judahites or Phoenician written by Tyrians—see Ahituv et al. 2012: 130; Lemaire 2013, respectively.

What Did Biblical Authors Know About the Southern Desert?

Let me start with the period of the latest redaction/s of the text by Priestly or post-Priestly scribes (Römer 2007) in the Persian period. The sparsely settled and demographically depleted province of Yehud stretched no further than Beth-zur in the south (Finkelstein 2010a). There was no Jewish presence at that time in the southern Hebron hills or the Beer-sheba Valley. And though several Persian period sites have been recorded in the Negev Highlands (Cohen and Cohen-Amin 2004: 159–201), activity at the key sites in the south was weak: Tell el-Qudeirat lost its importance, Tell el-Kheleifeh too seems to have declined, En Hazeva was not inhabited, there is no evidence for significant Persian period presence at Wadi Faynan and activity on the Edomite plateau was weak. Under these circumstances, Priestly author/s' knowledge of the southern desert must have been fragmentary at best.[5]

The toponyms that appear in the wandering narrative and itineraries can, then, hardly represent Persian period realities. This means that the biblical materials discussed here are based on earlier sources, which reflect earlier realities (from the text perspective see Noth 1968: 213, 1972: 224–227; Fritz 1970: 116–117; Davies 1983; 1995: 342). In the following paragraphs I wish to try to identify the background for these possible sources step by step, from late to early.

In the closing decades of its history, after the Assyrian pull-out from the region, Judah was still strongly present in the Beer-sheba Valley. Further to the southwest, the Iron IIC finds at Kadesh-Barnea indicate that the fort continued to function after Assyria's withdrawal. There are several clues that Judah was active there in the late seventh century: I refer to the Hebrew ostraca (Lemaire and Vernus 1980, 1983; Cohen 2007) that best-fit a date ca. 600 BCE, and to several contemporary Arad ostraca, which seem to refer to Judahite military units that moved in the desert (Ostracon 2 speaks about provisions for a four-day trip, which may fit a journey to Kadesh-Barnea: Aharoni 1981: 15, 145). There are enough clues, then, that Judahites were well-acquainted with the desert until the destruction of the kingdom. The importance of Judah as a participant in the Arabian trade network is manifested in the recently published, ca. 600 BCE Sabaean inscription, which refers to "the towns of Judah" (Bron and Lemaire 2009; Lemaire 2012).

The "Assyrian Century"—ca. 730-630 BCE—evidenced the strongest Judahite activity in the southern desert. Assyria shifted the main Arabian trade route from the difficult-to-control, arid, and isolated Dharb el-Ghazza to the Edomite plateau and the Beer-sheba Valley, which were dominated by the vassal kingdoms of Edom and Judah. This was the time of peak prosperity in the Beer-sheba Valley (Finkelstein 2012a contra Lipschits et al. 2011). The towns and forts there, and especially markets and khans such as the one unearthed at Aroer (Thareani-Sussely 2007: Thareani 2011: 301–307), were places where Judahite merchants and administrators met Edomites and Arabs from the desert (for Hebrew, Edomite, and south-Arabian inscriptions found at Aroer, see Thareani 2011: 223–228). Information about the south could also have been transmitted by Arab merchants who visited Jerusalem (Shiloh 1987; Lemaire 2012).

Beyond the Beer-sheba Valley, the Assyrians controlled the desert trade routes from four pivotal strongholds: Kadesh-Barnea in the east—a Judahite fortress that was probably commissioned by Assyria in order to guard the movement of people and commodities along the Dharb el-Gazza; the large Assyrian forts at En Hazeva and Tell el-Khelleifeh, which were probably manned by local people—Edomites, Arabs, and possibly also Judahites (for Assyria and the three forts see Na'aman 2001: 267–268); and the imperial center built on a large podium at Buseirah. At Kadesh-Barnea, and possibly also En Hazeva, Judahites could meet locals and collect

[5] The Yehud seal impression from Tell el-Qudeirat (Vanderhooft and Lipschits 2007: 27) should be viewed as a chance find, similar to an impression found in Babylon.

information about places and routes in the deeper desert. Judahites could have served in the Assyrian administration and hence travel also to more distant places such as Buseirah and Tell el-Kheleifeh.

This knowledge of the desert finds expression in a variety of biblical references. Ezekiel's description of the southern border of the Land of Canaan (47:19; 48:28) is based on two points of reference—Tamar in the east and Kadesh-Barnea in the west. Attention should also be given to the detailed geographical knowledge (including awareness of one of the desert roads) expressed in the description of the southern border of the tribe of Judah (Josh 15:2–3). Knowledge of the south is also reflected in Genesis 14. This chapter is made up of several layers and part of it is late in date, possibly as late as the Hellenistic period (Granerød 2010). But the military campaign itinerary, which mention El-paran (Eilat), Enmishpat (that is, Kadesh), and Hazazon-tamar (vv. 6–7), is based on the well-known Assyrian control points in the south.[6] This knowledge of the south is expressed in the wandering narrative, in which Kadesh plays a central role (the absence of Tamar is admittedly a problem). Needless to say, the story about the refusal of the king of Edom to let the Israelites cross his territory (Num 20:14–21) should also be anchored in the late eighth to early sixth centuries—the only time in the Iron Age and Persian period with a strong kingdom in this area.

This brings me to the years prior to 720 BCE and to what I consider the most tantalizing question—the origin of the strong Exodus-desert tradition in the Northern Kingdom, as expressed in the prophecies of Hosea and Amos (Hoffman 1983; 1989; Dozeman 2000).[7] The key site for addressing this issue is Kuntillet ʻAjrud, which dates to the first half of the eighth century BCE, that is, to the days of Jeroboam II (788–747 BCE).

Various pieces of information seem to indicate the existence of overland Arabian trade no later than the ninth century (Liverani 1992 for contacts with Mesopotamia; Sass 2005: 118; Jasmin 2005). In its northwestern branch this early Arabian trade could have passed to the Mediterranean coast in either of two routes: along the Edomite plateau (before the emergence of a territorial kingdom there) and the Dharb el-Ghazza. The latter was the shorter alternative, though more difficult to take for the paucity of water sources. In the ninth century the desert trade was probably dominated by Gath and Damascus (before and after ca. 830 BCE—Finkelstein forthcoming b).

This situation changed with the expansion of Adad-nirari III and the decline of Damascus in the closing years of the ninth century. The text of Adad-nirari that mentions Edom (Cogan 2008: 34–35) seems to indicate that he inherited the hegemony of Damascus in the south. Assyria achieved its interests in the region by promoting the power of the Northern Kingdom as an ally/vassal. Probably starting in the days of Joash (who is mentioned in the Tell el-Rimah stela as paying tribute to Adad-nirari) the Northern Kingdom controlled the territories that had previously been ruled by Damascus. And if one takes 2 Kings 14:8–14 as an historical account, Israel also subjugated Judah. The prosperity and domination of Israel in the south was strengthened in the days of Jeroboam II (see, e.g., 2 Kings 14:25).

The Kuntillet ʻAjrud finds indicate that in the first half of the eighth century the Northern Kingdom of Israel dominated not only the southern lowlands but also the desert trade route along the Dharb el-Ghazza and its outlet. Inscriptions (Na'aman 2012) and drawings (Ornan in press) unearthed at the site point to the strong involvement there of an Israelite monarch, most probably Jeroboam II, to the extent that Ornan sees the site as a royal trade station. Regardless of the nature of activity at the site (summaries in Na'aman 2012; Meshel 2012: 65–69), for the sake of this chapter the most important issue is the mention in the inscription of YHWH of

[6] For a possible connection between the routes of the Genesis 14 campaign and the desert wandering (in opposite directions) see Granerød 2010: 106–107.

[7] There is no clue for any importance of the Exodus-wandering tradition in Judah before 720 BCE; there are no references to this narrative in early Judahite prophetic works (Hoffman 1989: 181–182). This is especially significant in view of the possible presence of Judahites (under Israel) at Kuntiillet ʻAjrud.

Teman and YHWH of Samaria. Cult at Kuntillet 'Ajrud seems to have been devoted to YHWH of Teman, that is, YHWH of the southern arid zones and Asherah ("Asherat" according to Na'aman 2012). Teman is mentioned in the Hebrew Bible in relation to Edom, but also to Dedan in northwest Arabia (Jer 49:7–8). Noteworthy are Habakkuk's words, "God came from Teman, and the Holy One from Mount Paran" (Hab 3:3; for this and other reference that connect Teman, Paran, and Sinai, see Ahituv et al. 2012: 96, 130).

YHWH of Samaria should probably be understood as the protection deity of the capital of the Northern Kingdom (compared to YHWH of Jerusalem in the Beit Lei inscription: e.g., Lemaire 1984). The inscription may, in fact, refer to a temple of YHWH at Samaria (Keel and Uehlinger 1998: 228; Dijkstra 2001: 116; Schmid 2012: 53), which may also be hinted at in Hosea 8:6 (Zevit 2001: 391). The Northern Kingdom had two foundation myths, the Jacob Cycle (e.g., de Pury 1991; family tradition, according to Van der Toorn 1996: 287–315); and the Exodus-wandering narrative (state tradition, Van der Toorn 1996: 287–315).[8]

The early layer of the Jacob stories dealt with foundation of temples at Penuel and Bethel. This tradition was evidently promoted in these two cult places (for the importance of Bethel in the eighth century BCE see Finkelstein and Singer-Avitz 2009). Was there a cult-place directly related to the Exodus-wandering tradition? This too could have been venerated at Bethel (e.g., Van der Toorn 1996: 289, who sees the Exodus as an Ephraimite tradition; Dozeman 2000: 55; Mayes 2011: 136); the Temple of YHWH at Samaria is another option (for this temple having cultic literature of its own its own see Schmid 2010: 53). The strong connection of 'Ajrud to the king of Israel and the possible Exodus-related inscription found at the site may support this possibility.

Against this background, it is clear that people from the Northern Kingdom, including Samaria officials and merchants, frequented the site of Kuntillet 'Ajrud in particular and the Dharb el-Ghazza, including the head of the Gulf of Aqaba (above, Ezion-Geber), in general. There they must have met local nomads involved in the southern trade. From their own experience and from these contacts those Israelites must have learned about places and routes in the "deep" desert, mainly those located between the head of the Gulf of Aqaba and the Mediterranean coast.

This is the place to mention toponyms that appear in the Hebrew Bible only in the Numbers 33 list, that is, they are absent from the narrative in Exodus-Numbers and from the Deuteronomy itineraries, and are not mentioned in any other biblical text. They are Dophkah and Alush of vv. 12–14; the group of 12 places from Rithmah to Hashmonah in vv. 18–30, Abronah of vv. 34–35 and Zalmonah and Punon of vv. 41–43. None of them except for Punon can be identified. These place names probably come from a different, independent (Noth 1968: 243; Davies 1995), possibly northern eighth century BCE source. Whether they originally belonged to a pilgrimage itinerary (Noth 1940; 1968: 245–246; Knierim and Coats 2005: 309); whether such an itinerary was connected to the story of the journey of Elijah to Horeb in 1 Kings 19[9]; and whether such a pilgrimage route was related to Kuntillet 'Ajrud are impossible to say. One thing is clear—these sites were no longer relevant to Judahite scribes in the seventh century.

This is as far back as one can go with an answer to the question I posed at the beginning of this chapter: What, how, and when did biblical authors know about the southern desert?

[8] For a possible thematic relationship between the Moses and Jeroboam I stories see Albertz 2001; Schmid 2010: 83 and bibliography.

[9] Though the current text of 1 Kings 19 may represent late redactions (Schmid 2012: 60 and bibliography), and Horeb is a Deuteronomistic expression (Dozeman 1989: 67–68), the origin of the tradition may go back to the ninth century BCE (White 1997).

The Roots of the Exodus-Wandering Tradition

It should be clear that Hosea and Amos did not "invent" the Exodus-desert tradition. So what was the source of this tradition in the Northern Kingdom in the eighth century BCE? How far back can it be traced? Attempts to isolate a "moment in Egypt" in the thirteenth century BCE to fit the Exodus narrative (e.g., Halpern 1993; Kitchen 1998; Hoffmeier 1997; 2005) are doomed to failure (e.g., Finkelstein and Silberman 2001: 48–71), so a more nuanced explanation must be sought. With no clear evidence, neither in the biblical text or Egyptian sources nor in archaeology, one is forced to stride into the territory of historical speculation.

Redford (1987: 150–151; 1992: 412) suggested that the Exodus tradition may have originated from a memory of the expulsion of Canaanites from the Nile Delta in the sixteenth century BCE. Na'aman (2011; following Hendel 2001) recently proposed that the biblical story preserves a memory of oppression inflicted on the people of Canaan by the Egyptian administration in the Late Bronze II–III, in the thirteenth and twelfth centuries BCE. Bietak (1987) and Römer (2002: 54–67) too looked for roots of the Exodus-Moses tradition in the Late Bronze Age. The problem with these theories is that they do not explain why the memory was preserved and promoted in the Northern Kingdom. The southern lowlands—the Shephelah and southern coastal plain—would be a more reasonable place.

The reminiscence of an expulsion from the Delta at the end of the Middle Bronze should have been kept in the southern coastal plain and the area of the Besor, and hieratic inscriptions and other archaeological finds hint that economic oppression in the twelfth century BCE was probably the severest in the southern lowlands. As for the north, Egyptian rule should have been strongly felt mainly in the valleys, around Megiddo and Beth-shean (the main Egyptian stronghold in the area); not in the highlands. The weak holding of Egypt of the highlands is demonstrated by the maneuvers of Shechem of Labayu and his sons in the Amarna period. There is no hint of Egyptian economic pressure in this region; in fact, the highlands, including the north Samaria hills, were sparsely settled at that time. Finally, it is noteworthy that the Hebrew Bible expresses no knowledge of the situation in Canaan in the Late Bronze Age. To sum up this point, in order to get a full picture regarding the origin of the Exodus tradition, one also needs to look for a memory that is specifically connected to the northern highlands and its population, and which is preferably closer in time to the days of Hosea and Amos.

In several recent articles I discussed the role of Egypt of the 22nd dynasty, and more specifically Pharaoh Sheshonq I's campaign, in the decline of the first north Israelite territorial entity of the late Iron I, which was centered in the area of Gibeon-Gibeah to the north of Jerusalem (Finkelstein 2006; most recently Finkelstein and Fantalkin 2012). The Gibeon-Gibeah polity was replaced by the Northern Kingdom, which was centered in the area of Shechem-Tirzah. The rise of this entity too may have been related to the campaign of Sheshonq I (Finkelstein 2012b). Possible involvement of Egypt in the history of Jeroboam I, founder of the Northern Kingdom, is hinted in the LXX version of 1 Kings 12—the "Alternative Story" of the division of the United Monarchy. Scholars are divided regarding the importance of this text—whether reflecting a pre-Deuteronomistic source (Schenker 2000, 2008), or merely a late *midrash* (Talshir 1993; Sweeney 2007). Van der Toorn (1996: 287–315) and Albertz (2001) pointed to the possible function of the Exodus narrative as a charter myth or thanksgiving story in the days of Jeroboam I.

Memories of these events could have been preserved in the areas of both Bethel and Shechem; and they could have been embedded into earlier salvation from Egypt traditions from the lowlands, which were "imported" into the highlands when Israel expanded into the northern valleys (on early salvation from Egypt tradition see Dozeman 2000: 62, 69). And if indeed the "Alternative Story" is based on a pre-Deuteronomistic source, and had there been a

Moses figure already at this early stage (Smend 1995; Blum 2012), another motivation for the adaptation of the tradition could have been the thematic similarities between the biographies of Moses and Jeroboam I (Albertz 2001; Schmid 2010: 83; and bibliography).

The salvation from Egypt tradition thus became one of the two founding myths of Israel. In the early days of the Northern Kingdom it was still an oral tradition. There is no way to know if in this formative phase it included a component about wandering in the wilderness. And it is impossible to say whether at this early stage there was already a "connection" between the Jacob Cycle and the Egypt-Exodus narrative.

Conclusion

Let me summarize what I have proposed for the development of the Exodus-wandering tradition, now from early-to-late, with emphasis on long-term cultural memory (for this concept see, e.g., Assmann 1998; Hendel 2001) more than a specific, single event.

The beginning is vague and now untraceable. Memories of the stormy relationship between Egypt and the population of Canaan in the sixteenth to tenth centuries BCE could have accumulated gradually and developed into a strong tradition of salvation from Egyptian rule among the people of the region. The roots of this tradition were probably located in the lowlands; in the tenth century BCE it was "imported" into the northern part of the central highlands, where it became one of the two charter myths of the kingdom of Israel.

The first intimate acquaintance of people from the North with the southern desert came in the first half of the eighth century, with the strong royal Israelite activity along the Arabian trade route of the Dharb el-Ghazza. This period may provide the background for the first desert itineraries in the Hebrew Bible, as well as for the Exodus-desert materials in Hosea and Amos.

The Exodus-wandering tradition "migrated" to Judah after 720 BCE (Hoffman 1989: 181–182). Archaeology testifies to dramatic growth in Judah in the Iron IIB—in the number of new settlements, size of existing settlements, and population. This cannot be explained as natural growth and must reflect movement of people from Israel to the south after the fall of the Northern Kingdom (Finkelstein and Silberman 2006). These Israelites brought with them to Judah northern traditions, including the Exodus-wandering narrative. It was elaborated on and transformed in the period of Assyrian domination over Judah, when Judahites became intimately acquainted with places in the desert. It seems that elements in the life of Moses and strands of anti-Imperial ideology (Otto 2000: 51–67; Blanco Weissmann 2001; Römer 2002: 24–29; 2003: Schmid 2010: 81) also characterize this phase.

The withdrawal of Assyria from the region in the second half of the seventh century brought about a change in the geopolitical situation. Much of the area which had been dominated by the Assyrians was now ruled by the 26th dynasty in Egypt. Judah and Egypt—each with its own goals of territorial expansion and golden-age ideology—were now on a collision course. The Exodus story, especially the victory of YHWH over the pharaoh of Egypt, served well the ideology of Judah in the days of King Josiah, as a fable about the past and a prediction of the future. Realities of Egypt in the time of the 26th dynasty, when Judahites lived in the Delta, could also have influenced the continuing elaboration of the Exodus tradition (Redford 1987; 1992: 408–422; Van Seters 2001).

The geopolitical realities that facilitated knowledge of the desert in Judah disappeared at the end of the Iron Age. All the large desert fortresses were abandoned and Edom declined after 560 BCE. Priestly scribes who lived in Jerusalem in post-exilic times had no knowledge of the arid regions in the south. But the Exodus and wandering narrative continued to develop. The ongoing compilations, elaborations, and redactions in the Persian period were strictly literal. But they resonated well with the concerns of the time, mainly the "Exodus" from exile in Mesopotamia

(Hoffmann 1998). The work of these Priestly authors gave the Exodus tradition its final shape, and ultimate importance in Jewish and Western tradition, far beyond its modest beginning in the lowlands of Canaan and then the highlands of the Northern Kingdom of Israel.

The Exodus-wandering tradition is therefore the final product of many centuries of accumulation and growth, first oral and then written, with a complex history of redactions in the light of changing political and historical realities.

Acknowledgement The preparation of this chapter was supported by the Chaim Katzman Archaeology Fund, Tel Aviv University.

References

Aharoni, Y. 1963. Tamar and the roads to Elath. *Israel Exploration Journal* 13: 30–42.

———. 1981. *Arad Inscriptions*. Jerusalem: Israel Exploration Society.

Ahituv, S., E. Eshel, and Z. Meshel. 2012. The Inscriptions. In *Kuntillet ʻAjrud (Horvat Teman): An Iron Age II Religious Site on the Judah-Sinai Border*, ed. Z. Meshel, 73–142. Jerusalem: Israel Exploration Society.

Albertz, R. 2001. Exodus: Liberation History against Charter Myths. In *Religious Identity and the Invention of Tradition: Papers Read at a NOSTER Conference in Soesterberg, January 4-6, 1999*, ed. J.W. van Henten and A.W.J. Hautepen, 128–143. Assen: Van Gorcum.

Assmann, J. 1998. *Moses the Egyptian. The Memory of Egypt in Western Monotheism*. Cambridge: Harvard University Press.

Ayalon, E. 1995. The Iron Age II Pottery Assemblage from Horvat Teiman (Kuntillet ʻAjrud). *Tel Aviv* 22: 141–205.

Barker, G.W., R. Adams, O.H. Creighton, D.D. Gilbertson, J.P. Grattan, C.O. Hunt, D.J. Mattingly, S.J. McLaren, H.A. Mohamed, P. Newson, T.E.G. Reynolds, and D.C. Thomas. 1998. Environment and Land Use in the Wadi Faynan, Southern Jordan: the Second Season of Geoarchaeology and Landscape Archaeology (1997). *Levant* 30: 5–25.

Barker, G.W., O.H. Creighton, D.D. Gilbertson, C.O. Hunt, D.J. Mattingly, S.J. McLaren, D.C. Thomas, and G.C. Morgan. 1997. The Wadi Faynan Project, Southern Jordan: a Preliminary Report on Geomorphology and Landscape Archaeology. *Levant* 29: 19–40.

Bartlett, J.R. 1989. *Edom and the Edomites*. Sheffield: JSOT.

Beck, P. 1996. Horvat Qitmit Revisited via En Hazeva. *Tel Aviv* 23: 102–114.

———. 2000. The Art of Palestine during the Iron Age II: Local Traditions and External Influences (10th–8th Centuries BCE). In *Images as Media: Sources for the Culture History of the Near East and the Eastern Mediterranean (1st Millennium BCE)(OBO 175)*, ed. C. Uehlinger, 165–183. Fribourg: University Press.

Ben-Arieh, S. 2011. Temple Furniture from a Favissa at En Hazeva. *Atiqot* 68: 107–175.

Bienkowski, P. 2001a. Iron Age Settlement in Edom: A Revised Framework. In *The World of the Aramaeans II: Studies in History and Archaeology in Honour of Paul-Eugène Dion*, ed. P.M.M. Daviau, J.W. Wevers, and M. Weigl, 257–269. Sheffield: Academic.

———. 2001b. New Evidence on Edom in the Neo-Babylonian and Persian Periods. In *The Land that I will Show You. Essays in the History and Archaeology of the Ancient Near East in Honour of J. Maxwell Miller*, ed. J.A. Dearman and M.P. Graham, 198–213. Sheffield: Academic.

Bietak, M. 1987. Comments on the "Exodus". In *Egypt, Israel, Sinai: Archaeological and Historical Relationships in the Biblical Period*, ed. A.F. Rainey, 163–171. Tel Aviv: Tel Aviv University.

Blanco Weissmann, F. 2001. Sargon, Mose und die Gegner Salomos. Zur Frage vor-neuassyrischer Ursprünge der Mose-Erzählung. *BN* 110: 42–54.

Blum, E. 2012. Der historische Mose und die Frühgeschichte Israels. *HBAI* 1: 37–63.

Bron, F., and A. Lemaire. 2009. Nouvelle inscription sabéenne et le commerce en Transeuphratène. *Transeuphratène* 38: 11–29.

Carr, D.M. 2012. The Moses Story: Literary Historical Reflections. *HBAI* 1: 7–36.

Coats, G.W. 1972. Wilderness Itineraries. *CBQ* 34: 135–152.

Cogan, M. 2008. *The Raging Torrent: Historical Inscriptions from Assyria and Babylonia Relating to Ancient Israel*. Jerusalem: Carta.

Cohen, R. 2007. Inscriptions. In *Excavations at Kadesh Barnea (Tell el-Qudeirat) 1976–1982 (IAA Reports 34)*, ed. R. Cohen and H. Bernick-Greenberg, 245–254. Jerusalem: Israel Antiquities Authority.

Cohen, R., and H. Bernick-Greenberg. 2007. *Excavations at Kadesh Barnea (Tell el-Qudeirat) 1976–1982 (IAA Reports 34)*. Jerusalem: Israel Antiquities Authority.

Cohen, R., and R. Cohen-Amin. 2004. *Ancient Settlement of the Negev Highlands. Volume II: The Iron Age and the Persian Period (Israel Antiquities Authority Reports No. 20)*. Jerusalem: Israel Antiquities Authority.

Cohen, R., and Y. Yisrael. 1995. The Iron Age Fortresses at ʻEn Hazeva. *BA* 58: 223–235.

Cross, F.M. 1973. *Canaanite Myth and Hebrew Epic: Essays in the History of the Religion of Israel*. Cambridge: Harvard University Press.

Davies, G.I. 1974. The Wilderness Itinereries: A Comperative Study. *Tyndale Bulletin* 25: 46–81.

———. 1979. *The Way of the Wilderness: A Geographical Study of the Wilderness Itineraries in the Old Testament*. Cambridge: Cambridge University Press.

———. 1983. The Wilderness Itineraries and the Composition of the Pentateuch. *VT* 33: 1–13.

———. 1990. The Wilderness Itineraries and Recent Archaeological Research. In *Studies in the Pentateuch (Supplements to Vetus Testamentum 41)*, ed. J.A. Emerton, 161–175. Leiden: Brill.

Davies, E.W. 1995. *Numbers: Based on the Revised Standard Version*. London: M. Pickering.

Dijkstra, M. 2001. El, the God of Israel – Israel, the People of YHWH: On the Origins of Ancient Israelite Yahwism. In *Only One God? Monotheism in Ancient Israel and the Veneration of the Goddess Asherah*, ed. B. Becking, M. Dijkstra, M.C.A. Korpel, and J.H. Vriezen, 81–126. Sheffield: Academic.

DiVito, R. 1993. The Tell el-Kheleifeh Inscriptions. In *Nelson Glueck's 1938–1940 Excavations at Tell el-Kheleifeh: A Reappraisal*, ed. G.D. Pratico, 51–63. Atlanta: Scholars.

Dozeman, T.B. 1989. *God on the Mountain: A Study of Redaction, Theology and Canon in Exodus 19–24*. Atlanta: Scholars.

———. 2000. Hosea and the Wilderness Wandering Tradition. In *Rethinking the Foundations: Historiography in the Ancient World and in the Bible, Essays in Honour of John Van Seters*, ed. S.L. McKenzie and T. Römer, 55–70. Berlin: de Gruyter.

———. 2011. The Priestly Wilderness Itineraries and the Composition of the Pentateuch. In *The Pentateuch. International Perspectives on Current Research*, ed. T.B. Dozeman, K. Schmid, and B.J. Schwartz, 257–288. Mohr Siebeck: Tübingen.

Dozeman, T.B., and K. Schmid (eds.) 2006. *A Farewell to the Yahwist? The Composition of the Pentateuch in Recent European Interpretation*. Atlanta: Society of Biblical Literature.

Finkelstein, I. 1992. Edom in the Iron I. *Levant* 24: 159–166.

———. 1995. *Living on the Fringe: The Archaeology and History of the Negev, Sinai and Neighbouring Regions in the Bronze and Iron Ages*. Sheffield: Academic.

———. 2005. Khirbat en-Nahas, Edom and Biblical History. *Tel Aviv* 32: 119–125.

———. 2006. The Last Labayu: King Saul and the Expansion of the First North Israelite Territorial Entity. In *Essays on Ancient Israel in its Near Eastern Context, A Tribute to Nadav Na'aman*, ed. Y. Amit, E. Ben Zvi, I. Finkelstein, and O. Lipschits, 171–177. Eisenbrauns: Winona Lake.

———. 2010a. Kadesh Barnea: A Reevaluation of its Archaeology and History. *Tel Aviv* 37: 111–125.

———. 2010b. The Territorial Extent and Demography of Yehud/Judea in the Persian and Early Hellenistic Periods. *Revue Biblique* 117: 39–54.

———. 2012a. Comments on the Date of Late-Monarchic Judahite Seal Impressions. *Tel Aviv* 39: 203–211.

———. 2012b. Tell el-Farah (Tirzah) and the Early Days of the Northern Kingdom. *Revue Biblique* 119: 331–346.

———. Forthcoming a. The Archaeology of Tell el-Kheleifeh and the History of Ezion-geber/Elath. *Semitica*.

———. Forthcoming b. The Southern Steppe of the Levant ca. 1050-750 BCE: A Framework for A Territorial History. *Palestine Exploration Journal*.

Finkelstein, I., and A. Fantalkin. 2012. Khirbet Qeiyafa: An Unsensational Archaeological and Historical Interpretation. *Tel Aviv* 39: 38–63.

Finkelstein, I., and L. Singer-Avitz. 2009. Reevaluating Bethel. *Zeitschrift des Deutschen Palästina-Vereins* 125: 33–48.

Finkelstein, I., I. Koch, and O. Lipschits. 2012. The Biblical Gilead: Observations on Identifications, Geographic Divisions and Territorial History. *Ugarit-Forschungen* 43: 131–159.

Finkelstein, I., and E. Piasetzky. 2006. ^{14}C and the Iron Age Chronology Debate: Rehov, Khirbet en-Nahas, Dan and Megiddo. *Radiocarbon* 48: 373–386.

———. 2008. The Date of Kuntillet 'Ajrud: The ^{14}C Perspective. *Tel Aviv* 35: 175–185.

Finkelstein, I., and B. Sass. 2013. The West Semitic Alphabet from the Late Bronze to the Iron IIA Examined from an Archaeological Starting Point. *Hebrew Bible and Ancient Israel*. 149–220.

Finkelstein, I., and N.A. Silberman. 2001. *The Bible Unearthed: Archaeology's New Vision of Ancient Israel and the Origin of its Sacred Texts*. New York: The Free Press.

———. 2006. Temple and Dynasty: Hezekiah, the Remaking of Judah and the Rise of the Pan-Israelite Ideology. *Journal for the Study of the Old Testament* 30: 259–285.

Finkelstein, I., and L. Singer-Avitz. 2009. Reevaluating Bethel. *Zeitschrift des Deutschen Palästina-Vereins* 125: 33–48.

Freud, L. 2008. The Date of Kuntillet 'Ajrud: A Reply to Lily Singer-Avitz. *Tel Aviv* 35: 169–174.

Fritz, V. 1970. *Israel in der Wüste: Tradition-geschichtliche Untersuchung der Wüstenüberlieferung des Jahwisten*. Marburg: N. G. Elwert.

Gertz, J.C. 2000. *Tradition und Redaktion in der Exoduserzählung. Untersuchungen zur endredaktion des Pentateuch*. Göttingen: Vandenhoeck & Ruprecht.

Gilboa, A., A.J.T. Jull, I. Sharon, and E. Boaretto. 2009. ^{14}C Dates from Tell el-Qudeirat (Kadesh Barnea). *Tel Aviv* 36: 82–94.

Glueck, N. 1965. Ezion-geber. *BA* 28: 70–87.

Granerød, G. 2010. *Abraham and Melchizedek: Scribal Activity of Second Temple Times in Genesis 14 and Psalms 10*. Berlin: de Gruyter.

Gunneweg, J., I. Perlman, and Z. Meshel. 1985. The Origin of the Pottery of Kuntillet 'Ajrud. *Israel Exploration Journal* 35: 270–283.

Halpern, B. 1993. The Exodus and the Israelite Historian. *Eretz Israel* 24: 89*–96*.

Hauptmann, A. 2007. *The Archaeometallurgy of Copper: Evidence from Faynan, Jordan*. Berlin: Springer.

Hendel, R. 2001. The Exodus in Biblical Memory. *Journal of Biblical Literature* 120: 601–608.

Hoffman, Y. 1983. *The Doctrine of the Exodus in the Bible*. Tel Aviv: Tel Aviv University (Hebrew).

———. 1989. A North Israelite Typological Myth and a Judaean Historical Tradition: The Exodus in Hosea and Amos. *Vetus Testamentum* 39: 169–182.

———. 1998. The Exodus – Tradition and Reality The Status of the Exodus Tradition in Ancient Israel. In *Jerusalem Studies in Egyptology*, ed. I. Shirun-Grumach, 193–202. Harrassowitz: Wiesbaden.

Hoffmeier, J.K. 1997. *Israel in Egypt: The Evidence for the Authenticity of the Exodus Tradition*. New York: Oxford University Press.

———. 2005. *Ancient Israel in Sinai: The Evidence for the Authenticity of the Wilderness Tradition*. Oxford: Oxford University Press.

Jasmin, M. 2005. Les conditions d'émergence de la route de l'encens à la fin du IIe millénaire avant notre ère. *Syria* 82: 49–62.

Kallai, Z. 1998. The Wandering-Traditions from Kadesh-Barnea to Canaan: A Study in Biblical Historiography. In *Biblical Historiography and Historical Geography*, ed. Z. Kallai, 165–174. P. Lang: Collection of Studies, Frankfurt.

Keel, O., and C. Uehlinger. 1998. *Gods, Goddesses and Images of Gods in Ancient Israel*. Minneapolis: Fortress.

Kitchen, K.A. 1998. Egyptians and Hebrews, from Raamses to Jericho. In *The Origin of Early Israel – Current Debate (Beer-Sheva XII)*, ed. E. Oren, 65–131. Jerusalem: Ben Gurion University.

Knierim, R.P., and G.W. Coats. 2005. *Numbers*. Grand Rapids: Eerdmans.

Kratz, R.G. 2005. *The Composition of the Narrative Books of the Old Testament*. London: T. & T. Clark.

Lee, W.W. 2003. *Punishment and Forgiveness in Israel's Migratory Campaign*. Grand Rapids: Eerdmans.

———. 2008. The Concept of the Wilderness in the Pentateuch. In *Israel in the Wilderness. Interpretations of the Biblical Narratives in Jewish and Christian Traditions*, ed. K.E. Pomykala, 1–16. Leiden: Brill.

Lemaire, A. 1984. Date et origine des inscriptions hebraïques et pheniciennes de Kuntillet 'Ajrud. *Studi Epigrafici e Linguistici* 1: 131–143.

———. 2012. New Perspectives on the Trade between Judah and South Arabia. In *New Inscriptions and Seals Relating to the Biblical World*, ed. M. Lubetski, 93–110. Atlanta: Society of Biblical Literature.

———. 2013. Remarques sur les inscriptions phéniciennes de Kuntillet 'Ajrud. *Semitica* 55: 83–99.

Lemaire, A., and P. Vernus. 1980. Les ostraca paléo-hébreux de Qadesh-Barnéa. *Orientalia* 49: 341–345.

———. 1983. L'ostracon paléo-hébreux No. 6 de Tell Qudeirat (Qadesh-Barnéa). In *Fontes atque pontes. Fine Festgabe für Hellmut Brunner*, ed. M. Görg, 302–326. Harrassowitz: Wiesbaden.

Levy, T.E., M. Najjar, A.D. Gidding, I.W.N. Jones, K.A. Knabb, K. Bennallack, M. Vincent, A.N. Lamosco, A.M. Richter, C. Smitheram, L.D. Hahn, and S. Balaswaminathan. 2012. The 2011 Edom Lowlands Regional Archaeology Project (ELRAP): Excavations and Surveys in the Faynan Copper Ore District, Jordan. *Annual of the Department of Antiquities of Jordan* 56: 423–445.

Levy, T.E., M. Najjar, and E. Ben-Yosef. 2014. *New Insights into the Iron Age Archaeology of Edom, Southern Jordan: Surveys, Excavations and Research from the Edom Lowlands Regional Archaeology Project (ELRAP)*. Los Angeles: UCLA Cotsen Institute of Archaeology Press.

Lipschits, O., O. Sergi, and I. Koch. 2011. Judahite Stamped and Incised Jar Handles: A Tool for Studying the History of Late Monarchic Judah. *Tel Aviv* 38: 5–41.

Liverani, M. 1992. Early Caravan Trade between South-Arabia and Mesopotamia. *Yemen* 1: 111–115.

———. 2005. *Israel's History and the History of Israel*. London: Equinox.

MacDonald, B., L.G. Herr, M.P. Neeley, T. Gagos, K. Moumani, and M. Rockman. 2004. *The Tafila-Busayra Archaeological Survey 1999–2001, West-Central Jordan*. Boston: American Schools of Oriental Research.

Martin, M., and I. Finkelstein. 2013. Iron IIA Pottery from the Negev Highlands: Petrographic Investigation and Historical Implications. *Tel Aviv* 40: 6–45.

Mastin, B.A. 2004–2007. The Theophoric Elements *yw* and *yhw* in Proper Names in Eight-Century Hebrew Inscriptions and the Proper Names at Kuntillet 'Ajrud. *Zeitschrift für Althebräistic* 17–20: 109–135.

———. 2009. The Inscriptions Written on Plaster at Kuntillet 'Ajrud. *Vetus Testamentum* 59: 99–115.

———. 2011. Who Built and who Used the Buildings at Kuntillet 'Ajrud? In *On Stone and Scroll: Essays in Honour of Graham Ivor Davies*, ed. J.K. Aitkin, K.J. Dell, and B.A. Mastin, 69–85. Berlin: de Gruyter.

Mattingly, D., J.G. Newson, R. Tomber, G. Baker, D. Gilberston, and C. Hunt. 2007. The Making of Early States: The Iron Age and Nabataean Periods. In *Archaeology and Desertification, the Wadi Faynan Landscape Survey, Southern Jordan*, ed. G. Barker, D. Gilbertson, and D. Mattingly, 271–303. Oxford: Council for British Research in the Levant.

Mayes, A.D.H. 2011. Pharaoh Shishak's Invasion of Palestine and the Exodus from Egypt. In *Between Evidence and Ideology*, ed. B. Becking, 129–144. Brill: Leiden.

Meshel, Z. 2012. *Kuntillet 'Ajrud (Horvat Teman): An Iron Age II Religious Site on the Judah-Sinai Border*. Jerusalem: Israel Exploration Society.

Na'aman, N. 1991. The Kingdom of Judah under Josiah. *Tel Aviv* 18: 3–71.

———. 2001. An Assyrian Residency at Ramat Rahel? *Tel Aviv* 28: 260–280.

———. 2011. The Exodus Story: Between Historical Memory and Historiographical Composition. *Journal of Ancient Near Eastern Religions* 11: 39–69.

———. 2012. The Inscriptions of Kuntillet ‘Ajrud Through the Lens of Historical Research. *Ugarit-Forschungen* 43: 1–43.

North, R.S.J. 2001. Perspective of the Exodus author(s). *Zeitschrift für die Alttestamentliche Wissenschaft* 113: 481–504.

Noth, M. 1940. Der Wallfahrtsweg zum Sinai (Nu 33). *Palästina-Jahrbuch* 36: 5–28.

———. 1968. *Numbers: A Commentary*. London: SCM.

———. 1972. *A History of Pentateuchal Traditions*. Englewood Cliffs: Prentice-Hall.

Ornan, T. In press. The Drawings from Kuntillet ‘Ajrud Reconsidered. In *To Yahweh Teiman and his Ashera, the Inscriptions and Drawings from Kuntillet ‘Ajrud*. S. Ahituv, E. Eshel, Z. Meshel, and T. Ornan. Jerusalem *(Hebrew)*.

Otto, E. 2000. Mose und das Gesetz: Die Mose-Figur als Gegenentwurf Politischer Theologie zur neuassyrischen Königsideologie im 7 Jh. v. Chr. In *Mose: Ägypten und das Alte Testament*, ed. E. Otto, 43–83. Stuttgart: Verlag Katholisches Bibelwerk.

Pratico, G.D. 1993. *Nelson Glueck's 1938–1940 Excavations at Tell el-Kheleifeh: A Reappraisal*. Atlanta: Scholars.

de Pury, A. 1991. Le cycle de Jacob comme légende autonome des origines d'Israël. In *Congress Volume Leuven 1989*, ed. J.A. Emerton, 78–96. Brill: Leiden.

Redford, D.B. 1987. An Egyptological Perspective on the Exodus Narrative. In *Egypt, Israel, Sinai: Archaeological and Historical Relationships in the Biblical Period*, ed. A.F. Rainey, 137–161. Tel Aviv: Tel Aviv University.

———. 1992. *Egypt, Canaan, and Israel in Ancient Times*. Princeton: Princeton University Press.

Römer, T. 1990. *Israels Väter. Untersuchungen zur Väterthematik im Deuteronomium und in der Deuteronomistischen Tradition (OBO 99)*. Freiburg: Universitatsverlag Freiburg.

———. 2002. *Moïse: "lui que Yahvé a connu face à face"*. Paris: Gallimard Découvertes.

———. 2003. La construction d'une "vie de Moïse" dans la Bible hébraïque et chez quelques auteurs hellénistiques. *Revue de l'Institut Catholique de Paris* 85: 13–30.

———. 2007. Israel's Sojourn in the Wilderness and the Construction of the Book of Numbers. In *Reflection and Refraction. Studies in Biblical Historiography in Honour of A. Graeme Auld*, ed. R. Rezetko, T.H. Lim and W.B. Aucker, 419–445. Brill: Leiden.

Römer, T., and K. Schmid (eds.). 2007. *Les Dernières Rédactions du Pentateuque, de l'Hexateuque et de l'Ennéateuque*. Leuven: Uitgeverij Peeters.

Roskop, A.R. 2011. *The Wilderness Itineraries: Genre, Geography, and the Growth of Torah*. Winona Lake: Eisenbrauns.

Sass, B. 2005. *The Alphabet at the Turn of the Millennium*. Tel Aviv: Tel Aviv Univesity.

Schenker, A. 2000. Jeroboam and the Division of the Kingdom in the Ancient Septuagint: LXX 3 Kingdoms 12.24 A-Z, MT 1 Kings 11-12; 14 and the Deuteronomistic History. In *Israel Constructs its History: Deuteronomistic History in Recent Research*, ed. A. de Pury, T. Römer, and J.-D. Macchi, 214–257. Sheffield: Academic.

———. 2008. Jeroboam's Rise and Fall in the Hebrew and Greek Bible. *Journal for the Study of Judaism* 39: 367–373.

Schmid, K. 2010. *The Old Testament: A Literary History*. Minneapolis: Fortress.

———. 2012. Genesis and Exodus as Two Formerly Independent Traditions of Origins for Ancient Israel. *Biblica* 93: 187–208.

Shiloh, Y. 1987. South Arabian Inscriptions from the City of David, Jerusalem. *Palestine Exploration Quarterly* 119: 9–18.

Singer-Avitz, L. 2006. The Date of Kuntillet ‘Ajrud. *Tel Aviv* 33: 196–228.

———. 2008. The Earliest Settlement at Kadesh Barnea. *Tel Aviv* 35: 73–81.

Smend, R. 1995. Mose als geschichtliche Gestalt. *Historische Zeitschrift* 260: 1–19.

Sweeney, M.A. 2007. A Reassessment of the Masoretic and Septuagint Versions of the Jeroboam Narratives in 1 Kings/3 Kingdoms 11-14. *Journal for the Study of Judaism* 38: 165–195.

Talshir, Z. 1993. *The Alternative Story of the Division of the Kingdom (Jerusalem Biblical Studies 6)*. Jerusalem: Simor.

Thareani-Sussely, Y. 2007. Ancient Caravanserais: An Archaeological View from ‘Aroer. *Levant* 39: 123–141.

———. 2011. *Tel ‘Aroer: The Iron Age II Caravan Town and the Hellenistic Early Roman settlement*. Jerusalem: Hebrew Union College.

Ussishkin, D. 2010. En Ḥaṣeva: On the Gate of the Iron Age II Fortress. *Tel Aviv* 37: 246–253.

Van der Toorn, K. 1996. *Family Religion in Babylonia, Syria and Israel: Continuity and Change in the Forms of Religious Life*. Leiden: Brill.

Vanderhooft, D., and O. Lipschits. 2007. A New Typology of the Yehud Stamp Impressions. *Tel Aviv* 34: 12–37.

Van Seters. 2001. The Geography of the Exodus. In *The Land that I Will Show You: Essays on the History and Archaeology of the Ancient Near East in Honor of J. Maxwell Miller*, ed. J.A. Dearman and M.P. Graham, 255–276. Sheffield: Academic.

Walsh, J.T. 1977. From Egypt to Moab: A Source Critical Analysis of the Wilderness Itinerary. *Catholic Biblical Quarterly* 39: 1–19.

White, M. 1997. *The Elijah Legends and Jehu's Coup*. Atlanta: Scholars.

Zevit, Z. 2001. *The Religions of Ancient Israel: A Synthesis of Parallactic Approaches*. London: Continuum.

4 Exodus Dates and Theories

Lawrence T. Geraty

Abstract

This chapter surveys in a brief and introductory manner the scholarly views regarding the dating of the Exodus described in the Hebrew Bible, including especially the "traditional" 18th dynasty date (ca. 1450 BCE) and the current "consensus" 19th dynasty date (ca. 1250 BCE), but touching also on other Exodus dates advocated from ca. 2100 BCE through ca. 650 BCE. These are summarized in table form with the relevant bibliography. Theories of the date are usually accompanied by identifications of the pharaohs involved, and these are briefly surveyed as well.

While several scholars doubt there was any such thing as a historical exodus of Israelites from Egypt (such as Fleming, Halpern, Hendel and Propp), many others propose some kernel of historicity though details differ widely (Allen, Assmann, Batto, Benz, Berner, Bietak, Bloch, Dever, Faust, Finkelstein, Friedman, Galvin, Harris, Hoffmeier, Hollis, Maeir, Matthews, Moro, Moshier, Mullins, Na'aman, Noegel, Redford, Rendsburg, Russell, Schneider, Sparks). Others advocate a science-based approach (Bronk Ramsey, Dee, Grattan, Higham, Levy, McCoy, Salamon, Ward, Wiener). Most others recognize the theme's importance primarily in Israelite memory.

L.T. Geraty (✉)
La Sierra University, Riverside, CA, USA

American Schools of Oriental Research (ASOR), Boston, MA, USA
e-mail: lgeraty@lasierra.edu

As a member of the organizing committee for the conference, I want to thank each of the speakers for their participation, for honoring us with their presence, and for submitting their papers as chapters for this book. It has been a long time since so many eminent scholars got together to talk about the biblical Exodus and its context in history and religious thought, much less prepare their remarks for publication. Indeed one might say that we have a majority of the consensus-making scholars of the Exodus represented here, assembled all in one place for the first time in such numbers to discuss the Exodus.

Each of you gives a specialized treatment of some aspect of the topic of the Exodus. My assignment is to give a brief overview of theories and dates as a backdrop for the rest of the chapters. I apologize in advance if it seems too elementary. My purpose is not to break new ground but rather to summarize what I call "Exodus dates I have known!"

T.E. Levy et al. (eds.), *Israel's Exodus in Transdisciplinary Perspective*,
Quantitative Methods in the Humanities and Social Sciences, DOI 10.1007/978-3-319-04768-3_4,

I begin with the traditional date in the 18th dynasty, move to the current consensus date in the 19th dynasty, and then even more briefly touch on several other Exodus theories. Views that it never happened do not give us a date since it is a nonevent. Views that it did not happen in much like the way described in the Bible are not very datable either. In other words, this overview of dates necessarily pertains more to "centrist" rather than "minimalist" positions, as well, of course, the face value, somewhat literalist understanding of the Exodus traditions—namely, that there was a datable event, that we find the round numbers helpful, that Moses had a genuine relationship to the Exodus narrative, and the like.

Traditional Date ca. 1450 BCE, in the 18th Dynasty

This theory begins with the chronological statements in the Hebrew Bible. The key text is 1 Kings 6:1, "In the 480th year after the Israelites had come out of Egypt, in the 4th year of Solomon's reign over Israel, in the month of Ziv, the second month, he began to build the temple of the Lord."

Most scholars agree that the fourth year of Solomon is ca. 970 BCE, plus or minus a few years. Adding 480 years to that number, one gets 1450 BCE (some would argue that "480th" is an ordinal number and that 479 years should be added). That would take us to the New Kingdom's 18th dynasty, perhaps on the death of Thutmose III that year, depending on the chronology used, or some have suggested Amenhotep II. It seems too good to be true, doesn't it?

The usual critical response is that the 480-year figure is based on the theory that there were 12 generations of 40 years each between the Exodus and the United Monarchy. But modern scholarship suggests that we know that 40 years is too many for a generation, with the suggestion that 25 years would be closer to the mark. And if you multiply 12 times 25, that would be 300 years. When 300 is added to 970 BCE, that would take one back to ca. 1270 BCE in the 19th dynasty—but that is getting ahead of our story.

Traditionalists say that I Kings 6:1 is not the only chronological datum. There is Judges 11:26 where Jephthah, the Gileadite judge, is in debate with the Ammonites who are trying to retake the Mishor or the Madaba Plains from the Israelites. In response to the Ammonite argument that this territory really belongs to them, Jephthah says, "For 300 years Israel occupied Heshbon, Aroer, the surrounding settlements and all the towns along the Arnon. Why didn't you retake them during that time?"

Jephthah, as one of the so-called judges, is dated to the eleventh century BCE. If one adds Jephthah's round number of 300 years that the Israelites had been in the Transjordan, it takes you to roughly 1400 BCE for the Conquest and of course that fits the 1450 date for the Exodus. Traditionalists point out that 300 is not divisible by a generation of 40 years, though as a round number it could fit almost any length of a generation. In other words, the figures in 1 Kings 6:1 and Judges 11:26 work together to suggest an Exodus from Egypt ca. 1450 BCE.

Does such a date work within the context of what we know about the New Kingdom's 18th dynasty? Following William Shea, one spokesperson for the traditional date, the answer would be yes. Shea (1982: 233) writes:

> The pharaohs of this period must be dated as accurately as possible before the attempt is made to associate biblical events with them, because if they have been misdated then the correlations suggested by the biblical date for the Exodus will be incorrect. The chronology of the 18th dynasty has been established by using three types of data: Sothic cycle dates, new moon dates, and the highest-numbered regnal years attested for each of the kings who ruled during this period.

A recent paper by Doug Petrovich at the University of Toronto in the *Journal of Ancient Egyptian Interconnections* argues for a traditional dating of the Exodus based on archaeological evidence and inscriptional evidence as well as Egyptian dynastic chronology (Petrovich 2013; see Billington 2013 for a helpful summary and background).

Following the correlations suggested by scholars such as Shea, Petrovich, and others, a typical outline is given below, with apologies to those who have somewhat different alternatives on details.

Thutmose I would fit as the pharaoh who issued the death decree for all male Hebrew babies. It would have come early in his reign and thus not have affected Aaron who was 3 years older than Moses (Exod 7:7). He moved his court to Memphis where his daughter might have come in contact with baby Moses.

It is possible that Hatshepsut, then, would have been pharaoh's daughter who rescued the baby Moses, who would have grown up during the reign of her father, Thutmose I, and her husband, Thutmose II. If Moses fled Egypt when he was 40 (according to Acts 7:23), then it would have been late in Hatshepsut's reign when her co-regent, Thutmose III, would have begun to assert independence. She would also have been the pharaoh who died while Moses was in exile (Exod 2:23).

We know about the death of Thutmose III in March (i.e., the time of Passover) of ca. 1450 BCE from the tomb biography of Amenemhab who served in the Egyptian navy under several pharaohs. The cities of Pithom and what would later be called Ra'amses, cities built by the enslaved Israelites (Exod 1:11), would have been needed as store cities for the many expeditions Thutmose III led into Asia.

It is interesting that the mummy in the Cairo Museum labeled Thutmose III has been estimated to be between 35 and 40 years of age according to Harris and Wente in their *X-Ray Atlas of the Royal Mummies* (1980: table 6.4) or 45 years as per Harris and Weeks in their book *X-Raying the Pharaohs* (1973: 138)[1]; yet, we know that he reigned 54 years, so presumably he should have been at least 60 if not 70 when he died. Could another body have been substituted for Thutmose III when his was not recovered from the Reed Sea/Red Sea?

Bear with me now, I am merely describing the scenario proposed by current advocates of the traditional date of the Exodus, ca. 1450 BCE (and allowing for small shifts such as Shea 2003: 254, who abandoned his elaborate 1450 scenario and embraced 1446 BCE). Readers will have to judge for themselves the merits of the arguments and the evidence, considered by most of the scholars, I am sure, to be nonhistorical fantasy at worst and circumstantial at best.

There is good evidence for a 2-year 4-month, co-regency of Thutmose III and his son, Amenhotep II, who happened to be campaigning in Asia when his father died. He rushed back to Egypt to assume sole kingship at which time he executed the foreign chiefs he brought back with him as captives. From a biblical point of view, such an unusual action fits the actions of an enraged son of the pharaoh of the Exodus who returned to Egypt to find his father dead from circumstances caused by the Hebrew slaves. It is also interesting to note that the first contemporary Egyptian reference to 'Apiru outside of Egypt comes from this time when Amenhotep II brought back to Egypt from Syro-Palestine some 3,600 'Apiru from among his 90,000 captives. Was this to compensate for the loss of Hebrew slaves?

The so-called Dream Stele of Thutmose IV was taken by previous advocates of a fifteenth-century Exodus as indirect evidence for the tenth plague, death of firstborn, because he claimed that he was not a legitimate heir, i.e., not a firstborn, and thus not in line to succeed his father, Amenhotep II.[2] There was likely another

[1] My thanks to Thomas Schneider for calling my attention to this data on the X-raying of what is believed to be Thutmose III's mummy and for these references, though he adheres of course to the generally accepted dating of Thutmose III's life and reign and identification of his mummy.

[2] My thanks again to Thomas Schneider for a Betsy Bryan reference questioning Thutmose IV's own statement of his non-legitimacy as heir to the throne, i.e., alluding to the speculative possibility that maybe he was indeed a firstborn but was denying it in the Dream Stela for an obscure reason of gaining divine legitimation instead of relying on birthright. However, Bryan does not directly say this and merely suggests that it is *possible* that the Dream Stela is not partially historical but may be wholly fictitious, but she is not definite about that possibility either (Bryan 1998: 41). Aidan Dodson finds other evidence of an elder brother of Thutmose IV who was heir

son, a firstborn, who died earlier in the reign of Amenhotep II, thus not affecting the legitimacy of Thutmose IV's claim to succession on other, rare grounds.

The Amarna Letters mention the Habiru, some of whose activities could be consistent with what we know about the Hebrews in the early period of the Judges, though as is well known, while the Hebrews could have been Habiru, not all Habiru were Hebrews.

I can imagine many conference colleagues saying that this circumstantial evidence from the 18th dynasty is just that and very tenuous if not hypothetical in the absence of more direct statements. Indeed, many objections have been raised to this traditional date for the Exodus. I have already mentioned the objection that the traditional date is based on 12 generations of 40 years each.

The major objection, however, to a fifteenth century date for the Exodus (other than the nature of the biblical text) relates to the results of excavations at Palestinian cities mentioned in the biblical account—cities such as Heshbon, Jericho, Ai, Bethel, Debir, and Gibeon. I think that the key site of Hazor can be argued either way. Ad hoc explanations can be supplied for most of these sites, but there is no question that the archaeological evidence is inconclusive at best and problematic at worst.

Consensus Date in the Late Thirteenth Century BCE, in the 19th Dynasty

Pharaoh Merneptah's "Israel" Stele (ca. 1220 or 1208 BCE depending on the Egyptian chronology used) fixes the latest date before which the Exodus must have occurred since it mentions Israel as a people among names that otherwise refer to places in Palestine. We have already referred to the reinterpretation of the chronological datum in 1 Kings 6:1 as a symbolic, idealized interval. But it is primarily the archaeological evidence from excavated Palestinian sites that has been used to bolster the current scholarly consensus that if there was an Israelite Exodus from Egypt, it must have occurred sometime in the thirteenth century BCE, in other words, sometime during the 19th dynasty. Does evidence from Egypt in the thirteenth century support such a theory?

Exodus 1:11 indicates that the Israelites built the city of Ra'amses for the pharaoh of the oppression. Only two pharaohs before Merneptah bore the name Ramses (or Ramesses, among variant spellings). Ramses I was not very significant since he reigned less than 2 years. But Ramses II ruled Egypt from ca. 1290 through ca. 1224 BCE and is known as a great builder. (Again, there are several sets of competing chronologies of the New Kingdom and thus dates for Ramses II; another fairly commonly accepted dating is ca. 1279–1213 BCE.)

Ramses II's royal residence, Pi-Ramses (or Piramses), was located in the Eastern Nile Delta. At one time Tanis, on the Tanitic branch of the Nile, was thought to be Piramses. But the literary evidence relating to Piramses does not accord well with Tanis, and there is no architectural and stratigraphic evidence for its existence prior to the 20th dynasty (though there is an inscription of Ramses II mentioning Tanis).

Since 1930 various studies have suggested that Qantir, on the Pelusiac (easternmost) branch of the Nile, may be Piramses because of the fertility of its surrounding fields, its location on both the land and sea routes to Asia, the existence of a palace of Ramses II there, and the geographical divisions of the city and its surrounding regions that all correspond to the literary references to Piramses. Tell el-Dab'a covers quite a bit of territory just to the south of Qantir. As Manfred Bietak has shown, the occupation of this site under the Middle Kingdom's 12th and 13th dynasties was brought to an end with a violent destruction. Three Hyksos strata or

and firstborn to Amenhotep II, designated Amenhotep C, particularly another stela found near the Great Sphinx, not to be confused with the Dream Stela (Dodson and Hilton 2004: 132, 135, 137–8). Dodson notes the unusual "mutilation of the monuments" of Amenhotep C and seems to suggest that Thutmose IV's Dream Stela was intended to "justify Thutmose's seizure of power" from him (ibid.: 137). Rather than a political coup, perhaps the premature death of firstborn Amenhotep C gave an opportunity to Thutmose IV to seize the throne.

building sub-phases follow this destruction, and the city enlarged progressively through these three periods. The destruction of the third and last Hyksos stratum has been connected with the recapture of Lower Egypt by the early 18th dynasty.

At one time it was believed that Tell el-Dab'a was left unoccupied until it was rebuilt under the 19th dynasty, thus seeming to lend support to the theory that the Exodus must have taken place in the thirteenth century BCE, during the 19th dynasty. But again, Bietak's excavations have shown major activity in the 18th dynasty, specifically concentrated in the reigns of Thutmose III and Amenhotep II, when Tell el-Dab'a was a major palace district, port city, trade center, and primary naval base, with a "squatter settlement" surrounding the area. (Could the "squatters" have been "slaves"?) Professor Bietak suggests that the geography of the Exodus reflects the Ramesside period not to mention the 400 Year Stele, the Anastasi Papyrus, the Israel Stele, the four-room house in twelfth-century Thebes, the hill country settlements in Palestine, etc. (Bietak 2003, Bietak 2010a, b; Bietak, this volume, Chap. 2).

There are other reasons, too, that have been adduced to support such a theory:

1. Egyptian tomb reliefs depict 'Apiru or Habiru working in vineyards in the northeastern delta of Egypt but Albright (1968: 157) had pointed out that in the 19th dynasty there appears to have been a drastic reduction of wine-jar sealings (due to Exodus?).
2. Ramses II's military campaign into the Sinai, Negev, and Transjordan appears to have taken place in his 18th year; some have suggested that this could have been a "search and destroy" mission against escaped Israelites.
3. Ramses II's treaty with the Hittites contains an unusual stipulation, i.e., that it was incumbent on the Hittite king, in the event that Ramses' "own subjects" committed "another crime against him," that the Hittite king would come to his aid in suppressing such a disorder. Could this stipulation from Ramses' 21st year be a veiled reference to the Exodus?
4. George Mendenhall (1955: 30) suggested that the similarities between the Hittite covenant treaties and the form of the Mosaic covenant could be most easily explained if Moses had been in Egypt during the thirteenth century BCE.

Beyond the generalized nature of the preceding suggestions, some problems with dating the Exodus in the thirteenth century arise when the history of this period is examined from the biblical point of view. These problems have to do with the pharaohs involved: the pharaoh of the oppression, who died while Moses was in exile (Exod 1:23), and the pharaoh who died during the Exodus (Psalm 136:15; cf. Exod 14:4, 17–18, 26–28). It must be said that neither Seti I nor Ramses II works well for the pharaoh of the oppression, nor do Ramses II or Merneptah work that well for the pharaoh of the Exodus.

Other Theories for the Exodus

Having started this chapter with a brief discussion of the traditional early date of ca. 1450 BCE, and then moved on to the mainstream-consensus late date of ca. 1250 BCE, I would like now to proceed to briefly review the broad range of other proposed Exodus dates in chronological sequence from the earliest to the latest Exodus dates. This will be in summary form only because there is no space to explore any of these other theories in any depth, and it should be noted that some of the major advocates of these other theories have chapters in the present volume.

I want to emphasize that I try not to get bogged down in minute arguments over exact years that would be impossible to determine anyway based on our current state of knowledge despite what some advocates might claim—even to the exact day, month, and year! I round off dates to the nearest 50 or 100 years for clarity and simplicity of presentation wherever possible.

These proposed dates and theories of the Exodus can be easily seen in the tables that Brad Sparks has conveniently put together for me, which summarize chronologically the main

Table 4.1 Proposed dates and theories of the Exodus

Approximate date (±50 to ±100 years)	Theory/pharaoh identification	Scholars (non-exhaustive list)
ca. 2100 BC	MB I/EB IV (IBA) Exodus	R.Cohen (1983), Anati (1985), Neev and Emery (1995), Alon (1999); cf. Nigro (2014)
ca. 1700 BC		Josephus, Africanus
ca. 1600 BC	Thera eruption	Bruins and van der Plicht (1996)
ca. 1500–1450 BC	Thera eruption	Bennett (1963), Galanopoulos (1964), Vitaliano (1968, 1973), Goedicke (1981, 2004)
ca. 1550 BC	Hyksos expulsion as Exodus	Manetho (Josephus agrees; dated ca. 1700 BC), Hall (1927)
ca. 1350 BC	Leper expulsion as Exodus	Manetho (Josephus disputes; dated ca. 1500 BC)
ca. 1500 BC	Ramses II as pharaoh of oppression	Archbishop Ussher (1650) (dated Exodus 1491 BC)
ca. 1450 BC	Traditional early date Exodus (Thutmose III or Amenhotep II)	Watzinger (1913), Peet (1922), Horn (1977), Goedicke (1981, 2004), Petrovich (2013), Billington (2013), et al.
ca. 1400 BC (1840s dating of Ramses II Dyn.19—current dating ca. 1250 BC)	Ramses II as pharaoh of oppression	Lepsius (1849) (dated Exodus ca. 1314 BC)
ca. 1300 BC	Traditional Jewish date (ca. 1313 BC)	R. Yose b. Halafta, *Seder Olam* (ca. 160 AD)
ca. 1250 BC (Ramses II and/or Merneptah)	Late date Exodus	Current consensus
ca. 1170 BC	Sea Peoples-Philistine era Exodus	D.N.Freedman (1980, 2006), Rendsburg (1992), Gordon (1997)
ca. 650 BC	Egyptian Saite period Exodus	Redford (1992)

Exodus pharaoh identity theories together with their primary advocates (Tables 4.1, 4.2, and 4.3).

1. The theory with the earliest Exodus date centers in the Middle Bronze I or Early Bronze IV, now often called the Intermediate Bronze Age (IBA), ca. 2100 BCE (and there are various tugs on the dating of this period, upward and downward). Primary advocates include Rudolph Cohen in 1983 (and 1992, 1999), Emmanuel Anati in 1985 (1986, 1997, 2001, 2013: 74–81; see this volume, Chap. 35), David Neev and Kenneth Emery in 1995, and David Alon in 1999. Cohen, Anati, and Alon cite various Egyptian text parallels to the Exodus. Current excavator of Jericho, Lorenzo Nigro (2014) suggests that memory of the EB destruction of Jericho entered the Biblical Conquest narrative.

   This archaeological period is correlated with the late Old Kingdom and First Intermediate Period in Egypt, with some advocates connecting the Exodus specifically with pharaoh Pepi II Nefer-ka-re of the 6th Dynasty. Some have observed a connection with the report of Artapanus of Alexandria, ca. 200 BCE, who described the birth story of Moses as occurring during the reigns of pharaohs simultaneously ruling a country divided North and South, under pharaoh Nekheph-res or Kheneph-res in the South and Palmanoth-es in the North.
2. An early theory advocated by the chronographer Julius Africanus and the Roman era Jewish historian Josephus put the Exodus—in what we would now reckon at ca. 1700 BCE—based on the information known to them at the time about the Hyksos.
3. A more recent theory takes advantage of the volcanic eruption of Thera, now better known as Santorini, dated to the sixteenth century BCE, and what it did to the Eastern Mediterranean. The correlation of the Thera eruption to the Exodus event was first proposed by British mathematician John G. Bennett in 1962 (Bennett 1963), then

Table 4.2 Exodus pharaoh identifications and theories

Pharaohs of oppression and Exodus	Dynasty	Period	Date (approx. *current* dating)	Author or advocate (non-exhaustive list)
(Pepi II Nefer-ka-re implied)	Dynasty 6 end	Old Kingdom end	ca. 2100 BC	(R. Cohen 1983, Anati 1985, Neev and Emery 1995, Alon 1999)
Northern–Southern co-rulers Nekheph-res/Khe-neph-res (in S Egypt) (Nefer-ka-re Pepi II) (Khaneferre Sobekhotep?)	Dyn. 6/13?	Old Kingdom end/MK?	ca. 2100/1700? BC	Artapanus (ca. 200 BC)
Pal-manothes (in N Egypt) (Amenem-het III/ Amenm-hethes III) (Amenhotep II?)	Dyn. 12/18?	Middle Kingdom end/NK?	1800/1400? BC	(via Clement of Alexandria, ca. 200 AD, Eusebius, ca. 326 AD)
Sesostris or son/grandson (Sesostris III or Amenemhat/Ammenemes III/IV)	Dynasty 12	Middle Kingdom end	ca. 1800 BC	Whiston (1722)
Hyksos Expulsion and Leper Expulsion theories	Early to Mid-Dyn. 18	New Kingdom	ca. 1500 to 1300 BC	Manetho (ca. 260 BC) Josephus (ca. 100 AD) Africanus (ca. 222 AD) (dated ca. 1700 BC) Eusebius (ca. 326 AD) (dated ca. 1500 BC)
Ramses II Merneptah	Dynasty 19	New Kingdom	ca. 1250 BC	Archbishop Ussher (1650) (dated Ramses ca. 1500 BC)
Ramses III	Dynasty 20	New Kingdom	ca. 1170 BC	D.N.Freedman (1980), Rendsburg (1992), Gordon (1997)
Bokkhoris	Dynasty 24	Third Intermediate	ca. 700 BC	Lysimachus (ca. 100 BC)

Table 4.3 Hyksos and Leper Expulsion theories of the Exodus

Amoses/Tethmosis (Ahmose) or Akhenkheres (Akhenaten?)	Early Dyn. 18	New Kingdom (Hyksos expulsion)	ca. 1550 BC	Manetho (ca. 260 BC) Josephus (ca. 100 AD) Africanus (ca. 222 AD) (dated ca. 1700 BC)
	Mid-Dyn. 18	New Kingdom (Hyksos expulsion)	ca. 1350 BC	Eusebius (ca. 326 AD) (dated ca. 1500 BC)
Amenophis—Sethos—Rampses (Amenhotep IV Akhenaten-Seti I -Ramses II??)	Late Dyn. 18/ Early Dyn. 19	New Kingdom (Leper expulsion)	ca. 1300 BC	Manetho (ca. 260 BC) (Exodus No. 2)

elaborated upon by Greek archaeologist Angelos Galanopoulos in 1964, supported by geologist Dorothy Vitaliano in 1968 (14–16, and 1973: 254–256), and further developed by the Egyptologist Hans Goedicke in 1981 and updated at book length in 2004. They put the date of the eruption and the Exodus at around the traditional date ca. 1450 BCE based on early radiocarbon dating of organic material in Theran ash (Goedicke, in the 1470s). This dating has aroused great controversy due to later radiocarbon dating, as well as tree ring and ice core dates now largely abandoned, which point to Thera's eruption at ca. 1650–1600 BCE. The new lower Thera dating has in turn been connected with the Exodus by archaeologists Hendrik Bruins and Johannes van der Plicht in 1996, writing in *Nature*.

4. Manetho, the third century BCE priestly historian who divided the Egyptian pharaohs into dynasties, believed that the Exodus could be equated with the Hyksos expulsion from Egypt which dates to ca. 1550 BCE if rendered in our modern dating system. Josephus essentially endorsed Manetho's theory though with a chronology we would now reckon at about 1700 BCE or some 150 years earlier than modern reconstructions of Manetho's dates. In 1927, H. R. Hall supported Manetho's view (p. 5), and many others have followed since. Manetho also may have offered another theory that the Exodus was the expulsion of lepers from Egypt ca. 1350 BCE (though Erich Gruen has argued that this is a later scribal interpolation: 1998: 57–72). The leper correlation with the Exodus was hotly disputed by Josephus as anti-Semitism.
5. Archbishop Ussher, the same divine who advocated the date for Creation in 4004 BCE, in 1650 dated the Exodus to 1491 BCE, suggesting that Ramses Miamun (Ramses II) was the Pharaoh of the Oppression prior to the Exodus. That this was so, popular at the time and in succeeding centuries, appears totally forgotten today.
6. Already in 1849, Karl Lepsius had dated the Exodus to 1314 BCE. Lepsius suggested—like Ussher before him—that Ramses II was the Pharaoh of the Oppression, but roughly around 1400 BCE (Lepsius 1849: 360–364), whereas Ussher put him in the sixteenth century BCE. These very high dates for Ramses II would surprise us today. The dating of Ramses by other nineteenth-century Egyptologists was still higher than Lepsius' and led to an Exodus date in agreement with the traditional date of the Exodus around 1450 BCE. As the work of Egyptology progressed, Ramses II continued to be connected to the Exodus even as the date of Ramses came down to ca. 1250 BCE, until this date and identification of pharaoh became the current consensus. Thus the early date became the late date of the Exodus without anyone realizing this effect of the downward slide in dates of Egypt's dynasties.
7. We have already treated at a bit more length the so-called traditional early date for the Exodus ca. 1450 BCE—usually under Thutmose III or Amenhotep II. Notable modern advocates of that approximate date include archaeologist Carl Watzinger in 1913 (in Sellin and Watzinger 1913: 108, 111), Egyptologist T. Eric Peet in 1922 (p. 112), Egyptologist Siegfried Horn in 1977 (p. 23), and many of a more conservative bent than Watzinger and Peet, up to the present time as I have already mentioned. It is worth noting that most of these are Egyptologists/archaeologists rather than biblical scholars. In 2004, Egyptologist Hans Goedicke suggested specific dates of ca. 1473–1472 BCE for the Exodus occurring during the reign of female pharaoh Hatshepsut (Goedicke 2004: 104).
8. The traditional Jewish date of Rabbi Yose ben Halafta (in the Seder Olam, ca. 160 CE) put the Exodus ca. 1313 BCE.
9. Again, we have already treated at a bit more length the so-called current consensus late date Exodus of ca. 1250 BCE with Ramses II and/or Merneptah as pharaoh. Most scholars who give any credence at all to a historical Exodus from Egypt could be cited as advocates for this theory.
10. David Noel Freedman in 1980 (p. 131; and 2006: 297), as well as Gary Rendsburg (1992: 513–516) and Cyrus Gordon (1997: 151–152), suggested the possibility of an Exodus during the time of the Sea Peoples and the Philistines, ca. 1175–1170 BCE.
11. Egyptologist Donald Redford in 1992 suggested the possibility of an Egyptian Saite period Exodus, connected to a Hyksos or a leper expulsion, as late as ca. 650 BCE.

As one can see, theories regarding the Israelite Exodus from Egypt have in modern times been seriously proposed yielding Exodus dates across a millennium and a half of time. There are some additional useful studies in the bibliography for those wishing to see a general discussion of the Exodus (Aling, Bright, Finkelstein and Silberman, Gottwald, Hoffmeier, Humphreys, Kitchen, Shanks, and Wright).

References

Albright, William F. 1968. *Yahweh and the Gods of Canaan*. New York, NY: Doubleday.

Aling, Charles F. 1981. *Egypt and Bible History*. Grand Rapids, MI: Baker.

Alon, David. 1999. *New Light on the Exodus* [Hebrew]. Jerusalem: Ariel Publishing.

Anati, Emmanuel. 1985. Has Mt. Sinai Been Found? *Biblical Archaeology Review* 11(4, July/August): 42–57.

———. 1986. *Montagna di Dio*. Milan, Italy: Editoriale Jaca Book. English translation: Anati, Emmanuel. 1986. *The Mountain of God*. New York, NY: Rizzoli.

———. 1997. *Esodo tra mito e storia*. Valcamonica: Edizioni del Centro.

———. 2001. *The Riddle of Mount Sinai: Archaeological Discoveries at Har Karkom*, Studi Camuno, vol. 21. Valcamonica, Italy: Edizioni del Centro.

———. 2013. *Is Har Karkom the Biblical Mount Sinai?* Capo di Ponte, Italy: Atelier Monographs.

Bennett, John G. 1963. Geophysics and Human History. *Systematics* 1: 127–156.

Bietak, Manfred. 2003. Israelites Found In Egypt, Four-Room House Identified in Medinet Habu. *Biblical Archaeology Review* 29(Sept/Oct): 40–47, 49, 82–83.

———. 2010a. From Where Came the Hyksos and Where Did They Go? In *The Second Intermediate Period (Thirteenth-Seventeenth Dynasties)*, Orientalia Lovaniensia Analecta, vol. 192, ed. Marcel Marée, 139–182. Leuven: Peeters.

———. 2010b. Houses, Palaces And Development Of Social Structure in Avaris. In *Cities And Urbanism In Ancient Egypt* [Papers from a Workshop in November 2006 at the Austrian Academy of Sciences], eds. Manfred Bietak, et al. Österreichischen Akademie der Wissenschaften, Denkschriften Der Gesamtakademie, vol. 60, 11–68. Vienna: Verlag der Österreichischen Akademie der Wissenschaften.

Billington, Clyde. 2013. Exodus and the Abandonment of Avaris. *Artifax* (Summer):16–17.

Bright, John. 2000. *A History of Israel*, 4th ed. Louisville, KY: Westminster John Knox.

Bruins, Hendrik J., and Johannes van der Plicht. 1996. The Exodus Enigma. *Nature* 382(July 18): 213–214.

Bryan, Betsy M. 1998. Antecedents to Amenhotep III. In *Amenhotep III: Perspectives on His Reign*, ed. David B. O'Connor and Eric H. Cline, 27–62. Ann Arbor, MI: Univ. of Michigan Press.

Cohen, Rudolph. 1983. The Mysterious MB I People: Does the Exodus Tradition in the Bible Preserve the Memory of Their Entry into Canaan? *Biblical Archaeology Review* 9(4, July/August): 16–29.

———. 1992. The Nomadic or Semi-Nomadic Middle Bronze Age I Settlements in the Central Negev. In *Pastoralism in the Levant*, ed. O. Bar-Yosef and A. Khazanov, 105–131. Madison, WI: Prehistroy Press.

———, ed. 1999. *Ancient Settlement of the Central Negev*. Vol. 1: The Chalcolithic Period, the Early Bronze Age and the Middle Bronze Age I. Israel Antiquities Authority Reports, no. 6, Jerusalem: IAA [Hebrew with Eng. summary].

Dodson, Aidan, and Dyan Hilton. 2004. *The Complete Royal Families of Ancient Egypt*. London: Thames & Hudson.

Finkelstein, Israel, and Neil A. Silberman. 2001. *The Bible Unearthed*. New York, NY: Free Press.

Freedman, David Noel. 1980. *Pottery, Poetry and Prophecy: Studies in Early Hebrew Poetry*. Winona Lake, IN: Eisenbrauns.

Freedman, D.N., and Miano David. 2006. 'His Seed Is Not': 13th-Century Israel. In *Confronting the Past: Archaeological and Historical Essays on Ancient Israel in Honor of William G. Dever*, ed. Gitin Seymour et al., 295–301. Winona Lake, IN: Eisenbrauns.

Galanopoulos, Angelos G. 1964. Die Agyptischen Plagen und der Auszug Israels aus geologischer Sieht. *Das Altertum* 10(3): 131–137.

Goedicke, Hans. 1981. Goedicke Defends His Exodus Thesis, letter. In *BAR*, vol. 8(2, March/April), 12a–b.

———. 2004. *The Speos Artemidos Inscription of Hatshepsut and Related Discussions*. Oakville, CT: HALGO, Inc.

Gordon, Cyrus H., and Gary Rendsburg. 1997. *The Bible and the Ancient Near East*, 4th ed. New York, NY: W. W. Norton.

Gottwald, Norman K. 1979. *The Tribes of Yahweh*. Maryknoll, NY: Orbis Books.

Gruen, Erich S. 1998. *Heritage and Hellenism: The Reinvention of Jewish Tradition*. Berkeley, CA: Univ. of California Press.

Hall, H.R. 1927. "Édouard Naville" [obit]. *Journal of Egyptian Archaeology* 13(1/2, Apr): 1–6.

Harris, James E., and Kent R. Weeks. 1973. *X-Raying the Pharaohs*. New York, NY: Scribner's.

Harris, James E., and Edward Wente. 1980. *An X-Ray Atlas of the Royal Mummies*. Chicago, IL: Univ. of Chicago.

Hoffmeier, James K. 1999. *Israel in Egypt: The Evidence for the Authenticity of the Exodus Tradition*. Oxford: Oxford Univ. Press.

———. 2005. *Ancient Israel in Sinai*. Oxford: Oxford Univ. Press.

Horn, S.H. 1977. What We Don't Know About Moses and the Exodus. *Biblical Archaeology Review* 3(2, Jun): 22–31.

Humphreys, Colin J. 2004. *The Miracles of Exodus*. New York, NY: Harper.

Kitchen, Kenneth A. 1977. *The Bible in its World.* Nottingham: InterVarsity.

Lepsius, (Karl) Richard. 1849. *Die Chronologie der Aegypter.* Berlin: Nicolaische Buchhandlung.

Mendenhall, George E. 1955. *Law and Covenant in Israel and the Ancient Near East.* Pittsburg: Presbyterian Board of Colportage of Western Pennsylvania.

Neev, David, and Kenneth O. Emery. 1995. *The Destruction of Sodom, Gomorrah, and Jericho: Geological, Climatological, and Archaeological Background.* New York, NY: Oxford Univ. Press.

Nigro, Lorenzo. 2014. Tell Es-Sultan – A Pilot Project for Archaeology in Palestine. Ancient Near East Today [ASOR Blog online] 2(3).

Peet, T. Eric. 1922. *Egypt and the Old Testament.* London: Hodder and Stoughton.

Petrovich, Douglas. 2013. Toward Pinpointing the Timing of the Egyptian Abandonment of Avaris during the Middle of the 18th Dynasty. *Journal of Ancient Egyptian Interconnections* 5(2): 9–28.

Redford, Donald B. 1992. *Egypt, Canaan, and Israel in Ancient times.* Princeton, NJ: Princeton Univ. Press.

Rendsburg, Gary A. 1992. The Date of the Exodus and the Conquest/Settlement: The Case for the 1100s. *Vetus Testamentum* 42: 510–527.

Sellin, Ernst, and Carl Watzinger. 1913. *Jericho.* Leipzig: J. C. Hinrichs.

Shanks, Hershel, ed. 1999. *Ancient Israel,* rev. Washington, DC: Biblical Archaeology Society.

Shea, William H. 1982. Exodus, Date of. In *International Standard Bible Encyclopedia.* Rev. ed. Geoffrey W. Bromiley, vol. 2, 230–238. Grand Rapids, MI: Eerdmans.

———, 2003. The Date of the Exodus. In Giving the Sense. David M. Howard, Jr., and Michael A. Grisanti (eds.). Grand Rapids, MI: Kregel.

Ussher, James. 1650. *Annales Veteris et Novi Testamenti.* Vol. 1. London [Ann Arbor, MI: University Microfilms, 1979 microfilm] [Latin].

Vitaliano, Dorothy B. 1968. Geomythology: The impact of geologic events on history and legend, with special reference to Atlantis. *Journal of the Folklore Institute* (Indiana University) 5: 5–30.

———. 1973. *Legends of the Earth: Their Geologic Origins.* Bloomington, IN: Indiana Univ. Press.

Wright, G. Ernest. 1962. *Biblical Archaeology,* rev. Philadelphia, PA: Westminster.

——— (ed.). 1965. *The Bible and the Ancient Near East.* New York, NY: Doubleday Anchor.

The Exodus as Cultural Memory: Egyptian Bondage and the Song of the Sea

5

Ronald Hendel

Abstract

This essay situates the cultural memory of the Exodus in a dialectic between historical memory and ethnic self-fashioning. Memories of the Egyptian Empire in Canaan have been transformed into a memory of liberation from Egyptian bondage, with this political transition mapped onto the geographical space of Egypt and Canaan. The mnemohistory of the Exodus has roots in the LB/Iron Age transition, which has been narrativized as an ethnic myth of origins. The oldest expression of this ethnic myth, the Song of the Sea, transmutes the memory of Egyptian collapse into a song of the Divine Warrior, wherein Yahweh is the sole king and Pharaoh is chaos vanquished.

The biblical narrative of the exodus is approached as a production of cultural memory, not as an inerrant divine witness. This involves the task of what Jan Assmann calls mnemohistory, in which historical memory, folklore, ethnic self-fashioning, and literary artistry converge. The mnemohistory of the Exodus is an interdisciplinary inquiry. It is unnecessary to include geological events, such as volcanic eruptions, in this explanatory model.

Egyptian Bondage

The historical problem of the Exodus is aptly expressed by William Dever:

> The implications of the new picture of indigenous Late Bronze Age Canaanite origins for the majority of the early Israelite population is clear. Not only is there no archaeological evidence for an exodus, there is no need to posit such an event. We can account for Israelite origins, historically and archaeologically, without presuming any Egyptian background. As a Syro-Palestinian archaeologist, I regard the historicity of the Exodus as a dead issue. (1997: 81)

In terms of what Fernand Braudel (1980: 27–29) calls *l'histoire événementielle*, "the history of events," it does seem to be the case that the Exodus story is unhistorical (Frerichs and Lesko 1997). But in terms of what we may call cultural history, the prominence of the Exodus story requires explanation. Jan Assmann coined the term "mnemohistory" to describe inquiry into the history of the remembered past:

R. Hendel (✉)
University of California, Berkeley, CA, USA
e-mail: hendel@berkeley.edu

T.E. Levy et al. (eds.), *Israel's Exodus in Transdisciplinary Perspective*,
Quantitative Methods in the Humanities and Social Sciences, DOI 10.1007/978-3-319-04768-3_5,

> Unlike history proper, mnemohistory is concerned not with the past as such, but only with the past as it is remembered. It surveys the story-lines of tradition, the webs of intertextuality, the diachronic continuities and discontinuities of reading the past. Mnemohistory is not the opposite of history, but rather is one of its branches or subdisciplines, such as intellectual history, social history, the history of mentalities, or the history of ideas.... Mnemohistory is reception theory applied to history. (1997: 8–9)

The task of mnemohistory is not to confirm or disconfirm the events of the Exodus, but to trace its history as cultural memory—to inquire into how such memories are constituted, how they change over time, and how they are mobilized, contested, and transformed by various agents and groups. This is the history of cultural memory (Hendel 2010), in contrast to the history of events.

In my 2001 article, "The Exodus in Biblical Memory," I raised the question of why a community of indigenous Canaanites would embrace a story of deliverance from Egyptian bondage as a shared cultural memory. I proposed that the story of the Exodus derives, at least in part, from Canaanite memories of Egyptian oppression during the Egyptian Empire of the Late Bronze Age, when Canaan was an Egyptian province:

> The oppressive rule of Pharaoh and the enslavement of the ancestors—these are memories that could have been shared by many segments of the population of early Israel. It is plausible that some people in early Israel had indeed escaped from slavery in Egypt. The Egyptian names of Moses, Phineas, and Hophni are perhaps testimony to the Egyptian origin of some of the Levite lineages. But—and this is the important point—for the exodus story to take root in early Israel it was necessary for it to pertain to the remembered past of settlers who did *not* immigrate from Egypt. By leaving the name of Pharaoh a blank, the memory of Egyptian oppression could extend to all who had felt the oppression of Pharaoh at any time in the remembered past. This extension of reference extends broadly throughout Canaan during the Egyptian Empire of the Late Bronze Age. (2001: 605)

In a subsequent essay I emphasized how this memory served to constitute Israel's collective identity and ethnic boundaries:

> The cultural boundaries of early Israel were, at least in part, constructed by the dissemination of stories about the deliverance of Israel from Egyptian bondage and birth of a free people in the Promised Land. It is important to note that even Israelite settlers who had never been slaves in Egypt could easily participate in this narrative memory, for Egypt had been the overlord of Canaan for several centuries previously (ca. 1500–1150 B.C.E.). Egyptian rule during this period had often been harsh, including the regular export of Canaanite to Egypt to serve as slaves. With the waning of the Egyptian Empire in Canaan, the memory of oppression and slavery and the concomitant memory of deliverance to freedom would have resonated in the drama of the Exodus story. By adopting this story as their own, the villagers in the highlands became Israelites, and a mixed multitude crystallized its collective identity as the people of Yahweh. (2005: 9)

To reiterate some parts of this argument, many Canaanites during the period of Egyptian rule were slaves in Egypt. Some were prisoners of war, others were sent as tribute by Canaanite kings, some were sold into slavery, and at times entire Canaanite settlements were forcibly relocated to Egypt. All sectors of Canaanite society were familiar with the heavy hand of Pharaoh, and many had knowledge of those who were taken into slavery.

I would now emphasize that the *ideology* of slavery to Pharaoh—in addition to the proliferation of Canaanite slavery in Egypt—was a formative part of this cultural memory. According to Egyptian imperial ideology, the province of Canaan was the personal property of the Pharaoh. All of its inhabitants were his slaves, from kings to peasants. The Amarna letters are important evidence for this political ideology. In the formulaic diction of these letters, the Canaanite ruler proclaims his abject servitude to Pharaoh:

> [T]o the king, the Sun, my lord: [Mess]age of ʻAbdi-Ašratu, your [s]lave, the dirt under your feet. I fall at the feet of the king, my lord, seven times and seven times. As I am a slave of the king and a dog of his house, I guard all Amurru for the king, my lord. (EA 60; Moran 1992: 131–132)
>
> Say [to the ki]ng, my lord and my [Su]n: Message of Biridiya, the loyal slave of the king. I fall at the feet of the king, my lord and my Sun, seven times and seven times. May the king, my lord, take cognizance of his slave and his city. (EA 365; Moran 1992: 363)

Note the last phrase—Birdiya asks Pharaoh to take cognizance of *his* slave and *his* city. Pharaoh is the master, and the people of Canaan are his property. A Babylonian king summarizes this

Fig. 5.1 Tutankhamun's sandals with Canaanite and Nubian captives (Veldmeijer 2012: figs. 3.43)

relationship succinctly in a letter to Pharaoh: "[C]anaan is your country, and [its] king[s are your slaves]" (EA 8; Moran 1992: 16).

As master of Canaan, Pharaoh was the legitimate source of power and state violence. This power was expressed militarily and discursively. As Pharaoh proclaims in a letter to the Canaanite king of Gezer: "Amun has indeed put the Upper Land, the Lower Land, where the sun rises, where the sun sets, under the feet of the king" (EA 369; Moran 1992: 366). The metaphor of Pharaoh trampling on his Canaanite captives is literalized in a pair of royal sandals from the tomb of Tutankhamun, depicting bound Canaanite and Nubian captives on the soles (Fig. 5.1). With each step, Pharaoh tramples the foreign captive. Notice that these foreign captives who are "under the feet of the king" are in the imperial provinces, in "the Upper Land, the Lower Land." All the foreign subjugated peoples are slaves of Pharaoh, even when they are in their own land. This is the ideology of empire, which was conveyed by various media—speech, images, architecture, etc.—to the "captive" Canaanites.

Canaanite captives—a category that included every Canaanite—labored for Pharaoh in Egypt *and* in Canaan. According to the Amarna letters and other evidence, Pharaoh had vast agricultural holdings in the Jezreel valley (Na'aman 2005). Canaanite corveé workers did the agricultural labor, perhaps alongside Egyptian soldiers. The king of Megiddo describes his provision of corveé labor in a letter to Pharaoh:

> Only I am cultivating in Šunama, and only I am furnishing corvée workers. . . . Only I (by myself) furnish corvée workers. From Yapu [Joppa] they come, from [*my*] *resources* here, and from Nuribta. (EA 365; Moran 1992: 363)

Conscripted Canaanites worked the Pharaonic fields. Through various forms of subjection—including agricultural labor, building construction, annual and irregular tribute, and grain taxes—everyone, from king to peasant, was a slave to Pharaoh.

While the evidence of Egyptian imperial rule in Canaan is often sketchy and capable of multiple interpretations (see Bryan 1996), there is a consensus among historians that the last phase of imperial rule in Canaan during the nineteenth and twentieth dynasties grew very harsh. Nadav Na'aman has recently collected a variety of evidences to substantiate this view:

> The Egyptian occupation of Canaan—particularly in its southern districts—was intensified during the Nineteenth-Twentieth Dynasties. This heightened occupation is reflected in the strengthening of the main route leading from Egypt to Canaan; the fastening [*sic*] of the supervision on the quarry operations in the Arabah; the extensive annexation of Canaanite territories in south Canaan and northern valleys; the increasing demand of taxes and gifts and the delivery of income from the conquered and incorporated territories to the royal treasury and temples in Egypt; and the intensified supervision of the movements of the pastoral nomads and 'Apiru bands . . . The burden imposed on the city-state rulers, the farmers and nomads must have greatly increased, hence the many rebellions and the Egyptian campaigns aimed at suppressing them. (2011: 55)

As Ellen Morris observes, the end of the Egyptian occupation of Canaan was sometimes violent.

> Archaeological evidence suggests that the final end of Egyptian rule in the north was a short and bloody affair. Not only does evidence for Egyptian occupation cease abruptly in the reign of Ramesses VI, but nearly every Egyptian base in Canaan seems to have been torched, whether by enemy attackers, by garrison uprisings, or by the Egyptian themselves as they retreated homeward. (2005: 709)

It seems that the Egyptian masters were not well loved by their former slaves.

With the collapse of the Egyptian Empire in the middle of the twelfth century B.C.E., the heavy hand of Egypt on its colonial territories was no more. To the former Canaanite subjects, the deliverance from Egyptian bondage must have seemed a great wonder. They were no longer under the feet of Pharaoh, his "living captives" (*sqrw-'nḫ*). The memory of the house of bondage was nurtured by Canaanites who never went to Egypt, but who nonetheless lived under the heavy hand of Pharaoh during the years of the Egyptian Empire in Canaan. After generations of enslavement, the Canaanites were released from Egyptian bondage.

In his recent article, Na'aman agrees with my proposal about the background of this biblical memory. He writes: "The negative biblical description of Egypt as 'house of bondage' reflects well the Egyptian reality of the New Kingdom" (2011: 49). However, he argues that in my 2001 article I "did not try to explain the relation between the historical memory of the settlers in Canaan and the tradition of subjugation and delivery from bondage in Egypt" (2011: 62). I respectfully demur from Na'aman's assessment, since this is precisely in what my proposal consists. Nonetheless, Na'aman's article marshals considerable evidence for the plausibility of this proposal.

Na'aman rightly observes that "the figure of Moses, probably as leader of a tribal group, might have influenced the shift of the story from Canaan to Egypt" (2011: 66). I would add that other *exodoi* of Canaanite slaves from Egypt may have contributed as well. Tens of thousands of Canaanites were slaves in Egypt during the Empire. While many may have remained there and become naturalized Egyptians, it stands to reason, as Barry Kemp observes, that "some doubtless fled back to their homeland" (2006: 34). The movement of former Canaanite slaves from Egypt to the Israelite highlands—during and after the collapse of the Egyptian Empire—is arguably part of the historical landscape. In this respect, I agree with Abraham Malamat that the Exodus may have been a "'durative' event . . . a steady flow of . . . [Canaanite slaves] coming out of Egypt during a lengthy period, perhaps encompassing hundreds of years" (1997: 16). The point that I have emphasized, however, is that the memory of Egyptian slavery pertained to many or most of the early settlers in Israel, irrespective of whether they had dwelled in Egypt.

This historical memory was ultimately transposed into something more permanent: a narrative of ethnic origins, through which a "mixed multitude" of settlers in the highlands imagined itself into a new community. As recent works have emphasized (Killebrew 2005; Faust 2006), this multitude presumably included Canaanite peasant farmers and pastoralists, nomads (Shasu), peasant bandits ('Apiru), and other segments of Canaanite society. The story of the Exodus arguably served as a *mythomoteur* (Smith 1986: 15) that played a vital role in transforming this mixed multitude into a cohesive community.

Between the traumatic memory of the Egyptian Empire and its coalescence into a story of ethnic origins, the memories were transfigured. Most notably, the house of bondage was restricted to the land of Egypt, and Egyptian rule in the land of Canaan was forgotten. This is arguably a strategic forgetting. The ancestors went to Egypt due to a famine in the land precisely so that they could return as a newly constituted people. The geographical restriction of the house of bondage to Egypt streamlines the story and enables the extraterritorial creation of an ethnic boundary that distinguishes Israelites from Canaanites. As Yair Zakovitch observes, "The intimate relationship between the children of Israel and the Canaanites necessitated the reinforcement and intensification of the Exodus myth, the belief that the people of Israel were created in Egypt and received their law and culture in the wilderness" (1991: 133).

In this spatial displacement, the Exodus from Egypt becomes the narrative site for forging a new people, who enter Canaan with clearly marked external ethnic boundaries. In this sense the spatial movement overcomes the problem of the basic lack of cultural difference between early Israel and its Canaanite neighbors. Through the strategic processes of cultural memory, Israel projected its origins to "outside" in order to construct a distinctive identity "inside." As Peter Machinist emphasizes, "A story of outside entrance into Palestine . . . would have served as an important pole around which a collective identity could be segregated and consolidated" (1994: 52).

Coming from "elsewhere," as Nili Wazana observes, also connotes newness, reflecting a self-consciousness of being a recently formed people in the midst of older civilizations: "younger emerging societies, remembering their formation in 'historical times'... [translated] otherness to outsider origins, yet turned it to a mark of divine chosenness" (2005: 238–239). In other words, Israel's self-conception as a chosen people coming from elsewhere compensates for its awareness of its recent origins. As a consequence of historical belatedness, they "based their claim to the land on divine assignment rather than on ancestral right of possession" (2005: 239).

The movement from outside to inside is also a resumption of the ancestral migrations and divine promises in the patriarchal narratives. The Exodus narrative signals this linkage both in P (Exod 6:2–8) and E (Exod 3:6–10*). This intertextual relationship creates a doubled narrative of legitimation, as the promise of the land to the ancestors, given while they were in the land, is accomplished by the movement back to the land as a consequence of the Exodus. The origins from elsewhere are a fulfillment of ancestral promises made on the "inside." A previous layer of legitimation, causally related to the others, is the postdiluvian curse of Canaan (Gen 9:25). Israel's formation as a people is ultimately a consequence of the curse of the indigenous inhabitants. Outside and inside are related by means of these narrative intersections, such that the people and the land have an elective affinity over the *longue durée* from the Noachic era to the time of Moses.

The Exodus as a cultural memory constructs a passage from slavery to freedom, an escape from the house of bondage to a new life in the Promised Land. This symbolic rite of passage has narrative, ideological, religious, and historical dimensions. The "escape from Egypt" is literalized as a journey at the same time that it is a metaphor for transformation into a distinctive people and polity.

Excursus: Other Inquiries

Since my earlier essays I have learned that this proposal has been advanced in various forms by other scholars.[1] I take this to be independent confirmation of its cogency. In his 1993 book, *The End of the Bronze Age: Changes in Warfare and the Catastrophe ca. 1200 B.C.*, Robert Drews writes:

> Prior to the Catastrophe, the land of Israel had for almost four hundred years chafed under Egyptian hegemony, a condition so unthinkable in post-Catastrophe circumstances that tradition seems eventually to have transformed it into four hundred years of Israelite "bondage" in the land of Egypt. (1993: 173)

Drews does not develop this idea in detail, but he deserves credit for raising it.

In a 1996 article, "Egyptian Taskmasters and Heavy Burdens: Highland Exploitation and the Collared-Rim Pithos of the Bronze/Iron Age Levant," David Wengrow argues that the distribution of these pithoi is evidence of increased Egyptian exploitation of the southern Levantine

[1] My thanks to Stephen Russell and Konrad Schmid for bringing most of the following to my attention. See Russell (2009: 108–109).

highlands in the later years of the Egyptian Empire. He draws out some implications:

> The centuries-old Egyptian influence over the Levantine economy, previously evident only at elite levels of Canaanite society, was for the first time apparent to the populace at large. Villagers, previously fragmented by allegiances to rival city-states, were now able to conceive of themselves as united in opposition to an alien power, providing a focus for the crystallization of southern Levantine ethnic and ultimately national identity manifested in the territorial states of Israel and Judah which followed the end of Egyptian rule in Canaan. (1996: 323)

Wengrow does not mention the Exodus story in connection with the "crystallization of southern Levantine ethnic and . . . national identity . . . which followed the end of Egyptian rule," but the title of the article makes this connection clear. If the Egyptian Empire could be perceived by villagers as "Egyptian taskmasters," then their condition of collective bondage is a plausible agent in the crystallization of social identity. I am not concerned here with the correctness of Wengrow's analysis of the economic function of the collared-rim pithoi (see Faust 2006: 200), but with his description of how the villagers' responses to Egyptian bondage may have yielded social transformation and story.

In his 1998 book, *The Original Torah: The Political Intent of the Bible's Writers*, S. David Sperling formulated another version of this proposal, working out of the "internal revolt" model of Israelite origins developed by George Mendenhall and Norman Gottwald:

> As these farmers withdrew from a more stratified social order, which was economically and politically burdensome, they encountered hostility from the older order. This order had been imposed by the Egyptian empire which claimed Canaan as its own from the eighteenth through twentieth dynasties (ca. 1560–1080 B.C.E.) .... Given that the Amarna letters demonstrate local consciousness and strong opposition to the collusion between the Canaanite rulers and Egypt, we must interpret the Hebrew traditions of servitude *in* Egypt as allegories of servitude *to* Egypt. (1998: 52, 54)

In Sperling's formulation, the early Israelites were peasants in rebellion against the political hegemony of the Canaanite city-states. This theory has some unnecessary premises and has been subjected to serious criticism (e.g., Boer 2002; Faust 2006: 96–98). However, the insight that the narrative of "servitude *in* Egypt" derives substantially from "servitude *to* Egypt" is cogent. This relationship is, however, better regarded as a feature of cultural memory than as an "allegory," which is a literary mode that is foreign to the Exodus story.

In his 2003 book, *Oltre la Bibbia: Storia antica di Israele* (translated as *Israel's History and the History of Israel*), Mario Liverani advanced another independent version of this proposal. His deep knowledge of the political history of the Late Bronze Age informs his formulation:

> The basic idea was that Yahweh had delivered Israel from Egyptian power and had given them control—with full autonomy—of the land where they had already lived. There was an agreed 'memory' of the major political phenomenon that had marked the transition from submission to Egypt in the Late Bronze Age to autonomy in Iron Age I . . . [T]he terminology of 'bringing out' and 'bringing back' . . . had already been applied in the Late Bronze Age texts to indicate a shifting of sovereignty, without implying any physical displacement of the people concerned, but only a shift of the political border .... This is an idiomatic use of the code of movement (go in/go out) to describe a change in political dependence." (2005: 278)

While it seems implausible that the verbs of "bringing out" and "bringing back" are idioms for shifting sovereignty in Late Bronze Age or biblical texts (see Na'aman 2011: 62), his perception that the narrative's "code of movement" has political resonance is acute.

In the multi-authored 2006 book, *Grundinformation Altes Testament* (translated as *T&T Clark Handbook of the Old Testament*), Angelika Berlejung formulated another independent version of this idea:

> It is possible . . . that the Egypt of the exodus did not in fact refer to Egypt in the geographical sense, but rather to Egypt in a political sense, so that a migration from southern Palestinian city-states could be interpreted as an exodus from Egypt. In this case, there would be no need to look for the presence of Asian Semites in Egypt proper, which would otherwise have constituted the logical presupposition for an exodus. (Gertz et al. 2012: 108)

I would aver that we should not exclude the influence of "Asian Semites in Egypt proper" on the Exodus memory. Canaanites were slaves to Egypt both in Egypt and in Canaan; hence, we should imagine a synthesis of memories of geographical *and* political slavery rather than conceive of an either/or. Nonetheless, we see that this proposal, with variations, has been circulating in the byways of recent scholarship. Its independent derivation from the textual and historical evidence by scholars in different fields and intellectual cultures may indicate its plausibility.

The Song of the Sea and Egyptian Ideology

The Song of the Sea in Exodus 15 is arguably the oldest representation of the Exodus in the Hebrew Bible. The text has the greatest concentration of typologically archaic linguistic features of any text in the Hebrew Bible, and it has no linguistic features that are postclassical (see Bloch 2009, 2012). This distinguishes the song from other compositions that have a mixture of typologically archaic and postclassical linguistic features, such as Second Isaiah, to which it is sometimes compared. In terms of absolute date, it is unlikely to be later than the mid-eighth century B.C.E. (Bloch 2012: 164) and could certainly be earlier. We cannot specify an absolute date more precisely. However, the song is the most quoted text within the Hebrew Bible (e.g., Josh 2:9; Isa 12:2; Pss 78:13; 118:14; perhaps 1 Kgs 8:13; see Russell 2009: 135–142),[2] and as such it is at least relatively early. In view of its relative antiquity and influence in the Bible, it holds a key place in a mnemohistory of the Exodus.

The text is a victory hymn in praise of Yahweh (see Propp 1998: 507). It has thematic affinities with the divine enthronement psalms (Pss 47; 93; 96–99), but its focus on a particular battle places it in closer relationship to the Song of Deborah in Judges 5. The genre of victory hymn conditions the representational possibilities of the text. It is not a narrative genre, and hence we cannot expect a chronologically connected sequence. Such hymns consist of pastiches of images, including variations on traditional motifs and tropes.

As Frank Cross has demonstrated (1973), much of the poetic diction in the song is derived from Canaanite mythology, particularly what he calls the mythic pattern of the primordial victory of the Divine Warrior. In Cross's analysis, this pattern consists of three major themes:

1. The combat of the Divine Warrior and his victory at the sea
2. The building of a sanctuary on the "mount of possession" won in battle
3. The god's manifestation of "eternal" kingship

This mythic pattern is exemplified in the Ugaritic myth of Baal's battle with Yamm ("Sea") and the Babylonian myth of Marduk's battle with Tiamat ("Sea").

The oldest reference to this myth of divine combat is from a prophetic letter from Mari (ca. 1760 B.C.E.), in which the god Adad describes his past beneficence to the king:

> I restored you to the th[rone of your father's house],
> and the weapon[s] with which I fought with Sea [*Têmtim*] I handed you.
> I anointed you with the oil of my luminosity;
> nobody will offer resistance to you. (Nissinen 2003: 22)

Adad's gift of his weapons (which were displayed at a temple in Terqa; see Durand 1993: 53) conveys to the king the power to defeat chaos and to maintain order, both in the cosmos and the political world.

During the Egyptian Empire, the Canaanite myth of the Divine Warrior was absorbed into new forms of Egyptian royal ideology, in which the Pharaoh's victory over foreigners as instantiations of chaos was a central theme. As Anthony Spalinger observes, a focal point of Egyptian art and narrative in this period was the representation of the victorious king in battle against his chaotic adversaries:

[2] Propp (1998: 565) considers "the possibility that Psalm 78 constitutes our earliest commentary on the Song." On the rationale for dating Psalm 78 prior to the fall of the Northern Kingdom, see Day (2004: 237–238).

> The major theme of these ... forms of historical narration was that of the suppression of chaos. Anarchy was the attribute of the enemy. Egypt (Amun, king) represented Truth (*Maat*) and permanence .... The Egyptian king, as deputy and son of Amun, wars against the recalcitrant foes. Always in a chariot, Pharaoh shoots his arrows and fells the major opponent .... The actions of foe and king alike are presented in a rigid political-theological framework, one that views war as a personal contest of the Pharaoh against chaos. (2005: 77–78)

Pharaoh was prominently represented as a Divine Warrior, absorbing the attributes of the Egyptian god Montu and the Canaanite god Baal. A perspicuous example is the description of Ramesses III in the battle against Libya, from an inscription at his temple at Medinet Habu:

> Raging and stretching out (his) right arm, plunging into battle, and slaying myriads in their places under his horses. He looks on the swarming mass as (mere) grasshoppers, crushed, ground down and pulverized(?) like flour. Firm of horns, relying on his strong arm, (so that) millions and myriads are despised before him, whose form is like Montu when he goes forth. Every land writhes (in travail) because of him, (just) at the thought of him. (Kitchen, 2008: 23)

The ideology of the Pharaoh as Divine Warrior was also expressed by his Canaanite vassal kings, as in an Amarna letter that William Moran characterizes as "a hymn to Pharaoh":

> My lord is the Sun who comes forth over all lands day by day ... who establishes the entire land in peace, by the power of his arm; who gives forth his cry in the sky like Baal, and all the land is frightened at his cry. (EA 147; Moran 1992: 233)

From this letter and from the military and economic footprint of the Egyptian Empire in Canaan, we may infer that the general outlines of Egyptian ideology were familiar to many Canaanites. For instance, as Keel and Uehlinger observe, "[t]he dominant human form on Egyptian seal amulets dating to the Late Bronze Age II ... for southern Palestine ... is the *pharaoh*" (1998: 80). The image of the Pharaoh as a victorious warrior conveys the imperial ideology (Fig. 5.2).

The Song of the Sea draws on the widely known mythology of the Divine Warrior to portray Yahweh's victory over Pharaoh. As Cross comments: "The power of the mythic pattern was enormous. The Song of the Sea reveals this power as mythological themes shape its mode of presenting epic memories" (1973: 143–144). I would add that the mythological themes serve, in part, to present the memories in such a way that contests Egyptian claims of Pharaonic power. That is, the song's depiction of the defeat of the Egyptian army provides a countermemory to the Egyptian ideology of Pharaoh as Divine Warrior.

The sense of the Song of the Sea as a countermemory is evident whether or not it deliberately mimics Egyptian tropes. That is, it alludes in a general sense to the Egyptian discourse of Pharaonic power. The semiotic relationship exists at the level of cultural memory, in which the discourse about Egypt in the song defeats Egyptian claims about the invincibility of Pharaoh. The song depicts Pharaoh as the agent of chaos, whose troops are devastated by the Yahweh, the true Divine Warrior. The harsh rule of Pharaoh is replaced by the rule of Yahweh, who redeems his people and assumes eternal kingship. Yahweh's victory overcomes the memory of Pharaoh as the all-powerful master of the Hebrew slaves. Pharaoh is now the emblem of chaos, an instantiation or a metonym of sea, his troops cast into the deep waters. In the language of the Amarna letter, Yahweh—not Pharaoh—is now the triumphant Divine Warrior, "who establishes the entire land in peace, by the power of his arm."

The song introduces its countermemory by expressing a strong contrast between Yahweh and Pharaoh:

> I will sing to Yahweh, for he is highly exalted;
> Horse and its rider he hurled into the sea. (Exod 15:1)

As William Propp observes, this verse introduces a polarity of space and power between the two adversaries: Yahweh is the victor from "above" (גָּאֹה גָּאָה, "highly exalted") and later ascends to his mountain of kingship, while the enemy is "below," hurled into the sea, a passive object of Yahweh's wrath.

Fig. 5.2 Pharaoh smiting enemies: Rock relief of Ramesses II smiting Canaanite, Nahr el-Kelb, Lebanon (Weissbach 1922: fig. 6); scarabs from nineteenth-dynasty Canaan: (**a**) Beth-Shean, (**b**) Tell Beit Mirsim, (**c**) Tell el-Far'ah (Keel 1985: fig. 400)

Notably, the identity of the defeated enemy, referred to as סוּס וְרֹכְבוֹ ("horse and its rider"), is deliberately opaque. The initial implication seems to be that Pharaoh is the enemy, since he is the "rider" par excellence, characteristically represented driving his chariot alone into battle. As Mario Liverani characterizes this conventional Egyptian scene:

> On the one side there is Pharaoh alone, with his battle instruments, the chariot and the bow, both endowed with high symbolic values. On the other side there is the enemy 'coalition,' countless but

> inferior, foredoomed to defeat. Pharaoh just charges straight [at] the enemy and defeats the multitudes. (1990: 117)

The diction of the song in verse 1 suggests that Pharaoh alone is cast into the sea. In this introductory couplet, he seems to be the image of chaos defeated.

After some praise for the victorious Divine Warrior ("Yahweh is a man of war, Yahweh is his name . . ."), the song returns to the war scene:

> Pharaoh's chariots and his warriors he threw into the sea;
> And his choice officers sank into the Reed Sea.
>
> The deeps covered them;
> They went down into the depths like stone. (Exod 15:4–5)

The terse description of victory in the opening verse is now presented more fully. The grammatically singular "horse and its rider" in verse 1 is now explicated as a plural, "Pharaoh's chariots and his warriors." The enemy is not Pharaoh alone, but a myriad army. In retrospect, the grammatical singular is shown to be a collective. The identity of "the sea" in verse 1 is also more fully revealed. In the parallel sequence in verse 4, *yam* ("sea") expands into *yam suf* ("the Reed Sea"). The sea is further colored by cosmological terms in verse 5, *tehomot* ("the deeps") and *meṣulot* ("the depths"), which foreground the affective resonance of mythic waters. These cosmological terms echo the old mythology of Canaanite Yamm ("Sea") and the biblical mythic imagery of the descent to Sheol (e.g., 2 Sam 22:17; Ps 69:2–3,15-16; Jonah 2:3, 6–7; see Propp 1998: 530).

Yahweh's defeat of Pharaoh's troops at the sea is, by means of this diction, represented as the fall of Egyptian power into chaos and death. The mighty army falls by Yahweh's arm and is then covered by the mythic sea, sinking into its chaotic depths, powerless and inert ("like a stone"). Their downward path is an inverted mirror of Yahweh's exalted victory. Pharaoh's troops, the human enemy, are obliterated by Yahweh's power.

The song's representation of the battle scene is elaborated in three further verses, each separated by praise to the victorious Divine Warrior:

> At the breath of your nostrils, the waters piled up,
> They stood like flowing hills,
> The deeps congealed in the heart of the sea.
>
> . . . .
>
> You blew with your breath, and the sea covered them,
> They sank like lead in the mighty waters.
>
> . . . .
>
> You stretched out your right arm,
> And the earth swallowed them. (Exod 15:8, 10, 12)

Here are elaborated images of the roiling sea, whose recesses fuse by the force of Yahweh's powerful breath and into which the armies sink. Yahweh's mighty arm causes the earth (or underworld) to open its mouth in turn and swallow the enemy. The latter action echoes the violent appetite of Mot ("Death") in Canaanite myth, who swallows Baal in his gaping maw. This image recurs in the death of Korah and his co-conspirators in Num 16:32, "The earth opened its mouth and swallowed them."

James Hoffmeier and Manfred Görg have argued that the motif of Yahweh's mighty arm derives from the Egyptian motif of Pharaoh's mighty arm (see above, EA 147; Hoffmeier 1986; Görg 1986). According to Hoffmeier, this "was used as a deliberate play on the Egyptian concept of the victorious pharaoh who conquered his enemies with his powerful arm (*ḫpš*) and outstretched hand (*pr-ʿ*)" (2005: 78). I would note that the motif of the mighty arm of the Divine Warrior reaches back to older mythology (e.g., Ninurta is "lord whose powerful arm is fit to bear the mace"). Yet it is germane to our theme that in the song it is Yahweh's mighty arm that prevails, not Pharaoh's.

The several images of the battle at the sea make it clear that this is not a single connected narrative. It is a pastiche of parallel images that connote the total defeat of Pharaoh's forces by the incomparable power of Yahweh. As Cross emphasizes, the diction of the battle at the sea draws on the affective force of the mythic pattern, in which the Divine Warrior defeats the sea as an embodiment of chaos. In the song, the sea is naturalized and localized as the *Yam Suf* ("Reed Sea"), which is both a geographical locale and a mythic point of descent into the underworld.

Fig. 5.3 Seti I returning to Egypt on the "Way of Horus" in northern Sinai with Shasu captives (Gardiner 1920: pl. XI)

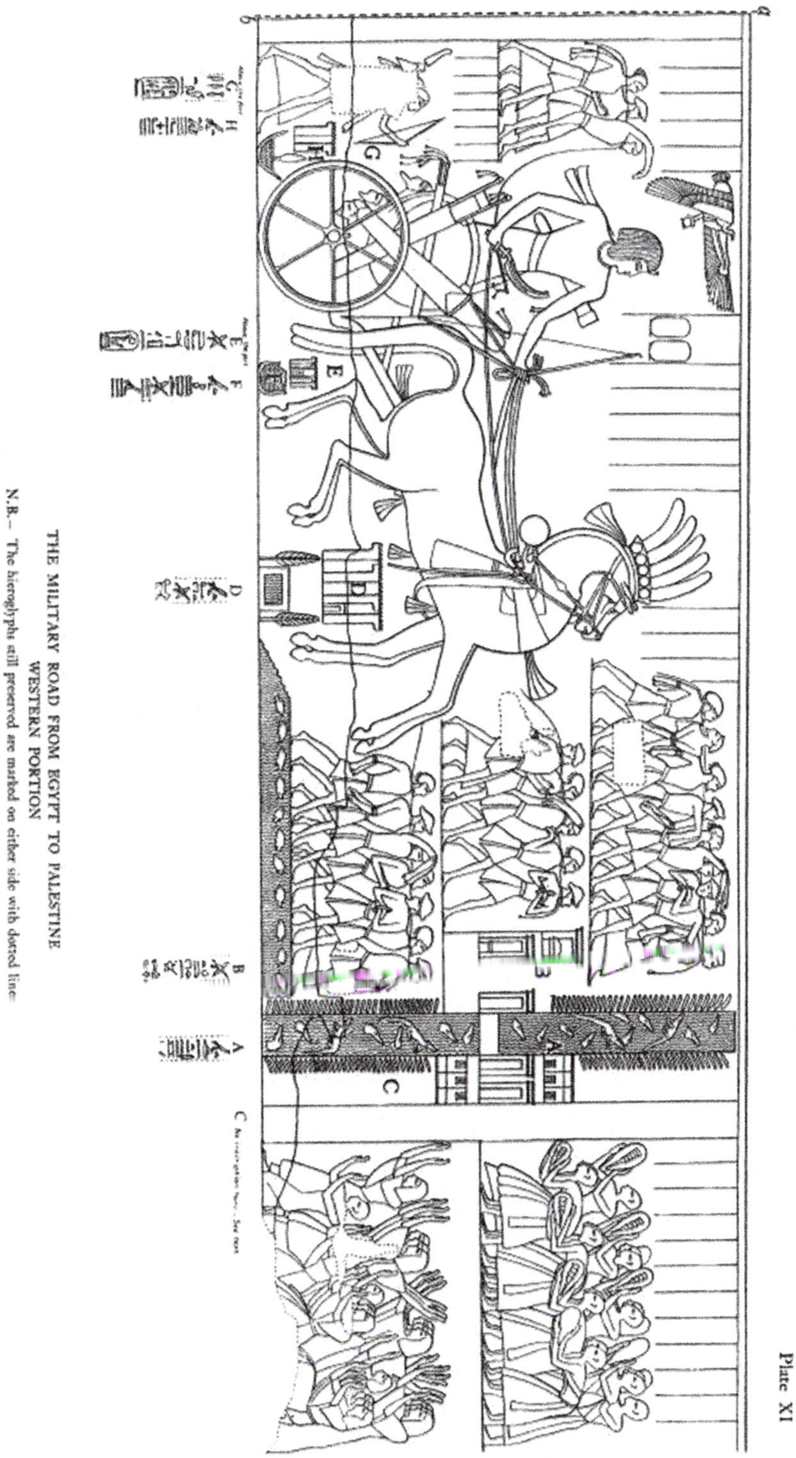

As commentators have long noted, *Yam Suf* is elsewhere a designation of the Red Sea, including its two northern gulfs (e.g., Exod 23:31; Num 14:25; Num 33:10–11; Deut 1:40; 1 Kgs 9:26; see Propp 2006: 752–753). In the song this body of water, literally the "Sea of Reeds," corresponds to the Egyptian trope of the marshy waters that serves as the boundary between the civilized land of Egypt and the outside world of chaos. In the song, this watery boundary is the point of divine victory and redemption of Israel from Egypt. Like the Egyptian marsh waters, it is a boundary between chaos and order, but now the geography of chaos and order is reversed. Egypt is the land of chaos and bondage, and the wilderness of Sinai is the path to redemption and the holy land. The Sea of Reeds is a geographical and a symbolic *limen* ("boundary") in Israel's rite of passage into a liberated people.

During the Egyptian Empire, the watery boundary between Egyptian truth (*Maat*) and foreign chaos is illustrated by a victory stele of Seti I (Fig. 5.3). Pharaoh is returning from his victories in Canaan via the "Way of Horus" through the northern Sinai, driving before him a myriad of Shasu captives. The bound captives face every direction with arms akimbo, representing the chaos that Pharaoh has subdued (Davies 2012). The victorious Pharaoh approaches the "dividing canal" (*tꜣ dnit*) at the frontier, which is filled with crocodiles and edged with reeds (see Hoffmeier 1996: 164–175). On the other side, in Egypt, are, according to the caption, "chief priests and mayors of Upper and Lower Egypt," paying obeisance to Pharaoh.

The victory of Yahweh at the Reed Sea is a countermemory of the imperial ideology that this visual image eloquently expresses. The marsh-edged "dividing canal" that Pharaoh crosses in triumph with captives in tow is replaced in Israelite memory with the Reed Sea where Yahweh defeated Egyptian chaos and from which he led the liberated slaves to the Promised Land. Yahweh's victory at the sea effects a liminal passage of geography and identity for Israel. Yahweh's victory over chaos is the foundation of a new collective identity.

The movement of the song from the watery chaos of Egypt's boundary to the holy pasture and mountain of Yahweh's enthronement is a move from periphery to center, from chaos to order, and from Pharaonic slavery to Yahweh's eternal rule. The mythic rite of passage is completed when Israel is planted in Yahweh's mountain sanctuary, surrounded by other peoples cowed by Yahweh's might, silent as stones. In the far periphery are the equally silent Egyptians, who have sunk like stones into the waters, into nonexistence. The memory of Egyptian bondage is countered by Yahweh's victory and the creation of a new people, the *'am zu ga'alta*, the "people that you redeemed" (Exod 15:13). In the song—a *mythomoteur* of Israel's cultural memory—Egypt's fall is a prelude to Israel's ascent.

References

Assmann, Jan. 1997. *Moses the Egyptian: The Memory of Egypt in Western Monotheism*. Cambridge, MA: Harvard University Press.

Bloch, Yigal. 2009. The Prefixed Perfective and the Dating of Early Hebrew Poetry: A Re-Evaluation. *Vetus Testamentum* 59: 34–70.

———. 2012. The Third-Person Masculine Plural Suffixed Pronoun *–mw* and Its Implications for the Dating of Biblical Hebrew Poetry. In *Diachrony in Biblical Hebrew*, ed. C.L. Miller-Naudé and Z. Zevit, 147–170. Winona Lake, IN: Eisenbrauns.

Boer, Roland (ed.). 2002. *Tracking "The Tribes of Yahweh": On the Trail of a Classic*. Sheffield: Sheffield Academic.

Braudel, Fernand. 1980. *On History*. Chicago: University of Chicago Press.

Bryan, Betsy M. 1996. Art, Empire, and the End of the Late Bronze Age. In *The Study of the Ancient Near East in the 21st Century*, ed. J.S. Cooper and G.M. Schwartz, 33–79. Winona Lake, IN: Eisenbrauns.

Cross, Frank Moore. 1973. The Song of the Sea and Canaanite Myth. *Canaanite Myth and Hebrew Epic: Essays in the History of the Religion of Israel*, 112–44. Cambridge, MA: Harvard University Press.

Davies, Vanessa. 2012. The Treatment of Foreigners in Seti's Battle Reliefs. *Journal of Egyptian Archaeology* 98: 73–85.

Day, John. 2004. How Many Pre-Exilic Psalms Are There? In *In Search of Pre-Exilic Israel*, ed. J. Day, 225–250. London: T&T Clark.

Dever, William G. 1997. Is There Any Archaeological Evidence for the Exodus? In *Exodus: The Egyptian Evidence*, ed. E.S. Frerichs and L.H. Lesko, 67–86. Winona Lake, IN: Eisenbrauns.

Drews, Robert. 1993. *The End of the Bronze Age: Changes in Warfare and the Catastrophe ca. 1200 B. C.* Princeton, NJ: Princeton University Press.

Durand, Jean-Marie. 1993. Le mythologème du combat entre le dieu de l'orage et la mer en Mésopotamie. *MARI* 7: 41–61.

Faust, Avraham. 2006. *Israel's Ethnogenesis: Settlement, Interaction, Expansion and Resistance*. London: Equinox.

Frerichs, Ernest S., and Leonard H. Lesko (eds.). 1997. *Exodus: The Egyptian Evidence*. Winona Lake, IN: Eisenbrauns.

Gardiner, Alan H. 1920. The Ancient Military Road between Egypt and Palestine. *Journal of Egyptian Archaeology* 6: 99–116.

Gertz, Jan Christian, Angelika Berlejung, Konrad Schmid, and Markus Witte. 2012. *T&T Clark Handbook of the Old Testament: An Introduction to the Literature, Religion and History of the Old Testament*. London: T&T Clark. German original. 2006.

Görg, Manfred. 1986. 'Der starke Arm Pharaos': Beobachtungen zum Belegspektrum einer Metapher in Palästina und Ägypten. In *Hommages à François Daumas*, ed. H. Altenmüller et al. Montpellier: Université Paul Valéry.

Hendel, Ronald. 2001. The Exodus in Biblical Memory. *Journal of Biblical Literature* 120: 601–622.

———. 2005. *Remembering Abraham: Culture, Memory, and History in the Hebrew Bible*. New York, NY: Oxford University Press.

———. 2010. Cultural Memory. In *Reading Genesis: Ten Methods*, ed. R. Hendel, 28–46. Cambridge: Cambridge University Press.

Hoffmeier, James K. 1986. The Arm of God Versus the Arm of Pharaoh in the Exodus Narratives. *Biblica* 67: 378–387.

Hoffmeier, James K. 1996. *Israel in Egypt: The evidence for the authenticity of the Exodus tradition*. New York: Oxford University Press.

Hoffmeier, James K. 2005. *Ancient Israel in Sinai: The Evidence for the Authenticity of the Wilderness Tradition*. New York, NY: Oxford University Press.

Keel, Othmar. 1985. *The Symbolism of the Biblical World: Ancient Near Eastern Iconography and the Book of Psalms*. New York, NY: Crossroad.

Keel, Othmar, and Christoph Uehlinger. 1998. *Gods, Goddesses, and Images of God in Ancient Israel*. Minneapolis: Fortress.

Kemp, Barry J. 2006. *Ancient Egypt: Anatomy of a Civilization*, 2nd ed. London: Routledge.

Killebrew, Ann E. 2005. *Biblical Peoples and Ethnicity: An Archaeological Study of Canaanites, Egyptians, Philistines, and Early Israel, 1300-1100 BCE*. Atlanta: Society of Biblical Literature.

Kitchen, Kenneth A. 2008. *Ramesside Inscriptions: Translated and Annotated Volume V: Setnakht, Ramesses III and Contemporaries*. Oxford: Blackwell.

Liverani, Mario. 1990. *Prestige and Interest: International Relations in the Near East ca. 1600-1100 B. C.*. Padua: Sargon.

Liverani, Mario. 2005. *Israel's History and the History of Israel*. London: Equinox. Italian original, 2003.

Machinist, Peter. 1994. Outsiders or Insiders: The Biblical View of Emergent Israel and Its Contexts. In *The Other in Jewish Thought and History: Constructions of Culture and Identity*, ed. J. Silberstein and R.L. Cohn, 35–60. New York: New York University Press.

Malamat, Abraham. 1997. The Exodus: Egyptian Analogies. In *Exodus: The Egyptian Evidence*, ed. Ernest S. Frerichs and Leonard H. Lesko, 15–26. Winona Lake, IN: Eisenbrauns.

Meyers, Carol. 2005. *Exodus*. Cambridge: Cambridge University Press.

Moran, William L., ed. and trans. 1992. *The Amarna Letters*. Baltimore, MD: Johns Hopkins University Press.

Morris, Ellen Fowles. 2005. *The Architecture of Imperialism: Military Bases and the Evolution of Foreign Policy in Egypt's New Kingdom*. Leiden: Brill.

Na'aman, Nadav. 2005. Pharaonic Lands in the Jezreel Valley in the Late Bronze Age. *Canaan in the Second Millennium B.C.E.: Collected Essays*, vol. 2, 232–241. Winona Lake, IN: Eisenbrauns.

———. 2011. The Exodus Story: Between Historical Memory and Historiographical Composition. *Journal of Ancient Near Eastern Religions* 11: 39–69.

Nissinen, Martti, ed. and trans. 2003. *Prophets and Prophecy in the Ancient Near East*. Atlanta: Society of Biblical Literature.

Propp, William H.C. 1998. *Exodus 1-18*. Anchor Bible. New York, NY: Doubleday.

———. 2006. *Exodus 19-40*. Anchor Bible. New York, NY: Doubleday.

Russell, Stephen C. 2009. *Images of Egypt in Early Biblical Literatures: Cisjordan-Israelite, Transjordan-Israelite, and Judahite Portrayals*. BZAW 403. Berlin: de Gruyter.

Smith, Anthony D. 1986. *The Ethnic Origins of Nations*. Oxford: Blackwell.

Spalinger, Anthony J. 2005. *War in Ancient Egypt: The New Kingdom*. Oxford: Blackwell.

Sperling, S. David. 1998. *The Original Torah: The Political Intent of the Bible's Writers*. New York: New York University Press.

Veldmeijer, André J. 2012. *Tutankhamun's Footwear: Studies of Ancient Egyptian Footwear*. Leiden: Sidestone.

Wazana, Nili. 2005. Natives, Immigrants and the Biblical Perception of Origins in Historical Times. *Tel-Aviv* 32: 220–244.

Weissbach, F.H. 1922. *Die Denkmäler und Inschriften an der Mündung des Nahr el-Kelb*. Berlin: de Gruyter.

Wengrow, David. 1996. Egyptian Taskmasters and Heavy Burdens: Highland Exploitation and the Collared-Rim Pithos of the Bronze/Iron Age Levant. *Oxford Journal of Archaeology* 15: 307–326.

Zakovitch, Yair. 1991. *"And You Shall Tell Your Son..." The Concept of Exodus in the Bible*. Jerusalem: Magnes.

Part II

Science-Based Approaches to the Exodus

Radiocarbon Dating and the Exodus Tradition

6

Michael W. Dee, Christopher Bronk Ramsey, and Thomas F.G. Higham

Abstract

The Exodus tradition relays a sequence of events for which no precise historical setting has ever been established. Documentary evidence from Egypt has failed to substantiate the Biblical account. Consequently, unraveling the relationship between the text and the material evidence has become the domain of Biblical scholarship and archaeological excavation. Radiocarbon dating can play a significant role in supporting or opposing archaeological hypotheses. The chronometric technique has long been employed across the Eastern Mediterranean region and is now capable of situating Bronze Age events on a decadal time scale. In this chapter, the radiocarbon evidence pertaining to the Exodus tradition is evaluated and the role of the method in future research considered.

The Exodus tradition is the subject of three interlocking fields of research. At the forefront is analysis of the Biblical texts and their historical, philological, and cultural interpretation. Indeed, scholars such as Jan Assmann and Ronald Hendel elegantly rendered the value of the Exodus tradition as a resource for understanding the origins of communal identity at the San Diego conference. Archaeological research, on the other hand, primarily concerns itself with the historicity of the texts by critical examination of the remnants of the Eastern Mediterranean Bronze Age. The degree of legitimacy afforded the texts *a priori* differentiates one archaeological approach from another. Indeed, such opposing viewpoints were plainly evident in San Diego (see Chaps. 15 and 3). Scientific analysis is somewhat further removed from the text and generally acts to support or oppose the various archaeological hypotheses. This chapter discusses the contribution of chronometry to the historicity of the Exodus tradition. It centers on the measurements obtained by radiocarbon dating, the most widely used of all scientific dating methods.

M.W. Dee (✉)
Oxford Radiocarbon Accelerator Unit, University of Oxford, Oxford, UK
e-mail: michael.dee@rlaha.ox.ac.uk

T.E. Levy et al. (eds.), *Israel's Exodus in Transdisciplinary Perspective*, Quantitative Methods in the Humanities and Social Sciences, DOI 10.1007/978-3-319-04768-3_6,

Introduction

The Contribution of Radiocarbon Dating to Research on the Exodus

Proving whether the Exodus tradition is founded on actual or mythological events, or a combination of the two, is of course beyond the capability of radiocarbon dating. Indeed, the method can only contribute to such debates by testing the narrative as it is interpreted by different scholars. The assumption is that independently derived dates for events described in the text may help to elucidate their authenticity. However, no material evidence has ever been uncovered that can be unequivocally attributed to the flight of the Israelites (or a group that might be identified as such) out of Egypt during the Bronze Age (Meyers 2005: 6). As a result, chronometry must help fix and refine dates for Bronze Age events that are found in the archaeological record to determine whether the information obtained coheres with the Exodus tradition.

Bronk Ramsey et al. (2010) published the first high-precision radiocarbon chronology for the dynastic period of ancient Egypt. The outputs of this study serve as a check on the calendar dates commonly assigned to the Egyptian historical chronology and hence provide information on the potential date of the Exodus and the identity of the pharaohs involved. Secondly, the conquest of the Canaanite cities described in the Book of Joshua is a specific episode of the tradition that should also be detectable by radiocarbon dating. A number of the relevant sites have now been excavated, and some contain destruction layers that contain datable organic remains. Finally, more than 100 radiocarbon dates already exist for the Minoan eruption of Thera. This cataclysmic event is often connected with the Exodus tradition (see Chaps. 7 and 9) because of its proposed coincidence with the end of the Hyksos period and its likely instigation of tsunami in the Eastern Mediterranean (see Chap. 9). Before examining these examples in detail, however, it is necessary to provide a brief overview of the radiocarbon technique.

Summary of the Radiocarbon Method

Radiocarbon dating involves measuring the concentration of the unstable isotope radiocarbon (^{14}C or C-14), which is approximately a trillion times less common than ordinary carbon (^{12}C). The precise value of this ratio ($^{14}C/^{12}C$) is what determines the date obtained (Libby et al. 1949). Radiocarbon is formed naturally in the upper atmosphere, and thereafter its chemical behavior is no different from the stable forms of carbon. It is rapidly oxidized to carbon dioxide, becomes dispersed around the globe, and enters the food chain by way of photosynthesis (see Aitken 1990: 57–75). Both plants and animals continually refresh their radiocarbon levels by turning over cellular tissue; however, when the organism dies, the ^{14}C atoms within the organism disappear exponentially due to nuclear decay. Thus the $^{14}C/^{12}C$ ratio of the sample reveals how much time has elapsed since the organism was alive. It is important to note that the sample material had to be once alive for a radiocarbon date to be obtained, and shorter lived materials, like flowers, seeds, and hair, usually provide the most accurate results (see Waterbolk 1971; Dee et al. 2012).

At the University of Oxford, the $^{14}C/^{12}C$ ratio is measured using accelerator mass spectrometry (AMS), an approach that requires samples 1/1,000 the size of the original decay-counting method. Typical AMS samples range from about 500 mg of bone to just 25 mg of plant material or charcoal (see Dee et al. 2012). Once the $^{14}C/^{12}C$ ratio has been obtained, a raw radiocarbon date can be calculated. However, for the purposes of absolute chronology it is imperative that this result is then *calibrated* against a reference curve of $^{14}C/^{12}C$ measurements made primarily on dendrochronologically dated wood. This produces a calibrated radiocarbon date, which is expressed as a probability function over calendar time (BC/AD or BCE/CE). Calibrated dates are most commonly quoted using their 68 or 95 % highest probability ranges. Such ranges often extend over 200–300 calendar years making them of limited use for high-precision chronology. However, computer programs have been developed that allow complex probability calculations to be made on groups of related radiocarbon dates. Such methods, based on Bayesian statistics, allow for final estimates to be produced of the order of a few decades (Buck et al. 1991; Bronk Ramsey 2009).

Radiocarbon Dating of New Kingdom Egypt

The period of time to which the Exodus relates has long been a source of controversy. In fact, scholars have proposed dates ranging from the late third millennium to the mid-first millennium BCE (Walton 2003: 259–272; Meyers 2005: 8). The uncertainty is mainly due to the absence of any Egyptian royal names in the text. The two most commonly proposed scenarios relate to the expulsion of the (Semitic) Hyksos from Egypt at the onset of the New Kingdom and the reign of Ramesses II in the later New Kingdom (Walton 2003: 259–272). These are also the two main possibilities considered in this chapter. A stele dating to the reign of king Merneptah of the late New Kingdom is generally regarded as a *terminus ante quem* for the Exodus because it describes the Israelites as a Levantine people (Petrie and Spiegelberg 1897; Hoffmeier 1996: 27). Furthermore, there is a temporality to the events described in the narrative (summarized by Walton 2003: 252–272), which offer a suite of chronological parameters that should be testable by means of archaeological analysis and radiocarbon dating.

During the last decades of the twentieth century, radiocarbon dating was often dismissed as unreliable in Egypt or inadequate for the chronological problems of the dynastic era (e.g., Shaw 1985). This notion was finally quashed by the study of Bronk Ramsey et al. (2010; see also Dee 2013). By applying Bayesian statistics to a new set of high-quality radiocarbon data, Bronk Ramsey et al. (2010) provided a science-based chronology for the Old, Middle, and New Kingdoms of Egypt of unprecedented precision. The accession dates produced for the rulers of the New Kingdom averaged just 24 calendar years (95 % probability). Importantly, the dates were almost precisely synchronous with current versions of the historical chronology (see Fig. 6.1). These results showed that radiocarbon could reach the levels of precision and accuracy necessary to differentiate events of the New Kingdom on a decadal time scale. Moreover, they illustrated the potential for testing the correspondence of material remains with the Egyptian historical chronology, even in the absence of any ceramic or written linkages. The most valuable insights obtained by this approach for the Exodus may still be to come, but there is already sufficient data available to examine the conquests of Joshua, and the timing of the Minoan eruption of Thera.

Dating the Conquest of the Cities of Canaan

The 6th Book of the Hebrew Bible recounts the story of the Israelites' entry into Canaan after the Exodus and their subsequent victories under the leadership of Joshua (see Dever 2003: 23–74; Hoffmeier 1996: 33–44). It describes their conquest of 31 Canaanite cities including Jericho, Meggido, and Jerusalem. Of the 31 sites, 26 have since been identified and 22 have been excavated (Walton 2003: 259–272). Archaeological analysis has already brought to light a number of discrepancies with the text, ranging from no sign of occupation during the Middle and Late Bronze Ages to no relevant evidence of destruction. Such observations are discussed in detail elsewhere (Hoffmeier 1996: 33–44; Dever 2003: 23–74; Walton 2003: 259–272). However, if the sites were indeed razed to the ground, they should yield material suitable for radiocarbon dating. Such remains might include burnt posts and beams, charred matting, seeds, or even burnt bone. However, scholars like Hoffmeier (1996: 35) point out that the Book of Joshua only explicitly describes three cities as having been *burnt*—Jericho, Ai, and Hazor. The remaining cities could well have been conquered without being razed. Nevertheless, charred remains should still be identifiable at these three sites.

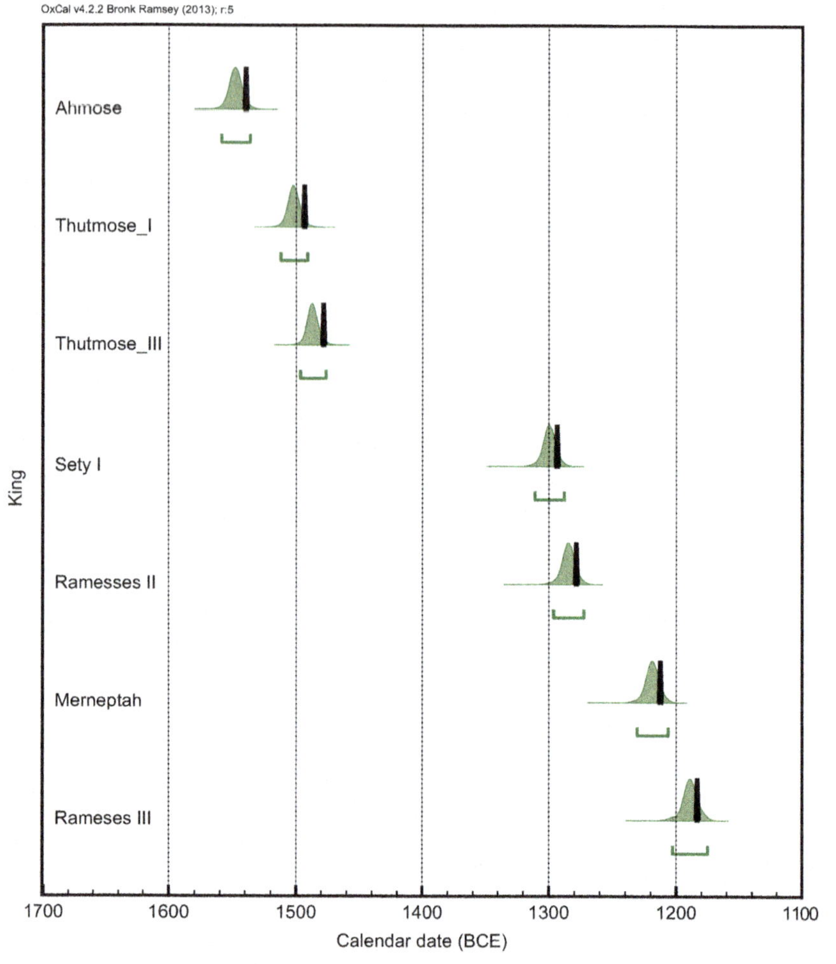

Fig. 6.1 Radiocarbon dates (*green distributions*, 95 % probability ranges beneath) for the accession years of selected kings of the New Kingdom (see Dee 2013, for data). The *black lines* indicate Kitchen's (2000) historical estimates for the same events

The Biblical city of Ai is most commonly associated with et-Tell (Albright 1924), although this attribution is not universally accepted (Wood 2000). Indeed, et-Tell is problematic from a Biblical perspective, as it shows no sign of occupation in the Middle Bronze Age and even in the Iron Age no evidence of destruction by fire. On the contrary, Hazor is unanimously linked with modern-day Tell el-Qedah and comprises occupation layers throughout the Middle and Late Bronze Ages. One such Late Bronze Age stratum has yielded a radiocarbon date on a sample of burnt wheat. The absolute date range was 1406–1273 BCE (95 % probability, UCIAMS-118997: 3065 ± 15 BP, Ben-Tor 2013), somewhat early for an Exodus in the reign of Ramesses II (1279–1213 BCE, Kitchen 2000), especially considering the 40 years spent "in the wilderness" prior to settlement in Canaan. However, no definitive conclusions should be drawn about the destruction layer until more radiocarbon results have been obtained.

Of all the sites featured in the Conquest narrative, however, Jericho (Tell es-Sultan) has received the most attention (see Watzinger 1926; Garstang and Garstang 1948; Kenyon 1957). The destruction and conclusion of the City IV occupation phase of Jericho have previously been assigned to the Biblical event (Garstang and Garstang 1948). However, since the excavations of Kathleen Kenyon 1952–1958, the consensus viewpoint has been that City IV was destroyed at the end of the Middle Bronze Age, *ca.* 1580 BCE (Kenyon 1957). This

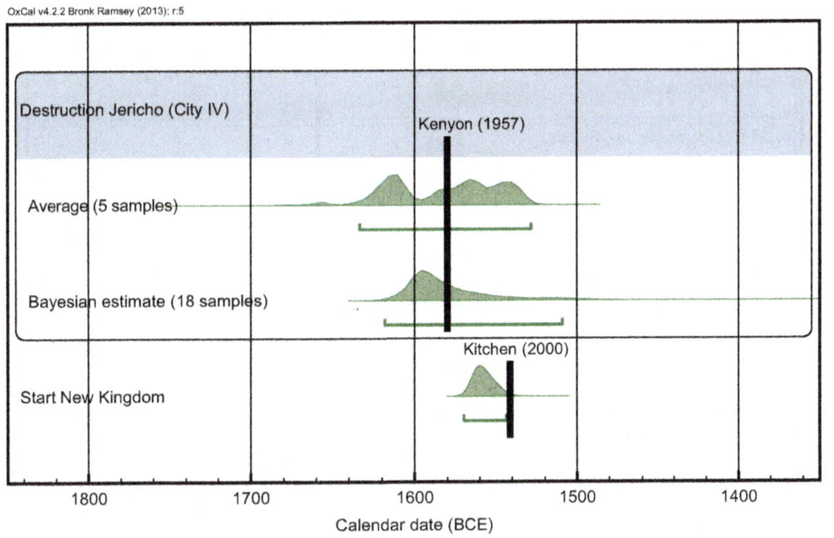

Fig. 6.2 Calibrated radiocarbon dates (*green*, 95 % probability ranges beneath) relating to the destruction of Jericho (City IV). At the *top* is the average obtained on 5 charred seed samples (1634–1529 BCE, 95 %) by Bruins and van der Plicht (1995), and in the *center* is an updated estimate using Bayesian modeling of all 18 dates from the same study (1619–1510 BCE, 95 %). The *black line* shows Kenyon's (1957) historical estimate for the destruction event. At the *bottom* is the radiocarbon date (1570–1544 BCE, 95 %) for the commencement of the New Kingdom, from Ramsey et al. (2010). The *black line* represents Kitchen's (2000) historical estimate for the beginning of the New Kingdom

interpretation excludes it from a Ramesside period Exodus, but does closely align the event with the expulsion of the Hyksos and the commencement of the New Kingdom 1540 BCE (Kitchen 2000) and 1570–1544 BCE (95 % range) (Bronk Ramsey et al. 2010).

In 1995, Bruins and van der Plicht published radiocarbon dates on 12 charcoal and 6 charred seed samples from the City IV destruction layer. The 12 charcoal samples returned slightly older ages than the seed samples, most probably because of inbuilt age (see Waterbolk 1971; Dee et al. 2012). To refine their estimate, Bruins and van der Plicht (1995) chose to average five of the short-lived seed measurements, excluding one as an outlier. In their opinion, the seeds were likely to have come from the last harvest before the conflagration. The calibrated date range for the estimate is given in Fig. 6.2, along with a new date calculated from all 18 samples using a Bayesian approach. The agreement between the radiocarbon results and Kenyon's interpretation of the archaeology is immediately apparent. Furthermore, the similarity of the destruction dates and estimates for the start of the New Kingdom is also evident. Indeed, this example demonstrates the potential for radiocarbon-based testing of the Exodus, as both the start of the New Kingdom and the putative event have been fixed in time by the chronometric method. In this case, the evidence clearly places the fall of City IV in the late Hyksos period. Once more, this date does not easily accommodate an extended period of time in the Sinai (the wilderness years) after flight from a pharaoh of the New Kingdom.

The Minoan Eruption of Thera

The eruption of Santorini (Classical Thera) was one of the largest volcanic events of the Holocene (Pyle 1997). Its likely effects on the civilizations of the Aegean Bronze Age are discussed in detail elsewhere (Marinatos 1939; Davis 1992; Warren 1995). Of particular interest for research on the Exodus is the possible impact of the eruption on the Nile Delta. Indeed, recent

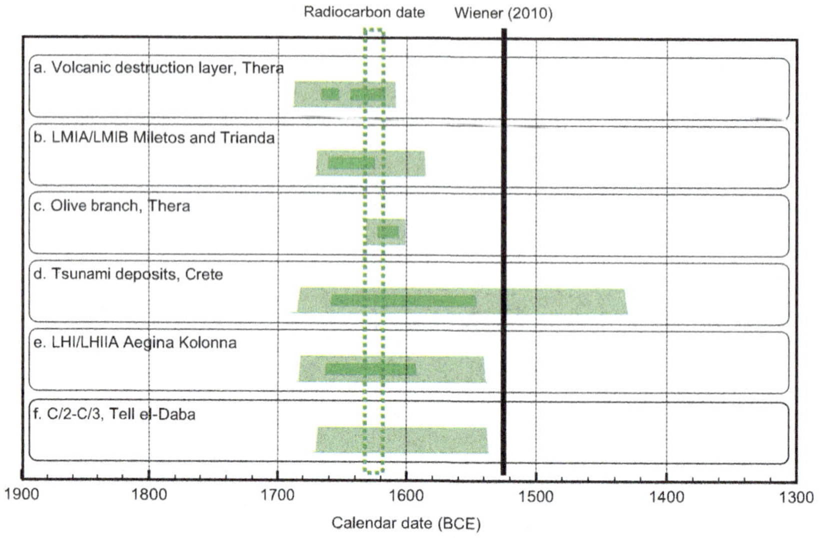

Fig. 6.3 The radiocarbon calibrated date ranges (*dark green*, 68 % probability; *light green*, 95 % probability) produced for the eruption of Thera at various locations across the Eastern Mediterranean (*a.* and *b.* Manning et al. 2006; *c.* Friedrich et al. 2006; *d.* Bruins et al. 2008; *e.* Wild et al. 2010; *f.* Kutschera et al. 2012). The *dotted box* approximates the most probable date on the basis of radiocarbon. A typical historical estimate, taken from Wiener (2010), is shown by the *black line*

scientific analysis suggests that the cataclysm probably dispersed tsunami around the Eastern Mediterranean (Sigurdsson et al. 2006). Debris consistent with such an occurrence has already been identified on Crete (Bruins et al. 2008). At the San Diego conference, Ward presented computer simulations for the event including some resulting in the inundation of the northeastern Egyptian coastline. The obvious parallels with the Exodus narrative mean that some comment on the controversy over the date of the eruption is required.

Because its geophysical impact was so marked, radiocarbon dating the eruption of Thera should have been straightforward. However, since the first radiocarbon results were published in the 1970s, a schism has developed between the historical and scientific communities over when the event actually took place (Betancourt and Weinstein 1976; Betancourt 1987; Manning et al. 2006; Friedrich et al. 2006; Bietak and Höflmayer 2007). In brief, the archaeo-historical estimates for the eruption are primarily based on ceramic synchronisms between the Aegean and Egypt (summarized by Höflmayer 2012). These linkages date the event to the late sixteenth century BCE or at least after the commencement of the New Kingdom. Radiocarbon dating, on the other hand, has consistently affirmed a late seventeenth century BCE date or during the Second Intermediate Period. Despite the vast quantity of literature submitted on both sides of the debate, no agreement has yet been reached (see Fig. 6.3; Bietak and Höflmayer 2007; Friedrich and Heinemeier 2009; Wiener 2010; Höflmayer 2012; Manning 2014). Nonetheless, some very recent archaeological finds may provide a route forward.

Crucial to the historical dating of the eruption is the Deltaic site of Tell el-Daba, the Hyksos capital, and nexus for cultural interchange with southwest Asia and the Aegean during the Second Intermediate Period. Theran pumice, acting as a *terminus ante quem* for the eruption, is first attested at the site in a workshop (Bietak and Höflmayer 2007) assigned to the reign of Thutmose III (ca. 1479–1425 BCE, Kitchen 2000). By contrast, samples from the same context produced radiocarbon data concurrent with other results for the eruption from across the region (Fig. 6.3). A crucial *terminus post quem* for the eruption at the site of Tell el-Daba is a seal of the Hyksos king, Khayan, found in Stratum D/3. His rule is thought to predate the eruption, largely because a ceramic lid bearing his name was found on Crete in phases prior to the eruption (Höflmayer 2012), although the integrity of this find has recently been challenged (Manning 2014). The chronology of the Hyksos kings is at best fragmentary, but Khayan has traditionally been dated to the late 15th Dynasty, only a matter of decades before the Hyksos were expelled from Egypt. Correspondingly, Stratum D/3 at Tell el-Daba is archaeologically dated to *ca.* 1590 BCE and the Minoan eruption is placed later in the same century. However, evidence has recently come to light that suggests that both the relative and absolute dating of Khayan are incorrect. Forstner-Müller et al. (2013) reported the finding of another seal with the name of Khayan at Tell el-Daba, this time in Stratum E/1, that is, beneath D/3. By the general chronology of the site, this shifts Khayan's reign back to at least 1620 BCE (see Kutschera et al. 2012). Even more significant was a discovery made recently at Tell Edfu in Upper Egypt by Moeller and Marouard (2011). They describe the excavation of a closed context containing dozens of clay seals impressed with royal names. The find strongly suggests that Khayan's rule was [illegible] very soon after or even contemporaneous with Sobekhotep IV of the 13th Dynasty. The implication is that Khayan's reign might actually belong in the early seventeenth century BCE. If further corroborated, this new *terminus post quem* for the eruption of Thera would easily accommodate the late seventeenth century dates provided by radiocarbon (Fig. 6.3).

Setting aside the dispute with the historical chronology, radiocarbon dates for the eruption of Thera and for the expulsion of the Hyksos have consistently been separated by 50 years or more. From the perspective of the Exodus, therefore, the eruption remains an anachronism, which implies that its relationship with the tradition is complex one and certainly not resolvable by chronometry alone (see Chap. 7).

Conclusion: Radiocarbon Dating and the Exodus Tradition

Radiocarbon dating is only able to play a supporting role in furthering the understanding of the Exodus tradition. Too little material evidence is available for the technique to testify for or against its historicity. The few events of the second millennium BCE that even tangentially relate to the narrative, such as the destruction of Jericho and the eruption of Thera, still present an incoherent picture and one which fails to provide a recognizable sequence of events. If highly compelling new archaeological or textual evidence were to come to light, scientific analysis would soon occupy the focus of attention. This chapter has shown that radiocarbon dating is already capable of addressing such a challenge, should the occasion arise. However, the more likely scenario is that incremental contributions will be made by chronometry over time, which will help textual and archaeological scholars gain a clearer understanding of this enigmatic and multifaceted tradition.

References

Aitken, M.J. 1990. *Science-based Dating in Archaeology*. London: Longman.

Albright, W.F. 1924. The Jordan Valley in the Bronze Age. *The Annual of the American Schools of Oriental Research* 6: 13–74.

Ben-Tor, A. 2013. Who Destroyed Canaanite Hazor? *Biblical Archaeology Review* 39(4): 26–36.

Betancourt, P.P. 1987. Dating the Aegean Late Bronze Age with radiocarbon. *Archaeometry* 29: 45–49.

Betancourt, P.P., and G.A. Weinstein. 1976. C-14 and the Beginning of the Late Bronze Age in the Aegean. *American Journal of Archaeology* 80: 329–348.

Bietak, M., and F. Höflmayer. 2007. Introduction: High and Low Chronology. The Synchronisation of Civilisations in the Eastern Mediterranean in the Second Millennium B.C. III. *Proceedings of the SCIEM 2000-3rd Euroconference, Vienna* (28 May to 1 June 2003), 13–23. ed. M. Bietak, and E. Czerny. Denkschriften der Gesamtakademie 37; Contributions to the Chronology of the Eastern Mediterranean 9. Vienna: Österreichische Akademie der Wissenschaften.

Bruins, H.J., and J. van der Plicht. 1995. Tell es-Sultan (Jericho): Radiocarbon Results of Short-lived Cereal and Multiyear Charcoal Samples from the End of the Middle Bronze Age. *Radiocarbon* 37(2): 213–220.

Bruins, H.J., A. MacGillivray, C.E. Synolakis, C. Benjamini, J. Keller, H.J. Kisch, A. Klügel, and J. van der Plicht. 2008. Geoarchaeological Tsunami Deposits at Palaikastro (Crete) and the Late Minoan IA Eruption of Santorini. *Journal of Archaeological Science* 35(1): 191–212.

Buck, C.E., C.D. Litton, and A.F.M. Smith. 1991. Calibration of Radiocarbon Results Pertaining to Related Archaeological Events. *Journal of Archaeological Science* 19: 497–512.

Davis, J.L. 1992. Review of Aegean Prehistory I: The Islands of the Aegean. *American Journal of Archaeology* 96(4): 699–756.

Dee, M.W. 2013. A Radiocarbon-based Chronology for the New Kingdom. In *Radiocarbon and the Chronologies of Ancient Egypt*, ed. A.J. Shortland and C. Bronk Ramsey, 65–75. Oxford: Oxbow.

Dee, M.W., J.M. Rowland, T.F.G. Higham, A.J. Shortland, F. Brock, S.A. Harris, and C. Bronk Ramsey. 2012. Synchronizing Radiocarbon Dating and the Egyptian Historical Chronology Through Improved Sample Selection. *Antiquity* 86: 868–883.

Dever, W.G. 2003. *Who Were the Early Israelites, and Where Did They Come From?* Grand Rapids, MI: Eerdmans.

Forstner Müller, I., P. Rose, C. Reali, and H. Tronchère. 2013. *Preliminary Report on the Season 2012 at Tell el-Dab'a/Sharqeya.* http://www.auaris.at/downloads/Report_SCA_engl_arabic_small.pdf

Friedrich, W.L., and J. Heinemeier. 2009. The Minoan Eruption of Santorini 2009 Radiocarbon Dated to 1613 ± 13 B.C. In *Time's Up! Dating the Minoan Eruption of Santorini. Acts of the Minoan Eruption Chronology Workshop, Sandbjerg, November 2007*, ed. D. Warbuton, 285–293. Aarhus: Aarhus University Press.

Friedrich, W.L., B. Kromer, M. Friedrich, J. Heinemeier, T. Pfeiffer, and S. Talamo. 2006. Santorini Eruption Radiocarbon Dated to 1627–1600 B.C. *Science* 312 (5773): 548.

Garstang, J., and J.B.E. Garstang. 1948. *The Story of Jericho*. London: Marshall, Morgan and Scott.

Hoffmeier, J.K. 1996. *Israel In Egypt: The Evidence for the Authenticity of the Exodus Tradition*. Oxford: Oxford University Press.

Höflmayer, F. 2012. The Date of the Minoan Santorini Eruption: Quantifying the "Offset". *Radiocarbon* 54 (3–4): 435–448.

Kenyon, K.M. 1957. *Digging Up Jericho*. London: Ernst Benn.

Kitchen, K.A. 2000. Regnal and Genealogical Data of Ancient Egypt (Absolute Chronology I) the Historical Chronology of Ancient Egypt, a Current Assessment. ed. M. Bietak. *The Synchronisation of Civilisations in the Eastern Mediterranean in the Second Millennium BC I. Proceedings of an International Symposium, Handorf* (15–17 November 1996), *and Vienna* (11‾12 May 1998), 39–52. Denkschriften der Gesamtakademie 19. Vienna: Österreichische Akademie Der Wissenschaften.

Kutschera, W., M. Bietak, E.M. Wild, C. Bronk Ramsey, M.W. Dee, R. Golser, K. Kopetzky, P. Stadler, P. Steier, U. Thanheiser, and F. Weninger. 2012. *The Chronology of Tell el-Daba: A Crucial Meeting Point of 14C Dating, Archaeology, and Egyptology in the 2nd Millennium BC* 54(3–4): 407–422.

Libby, W.F., E.C. Anderson, and J.R. Arnold. 1949. Age Determination by Radiocarbon Content: World-wide Assay of Natural Radiocarbon. *Science* 109(2827): 227–228.

Manning, S.W. 2014. *Test of Time: The Volcano of Thera and the Chronology and History of the Aegean and East Mediterranean in the Mid Second Millennium BC*, 2nd ed. Oxford: Oxbow Books.

Manning, S.W., C. Bronk Ramsey, W. Kutschera, T.F.G. Higham, B. Kromer, P. Steier, and E.M. Wild. 2006. Chronology for the Aegean Late Bronze Age 1700–1400 B.C. *Science* 312(5773): 565–569.

Marinatos, S. 1939. The Volcanic Destruction of Minoan Crete. *Antiquity* 13(52): 425–439.

Meyers, C.L. 2005. *Exodus*. Cambridge: Cambridge University Press.

Moeller, N., and G. Marouard (with a contribution by N. Ayers). 2011. Discussion of Late Middle Kingdom and Early Second Intermediate Period History and Chronology in Relation to the Khayan Sealings from Tell Edfu. *Egypt and the Levant* 21: 87–121.

Petrie, W.M.F., and W. Spiegelberg. 1897. *Six Temples at Thebes, 1896*. London: Quaritch.

Pyle, D. 1997. The Global Impact of the Minoan Eruption of Santorini, Greece. *Environmental Geology* 30(1–2): 59–61.

Ramsey, C. Bronk. 2009. Bayesian Analysis of Radiocarbon Dates. *Radiocarbon* 51(1): 337–360.

Ramsey, C. Bronk, M.W. Dee, J.M. Rowland, T.F.G. Higham, S.A. Harris, F. Brock, A. Quiles, E.M. Wild, E.S. Marcus, and A.J. Shortland. 2010. Radiocarbon-based Chronology for Dynastic Egypt. *Science* 328(5985): 1554–1559.

Shaw, I.M.E. 1985. Egyptian Chronology and the Irish Oak Calibration. *Journal of Near Eastern Studies* 44 (4): 295–317.

Sigurdsson, H., S. Carey, M. Alexandri, G. Vougioukalakis, K. Croff, C. Roman, D. Sakellariou, C. Anagnostou, G. Rousakis, C. Ioakim, A. Gogou, D. Ballas, T. Misaridis, and P. Nomikou. 2006. Marine Investigations of Greece's Santorini Volcanic Field. *EOS, Transactions American Geophysical Union* 87(34): 337–339.

Walton, J.H. 2003. Exodus, Date Of. In *Dictionary of the Old Testament: Pentateuch*, ed. T.D. Alexander and D.W. Baker, 258–272. Leicester: InterVarsity.

Warren, P.M. 1995. Minoan Crete and Pharaonic Egypt Interconnections in the Second Millennium BC. In *Egypt, the Aegean and the Levant:*

Interconnections in the Second Millennium BC, ed. W. V. Davies and L. Schofield, 1–18. London: British Museum Press.

Waterbolk, H.T. 1971. Working with Radiocarbon Dates. *Proceedings of the Prehistoric Society* 37: 15–33.

Watzinger, C. 1926. Zur Chronologie der Schichten von Jericho. *Zeitschrift der Deutschen Morgenlandischen Gessellschaft* 80: 131–136.

Wiener, M.H. 2010. A Point in Time. In *Cretan Offerings: Studies in Honour of Peter Warren*, ed. O. Krzyszkowska, 367–394. London: The British School at Athens.

Wild, E.M., W. Gauß, G. Fortenpointner, M. Lindblom, R. Smetana, P. Steier, U. Thanheiser, and F. Weninger. 2010. 14C Dating of the Early to Late Bronze Age Stratigraphic Sequence of Aegina Kolonna, Greece. *Nuclear Instruments and Methods in Physics Research B* 268(7–8): 1013–1021.

Wood, B.G. 2000. Khirbet el-Maqatir, 1995-1998. *Israel Exploration Journal* 50(1–2): 123–130.

The Thera Theories: Science and the Modern Reception History of the Exodus

7

Mark Harris

Abstract

While scholars continue to debate the elusive historical and critical issues behind the Exodus narrative, a steady stream of interpretations inspired by the natural sciences has been appearing. These interpretations take naturalistic portions of the text at face value and insist that they are to be understood in terms of natural catastrophes. Biblical scholars and archaeologists may highlight the complex *human* factors behind the genesis and evolution of the exodus traditions, but these *naturalistic* interpretations paint more-or-less apocalyptic scenarios informed by scientific research into volcanoes, earthquakes, and other spectacular natural phenomena. The eruption of Thera (Santorini) in the seventeenth century BC has featured heavily, not least because it can also be invoked as an explanation for the myth of Atlantis. This chapter presents an overview of these "Thera theories" in order to investigate their main interpretative strategies. While serious difficulties are identified with the Thera theories, it will be argued that, as acts of reading, they possess a strong imaginative appeal.

The chapters and presentations by Stephen Moshier, John Grattan, Michael Dee, Steven Ward, and Amos Salamon all bring the natural sciences to the text of Exodus in order to examine its naturalistic context, an approach that was memorably challenged in one of the discussion sessions at the conference as being "intellectually indefensible." Grattan's presentation outlines the effects of volcanoes on human culture, and this chapter looks at one particular volcano, Thera. In this way, the chapter examines a recurring motif in naturalistic studies of Exodus, the idea that the miracle traditions might be explained by natural catastrophes from Thera such as tsunamis (Ward, Salamon). While the chapter is critical of naturalistic readings, it nevertheless suggests an "intellectual defense" within the context of post-modernist biblical criticism.

M. Harris (✉)
School of Divinity, University of Edinburgh,
New College, Mound Place, Edinburgh, EH1 2LX,
United Kingdom
e-mail: Mark.Harris@ed.ac.uk

Introduction

Naturalistic explanations of the plagues and sea crossing traditions of Exodus have had a

T.E. Levy et al. (eds.), *Israel's Exodus in Transdisciplinary Perspective*,
Quantitative Methods in the Humanities and Social Sciences, DOI 10.1007/978-3-319-04768-3_7,

checkered reception. For while there are numerous books (many at the popular level), TV documentaries, Web sites, and scientific articles that claim to provide the definitive explanation for "what really happened"—many of them using the legendary Minoan eruption of the volcano Thera—the professional guild of biblical scholarship has tended to exercise caution. The weighty scholarly commentaries on Exodus, for instance, conspicuously overlook these "Thera theories," preferring to emphasize the complex human (social, political, and religious) factors standing behind the text. In some cases, this scholarly caution extends to outright skepticism. Thus, Miller and Hayes (1986: 65), in their classic text *A History of Ancient Israel and Judah*, indict the Thera theories with serious credibility issues:

> Theories of this sort attempt to give naturalistic and scientifically acceptable explanations for the more fantastic and miraculous biblical claims. In our opinion, however, these theories presuppose such hypothetical scenarios, such a catastrophic view of history, and such marvelous correlations of coincidental factors that they create more credibility problems of their own than the ones they are intended to solve.

Although naturalistic approaches are rarely denounced as explicitly as this, the general disinterest displayed by the scholarly guild indicates that the credibility issues are felt widely, even if they are not articulated so forcefully. At best, naturalistic explanations appear to be of limited interest in professional biblical interpretation, in spite of the enthusiastic reception such explanations often receive in the wider Bible-reading world.

If there is a silent hermeneutical debate going on over the relevance of naturalistic explanations, then it is the purpose of this paper to bring the debate into the open by examining the Thera theories and their main hermeneutical tactics. It will be argued that the theories work by means of a selective literalism which neglects many of the crucial critical questions raised by biblical scholarship. Nevertheless, it will also be suggested that the Thera theories may be seen as creative acts of reading in their own right. This point allows a further factor to come into focus, that of "imagination," a keyword in the conference subtitle, *Israel's Exodus Between Text and Memory, History and Imagination*, and a useful term for appreciating the popular apocalyptic appeal of the Thera theories.

The Minoan Eruption

First, it is important to lay out what we can say confidently about the effects of the Minoan eruption on Bronze Age civilization; unfortunately it is rather little. It is widely recognized that "the great and archaeologically pivotal Thera eruption" (Ramsey et al. 2004: 325) offers the unique possibility of providing a common archaeological marker across eastern Mediterranean Bronze Age chronologies (e.g., Friedrich et al. 2006). And yet, despite this potential, there are unresolved problems with reconciling the radiocarbon date of the eruption (ca. 1610 BC) with the date obtained by more traditional archaeological methods (ca. 1500 BC), and this discrepancy is still, after several decades, the subject of intense debate. Michael Dee's paper in this volume addresses the debate directly, and so it will not be rehearsed here; suffice it to say that the date of the Minoan eruption, and how it relates to Egyptian chronologies, is of crucial significance in assessing the Thera theories.

A further point to note about the Minoan eruption is an intriguing but resounding silence in human history. No textual reports are known that unambiguously report the eruption or its effects,[1] and neither is there archaeological evidence to suggest that surrounding civilizations were adversely affected. The only victims appear to be the people of Akrotiri, the abandoned Minoan habitation on Thera, but in spite of extensive excavations here no human casualties of the eruption have been found, so it seems that the people received sufficient warning to flee. On the one hand, then, abundant geological evidence exists for the Minoan eruption—the very visible stratigraphy of the island, together with ash

[1] Notwithstanding the controversy over texts such as the Ahmose Tempest Stela (Foster et al. 1996; Wiener and Allen 1998).

and pumice deposits elsewhere in the Eastern Mediterranean—but on the other hand, no evidence exists to indicate that any human beings were seriously affected beyond those who were forced to abandon their homes on Thera. This point underscores what might be called the *paradox* of Thera, that a very sizeable volcanic eruption occurred in the midst of Bronze Age civilizations, but left no unambiguous mark on them (Ellis 1998: 161). This is an important point to bear in mind as we examine the Thera theories, since they claim that this volcano was decisive in causing large-scale cultural cataclysms, preeminently the demise of Minoan civilization on Crete, and that it is reflected in stories such as Plato's tale of Atlantis, and in Exodus. However, the nature of the evidence available means that such theories must by necessity take the form of imaginative conjecture.

All of this is not to imply that nothing is known of the Minoan eruption; on the contrary, Thera is probably the best-studied volcano of the ancient world. The Minoan eruption took place over three or four stages, spread out over days or weeks (Friedrich 2000: 72–7; de Boer and Sanders 2002: 54–5). The third, explosive stage, was the most spectacular, with a Volcanic Explosivity Index (VEI) of perhaps 7.[2] Like the Richter scale, the VEI is logarithmic, and for comparison, the famous eruption of Krakatau in 1883, with which Thera is invariably compared, had a VEI of 6, and Mount St Helens in 1980 a VEI of 5. This puts Thera towards the "super-colossal" end of the spectrum: unusually large, but not uniquely so.

Potentially the most devastating component of the Minoan eruption was its tsunamis; they are, however, also the most difficult component to quantify and trace. This is why a comparison is so often made with Krakatau. Both Thera and Krakatau involved very explosive eruptions of the Plinian type, and both were island volcanoes that collapsed when the magma chamber emptied, leaving a caldera in the middle of the sea, and producing potentially huge waves in the process (Novikova et al. 2011). The eruption of Krakatau is well-documented, including eyewitness testimony. Krakatau is said to have destroyed some 160 towns and villages, and to have killed as many as 40,000 people (de Boer and Sanders 2002: 157), largely on account of the enormous tsunamis generated. Thera also collapsed—the present ring-shape of the island bears this out—and it seems to have generated tsunamis, since evidence of Theran tsunami deposits has been found in several places on the coast of Turkey (Minoura et al. 2000) and at Palaikastro in Crete (Bruins et al. 2008, 2009).

However, at this point, we must be careful that the analogy with Krakatau does not lead to unwarranted claims. For one, it is known that caldera collapse is just one of the factors in generating tsunamis in this kind of eruption. Furthermore, work of the last two decades has shown that Thera *already* had a caldera from a previous eruption, which means that any collapse was not as great as had been previously supposed (Dominey-Howes 2004: 120). It is also possible that what caldera collapse there was took place rather slowly, over hours rather than minutes, meaning that the tsunamis from the collapse would be correspondingly smaller. These points have led to great uncertainty being expressed in the recent scientific literature about the extent and magnitude of the tsunamis from Thera, and about the value of making a close analogy with Krakatau. Added to this uncertainty is the fact that, unlike Krakatau, we know of no human casualties from the Theran tsunamis, nor of the effect they might have had, for instance, on the Minoan ports. Tsunamis are frequently invoked in the Thera theories, suggesting that they played a crucial role in the end of Minoan civilization, and in the sea crossing of Exod 14–15. However, it appears that caution is in order. Recent studies of the Theran tsunamis have cast strong doubts on the likelihood that tsunami impact occurred beyond the Aegean (Minoura et al. 2000; Dominey-Howes 2004; Pareschi et al. 2006; Salamon et al. 2007). In other words, it is

[2] VEI data from the database of the Smithsonian Institution Global Volcanism Program (http://www.volcano.si.edu/, accessed on 7th August 2013).

appearing increasingly doubtful that a significant tsunami from Thera ever struck the Egyptian coastline.

A closely related issue is that of the ash cloud produced by Thera. The volcano ejected perhaps 60 m^3 of rock into the sea and air, creating a plume of volcanic tephra into the upper atmosphere (Sivertsen 2009: 25). Like the tsunamis, the ash cloud is crucial for the Thera theories. Did the ash fallout make agriculture on Crete impossible? Did the ash turn the Nile red, and cast darkness over Egypt, thus explaining the plague narratives? As with the tsunamis, recent work suggests caution. The ash cloud seems to have extended mostly due east of Thera. Thus Rhodes suffered from extensive ash fall,[3] but Crete, to the south, was much less affected (Friedrich 2000: 80; Johnston et al. 2012), and Egypt perhaps not at all, although some ash must have found its way there, since minute traces of it have been found in the Nile Delta. Bietak (2006: 63), though, considers it "highly unlikely. . .that Egypt was shrouded in a cloud of volcanic ash." If so, those Thera theories that call upon the ash cloud to explain the plagues of Egypt are looking increasingly dubious.

Gathering the evidence together, it is clear that many of the key parameters of the Minoan eruption are well defined. However, those parameters that are key to the Thera theories—the date, and the extent of the tsunamis and ash cloud—are uncertain. It stands to reason that the eruption must have affected human settlements in the Aegean (if not further afield), but with no substantiating evidence it is almost impossible to say how. That there was significant cultural disruption towards the end of the Middle Bronze Age is indisputable, and the Minoan eruption offers a convenient naturalistic explanation. However, caution is once again in order, since the cultural response to disaster is known to be highly complex (Torrence and Grattan 2002). Moreover, the almost complete absence of evidence for such a causal link between Thera and cultural disruption means that the Thera theories must be considered as highly conjectural by definition.[4] Nonetheless, they remain a rich resource for the imagination.

The Thera Theories

There are many Thera theories, and space permits only a small number of representatives to be discussed. Perhaps the most significant is that due to Spyridon Marinatos, who proposed in 1939 that Thera was responsible for the demise of Minoan Crete. Signs of significant earthquake damage to Minoan palaces could be explained, he thought, by the eruption of Thera. And by making the analogy with Krakatau explicit, Marinatos (1939: 431) filled in the very substantial blanks in our knowledge of how Thera affected humans:

> No historical account survives of this great earthquake, but fortunately we have an excellent means of reconstructing all the phenomena which accompanied the disaster, in the eruption of Krakatau.

In this way, Marinatos described an apocalyptic scene of destruction in Crete, as giant waves obliterated the coastal towns, while earthquakes, lightning, and a rain of red-hot ash and pumice worked devastation inland. After such an "irreparable blow" (*ibid.* 437) Minoan civilization sank into decline before dying out completely. Such a picture—based on Krakatau but displaced in time and space to Crete—clearly bears a great deal of similarity to the genre of apocalypse, and herein lies its imaginative appeal, as witnessed by the enduring appeal of the Thera theories. Ominously, the editors added a caveat at the end of Marinatos' article, suggesting the need for further supporting evidence, evidence which is still largely outstanding.

[3] However, there is no sign that it had a lasting effect on Minoan civilization there (Doumas and Papazoglou 1980).

[4] Note, for instance, the elaborate ongoing debate over whether Thera played any part at all in the demise of the Minoans. The evidence, such as it is, points to human factors as decisive, but Thera may have been a catalyst. See, for instance, Driessen (2002) for a positive view, and Manning and Sewell (2002) for a skeptical view.

Marinatos's apocalypse soon became linked with Plato's story of the destruction of Atlantis, although Marinatos himself was careful not to make too much of this.[5] Galanopoulos and Bacon though, in their well-known book of 1969, *Atlantis: The truth behind the legend*, made the links between Thera, Atlantis, and the exodus explicit. They explained that the circular caldera of modern-day Santorini possesses a ring-like geography similar to Plato's description for Atlantis. Hence, they made the case that Thera/Santorini/Crete was the home of the legendary civilization of Atlantis, what we know as the Minoans, destroyed by the gigantic volcanic eruption of Thera.

Since Galanopoulos and Bacon were working with a fifteenth century BC eruption date, close to the traditional date of the exodus adduced from 1 Kings 6:1,[6] they also argued that Thera was decisive in allowing the Israelites to escape from Egypt (Galanopoulos and Bacon 1969: 193–9). The first nine of the ten plagues was the sequence of natural events that might be expected from the storms, earthquakes, and fallout of volcanic tephra that would accompany a large volcanic eruption. And the sea crossing could also be explained by Thera, if the *yam suph* was Lake Sirbonis, a shallow lagoon currently separated from the Mediterranean by a narrow spit of land. Galanopoulos and Bacon saw the Israelites standing there at the end of the eruption, as Thera's magma chamber filled with seawater. The sea would have ebbed away, giving the Israelites time to cross the lagoon before the tsunamis came crashing in, taking the pursuing Egyptians by surprise.

Galanopoulos's and Bacon's hypothesized role for Thera has taken firm hold, and we find it repeated in various guises in later Thera theories.[7] Their hypothesis is not without its problems though, one of the most relevant being the selective way that the texts are treated in order to obtain a fit with the hypothesis, a perennial feature of naturalistic explanations.[8] But the model's most significant problem is its fifteenth century BC eruption date. On the one hand, it is two centuries earlier than the widespread scholarly view that the exodus should be located in the time frame of Ramesses II. On the other hand, it is nearly two centuries later than the now widely accepted radiocarbon date for the Minoan eruption. More recent Thera theories have responded to this problem with ingenuity. Sivertsen and Booysen offer two such recent theories, taking the revised radiocarbon date seriously (which means bringing the Hyksos into focus as candidate Israelites), but simultaneously proposing additional exoduses in other time frames.

The geologist Barbara Sivertsen published a comprehensive study of the exodus, *The Parting of the Sea*, in 2009. Her sub-title explains her approach well: *How Volcanoes, Earthquakes, and Plagues Shaped the Story of Exodus*. Sivertsen suggests that the text of Exodus may be explained by a whole portfolio of natural disasters, not just the Minoan eruption. Three different volcanoes feature, plus earthquakes, tsunamis, a hurricane wind, climate change, and various diseases. In fact, disasters take center stage in Sivertsen's telling of the Exodus narrative, and there are relatively few human actions described that are not a direct response to natural catastrophe. Such a reading appeals greatly to the imagination, and like Marinatos's 1939 paper, it is perhaps best described as an apocalypse informed by modern science.

In common with earlier Thera theories, Sivertsen tends to read the text quite literally when a naturalistic explanation presents itself, but otherwise treats the text relatively loosely.

[5] Atlantis does not appear, for instance, in Marinatos' 1939 paper, but in 1950 he speculated about the origins of the myth (Ellis 1998: 232–5).

[6] The various exodus chronologies are discussed in this volume by Lawrence Geraty.

[7] It is, for instance, the basic picture given by Ian Wilson, supported by Hans Goedicke's argument that the sea crossing is described in the Speos Artemidos inscription (Wilson 1985: 130–7).

[8] For example the "strong east wind" that drove the sea back (Exod 14:21) is dismissed as insufficient to move the waters (Galanopoulos and Bacon 1969: 197), in spite of the fact that this wind has been highlighted as the most credibly naturalistic component of the sea crossing by scholars (Harris 2007: 8–11).

So, for example, Sivertsen revives the argument that Mount Sinai was one of the Arabian volcanoes. In support, she suggests a highly literal and naturalistic reading of the theophany of Exodus 24, where God is described as having "under his feet…something like a pavement of sapphire stone, like the heaven for clearness" (24:10). Sivertsen connects this text with the Harrat ar-Raha volcano in Arabia, which is unusually rich in olivine crystals. Usually green, Sivertsen (2009: 74) explains that olivines can appear blue like sapphires. Such a strongly literalistic approach is in marked contrast with that taken by professional biblical scholars, who would be more likely to see the reference to sapphires in terms of a complex of stock symbols and metaphors used in biblical theophanies and apocalypses (e.g., Ezek 1:26; Rev 9:17).

On the other hand, Sivertsen takes a less literal approach to the text when she accommodates de Vaux's idea that the Exodus text might preserve two presentations of the story merged together, the "exodus-flight" and "exodus-expulsion" (*ibid.* 42–3). Sivertsen interprets this idea as indicating two historical exoduses that occurred nearly two centuries apart, each precipitated by a different volcano. The first exodus is driven by Thera, erupting in the late seventeenth century, and causing the first nine plagues. This terrifies the Asiatics of Avaris to flee (from natural catastrophe, not oppression). They cross over one of the seas of reeds into Sinai by means of a land bridge, which is exposed by a strong wind driving back the waters (*ibid.* 45). They come to the holy mountain, a volcano in Arabia, and so on to Canaan by way of Jericho. This is the first exodus. The second exodus takes place in the reign of Tuthmosis III, around 1450 BC, and is prompted by an outbreak of food poisoning among Egyptian children. This, the tenth plague, allows Shasu/Israelite slaves to escape, and they flee to the Mediterranean coast pursued by the Egyptian army. Coincidentally, another volcano erupts in the distant Aegean (this time from an underwater volcano), producing darkness through its tephra cloud, and a tsunami that destroys the Egyptians while the slaves are encamped a little way inland (*ibid.* 134–6).

Sivertsen's approach is open to similar criticisms as those of Galanopoulos and Bacon. Hermeneutically, it shows inconsistency in reading the text literally when a suitable naturalistic hypothesis presents itself, but loosely at other times, sometimes in order to incorporate a naturalistic hypothesis. However, in its favor, Sivertsen offers a memorable and imaginative reading which, because of its heavy use of naturalistic and catastrophic components, possesses many qualities of the traditional apocalypse; in such a way, modern science recreates the ancient genre.

The final Thera theory we shall examine is the most recent, and also the most elaborate: the engineer Riaan Booysen's study of 2012, *Thera and the Exodus*. Booysen makes it clear from the beginning that he does not believe in miracles (Booysen 2012: 2). Everything in the Exodus text must be explained naturalistically, or else be rejected. Ironically, Booysen is able to explain the plagues and the sea crossing entirely naturalistically, and he multiplies them. There is "not one but possibly up to three eruptions of Thera" (*ibid.*) spread over several centuries in his naturalistic models, a real intensification over Sivertsen. In this way, Booysen sets up three exodus-style events. He describes an *eisodos* (the entry of the Hyksos into Egypt), and two Hyksos exoduses, one in the reign of Ahmose, and the other in the reign of Amenhotep III. Each of these three events is facilitated by a different Theran eruption[9] and associated flooding (*ibid.* 281, 288–9, 309). The plagues are associated with the final eruption, but the sea event occurs on two or perhaps three different historical occasions, with armies being swept away repeatedly.

Putting aside the question of whether there is material evidence for so many eruptions of Thera (*ibid.* 320–28), we may note that Booysen's reading of the Exodus text is similar to previous Thera theories: it is literalistic when a naturalistic explanation is in view, but looser otherwise. And

[9] The Minoan eruption is the second of Booysen's three Theran eruptions.

as with Sivertsen, the overall Exodus narrative proliferates into multiple exoduses that bear only a loose relationship with the original story, while naturalistic elements (especially the plagues and sea crossing) are carefully preserved. One thing is clear: the Thera theories are following a trajectory towards increasingly complex naturalistic scenarios while the historical, theological and textual questions raised by critical scholarship are largely overlooked. A careful discussion of the hermeneutical implications and limits of such naturalistic explanations is long overdue.

The Hermeneutics of Naturalistic Explanations

The Thera theories operate on the principle that a remarkable event requires a remarkable explanation, and they demonstrate that, in this instance at least, science does not disappoint. Science may habitually explain the remarkable in mundane terms (e.g., the laws of physics), but when applied to the stories of the plagues and sea crossing, science is clearly capable of heightening the sense of the remarkable by translating these outlandish stories into vivid modern terms which appeal to the imagination through their apocalyptic character.[10] Indeed, the breadth and scale of the Thera theories is remarkable in itself, as is their suggestion of remarkable coincidences; the tsunamis that deliver the Israelites in the nick of time but destroy the Egyptians are probably the best examples. While these coincidences and heightened remarkabilities do not appear to be problematic for those natural scientists and others who formulate Thera theories, they raise credibility issues for biblical scholars such as Miller and Hayes, quoted at the beginning of this article.

It is worth pursuing the credibility issues further. Partly they arise from the lack of evidence that the Minoan eruption had any serious effect on Egyptian life ("The Minoan Eruption"). However, there is a more significant hermeneutical issue at stake.

In his conference lecture on geophysical perspectives of the exodus, Steven Ward presented several possible scenarios by which tsunamis and floods could be generated in order to explain the sea crossing, making the point that these were reasonable possibilities for what might have happened. My approach is quite different. I do not dispute that it is reasonable to suppose that tsunamis may have struck the Egyptian coast from time to time, but I question *whether such a model provides a good reading of the text* of Exodus 14–15. Four factors must be considered:

1. There are many possible naturalistic explanations for the sea event, not just tsunamis and floods (Harris 2007), and these should all be weighed in the balance before one is considered more reasonable than the others.
2. Source critical approaches distinguish three or perhaps even four different sea traditions/elements woven together in the text,[11] several of which bear no relation to tsunamis. Significantly, all of these traditions are *naturalistic to some degree in their own right.* In other words, the text itself possesses a degree of naturalism. It is ironic then, that modern naturalistic explanations so often focus on one

[10] Note that theological reservations are often expressed about naturalistic explanations like the Thera theories. The complaint is that, although such theories may heighten the sense of the remarkable, they also strip the biblical text of its theological depth by explaining away the text's notion of divine action in mundane (scientific) terms. Prof Dever made such a point in his lecture at the conference, saying that "Scientific explanations of the miracles miss the point." More forceful statements of such a view are made, for instance, by John Durham in his commentary, writing of Hort's explanation of the fifth plague: "This kind of wild speculation is also misleading, and not alone because of its absurdity. Worse still is its discrediting of the theological tenor of the biblical narrative, which will admit no naturalistic and hence nonmiraculous 'explanations'" (Durham 1987: 118). Durham assumes here that science and the miraculous are mutually exclusive, a controversial view. Addressing this point adequately, however, requires a wide-ranging philosophical discussion of the concept of miracle which is beyond the scope of this article.

[11] A strong east wind (Exod 14:21); mud clogging the Egyptians' chariot wheels (14:25); walls of water (14:22); a meteorological storm during a sea crossing (15:8–10).

element/tradition to the exclusion of others; there is a tendency towards selective and literalistic reading. Such explanations are unable to respect the diversity of traditions in the text, and they offer little help in the critical task of plotting the evolutionary process by which the traditions came together.
3. All other things being equal, an explanation such as the tsunami model of the sea crossing, which works through incredible coincidences and tiny probabilities, is inherently unfavorable compared to one that works with *higher* degrees of probability.
4. Biblical scholarship routinely works with such *higher* probability explanations, aware that the everyday human phenomena of story-telling and retelling—as identified in source, form, redaction, and textual criticism, including recognition of mythological motifs—account for much of what we read in the text before we ever try to read it terms of putative natural phenomena.

All of this means, I believe, that Miller and Hayes are right to complain of serious credibility issues in the Thera theories; there are many other factors beyond the naturalistic to consider in interpreting the text, not least human factors. Propp's assessment (1999: 347–8) is uncompromising but insightful:

> Rationalistic explanations for miracles...are anachronistic today. To believe that the Bible faithfully records a concatenation of improbable events, as interpreted by a prescientific society, demands a perverse fundamentalism that blindly accepts the antiquity and accuracy of biblical tradition while denying its theory of supernatural intervention.

Propp is quite right: the selective and literalistic readings that many naturalistic explanations perform are related to "fundamentalism" in that they assume the ancient text should be read like a modern scientific account; they are "perverse" because they largely neglect the theological, critical, and contextual questions that surround the ancient text.

However, if naturalistic approaches are often uncritical, it is important not to reject them uncritically, ruling them out of court altogether. Aside from the possibility that they may suggest helpful contextual insights into the text, in the case of the Thera theories at least it is also worth considering their patent popular appeal as *imaginative* acts of reading. The Thera theories add modern color and vitality to the text, and in visualizing the story of the plagues and sea crossing—a story with many extraordinary and implausible components—their apocalyptic character is a help, not a hindrance, in this imaginative task. From that point of view, the Thera theories complement, if not improve upon, Greta Hort's well-known naturalistic model of the plagues (Hort 1957–58), where they are explained as intensifications of the perennial natural phenomena of Egypt, like the *khamsin* desert sandstorm for the plague of darkness. Hort's model has a tendency to downplay the spectacular and apocalyptic elements of the narrative, while the Thera theories amplify them. It is at this level then—that of imaginative reading—that the Thera theories are most effective, explaining their popularity beyond the guild. There is even a sense in which the Thera theories stand alongside other forms of Bible reading in the postmodern proliferation.[12] In short, the Thera theories give us *scientific* readings that challenge or inspire the imagination, even if there are scholarly difficulties surrounding their hermeneutical treatment of the text, and their ability to say what really happened.

The conference was highly significant in bringing together natural scientists and critical scholars, historians, and archaeologists. The phoney war over naturalistic interpretations—reflecting to some degree the parallel (but not so phoney) war between science and religion—will hopefully be played out rather more openly through such conversations, and, one might hope, work to the mutual benefit of science and critical scholarship in understanding these normative texts.

[12] For example liberation readings, or post-colonial readings.

References

Bietak, Manfred. 2006. The Volcano Explains Everything – Or Does It? *Biblical Archaeology Review* 2006(Nov/Dec): 60–65.

Booysen, Rian. 2012. *Thera and the Exodus: The Exodus explained in terms of natural phenomena and the human response to it*. Winchester, UK: O-Books.

Bruins, Hendrik J., J. Alexander MacGillivray, Costas E. Synolakis, Chaim Benjamini, Jörg Keller, Hanan J. Kisch, Andreas Klügel, and Johannes van der Plicht. 2008. Geoarchaeological tsunami deposits at Palaikastro (Crete) and the Late Minoan IA eruption of Santorini. *Journal of Archaeological Science* 35: 191–212.

Bruins, Hendrik J., Johannes van der Plicht, and Alexander J. MacGillivray. 2009. The Minoan Santorini eruption and tsunami deposits in Palaikastro (Crete): Dating by geology, archaeology, ^{14}C, and Egyptian chronology. *Radiocarbon* 51: 397–411.

de Boer, Jelle Zeilinga, and Donald Theodore Sanders. 2002. *Volcanoes in Human History: The Far-Reaching Effects of Major Eruptions*. Princeton, NJ: Princeton University Press.

Dominey-Howes, Dale. 2004. A re-analysis of the Late Bronze Age eruption and tsunami of Santorini, Greece, and the implications for the volcano-tsunami hazard. *Journal of Volcanology and Geothermal Research* 130: 107–132.

Doumas, C., and L. Papazoglou. 1980. Santorini tephra from Rhodes. *Nature* 287: 322–324.

Driessen, Jan. 2002. Towards an archaeology of crisis: Defining the long-term impact of the Bronze Age Santorini eruption. In *Natural Disasters and Cultural Change*, ed. R. Torrence and J. Grattan, 250–263. London: Routledge.

Durham, John I. 1987. *Exodus*, Word biblical commentary, vol. 3. Waco, TX: Word.

Ellis, Richard. 1998. *Imagining Atlantis*. New York, NY: Knopf.

Foster, Karen Polinger, Robert K. Ritner, and Benjamin R. Foster. 1996. Texts, Storms, and the Thera Eruption. *Journal of Near Eastern Studies* 55: 1–14.

Friedrich, Walter L., Bernd Kromer, Michael Friedrich, Jan Heinemeier, Tom Pfeiffer, and Sahra Talamo. 2006. Santorini Eruption Radiocarbon Dated 1627–1600 B.C. *Science* 312: 548.

Friedrich, Walter L. 2000. *Fire in the Sea: The Santorini Volcano: Natural History and the Legend of Atlantis* (Trans. A. R. McBirney in accordance with 1994 German ed). Cambridge, UK: Cambridge University Press.

Galanopoulos, A.G., and Edward Bacon. 1969. *Atlantis: The Truth Behind the Legend*. London: Nelson.

Harris, Mark J. 2007. How did Moses part the Red Sea? Science as salvation in the Exodus Tradition. In *Moses in Biblical and Extra-Biblical Traditions*, ed. A. Graupner and M. Wolter, 5–31. Berlin: de Gruyter.

Hort, Greta. 1957–58. The Plagues of Egypt. *Vetus Testamentum* 69: 84–103; 70: 48–59.

Johnston, E.N., J.C. Phillips, C. Bonadonna, and I.M. Watson. 2012. Reconstructing the tephra dispersal pattern from the Bronze Age eruption of Santorini using an advection–diffusion model. *Bulletin of Volcanology* 74: 1485–1507.

Manning, S.W., and D.A. Sewell. 2002. Volcanoes and history: A significant relationship? The case of Santorini. In *Natural Disasters and Cultural Change*, ed. R. Torrence and J. Grattan, 264–291. London: Routledge.

Marinatos, Sp. 1939. The Volcanic Destruction of Minoan Crete. *Antiquity* 13: 425–439.

Miller, J. Maxwell, and John H. Hayes. 1986. *A History of Ancient Israel and Judah*. London: SCM.

Minoura, K., F. Imamura, U. Kuran, T. Nakamura, G.A. Papadopoulos, T. Takahashi, and A.C. Yalciner. 2000. Discovery of Minoan tsunami deposits. *Geology* 28: 59–62.

Novikova, T., G.A. Papadopoulos, and F.W. McCoy. 2011. Modelling of tsunami generated by the giant Late Bronze Age eruption of Thera, South Aegean Sea, Greece. *Geophysical Journal International* 186: 665–680.

Pareschi, Maria Teresa, Massimiliano Favalli, and Enzo Boschi. 2006. Impact of the Minoan tsunami of Santorini: Simulated scenarios in the eastern Mediterranean. *Geophysical Research Letters* 33: L18607.

Propp, William H.C. 1999. *Exodus 1–18: A New Translation with Introduction and Commentary*, The Anchor Yale Bible. New Haven, CT: Yale University Press.

Ramsey, Christopher Bronk, Sturt W. Manning, and Mariagrazia Galimberti. 2004. Dating the volcanic eruption at Thera. *Radiocarbon* 46: 325–344.

Salamon, Amos, Thomas Rockwell, Steven N. Ward, Emanuela Guidoboni, and Alberto Comastri. 2007. Tsunami Hazard Evaluation of the Eastern Mediterranean: Historical Analysis and Selected Modeling. *Bulletin of the Seismological Society of America* 97: 705–724.

Sivertsen, Barbara J. 2009. *The Parting of the Sea: How Volcanoes, Earthquakes, and Plagues Shaped the Story of Exodus*. Princeton, NJ: Princeton University Press.

Torrence, Robin, and John Grattan. 2002. The archaeology of disasters: Past and future trends. In *Natural Disasters and Cultural Change*, ed. R. Torrence and J. Grattan, 1–18. London: Routledge.

Wiener, Malcolm H., and James P. Allen. 1998. Separate Lives: The Ahmose Tempest Stela and the Theran Eruption. *Journal of Near Eastern Studies* 57: 1–28.

Wilson, Ian. 1985. *The Exodus Enigma*. London: Weidenfeld and Nicolson.

Which Way Out of Egypt? Physical Geography Related to the Exodus Itinerary

8

Stephen O. Moshier and James K. Hoffmeier

Abstract

The Exodus narrative is rich with geographic references that are properly understood in the context of the ancient landscape of the eastern Nile Delta and adjacent Sinai Peninsula. Changes in the physical geography of the region reflect dynamic interactions between the Nile river system, the Mediterranean Sea, and tectonics of the Red Sea rift system. Coordinating field geology, archaeological sites, digital topography, and satellite imagery with Geographic Information Technology resulted in a map depicting the physical geography of the area of interest during the Bronze Age. The map reveals different positions of the Mediterranean coastline with associated lagoons and the existence of Pelusiac Nile distributaries, lakes, and wetlands. The restored geography constrains the path of the ancient "Ways of Horus," the militarized coastal road between Egypt and the land of the Philistines, but also provides a plausible map of the region that is described in the Exodus texts.

The relationship of the Exodus narrative to the natural geography in which it is set is no less important for understanding its meaning than historical, cultural, religious, linguistic, and literary contexts that are the subject of other contributions in this volume. Like a cryptic text in a palimpsest manuscript, ancient landscapes of the Near East are hidden beneath the veneer of desert sand and the infrastructure of human development. The purpose of this project is to create a map that depicts the natural geography of the eastern Nile Delta and northwestern Sinai during the Late Bronze Age, some 3,500–3,200 years before present. This map provides a geographic context for the history and archaeology of ancient Egypt with implications for the setting of the biblical Exodus.

Introduction

Many editions of the Hebrew Tanakh and Christian Holy Bible feature a map showing one or multiple alternative Exodus routes out of the Nile Delta into the Sinai Peninsula. The routes are

S.O. Moshier (✉)
Wheaton College, Wheaton, IL, USA
e-mail: Stephen.Moshier@wheaton.edu

T.E. Levy et al. (eds.), *Israel's Exodus in Transdisciplinary Perspective*,
Quantitative Methods in the Humanities and Social Sciences, DOI 10.1007/978-3-319-04768-3_8,

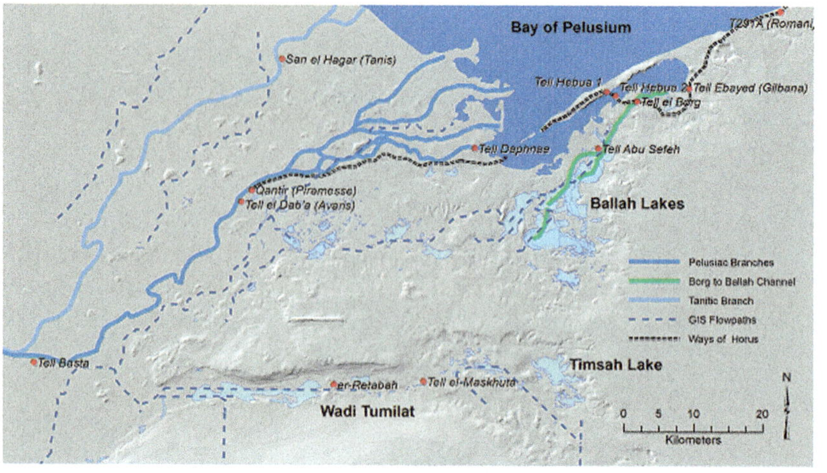

Fig. 8.1 Physical geography of the eastern Nile Delta and northwest Sinai during the Late Bronze Age. This illustration shows relief with shading derived from digitized topographic map contours. See text for sources of other geographic information

based upon various interpretations of the itineraries contained in the scriptures (Exod 12–19; Num 33). Archaeological excavations and studies of ancient texts during the past century contribute information relevant to the Exodus itinerary. For example, Piramesse, Sukkoth, and Migdol of the Exodus narrative are reasonably identified with locations known from ancient Egyptian archaeology and epigraphy (Bietak, Chap. 2). Other locations in the itinerary, such as Etham, Pi Hahiroth, Baal Zephon, and especially the Re(e)d Sea, remain ambiguous or undiscovered. Bible maps generally show the *modern* geography of settlements, river courses, lakes, and coastlines. However, geologic studies reveal changes in the land that have implications for some of these problematic locations and overland routes traveled by ancient people. In particular, surveys in the region over the past 40 years have identified and delineated abandoned Nile distributaries, significant ancient inland lakes (now dry or changed), the migrating Mediterranean coastline, and overall evolution of the Nile Delta plain. This chapter presents a map of the natural geography of the region during the Bronze Age based upon multiple sources from cartography, archaeology, and geology (Fig. 8.1).

Methods: Cartography

Figure 8.1 was complied with the ArcMap (version 10) Geographic Information System (GIS). GIS uses computing to organize and analyze spatial data coordinated to a particular geographic grid. With GIS, physical terrain may be depicted effectively from digitized land surface elevation data. High-resolution surface elevation data, publically available for many countries, was not available for Egypt at the time of this study. However, high-resolution surface elevation data contain the "overprint" of human infrastructure, such as roads, canals, bridges, and buildings. These structures and modifications of the terrain can influence GIS applications used to predict natural water flow or most effective overland travel routes. Such GIS applications are now routinely employed in the study of ancient cultures (Wilkinson 2003). A collection of vintage Survey of Egypt topographic maps at various scales was used to create the terrain model for this study.[1] The region of interest was relatively uninhabited when the maps were produced between 1900 and 1945. Features added to the GIS project included elevation data (digitized contours), outlines of inland water bodies, and locations of archaeological sites (tells). The Suez Canal (originally completed in 1869) is the most significant feature on the Survey of Egypt maps that was not included in Fig. 8.1.

[1] Maps used to create the terrain model in Fig. 8.1 are from the Survey of Egypt collection (British War Office, Intelligence Division) at the Center for Ancient Middle Eastern Landscapes, Oriental Institute, University of Chicago.

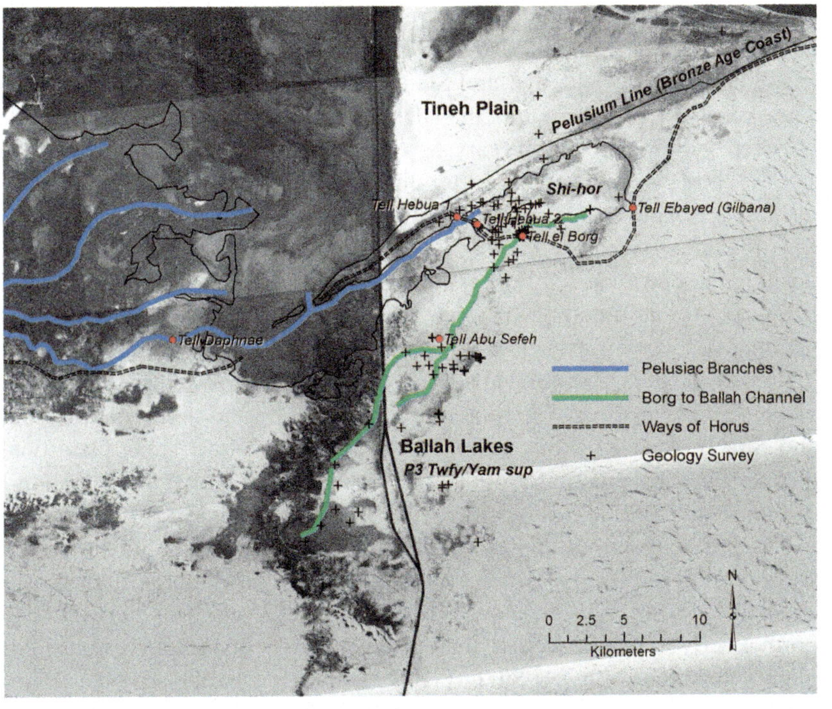

Fig. 8.2 Georectified CORONA satellite photographs (December 1968), showing the Mediterranean coast during the Late Bronze Age (*single black trace*), Pelusiac Nile branches (*blue traces*), flowpath of a channel between Tell el-Borg and the area of the former Ballah Lakes (*green traces*), the Ways of Horus road (*double, dashed black trace*), and field study locations (*crosses*). The Suez Canal is the linear feature running north–south in the center of the photographs. CORONA images acquired from US Geological Survey, EROS Data Center, Sioux Falls, SD

Methods: Landscape and Surface Geology

The depiction of ancient physical geography in this project is informed by a general understanding of landscape evolution in the region over the past 5,000 years (summarized below). Locations of ancient Nile branches and the Mediterranean coastline in Fig. 8.1 were derived from previous studies by Bietak (1975), Marcolongo (1992), Coutellier and Stanley (1987), Stanley (2002), Hoffmeier and Moshier (2006), and Moshier and El-Kalani (2008). Satellite imagery reveals large-scale surface features and patterns that are difficult to recognize on the ground. Contemporary digital space imagery offers remarkable resolution (size of visible objects) and records a range of emitted or reflected bands on the electromagnetic spectrum. However, many areas of archaeological and historical interest have been altered or covered in recent years by agricultural projects, urbanization, and other human modifications of the land. We used declassified spy satellite photos from the CORONA mission (1960–1972) to identify surface features, such as ancient coastlines and river channels. Figure 8.2 is a mosaic of CORONA photos rectified to the geographic grid, showing the locations of archaeological sites and interpretations of ancient surface features.

Methods: Modeling Surface Hydrology

Using the DEM as a representation of the natural landscape, GIS subroutines are available to determine where surface water should accumulate and flow along channel pathways (Spatial Analyst Hydrology tools in ArcMap). Pathways with the strongest "flow accumulation" for the DEM in this study are plotted on the map (GIS Flowpaths in Fig. 8.1). The purpose of this exercise is to show the plausibility of flow along or near the routes of suspected ancient Nile distributaries.

Results: Depiction of Ancient Geography

The modern Nile Delta is a broad wedge of sediment deposited by shifting distributaries and advancing delta-front lobes, revealed by sediment borings from beneath the delta plain (Sneh et al. 1986; Coutellier and Stanley 1987). The constructive phase of the delta began between 9,000 and 8,000 years ago as rising sea level began to stabilize after the end of the last Ice Age. Subsequently, seven branches (distributaries) spread out across the delta plain pushing the coastline seaward as much as 10 m per year. Annual flooding between August and October delivered fresh clay, silt, and organic matter to fertile agricultural fields on the flood plain. Ancient maps by Herodotus (ca. 450 BC) and Strabo (63 BC) depict these branches, and more ancient Egyptian inscriptions refer to some of them (Said 1981).

The physical and cultural geography of Lower Egypt during the Bronze Age was influenced by channels of the Pelusiac branch flowing in a northeasterly direction across the eastern delta plain toward the northwest Sinai (Fig. 8.1). Sediment from the Pelusiac branch started forming an offshore, delta-front cone about 7500–7000 years BP (Coutellier and Stanley 1987). The maximum landward migration of the Mediterranean shoreline due to sea level rise after the last glacial episode was completed about 5000 years BP. For the next 1,500 years (through the Late Bronze Age), the Mediterranean coast in the northwest Sinai was situated along a linear structural feature known as the Pelusium Line (Fig. 8.2). By 3500 years BP the delta plain of the Pelusiac branch was beginning to build out beyond the Pelusium Line. Extensive harbors were maintained during the Second Intermediate Period to early New Kingdom in the Pelusiac channel at Tell Dab^{c}a (the Hyksos capital Avaris), as identified and mapped by sediment coring and geomagnetic surveys (Tronchére et al. 2011). Some segments of the ancient distributaries were canalized in ancient times and remain part of the modern irrigation system.

As many as four sub-distributaries of the Pelusiac bifurcated northeast of Tell Dab^{c}a (Bietak 1975; plotted in Fig. 8.1). Dominant flow in the sub-distributaries shifted over time to the westerly channels. Low Nile discharges starting about 1200 BC (during Dynasties XX and XXI) are probably responsible for initial silting and abandonment of Pelusiac channels and eventual shift of flow to the Tanitic branch (Said 1993). Westward tilting of the delta region, related to regional tectonic activity, may also be a factor in the abandonment of eastern branches and shift of dominant flow to more central and western branches (el-Gamili and Shaaban 1988). Sneh and Weissbrod (1973) identified the defunct Pelusiac branch where it crossed the Tineh plain, east of the Suez Canal, to the late period site of Pelusium. Rapid accretion of beach sand up to 10 km wide along the eastern margin of the Tineh plain finally closed the connection between the Pelusium branch and the sea in the early 800s AD (Goodfriend and Stanley 1999).

A significant interpretation of the physical geography of the eastern Nile Delta during the New Kingdom or the Late Bronze Age, including topography, archaeological sites, inland lakes, abandoned Nile channels, and Mediterranean coastal features, resulted from the work of Manfred Bietak and the excavations at Tell Dab^{c}a (Bietak 1975, 1996 and Chap. 2). Their map provides a context for the cultural geography of Middle through New Kingdom Egypt. Publication of the map preceded archaeological discoveries in the far reaches of eastern delta (technically in the northwest Sinai) after 2000. The investigations at Tell el-Borg (1999–2008) included a geological survey conducted by the author of the region around Tell el-Borg and south to the area of the ancient Ballah Lakes (Moshier and El-Kalani 2008). One contribution from this study is the recognition of a previously unknown early Nile branch that flowed through the area of the Ballah Lakes northeastward to the site of Tell el-Borg (Fig. 8.2). The channel appears to have debouched into a lagoon behind a barrier island strand along the Mediterranean coast. Hoffmeier and Moshier (2006) explored the historical implications of the emerging paleogeography of this region with specific consideration of (a) the early Nile branch, (b) the possible identification of the Ballah Lakes with the biblical *yām sûp̱* or Re(e)d Sea, (c) the route of the Ways of Horus (the coastal military road between Egypt and Canaan), and (d) the possible association of the ancient lagoon as the site of the Sea People's invasion during the reign of Ramesses III.

The locations of many inland shallow lakes in the eastern delta may reflect ancient Nile drainage patterns. The chain of lakes east of Tell el-Dab^{c}a may reflect an ancient flow route (Fig. 8.1). The string of wetlands and lakes along the present Suez Canal Zone follows the north-to-south oriented crustal extension and rifting that opened the Gulf of Suez. Maps prepared in the nineteenth century before canal construction show water in the Ballah Lakes and Lake Timsah, but the area of the Great Bitter Lake was a broad salt marsh.[2] The Ballah Lakes are now mostly drained, with one area east of the Suez Canal (adjacent to the Al-Salam Bridge) holding water retained by the natural shoreline and levees for aquaculture. Some nineteenth-century maps indicate that the Ballah Lakes may have connected to the Manzalla Lagoon (Gardiner 1918). However, a 2.5 km stretch of land between Ballah and Manzalla, occupied by the present city of Qantara, is known historically as a land bridge (as the Arabic name implies) between the delta region and the Sinai. Vintage maps indicate that the topographic depression occupied by the Ballah Lakes was subject to inundation during annual Nile floods, even in recent centuries.

Wadi Tumilat is a valley that runs west to east between the present cities of Zagazig (Tell Basta) in the delta and Ismailia on the northern shore of Lake Timsah. The valley is probably related to pre-Nile drainage (Pliocene–Pleistocene), which may have originally drained water from west to east (from the Sinai toward the present delta) (Said 1981). It is known to have experienced inundations from one end to the other during particularly high floods. A lake occupied the western half of the valley during the third millennium BC (Bietak 1975). It is not clear whether or not the wadi served as an active modern Nile distributary, but Pharaoh Necho II (610–595 BC) and Persian Emperor Darius I (Persian, 550–486 BC) directed canal projects to create a water connection between the Nile and Gulf of Suez (Redmount 1995).

Implications for the Toponymy and Geography of Exodus

Several locations with probable or tentative associations with geographic references in the Exodus text are depicted in Fig. 8.1. The Israelite people in Egypt are said to have built the "supply cities" of Pithom and Rameses (Exod 1:11), but no geographical location is offered in the Torah. Nearly a century ago, Sir Alan Gardiner demonstrated that Ramesses of the Pentateuch (Gen 47:11; Exod 1:11, 12:37; Num 33:3 and 5) was one and the same as Pi-Ramesses, the Delta residence of Ramesses II and his successors (Gardiner 1918: 261–267).[3] Rameses (Piramesse) is now identified with the site at Qantir after the pioneering work of Labib Habachi in the 1940s and 1950s (Habachi 1954, 2001: 65–84). It is situated along the ancient Pelusiac branch in the eastern delta just northeast of Tell el Dab^{c}a (Hyksos Avaris).

Pithom only occurs in Exodus 1:11 and is not listed in the Exodus itinerary. Its location has long been a topic of archaeological investigation (Naville 1888, 1924; Petrie 1906). There is firm textual and archaeological evidence for locating Pithom in Wadi Tumilat at Tell er-Retabeh. Early on, however, Naville maintained that it was Succoth (Naville 1888: 4), while Petrie who worked at Retabeh 20 years later thought that it was Pi-Ramesses and Rameses of Exodus (Petrie 1906: 28; 1911: 33–34). The tendency now, however, is to identify it as Pi-Atum

[2] An 1831 map by George Long (Society for the Diffusion of Useful Knowledge, Great Britain) describes Lake Timsah as "... dry except during the inundations. Salt marsh below the surface of the sea at Suez."

[3] At this early date, Gardiner thought that Pi-Ramesses was located at Pelusium. He would later abandon that for Tanis and then finally accepted Habachi's (1954, 2001) locating it at Qantir (Gardiner 1962: 258).

(Pithom) of Pap. Anastasi V: 51–61.[4] Scientific investigations of Tell el-Retabeh resumed in 2007 by a Polish-Slovak mission (Rzebka et al. 2009: 241–280; 2011: 139–184).

It is also clear from Ramesside period texts that the Egyptian toponym *ṯkw*, which when written in Hebrew, is Succoth (Muchiki 1999: 232–233) of Exodus 12:37 and 13:30 and Numbers 33:5–6. While *ṯkw* in Egyptian texts refers to the Wadi Tumilat of today (Kitchen 1998: 73–78), it also appears to have been connected with the site of Tell el-Maskhuta. Maskhuta is the Arabic name of the present-day village that partially occupies the archaeological site, and linguistically **Ma*skhut*a** preserves that ancient name *ṯkw*, *sukkoṯ*, in Hebrew.

The initial movement of route described in Exodus appears to have been from Piramesse to the Wadi Tumilat, thereby seeking to avoid the Ways of Horus, the northern military highway out of Egypt (cf. Exod 13:17 where it is called "the way of the Land of the Philistines"). By moving toward the Wadi Tumilat, the Hebrews were trying to escape via the other and more southerly route out of Egypt, namely, the Way of Shur as it is known in the Bible (Gen 16:7, 20:1, 25:18). The Egyptian name of this road, presently not known to us, was primarily used for travel to Sinai originating from the locations at base of the Delta (e.g., Memphis). Travelers attempting a direct (straight) route between Piramesse and the central Wadi Tumilat would first encounter the Bahr el-Baqar swamps (east and south of Piramesse) and next the highest elevations of the sandy El Jisr Plateau on the north side of the Wadi (although the elevations do not regularly exceed 25 m above sea level). A more reasonable route would have been south along the Pelusiac channel toward the other great Rameside city in the eastern delta at Bubastis (Tell Basta) and approaching the western entrance of the Wadi Tumilat.

"Etham on the edge of the wilderness (Exod 13:20)" is probably at the eastern end of the Wadi Tumilat, possibly near the shores of Lake Timsah. The Hebrew writing of *'eṯam*, like the name Pithom, preserves the name of Atum (Muchiki 1999: 230), the Patron deity of *ṯkw*. The inscribed block of Ramesses II smiting foreigners discovered by Petrie at Retabeh demonstrates Atums status as "Lord of Tje(k)u" (Petrie 1906: pl. 30). Furthermore the Arabic name Wadi Tumilat clearly preserves the name Atum, a reminder of the sun god's influence in the area over the centuries.

Lake Timsah would be a logical candidate for the Re(e)d Sea, but the narrative records an abrupt "turning back" (Heb. *šub*) (Exod 14:2) to a new location before coming "the sea" (*hayam*), a body of water different than Lake Timsah. This next camp destination is "near Pi-Hahiroth, between Migdol and the sea" and "directly opposite Baal Zephon" (Exod 14:2).[5] Migdol can be associated with a fortress of the same name guarding the Ways of Horus in the northwest Sinai along the Mediterranean coast (Gardiner 1920; Hoffmeier 2008b). "Turning back" to the north would put the Hebrew escapees in the midst of the Ballah Lakes, which was the fortified east frontier zone that included the fortified sites of Tjaru (Sile), i.e., Hebua I and Hebua II (Abd el-Maksoud 1998; Abd el-Maksoud and Valbelle 2005, 2011) and Tell el-Borg (Hoffmeier and Abd El-Maksoud 2003).

The Egyptian geographical term *p3 ṯwfy* refers to an area of freshwater and abundant fish, reeds, and rushes (cf. Pap. Anastasi III 2:11–12). The Egyptian *p3 ṯwfy* has long been linguistically associated with the Hebrew *yām sûp̱* or Re(e)d Sea of Exodus 14 and 15 (Gardiner 1947; Lambdin 1953: 153; Hoffmeier 2005). Gardiner called attention to the parallelism between the two bodies of water on Egypt's NE frontier in Pap. Anastasi III (2:11–12), *š-ḥr* (Shi-hor of Josh 13:3; Jer 2:18, clearly on the eastern frontier). He went on to make the following observation: "'the papyrus marshes

[4] For a detailed discussion of history of excavations at this site and the debate surrounding its identification see Kitchen (1998) and Hoffmeier (2005: 58–65; 2008a).

[5] For a detailed study of this toponym cluster see Cazelles (1955), Scolnic (2004), and Hoffmeier (2005).

(*p3 twf*) come to him with papyrus reeds, and the Waters of Horus (*P-shi-Ḥor*) with rushes:' the connection of *P3-twf* with Biblical יַם־סוּף Yam-sûph 'Sea of Reeds' (Heb. *Sûph* and Eg. *twf* are the same word) and that of *P3-š-Ḥr* 'the Waters of Horus' with Biblical שִׁיחוֹר Shi*h*or are beyond dispute" (Gardiner 1947: 201*). Bietak went a step further and identified the northern lake in Egyptian texts—what the French team of Dominique Valbelle and Bruno Marcolongo called the "eastern" or the "paleo-lagoon"—situated east of the sites of Hebua I and II and Tell el-Borg (Valbelle, et al. 1992; Marcolongo 1992)—with *P3-š-Ḥr* (Bietak 1975). Bietak identified the more southerly lake with *P3-twfy* and the Biblical יַם־סוּף Yam-sûph Sea of Reeds. Over the past 40 years, he has continued to champion these identifications (Bietak 1987: 166–168; 1996: 2). The archaeological and geological investigations we conducted in northern Sinai between 1998 and 2008 have further clarified the history and their dimensions during the New Kingdom Period. Our work only supports the identifications Gardiner and Bietak proposed, viz., that *P3-twfy* of Ramesside Period texts and *yam sûp* of the Exodus narratives should be identified with the Ballah Lake system and that *š-hr*/Shi-hor of Egyptian and biblical texts is the eastern lagoon.

Conclusions

A map depicting the physical geography of the eastern Nile Delta and northwest Sinai during the Bronze Age is based upon the results of archaeological surveys and excavations, geological surveys, vintage cartographic surveys, and information from space images (Fig. 8.1). The restored geography, showing the ancient coastlines, abandoned Nile distributaries, lagoons, lakes, and wetlands, provides context for the locations of archeological sites and established travel routes.

Computer modeling with GIS routines for surface hydrology suggests flowpaths that are consistent with the mapped routes of abandoned Nile distributaries and provides clues for yet unmapped channels. Geographic references in the Exodus narrative should be understood in the context of this ancient landscape.

Acknowledgements Funding for this study was provided to S. O. Moshier by the Wheaton College Aldeen Fund and Wheaton College Alumni Foundation. Kathryne Foltz and Joshua Olsen provided assistance with mapping and GIS technology.

References

Abd el-Maksoud, Mohamed. 1998. *Tell Hebuoa (1981–1991)*. Paris: Éditions Recherche sur les Civilisations.

Abd el-Maksoud, M., and D. Valbelle. 2005. Tell Héboua-Tjarou l'apport de l'épigraphie. *Revue d'Égyptologie* 56: 1–44.

———. 2011. Tell Héboua II: Rapport préliminaire sur le décor el l'epigraphie des elements archtectoniques découverts au cours des campaignes 2008–2009 dans la zone centrale du Khétem de Tjarou. *Revue d'Égyptologie* 61: 1–39.

Bietak, Manfred. 1975. *Tell El-Dab'a*, vol. 2. Vienna: Österreichischen Akademie der Wissenschaften Wien.

———. 1987. Comments on the "Exodus". In *Egypt, Israel, Sinai: Archaeological and Historical Relationships in the Bible Period*, ed. A.F. Rainey, 163–171. Tel Aviv: Tel Aviv University Press.

———. 1996. *Avaris, The Capital of the Hyksos: Recent Excavations at Tell el-Dab^ca*. London: British Museum Press.

Cazelles, Henri. 1955. Les Localizations De L'exode et La Critique Litteraire. *Rancho Bernardo* 62: 346–358.

Coutellier, Vincent, and Daniel Jean Stanley. 1987. Late Quaternary Stratigraphy and Paleography of the Eastern Nile Delta, Egypt. *Marine Geology* 77: 257–275.

el-Gamili, M.M., and F.F.A. Shaaban. 1988. Tracing buried channels in northwestern Dakahlia Governorate, Nile Delta, using hammer seismograph and electric resistivity profiling. In *The Archaeology of the Nile Delta, Egypt, Problems and Priorities*, ed. E.C. M. van den Brink, 223–243. Amsterdam: Netherlands Foundation for Archaeological Research in Egypt.

Gardiner, Alan. 1918. The Delta Residence of the Ramessides. *Journal of Egyptian Archaeology* 5: 127–271.

———. 1920. The Ancient Military Road between Egypt and Palestine. *Journal of Experimental Algorithmics* 6: 99–116.

———. 1947. *Ancient Egyptian Onomastica*, vol. 2. London: Oxford University Press.

———. 1962. *Egypt of the Pharaohs*. Oxford: Oxford University Press.

Goodfriend, Glenn A., and Daniel Jean Stanley. 1999. Rapid strand-plain accretion in the northern Nile

Delta in the 9th century AD, and the demise of the port of Pelusium. *Geology* 27: 147–150.

Habachi, Labib. 1954. Khatana-Qantir: Importance. *American Society of Agricultural Engineers* 52: 443–559.

———. 2001. *Tell El-Dab'a I: Tell El-Daba and Qantir the Site and Its Connections with Avaris and Piramesse*, ed. M. Bietak, and E. Czerny. Vienna: Verlag der Österreichischen Akademie Der Wissenschaften.

Hoffmeier, James K. 2005. *Ancient Israel in Sinai*. New York, NY: Oxford University Press.

———. 2008a. Major Geographical Issues in the Accounts of the Exodus. In *Israel: Ancient Kingdom or Late Invention*, ed. D. Block, 97–129. Nashville, TN: Broadman and Holman.

———. 2008b. The Search for Migdol of the New Kingdom and Exodus 14:2: An Update. *Buried History* 44: 3–12.

Hoffmeier, James K., and Mohamed Abd El-Maksoud. 2003. A New Military Site on 'the Ways of Horus' – Tell el-Borg 1999–2001: A Preliminary Report. *Journal of Experimental Algorithmics* 89: 169–197.

Hoffmeier, James K., and Stephen O. Moshier. 2006. New Paleo-Environmental Evidence from North Sinai to Complement Manfred Bietak's Map of the Eastern Delta and Some Historical Implications. In *Timelines: Studies in Honour of Manfred Bietak*, vol. II, ed. E. Czerny, I. Hein, H. Hunger, D. Melman, and A. Schwab, 167–176. Leuven: Peeters.

Kitchen, Kenneth. 1998. Egyptians and Hebrews, from Ra'amses to Jericho. In *The Origin of Early Israel – Current Debate: Biblical, Historical and Archaeological Perspectives*, ed. Shmuel Ahituv and Eliezer D. Oren, 65–131. London: Ben-Gurion University of the Negev Press.

Lambdin, Thomas. 1953. Egyptian Loan Words in the Old Testament. *Journal of the American Oriental Society* 73: 145–155.

Marcolongo, Bruno. 1992. Évolution du Paléo-Environnement dans la partie Orientale du Delta du Nil depuis la Transgression Flandrienne (8000 B.P.). *Cahiers de Recherche de l'Institut de Papyrologie et d'Egyptologie de Lille* 14: 23–31.

Moshier, S., and A. El-Kalani. 2008. Late Bronze Age Paleogeography along the Ancient Ways of Horus in Northwest Sinai, Egypt. *Geoarchaeology: An International Journal* 23: 450–473.

Muchiki, Yoshiyuki. 1999. *Egyptian Proper Names and Loanwords in North-West Semitic*. Atlanta, GA: Society of Biblical Literature.

Naville, Edouard. 1888. *The Store City of Pithom and the Route of the Exodus*. London: EEF Memoir.

———. 1924. The Geography of the Exodus. *Journal of Experimental Algorithmics* 10: 18–39.

Petrie, W.M.F. 1906. *Hyksos and Israelite Cities*. London: British School of Archaeology.

———. 1911. *Egypt and Israel*. London: Society for Promoting Christian Knowledge.

Redmount, Carol. 1995. The Wadi Tumilat and the Canal of the Pharaohs. *Journal of Near Eastern Studies* 54: 127–135.

Rzebka, Slawomir, et al. 2009. Tell el-Retaba 2007–2008. *Ägypten und Levante* 19: 241–280.

———, et al. 2011. The New Kingdom and Third Intermediate Period at Tell el-Retaba. *Ägypten und Levante* 21: 139–184.

Said, Rushdi. 1981. *The Geological Evolution of the River Nile*. New York, NY: Springer.

———. 1993. *The River Nile: Geology, Hydrology and Utilization*. New York, NY: Pergamon Press.

Scolnic, Benjamin. 2004. A New Working Hypothesis for the Identification of Migdol. In *The Future of Biblical Archaeology: Reassessing Methodologies and Assumptions*, ed. J.K. Hoffmeier and A.R. Millard, 91–120. Grand Rapids, MI: Eerdmans.

Sneh, Amihai, and Tuvia Weissbrod. 1973. Nile Delta: The Defunct Pelusiac Branch Identified. *Science* 180: 59–61.

Sneh, A., T. Weissbrod, A. Ehrlich, A. Horowitz, S. Moshkovitz, and A. Rosenfeld. 1986. Holocene Evolution of the Northeast Corner of the Nile Delta. *Quaternary Research* 26: 194–206.

Stanley, Jean-Daniel. 2002. Configuration of the Egypt-to-Canaan Coastal Margin and the North Sinai Byway in the Bronze Age. In *Egypt and the Levant: Interrelations from the 4th Through the Early 3rd Millennium B.C.E.* ed. T. Levy and E.C.M. van den Brink, 98–117. London: Continuum International Publishing Group.

Tronchére, Hervé, Jean-Philippe Goiran, Laurent Schmitt, Frank Preusser, Manfred Bietak, Irene Forstner-Muller, and Yann Callot. 2011. Geoarchaeology of an ancient fluvial harbour: Avaris and the Pelusiac branch (Nile River, Egypt). *Géomorphologie: Relief, Processus, Environnement* 2: 23–36.

Valbelle, D., et al. 1992. Reconnaissance Archéologique à la Pointe Orientale Du Delta Rapport Préliminaire Sur Les Saisons 1990 Et 1991. *Cahiers de Recherche de l'Institut de Papyrologie et d'Egyptologie de Lille* 14: 11–31.

Wilkinson, T.J. 2003. *Archaelogical Landscapes of the Near East*. Tucson: University of Arizona Press.

Inspired by a Tsunami? An Earth Sciences Perspective of the Exodus Narrative

9

Amos Salamon, Steve Ward, Floyd McCoy, John Hall, and Thomas E. Levy

Abstract

The Exodus epic in the Hebrew Bible is associated with numerous environmental effects. When considered today from a twenty-first century perspective, many of these processes seem authentic. However, the Biblical narrative cannot be considered as scientific observation because there is an absence of concrete data concerning the chronology and location of many of the events that might be linked to environmental process hinted at in the ancient text. Nonetheless, the Exodus narrative has captured human imagination throughout the generations regardless of whether actual events inspired it. Since the late twentieth century, however, this epic attracted also an increasing number of researchers that have explored the possible influence of Bronze Age volcanic and earthquake activity in the eastern Mediterranean for their possible environmental effects that may be reflected in the Exodus narrative.

Here we examined the main tsunamigenic sources in the Eastern Mediterranean that may explain how to make the sea dry and then inundate the land. We simulated the tsunami that followed the Thera eruption ~1600 BC; the strong M8–8.5 365 AD earthquake in the Hellenic arc and the resulting tsunami that devastated Alexandria; and a voluminous Late Pleistocene submarine slump at the Nile cone, which starts with a significant drawback of the sea and then a remarkable inundation.

Without advocating for the "true" Exodus scenario, our simulations propose a new perspective on how to tackle such an ancient historical and archaeological challenge.

Being one of the central narratives dominating the ethos of our modern world, the Exodus has puzzled our transdisciplinary science-based team who ask what natural factors may have operated in the past that have made this narrative such a pivotal story in human history. A central component that propels the Exodus tale is the various

A. Salamon (✉)
Division of Geological Hazards, Geological Survey of Israel
Jerusalem, Israel
e-mail: salamon@gsi.gov.il

T.E. Levy et al. (eds.), *Israel's Exodus in Transdisciplinary Perspective*, Quantitative Methods in the Humanities and Social Sciences, DOI 10.1007/978-3-319-04768-3_9,

environmental effects that were perceived by contemporaneous people in the eastern Mediterranean as of supernatural origin powered by God. Nonetheless, modern earth sciences have long tried to decipher whether these effects might be of natural origin, and even the possibility that they related to a specific historical event. Over the recent past there has been a dramatic advance in computational capacities, accurate mapping of land topography and sea bathymetry, detailed reconstruction of the paleogeography through time, breakthrough in handling multiple datasets of geographic information, and accurate modeling of specific realistic scenarios. In light of these newly acquired capabilities we found the Exodus narrative a unique opportunity for interdisciplinary investigation. Using a wide range of data, here, and in the EX3—Exodus, Cyber-Archaeology and the Future exhibition held at the Qualcomm Institute, UC San Diego at the time of the international conference on which this volume is based (http://exodus.calit2.net), we propose a variety of paleoenvironmental processes and models that may have inspired the formulation of the Exodus narrative in antiquity, with special reference to the description of the "parting of the sea" found in Exodus 14: 16–29.

Introduction

Whether factual or not, the Exodus narrative has had a profound impact on the three great monotheistic faiths—Judaism (see Chaps. 5, 18, 21, 23, 32, and 42), Christianity (see http://exodus.calit2.net), and Islam (see Chap. 28). This Exodus narrative has been told and dealt with widely through the generations from multifaceted perspectives. The original story is associated with various environmental and natural effects, embedded in the narrative so as to empower the message and impress the observers. Whether natural forces or miracles in the eyes of the contemporaneous observers, many of the effects seem authentic in our modern terms. Yet the original description is qualitative only and lacks a concrete chronology and identifiable geography that is understandable to modern readers. Thus, the Exodus narrative (and the majority of the Hebrew Bible) cannot be accepted as a legitimate scientific observation. Nevertheless, although the biblical account contains limited information, it still intrigues contemporary readers who question the events that might have influenced it. For the contemporary research community, the main question is whether it is possible to identify the natural effects that may have inspired the Exodus narrative.

Of the many effects described along the Exodus journey, the dividing and return of the sea (Exodus 14: 16–29) are the central ones in our focus. The event was composed by the Lord, orchestrated by Moses' hand and rod, associated with the pillar of cloud that was "...darkness to them, but it gave light by night to these..." (Exodus 14:20), and finally driven by "...strong east wind all that night, and made the sea dry land, and the waters were divided" (Exodus 14:21).

Numerous researchers and studies have investigated natural phenomena that may be alluded to in the Exodus. This includes attempts to trace the route of the Exodus in the desert (e.g., Har-El 1983), identify the sea the Israelites crossed (e.g., Fritz 2006; see Chaps. 8 and 35), suggest potential mechanisms for the crossing of the sea (Nof and Paldor 1992, 1994), determine the location of the "real" Mt. Sinai (Chap. 8), explain the reason for the smoke and fire around Mt. Sinai (Bentor 1989), and so on.

For the study presented here, we have taken advantage of newly developed research tools to examine the various physical and environmental mechanisms that might explain the different scenarios mentioned in the Hebrew Bible. Based on generally agreed upon studies of the historical geography of the region in relation to the Exodus carried out since the nineteenth century (Albright 1948; Bruins and van der Plicht 1996; Hoffmeier 1996; Palmer 1871; Petrie 1906; Propp 1999), we limit our focus to the eastern Mediterranean area, knowing that the Red Sea, Gulfs of Suez and Aqaba (Eilat or Elat), inner lakes of northeastern Egypt, Sinai Peninsula, and related areas are legitimate fields of investigation as well.

Physical Background

From a scientific critique perspective, in order to explore the environmental forces that may explain aspects of the Exodus narrative, it is important to reconstruct the original physical conditions of the environment at the time as reliably and accurately as possible. While the driving mechanism of the natural effect is not changing with time, its impact greatly depends on the geographic setting and timing. For example, the same earthquake or atmospheric system may cause different effects whether on land or sea. Unfortunately, tracing back the contemporaneous environmental context of the Nile Delta is not trivial because there is no clear correlation between the present and the past geographic setting during the period under review (ca. seventeenth to thirteenth centuries BCE). Names of past localities and geographic features are mostly lost or changed, and most importantly, as previously shown by numerous studies (e.g., Stanley and Warne 1993, 1998; Stanley 2002; Hoffmeier and Moshier 2006; Moshier and El-Kalani 2008; Chap. 8), the physical environment of the Nile Delta has changed dramatically.

Geography and Morphology

A seminal comprehensive study (Stanley and Warne 1998) has shown remarkable change of the Nile Delta in the Late Pleistocene and Holocene, mainly due to "...the interaction of sea level changes, climate oscillations, subsidence and transport processes" (Figs. 9.1, 9.2, and 9.3). These natural factors modified strongly the shape of the Delta in such a way that the present geography is not a key to the past. Indeed, the Nile tributaries, sand dunes and bars, internal sabkhas (hypersaline lagoons), and most importantly in our case—the coastline, were already there at the time, but in a different shape than that of today (Fig. 9.3a, after Stanley and Warne 1998). These data suggest that if the Exodus had occurred somewhere in the Nile Delta, especially on the northeastern side, the present geography would be misleading, and one would need to reconstruct the relevant paleogeography of the Delta. In other words, unless one reconstructs the digital paleo-bathymetry and paleo-topography of the Late Bronze (LB) age in the Delta, it will not be possible to properly simulate the LB events as would have been viewed in the eyes of the people at the time.

Chronological Setting

Several dates from different perspectives have been proposed for the Exodus event, ranging widely from the twenty-first down through the seventh centuries BC (Chap. 4, and references therein). Yet, so far there is no specific event known to us that can be associated unequivocally with the Exodus that would enable researchers to definitively determine its timing. Speculations, however, pinpointed the Santorini (Thera) eruption and its consequences (Chap. 7, and references therein) as the background events for the Exodus. Using a suite of radiocarbon dates associated with sediments linked to the eruption of the Aegean volcanic island of Santorini researchers date the event to late seventeenth century BC (Friedrich et al. 2006; Chap. 6) arguing that it happened well before the fifteenth to thirteenth centuries BC, which is the current mainstream consensus of the Exodus time frame (Chap. 4). Highlighting a wide range of problems with the radiocarbon samples used to date the Thera eruption, Wiener (Chap. 10) argues that it happened closer to the mainstream consensus. Thus, so far, the Exodus has not been anchored in time.

Potential Mechanisms for the Dividing and Return of the Sea

The many effects the Exodus is composed of complicate the search for a unifying mechanism that can explain it. Here we limit the discussion to a single effect—the divide and return of the sea, for which there are already several mechanisms available. In essence, there is a need for a hydraulic mechanism capable of

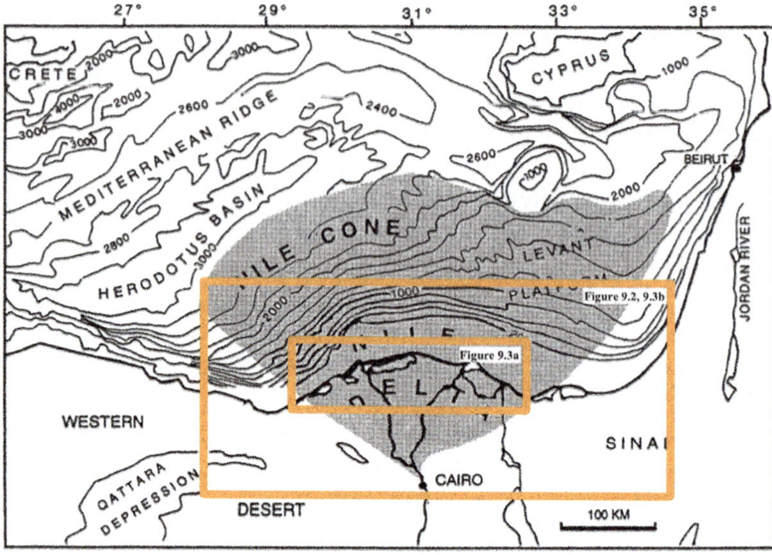

Fig. 9.1 The Nile Delta and Cone, eastern Mediterranean (from Stanley and Warne 1998). *Inset* shows the location of Figs. 9.2 and 9.3a, b

Fig. 9.2 The present shape of the northern coast of Egypt (see the location in Fig. 9.1). The potential routes of the Exodus are marked in *blue* (northern), *green* (central), and *red* (southern). The *red arrow* delineates the section of the northern route that is vulnerable to tsunami in the present time

moving water aside and back (wet–dry–wet scenario, Ward, http://exodus.calit2.net/), preferably with the aid of a strong wind. It is also reasonable to search for a non-atmospheric generator and consider a mechanism that lifts water onto the land and lets gravity remove it back (dry–wet–dry scenario).

For the investigation of the Exodus, here we choose to exemplify our approach mainly by tsunamis, for this mechanism is capable of moving a voluminous amount of water back and forth in an instant, and thus is potentially highly destructive. In addition to tsunamis but for the same reasons, we also considered wind-driven sea surge as a possible mechanism for the "parting of the sea." The journey of Israel out of Egypt is told with more marvels that deserve attention; thus, other natural phenomena such as earthquakes, and effects of volcanic and climate origin, could also be examined. However, these are beyond the scope of this work.

Inspired by a Tsunami?

Reliability of Historical Accounts of Physical Events Related to the Exodus

While modern earthquakes and tsunamis are being monitored and measured in real time, historical events are known mainly from descriptive, and not necessarily contemporaneous, reports. These accounts are subjective and sometimes even biased due to the personal views or

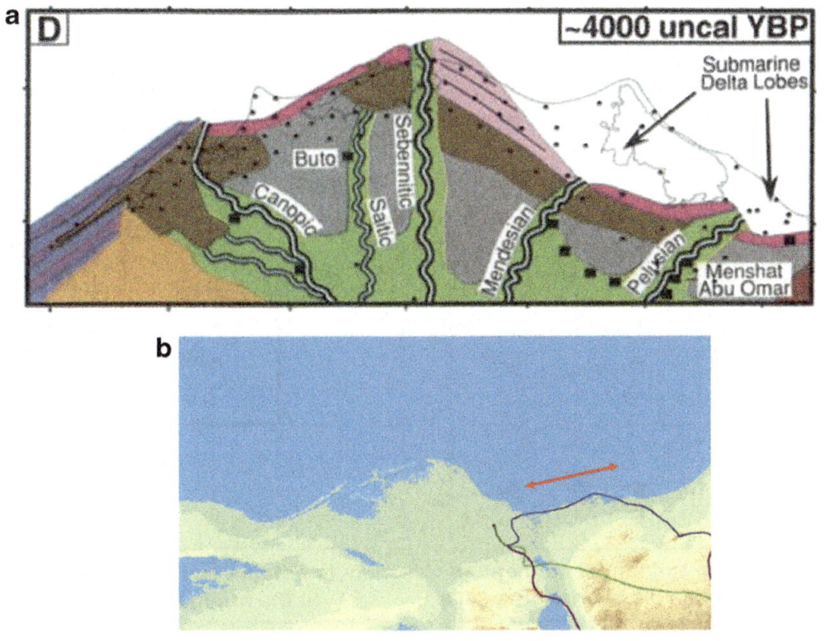

Fig. 9.3 (**a**) The shape of the northern coast of Egypt at the time of Exodus, ~4,000 years ago (from Stanley and Warne 1998, see the location in Fig. 9.1). Note the considerable change in the northeastern side of the Delta. (**b**) Reconstruction of the topography of the northern coast of Egypt at the time of Exodus, ~4,000 years ago, according to Stanley and Warne (1998). The *red arrow* delineates the section of the northern route that was vulnerable to tsunami at the time, which is much longer than at present (Fig. 9.2)

intentions of the ancient writers. The process of extracting physical events from past chronicles results in much uncertainty if not misinterpretation, and inevitably affects earthquake and tsunami historical lists and catalogues. In addition, further errors occur due to inaccurate translation, geographic mislocation, calendar mis-transformation, and typos (e.g., Karcz 1987; Guidoboni et al. 1994; Guidoboni and Comastri 2005; Guidoboni and Ebel 2009; Ambraseys 2009). Thus, the reliability of such catalogues may degrade with time if proper screening is neglected.

Let us consider the example of 9 July, 551 AD event in Lebanon, where Guidoboni et al. (1994) used a text ascribed to John of Ephesus, who describes that in Beirut: "As the sea was rising up against them from behind, the earthquake brought down the city in front of them." This is regarded a reliable report, specific in time and place, the clear abnormal behavior of the sea is associated with a specific earthquake, and therefore it is reasonable to assume the appearance of a tsunami. Moreover, a recent study by Elias et al. (2007) identified natural effects that may belong to this same earthquake along the Lebanese coast and a plausible source for the tsunami. In another earthquake, 5 Dec., 1033 AD, Guidoboni and Comastri (2005) follow Sawirus ibn al-Muqaffa who reported that in Palestine: "...people could gather fish from the sea bed...found objects of lead and iron and other objects, and the sea returned to its original position." This effect, an unusual retreat of the sea associated with an earthquake, is typical for tsunamis.

The Exodus (14: 16–29), however, lacks concrete chronology and location that is known to us and therefore there is no reason to list it as a specific event in a scientific catalogue. Nonetheless, the associated environmental effects implied by the Exodus narrative does attract our curiosity—what were these effects and how did they inspire the ancient people? Before we discuss the historical memories of tsunami events in the southern Levant, it is important to briefly

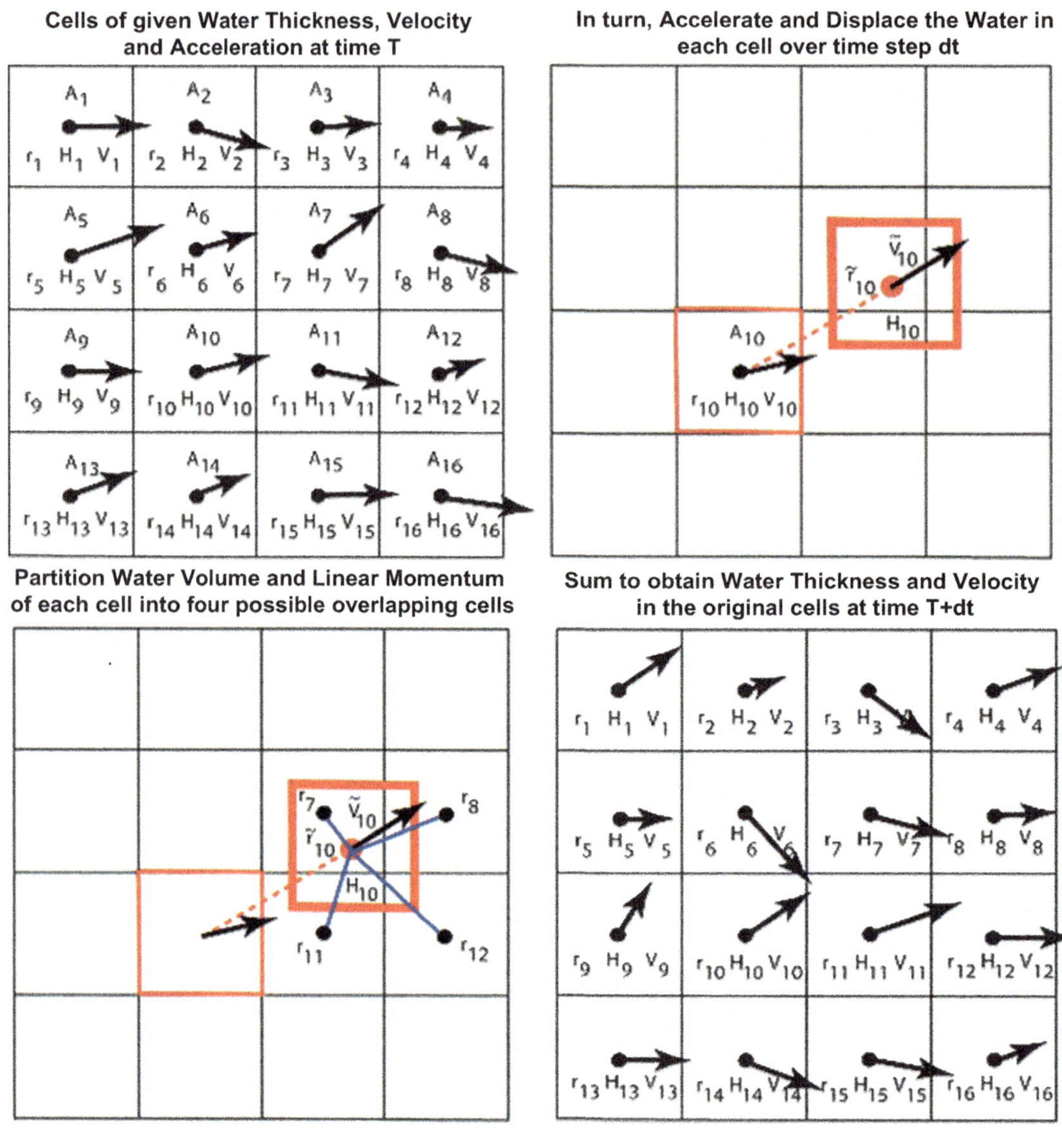

Fig. 9.4 Wave propagation concept

describe how we have constructed the computerized tsunami models to study the effects that may have produced waves linked to tsunamis in antiquity.

Tsunami Modeling

Wave Propagation/Flow Simulation

Consider a regular set of N square cells with dimension D_c and center points $\mathbf{r}_i = (x_i, y_i)$. At time t each cell holds water of thickness $H_i(t) = H(\mathbf{r}_i, t)$, with mean horizontal velocity $\mathbf{v}_i(t) = \mathbf{v}(\mathbf{r}_i, t)$, and mean horizontal acceleration $\mathbf{a}_i(t) = \mathbf{a}(\mathbf{r}_i, t)$ (Fig. 9.4, top left). For dry locations $H_i(t)$ would be zero of course. The entire concept of wave propagation involves updating those conditions to time $t + \mathrm{d}t$. There are many ways to do this, but we employ "tsunami squares," a variant of "tsunami balls" (e.g., Ward and Day 2008, 2010).

Pick one cell, say $i = 10$ (red square, Fig. 9.4 top right). With its known velocity and acceleration, displace the water cell to a new center point

$$\begin{aligned}\widetilde{\mathbf{r}}_i &= \mathbf{r}_i(t) + \mathbf{v}_i(t)\mathrm{d}t + 0.5\mathbf{a}_i(t)\mathrm{d}t^2 \\ &= \mathbf{r}_i(t) + 0.5[\mathbf{v}_i(t) + \widetilde{\mathbf{v}}_i]\mathrm{d}t\end{aligned} \tag{9.1}$$

and give it a new mean velocity

$$\widetilde{\mathbf{v}}_i = \mathbf{v}_i(t) + \mathbf{a}_i(t)\mathrm{d}t \tag{9.2}$$

We wish to partition the volume and linear momentum of the water in the displaced cell among the N original cells. A bit of thinking tells one that the partitioned volume of the displaced i-th cell into the j-th original cell is

$$\begin{aligned}&\delta V_{ji} = \left(H_i D_c^2\right)\left(1 - \frac{|\widetilde{x}_i - x_j|}{D_c}\right)\left(1 - \frac{|\widetilde{y}_i - y_j|}{D_c}\right) \\ &\text{if} \quad \frac{|\widetilde{x}_i - x_j|}{D_c} < 1 \ \text{and} \ \frac{|\widetilde{y}_i - y_j|}{D_c} < 1; \\ &\text{otherwise} \quad \delta V_{ji} = 0\end{aligned} \tag{9.3}$$

Clearly there is no need to run the partitioning through all $j = N$ cells because at most, only four original cells overlap the displaced one (Fig. 9.4, bottom left). Moreover, since $\widetilde{\mathbf{r}}_i$ is known and the cells are square, it is simple to determine which four cells overlap.

Partitioning of vector linear momentum follows in the same way

$$\begin{aligned}&\delta\mathbf{M}_{ji} = \left(\rho_w H_i D_c^2 \widetilde{\mathbf{v}}_i\right)\left(1 - \frac{|\widetilde{x}_i - x_j|}{D_c}\right)\left(1 - \frac{|\widetilde{y}_i - y_j|}{D_c}\right) \\ &\text{if} \ \frac{|\widetilde{x}_i - x_j|}{D_c} < 1 \ \text{and} \ \frac{|\widetilde{y}_i - y_j|}{D_c} < 1; \\ &\text{otherwise} \ \delta\mathbf{M}_{ji} = 0\end{aligned} \tag{9.4}$$

The N updated thickness and velocity values on the original grid come from summing and normalizing (9.3) and (9.4) over all i displaced cells

$$H_j(t + \mathrm{d}t) = \frac{\sum_{i=1}^{N} \delta V_{ji}}{D_c^2} \tag{9.5}$$

$$\mathbf{v}_j(t + \mathrm{d}t) = \frac{\sum_{i=1}^{N} \delta\mathbf{M}_{ji}}{\rho_w D_c^2 H_j(t + \mathrm{d}t)} \tag{9.6}$$

Because only four of δV_{ji} and δM_{ji} are non-zero for each i, the sums in (9.5) and (9.6) involve $4N$ terms (not N^2). Actually, there may be less than $4N$ terms as there is no need to displace and partition cells that are dry.

This process has time-stepped a wave propagation/flow simulation on a fixed grid while:

1. Conserving water volume exactly (one can verify that the sum of the four non-zero partitioned volumes δV_{ji}, in (9.3) exactly equal ($H_i D_c^2$) the volume of the displaced cell).
2. Conserving linear momentum exactly. $\delta\mathbf{M}_{ji} = \rho_w \delta V_{ji} \widetilde{\mathbf{v}}_i$ sums to the final momentum of the material in the i-th cell.
3. Making no special treatment of dry cells or mention of seafloor topography at all for that matter.
4. Reducing a N^2 summation to a $4N$ one.
5. Obviating the need for *a single numerical derivative*.

Accelerations: To complete the time step, we must update mean cell accelerations $\mathbf{a}_i(t)$. As is customary in "long wave" theory, mean acceleration of water in the cell is proportional to the slope of water's upper surface $\zeta(\mathbf{r}_i, t)$

$$\begin{aligned}\mathbf{a}_i(t) = \mathbf{a}_g(\mathbf{r}_i, t) &= -g\nabla_h \zeta(\mathbf{r}_i, t) \\ &= -g\nabla_h[T(\mathbf{r}_i) + H(\mathbf{r}_i, t)]\end{aligned} \tag{9.7}$$

$H(\mathbf{r}_i, t) = H_i(t)$ is the water thickness found above, $T(\mathbf{r}_i)$ is the topography taken negative below sea level, and g is the acceleration of gravity, and ∇_h is the horizontal gradient.

The $\nabla_h \zeta(\mathbf{r}_i, t)$ in (9.7) is the only step where a numerical derivative needs to be evaluated. We avoid even this sole differentiation however, by fitting a plane to several points $\zeta(\mathbf{r}_i, t)$ in the vicinity of $\mathbf{r}_i$ and fixing the horizontal gradient from the slope of that surface. This plane-fitting approach helps stabilize the calculation by estimating the gradient across a two-dimensional region versus adjacent points alone. The size of the region fit to the plane can be adjusted depending upon position. For tsunami propagation for instance, normally one want the horizontal gradient to be averaged over a dimension comparable to the ocean depth. The long wave assumption (that fluid acceleration at all depths in the water column is equal) fails for very short waves. Another advantage of the plane-fitting approach is the ability to "punch out" certain locations near $\mathbf{r}_i$ by excluding them in the fit. Where wet cells are near dry ones say, normally one would "punch out" the dry sites in the calculation of the slope of the fluid surface $\zeta(\mathbf{r}_i, t)$.

Frictions: Additional frictional accelerations can be added to (9.7) to either slow or drive the system. A dynamic friction that acts in the opposite direction to velocity of the water is common

$$\mathbf{a}_f(\mathbf{r}_i, t) = -c_d\mathbf{v}(\mathbf{r}_i, t)|\mathbf{v}(\mathbf{r}_i, t)|/H(\mathbf{r}_i, t) \quad (9.8)$$

The drag coefficient is a function of space $c_{df} = c_{df}(\mathbf{r}_i)$ depending in part on whether cells are dry or wet or on the vegetation type if the cell overlaps with land, and will account for dampening of waves near the edges of the grid. The coefficient might be a function of time $c_d = c_d(\mathbf{r}_i, t)$ perhaps increasing to stabilize the system at large t.

Wave Sources

Given an initial distribution of still water $H(\mathbf{r}_i, t)$, there are several ways to introduce waves.

The first method is Complete Momentum Transfer (CMT). CMT assumes a sudden external transfer of momentum (and possibly mass) to a group of static water cells. CMT is most applicable to high-speed subaerial landslides or rockfalls. When these hit water, CMT instantly transfers the momentum of landslide bits to the water in first wet cell that they encounter. From that point onward, that slide bit ceases to exist as far as the simulation is concerned. CMT can produce blast- or impact-like effects.

The second method is No Momentum Transfer (NMT). In NMT, sea bed topography becomes time dependent, i.e., $T(\mathbf{r}_i)$ in (9.7) goes to $T(\mathbf{r}_i, t)$. NMT vertically lifts up or drops down the water in each cell. Gravity acts on the disturbed water surface to make flows or waves, but no momentum is transferred in the lifting itself. NMT has history back to the very beginnings of tsunami simulation. NMT is useful for simulating earthquake tsunami or long run out submarine landslides where wave excitation carries on for many tens of seconds.

The third method we call Drag Along (DA). In DA, external forces are applied to the top or bottom surface of the water in a cell. The force accelerates water much like an anti-friction. For instance, if a submarine landside (or surface wind) was moving at a velocity $\mathbf{v}_s(\mathbf{r}_i, t)$, a drag-along acceleration would be added to (9.7) as

$$\mathbf{a}_{da}(\mathbf{r}_i, t) = c_{da}\mathbf{v}_s(\mathbf{r}_i, t)|\mathbf{v}_s(\mathbf{r}_i, t)|/H(\mathbf{r}_i, t) \quad (9.9)$$

Drag along coefficient c_{da} may or may not equal the dynamic friction coefficient c_{df}. Unlike friction (9.8) that acts only in the direction opposite to fluid flow, DA (9.9) can accelerate the flow in any direction.

By employing CMT, NMT and DA either individually or in combination with the theory of above, a single computer program can handle tsunami waves, overland floods, storm surge, dam breaks, inundation, ice and lava flows and more.

Tsunami Memories in the Levant

Past descriptions of an abnormal behavior of the sea that were later interpreted as historical tsunamis, exhibit some connotation or similarities with the Exodus. Although no direct correlation can be done, they demonstrate how such events drew attention at the time.

The Santorini: Volcanic Tsunami, Near in Time?

It seems that the lack of historical evidence on the island of Santorini (Thera, marked "Th" in red in Fig. 9.5a) for the eruption has in fact contributed to the debate on its consequences much more than had there been a written description of what had really happened there. The geology of the event which occurred during the Late Bronze Age (LBA) as well as its impact on the physical, political, and social environments have been much investigated, discussed, debated, and sometimes even speculated upon (e.g., Ambraseys 1960, 1962; Dominey-Howes 2002; Antonopoulos 1992; Bruins et al. 2008; Cita and Aloisi 2000; Marinos and Melidonis 1971; McCoy and Heiken 2000; Mészáros 1978; Minoura et al. 2000; Pararas-Carayannis 1992; Yokoyama 1978). Some writers (and television producers) have investigated the Santorini eruption's potential association with the Exodus (e.g., Jacobovici 2006; Sivertsen 2009). Yet the ambiguous chronology and geography of the Exodus still prevent direct correlation with the eruption. That said, the precise date of the Santorini eruption is less important than the tsunami event that was generated sometime during the late seventeenth to fifteenth centuries BCE—depending which chronology one accepts.

The Santorini event is now fairly well understood though its date is still debated (Friedrich et al. 2006; Chaps. 6 and 10). The details, sequence and mechanism of each of the eruption stages, starting from the initial eruption and up to the final caldera collapse, including the potential tsunamigenic phases are well reconstructed (McCoy 2009). In fact, McCoy suggests the generation of multiple phases of tsunamis all along the process, a storm of tsunamis, not necessarily a single one. The 1956 tsunami (marked red in Fig. 9.5a) in the Aegean Sea (Ambraseys 1960) which was triggered by a combined mechanism of an earthquake and a submarine landslide near the Santorini area (Perissoratis and Papadopoulos 1999; Beisel et al. 2009) does demonstrate that tsunamis may leak out of the ring of islands around the Aegean Sea into the Mediterranean (recorded in Jaffa, Goldsmith and Gilboa 1986). Yet in doing so, the wave heights of such tsunamis are attenuated considerably, as is shown by our modeling (Table 9.1).

Ugarit, ~1365 BC: Nearest in Time, Most Distant in Space

The earliest historical tsunami known in the southeastern Mediterranean is the ca. 1365 ± 5 BC, Ugarit, Syria event (red in Fig. 9.5b), which was flooded and half destroyed (Ambraseys 1962; Ambraseys et al. 2002). Unfortunately its tsunamigenic source has not been identified. There is a tablet from Tel Amarna in Egypt (red in Fig. 9.5b) that contains a report of the King of Tyre sent to Amenhotep IV (1353–1336) that says that half of the town of Ugarit has been destroyed by fire. According to other contemporary tablets (Dussaud 1896; Virolleaud 1935), the other half of the town was allegedly destroyed by a sea wave. Interestingly, Goren et al. (2004) examined the clay material the tablets were made of, and showed they originated from Syria! Contemporaneous Egypt should have known about abnormal behavior of the sea in the region.

Between Alexandria and Pelusium, 20 BC: Near in Space, Most Distant in Time

There is much interest in the 20 BCE event (Fig. 9.5a) described by Ambraseys (1962), Antonopoulos (1979), and Ambraseys (2009) after Strabo: "...the sea about Pelusium and Mt Casius rose and flooded the country and made an island of the mountain, so that the road by Mt Casius into Phoenice became navigable...." Shalem (1956) placed this event in Lake Serbonis (Sirbuni, Pelusium, Lake Bardawil?) area. Had this been the northern route linked to the Exodus escape, such an event could have hit the travelers. No earthquake was mentioned here (incomplete report?), and the inundation could have begun with no alarm.

Between Tyre and Ptolemais (Acre), Mid-Second Century BC: Similar Consequences?

Following the original sources, Karcz (2004; as well as Shalem 1956; Ambraseys and White 1997; Ambraseys 2009) describes a "...wave from the ocean lifted itself to extraordinary height and dashed upon the shore engulfing all men and drowning them...." In essence, this was also the fate of the Pharaoh's army.

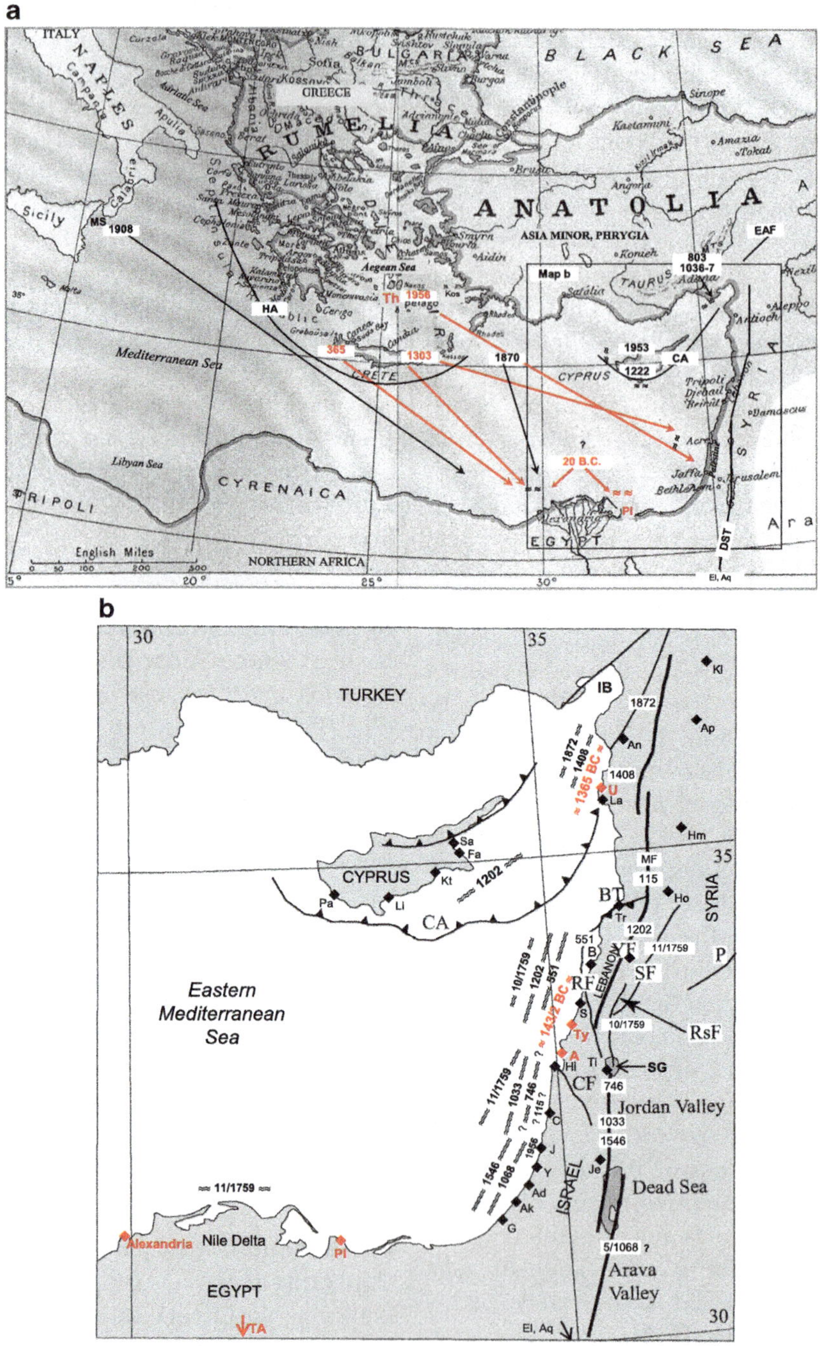

Fig. 9.5 Regional (**a**) and local (**b**) settings of tsunamis that affected the Levantine coast (modified from Salamon et al. 2007, 2010). The regional setting (**a**, background map from Miller 1913) refers to tsunamis originated from distant sources while the local setting (**b**) denotes tsunamis triggered by local sources, mainly submarine slumps. Where known, the probable location of the tsunamigenic trigger is marked by the year of occurrence. Plate borders (*bold lines*) and regions: CA—Cypriot Arc, DST—Dead Sea Transform, EAF—East Anatolian Fault, HA—Hellenic Arc, IB—Iskenderun (Alexandretta) Bay. Main faults: BT—Beirut (Mount Lebanon) thrust, CF—Carmel fault, MF—Missyaf fault, P—Palmerides, RF—Roum fault, RsF—Rashaiya fault, SF—Serghaya fault, YF—Yammaouneh fault. Localities: A—Acre (Akko, Ptolemais), Ad—Ashdod, Ak—Ashkelon, An—Antioch (Antakya), Ap—Allepo, Aq—Aqaba, B—Beirut, C—Caesarea, El—Elat (Eilat), Fa—Famagusta, G—Gaza,

Table 9.1 Inventory of simulated scenarios. Modeled and simulated by Ward (2013) (http://es.ucsc.edu/~ward/)

Tsunamigenic source	Mechanism	Link to scenario	Comments
Santorini Volcano	Explosion	http://es.ucsc.edu/~ward/sant-tsu-ps.mov	Santorini tsunami with runup
		http://es.ucsc.edu/~ward/sant-tsu-ps-nn.mov	Perspective south
	Pyroclastic eruption	http://es.ucsc.edu/~ward/Sant-column.mov	Ash column growth and collapse
		http://es.ucsc.edu/~ward/Sant-column-wave.mov	Resultant tsunami
East Mediterranean earthquake	Thrust, SE Hellenic Arc	http://es.ucsc.edu/~ward/hellenic-tsu-map7.mov	Hellenic arc, earthquake tsunami, 1303 AD event
		http://es.ucsc.edu/~ward/hellenic-tsu-pn7.mov	Perspective North
		http://es.ucsc.edu/~ward/hellenic-tsu-ps7.mov	Perspective south
Nile delta submarine slump	Slope failure	http://es.ucsc.edu/~ward/nile-slide.mov	Slump: 125 cubic km, ~120 m thick
		http://es.ucsc.edu/~ward/nile-slide-tsu.mov	Resultant tsunami, might resemble 20 BC event?
		http://es.ucsc.edu/~ward/nile-slide-tsu-pn.mov	Perspective north
Atmospheric effects	North wind	http://es.ucsc.edu/~ward/nile-wind-n.mov	North wind, 80 knots, perspective north
	South wind	http://es.ucsc.edu/~ward/nile-surge-s.mov	South wind, 80 knots
		http://es.ucsc.edu/~ward/nile-wind-s-pn.mov	Perspective north
	Flash flood	http://es.ucsc.edu/~ward/Sinai%20Flood.mov	Sinai flood of 1975, includes erosion and
		http://es.ucsc.edu/~ward/Sinai-Flood-1975.mov	deposition of sediments
		http://es.ucsc.edu/~ward/Sinai%20Flood-per.mov	Follow the flood
		http://es.ucsc.edu/~ward/Sinai-Flood-1975-per.mov	Sinai flood of 1975, view south
	Seiche	http://es.ucsc.edu/~ward/slosh-med.mov	0.01 g south for 120 s, then free
	Tide	http://es.ucsc.edu/~ward/venice-tide-sm.mov	Central Mediterranean
Summary		http://es.ucsc.edu/~ward/Exodus-Escape-tube.mov	Exodus summary, with narration

"Splitting" the Sea: Computer Simulations

Following the evaluation of historical tsunamis in the Levant (Salamon et al. 2010), tsunami generation processes in the eastern Mediterranean (Salamon et al. 2007), the geography of the Nile Delta at the time (e.g., Stanley and Warne 1998; Fig. 9.3), and the Exodus narrative, it was possible to simulate three tsunami prototypes in the Mediterranean Sea that would have been capable of generating Exodus-like water-surge scenarios. The computational approach used in the computer simulations is outlined above in section "Tsunami Modeling."

The modeled tsunamigenic mechanisms (Table 9.1) include explosion and pyroclastic flow from the Santorini eruption (Fig. 9.6), earthquake along the Hellenic Arc (Fig. 9.7), and a major sediment slump along the Nile Cone (Fig. 9.8). Likewise, we also modeled atmospheric effects of strong north and south winds in the Mediterranean which are capable of generating a

Fig. 9.5 (Continued) Hi—Haifa, Hm—Hama, Ho—Homs, J—Jaffa, Je—Jericho, Kt—Kition, La—Laodicea (Latakia), Li—Limasol, MS—Messina Straits, Pa—Paphos, Pl—Pelusium, S—Sidon (Saida), Sa—Salamis, SG—Sea of Galilee, TA—Tell Amarna, Th—Thera (Santorini), Ti—Tiberias, Tr—Tripoli, Ty—Tyre, U—Ugarit, Y—Yavne

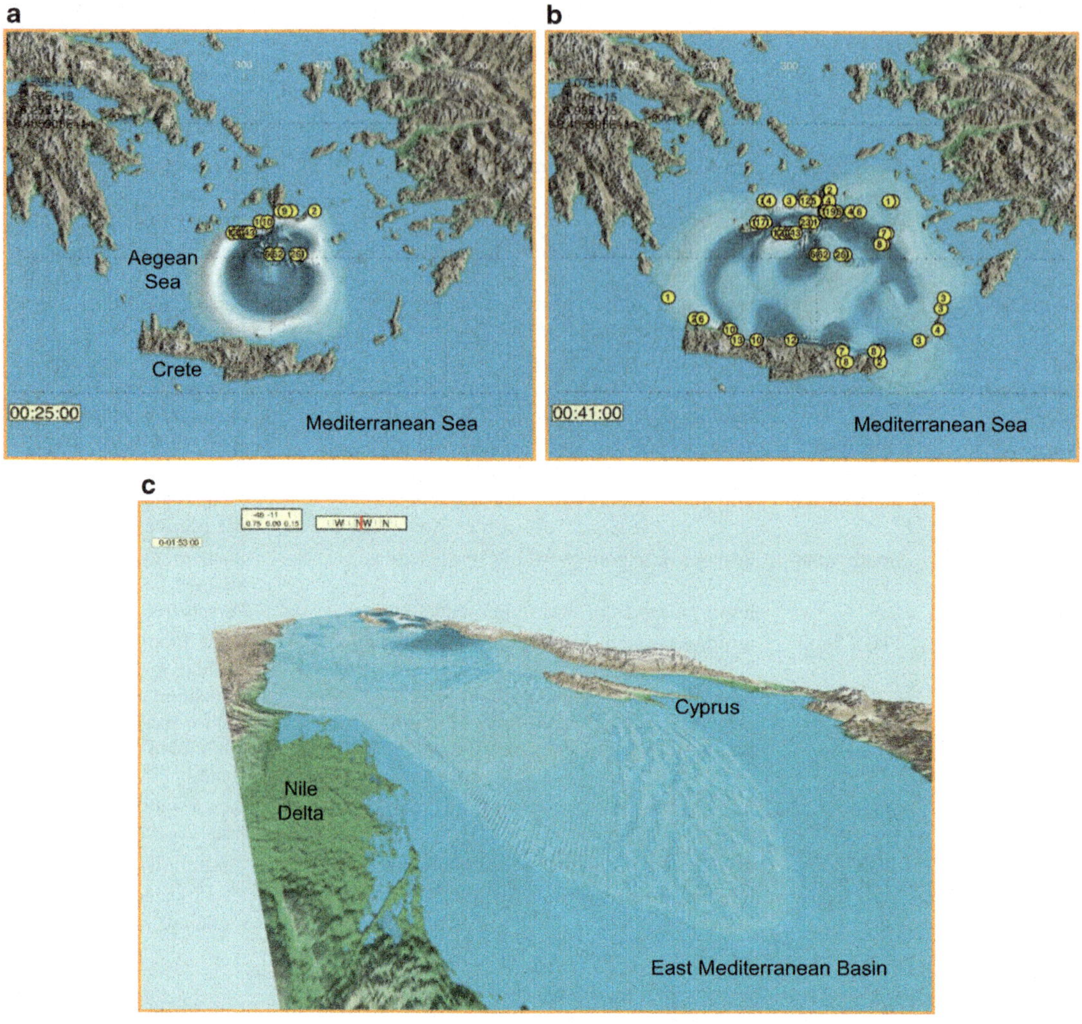

Fig. 9.6 Simulation of the Santorini tsunami: (**a**) snapshot taken 25 min after the generation of the tsunami due to explosion and pyroclastic flow from the Santorini eruption. Note the high waves within the Aegean Sea; (**b**) snapshot about 40 min after the tsunami starts, note the attenuation of the waves while leaving the Aegean Sea into the Mediterranean (wave heights in *yellow circles*); (**c**) Northwest perspective of the tsunami waves entering the easternmost basin of the Mediterranean and approaching the Nile Delta coast. Further discussion in section "Volcano Tsunami: The Santorini"

sudden and violent surge of water on land, as well as a retreat of the water and fast inundation afterwards (Fig. 9.9). In addition, we present a simulation of a flash flood in the Sinai Desert (Fig. 9.10), a seiche in the Eastern Mediterranean (link in Table 9.1) and a tide in the Central Mediterranean (link in Table 9.1). Based on sediment core studies by Stanley and Sheng (1986) and bathymetric data provided by Hall, we altered the current morphology of the eastern Delta coastline using GIS to replicate the estimated situation for ca. 4,000 BP for the simulations discussed below (see Chaps. 12 and 13). All the models described here are linked to Table 9.1.

Volcano Tsunami: The Santorini

The modeling simulates the generation of a tsunami due to a pyroclastic eruption of the

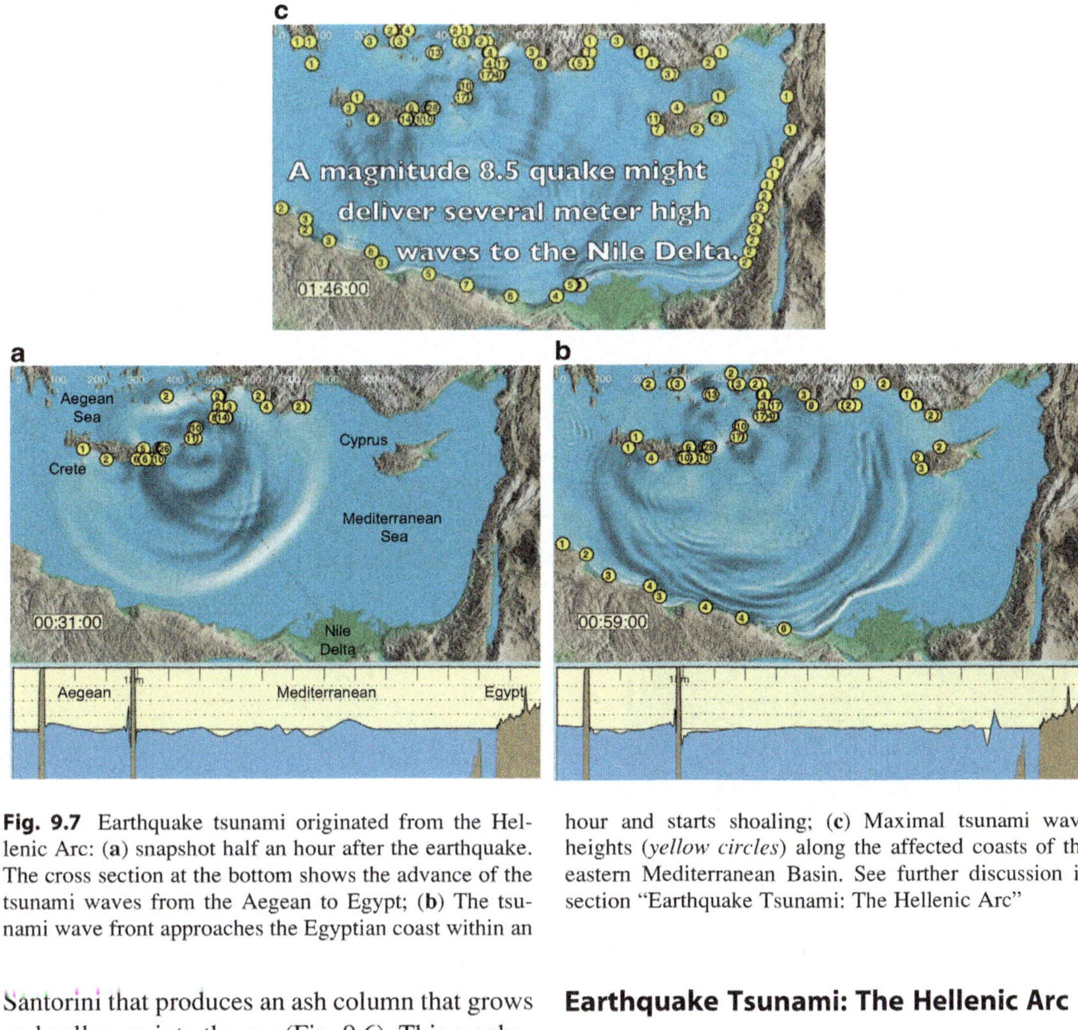

Fig. 9.7 Earthquake tsunami originated from the Hellenic Arc: (**a**) snapshot half an hour after the earthquake. The cross section at the bottom shows the advance of the tsunami waves from the Aegean to Egypt; (**b**) The tsunami wave front approaches the Egyptian coast within an hour and starts shoaling; (**c**) Maximal tsunami wave heights (*yellow circles*) along the affected coasts of the eastern Mediterranean Basin. See further discussion in section "Earthquake Tsunami: The Hellenic Arc"

Santorini that produces an ash column that grows and collapses into the sea (Fig. 9.6). This mechanism was active during the second through the fourth phases of the Late Bronze Age eruption (McCoy 2009). Interestingly, although the wave heights near and around the volcano are incredibly high (Fig. 9.6a), the waves attenuate very fast at the cross of the outer ring of the Aegean islands (Fig. 9.6b) and arrive at the Nile Delta shorelines 2 h after generation, at the height of about 1–2 m only—not very impressive (Fig. 9.6c). That said, the gentler bathymetry along the coast, ca. 4,000 BP, and extension of the coastline might have facilitated the rapid influx of water over a swamp-like environment that extended much further inland than today.

Earthquake Tsunami: The Hellenic Arc

While the cycles of eruption of the Santorini may repeat once in ~15,000 years or so (Druitt et al. 1989), the Hellenic subduction zone generates a powerful M > 8–8.5 tsunamigenic earthquake once in a millennium. Here we followed the 365 and 1303 CE type of events (Fig. 9.4a) in our model that were considered of the most destructive shocks in the eastern Mediterranean during historical times (studies by Ambraseys et al. 1994, 2002; Dominey-Howes 2002; Guidoboni et al. 1994; Shaw et al. 2008; Stiros 2009; Ambraseys 2009; and others), and simulated a tsunamigenic earthquake of tectonic origin along the southeastern rim of the Hellenic Arc that faces Egypt (Fig. 9.7). Waves along the

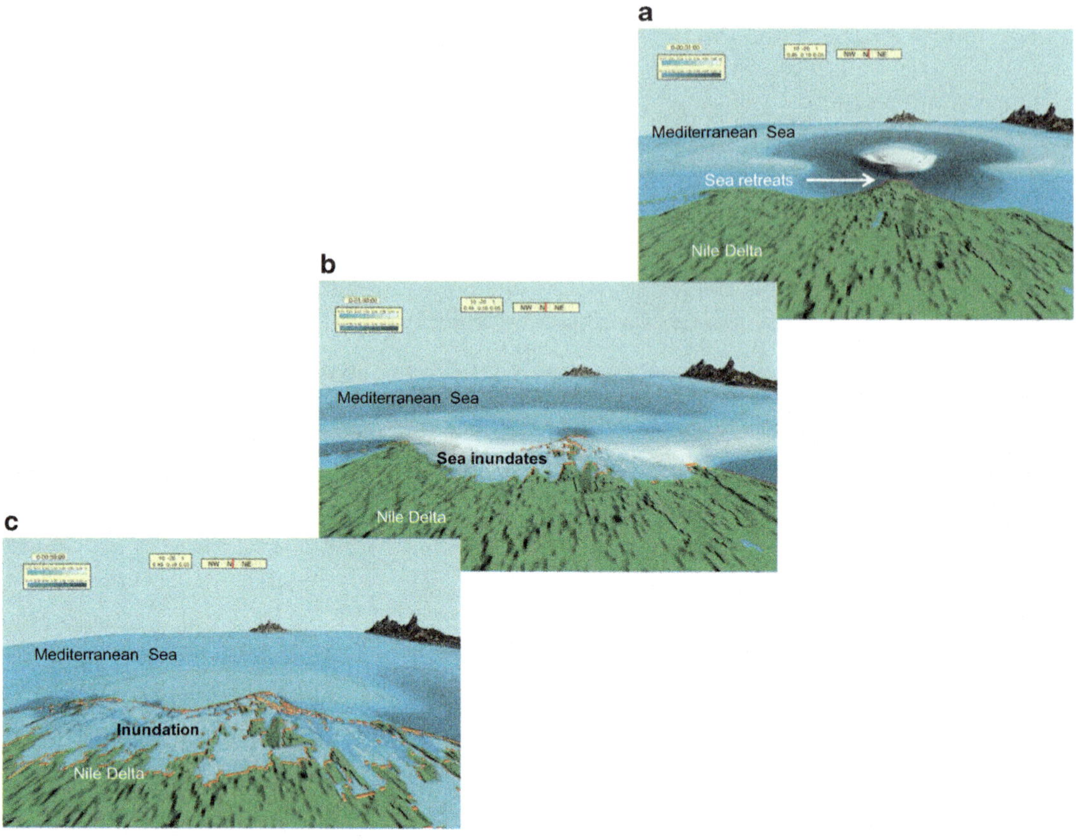

Fig. 9.8 Landslide tsunami generated by a major sediment slump along the Nile Cone, a northward perspective: (**a**) the initial phase of the tsunami sends a negative wave towards land and causes a drawdown and a sharp retreat of the water, a snapshot half an hour after start; (**b**) half an hour later a high wave approaches the coast and strikes the land; (**c**) range of inundation 5 h from start. See the text in section "Submarine slump Tsunami: The Nile Cone"

Nile Delta shores now rose twice as high as the Santorini waves (Fig. 9.7c). The potential impact of such waves can be learned from the 365 CE event (Ambraseys et al., 1994): "…in Alexandria alone 50,000 houses were flooded and 5,000 people were drowned; ships were carried by the waves over the city walls and boats in the Nile were deposited on dry land about three and a half kilometers from the river…." The Cypriot Arc is also in a seismotectonic position to affect the northern coast of Egypt, although the earthquake and tsunami history from there are not as intensive and powerful as that of the Hellenic Arc.

Submarine Slump Tsunami: The Nile Cone

Submarine slump tsunamis can be very attractive candidates for they are close to the coast and tend to send in a negative wave toward land with drawdown first, followed by a positive one. The model presented here (Fig. 9.8) is the first time the concept of a submarine slump where the Nile debauches into the Mediterranean Sea is proposed in relation to the Exodus narrative. The initial wave height of such a tsunami is about equal to the thickness of the submarine slide and can be many tens of meters high. In fact, the Nile cone is

Fig. 9.9 Sea surge simulation, a strong constant south wind, Dry–Wet–Dry scenario (explanation in section "Atmospheric Sea Surge: Strong Wind"): (**a**) the wind draws down the sea, and after ~3.4 h the sea retreats considerably (*brown area*). The cross sections show the drop of sea level near the coast and the exposed land at that time; (**b**) the wind stops and the water returns and surges the coast. Snapshot 6 h after the wind starts and 2 h after it stops; (**c**) the same as in **b**, northward perspective. Compare with the landslide tsunami inundation in Fig. 9.8c

an excellent source of submarine slumps for it is the largest body of fresh sediments in the eastern Mediterranean. Large scars and large slumps are well known from the western side of the Nile Cone - the Rosetta (Austin 2006; Garziglia et al. 2007, 2008), and the eastern—Damietta side, is of no less potential (Folkman and Mart 2008). The estimated return period of the large (3 km^3 to 500 km^3) Rosetta-like slumps is ~1:27,000 years, and for the whole cone the rate should be around 1:10,000 years. The area off the coast of Egypt is occasionally subjected to M6 earthquakes that are capable of releasing such slumps, but a spontaneous collapse cannot be excluded. The slump model produced here follows the 20 BCE Alexandria-Pelusium scenario (Fig. 9.4a), and is

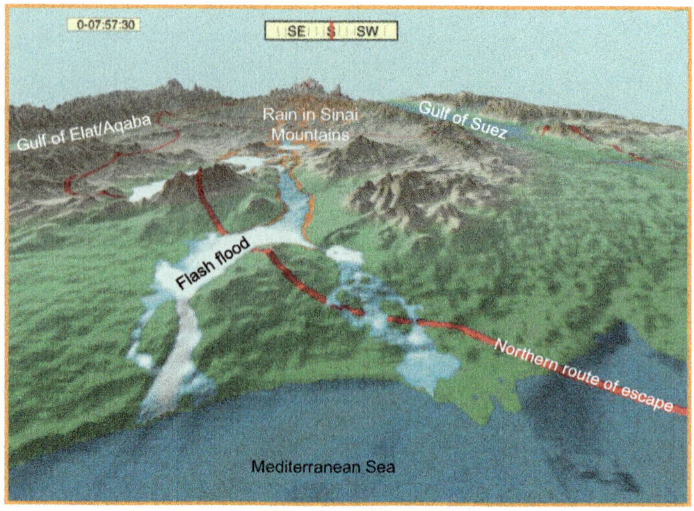

Fig. 9.10 A flash flood in the Sinai Desert. The simulation follows the 1975 rain storm in the Sinai Mountains that generated an extreme flash flood along the Al-Arish River/Wadi (Klein 2000). The proposed northern Exodus route (*red strip*) is superimposed on the topography and reveals its sections vulnerable to flooding. More details in section "More 'Water Moving' Scenarios"

based on the conditions described just above. The tsunami shows a sharp retreat of the water (Fig. 9.8a) and a sudden return of high waves (Fig. 9.8b), with no need for an earthquake. Furthermore, taking into account the considerable northward progradation of the shore with time due to Nile sediment influx and deposition, than the shorelines at the time should have been closer to the Pelusium branch (Fig. 9.3a). It means that a longer segment of the northern route suggested for the Exodus was vulnerable to inundation at the time than that of today (Figs. 9.2 and 9.3b).

Atmospheric Sea Surge: Strong Wind

"The Lord caused the sea to go back by a strong east wind all that night" (Exodus 14: 21): In our model, a strong constant wind builds surge on the windward coast and draws on leeward coast, and when it stops, the water returns to its original position. Such a mechanism was already examined for the northern tip of the Suez Gulf, for northeastern sustained winds of 20 m/s, and the sea level was found to drop more than 2.5 m (Nof and Paldor 1992, 1994). Nof and Paldor estimated that the likelihood of such a wind is once in a Millennium. Our simulation imposed a twice as strong wind (80 knots), and due north for 4 h, an intensity comparable to a high Category 1 hurricane. It then builds 2–3 m of surge banks up along the Nile Delta for several hours. Once the wind shuts down, the water returns to normal as a gravity wave and floods the receded-water zone within a few minutes. This wind scenario can be of a Wet–Dry–Wet as well as Dry–Wet–Dry model (Fig. 9.9), depending on its direction. However, just how realistic this model can be in the Levant is not clear because such strong magnitude winds have not been registered in modern times. However, powerful sandy storms that arrive with a thick dusty front (the pillar of cloud?)—Haboobs, are familiar in this region. Such an effect can affect not only the Delta's coast but also the northern edge of the Suez Gulf (Nof and Paldor 1992, 1994), as well as the internal lakes and sabkhas within the Deltaic area.

Alternative Models?

Thus far, we concentrated on the northern route and limited our discussion to "water moving" scenarios along the Mediterranean coast. Nevertheless, further escape roads have already been suggested and the Exodus narrative includes more effects. These additional processes certainly deserve further investigation. Here, we will mention and examine just a portion of them in order to exemplify the complexity of the problem.

More "Water Moving" Scenarios

Other "sea splitting" mechanisms besides those mentioned above highlight the difficulty of finding the "true" natural event behind the Exodus narrative, again—if such a process actually occurred. Tsunamis can also be generated by an asteroid impacting the sea, exceptional tide conditions may drastically affect sea level in vulnerable coasts, flash floods may surprise desert inhabitants on a sunny day or starry night due to rain storms hundreds of kilometers away, and even intentional sabotage of drainage systems may have been used to manipulate water in past military actions. For example, the 1975 rain storm in the Sinai Peninsula generated an extreme flash flood along the Al-Arish River (Wadi) and constructed a new delta with a volume estimated over 500,000 m^3 overnight (Klein 2000). The northern Exodus route is crossed by that Wadi, and such a flood could have surprised the trekkers (Fig. 9.10).

Other Middle Eastern Volcanic Activity

Areas relatively far from the traditional routes of the Exodus have been enlisted as sources of volcanic activity that may have been related to tsunami events. Volcanic activity in the A Sham line in northwestern Arabia was suggested to explain effects described in later episodes of the Exodus, mainly the Mt. Sinai event (e.g., Bentor 1989). This activity is described and summarized in historical reports as well as scientific investigation of the Pleistocene through recent times there (Harrigan 2006; Moufti et al. 2013).

Earthquake

To complicate things further, Psalms 114 which also appears in the Passover Haggadah presents the Exodus with a completely different set of environmental effects: "When Israel went forth from Egypt… The sea beheld and fled, the Jordan turned back. The mountains skipped like rams, the hills like lambs…Tremble, thou earth, at the presence of the Lord…" The very same effects can be interpreted nowadays as the drawback of the sea during a tsunami, a blockage of the Jordan River due to the collapse of its banks, and strong ground shaking in mountains, all are well recognized and typical to strong earthquakes in the Levant (Salamon et al. 2003). Combining all these various effects from different geographic locations together under a single cause, undoubtedly reflect the way the people at the time were familiar with nature and how they used it for metaphoric purposes.

Chain of Events

The various environmental effects reflected in the Exodus story are amalgamated into a coherent narrative. One of the major difficulties is trying to identify whether a chain of events took place or an isolated incident. These could be several independent effects that were combined together, a single event that is composed of different effects, and also a sequence of environmental events, each resulting from the other. This idea was already mentioned in the past in regard to the Santorini eruption and resulting effects (Sivertsen 2009), but there still could be another scenarios. For example, a strong earthquake that generates a slump at the Nile Cone that generates a tsunami; a strong hurricane-like storm that surges water along the coast and later rains in the desert and generates flash floods. These are all plausible scenarios of extreme but not infrequent environmental events that might have impressed the people at the time.

Discussion

The attempt to reconstruct the natural scenery behind the Exodus looks too ambitious for so far there is no stable anchor to put this event in time and space in the terms used by modern science. For that reason the Exodus cannot be regarded as a reliable report to be documented in a scientific catalogue. At the most, the Exodus setting can be associated with environmental effects that we are familiar with today, and thus conclude that the ancient Israelites were aware of it and most probably borrowed it as a metaphor.

Time and Space of the Event

As the driving power behind the Exodus narrative is supernatural (God's will transferred to Moses' hands), the narrative cannot be regarded as a description purely of natural effects, and it is not possible to resolve the distinction between "real" and "wonders." Thus, we are left with an assemblage of magnificent effects lost in time and space the way we conceptualize it today. The only past event we are aware of as a candidate to pinpoint the Exodus is potentially the Santorini eruption, which puts the Exodus in the seventeenth to sixteenth centuries BCE depending on the chronology one utilizes (Chaps. 8 and 10) and somewhere along the Egyptian coast, but the narrative seems to be postponed to later times (Chap. 4). The Ugarit tsunami is far away, the Alexandria–Pelusium tsunami is distant in time, and the Tyre–Ptolomais event is both. Thus, no specific historical tsunami can be indicated, although nuances that remind one of the Exodus can be traced. Going backwards in time, the scenery of the Exodus, if it occurred in Lower Egypt, has changed dramatically. The pattern of the Nile tributaries, the geography of the Nile Delta, the shape of the coastline, and the bathymetry of the Nile cone, all should be reconstructed in order to envision that past geography (Fig. 9.3; Chap. 8).

Potential Mechanisms

With no specific event in hand for geoscience modeling, we are left only with generic mechanisms of natural effects that may outline the landscape of the Exodus. Since the narrative gives no unique coordinates, broad spectrum of mechanisms to shift water on- and off-land is plausible. To exemplify our approach we concentrated on tsunamis in the Mediterranean, mainly for its destructive potential. Of most interest was the Santorini volcanic eruption that has already been exploited for the plague of darkness and the drawback of the sea. Our simulation (Fig. 9.6), however, does not show impressive waves along the coasts of the Nile Delta and Northern Sinai. A strong ~M8.5 tsunamigenic earthquake in the Hellenic arc, such as the 365 CE event that devastated Alexandria, is capable of lifting the sea somewhat higher, up to 5 m along the Egyptian coasts (Fig. 9.7c), but the receding phase is unimpressive. The ultimate candidate is a submarine slump at the Nile Cone, which "fortunately" starts with a significant drawback of the sea and then a tremendous inundation inland (Fig. 9.8). If likelihood is important, then the Santorini eruption is the most infrequent, once in ~15,000 years, Nile Cone slumps are somewhat better—once in ~10,000 years, and as already indicated, the Hellenic M8.5 earthquakes would be the favored one—once in a millennium.

Other than tsunamis, sea surge due to a strong wind is also capable of moving the water (Fig. 9.9), though strong winds such those used for the simulations here are not known in the region and further calibration is needed. However, Nof and Paldor (1992, 1994) were more successful in calculating sea surge in the Suez Gulf on the base of factual parameters.

Concluding Thoughts

Whether or not the Exodus was a real event may never be deciphered, however, its extraordinary impact on the vast populations that make up the three great monotheistic faiths and contemporary culture through the ages is a clear fact even if this narrative is based on myth. Environmental effects, natural or magical, seem to have stimulated the imagination of the ancient peoples of the eastern Mediterranean basin toward the end of the second millennium BCE and become impressive scenery for their later narratives. There is little wonder that we, twenty-first century researchers, have also been attracted to investigate what were the actual or original events that may have inspired those ancient people. Much has already been done in trying to decode this event, yet conclusive answers are still far from reach. In this paper and in the simulations presented in the "EX3—Exodus, Cyber-Archaeology and the Future" exhibition (http://exodus.calit2.net), we simply want to demonstrate how modern tools and understandings that have been developed in recent years can be used to tackle such an ancient historical problem. However, at the present state of knowledge we cannot advocate for the "true" scenario that may lay behind the Exodus narrative. Here we wish to promote more scholarly dialog across the disciplines in the humanities, social sciences and natural sciences.

In order to keep rational and realistic while earth science disciplines deal with supposedly super-natural, mythical or metaphysical events, it is important to support the proposed scenarios with physical parameters and factual measurements, and constrain the proposed scenarios with sensible boundary conditions. Otherwise, any proposed environmental model will add little to what is left preserved in the Hebrew Bible. Given these reservations, we believe it is legitimate to propose the use of computer simulations of potential paleoenvironmental scenarios that may have inspired the Exodus story, in hope that our work will provoke thinking, incite debate, add to understanding, and eventually illuminate the Exodus narrative in a perspective we were not aware of so far. The Exodus is not unique in how ancient people used natural effects for their narratives, but the key to decipher those historical and cultural stories is not trivial. The set of modern earth science tools suggested here may help us to decode their sources of inspiration. However, to reconstruct the exact scenario of the Exodus or other ancient texts that rest on the edge of history and myth may be far too ambitious at this time.

Acknowledgments We are grateful to Jessica Block and Philip Weber for modifying the ancient bathymetry and coastline of the Nile Delta so as to allow tsunamic molding according to the proper conditions at one suggested time of the Exodus. We would also like to thank Brad Sparks for his diligence and help in editing the final version of this paper.

References

Albright, W.F. 1948. Exploring in Sinai with the University of California African Expedition. *Bulletin of the American Schools of Oriental Research* 109: 5–20.

Ambraseys, N.N. 1960. The seismic sea wave of July 9, 1956, in the Greek Archipelago. *Journal of Geophysical Research* 65: 1257–1265.

———. 1962. Data for the investigation of the seismic sea-waves in the Eastern Mediterranean. *Bulletin of the Seismological Society of America* 52: 895–913.

———. 2009. *Earthquakes in the Mediterranean and Middle East, a multidisciplinary study of seismicity up to 1900*. Cambridge: Cambridge University Press.

Ambraseys, N.N., C.P. Melville, and R.D. Adams. 1994. *The seismicity of Egypt, Arabia and the Red Sea: A historical review*, xix + 181 pp. Cambridge, UK: Cambridge University Press.

Ambraseys, N.N., and D. White. 1997. The seismicity of the eastern Mediterranean region 550–1 BC: A reappraisal. *Journal of Earthquake Engineering* 1: 603–632.

Ambraseys, N.N., J.A. Jackson, and C.P. Melville. 2002. Historical seismicity and tectonics: The case of the eastern Mediterranean and the Middle East. In *International handbook of earthquake and engineering seismology*, ed. W.H.K. Lee, H. Kanamori, P.C. Jennings, and C. Kisslinger, 747–763. Amsterdam: Academic Press.

Antonopoulos, J. 1979. Catalogue of tsunamis in the eastern Mediterranean from antiquity to present times. *Annali di Geofisica* 32: 113–130.

———. 1992. The great eruption of Thera volcano and the ensuing tsunami in the Greek Archipelago. *Natural Hazards* 5: 153–168.

Austin, B.J. 2006. Late Neogene Mass Transport Deposits of the West Nile Delta, Egypt. Abstracts, *American Association of Petroleum Geologists*, Annual Meeting, Houston, Texas. http://www.searchanddiscovery.net:16080/documents/2006/06088houston_abs/abstracts/austin.htm.

Beisel, S., L. Chubarov, I. Didenkulova, E. Kit, A. Levin, E. Pelinovsky, Y. Shokin, and M. Sladkevich. 2009. The 1956 Greek tsunami recorded at Yafo, Israel, and its numerical modeling. *Journal of Geophysical Research* 114, C09002. doi:10.1029/2008JC005262.

Bentor, Y.K. 1989. Geological events in the Bible. *Terra Nova* 1: 326–338.

Bruins, H.J., J.A. MacGillivray, C.E. Synolakis, C. Benjamini, J. Keller, H.J. Kisch, A. Klügel, and J. van der Plicht. 2008. Geoarchaeological tsunami deposits at Palaikastro (Crete) and the Late Minoan IA eruption of Santorini. *Journal of Archaeological Science* 35: 191–212.

Bruins, H.J., and J. van der Plicht. 1996. The Exodus enigma. *Nature* 382(6588): 213–214.

Cita, M.B., and G. Aloisi. 2000. Deep-sea tsunami deposits triggered by the explosion of Santorini

(3500 y BP), eastern Mediterranean. *Sedimentary Geology* 135: 181–203.

Dominey-Howes, D. 2002. Documentary and geological records of tsunami in the Aegean Sea region of Greece and their potential value to risk assessment and disaster management. *Natural Hazards* 25: 195–224.

Druitt, T.H., R.A. Mellors, D.M. Pyle, and R.S.J. Sparks. 1989. Explosive volcanism on Santorini. *Greece, Geological Magazine* 126: 95–126.

Dussaud, R. 1896. Voyage en Syrie, octobre–novembre 1895. *Notes: Archaeology Review* 3(28): 299–336.

Elias, A., P. Tapponnier, S.C. Singh, G.C.P. King, A. Briais, M. Daëron, H. Carton, A. Sursock, E. Jacques, R. Jomaa, and Y. Klinger. 2007. Active thrusting offshore Mount Lebanon: Source of the tsunamigenic A.D. 551 Beirut-Tripoli earthquake. *Geology* 35: 755–758. doi:10.1130/G23631A.1.

Folkman, Y., and Y. Mart. 2008. Newly recognized eastern extension of the Nile deep-sea fan. *Geology* 12: 939–942. doi:10.1130/G24995A.1.

Friedrich, W.L., B. Kromer, M. Friedrich, J. Heinemeier, T. Pfeiffer, and S. Talamo. 2006. Santorini Eruption Radiocarbon Dated to 1627–1600 BC. *Science* 312: 548. doi:10.1126/Science.1125087.

Fritz, G.A. 2006. *The Lost Sea of the Exodus. USA: Instantpublisher.com.* Collierville, TN: Instantpublisher.

Garziglia, S., M. Ioulalen, S. Migeon, E. Ducassou, J. Mascle, O. Sardou, and L. Brosolo. 2007. Triggering factors and tsunamigenic potential of a large submarine mass failure on the Western Nile margin (Roseta area, Egypt). In *Submarine Mass Movements and Their Consequences*, eds. Lykousis, V., D. Sakellariou, J. Locat, pp 347–355.

Garziglia, S., S. Migeon, E. Ducassou, L. Loncke, and J. Mascle. 2008. Mass-transport deposits on the Rosetta province (NW Nile deep-sea turbidite system, Egyptian margin): Characteristics, distribution, and potential causal processes. *Marine Geology* 250: 180–198.

Goldsmith, V., and M. Gilboa. 1986. Tide and low in Israel. *Ofakim Be'Geographia* 15: 21–47 (in Hebrew).

Goren, Y., I. Finkelstein, and N. Na'aman. 2004. Inscribed in Clay: Provenance Study of the Amarna Tablets and other Ancient Near Eastern Texts. Monograph Series of the Institute of Archaeology, Tel Aviv University No. 23. Tel Aviv.

Guidoboni, E., and A. Comastri. 2005. *Catalogue of earthquakes and tsunamis in the Mediterranean area from the 11th to the 15th century*, 1037. Rome-Bologna: INGV-SGA.

Guidoboni, E., A. Comastri, and G. Traina. 1994. *Catalogue of Ancient Earthquakes in the Mediterranean Area up to the 10th Century*. Bologna, Italy: ING-SGA. 504 pp.

Guidoboni, E., and J.E. Ebel. 2009. *Earthquakes and Tsunamis in the Past. A Guide to Techniques in Historical Seismology*. Cambridge: Cambridge University Press. xi+590 pp.

Har-El, M. 1983. *The Sinai Journeys*. San Diego, CA: Ridgefield Publishing Company.

Harrigan, P. 2006. Volcanic Arabia. Saudi Aramco World, MA06, 57, No. 2. http://www.saudiaramcoworld.com/issue/200602/volcanic.arabia.htm.

Hoffmeier, J.K. 1996. *Israel in Egypt – The Evidence for the Authenticity of the Exodus Tradition*. New York and others: Oxford University Press.

Hoffmeier, J.K., and S.O. Moshier. 2006. New paleo-environmental evidence from North Sinai to complement Manfred Bietak's map of the Eastern Delta and some historical implications. In *Timelines: Studies in honour of Manfred Bietak, Vol. II*, Orientalia Lovaniensia Analecta, vol. 149, ed. E. Czerny, I. Hein, H. Hunger, D. Melman, and A. Schwab, 167–176. Leuven: Peeters Publishers.

Jacobovici, S. 2006. *The Exodus Decoded*. The History Channel.

Karcz, I. 1987. Bibliographic reliability of catalogues of historic earthquakes in and around Israel, II Catalogue of Turcott and Arie (1986). Geological Survey of Israel, Report GSI/10/87.

Karcz, I. 2004. Implications of some early Jewish sources for estimates of earthquake hazard in the Holy Land. *Annals of Geophysics* 47: 759–792.

Klein, M. 2000. The formation and disappearance of a delta at the El-Arish river mouth. The Hydrology-Geomorphology Interface: Rainfall, Floods, LeVeque, R. J., 2006. *Clawpack software*. http://www.clawpack.org.

Marinos, G., and N. Melidonis. 1971. On the strength of seaquakes (tsunami) during the prehistoric eruption of Santorini. In *Acta of the 1st International Scientific Congress on the Volcano of Thera, held in Greece, 15th–23rd September, 1969*, General Direction of Antiquities and Restoration, ed. A. Kaloyeropoyloy, 277–282. Athens: Archaeological Services of Greece.

McCoy, F.W. 2009. The Eruption in the Debate About the Date. In *Time's Up! Dating the Minoan Eruption of Santorini [Acts of the Minoan Eruption Chronology Workshop, Sandbjerg, Denmark, Nov. 2007]*, ed. Warburton, D.A. *Monographs, Danish Institute Athens* 10: 73–90.

McCoy, F.W., and G. Heiken. 2000. The Late-Bronze Age explosive eruption of Thera (Santorini), Greece: Regional and local effects. *Geological Society of America Special Papers* 345: 43–70. doi:10.1130/0-8137-2345-0.43.

Mészáros, S. 1978. Some words on the Minoan tsunami of Santorini. In *Thera and the Aegean World II*, ed. Doumas, C. Papers presented at the Second International Scientific Congress, Santorini, Greece, Aug. 1978. London: Thera Foundation 1: 257–261.

Miller, W. 1913. The Ottoman Empire, 1801–1913. Cambridge University Press, 1913. Perry-Castañeda Library, Map Collection, The University of Texas at Austin. http://www.lib.utexas.edu/maps/historical/ottoman_empire_1801.jpg.

Minoura, K., F. Imamura, U. Kuran, T. Nakamura, G.A. Papadopoulos, T. Takahashi, and A.C. Yalciner. 2000. Discovery of Minoan tsunami deposits. *Geology* 28: 59–62.

Moshier, S.O., and A. El-Kalani. 2008. Late Bronze Age Paleogeography along the Ancient Ways of Horus in Northwest Sinai, Egypt. *Geoarchaeology: An International Journal* 23(4): 450–473.

Moufti, M.R., A.M. Moghazi, and K.A. Ali. 2013. 40Ar/39Ar geochronology of the Neogene-Quaternary Harrat Al-Madinah intercontinental volcanic field, Saudi Arabia: Implications for duration and migration of volcanic activity. *Journal of Asian Earth Sciences* 62: 253–268. http://dx.doi.org/10.1016/j.jseaes.2012.09.027.

Nof, D., and N. Paldor. 1992. Are There Oceanographic Explanations for the Israelites' Crossing of the Red Sea? *Bulletin American Meteorological Society* 73(3): 305–314.

———. 1994. Statistics of Wind over the Red Sea with Application to the Exodus Question. *Journal of Applied Meteorology* 33(8): 1017–1025.

Palmer, E.H. 1871. *The Desert of the Exodus*. Cambridge: Deighton, Bell and Co.

Pararas-Carayannis, G. 1992. The tsunami generated from the eruption of the volcano of Santorin in the Bronze Age. *Natural Hazards* 5: 115–123. http://www.drgeorgepc.com/TsunamiSantorin.html.

Perissoratis, C., and G.A. Papadopoulos. 1999. Sediment instability and slumping in the southern Aegean Sea and the case history of the 1956 tsunami. *Marine Geology* 161: 287–305.

Petrie, W.M.F. 1906. Researches in Sinai. London.

Propp, W.H.C. 1999. *Exodus 1–18: A new translation with introduction and commentary*. New York, NY: Doubleday.

Salamon, A., A. Hofstetter, Z. Garfunkel, and H. Ron. 2003. Seismotectonics of the Sinai subplate – The eastern Mediterranean region. *Geophysical Journal International* 155: 149–173.

Salamon, A., T. Rockwell, S.N. Ward, E. Guidoboni, and A. Comastri. 2007. Tsunami hazard evaluation of the eastern Mediterranean: Historical analysis and selected modeling. *Bulletin of the Seismological Society of America* 97(3): 705–724. doi:10.1785/0120060147.

———. 2010. A critical evaluation of tsunami records reported for the Levant coast from the second millennium BCE to the present, Israel. *Israel Journal of Earth Sciences* 58(3–4): 327–354.

Shalem, N. 1956. Seismic tidal waves (tsunamis) in the Eastern Mediterranean. *Society for Exploration of Eretz Israel* 20: 159–170 (in Hebrew).

Shaw, B., N.N. Ambraseys, P.C. England, M.A. Floyd, G. J. Gorman, T.F.G. Higham, J.A. Jackson, J.-M. Nocquet, C.C. Pain, and M.D. Piggott. 2008. Eastern Mediterranean tectonics and tsunami hazard inferred from the AD 365 earthquake. *Nature Geoscience* 1: 268–276.

Sivertsen, B.J. 2009. *The Parting of the Sea: How Volcanoes, Earthquakes, and Plagues Shaped the Story of the Exodus*. Princeton, NJ: Princeton University Press. ISBN 978-0-691-13770-4.

Stanley, D.J. 2002. Configuration of the Egypt-to-Canaan coastal margin and the north Sinai byway in the Bronze Age. In *Egypt and the Levant: Interrelations from the 4th through the early 3rd millennium BCE*, ed. T. Levy and E.C.M. van den Brink, 98–117. London: Continuum International Publishing Group.

Stanley, D.J., and H. Sheng. 1986. Volcanic shards from Santorini (Minoan ash) in the Nile Delta, Egypt. *Nature* 320: 733–735.

Stanley, D.J., and A.G. Warne. 1993. Sea level and initiation of Predynastic culture in the Nile delta. *Nature* 363: 435–438.

———. 1998. Nile delta in its destruction phase. *Journal of Coastal Research* 14: 794–825.

Stiros, S.C. 2009. The 8.5+ magnitude, AD 365 earthquake in Crete: Coastal uplift, topography changes, archaeological and historical signature. *Quaternary International* 216: 54–63.

Virolleaud, C. 1935. La revolte de Koser contre Baal. *Syria* 16: 29–45.

Ward S.N. 2013. http://es.ucsc.edu/~ward/.

Ward, S.N., and S. Day. 2008. Tsunami balls: A granular approach to tsunami run up and inundation. *Communications in Computational Physics* 3(1): 222–249.

———. 2010. The Lituya bay landslide and tsunami – A tsunami ball approach. *Journal of Earthquake and Tsunami* 4(4): 285–319.

Yokoyama, I. 1978. The tsunami caused by the prehistoric eruption of Thera. In: Doumas C (ed) Thera and the Aegean World I. Papers presented at the Second International Scientific Congress, Santorini, Greece, August 1978. London: Thera Foundation 1: 277–283.

10 Dating the Theran Eruption: Archaeological Science Versus Nonsense Science

Malcolm H. Wiener

Abstract

Various attempts have been made to ascribe the story of the Exodus to a particular time and place. The most common proposal is to place the Exodus just before the period of the Judges and the United Monarchy at the time of the collapse of Egyptian power around 1130 BC, with Moses and his followers taking the opportunity to flee Egypt then (or in one version, leave the Egyptian controlled area of Canaan to move up to the central highlands, pausing en route to receive the Ten Commandments). A less common but still noteworthy attempt has placed the Exodus at the time of the expulsion of the Hyksos from Egypt around 1525 BC, occasionally accompanied by the suggestion that the parting of the Red Sea described in the biblical accounts is a reflection of the tsunami (giant waves) which followed the massive eruption of the volcano on the island of Thera (Santorini) at that time. This interpretation has been buttressed by an Egyptian inscription known as the Ahmose Tempest Stele of this date. The purpose of this paper is to assess the arguments for and against placing the eruption of the Theran volcano at c. 1525 BC and for associating the Ahmose Tempest Stele with the eruption.

Introduction

Various authors have sought to connect the tsunami following the early Late Bronze Age eruption of the volcano on the Aegean island of Thera with the parting of the Sea of Reeds described in the biblical story of the Exodus (e.g., Goedicke 1992, 2004). A connection between the "plague of darkness" which could be felt (Ex. 10:21) and the fall of tephra (ash) resulting from the Theran eruption has also been suggested (Stanley and Sheng 1986). (But cf. Bietak, Chap. 2, regarding other Biblical references to the Exodus which reflect developments in the thirteenth–twelfth centuries BC.) Others have proposed that the storm and destruction described in the Ahmose Tempest Stele (Fig. 10.1) should be connected to the Theran eruption (Davis 1990; Foster and Ritner

I am grateful to Tom Levy, Peter Warren, Steven Soter, Doug Keenan, Daphna Ben-Tor, and Chris Bennett for helpful comments; and to Jayne Warner, Erin Hayes, Jason Earle, and Rebecca Hahn for research assistance and editorial advice.

M.H. Wiener (✉)
Institute for Aegean Prehistory (INSTAP)
Greenwich, CT, USA
e-mail: mhwiener@villacandia.com

T.E. Levy et al. (eds.), *Israel's Exodus in Transdisciplinary Perspective*,
Quantitative Methods in the Humanities and Social Sciences, DOI 10.1007/978-3-319-04768-3_10,

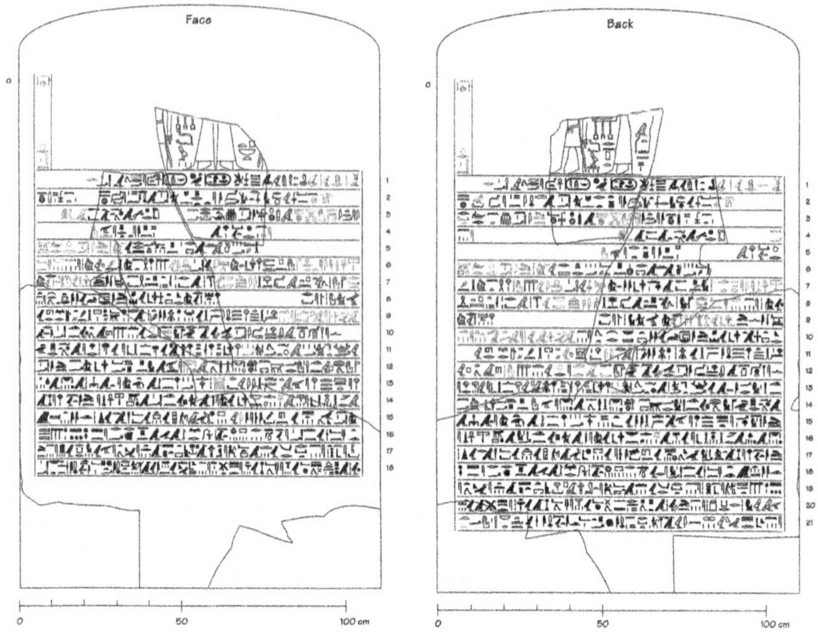

Fig. 10.1 The Tempest Stele of Ahmose. Reconstruction of the face and back (after Wiener and Allen 1998: Figs. 1a and 1b)

1996), thus approximately connecting in time the Theran eruption to the expulsion of the Hyksos. This proposal was questioned in a paper by Wiener and Allen (1998), the latter arguing that the stele was a typical "restoration text" composed to commemorate the restoration by Ahmose of temples and shrines destroyed by the Hyksos (rather than by a storm), and the former questioning whether the eruption could have had the impact in Upper Egypt described in the text. The Tempest Stele describes a great storm coming from the west, but Thera is located over 1,350 km north of Thebes, and the tephra from the eruption covered a wide arc to the east of Thera (McCoy and Heiken 2000: 57–59). Douglas Keenan has noted, however, that great volcanic eruptions can cause extreme storms in distant places in the years following via the action of atmospheric aerosols (pers. comm. of 15 August 2013, for which I am most grateful). The likely general contemporaneity in date between the Ahmose Stele and the eruption, properly dated by massive archaeological and now by new scientific evidence, strengthens the possible association. Finally, if the accession date for Ahmose is 1540 BC and the conquest of Avaris occurs in the 18th year (1522 BC) as some have proposed (Franke 1988: 264; Bourriau 2000: 185), the question arises as to whether an eruption plus tsunami in 1525 BC as suggested by the tree-ring plus ice-core evidence cited below might have inflicted significant damage on Hyksos shipping and society just prior to the conquest. It is the question of the date of the Theran eruption that is the subject of this paper.

The Date of the Eruption

The Archaeological Evidence for a Date c. 1525 BC

The massive archaeological evidence for a Theran eruption date after c. 1530 BC has been presented in detail by Manfred Bietak (2004, Chap. 2), Peter Warren (2006, 2009, 2010), various scholars working in Cyprus (Åström 1972: 758–760, 762, 765 n. 8; 2001: 50; Eriksson 1992; 2001; 2007: Table 1b), and the author (Wiener 2003, 2006a, b, 2007, 2009a, b, 2010). The evidence includes objects of New Kingdom type found in Late Helladic I contexts in the Shaft Graves of Mycenae; stratigraphical interconnections between Tell Dab'a, sites in the Near East, and especially Cyprus, with further links to the Aegean world; deposits of Theran pumice at 15 sites in the eastern Mediterranean no earlier than Egyptian New Kingdom contexts; and a Cypriot White Slip (WS) I vessel found in the Volcanic Destruction Level at Thera (Wiener 2001).

The WS I bowl deserves special attention, particularly since it was mentioned at the conference—in the discussion following the Dee et al. paper—by William Dever, who stated that WS I had been found in Near Eastern contexts as early as 1600 BC (if only according to the high Near Eastern chronology as set forth in Dever 1991, 1992a, b). That chronology is no longer accepted by the majority of scholars in view of the mounting evidence favoring the Middle to Low (but not the Ultra-Low) chronology (Barjamovic et al. 2012; Bietak 2013; Pfälzner 2004; compare Gasche et al. 1998). Even if one accepts the appearance of the earliest WS I by 1600 BC, it remains the case that the piece found in the eruption level on Thera is regarded by many leading specialists in Cypriot pottery as produced late in the WS I series and on the south coast of Cyprus, where the interconnections between Cypriot pottery and Egyptian datable objects are clear. Moreover, the piece found in the Theran eruption deposit was worn, damaged in antiquity, and then repaired. The odds against the bowl having been one of the first of its kind ever made, then carried to Thera immediately after its creation, acquiring signs of wear rapidly, being repaired after breakage, and used again all in a few years before its burial in the eruption, are enormous. Most importantly, Manfred Bietak (2013) has presented compelling evidence from his examination of sites in Egypt, the Levant, and Cyprus that WS I pottery does not arrive before Thutmoside levels c. 1500 BC, and is part of a well-understood sequence of Cypriot pottery involving thousands of pieces, not all of which can arrive with unexplained delays of centuries. Theran pumice from the eruption also does not appear before Thutmoside levels, suggesting the possibility of an eruption date still later than 1525 BC.

It is important to note that the chronological question posed by the Cypriot pottery comparison is not just a matter of one pot, nor a question of just one type of pottery allegedly arriving at Thera before its arrival in the Near East and Egypt (Manning 1999: 119–129), but a number of successive pottery classes stretching over centuries, involving many hundreds of examples. Surely not all could have seen the elapse of many decades between their manufacture in Cyprus and deposition abroad. The linkages connect sites in Egypt, the Near East, Cyprus, and the Aegean. The evidence from Tell el-'Ajjul near Gaza is of particular relevance, for it includes in the same stratum 23 pieces of Cypriot WS I pottery, Egyptian New Kingdom material produced no earlier than 1550 BC, pumice from the Theran eruption, and radiocarbon dates centered around c. 1525 BC (Fischer 2009: Fig. 4).

Peter Fischer, recent excavator of Tell el-'Ajjul, in response to the assertion that radiocarbon evidence proved the correctness of the Aegean Long Chronology, noted that at the outer boundary of the two-sigma margin of the radiocarbon distribution, the radiocarbon data could support the Long Chronology. The correct position in this regard will be clarified in a forthcoming publication (Fischer, pers. comm. of 6 April 2010, for which I am most grateful). Fischer (2009: 265) concludes by stating that the stratum in which the pumice first appears should be dated c. 1560–1530 BC, which is far later than the dates proposed for the eruption by proponents of the Aegean Long Chronology.

Fischer's comments with respect to the pumice are of particular interest in light of recent developments. He notes:

> During the Minoan eruption of Santorini, a large volume of magma was ejected in the short time span of not more than a few days and the eruption products were distributed over a large area.... For a long time, the estimates of the volume ranged from 16 to 35 km^3 of magma, but recent investigations suggest that a range of 100 km^3 of magma is more probable. It is obvious that the major part of the material erupted was deposited directly into the sea. As it consists mainly of highly vesicular silicate glass, pumice floats on the water. Depending on the size of the pumice lumps some time elapses before the fine glassy bubble walls

> break and the pumice becomes soaked and finally sinks. Experiments have shown that even relatively small samples of pumice can float on sea water for more than 1.5 years. Pumice may therefore be expected to be transported by marine currents and wind over large distances all over the Eastern Mediterranean region. It can be assumed that within weeks after the eruption large amounts of pumice accumulated along the shorelines (Fischer 2009: 262).

A subsequent detailed study examined over 400 samples of pumice found in 15 archaeological excavations at sites in Egypt, the Levant, Cyprus, and the Dodecanese. Over 90 % of the samples came from the LM IA eruption of Thera. Every one of these was found in a New Kingdom context dating no earlier than c. 1540 BC. The 30 pieces of pumice from earlier stratigraphic contexts all come from earlier eruptions at Nisyros and Gyali in the Dodecanese, with the exception of one piece sourced to an earlier eruption of Mt. Etna in Sicily (Bichler in press; see also Sterba et al. 2009). A prior publication, based upon an examination of about 300 of the pumice samples, noted that if the seventeenth century BC date were correct, "it would indeed be most peculiar a phenomenon that pumice from the Minoan Santorini eruption were abundantly available along the shores of the Eastern Mediterranean, yet for some reason had been left unnoticed and unused by the local inhabitants for 100–150 years" (Steinhauser et al. 2010: 408). The recent excavation by Dr. Abd el-Maksoud at Tell Hebwa in the eastern Nile Delta, a major fortress established by the Hyksos and greatly expanded by Thutmosis III and his successors, provides further confirming information in this regard—no Theran pumice in the Hyksos levels, but much found in the New Kingdom strata.[1] Fourteen separate uses of pumice in antiquity have been studied and its importance with regard to metallurgy noted in particular (Wiener in Wiener and Allen 1998: 26). It seems inconceivable that some of the vast amount of pumice ejected would not have floated to the sites in question, or that it would have lain unused in 15 different places for most of a century.

The archaeological evidence provides a date for the eruption between c. 1550–1480 BC. A date of 1525 BC would match tree-ring indications from the following year found in Arizona, California, and Nevada in the USA plus a weaker signal in trees from Yamal, Siberia. An acid spike in a Greenland ice core also indicates a volcanic event somewhere in that year (Wiener 2006a: 320–323; Salzer and Hughes 2007; Salzer and Hughes also refer to an event in 1544 BC, so far observed only in trees in Arizona, California, and Nevada). Against this mass of evidence for the traditional date of the Theran eruption, Dee et al. (Chap. 6) assert that the proposed Long Chronology is supported archaeologically by the context of the Khyan Lid (an alabaster lid of a vessel with the cartouche of the Hyksos ruler Khyan) found at Knossos by Arthur Evans a century ago (Evans 1921: 418–422). Dee et al.'s claim was based on recent evidence of sealings of the late 13th Dynasty pharaoh Sobekhotep IV and of the Hyksos ruler Khyan found together (Moeller et al. 2011) and on evidence of a Khyan sealing found in an early Second Intermediate Period context at Tell el-Dab'a (Forstner Müller et al. 2012: 4). These discoveries may tend to support the view of Ward (1984), Tufnell (1984), Warren and Hankey (1989: 136), and Ben-Tor (2010) that Khyan belongs at the beginning of the chain of Hyksos rulers rather than near the middle, as once suggested by Bietak (2001: 139; 2010: 102), although as Ben-Tor has noted (2010: 95), seals are often reused or appear in later deposits, as in the case of both late Middle Kingdom and early Second Intermediate Period seals found in a Thutmosis III context at Tell el-Dab'a. Warren and Hankey had already assumed that Khyan was the first of the Hyksos rulers in presenting the Aegean Short Chronology, but even moving the reign of Khyan back to the late eighteenth century BC would present no problem for the Aegean Short Chronology for the reason stated below. Unfortunately, Dee et al. invent a totally imaginary Knossian context for the Khyan Lid from Knossos by asserting that it

[1] I thank Prof. Manfred Bietak for this information provided via pers. comms. of 25 and 26 March 2013.

was found in a deposit buried by Theran ash (!). There is no such context, and in fact no Theran tephra (ash) at all reported from Knossos, which lies to the west of the tephra fall zone. Indeed, there is no basis whatever for assuming that the lid arrived in Crete during Khyan's reign. No trace of a stone vessel for the lid was found. It is highly likely that the lid, like all other known pieces of Egyptian stone found in Crete, was imported because of the acute Minoan interest in acquiring stone of various types for reworking. Indeed, there are several examples of Egyptian Old Kingdom stone vessels reworked into stone vessels of Minoan shape in Late Minoan I, around a thousand years after their original creation in Egypt. The dates of Khyan's reign are also not critical to the detailed sequence of imports and exports of pottery and other objects found at Dab'a and at sites in the Near East and Cyprus that support the standard Aegean Short Chronology published in this volume and elsewhere in great detail by Bietak (see also, e.g., Warren 2006, 2009, 2010; Wiener 2003, 2006a, b, 2007, 2009a, b, 2010).

The Radiocarbon Evidence

Dee et al. (Chap. 6) state emphatically that radiocarbon measurements place the date of the Theran eruption so firmly in the seventeenth century BC as to foreclose any further discussion of the subject. Dee in his presentation at the conference made special reference to a paper in the journal *Science* (Manning et al. 2006a), which he described as "a very big deal." Indeed the paper in question, because it was published in *Science*, has had a major impact on the field of Aegean prehistory. Accordingly it is all the more unfortunate that the paper is without scientific or statistical validity.

To understand why this is so, we must first examine the actual radiocarbon measurements that comprise the Theran 28-measurement data set. This is no easy task, as the measurements were published only in the Supporting Online Material (Manning et al. 2006b), and in a format extremely difficult to decipher. For example, two measurements of the same barley seed cluster with the same sample number (M10/23A N012) gave discordant readings of 3400 BP ± 31 and 3318 BP ± 28. The two measurements were presented apart from one another on separate pages, and with no statement that these were measurements from the same sample, thus making it difficult for even a diligent reader of the voluminous Supporting Online Material to grasp the critical underlining facts. No reference was made to the fact that a number of the measurements were holdovers from years past before the development of modern methods of pretreatment. Among the seeds included in the data set were seeds of the same species with central radiocarbon range measurements 215 years apart in the case of peas and 97 years apart in the case of barley. Samples of grain seeds shaded in gray indicating they had been excluded as outliers gave central ages 350 radiocarbon years apart (with one-sigma error ranges of ± 80 and ± 70 radiocarbon years, producing a combined error range at the margin of 500 [350 + 150] radiocarbon years). Initial inquiries concerning the reasons for exclusion produced no response, but in 2009 an article by Manning et al. (2009: Fig. 2) appeared which seemed to include these measurements in the 28 measurement data set and Christopher Bronk Ramsey at the Oxford Radiocarbon Accelerator Unit Research Laboratory for Archaeology has kindly confirmed that they had been included at the outset; the Supporting Online Material was modified accordingly on 24 April 2013 (pers. comm. of 23 May 2013).

Astonishingly, these wholly disparate measurements were then combined via the OxCal 3.1 program to produce an error range of ± 7.5 radiocarbon years for dating the Theran eruption! The claimed precision reflects the unfortunate fact that the OxCal, Calib, and other calibration programs reduce stated error bands in response to the number of measurements made, largely irrespective of how consistent, inconsistent, or even grossly incommensurate the measurements themselves are, except for the exclusion of what are deemed "outliers." Timothy Jull, Director of the NSF-Arizona AMS Facility, Elisabetta Boaretto of the Rehovot lab, and the author have each independently noted this problem (Boaretto et al. 2005: 43–44; Boaretto 2007: 215; 2009: 280; Jull in Sharon et al. 2007: 9; Wiener 2010). Some Bayesian calibration programs give extra weight to the central range when dealing with widely disparate measurements (rather than greater weight to the more recent measurement ages as has been suggested when reservoir effects are likely given the known environment). It is worth recalling that Bayesian procedures were first applied to radiocarbon measurements in order to reconcile differing measurements of the same sample. Combining measurements of seed samples 350–500 ^{14}C years apart (a difference likely resulting from different reservoir conditions as described below) violates basic principles of statistical analysis (Ward and Wilson 1978; Keenan 2012). The purported overall accuracy of ± 7.5 ^{14}C years for the date of the eruption is a consequence of the computer program, not a reflection of reality.

Claims equally lacking in scientific validity have been made with respect to radiocarbon measurements from a branch of an olive tree found on Thera (Wiener 2009a, b, denying claims made in Friedrich et al. 2006, 2009 and repeated by Dee et al., Chap. 6). Olive trees do not make annual rings as claimed by Friedrich et al. (Cherubini et al. 2013); absent that claim, there is no valid basis for asserting a radiocarbon wiggle-match to the calibration curve. There is no way of determining where the tree grew in relation to pre-eruption gas emission sources on Thera. The carbon from these vents of course lacked the ^{14}C isotope created in the atmosphere via the interaction of galactic cosmic rays with nitrogen atoms. Moreover, the final eruption of the Thera volcano was preceded by an earthquake and other precursor events which caused the populace to flee, with some then returning and attempting repairs and the removal of valuables before the final cataclysm. Trees growing in close proximity to one another may be differently affected by carbon releases in such circumstances. At Yellowstone caldera in the western United States a gas emission in 1978 resulted in a c. 25 % drop in ^{14}C in rings of some trees (Evans et al. 2010). Finally, there is no way of knowing whether the

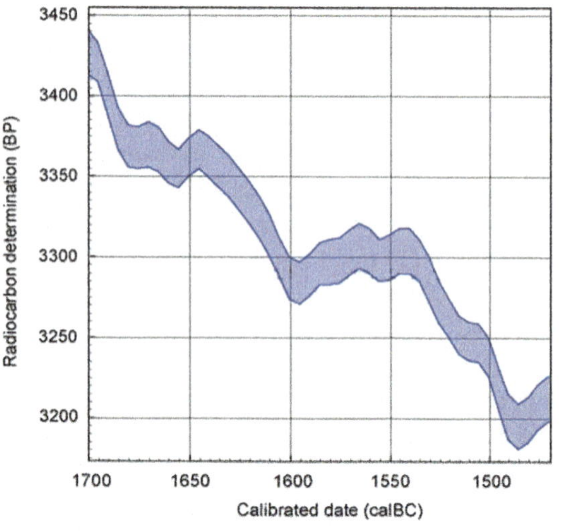

Fig. 10.2 INTCAL09 calibration curve. Detail of the period 1700–1470 BC

branch was living at the time of the eruption; olive trees often display dead branches (Rackham, pers. comm. of 11 May 2008; Blitzer, pers. comm. of 23 July 2008; see also Blitzer in press), whose removal may risk the health of the tree, particularly when metal saws are not routinely or readily available.

The claim of precision of ± 7.5 radiocarbon years is essential to the argument for an early eruption date because of the oscillating calibration curve at the critical period (Fig. 10.2) which provides similar radiocarbon ages at 1610 and 1530 BC. Were the error range stated at ± 20, for example, there would be no basis for distinguishing between the claimed eruption date in the seventeenth century BC and the archaeological interconnections plus Theran pumice contexts derived date of 1530–1525 BC. The position remains as stated by the former director of the Oxford Radiocarbon Accelerator Unit:

> Samples grown anytime within the span of calendar years 1620–1530 BC will have radiocarbon ages that are the same to within ± 20 years and hence be indistinguishable even if measured in a high precision laboratory. It would be necessary for the date to be as early as c. 1650 BC in order for the radiocarbon age to be distinguishable from that corresponding to 1620–1530 BC; this is a fact of nature—of the radiocarbon content of the atmosphere (Aitken 1988: 21–22).

In this connection, the Oxford laboratory report of a test on 96 decadal segments of wood of known date which showed an overall bias of 8.9 ± 3 years toward earlier dates (Bronk Ramsey et al. 2004) is of special interest, since such an offset would be sufficient in itself to bring the 28 data set of Theran measurements within the area of the oscillating calibration curve even if the combining of such widely disparate measurements were statistically acceptable. Moreover, analysis of 66 short-lived seed samples of known dates between AD 1700 and 1900 collected in Egypt produced a 19.5 ± 5 ^{14}C year offset from the radiocarbon measurements of tree rings of known date, mostly from the Rhine River Valley, that constitute the calibration curve (Dee 2010). The result confirms the rough estimate of a 20-year regional/seasonal offset previously proposed on the basis of the differences in growing seasons (Wiener 2003: 387). Within a single tree-ring calendar year, radiocarbon measurements differ by between 8 and 32 radiocarbon years (and may exceed this range) between the summer high of seasonal exchange of stratospheric radiocarbon with the atmosphere and the winter low (Levin et al. 1992; Housley et al. 1999: 167; Levin and Hesshaimer 2000; Keenan 2004: 102–103). Such regional/seasonal offsets should serve as a further caution with respect to claims of extremely high precision resulting from probabilistic analysis of disparate measurements as in the Thera dataset.

The problems for radiocarbon dating resulting from reservoir effects require special attention. The ^{14}C isotope is formed by the interaction of galactic cosmic rays with atmospheric nitrogen atoms. Solar wind deflects some of the galactic cosmic rays. Pronounced fluctuations in the solar wind modulate the production of ^{14}C over time, resulting in oscillations in the absorption of ^{14}C. ^{14}C is absorbed by living organisms until they die, at which point decay begins at a measureable rate (Reimer 2001: 2495; Soter, pers. comm. of 23 August 2013). Carbon within the earth is CO_2 in which there is no ^{14}C isotope. For every 1 % of earth carbon lacking the ^{14}C isotope in a sample submitted for radiocarbon measurement the reported radiocarbon age will be about 80 years older than the true date. The offsets from true ages caused by the various sources of CO_2 lacking ^{14}C are called "reservoir effects."

The critical sources of ^{14}C-absent carbon include volcanic and non-volcanic gas vents, geothermal fields, and general soil degassing. For example, in Italy the area of terrestrial CO_2 emissions stretches from Tuscany to Sicily, and from the Tyrrhenian Sea to the Apennines (Saupé et al. 1980; Rogie 1996; Minissale et al. 1997; Chiodini et al. 1999; 2004; Rogie et al. 2000; Cardellini et al. 2003; Gambardella et al. 2004; Marzaioli et al. 2005; Frezzotti et al. 2009: 109). Reported radiocarbon dates from Italian sites whose historical contexts are clear are frequently 100–300 years too early (Wiener 2010: 371). A similar phenomenon is reported with regard to Iceland, where volcanic/geothermal effects are thought to be the cause of radiocarbon dates 100–200 years earlier than presumed historical dates for the earliest European occupation levels (Sveinbjörnsdóttir and Heinemeier 2011). At Sulphur Banks on Hawaii, three living tree ferns and one living Ohia leaf plant growing 1–5 mile from a volcanic vent (whose true age was of course zero) produced radiocarbon age measurements between 81 and 303 years (Chatters et al. 1969: Table 2).

Oceans and seas also act as reservoirs of ^{14}C-deficient carbon, with the result that radiocarbon measurements of short-lived mollusks produce ages of more than 400 years older than their actual age (Stuiver et al. 1998: 1131–1135). The average Mediterranean offset has been estimated at 458 ± 85 years (Reimer and McCormac 2002). A recent study of 5-year Japanese tree-ring segments of known dendrochronological date from 1060 BC to AD 400 found that for some periods the radiocarbon dates obtained differed significantly from the calibration curve dates based on measurements from segments of long-lived European trees believed to be of known date. The authors suggest an "island effect" from surrounding marine carbon reservoirs as a possible cause (Imamura et al. 2007. See also Ozaki et al. 2007; Ozaki et al. 2009). Sakamoto et al. (2009) note that with respect to radiocarbon dates from the Japanese archipelago "possible local offsets of the curve cannot be ignored." Stuiver and Braziunas (1993) describe how irregular water-circulation oscillations of ^{14}C-deficient water, some with a periodicity of 40–50 years, operate globally. Several studies suggest the possibility of upwelling of ^{14}C-deficient carbon from the Aegean, either via the periodic exchange of water with the Black Sea which is rich in ^{14}C-deficient carbon (Keenan 2002), or the release of ^{14}C-deficient carbon from underwater volcanic vents, one of which, the submarine volcano Kolombo, is located 7 km north-northeast of Thera. Kolombo degasses bubbles of nearly pure CO_2 and is the largest of a chain of about 20 underwater volcanic craters (Carey et al. 2013). Given the prevalence of volcanic degassing in the region it would be foolhardy to assume that it was infrequent on pre-eruption Thera. The reservoir effect in the Mediterranean varies in time due to upwelling instabilities. Rapp and Hill (2006: 153) note that "upwelling of deep water occurs near many coastlines" and that it "is affected by the shape of the coastline and the bottom topography, local climate, and wind and current patterns."

An "estuary effect" can occur when salt water mixes with freshwater in river deltas. The Nile Delta is particularly vulnerable, for when the Nile floods recede annually, the salt water rushes in (Bietak 2013: 99). This furnishes a highly plausible explanation for the fact that a Thutmoside-era measurement from Avaris provided a radiocarbon date 120 years older than a number of radiocarbon dates for the same reign from Upper Egypt, together with four measurements from Hyksos-era contexts each also c. 120 years earlier than their archaeological contexts (Bietak and Höflmayer 2007; cf. Bronk Ramsey et al. 2010: 1556–1557).

Freshwater also poses problems. A recent article in *Radiocarbon* (Wood et al. 2013: 163) begins with the statement: "If ancient carbon is incorporated into lakes and rivers, it can be transferred along the foodchain where it can cause radiocarbon dates to appear erroneously old. This effect is known as the ^{14}C freshwater reservoir effect (FRE)." Freshwater rivers and lakes contain two main carbon reservoirs: CO_2 from the atmosphere and dissolved inorganic carbon from groundwater. Groundwater flowing through limestone bedrock or geothermal areas will appear depleted in ^{14}C (Wood et al. 2013: 163). When such groundwater is absorbed by plants or consumed by living creatures, radiocarbon measurements of their seeds or bones provide dates that are older than true dates (Rapp and Hill 2006: 149–150).

It is troubling to note that the reports of radiocarbon laboratories seldom if ever refer to the possibility of reservoir effects affecting the dates they provide. (Reports from laboratories similarly fail generally to note the inherent problems in comparing measurements of seeds with growing seasons of weeks or months with the 5- or 10-year calibration curve segments taken from long-lived trees, the possible effects of the 11-year sunspot cycle, potential seasonal/regional offsets, comparative pre-measurement treatment regimes, or inter-lab differences in calibration programs.) The paper by Dee et al. in this volume is typical of the silence concerning these critical questions. With respect to the Theran Volcanic Destruction Level seed measurements in particular, nothing was said in the *Science* paper cited about the likelihood that the gross differences in radiocarbon measurements (up to 350 ^{14}C years even without the addition of error ranges) were due to differences in the locations of the fields where the seeds were collected in relation to the areas of gas emissions.

Christopher Bronk Ramsey, the Director of the Oxford Radiocarbon Accelerator Unit, has kindly responded to my inquiries by stating that there are two potential methods of addressing the problem of reservoir effects within the OxCal program. The first is to insert what the investigator—for example, the archaeologist submitting the samples—regards as a minimum acceptable error range, for a single sample, group of samples, or the submitted material as a whole, based on the excavator's understanding of the nature of the sample database. In practice this is never done, given the intrinsic unquantifiability of such a minimum and the fact that archaeologists seldom understand the complex geophysical and statistical issues involved. The second method is to introduce an asymmetric adjustment to the error range (e.g., +20/−200) to reflect the excavator's understanding of the potential reservoir effects at the site. Such putative effects are impossible to quantify and few excavators would feel qualified to estimate them. In short, the radiocarbon laboratories believe that allowing for reservoir effects is the job of the excavator submitting the samples, while excavators assume that all matters relevant to radiocarbon dating have been considered by the laboratory engaged to provide radiocarbon dates and that standard canons of statistical inference will not have been violated.

Lastly, it is important to note that the term "probability" has a very different meaning in statistics than the meaning in general discourse. In radiocarbon discourse, "probability" (as in 68 % one-sigma probability or 95 % two-sigma probability) refers to the likelihood that the measurements (whatever their uncertainties) intersect the calibration curve (with its inherent problems) in a particular area, whereas in normal discourse the term "probability" implies that all contrary information, sources of uncertainty, and areas of insufficient knowledge have been considered (see Wiener 2010: 373 and n. 58). The aphorism attributed to the economist Aaron Levenstein seems apposite: "Statistics are like a bikini.... What they reveal is suggestive, but what they conceal is vital" (Lyons 1951).

Conclusion

Let us return to the Exodus. A tsunami or storms caused by the eruption of Thera have been proposed as causing the natural phenomena described in biblical accounts of the Exodus which are further said to retain, inter alia, a distant recollection of the expulsion of the Semitic Hyksos from Egypt and/or an escape from a Hyksos ruler/pharaoh. Separately, storms caused by the Theran eruption have been proposed as the source of the destruction described in the Ahmose Tempest Stele. Accordingly, it is relevant to state that however doubtful the proposed connections may appear on other grounds (see, e.g., papers by M. Harris, S. Ward, and A. Salamon in this volume; Wiener and Allen 1998), the assertion that the Theran eruption must be separated in time from these events by virtue of radiocarbon measurements is without scientific or statistical validity.

References

Aitken, M.J. 1988. The Minoan Eruption of Thera, Santorini: A Reassessment of the Radiocarbon Dates. In *New Aspects of Archaeological Science in Greece*, ed. R.E. Jones and H.W. Catling, 19–24. Athens: British School at Athens.

Åström, P. 1972. *The Swedish Cyprus Expedition*, Vol. 4, Part 1D: *The Late Cypriote Bronze Age: Relative and Absolute Chronology, Foreign Relations, Historical Conclusions*. Stockholm: The Swedish Cyprus Expedition.

———. 2001. The Relative and Absolute Chronology of Proto White Slip Ware. In *The White Slip Ware of Late Bronze Age Cyprus. Proceedings of an International Conference Organized by the Anastasios G. Leventis Foundation, Nicosia, in Honour of Malcolm Wiener, 29–30 October 1998*, ed. V. Karageorghis, 49–50. Vienna: Österreichische Akademie der Wissenschaften.

Barjamovic, G., T. Hertel, and M. Trolle Larsen. 2012. *Ups and Downs at Kanesh: Chronology, History and Society in the Old Assyrian Period*, Publications de l'Institut Historique et Archéologique Néerlandais à Stamboul, vol. 117. Leiden: Nederlands Instituut voor het Nabije Oosten.

Ben-Tor, D. 2010. Sequences and Chronology of Second Intermediate Period Royal-Name Scarabs, Based on Excavated Series from Egypt and the Levant. In *The Second Intermediate Period (Thirteenth -Seventeenth Dynasties): Current Research, Future Prospects*, Orientalia Lovaniensia Analecta, vol. 192, ed. M. Marée, 91–108. Leuven: Uitgeverij Peeters and Department of Oosterse Studies.

Bichler, M. In press. Thera Ashes. In *The Synchronisation of Civilisations in the Eastern Mediterranean in the Second Millennium B.C. IV. Proceedings of the SCIEM 2000 Closing Conference, Vienna, 26–28 January 2011*.

Bietak, M. 2001. Hyksos. In *The Oxford Encyclopedia of Ancient Egypt*, Vol. 2, ed. D.B. Redford, 136–143. Oxford: Oxford University.

———. 2004. Review of *A Test of Time*, by S.W. Manning. *Bibliotheca Orientalis* 61: 199–222.

———. 2010. A Palace of the Hyksos Khayan at Avaris. In *6 ICAANE. Proceedings of the 6th International Congress of the Archaeology of the Ancient Near East, 5 May–10 May 2009, "Sapienza", Università di Roma*, Vol. 2: *Excavations, Surveys and Restorations: Reports on Recent Field Archaeology in the Near*

East, eds. P. Matthiae, F. Pinnock, L. Nigro, and N. Marchetti, 99–109. Wiesbaden: Harrasowitz

———. 2013. Antagonisms in Historical and Radiocarbon Chronology. In *Radiocarbon and the Chronologies of Ancient Egypt*, ed. A.J. Shortland and C. Bronk Ramsey, 76–109. Oxford: Oxbow.

Bietak, M., and F. Höflmayer. 2007. Introduction: High and Low Chronology. In *The Synchronisation of Civilizations in the Eastern Mediterranean in the Second Millenium B.C. III. Proceedings of the SCIEM 2000 – 2nd EuroConference, Vienna, 28 May–1 June 2003*, ed. M. Bietak and E. Czerny, 13–23. Vienna: Österreichische Akademie der Wissenschaften.

Blitzer, H. In press. Olive Domestication and Cultivation in the Aegean. *Hesperia*.

Boaretto, E. 2007. Determining the Chronology of an Archaeological Site Using Radiocarbon: Minimizing Uncertainty. *Israel Journal of Earth Science* 56(2–4): 207–216.

———. 2009. Dating Material in Good Archaeological Contexts: The Next Challenge for Radiocarbon. *Radiocarbon* 51: 275–281.

Boaretto, E., A.J.T. Jull, A. Gilboa, and I. Sharon. 2005. Dating the Iron Age I/II Transition in Israel: First Intercomparison Results. *Radiocarbon* 47: 39–55.

Bourriau, J. 2000. The Second Intermediate Period (*c.* 1650–1550 BC). In *The Oxford History of Ancient Egypt*, ed. I. Shaw, 185–217. Oxford: Oxford University.

Bronk Ramsey, C., M.W. Dee, J.M. Rowland, T.F.G. Higham, S.A. Harris, F. Brock, A. Quiles, E.M. Wild, E.S. Marcus, and A.J. Shortland. 2010. Radiocarbon-Based Chronology for Dynastic Egypt. *Science* 328: 1554–1557.

Bronk Ramsey, C., T. Higham, and P. Leach. 2004. Towards High-Precision AMS: Progress and Limitations. *Radiocarbon* 46: 17–24.

Cardellini, C., G. Chiodini, F. Frondini, S. Giaquinto, S. Caliro, and F. Parello. 2003. Input of Deeply Derived Carbon Dioxide in Southern Apennine Regional Aquifers (Italy). *Geophysical Research Abstracts* 5. www.cosis.net/abstracts/EGS02/05731/EGS02-A-05731.pdf

Carey, S., P. Nomikou, K. Croff Bell, M. Lilley, J. Lupton, C. Roman, E. Stathopoulou, K. Bejelou, and R. Ballard. 2013. CO_2 Degassing from Hydrothermal Vents at Kolumbo Submarine Volcano, Greece, and the Accumulation of Acidic Crater Water. *Geology* 41: 1035–1038.

Chatters, R.M., J.W. Crosby III, and L.G. Engstrand. 1969. Fumarole Gaseous Emanations: Their Influence on Carbon-14 Dates, Washington State University *Technical Extension Service Circular*, vol. 32. Pullman, WA: Washington State University.

Cherubini, P., T. Humbel, H. Beeckman, H. Gärtner, D. Mannes, C. Pearson, W. Schoch, R. Tognetti, and S. Lev-Yadun. 2013. Olive Tree-Ring Problematic Dating: A Comparative Analysis on Santorini (Greece). *PLoS One* 8: e54730.

Chiodini, G., C. Cardellini, A. Amato, E. Boschi, S. Caliro, F. Frondini, and G. Ventura. 2004. Carbon Dioxide Earth Degassing and Seismogenesis in Central and Southern Italy. *Geophysical Research Letters* 31: L07615.

Chiodini, G., F. Frondini, D.M. Kerrick, J. Rogie, F. Parello, L. Peruzzi, and A.R. Zanzari. 1999. Quantification of Deep CO2 Fluxes from Central Italy. Examples of Carbon Balance for Regional Aquifers and of Soil Diffuse Degassing. *Chemical Geology* 159: 205–222.

Davis, E.N. 1990. A Storm in Egypt during the reign of Ahmose. In *Thera and the Aegean World III*, Vol. 3: *Chronology*, ed. D.A. Hardy and A.C. Renfrew, 232–235. London: The Thera Foundation.

Dee, M. 2010. Investigating the Accuracy of Radiocarbon Dating in Egypt: Checks with Samples of Known Age. Paper presented at Radiocarbon Dating & Egyptian Chronology Symposium, Oxford, England, 17–18 March 2010.

Dever, W.G. 1991. Tell el-Dab‘a and Levantine Middle Bronze Age Chronology: A Rejoinder to Manfred Bietak. *Bulletin of the American School of Oriental Research* 281: 73–79.

———. 1992a. The Chronology of Syria-Palestine in the Second Millennium B.C.E.: A Review of Current Issues. *Bulletin of the American School of Oriental Research* 288: 1–25.

———. 1992b. The Chronology of Syria-Palestine in the Second Millennium B.C. *Ägypten und Levante* 3: 39–51.

Eriksson, K.O. 1992. Late Cypriote I and Thera: Relative Chronology in the Eastern Mediterranean. In *Acta Cypria: Acts of an International Congress on Cypriote Archaeology Held in Göteborg on 22–24 August, 1991*, Part 3, ed. P. Åström, 152–223. Jonsered: Åström.

———. 2001. Cypriote Proto White Slip and White Slip I: Chronological Beacons on Relations between Late Cypriote I Cyprus and Contemporary Societies of the Eastern Mediterranean. In *The White Slip Ware of Late Bronze Age Cyprus. Proceedings of an International Conference Organized by the Anastasios G. Leventis Foundation, Nicosia, in Honour of Malcolm Wiener, 29–30 October 1998*, ed. V. Karageorghis, 51–64. Vienna: Österreichische Akademie der Wissenschaften.

———. 2007. *The Creative Independence of Late Bronze Age Cyprus. An Account of the Archaeological Importance of White Slip Ware*. Vienna: Österreichische Akademie der Wissenschaften.

Evans, A.J. 1921. *The Palace of Minos: A Comparative Account of the Successive Stages of the Early Cretan Civilization as Illustrated by the Discoveries at Knossos*, Vol. 1. London: Macmillan.

Evans, W.C., D. Bergfeld, J.P. McGeehin, J.C. King, and H. Heasler. 2010. Tree-Ring ^{14}C Links Seismic Swarm to CO_2 Spike at Yellowstone, USA. *Geology* 38: 1075–1078.

Fischer, P.M. 2009. The Chronology of Tell el-'Ajjul, Gaza: Stratigraphy, Thera, Pumice and Radiocarbon Dating. In *Time's Up! Dating the Minoan Eruption of Santorini. Acts of the Minoan Eruption Chronology Workshop, Sandbjerg, November 2007*, Monographs of the Danish Institute at Athens, vol. 10, ed. D.A. Warburton, 253–265. Aarhus: Aarhus University.

Forstner Müller, I., P. Rose, C. Reali, and H. Tronchère. 2012. Preliminary Report on the Season 2012 at Tell el-Dab'a/Sharqeya. www.auaris.at/downloads/Report_SCA_engl_arabic_small.pdf

Foster, K.P., and R.K. Ritner. 1996. Texts, Storms, and the Thera Eruption. *Journal of Near Eastern Studies* 55: 1–14.

Franke, D. 1988. Zur Chronologie des Mittleren Reiches. Teil II: Die sogenannte 'Zweite Zwischenzeit' Altägyptens. *Orientalia* 57: 245–274.

Frezzotti, M.L., A. Peccerillo, and G. Panza. 2009. Carbonate Metasomatism and CO_2 Lithosphere–asthenosphere Degassing beneath the Western Mediterranean: An Integrated Model Arising from Petrological and Geophysical Data. *Chemical Geology* 262: 108–120.

Friedrich, W.L., B. Kromer, M. Friedrich, J. Heinemeier, T. Pfeiffer, and S. Talamo. 2006. Santorini Eruption Radiocarbon Dated to 1627–1600 B.C. *Science* 312: 548.

———. 2009. Santorini Eruption Radiocarbon Dated to 1600 B.C.: Further Discussion. In *Tree-Rings, Kings, and Old World Archaeology and Environment: Papers Presented in Honor of Peter Ian Kuniholm*, ed. S.W. Manning and M.J. Bruce, 293–298. Oxford: Oxbow.

Gambardella, B., C. Cardellini, G. Chiodini, F. Frondini, L. Marini, G. Ottonello, and M.V. Zuccolini. 2004. Fluxes of Deep CO2 in Volcanic Areas of Central-Southern Italy. *Journal of Volcanology and Geothermal Research* 136: 31–52.

Gasche, H., J.A. Armstrong, S.W. Cole, and V.G. Gurzadyan. 1998. *Dating the Fall of Babylon: A Reappraisal of Second-Millennium Chronology*. Ghent: University of Ghent.

Goedicke, H. 1992. The Chronology of the Thera/Santorin Explosion. In *High, Middle or Low? Acts of the Second International Colloquium on Absolute Chronology, Schloß Haindorf/Langlois, 12–15 August 1990. Ägypten und Levante* 3: 57–62. Vienna: Austrian Academy of Sciences Press.

———. 2004. *The Speos Artemidos Inscription of Hatshepsut and Related Discussions*. Oakville, CT: Halgo.

Housley, R.A., S.W. Manning, G. Cadogan, R.E. Jones, and R.E.M. Hedges. 1999. Radiocarbon, Calibration and the Chronology of the Late Minoan IB phase. *Journal of Archaeological Science* 26: 159–171.

Imamura, M., H. Ozaki, T. Mitsutani, E. Niu, and S. Itoh. 2007. Radiocarbon Wiggle-Matching of Japanese Historical Materials with a Possible Systematic Age Offset. *Radiocarbon* 49: 331–337.

Keenan, D.J. 2002. Why Early-Historical Radiocarbon Dates Downwind from the Mediterranean Are Too Early. *Radiocarbon* 44: 225–237.

———. 2004. Radiocarbon Dates from Iron Age Gordion Are Confounded. *Ancient West & East* 3: 100–103.

———. 2012. Calibration of a Radiocarbon Age. *Nonlinear Processes in Geophysics* 19: 345–350.

Levin, I., R. Bösinger, G. Bonani, R.J. Francey, B. Kromer, K.O. Münnich, M. Suter, N.B.A. Trivett, and W. Wölfli. 1992. Radiocarbon in Atmospheric Carbon Dioxide and Methane: Global Distribution and Trends. In *Radiocarbon after Four Decades: An Interdisciplinary Perspective*, ed. R.E. Taylor, A. Long, and R.S. Kra, 503–518. New York, NY: Springer.

Levin, I., and V. Hesshaimer. 2000. Radiocarbon—A Unique Tracer of Global Carbon Cycle Dynamics. *Radiocarbon* 42: 69–80.

Lyons, L. 1951. *Evening Standard* (Uniontown, PA), 10 November, p. 4, col. 6.

Manning, S.W. 1999. *A Test of Time*. Oxford: Oxbow.

Manning, S.W., C. Bronk Ramsey, W. Kutschera, T. Higham, B. Kromer, P. Steier, and E.M. Wild. 2006a. Chronology for the Aegean Late Bronze Age 1700–1400 B.C. *Science* 312: 565–569.

———. 2006b. Supporting Online Material for Chronology for the Aegean Late Bronze Age 1700–1400 B.C. *Science* 312: 565. www.sciencemag.org/cgi/content/full/312/5773/565/DC1

———. 2009. Dating the Santorini/Thera Eruption by Radiocarbon: Further Discussion (AD 2006–2007). In *Tree-Rings, Kings and Old World Archaeology and Environment: Papers Presented in Honor of Peter Ian Kuniholm*, ed. S.W. Manning and M.J. Bruce, 299–316. Oxford: Oxbow.

Marzaioli, F., C. Lubritto, G. Battipaglia, I. Passariello, M. Rubino, D. Rogalla, S. Strumina, F. Miglietta, A. D'Onofrio, M.F. Cotrufo, and F. Terrasi. 2005. Reconstruction of past CO_2 Concentration at a Natural CO_2 Vent Site Using Radiocarbon Dating of Tree Rings. *Radiocarbon* 47: 257–263.

McCoy, F.W., and G. Heiken. 2000. The Late-Bronze Age Explosive Eruption of Thera (Santorini), Greece: Regional and Local Effects. In *Volcanic Hazards and Disasters in Human Antiquity*, Geological Society of America Special Paper, vol. 345, ed. F.W. McCoy and G. Heiken, 43–70. Boulder: Geological Society of America.

Minissale, A., G. Magro, O. Vaselli, C. Verrucchi, and I. Perticone. 1997. Geochemistry of Water and Gas Discharges from the Mt. Amiata Silicic Complex and Surrounding Areas (Central Italy). *Journal of Volcanology and Geothermal Research* 79: 223–251.

Moeller, N., G. Marouard, and N. Ayers. 2011. Discussion of Late Middle Kingdom and Early Second Intermediate Period History and Chronology in Relation to the Khayan Sealings from Tell Edfu. *Ägypten und Levante* 21: 87–121.

Ozaki, H., M. Imamura, H. Matsuzaki, and T. Mitsutani. 2007. Radiocarbon in 9th to 5th century B.C. Tree-Ring Samples from the Ouban 1 Archaeological Site, Hiroshima, Japan. *Radiocarbon* 49: 473–479.

Ozaki, H., M. Sakamoto, M. Imamura, H. Matsuzaki, T. Nakamura, K. Kobayashi, S. Itoh, E. Niu, and T. Mitsutani. 2009. Radiocarbon Dates of Japanese Tree Rings for 1060 BC–AD 400. Paper presented at the 20th International Radiocarbon Conference, Big Island, Hawaii, 31 May–5 June.

Pfälzner, P. 2004. Minoan Style in Inland Syria: The Evidence from the Palace at Qatna. Paper presented at the New York Aegean Bronze Age Colloquium, 28 May 2004.

Rapp, G.R., and C.L. Hill. 2006. *Geoarchaeology: The Earth Science Approach to Archaeological Interpretation*. New Haven, CT: Yale University.

Reimer, P.J. 2001. A New Twist in the Radiocarbon Tale. *Science* 294: 2494–2495.

Reimer, P.J., and F.G. McCormac. 2002. Marine Radiocarbon Reservoir Corrections for the Mediterranean and Aegean Seas. *Radiocarbon* 44: 159–166.

Rogie, J.D. 1996. Lethal Italian Carbon Dioxide Springs Key to Atmospheric CO_2 Levels. *Penn State Earth and Environmental Systems Institute. News and Events: News Archives.* www.eesi.psu.edu/news_events/archives/Lethal.shtml

Rogie, J.D., D.M. Kerrick, G. Chiodini, and F. Frondini. 2000. Flux Measurements of Nonvolcanic CO_2 Emission from Some Vents in Central Italy. *Journal of Geophysical Research* 105(B4): 8435–8445.

Sakamoto, M., S. Fujio, K. Kobayashi, H. Ozaki, and M. Imamura. 2009. Precise Chronology of Prehistoric Japan by Combination of Radiocarbon Dating and Pottery Typology. Paper presented at the 20th International Radiocarbon Conference, Big Island, Hawaii, 31 May–5 June.

Salzer, M.W., and M.K. Hughes. 2007. Bristlecone Pine Tree Rings and Volcanic Eruptions over the Last 5000 Yr. *Quaternary Research* 67: 57–68.

Saupé, F., O. Strappa, R. Coppens, B. Guillet, and R. Jaegy. 1980. A Possible Source of Error in ^{14}C Dates: Volcanic Emanations (Examples from the Monte Amiata District, Provinces of Grosseto and Siena, Italy). *Radiocarbon* 22: 525–531.

Sharon, I., A. Gilboa, A.J.T. Jull, and E. Boaretto. 2007. Report on the First Stage of the Iron Age Dating Project in Israel: Supporting a Low Chronology. *Radiocarbon* 49: 1–46.

Stanley, D.J., and H. Sheng. 1986. Volcanic Shards from Santorini (Upper Minoan Ash) in the Nile Delta, Egypt. *Nature* 320: 733–735.

Steinhauser, G., J.H. Sterba, E. Oren, M. Foster, and M. Bichler. 2010. Provenancing of Archaeological Pumice Finds from North Sinai. *Naturwissenschaften* 97: 403–410.

Sterba, J.H., K.P. Foster, G. Steinhauser, and M. Bichler. 2009. New Light on Old Pumice: The Origins of Mediterranean Volcanic Material from Ancient Egypt. *Journal of Archaeological Science* 36: 1738–1744.

Stuiver, M., and T.F. Braziunas. 1993. Sun, Ocean, Climate and Atmospheric $^{14}CO_2$: An Evaluation of Causal and Spectral Relationships. *The Holocene* 3: 289–305.

Stuiver, M., P.J. Reimer, and T.F. Braziunas. 1998. Precision Radiocarbon Age Calibration for Terrestrial and Marine Samples. *Radiocarbon* 40: 1127–1151.

Sveinbjörnsdóttir, Á.E., and J. Heinemeier. 2011. The Controversy of the Settlement Time of Iceland. Paper presented at Radiocarbon and Archaeology, 6th International Symposium, Pafos, Cyprus, 10–15 April.

Tufnell, O. 1984. *Studies on Scarab Seals*, Vol. 2: *Scarab Seals and their Contribution to History in the Early Second Millennium B.C.* Warminster: Aris & Phillips.

Ward, G.K., and S.R. Wilson. 1978. Procedures for Comparing and Combining Radiocarbon Age Determinations: A Critique. *Achaeometry* 20: 19–31.

Ward, W.A. 1984. Royal-Name Scarabs. In *Scarab Seals and their Contribution to History in the Early Second Millennium B.C*, Studies on scarab seals, vol. 2, ed. O. Tufnell, 151–192. Warminster: Aris & Phillips.

Warren, P.M. 2006. The Date of the Thera Eruption in Relation to Aegean-Egyptian Interconnections and the Egyptian Historical Chronology. In *Timelines: Studies in Honour of Manfred Bietak*, Vol. 2, Orientalia Lovaniensia Analecta, vol. 149, ed. E. Czerny, I. Hein, H. Hunger, D. Melman, and A. Schwab, 305–321. Leuven: Peeters.

———. 2009. The Date of the Late Bronze Age Eruption of Santorini. In *Time's Up! Dating the Minoan eruption of Santorini. Acts of the Minoan Eruption Chronology Workshop, Sandbjerg, November 2007*, Monographs of the Danish Institute at Athens, vol. 10, ed. D.A. Warburton, 181–186. Aarhus: Aarhus University.

———. 2010. The Absolute Chronology of the Aegean circa 2000 B.C.–1400 B.C. A Summary. In *Die Bedeutung der minoischen und mykenischen Glyptik: VI. Internationales Siegel-Symposium aus Anlass des 50 jährigen Bestehens des CMS, Marburg, 9.–12. Oktober 2008*, CMS Beiheft 8, ed. W. Müller, 383–394. Mainz am Rhein: Philipp von Zabern.

Warren, P.M., and V. Hankey. 1989. *Aegean Bronze Age Chronology*. Bristol: Bristol Classical.

Wiener, M.H. 2001. The White Slip I of Tell el-Dab'a and Thera: Critical Challenge for the Aegean Long Chronology. In *The White Slip Ware of Late Bronze Age Cyprus. Proceedings of an International Conference Organized by the Anastasios G. Leventis Foundation, Nicosia, in Honour of Malcolm Wiener, 29–30 October 1998*, ed. V. Karageorghis, 195–244. Vienna: Österreichische Akademie der Wissenschaften.

———. 2003. Time Out: The Current Impasse in Bronze Age Archaeological Dating. In *Metron: Measuring the Aegean Bronze Age. Proceedings of the 9th International Aegean Conference, New Haven, Yale University, 18–21 April 2002*, Aegaeum, vol. 24, ed. K.P. Foster and R. Laffineur, 363–399. Liège: Universitè de Liège Histoire de l'art et archéologie de la Grèce antique; Austin: University of Texas at Austin Program in Aegean Scripts and Prehistory.

———. 2006a. Chronology Going Forward (With a Query about 1525/4 B.C.). In *Timelines: Studies in Honour of Manfred Bietak*, Vol. 2, Orientalia Lovaniensia Analecta, vol. 149, ed. E. Czerny, I. Hein, H. Hunger, D. Melman, and A. Schwab, 317–328. Leuven: Peeters.

———. 2006b. Egypt & Time. *Ägypten und Levante* 16: 325–339.

———. 2007. Times Change: The Current State of the Debate in Old World Archaeology. In *The Synchronisation of Civilizations in the Eastern Mediterranean in the Second Millenium B.C. III. Proceedings of the SCIEM 2000 – 2nd Euro-Conference, Vienna, 28 May–1 June 2003*, ed. M. Bietak and E. Czerny, 25–47. Vienna: Österreichische Akademie der Wissenschaften.

———. 2009a. Cold Fusion: The Uneasy Alliance of History and Science. In *Tree-Rings, Kings, and Old World Archaeology and Environment: Papers Presented in Honor of Peter Ian Kuniholm*, ed. S.W. Manning and M.J. Bruce, 277–292. Oxford: Oxbow.

———. 2009b. Thera Discussion. In *Tree-Rings, Kings, and Old World Archaeology and Environment: Papers Presented in Honor of Peter Ian Kuniholm*, ed. S.W. Manning and M.J. Bruce, 317–332. Oxford: Oxbow.

———. 2010. A Point in Time. In *Cretan Offerings: Studies in Honour of Peter Warren*, Annual of the British School at Athens Supplement, vol. 18, ed. O. Krzyszkowska, 367–394. London: British School at Athens.

Wiener, M.H., and J.P. Allen. 1998. Separate Lives: The Ahmose Tempest Stele and the Thera Eruption. *Journal of Near Eastern Studies* 57: 1–28.

Wood, R.E., T.F.G. Higham, A. Buzilhova, A. Suvorov, J. Heinemeier, and J. Olsen. 2013. Freshwater Radiocarbon Reservoir Effects at the Burial Ground of Minino, Northwest Russia. *Radiocarbon* 55: 163–177.

Part III

Cyber-Archaeology and Exodus

The Sound of Exodus (EX3), "World Building" and the Museum of the Future: Adaptive, Listener-Centered Auditory Display for Multimodal Narrative in Advanced Visualization Spaces

11

Zachary Seldess, Peter Otto, Eric Hamdan, and Thomas E. Levy

Abstract

In the transdisciplinary world of cyber-archaeology, the "EX3 – Exodus, Cyber-archaeology and the Future" (EX3) exhibition offered a unique opportunity for researchers from many academic disciplines to work together in a collaborative space to help create "the museum of the future." In creating a narrative for the exhibition, the team happened to independently develop the same method of "World Building" advocated by the 5D Institute (http://5dinstitute.org/events/science-of-fiction) in order to create the world of the Exodus. One of the most innovative contributions to this effort comes from the Sonic Arts. In this chapter, a new approach to enriching the experience of cyber-archaeology and other advanced museum and visualization efforts is discussed. Using advanced, adaptive, listener-centered audio systems in development at UCSD, the authors experimented with computer-audio digital signal processing and acoustical design strategies to provide a well-controlled listening environment that complemented the advanced display technologies of EX3, with the goal of improving the overall audience experience of the show. Several supporting technologies and systems were integrated to control and deliver audio in close synchronization with visual content. Primary hardware and software components included custom control and signal processing software, unique content creation rendered by highly directional small format speaker arrays for optimal auditory intelligibility, and human image tracking that drives the auditory content rendering. Acoustical engineering was also employed in service of an extremely media-rich environment comfortably serving visiting audiences of up to 100 people, concentrated in a relatively small physical space. Specific techniques and design/engineering challenges are discussed; prospects for future work are noted.

Z. Seldess (✉)
Qualcomm Institute/California Institute of Telecommunications and Information Technology (Calit2), University of California, San Diego (UCSD)
La Jolla, CA, USA
e-mail: zseldess@eng.ucsd.edu

T.E. Levy et al. (eds.), *Israel's Exodus in Transdisciplinary Perspective*,
Quantitative Methods in the Humanities and Social Sciences, DOI 10.1007/978-3-319-04768-3_11,

Introduction

When the opportunity to host the Exodus conference (http://exodus.calit2.net) at UC San Diego's Qualcomm Institute (QI) arose, T.E. Levy suggested that the occasion could serve as a transdisciplinary experiment where many of the information technology methods practiced at the QI could be marshaled together to help tell the story of the Exodus in light of twenty-first century research. In planning the EX3 exhibition, we brought together (both physically and virtually) over 40 Biblical scholars, archaeologists, Egyptologists, computer scientists, engineers, ancient literature specialists, communications specialists, graphic designers, and geo-scientists to participate in a "World Building" process to help tell the narrative of the ancient Hebrews' Exodus from Egypt in a way that was meaningful for twenty-first century viewers, in particular university undergraduate students. The driving force for the narrative was an ecumenical anthropological archaeology view of the Israelite cultural system (Fig. 11.1) made up of five subsystems (social organization, subsistence, technology, economy/exchange, belief/religion) that made up the cultural fabric of the Israelites and how they interacted with their environment during the late second millennium BCE. For the EX3 Exodus narrative, the primary stories we wanted to tell concerned (1) the cyber-archaeology methodologies that field archaeologists can employ today to gather objective scientific data to engage in scientific story telling (Levy 2013); and (2) how geo-scientists deal with the numerous environmental processes reflected in Book of Exodus in the Hebrew Bible such as the famous "Parting of Sea" narrative (Exodus 14:16–29) when the waters of the Reed Sea open, and the Israelites miraculously escape from Egyptian military forces (see Salamon et al.; Wiener; Dee et al.; Moshier and Hoffmeier; Harris, this volume, Chaps. 6–10). The methodology we employed to construct a twenty-first century narrative of the Late Bronze/early Iron Age (ca. sixteenth to thirteenth century BCE, see Geraty, this volume, Chap. 4) Exodus story relied on production designer Alex McDowell's model of World Building for science fiction narrative in the cinematic arts industry (http://cinema.usc.edu/news/article.cfm?id=13503; http://5dinstitute.org/events/science-of-fiction). A wide range of professional researchers and graduate students were "locked" in a room once a week for several months at the QI to bring the insights of ancient literatures, archaeology, geo-science, engineering, computer science, and other fields together to create what McDowell calls "A thread of logic [that] evolves in a coherent and comprehensive way. . . offering new workflows, platforms, and paradigms. . ."

With only about 3 months of planning time for the EX3 exhibition, our use of the "world building" model to establish a twenty-first century narrative for EX3 was at a preliminary stage. McDowell's team has applied world building for low budget Hollywood films that cost around $45 million (Fig. 11.2; http://www.cinegrid.org) and high budget films; our EX3 budget was $45 K. We were able to develop a state-of-the-art exhibition because over $400 K worth of high definition 2D and 3D hardware, along with a multipurpose Information Technology room (Fig. 11.3) was made available for the project by the QI, along with volunteered time by faculty and graduate students. In this chapter, we use the Sonic Arts contribution to EX3 to highlight how by using the world building methodology it is possible to move beyond twentieth century linear narrative construction for exhibitions to a more holistic approach. One of the unique aspects of EX3 was reliance on a single artifact—a "standing stone" or "Matzeva" (Hebrew). As nomadism underlies the Exodus narrative and the *Matzeva* used in EX3 came from a cemetery that belonged to a nomadic population linked to the Biblical world that surrounds the Exodus narrative, this single object could serve as a "touchstone" for entering the ancient world of the Exodus. This enabled the EX3 design team to take full advantage of the state-of-the-art scientific visualization assets at the QI to create a new kind of narrative of the past.

We are multimodal creatures that use all of our senses to understand our surroundings.

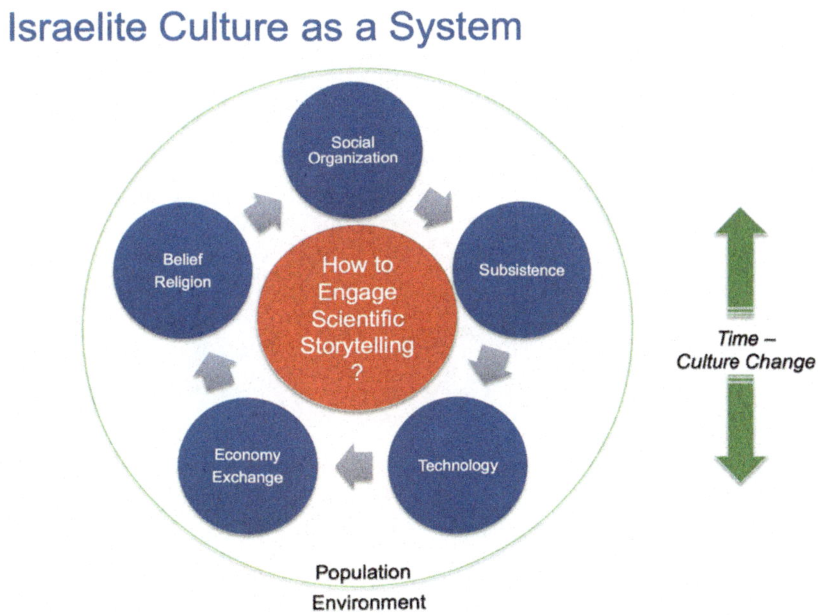

Fig. 11.1 Israelite culture as a system that interacts with its environment and changes through time

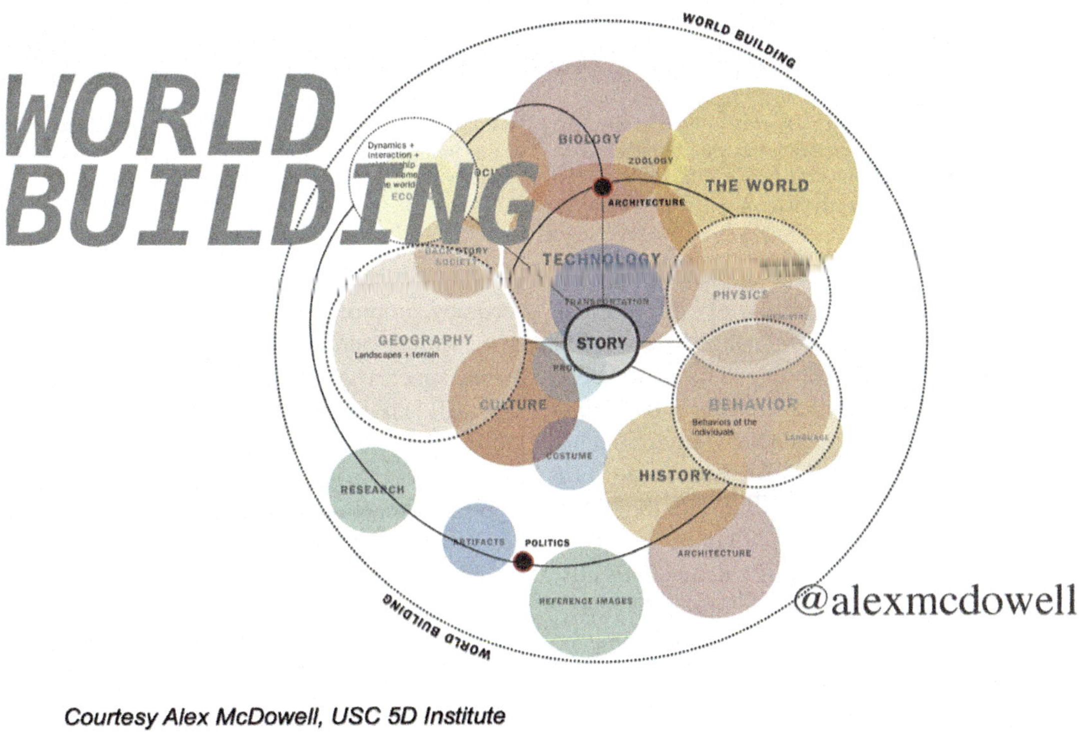

Fig. 11.2 McDowell's World Building model to create narrative for cinematic arts (http://5dinstitute.org/events/science-of-fiction; image courtesy Alex McDowell)

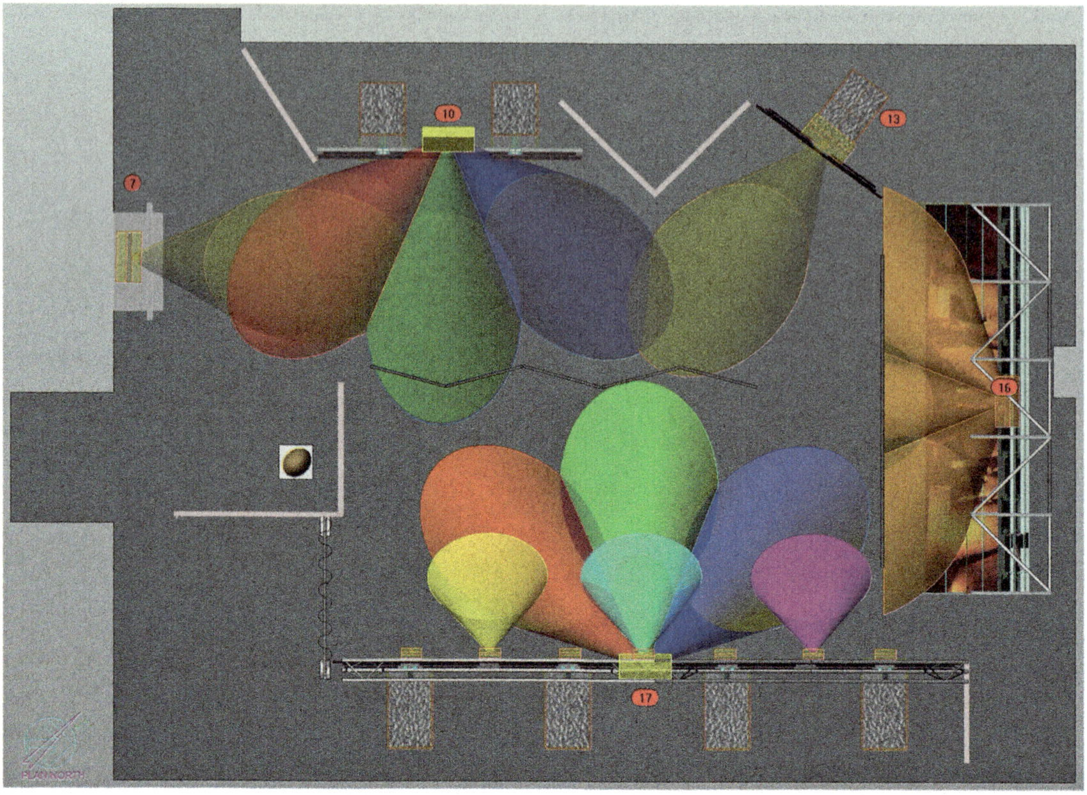

Fig. 11.3 (Diagram by Greg Dawe) Bird's-eye view of the Exodus EX3 event, illustrating the use of loudspeaker array beamforming throughout the space (at zones *7*, *10*, *13*, *16*, *17*). Traditional distributed loudspeaker rendering was also used on the large 8 × 4 tiled-display wall (zone *17*), as indicated by the smaller *yellow*, *cyan*, and *magenta* cones

Together the senses combine to provide us with continuous data-rich streams of information about our immediate environment and body. Of all the senses, hearing is unique in its remarkable ability to inform us, within milliseconds, about many aspects of our immediate environment, particularly those parts that we cannot see. For example, many of us are familiar with the experience of walking into a room full of people, such as a café, and by listening alone faithfully estimating the relative size of the space, its material properties, the presence of inanimate objects (e.g., the sound of chairs moving, the barista scooping ice), and even perhaps how many people are present by counting the number of voices heard and noticing their aggregate volume. In addition to using our hearing to inform us about the contents of our surroundings, we can localize individual sound sources with great speed and accuracy, improving our overall spatial awareness, search, and wayfinding abilities within a media space (Lokki and Grohn 2005; Seldess et al. 2011). Furthermore, the fine temporal resolution of our hearing allows us to perceive small fluctuations of sounds over time, both in amplitude and spectra, allowing for perception of detailed patterns and pitches. The refined temporal resolution of our hearing allows us to perceive notes and rhythms in music, differentiate the subtle timbral and temporal characteristics of various sound sources (such as the same note played on a piano and a cello, or two people singing the same melody), even allowing us to identify states within complex multidimensional dynamic systems (Hermann and Ritter 1999), such as the readiness of brewing coffee as the percolator begins to bubble and pop in a distinctly recognizable pattern.[1]

Augmenting modern high-resolution visual display environments with advanced sound rendering technology provides a powerful way for cyber-archaeologists to transform the processes and outcomes of their research into captivating storytelling experiences for specialists and general audiences alike. For example, with the appropriate hardware and training, cyber-archaeologists can capture and later reproduce the sounds heard at dig sites in order to share the original auditory environment with others, or record on-site impulse responses to be used in reconstructing the reverberant characteristics of the space, simulating the acoustic properties of the original site via off-site audio-visual display environments. Using these and other approaches, in concert with modern data visualization and virtual reality environments (Defanti et al. 2011), cyber-archaeologists can vividly recreate the multimodal experience of being at the site, perhaps even providing insights into the original uses of ancient spaces that exist today only in ruins. These audio experiences can also help model environmental effects that impacted ancient people in different times and places.

For the EX3 exhibition, in addition to traditional graphic poster display presentations and the display of a single physical artifact, as is typical within museum exhibits, five high-resolution visual display "exhibits" were presented, each providing a large amount of Exodus-related information in the form of high-resolution still images, videos, synthetic animations, 3D models, and 3D immersive virtual reality environments, rendered both in isolation and collage formats. Considering the impressive scale and density of the data presented throughout the EX3 exhibition (only 10 days in length), audio provided a much needed vehicle for leveraging additional modalities, widening the potential perceptual bandwidth for data reception (Hermann et al. 2011),[2] while at the same time not diminishing visitors' abilities to fully engage with the visual data. Our use of audio therefore served the diverse purposes of providing a greater sense of context and immersion through the more abstract domains of sound design, environmental soundscape, and musical accompaniment, as well as greatly enriching non-abstract data reception via voice-over narratives about the visual content, rendered together or in isolation as deemed appropriate for each visual display environment. The augmentation of the visual with the aural allowed us to combine the cinematic and immersive experience of virtual reality and film display environments with the traditional information-rich experience of a docent-led museum encounter.

Audio for EX3: In Context

Certain aspects of the EX3 exhibit posed significant impediments for designing and implementing a coherent and effective sonic counterpart to the rendered visual display. Most notably, we needed to find effective methods to overcome the challenges presented by the highly reflective acoustical properties of the display materials within the space (including poster-board and flat-panel graphic displays), as well as the dense and complex spatial relationships of "streams" of audio and video data within and between exhibits in the event.

Figures 11.3 and 11.4 provide a bird's-eye view of the EX3 layout as well as a front view representation of the content layout for one of the

[1] For a compelling real-world example of the use of audio as a way to identify subtle, and most often invisible, states within a complex system (in this case automotive mechanical issues), see the fascinating database of sound/symptom relationships available on the "Car Noise Emporium" page of NPR Car Talk's website (Magliozzi and Magliozzi 1999).

[2] *The Sonification Handbook*, edited by Thomas Hermann et al., currently provides the single most comprehensive overview of the state of the art in the fields of Auditory Display and Sonification. This book is a must read for anyone interested in past and present work using audio within the context of multimodal data analysis and exploration.

Fig. 11.4 (Photograph by Tom Defanti) The 4 × 2 tiled-display wall at EX3. Using custom depth camera tracking and speaker array beamforming, visitors standing in colored regions in front of the visual display received highly directional audio streams along with visual content at spatially correlated zones on the wall

large tiled-display environments featured at the event. As shown in Fig. 11.3, we positioned acoustically absorptive panels at key locations throughout the exhibit. These barriers, in addition to physically and visually compartmentalizing the various content areas within the exhibit,[3] also served the important role of sonically "deadening" the space by reducing the amount of globally diffuse sound propagation. However, in addition to traditional acoustic treatment of the space, given the suboptimal conditions (as is typical in the vast majority of real-world implementations), in order to achieve an acceptably intelligible and compelling audio experience we required creative and unconventional audio hardware and software solutions. We therefore chose to utilize alternative audio rendering strategies to properly address the issue of ambient noisiness and sonic crosstalk. These techniques included a novel optimized beamforming algorithm via small form-factor linear loudspeaker arrays (see section "Speaker Array Beamforming: Sonic Spotlights on the Listener"), as well as panning and audio decorrelation solutions for rendering across large areas of spatially distributed loudspeakers. Careful hardware placement and software design, further refined by use of custom depth camera tracking software (see section "Depth Camera Tracking for Real-time Audio Control"), which applied real-time control over the audio software based on user presence and location, resulted in a unique, high fidelity, and interactive audio experience throughout the event.

The presentation of information within and amongst areas of the EX3 exhibit was quite dense, much more so than a typical museum exhibit. A brief description of the presentation of content on the two display environments shown in Figs. 11.3 and 11.4 will help to frame our need for a novel approach to audio rendering:

The 4 × 2 tiled-display wall (see Figs. 11.3 (zone 10) and 11.5) presented information in three sections, each section corresponding to a sequential phase in the overall cyber-archaeology

[3] Acoustic barriers also served the important role of minimizing the amount of light bleeding between different display environments.

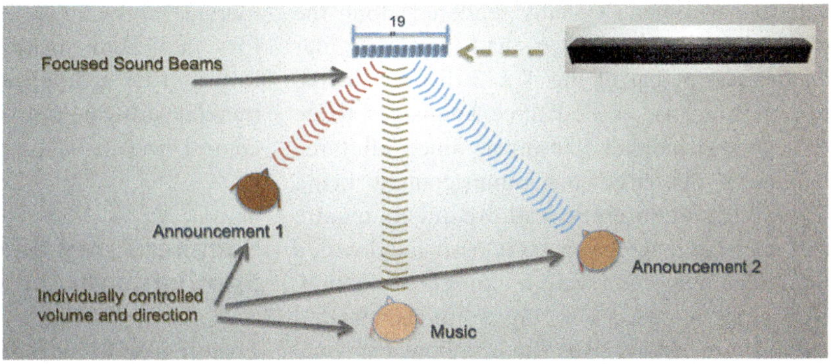

Fig. 11.5 Small speaker array emitting narrowly focused beams of sound, targeted at specific listener zones

workflow. The left side of the display wall presented data acquisition processes used in the field (laser scanning, GPS, aerial photography, ground penetrating radar, etc.). The middle section presented how the researchers processed and analyzed the data to create "layered realities," combining disparate data sets into an integrated virtual model of the entire site at various resolutions. This section contained pictures of the researchers working both in the field and back in their labs, as well as videos of the resultant virtual models, and diagrams of some of the team's data repositories and infrastructure systems. The right side focused on the research team's efforts to disseminate this information to the public, featuring video clips of animated flythroughs of the 3D site models. Visual content for the three regions looped continuously and in parallel, allowing visitors to fluidly shift focus between the different sequential steps in the cyber-archaeology process.

Audio for the 4 × 2 display wall featured voice-over narratives for each of the three sections, describing the content on visual display, and also presenting relevant information on each phase of the cyber-archaeology process that would otherwise be difficult to present visually within the physical and temporal constraints of the exhibit. Since the three areas of visual content were presented in parallel, we needed to find a way to allow three associated "streams" of audio to coexist without producing a cacophonous global sound space. Additionally, we needed to achieve a clear collocation of the visual and aural data while also making it effortless for listeners to perceptually separate audio streams, focusing their ears in on the stream associated with the relevant area of the wall.

Content on the 8 × 4 tiled-display wall (Fig. 11.3, zone 17) was divided temporally into two sections. The first section used the entire display wall to present transdisciplinary approaches towards the Exodus, highlighting various geophysical efforts to find new data to tie archaeological, historical, and textual evidence together. The second section split the wall into three areas presenting three hypotheses for natural events that may have been the basis for the "parting of the sea" narrative in the Hebrew Bible. The left area presented animated models reconstructing possible tsunamis resulting from the Thera-Santorini volcanic eruption.[4] The middle section presented animated models reconstructing [illegible] tsunamis that would have resulted from a submarine avalanche off the coast of Egypt. The right section presented reconstructions of tidal zones that could have been caused by large storm systems.

Audio for the 8 × 4 display wall featured both verbal and nonverbal sound cues. The first half of the presentation provided a voice-over narrative explanation of the visual content simultaneously in three languages (English, Arabic, and Hebrew), fitting the content and context of the data presented. The second half of the presentation featured reconstructed sounds of the

[4] Volcanic glass/ash traced to the Thera-Santorini eruption was found in Nile Delta cores extracted by the Smithsonian. For more on this, see Salamon et al., this volume, Chap. 9.

natural phenomena visually rendered along the display wall. Here we had very similar challenges to that of the 4 × 2 display. In the first section, we needed three languages to be rendered in parallel. However, since all three languages described the same content being displayed over the entire wall, we did not require that each language be spatially associated with a particular area of the wall. In the second section, we again needed to create clear spatial associations between the three regions of visual content presented on the wall and each region's associated nonverbal aural content. However, due to the nature of the sound presented (i.e., simulated environmental sounds), and in dealing with the large physical dimensions of the display wall, we needed these sounds to wash over larger areas of the space, rather than resolve into discrete "points" in front of the wall.

In both of the above display environments, and others presented within the EX3 exhibit, language-based audio streams simulated the presence of a virtual "docent," guiding visitors through otherwise overwhelming data-rich visual displays, while also providing the possibility of simultaneous space and listener-appropriate information delivery. Nonverbal audio cues helped to contextualize the visual content, providing greater immersion and overall sensory connection to the subject matter. To implement the above audio behavior, we needed to design graphics-driven audio software capable of employing a variety of audio rendering methods based on the needs of each exhibit, which could be controlled remotely both by the visual display software[5] and other remote drivers such as computer vision tracking software. In the following section we discuss our method for producing highly directional listener-centered audio using an optimized beamforming filter approach. In sections "'EX3_Audio' Audio Rendering Software" and "Depth Camera Tracking for Real-time Audio Control," we then discuss the audio "server" software designed to handle rendering for the above-mentioned audio use-cases, followed by a description of the depth camera tracking software designed to provide hands-free control over the audio rendering software.

Speaker Array Beamforming: Sonic Spotlights on the Listener

Too often audio systems, especially those used in public "multimedia" spaces, perform poorly. Museums, airports, transportation hubs, shopping areas, restaurants, and other public spaces using audio systems to convey information or entertainment share the same problematic tendencies: you hear them well when you don't want to, and you don't hear them well enough when you do want to. The EX3 event, with its concentration on advanced visualization systems seemed to create a scenario for extremely ill-behaved multimedia audio! How could we avoid a cacophony of intrusive and unintelligible sounds, and, rather, present highly nuanced sonic environments that matched the sophistication, depth, and creativity represented in EX3's visualization technology and content?

The solution would not be found in traditional loudspeaker technologies, which tend to diffuse sound in a roughly spherical pattern in all directions around the speaker. If traditional speakers are analogous to a floodlight, we needed a speaker that performed more analogous to a spotlight. To be precise, we needed more tightly controlled speaker diffusion, coupled with active sensing, to reverse the aforementioned problematic "leakage" tendency, activating audio only when and where needed. A beamforming speaker array system, developed in our labs, offered a reasonable solution, and in fact presented new possibilities for highly integrated audio-video interactions in transparent support of the intent of EX3.

Based on science related to the physics of radar and waveform synthesis, beamforming speaker arrays use constructive and deconstructive waveform phasing to reinforce acoustic waveforms in a desired direction and cancel or

[5] For a detailed discussion of the visualization software used in EX3, please refer to Schulze et al. and Srour et al., this volume, Chaps. 12 and 13.

minimize waveforms where they are not desired. Tightly integrating small beamforming speaker arrays in three of the EX3 installation areas solved much of the problem of "acoustical leakage" from one exhibit to the next. Further engineering allowed for the use of multilingual "zones" or beamed listening areas, as was used in the case of the largest 8 × 4 tiled-display featuring Christian, Islamic, and Hebraic interpretations of the Exodus. Here automated, simultaneous translations were isolated in each appropriate physical area of the exhibit. The audio design team devoted considerable attention to scaling and localizing sound fields to exactly match the physical location of displayed information and images, particularly on the larger 4 × 2 and 8 × 4 tiled-display walls. Observers reported a strong sense of "immersion" where audio was tightly coupled with each display's content and physical space.

Algorithm for Optimized Beamforming Filters

US Patent application WO 2012/068174 A2 describes *A Signal Processing Routine for Controlling a Speaker Array to Provide Spatialized, Localized, and Binaural Virtual Surround Sound*, jointly held by UCSD and the University of Southampton (Otto, Kamdar, Yamada, and Fazi). The patent is based on an optimization method for inverse filter design, applied to small speaker arrays (Olivieri et al. 2013) to achieve the beamforming effects described in the present paper.

"EX3_Audio" Audio Rendering Software

In designing the audio rendering software for the EX3 event we needed to provide an integrated solution utilizing the various rendering strategies required to support the above use-cases, while providing a simple method for concurrent graphics and camera tracking software to send network commands controlling and synchronizing audio content with the visual display. During the EX3 event, graphics software driving content on various display environments sent simple integer "cue" IDs, for real-time triggering and synchronization of audio content, to the appropriate audio server via UDP, TCP/IP, or Open Sound Control network protocols. Software settings could be initialized prior to runtime via a simple JSON configuration file or during runtime via the software graphical user interface. In an effort to support rapid prototyping (as is often required during preparation for complex multimedia productions) as well as to future-proof our work for use in other "museum of the future" events, we designed the software to support an arbitrary number of custom "renderer" modules inserted into the signal and control flow as plug-ins. These plug-in renderer modules, designed as custom patches using Cycling 74's Max programming environment (Zicarelli et al. 2013), can range from idiosyncratic rendering solutions intended to solve for a particular use-case, to sophisticated and generalized rendering algorithms intended to work across a variety of contexts. Plug-ins created for the EX3 event included support for linear array beamforming (as discussed in section "Speaker Array Beamforming: Sonic Spotlights on the Listener"), as well as more traditional matrix-mixing and 3D sound rendering approaches typically used on spatially distributed loudspeaker configurations for cinema and VR.

Having defined the relevant renderer modules for a given exhibit in the configuration file, a user then creates linkages between incoming cue IDs, the audio samples to be triggered by the cues, and the rendering modules to be utilized. Attributes relating to the rendering strategies employed are also defined, such as the target zone(s) for rendering and envelope to be applied to the signal during network-triggered on/off state changes. The software also allows for the definition, prior to runtime, of one or more "voices." Voices within the EX3_Audio software, analogous to the individual voices within a choir, define the maximum amount of simultaneous audio streams that can be rendered at any given time. Each cue ID is assigned either to its own unique voice, or to a voice shared by other cues, based on the particular rendering use-case. By assigning each cue ID to its own voice, we guarantee that the audio streams will be capable of rendering in parallel. Conversely, by assigning two or more cue IDs to the same voice, we safeguard against inappropriate simultaneities, guaranteeing that the last-most cue ID received will interrupt previous cues assigned to the same voice.

Depth Camera Tracking for Real-Time Audio Control

In an effort to minimize inter-zone crosstalk and listener fatigue resulting from an overabundance of unrelated auditory stimulus directed at participants within the EX3 exhibits, we created a custom depth-sensitive triggering system (EX3_VolumeTracker) using a Microsoft Kinect™ depth camera. By implementing a hands-free adaptive triggering system to dynamically enable and disable zone-based and listener-directed audio rendering, we were able to both more clearly establish the relationship between a given visual and auditory display,[6] as well as increase the intelligibility of audio content delivered to participants examining a given visual display, while also making it possible for visitors standing between zones or sitting in common areas within the exhibit to more easily converse. Our tracking approach can be summarized as follows:

1. Carve out a tracked real-space volume in front of the Kinect™ camera for use in search and analysis.
2. Define azimuthal zones within the camera's viewport that, when occupied by visitors, will enable appropriate audio content rendering.
3. Search for the presence of depth data within tracked azimuthal zones, alerting the audio rendering software accordingly via network messages in order to dynamically enable/disable audio content delivery.

[6] As mentioned above, in the case of EX3, this involved either voice-over narrative recordings or ambient sounds intended to accompany the visual data.

Of course, it is possible to utilize cameras without depth sensors in creating coarse zone-based triggering systems, but the addition of depth data allows us a more robust method for ignoring areas of 3-dimensional space within the camera's viewport via the scene culling described in step 1 above. Also, the minimal extra cost of a Kinect sensor with respect to good resolution webcams makes the effort worthwhile.

Method Description

For the description of our algorithm we only need assume access to the Kinect™ camera's depth image. Our method uses a right-handed coordinate system with z pointing directly out from the camera, y pointing out of the top of the camera, and x pointing to the right of the camera, when facing it. No, it does describe a right-handed system. Consider the "x pointing to the right of the camera, WHEN FACING IT" (http://www.evl.uic.edu/ralph/508S98/coordinates.html). We perform analysis on a given depth frame in two main phases:

1. Pre-processing of depth images.
2. Search for the presence of depth data within target azimuthal zones.

Depth Image Pre-processing

Before deriving the 3D locations of objects present within the Kinect™ camera's viewport, we first need to prepare the depth images. We convert the camera's raw depth values $r \in (0 - 2{,}047)$ into meters using the following formula (OpenKinect 2011):

$$d = 0.1236 \times \tan((r/2,842.5) + 1.1863)$$

where d is the distance in meters (and equivalent to location on the camera's real-space z axis).

Given the horizontal and vertical lens angles of the Kinect's infrared camera (used to acquire the depth image), we can now derive the real-space x and y coordinates of the depth data within the camera's viewport, in preparation for culling unwanted data prior to evaluating for physical presence within tracked zones. Real-space Cartesian (x, y) values in meters are derived as follows:

- Map the (x, y) pixel coordinates of the depth image to the range $(-1.0, 1.0)$ (left to right, bottom to top).
- Calculate the azimuth (az) and elevation (el) of each cell of the depth image using the following formulas:
  $az = \arctan(blob\_pixel_x, (1.0/\tan(horizontal\_lens\_angle)))$
  $el = \arctan(blob\_pixel_y, (1.0/\tan(vertical\_lens\_angle)))$
- Finally, calculate the real-space Cartesian (x, y) values of each cell in the depth image using the following formulas:
  $x = \tan(az) \times z$
  $y = \tan(el) \times -z$

Now that we have the Cartesian (x,y,z) coordinates for all cells in the depth image, we can easily remove unwanted regions within the camera's viewport prior to analysis by evaluating the location of the depth data in reference to user-defined clipping planes. We determine whether or not to consider a cell of depth data in our later analysis as follows:

Given the (x,y,z) Cartesian representation of a depth data cell (point Z) and clipping plane defined by three points: A, B, C

- Calculate the cross product n of AB and AC
- Calculate the dot product d of n and $Z–A$
- If the sign of d is negative, ignore the depth data in later analysis

Zone-Based Depth Data Evaluation

Based on our practical assessment of the audio requirements for the Exodus 3 exhibit, we determined that testing for the physical presence of objects only along the camera's azimuth plane (x/z Cartesian plane) would be sufficient in driving the on/off states of the various layers of directional audio accompanying visual data at relevant exhibits. It is worth noting, however, that since we do obtain the real-space Cartesian representation of the camera's depth data, it would be trivial to evaluate for physical presence within defined 3D Spherical or Cartesian space, rather than simply on the azimuth plane.

We first define azimuth ranges as discrete zones within the Kinect™ camera's viewport. For our software implementation, we allow the definition of 3 azimuth ranges, matching the maximum number of possible simultaneous audio streams rendered at the relevant exhibits within the event. The amount of possible tracked ranges within which we search for physical presence, however, is limited only by the resolution of the camera itself, with a depth image resolution of 640 by 480 pixels (providing a maximum of 640 angular subdivisions of the camera's horizontal lens angle). Given our knowledge of each cell of the depth image, as defined in section "Depth Image Pre-processing," we prepare our depth image for evaluation of physical presence as follows:

- Convert the camera's depth image to binary, setting all "valid" cells (determined as a result of the processes defined in section "Depth Image Pre-processing") to 1, and all invalid cells to 0.
- Multiply each cell of the resulting binary image with its corresponding azimuth value.
- All invalid cells within the camera's depth image will now be set to 0. We change those to a nonzero value outside the possible range of azimuth angles, such as 1,000.

As a result of the above processes, we have converted the original depth image into a map of azimuth values at each cell. We now simply search for cells in the converted depth image that fall within one or more of the defined azimuth ranges. If one or more cells of the image contain an azimuth value within a targeted range, we enable the directional audio for that zone (sending the appropriate network command to the audio rendering software). Conversely, if no valid depth data is found we disable the audio for that section. Since we only care if the defined azimuth ranges are either fully empty or contain an arbitrary amount of depth data, we need not continue our search within a zone for a given camera frame after the first valid pixel is found.

Method Implementation: "EX3_VolumeTracker"

We have designed a Kinect™ camera tracking server (EX3_VolumeTracker) that implements the above-described tracking method using the Max programming environment (Zicarelli et al.

2013). The server is set up via a simple JSON configuration file, and provides an intuitive GUI for real-time control over a variety of application parameters, depth and RGB image visualization, tracker logging, clipping plane definition, and network I/O connectivity. It receives remote commands via UDP or Open Sound Control at user-defined ports, and serves the tracking results as a three-element array of integers (0 or 1), representing physical presence within user-defined azimuth ranges, to an arbitrary number of listening clients. For the EX3 event, tracking software running at a given audio/visual display area would send the results of its analysis to the zone's corresponding custom audio rendering software (EX3_Audio), as discussed in section "'EX3_Audio' Audio Rendering Software."

Kinect RGB and Depth images are acquired using *jit.freenect.grab* (Pelletier 2012), an open-source third party Max object using the OpenKinect project's libfreenect library (OpenKinect 2012), that retrieves the camera's RGB, Infrared, and depth images, and accelerometer readings, allows control over the camera's tilt motor, and allows for the use of multiple cameras simultaneously. We decided to use libfreenect because the library provides all the raw data from the Kinect™ required for our use-cases, while also allowing our implementation to work out of the box, without driver installation requirements or other library dependencies (compared to other more feature-rich Kinect camera libraries and software tools, such as OpenNI 2013).

Conclusion and Future Work

By placing a systemic view of Israelite culture at the center of the ancient "world building" endeavor (Fig. 11.6), we have been able to engage in transdisciplinary scientific "storytelling" for the twenty-first century observer of one of the oldest written narratives in human history. This was achieved by engaging over 50 researchers from a wide range of fields, including the "sonic arts"—an area that until now has rarely engaged in meaningful research interaction with archaeologists, geo-scientists, Egyptologists, and other scholars interested in the material world. Creating advanced audio-visual displays for sharing information is not without its challenges. In museum and exhibit settings, the largest problem we face is isolating the audio to the display from which it originates, so it does not leak into and interfere with nearby listening positions. In addition, the ability to scale the audio image to the size of the display and even localize it to specific content on the display wall presents its own unique set of challenges. Luckily, new technology in the areas of audio beamforming, computer vision, and signal processing have made it possible to substantially isolate each exhibit (reducing sound leakage and distraction), create language zones, dynamically trigger audio streams only when listeners are present, personalize audio feeds while navigating the space, and provide immersive audio environments and simulations. Any combination of these new opportunities can be further developed to create entirely new audio-visual displays never experienced before.

In the EX3 exhibits, we demonstrated how the aforementioned advances in audio technology and computer vision could be combined and implemented in a "museum of the future." We demonstrated how multiple adjacent streams of sonic information could be clearly interpreted while centered about a large or shared visual display using beamforming. We also demonstrated how beamforming and distributed loudspeakers could be used to spatially link audio streams with sectioned visual content on tile-displays of several sizes. Finally, we demonstrated how a tracking system could be used to implement a "smart gate" on audio streams whose activation is dependent on the listener's position in the physical exhibit space.

In the future, cyber-archaeologists can tap into more sophisticated forms of audio acquisition and even more immersive, 3D audio reproduction techniques to enrich data sets, help interpret them, and augment media presentations with digital narratives while

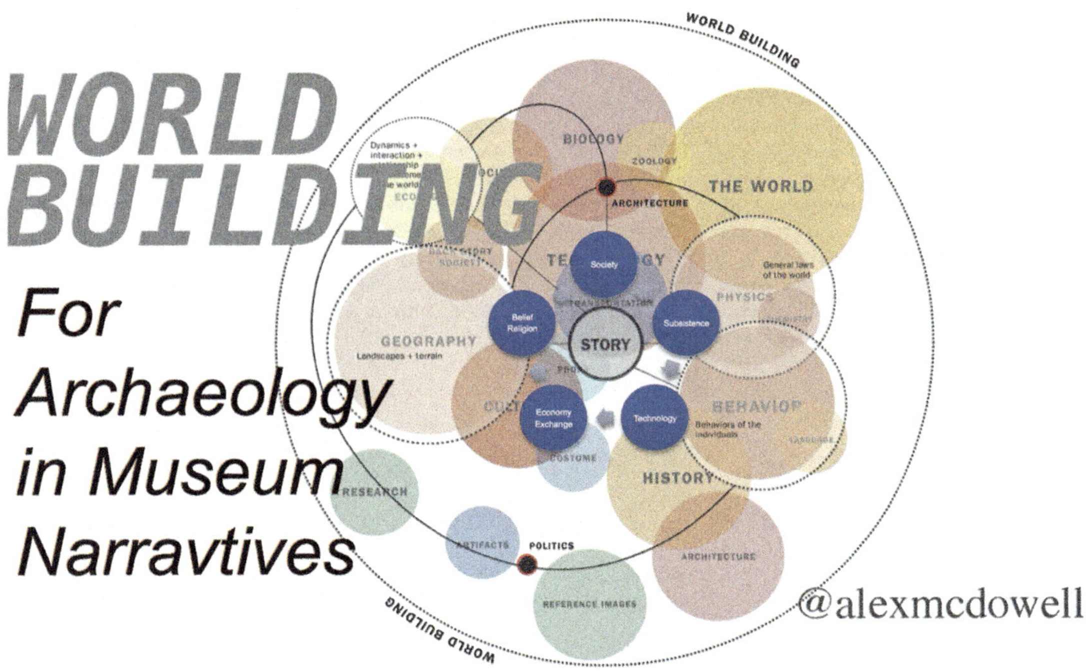

Courtesy Alex McDowell

Fig. 11.6 By placing the systemic model of human culture at the center of ancient world building, compelling new twenty-first century science-based narratives of the ancient world are possible

communicating information and concepts in user friendly, high resolution, experientially based learning and communication environments. It is easy to vividly imagine museum systems in the near future rivaling the most sophisticated media, VR, and entertainment venues. More difficult to manage are the myriad of details and subsystems that must be aggregated to bring these environments to life. Resources and design/engineering effort must be dedicated to unglamorous considerations such as acoustical treatment, power and cooling requirements, content creation, hardware, custom mounting and installation hardware, training media savvy docents, and handling the expense of computer-driven audio-video equipment. The next generation of archaeologists will need to be trained in a wider range of digital media skills, to allow them to plan site visits, equip their teams, and effectively gather, process, interpret, and present new types of data in a variety of new digital media forms. For the "museum of the future" our next challenge will be to create an "immersive" transmedia world building experiment, where we will take an ancient narrative, such as the Exodus from Egypt, and employ in-person scholars and online researchers and audiences to create enhanced science-based narratives of the past.

References

DeFanti, T.A. et al. 2011. The Future of the CAVE. *Central European Journal of Engineering* 1:16–37. Versita with Springer-Verlag.

Hermann, T., and H. Ritter. 1999. Listen to Your Data: Model-Based Sonification for Data Analysis. *Advances in Intelligent Computing and Multimedia Systems*: 189–194

Hermann, T. et al., eds. 2011. *The Sonification Handbook*. Berlin, Germany: Logos Verlag. http://sonification.de/handbook/.

Levy, T.E. 2013. Cyber-Archaeology and World Cultural Heritage: Insights from the Holy Land. *Bulletin of the American Academy of Arts & Sciences* LXVI:26–33.

Lokki, T., and M. Grohn. 2005. Navigation with Auditory Cues in a Virtual Environment. *IEEE Multimedia* 12: 80–86.

Magliozzi, R., and T. Magliozzi. 1999. Car Noise Emporium. *Car Talk*. http://www.cartalk.com/content/car-noise-emporium-27 (accessed August 15, 2013).

Olivieri, F. et al. 2013. Loudspeaker array processing for multi-zone audio reproduction based on analytical and measured electroacoustical transfer functions. *AES 52nd International Conference*. Guilford, UK.

OpenKinect. 2011. *OpenKinect.org wiki Imaging Information page*. http://openkinect.org/wiki/Imaging_Information (accessed August 15, 2013).

OpenKinect. 2012. *OpenKinect.org wiki main page*. http://openkinect.org. (accessed August 15, 2013).

OpenNI. 2013. *OpenNI.org main page*. http://www.openni.org (accessed August 15, 2013).

Pelletier, J.M. 2012. *jit.freenect.grab object for Max*. http://jmpelletier.com/freenect/ (accessed August 15, 2013).

Seldess, Z. et al. 2011. Sonnotile: Audio Annotation and Sonification for Large Tiled Audio/Visual Display Environments. *Proceedings of the 17th International Conference on Auditory*. Budapest, Hungary.

Zicarelli et al. 2013. *Max Visual Programming Environment*. http://cycling74.com/producst/max (accessed August 15, 2013).

12 The WAVE and 3D: How the Waters Might Have Parted—Visualizing Evidence for a Major Volcanic Eruption in the Mediterranean and Its Impact on Exodus Models

Jürgen P. Schulze, Jessica Block, Philip Weber, Thomas E. Levy, Gregory L. Dawe, Brad C. Sparks, and Thomas A. DeFanti

Abstract

To fully engage in Late Bronze Age "world building" and the Exodus narrative for the EX3 exhibition (see Chap. 11), transdisciplinary research in archaeology, geology, and computer graphics were integrated in a new 3D immersive Wide Angle Virtual Environment (WAVE). The goal was to marshal geological evidence for a hypothesis that might explain the "Parting of the Sea" narrative in the Book of Exodus. The research explores the possibility of a connection to the Santorini island (Thera) volcanic eruption of the Late Bronze Age inducing a tsunami that would first draw the water away from the shore before surging back into a large wave. We collected data from various sources and geo-located it on a 3D map of the Mediterranean region. Combined with an automated presentation sequence and narration, the resulting virtual reality application presents the data in a novel way, which allows for a more intuitive approach for its interpretation. This chapter introduces the new WAVE and describes how we created a real-time virtual reality demonstration to present archaeological and geological data that may inform elements of the Exodus story. We explain how the data was acquired, how it was fused onto a 3D terrain map, and how an automated demonstration was created with narration for the Exodus exhibition. The chapter examines the scientific features of the visualized data, as well as the implementation of the visualization software.

Introduction

By utilizing the power of 3D scientific visualization, ancient "world building" of the ancient Hebrew Exodus from Egypt was empowered at an unprecedented level. EX3 researchers explored environmental hypotheses linking Late Bronze Age tsunami events to the "Parting of the Sea" narrative in the Book of Exodus. Earlier researchers have studied geological influences on a range of events mentioned in ancient texts and/or observed in the archaeological record. Manfred Bietak (1996) excavated the ancient city site at Tell el-Dab'a (Avaris) along the now extinct Pelusiac branch of the Nile River, identifying the city as a harbor town. Daniel Jean Stanley (Stanley

J.P. Schulze (✉)
Qualcomm Institute, UC San Diego,
La Jolla, CA 92093, USA
e-mail: jschulze@eng.ucsd.edu

T.E. Levy et al. (eds.), *Israel's Exodus in Transdisciplinary Perspective*,
Quantitative Methods in the Humanities and Social Sciences, DOI 10.1007/978-3-319-04768-3_12,

et al. 1996; Stanley and Warne 1993a, b) collected and evaluated sediment cores in the Nile Delta to reconstruct the location of the Nile paleo-coastline of the Late Bronze Age, putting it farther south along its eastern flank. Stephen Moshier (Moshier and El-Kalani 2008) describes the geomorphic evidence for a northern Exodus route. Stanley also evaluated sediment cores in Lake Manzala containing volcanic ash deposits from the Thera volcanic eruption of the Late Bronze Age (Stanley and Sheng 1986). And Floyd McCoy, B. Goodman-Tchernov, Steven Ward, and T. Novikova have each collected and modeled evidence for a resulting tsunami from the Santorini (Thera) eruption that should have reached the ancient Nile coastline (Goodman-Tchernov et al. 2009; McCoy and Heiken 2000; Novikova et al. 2011). Each of these pieces of evidence has been integrated with to archaeological and biblical research on the Exodus (Sivertsen 2009), but never before have they been united into one scientific data visualization. Only Moshier had created a geospatial digital database of his work. Most other research results exist in scientific papers, but not in any digital or geospatial database. The WAVE immersive visualization environment provides a discipline-neutral platform for combining these data into a space that allows several people of different disciplines to collaboratively evaluate the data. Doing so has illuminated correlations that had not been made before.

The following sections of this chapter provide an overview of work related to our project, explain the various data types that went into our visualization application, how we processed them, and how we fused them into one application. We then report on how we turned the visualization application into an exhibit-ready demonstration, and we summarize what we learned from the experience.

Wave Construction and Geometry

The concept of an LCD-based virtual reality system such as the WAVE is described in a publication by DeFanti et al. (2011), which compares LCD-based systems to the traditional projector-based ones and finds that LCDs have significantly higher contrast and brightness, are easier to maintain, and have much smaller overall space requirements (Figs. 12.1, 12.2, 12.3, and 12.4).

Fig. 12.1 WAVE with researchers Chris McFarland, Jürgen Schulze, Greg Dawe, and Falko Kuester. Photo by Tom DeFanti

The University of California – San Diego's new WAVE display, true to its name, is shaped like an ocean wave, with a curved wall array of 35 55″ LG commercial LCD monitors that end in a "crest" above the viewer's head and a "trough" at his or her feet. It was fitting that the WAVE display was inaugurated at the EX3 exhibition featuring world building around the routes of the Exodus and the "Parting of the Sea" (Exodus 14) and the "museum of the future." The WAVE was designed by Tom DeFanti, Falko Kuester, and Greg Dawe. The WAVE, a 5 × 7 array of HDTVs, is now 20′ long by nearly 12′ high and can accommodate up to 20 people. Its curvature makes it appear like the department store windows of yore, or museum dioramas, in which the glass is positioned away from you and you cannot touch it, so it does not feel like it is there. The WAVE achieves that illusion in ultrahigh resolution (35 times 3D HDTV). Its curved aluminum structure is also a technical solution for the problem of the upper and lower parts of images on 3D passively-polarized screens ghosting as double images when simply mounted flat on a wall.

High-resolution computerized displays have evolved over the past decade from 2D to 3D panels and from one monitor to arrays of many monitors. They have transitioned from thick bezels (the rim that holds the glass display) to ultra-narrow bezels. Such technology is now widely used in television newsrooms, airports,

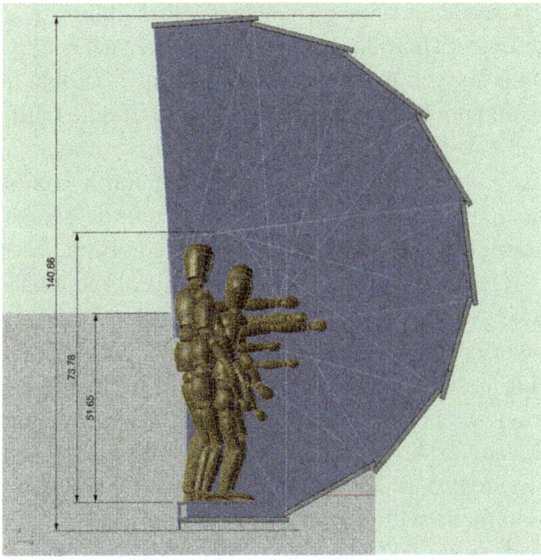

Fig. 12.2 CAD drawing of the WAVE design by Greg Dawe

Fig. 12.3 WAVE under construction for Exodus with Chris McFarland, Greg Dawe, and Andrew Prudhomme. Photo by Tom DeFanti

Fig. 12.4 WAVE computers as built by Joe Keefe, Chris McFarland, and Eric Lo. Photo by Tom DeFanti

and even retail stores, but not commonly in 3D like the WAVE.

With the creation of the WAVE, we wanted to give people an experience of looking over the edge, of hanging off a railing like you might do at the Grand Canyon. To do that, we had to provide an image above and below the viewer. When the data comes up and the ground plane disappears underneath you, it really feels like you are "flying over the data."

Related Work

The software we use as the basis for the development of our visualization application is built on top of our own CalVR (Schulze et al. 2013), which in turn uses the OpenSceneGraph API (OSG 2013) as its graphics engine and osgEarth (OSGEarth 2013) as its terrain rendering subsystem.

Our team has many years of experience with visualizing archaeological data in virtual reality. Notable publications include our *PNAS* paper by Levy et al. (2008), which mentions the use of our StarCAVE for a visualization of an excavation site in Jordan. More recently, Lin et al. (2011) have presented a virtual reality visualization application for an archaeological survey site in Mongolia.

Related work by other groups includes Vote et al.'s article (2000) on ARCHAVE, which is to our knowledge the first successful data visualization project for virtual reality in archaeology, using excavation data from the city of Petra in Jordan.

The proceedings of the annual International Symposium on Virtual Reality, Archaeology, and Cultural Heritage (VAST) provide a good source of a variety of case studies in the area of archaeology visualization projects. However, the projects presented there are typically on specific sites of archaeological interest, or technology related to data acquisition.

Research in landscape visualization has traditionally focused on policy and decision-making for environmental issues and urban planning (Dockerty et al. 2005; Block 2007; Sheppard 2005), which has been limited to static landscape design under different scenarios. More recently, with the broad use of Google Earth and ESRI's ArcGlobe, geospatial research has transitioned to the use of virtual globes to display and analyze geospatial data at various scales (Tomaszewski 2011; Tiede and Lang 2010; Smith and Lakshmanan 2011). However, these platforms have their limitations in data manipulation and accessibility. Most of this past research has focused on one geospatial scale, whereas our project took advantage of both local and regional data. Rarely have virtual globes been used to compare past and present geographies (Yano et al. 2012). The Kyoto case by Yano uses the virtual globe to show static images of different years. To our knowledge, we are the first to ever create a virtual reality demonstration of possible Exodus scenarios. The advantages of the osgEarth platform are its open source availability and functionality to manipulate all aspects of the geospatial data including topographic resolution and animation. With these functions, we were able to uniquely display information from a local to regional scale while adding 4D animations, and automated flight paths with narration.

The Exodus Data Fusion System

The platform in which the Exodus immersive visualization was developed was Qualcomm Institute's own virtual reality framework, named CalVR. CalVR is middleware software and facilitates writing applications for virtual reality systems by logically separating display configuration, rendering cluster and input devices from the application: the programmer can develop the application on a desktop computer, and later run it, normally without further modifications, in a large, graphics-cluster driven virtual reality (VR) system. CalVR is entirely written in the programming language C++ and uses the OpenSceneGraph API (OSG) as its graphics interface.

The terrain rendering engine, osgEarth is used to display the Mediterranean region and all geospatial data. osgEarth takes care of dynamically paging in the levels of detail of the terrain needed for a certain camera view: things in the foreground are rendered with greater detail than things at a distance. This allows us to maximize visual fidelity with a given hardware configuration. Note that it would not have been possible to use Google Earth for this application. While it is very similar to osgEarth in many aspects, Google Earth would not have run on the WAVE with its unusual screen configuration and head-tracked stereo vision.

Our goal was to merge (fuse) all of the available geographical data regarding the Exodus onto a very detailed 3D map of the region. We targeted this application for our newest walk-in virtual environment, the WAVE. Because the application was to be shown as part of the Exodus exhibition, we created an automatic, prerecorded demonstration mode with a voice-over, so that visitors only had to put 3D glasses on and watch the 8-min presentation. But even in demonstration mode, the application renders all the graphics in real-time and could be interrupted at any point to manually take over the camera controls—similar to turning off the autopilot in an airplane and flying it manually. The manual mode is useful for archaeologists to study the fused data in greater detail than is possible in the automated presentation, and also at their own pace.

Regional Geographic Base Map: Terrain and Imagery

Global terrain data was downloaded and merged from the US Geological Survey (USGS) and National Geospatial Intelligence Agency (NGA) Global Multi-resolution Terrain Elevation Data (GMTED) at a resolution of approximately 30 m per pixel. To make the topography complete, Mediterranean seafloor bathymetry was downloaded from the European Marine Observation and Data Network (EMODnet) (http://portal.emodnet-hydrography.ed/#) and merged with the land surface topography. These data provide the topographic basis for all other geospatial data to be evaluated. It provides depth to the seafloor, distance from the volcanic eruption to the Nile coast, and shows how and where a volcanic eruption and resulting tsunami could have traveled with topographic constraints. Global satellite imagery was draped over this topography in osgEarth using a Web map service (WMS) from readymap.com.

Four tiles of the USGS dataset comprised the entire land surface topography including the Mediterranean, Europe, and the Middle East, which we into merged a single regional file and then trimmed to our area of interest. Of the bathymetry data, each subset of the Mediterranean bathymetry was downloaded from the EMODnet data portal as an ESRI GRID format, reprojected to WGS84, resampled to a spatial resolution 30 m per pixel to match the terrain, and merged to the regional file. The completed terrain with all files were processed and merged, then exported into a 16-bit floating GeoTIFF-formatted file. This preprocessing workflow was performed using ESRI's ArcMap to prepare the data for the osgEarth software in the WAVE. Challenges of file size were encountered on the desktop PC used to run ArcMap. Solutions around this were made by cutting areas surrounding the study area before merging the complete file.

Modern satellite imagery was draped over the regional topography using a web map service to provide context for the regional fly-through. The specific data chosen from ReadyMap has been color-matched globally so as to provide visual continuity from the global, zoomed-out view to the zoomed-in, local views across the Nile Delta and the Sinai Peninsula. The compromise for choosing this coarse resolution dataset is that we do not see detail of current cities or urban infrastructure in the imagery as we zoom in. However, for this project that works in our favor. We successfully used present-day imagery to present Bronze Age information without interference of present-day anachronisms. An additional advantage to using the WMS is to save local memory for other data (Figs. 12.5 and 12.6).

Fig. 12.5 WAVE with researchers and audience during Exodus exhibition. Photo by Tom DeFanti

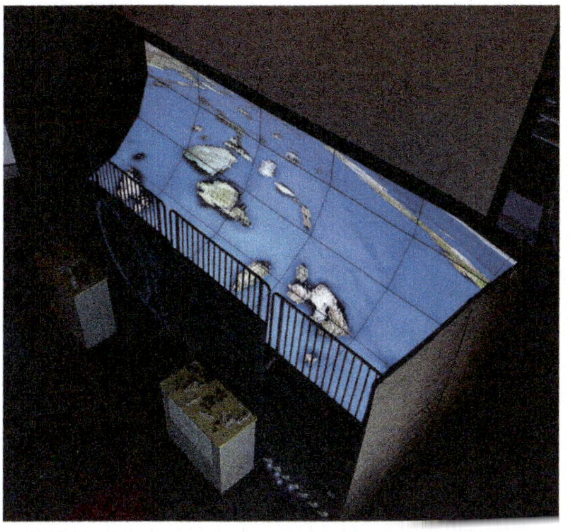

Fig. 12.6 WAVE from above. Photo by Tom DeFanti

Geological Data: Nile Sediment Drill Cores

Research published by Daniel Stanley was a result of the Smithsonian Institute's Nile Delta Drill Core and Sample Database (Stanley et al. 1996) to collect and log 87 sediment cores along the northern Nile Delta plain. The cores were drilled between 1985 and 1990, and subsequent lithologic logs (sediment description throughout each core) were created. Data for each core was collected including core number, core length, date of core recovery, approximate location description, and latitude and longitude with each lithologic log. These data provide the rock record to place the location and evolution of the Nile coast over the last 4,000 years. The coast evolution was reconstructed by Stanley (Stanley and Warne 1993b) and then digitized in ArcMap by the authors of this chapter using the figures in Stanley's paper. Reconstructing and visualizing the ancient topography is critical to addressing reasonable escape routes for the Hebrews (see Figs. 12.9 and 12.10).

Fig. 12.7 Sediment cores along northern flank of Nile Delta displayed in osgEarth

The core latitude, longitude and core length of each location were transcribed to Excel and then imported to ArcMap as a shapefile. That shapefile was then imported to osgEarth to be represented as 3D cylinders with uniform radii and varying depths depending on the length of the core. The 87 drill cores in the Nile Delta plain were colored bright orange (Fig. 12.7), and the 5 drill cores in Lake Manzala were colored a yellowish orange. For the purpose of being able to see the variation in core depths from the Earth surface, we represented these variations by extruding them above ground using their depth value (rather than representing them in the subsurface).

Stanley collected an additional five drill cores in Lake Manzala just east of the Nile Delta plain. These core locations were digitized using Stanley's map in ArcMap, and the exported shapefile was added to osgEarth as a separate file. These five locations contain the volcanic ash with characteristics matching the Thera volcanic eruption of the Late Bronze Age (Stanley and Sheng 1986). The Lake Manzala cores provide correlative evidence for how and where the volcanic ash from the Thera eruption affected the Nile coast.

Geophysical Data: Volcanic Eruption Simulation

Geophysicist Steven Ward of UC Santa Cruz used computational fluid dynamics modeling to create hypothetical, but geologically plausible, scenarios that could induce a tsunami in the Mediterranean Sea. He provided a scenario where a plinian volcanic eruption on Santorini induces a wave to propagate from the island southeastward, through the narrow outlet between the Greek islands of Crete, Karpathos, and Rhodes; ultimately arriving at the Egyptian and Israeli coasts. This simulation, shown on a virtual Earth, sheds light on the physical possibility of a wave in the Aegean Sea reaching 800 km across the Mediterranean to the Nile Delta. Although several geological processes could induce a tsunami in the Mediterranean Sea including a submarine landslide, a storm surge, a Hellenic subduction zone earthquake, and a Theran eruption, we chose to display and animate only the Theran eruption to represent the maximum amount of research discussed at this meeting.

To enhance the animation of the tsunami, we added a schematic ash plume at the location of Santorini island to initiate Ward's simulated wave animation. The height of the plume was animated to represent 35 km above sea level to correlate with prior calculated estimations (Booysen 2013). Figure 12.8 (image on right) shows ash plume and our visualization of the wave propagation. The plume height is relevant with respect to whether the Hebrews could have seen the plume from the Egyptian coast. According to Booysen, the 36-km ash plume height, estimated to be the possible height of Thera's ash plume, could not have been seen by the Hebrews considering the distance and the curvature of the Earth. However, Booysen estimated that the top of the Thera plume could

Fig. 12.8 Eruption of volcano on Santorini island with ash plume displayed in osgEarth

have reached approximately 58 km altitude if 100 km^3 of magma had been ejected. The authors of this chapter performed their own calculation to determine minimum height of the plume to be visible at the Nile Delta. The radius of the earth between Thera and Tel el-Dab'a (at midpoint latitude 33.6° N) is 6,372 km (rather than the equatorial radius of 6,378 km). The geocentric angle between Thera and Tell el-Dab'a is then 7.75° (1° is 69 statute miles = 60 nautical miles). Standard refraction effectively reduces this angle by about 0.57° to about 7.18°. This is approximately the angle below the horizon of the surface at Thera from Tell el-Dab'a. Using these corrections, we calculated the required height of the ash plume to be seen at Tell el-Dab'a to be $([1/\cos 7.18°] - 1)(6{,}372 \text{ km}) \approx 50$ km. Booysen cites the maximum possible plume height of a terrestrial volcano to be approximately 55 km. Therefore, it might have been possible for this eruption to be seen during the Hebrews' escape. The visualization of this data in the WAVE helped define and illustrate this debate.

Fluid dynamics simulations of a volcanically induced tsunami were provided by Steven Ward. 1,474 × 840 point ASCII Grid files were provided for each time step of the simulation, which were imported to ArcMap. Experiments were performed to determine the best way to import and process each file into an animation along the sea surface in osgEarth. We wrote a script to convert each ASCII Grid file to a 2,948 × 1,680 RGBA GeoTIFF file, with wave height mapped to a color gradient from light to dark blue, and a translucent alpha channel, so that the sea floor would be visible through the wave texture. Then each GeoTIFF was georectified in ArcMap using manually selected control points using the topography as the guide. Once each GeoTIFF was warped to fit the Earth surface by tessellating it into a mesh of 24 × 13 rectangles, a script was written to turn the 240 time steps into an animation, which represented a 4-h simulation time frame. Each time step was compressed using OpenSceneGraph's native binary compression into a 20 MB file, so that the animation occupied a total of 4.8 GB on disk. During rendering, multiple CPU threads were created to allow for a smooth rendering experience by asynchronously loading and buffering textures.

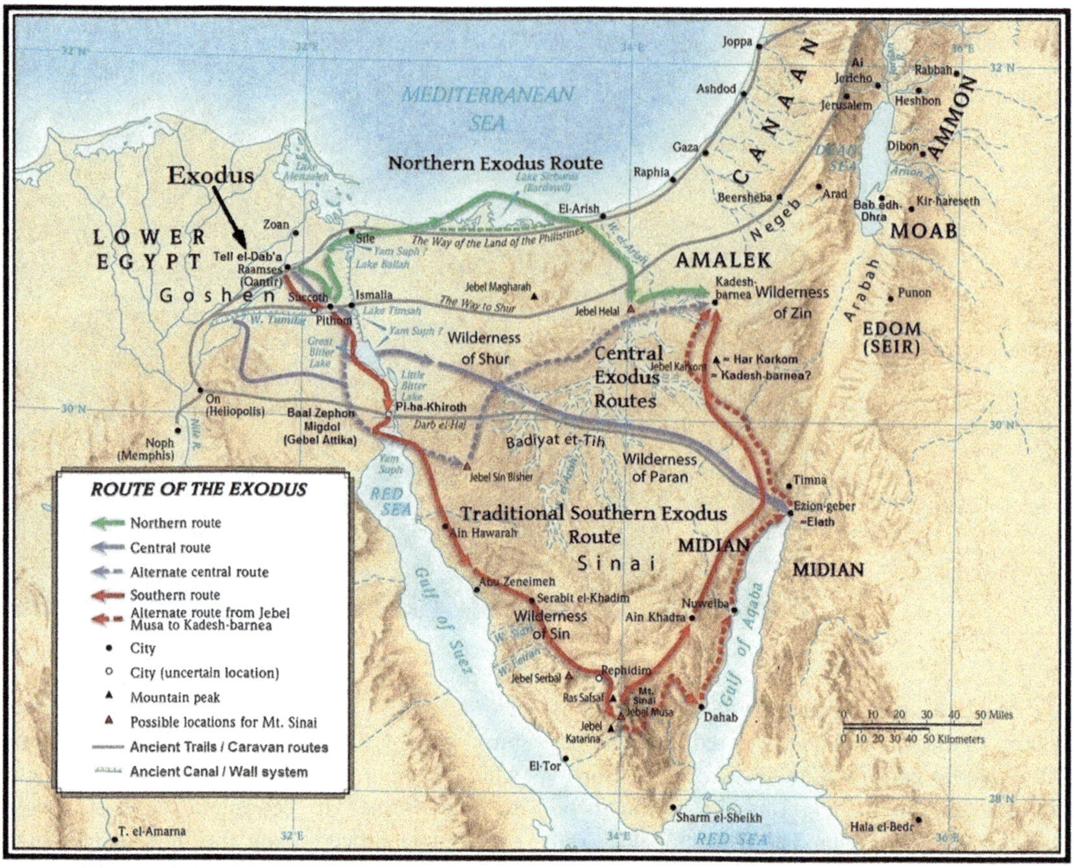

Fig. 12.9 Northern, Central, and Southern proposed Exodus routes (Ellis Smith 1993, as modified). (***Note: Route lines are in schematic outline only, not exact trail routes.***) Modern coastline is shown. See Fig. 12.10 below, and Chap. 9, Fig. 9.3b, for reconstruction of ancient coastline of the Nile Delta, ca. 2000–1000 BCE

Archaeological and Theological Data: Travel Routes

There are three main Exodus routes presented by which the Hebrews are argued to have escaped Egypt to Israel, see Figs. 12.9 and 12.10.

Moshier (Moshier and El-Kalani 2008), Bietak (Bietak 1996) and Stanley (Stanley and Warne 1993a) provide geomorphic and archaeological evidence for a northernmost route of the Exodus, also supported by archaeological evidence for the Biblical place names, such as *Yam Suph* (sometimes translated "Sea of Reeds" from the Hebrew), suggesting that the crossing of the Sea occurred in the salt marshes and shallow lakes between the Mediterranean and Red Seas (Fig. 12.9). The central route is based on the covenant experienced in Exodus 19:16–25 and also goes through the home of the Midianites, where Moses married his wife, Zipporah. The southern route is traditionally supported as the Exodus route placing Mount Sinai in the southern Sinai Peninsula. This route has been supported by the identification of *Yam Suph* as the "Red Sea" in the Greek Septuagint (a geographic site *identification*, not a Hebrew-to-Greek language *mistranslation*) (arguing for "Reed Sea" and claiming "Red Sea" is a mistranslation: Kitchen 2003: 261–3; Hoffmeier 2005: 81–85, 163–4; Hoffmeier 1999: 199–222; refuted by Batto, this volume, by Propp 2006: 752, Houtman 1993: I:128, Vervenne 1995: 424, et al.). The routes described above were georectified in ArcMap in order to digitize the paths of these Exodus routes into line shapefiles, and then translated into 3D tubes in osgEarth. It was necessary to draw

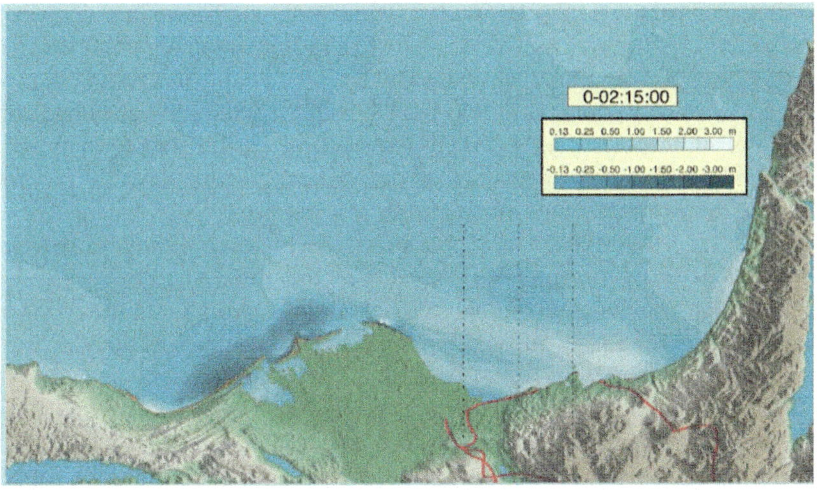

Fig. 12.10 Approaching tsunami wave along ancient coastline showing northern Exodus route affected by the oncoming wave (model by Steven Ward)

tubes instead of vectorized lines to make the travel routes more easily visible, and to be able to shade them to look properly three-dimensional. The travel paths were positioned 250 m above ground, so that we could avoid intersections with the ground.

Archaeological Data: Egyptian Forts

Bietak and Hoffmeier have excavated Egyptian forts in the northeastern Delta near the coast of the Sinai Peninsula that may be alluded to in Biblical texts (see also Moshier and Hoffmeier, Chap. 8). Due to the relevance of these locations to influence on the possible Exodus paths, the point locations were represented on the osgEarth using colored spheres with place names as labels. Key locations include Pi-Ramesses, Tell el-Dab'a (Avaris), Kadesh Barnea, Aqaba, and Jebel Musa. Because each of these landmarks provides evidence for different Exodus routes, we triggered each location to appear on the Earth when they were relevant to the narrated story, as it was told. The eight total site markers were imported to osgEarth as point shapefiles, similarly to the routes, and rendered as 3D spheres, hovering above the regional terrain. They vary in color and size depending on their relevance.

Telling the Story: The Demonstration Application

Merely fusing all of the above data into one large map would not have effectively told the fascinating story of the latest research results about the Exodus. We wanted the visitor experience to be such that the visitors could stand in front of the WAVE and watch the story unfold. We designed an automated flight path through the areas of interest to synchronize with a scripted narration written by UCSD archaeologist Prof. Thomas Levy. The script was written to concisely describe and view all of the data elements in this chapter in order to present the viewer with scientific evidence for the relationship between the Theran eruption and the Exodus of the Hebrews. Once the script was written, it was narrated by Tiffany Fox, one of our Institute's voiceover experts. We ended up with 12 separate parts of the narration, which we stored as separate audio clips on disk.

Using the timeline of the script, we experimented with ways to best choreograph the flight path to match the story. We tested flight speeds to approach the relevant regions synchronized with the narration. Knowing that the narration tries to convey the relevance of the regional geography, we took care to create the flight path from an oblique bird's eye view, minimizing turning motions to prevent motion sickness of the audience. Although the WAVE was designed as an immersive environment, zooming down to the ground level would have defeated the purpose of merging Mediterranean region-level data. We maintained regional views while focusing on key elements throughout the narration. The WAVE construction supported this nicely in that you could take advantage of the lower curvature of the tile display to simulate flying over the earth.

We recorded the flight path with OSG's path recorder, which we integrated into CalVR within the osgPathRecorder plugin. When activated, it recorded position and orientation of the viewer for every rendered frame, along with a time stamp. At playback time, positions and orientations were interpolated based on the elapsed time, so that the path played back at the same pace as originally recorded, despite likely differences in frame rate.

We also recorded the time stamps to trigger the audio clips and when to display relevant data in the story. During playback, at the appropriate times during the narration, the CalVR plugin triggered the playback of the audio clips by sending the clip's ID number to our audio server via a custom TCP protocol. The same plugin also turned on and off the various data types, such as drill cores, travel routes, and fort names.

We instrumented the WAVE with a Kinect device, which senses the presence of a person in the WAVE. If someone was there, CalVR automatically started the flight path and synchronized the audio clip. This allowed us to keep the WAVE quiet when nobody was watching, so as to avoid unnecessary distraction of visitors of other parts of the exhibition.

Discussion

Overall, the high-resolution, wide field of view display system of the WAVE was a very effective platform for the 3D virtual globe platform, particularly the automated flight over the terrain. Visitors found the narration matched with the automated flight path to be intuitive and insightful to follow in the fusion of data types for the telling of the story. These datasets had never been seen together before, and the product of this effort is exemplary interdisciplinary collaboration and research. Insights to where, when, and how the Late Bronze Age environment could have influenced the story of the Exodus has been revealed to researchers and lay audiences alike.

This paper describes an experiment in combining temporal geospatial data, prototyped on a hardware system that was built only days before exhibition opening. During development, the disparate data formats and sources posed challenges for the amount of work required to reduce data sizes to balance rendering ability and the required level of detail. Much time-consuming trial and error was necessary to converge on an overall real-time rendering rate. A compounding factor was that the WAVE was not available for testing until a few days before the Exodus exhibit opened, so we had to test on comparable, but not identical systems. Hence, some of the final tweaking had to happen in the days and hours before show time.

Other challenges were found when blending data across multiple software platforms. To solve this problem, the team collaborated to optimize functionality among the software available. This issue is particularly evident when converting geospatial data from a traditional 2D desktop application to a 3D environment. 3D geospatial visualization does not display vector data easily. In order to look good in a 3D environment, points must be represented as 3D objects, lines must be represented as tubes, and areas represented as rasters, thus increasing the volume of data to be rendered. By taking advantage of the new WAVE 3D immersive environment, and integrating it with the sonic arts (Seldess et al., Chap. 11) and MediaCommons Framework, we have been able to build the world of the Exodus in a meaningful way for twenty-first century audiences.

Conclusion

For the first time, ancient world building applied to the Hebrew Exodus from Egypt was taken to an immersive level by employing the new 3D WAVE experience. We successfully used CalVR, ArcGIS, GDAL, OpenSceneGraph, and other software tools to create a real-time rendered 3D demonstration of a possible scenario for the Parting of the Sea in the story of the Exodus. While we built the application on top of existing components, a large amount of custom work was necessary to bring all the pieces together seamlessly. In the future, it would be desirable that authoring tools be developed to simplify and streamline the many aspects of this project. But we were able to show that for an experienced team of content creators and programmers it is possible to pull together a complex 3D application in the relatively short time frame of about 2 months.

Acknowledgements We would like to thank the following researchers for contributions to this project: Stephen Moshier for his digital geospatial database (see Moshier and Hoffmeier, Chap. 8), Steven Ward for his collaboration and contributions of tsunami simulations, Daniel Jean Stanley for his Nile Delta data and intellectual contributions, Falko Kuester for WAVE leadership, Tiffany Fox for press release text concerning the WAVE, and last but not least, Brad C. Sparks for his Thera calculations and Exodus route maps and mapping data.

References

Bietak, M. 1996. *Avaris the Capital of the Hyksos: Recent excavations at Tell el-Dab'a*. London: The British Museum Press.

Block, J. 2007. *Visualization for Water Resource Management and for Salt River Paleo-Geomorphology in Central Arizona*. Unpublished M.S. thesis. Arizona State University.

Booysen, R. 2013. *Thera and the Exodus*. Alresford, Hants: O-Books.

DeFanti, T.A., D. Acevedo, R.A. Ainsworth, M.D. Brown, S. Cutchin, G. Dawe, K.-U. Doerr, A. Johnson, C. Knox, R. Kooima, F. Kuester, J. Leigh, L. Long, P. Otto, V. Petrovic, K. Ponto, A. Prudhomme, R. Rao, L. Renambot, D.J. Sandin, J.P. Schulze, L. Smarr, M. Srinivasan, P. Weber, and G. Wickham. 2011. The Future of the CAVE. *Central European Journal of Engineering* 1(1): 16–37. ISSN 1896-1541.

Dockerty, T., A. Lovett, G. Sunnenberg, K. Appleton, and M. Parry. 2005. Visualising the potential impacts of climate change on rural landscapes. *Computers, Environment and Urban Systems* 29(3): 297–320.

Ellis Smith, M.A. (ed.). 1993. *Holman Book of Biblical Charts, Maps, and Reconstructions*. Nashville, TN: Holman Bible.

Goodman-Tchernov, B.N., H.W. Dey, E.G. Reinhardt, F.W. McCoy, and Y. Mart. 2009. Tsunami waves generated by the Santorini eruption reached Eastern Mediterranean shores. *Geology* 37(10): 943–946.

Hoffmeier, J.K. 1999. *Israel in Egypt: The evidence for the authenticity of the Exodus tradition*. Oxford: Oxford University Press.

———. 2005. *Ancient Israel in Sinai*. Oxford: Oxford University Press.

Houtman, C. 1993. *Exodus [historical commentary on the Old Testament]*. Kok: Kampen. Vol. 1 of 4 vols. (1993–2002).

Kitchen, K. 2003. *On the reliability of the Old Testament*. Grand Rapids, MI: Eerdmans.

Levy, T.E., T. Higham, C. Bronk Ramsey, N.G. Smith, E. Ben-Yosef, M. Robinson, S. Münger, K. Knabb, J. P. Schulze, M. Najjar, and L. Tauxe. 2008. High Precision Radiocarbon Dating and Historical Biblical Archaeology in Southern Jordan. *Proceedings of the National Academy of Sciences* 105(43): 16460–16465.

Lin, A.Y.-M., A. Novo, P.P. Weber, G. Morelli, D. Goodman, and J.P. Schulze. 2011. A Virtual Excavation: Combining 3D Immersive Virtual Reality and Geophysical Surveying. In *Proceedings of ISVC 2011, Part II*, LNCS, vol. 6939, ed. G. Bebis et al., 229–238. Heidelberg: Springer.

McCoy, F.W., and G. Heiken. 2000. The Late Bronze Age explosive eruption of Thera (Santorini), Greece: Regional and local effects. *Geological Society of America*, Special Paper, 43–70.

Moshier, S.O., and A. El-Kalani. 2008. Late Bronze Age paleogeography along the ancient ways of Horus in Northwest Sinai, Egypt. *Geoarchaeology* 23(4): 450–473.

Novikova, T., G.A. Papadopoulos, and F.W. McCoy. 2011. Modelling of tsunami generated by the giant Late Bronze Age eruption of Thera, South Aegean Sea, Greece. *Geophysical Journal International* 186(2): 665–680.

OpenSceneGraph. 2013. OpenSceneGraph API: http://www.openscenegraph.org

osgEarth. 2013. osgEarth—Terrain on Demand: http://osgearth.org

Propp, W.H.C. 2006. *Exodus 19–40 [Anchor Bible 2.2]*. New York, NY: Doubleday.

Schulze, J.P., A. Prudhomme, P. Weber, and T.A. DeFanti. 2013. CalVR: An Advanced Open Source Virtual Reality Software Framework. In *Proceedings of IS&T/SPIE electronic imaging, the engineering reality of virtual reality*, San Francisco, CA, February 4, 2013, ISBN: 9780819494221

Sheppard, S.R.J. 2005. Landscape visualisation and climate change: The potential for influencing perceptions and behaviour. *Environmental Science & Policy* 8(6): 637–654.

Sivertsen, B.J. 2009. *The parting of the sea: How volcanoes, earthquakes, and plagues shaped the story of Exodus*. Princeton, NJ: Princeton University Press.

Smith, T.M., and V. Lakshmanan. 2011. Real-time, rapidly updating severe weather products for virtual globes. *Computers & Geosciences* 37(1): 3–12.

Stanley, D.J., J.E. McRea Jr., and J.C. Waldron. 1996. *Nile Delta Drill Core and Sample Database for 1985–1994: Mediterranean Basin (MEDIBA) Program, Smithsonian Contributions to the Marine Sciences*, 37th ed. Washington, DC: Smithsonian Institution Press.

Stanley, D.J., and H. Sheng. 1986. Volcanic shards from Santorini (Upper Minoan ash) in the Nile Delta, Egypt. *Nature* 320: 733–735.

Stanley, D.J., and A.G. Warne. 1993a. Sea level and initiation of Predynastic culture in the Nile delta. *Nature* 363: 435–438.

———. 1993b. Nile Delta: Recent Geological Evolution and Human Impact. *Science* 260: 628–634.

Tiede, D., and S. Lang. 2010. Analytical 3D views and virtual globes—Scientific results in a familiar spatial context. *ISPRS Journal of Photogrammetry and Remote Sensing* 65(3): 300–307.

Tomaszewski, B. 2011. Situation awareness and virtual globes: Applications for disaster management. *Computers & Geosciences* 37(1): 86–92.

Vervenne, M. 1995. The lexeme סוף (sûph) and the phrase ים־סוף (yam sûph). In *Immigration and emigration within the ancient Near East: Festschrift E. Lipinski [Orientalia Lovaniensia Analecta, 65]*, ed. K. Van Lerberghe and A. Schoors, 403–429. Peeters: Leuven.

Vote, E., D. Acevedo, M.S. Joukowsky, and D.H. Laidlaw. 2000. ARCHAVE - A Three Dimensional GIS for a CAVE Environment. In *CAA 2000: Computing Archaeology for Understanding the Past* [*Proceedings of Computer Applications for Archaeology (CAA) 28th Ann. Conf.*, Ljubljana, Slovenia, April 18–21, 2000]. Oxford: BAR (British Archaeological Reports) International Series, Archaeopress. http://graphics.cs.brown.edu/research/sciviz/archaeology/archave/

Yano, K., Y. Isoda, and T. Kawasumi. 2012. Virtual Kyoto as 4D-GIS. In *Virtual Geographic Environments*, ed. H. Lin and M. Batty, 349. Redlands, CA: ESRI Press.

MediaCommons Framework: An Immersive Storytelling Platform and Exodus

13

David Srour, John Mangan, Aliya Hoff, Todd Margolis, Jessica Block, Matthew L. Vincent, Thomas A. DeFanti, Thomas E. Levy, and Falko Kuester

Abstract

Scientific storytelling can be utilized to effectively convey research by synthesizing media-based data sets within a narrative frame. The "MediaCommons Framework" (MCF) was developed to efficiently integrate high-resolution media via cluster display systems for immersive, collaborative visualization. By incorporating temporal, spatial, and audio localization components into a wide-range of high-resolution media types, the MediaCommons Framework provides an ideal platform for scientific storytelling as it offers a coherent view of contextualized data, imparting a more engaging and intelligible experience to the public. As a case study, we describe our experiences using the framework to develop a storytelling application for the 2013 *EX3: Exodus, Cyber-archaeology, and the Future* digital museum exhibit. The use of the MCF within EX3 demonstrates the significance of advanced visualization in archaeology and the exigency to advance cyber-archaeological and other transdisciplinary research endeavors. This chapter's focus is on the technology used during the Exodus exhibit to convey stories to an audience within a museum-like environment. It includes high-level technical details about the framework used, our experiences developing and using storytelling applications for tiled-display walls, and the outcome of the public events these applications were used in. The experiences mentioned in this chapter are from the points of view of the framework's developers, exhibit content creators and managers, as well as archaeologists who have had the chance to use the applications to tell their stories to their targeted audience. This research embodies the ancient World Building methodology outlined in Seldess et al. (Chap. 11).

D. Srour (✉)
Qualcomm Institute, University of California, San Diego (UCSD), La Jolla, CA, USA
e-mail: dsrour@ucsd.edu

T.E. Levy et al. (eds.), *Israel's Exodus in Transdisciplinary Perspective*, Quantitative Methods in the Humanities and Social Sciences, DOI 10.1007/978-3-319-04768-3_13,

Introduction

With evolving technology, the amount of media that can be visualized has greatly increased in terms of both data resolution and the multitude of data types. Much effort has been put into the hardware aspect to reproduce scalable visualization environments which can effectively immerse participants as desired. Technology such as the OptiPortal (DeFanti et al. 2009: 114–123) has enabled scalability, portability, and ease of deployment for immersive environments. The ability to present larger data sets through tiled-display walls has led to novel ways in how information is conveyed to the end-user. Many software platforms have been created to help in visualizing media on clustered environments.

The Scalable Adaptive Graphics Environment (SAGE) (Renambot et al. 2004: 2004–2009), developed at the University of Illinois, generally processes the rendering needed for each frame on a head node and streams the distribution of pixels to tiled-display walls. This is a great technique as any visual application can be run on a tiled-display as long as it can redirect its pixel buffers to SAGE. On the other hand, this method is bounded by the resolution and complexity at which the head-node renders. The higher the resolution, the higher the resultant data bandwidth; the greater the complexity of the application, the more processing the head node must accomplish. This removes the parallelism that could be used from a cluster of rendering machines, making it harder to efficiently display larger data sets.

The Cross-Platform Cluster Graphic Library (CGLX) (Doerr and Kuester 2011: 320–332), developed at the University of California, San Diego solves these problems by taking advantage of the rendering parallelism of the cluster. By providing a simple abstraction layer for cluster synchronization and communication, it allows for OpenGL applications to be developed for scalable environments. One problem this presents is that CGLX is essentially as low-level as OpenGL, increasing the development time and efforts required to develop or reproduce complex collaborative visualization applications.

With both software and hardware solutions available, the needed lower level tools are available to display larger content on scalable display areas. CGLX being the most commonly used framework within our lab to display large content on planar tiled walls, developers noticed a lack of reusability between applications which led to impractical code maintenance tasks as many applications reinvented the wheel for commonly used features. It is for this reason that the creation of the MediaCommons Framework (MCF) came about. The MCF aims to abstract away the complexities of visualization on tiled-display walls. It allows developers to design simpler programs which they can develop and test on a standard desktop workstation. The application can then be executed as is within larger immersive environments.

With a flexible framework facilitating the development of tiled-display wall applications, advanced visualization techniques can be utilized for research dissemination by a wide spectrum of disciplines that have not previously exploited such technologies. Visualization is an essential process within the field of archaeology for analyzing the complex spatial relationships of sites from which inferences are drawn. While efforts are being made to digitally document cultural heritage sites for virtual preservation, the utility of cyber-archaeology practices are dependent upon the available methods for visual analytics and public outreach. The MCF serves as an excellent platform for these transdisciplinary enterprises as it permits comprehensive data sets to be displayed contextually and also supports "scientific storytelling," a technique that can increase public engagement with and interest in the content. To help tell the scientific story of twenty-first century Exodus research, we used Alex McDowell's 5D Institute methodology of "World Building" for the cinematic arts (http://5dinstitute.org/about).

We will discuss the history and goals behind the MCF, briefly review its technical aspects, and detail our experiences using it as a storytelling platform for the *EX3: Exodus, Cyber-*

archaeology, and the Future exhibit. Finally, we will discuss the role of visualization and storytelling specifically within the field of archaeology and the significance of transdisciplinary research in the study of cultural heritage.

History of MCF

CGLX was developed in house at UCSD and became one of the favored frameworks to display high resolution data. When installed, CGLX comes with utility applications to open and visualize various file formats on tiled-displays. Each application specialized in visualizing certain data types such as videos, flat and pyramidal tiffs (Yamaoka et al. 2011: 498–505) and 3D models. Unfortunately, these applications were crude, as many research projects are when they begin, and lacked the ability to open content dynamically at run time or to visualize different data types at once, restricting the way the desired message was presented to the audience. For example, being able to display a video and an image at the same time was impossible. This broke the story's flow since an application had to be shut down and another executed in order to visualize differing forms of media. It also greatly diminished the amount of data that could be provided at once, leaving critical information about the overall story out. As an example, an audience would grasp more information by being able to visualize an artifact with its 3D model, an image representation, and textual metadata all displayed at once in proximity of each other. From this desired functionality, MediaCommons was conceived to fully take advantage of the potential of scalable tiled-display walls.

The MCF started off as MediaCommons the application. The MediaCommons application was developed using CGLX to provide users with a generic and intuitive way to display and manipulate media, regrouping the default utility applications into a single module. It allowed users to display and manipulate data sets of mixed digital media. It also provided the ability to open and close media at run time, creating opportunities to use the display walls to tell stories in a more natural manner.

MediaCommons has been used around the world, for conferences, data analysis, collaboration, and presentations. There were times, however, for which MediaCommons failed to fit the bill. The format of how stories are told varies considerably depending on the message. The requirements on how such a message is delivered cannot be generalized into one application. Since the interaction in MediaCommons was built for the general use-case scenario, being able to tell a story in a specialized manner and with specific logic was impossible.

It became noticeable that the mechanisms needed to tell stories remained the same, with the only differences being the user interaction and the way these mechanisms were used. The key failing point of MediaCommons was the requirement of a human end-user driving the demonstration. While this user interaction worked well for presentations and unscripted stories, it was not feasible for cyclical content which was meant to run for hours, or even days, on end. To tackle this new set of requirements, the core components of MediaCommons were abstracted out of the application to create a high-level application programming interface (API) built on top of CGLX, OpenSceneGraph (Burns and Osfield 2004: 203) and smaller utility libraries (Fig. 13.1). This refactoring resulted in a suite of modules, collectively the MCF, which has been used to drive the general MediaCommons application as well as many different specialized storytelling applications, two of which will be explained in more detail.

MediaCommons Today

The modules offered in the MCF comprise intuitive user interface utilities, the ability to efficiently load and display large data sets, extend new data types, provide collaboration between both collocated users in a single environment and users in separated environments, and require minimal effort to reproduce features already used in MediaCommons. It also provides networking

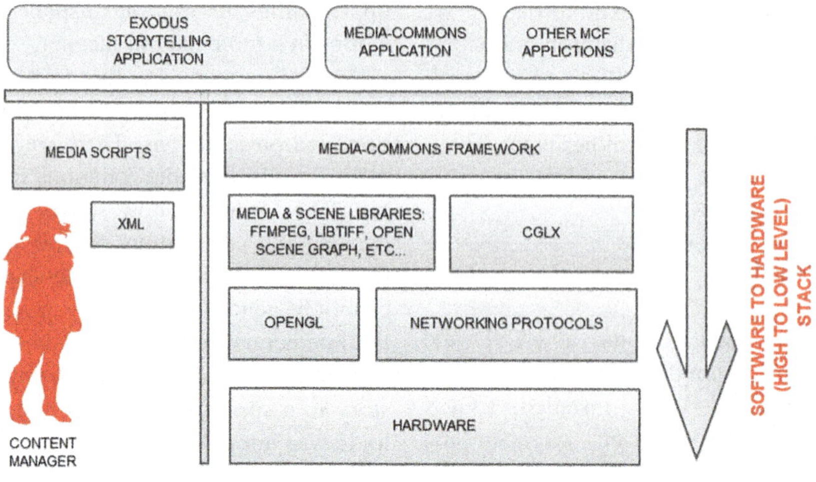

Fig. 13.1 Construction of Exodus Storytelling Application

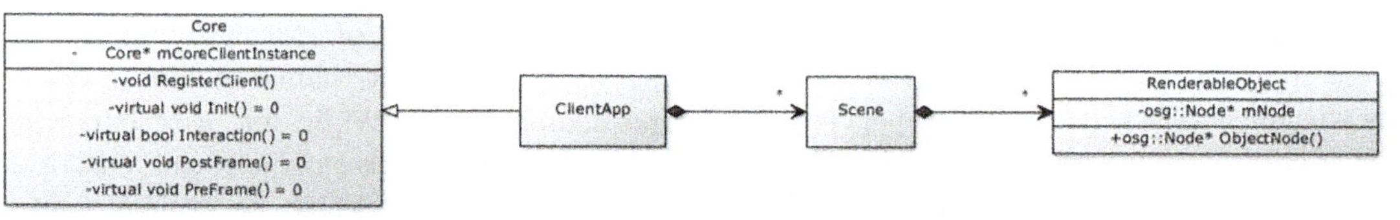

Fig. 13.2 UML class diagram of relationship between client class and MCF (MediaCommons Framework) core class

and profiling tools which may be used by developers to ensure optimal performance. The modules are extensible as developers are able to add new data types and features.

The following UML class diagram shows the relationship between the client class and the MCF's core class (Fig. 13.2). A developer would inherit from a core singleton instance. As seen, the core class offers a multitude of virtual functions within which specialized logic can be implemented. The client class may also hold one or more scenes which represent the visual components of the application. All media types are implemented (or extended) as RenderableObject classes and belong to each scene. These RenderableObject classes hide the complexity away from the developer and present an easy to use interface to manipulate each data type within a scene.

Equipped with these core features, a developer is able to quickly generate a visualization application that is scalable to any suitable tiled-display wall, specific to its own requirements on how a story is to be told to its audience. With data manipulation functionalities in place, all that is required to be developed is the specific application logic and desired user interfaces. The following use-case will demonstrate how crucial the MCF was in allowing developers to create specific storytelling applications within opportunistic deadlines, partially by enabling content managers to configure the media externally via scripting.

EX3: Exodus, Cyber-archaeology, and the Future

As our institute was responsible for hosting the *Exodus: Out of Egypt* conference and associated exhibition *EX3*, the requirements behind the content to be shown on our tiled display walls became more involved. Concrete stories were to be told using different strategies. The decision

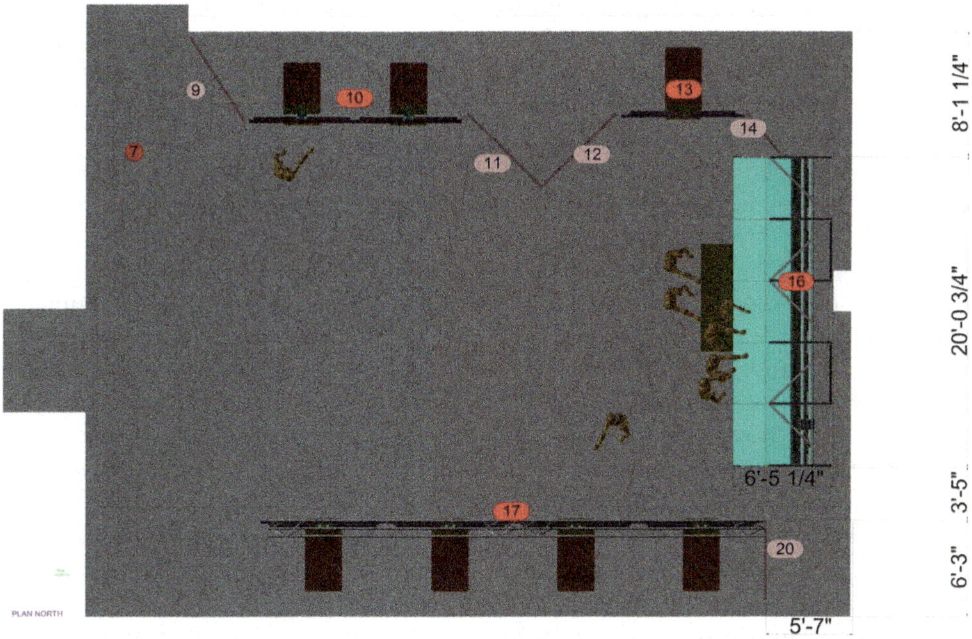

Fig. 13.3 Layout of EX3 Exodus multimedia environment

was also made to create a single application that would be executable on walls of differing resolutions, to ensure that its lifetime would extend that of a single conference or setup. The following list describes the given hardware and software requirements:

- Walls of differing resolutions and screen quantities (4 × 2, 2 × 2, 8 × 4).
- Different stories and content on each wall.
- Ability to specify the position and size of the media.
- Ability to display a wide range of high-resolution data types.
- Ability to display multiple media items at once.
- Fully automated.
- Specific audio localization cues.
- Configurable duration time for each media asset.
- Infinite looping playback.

The goal was to create an application that could be used in a museum environment to tell a complete story through immersive multimedia (Fig. 13.3). Some of the requirements were similar to those for earlier projects, but others required further implementation. All the core mechanics for media manipulation were readily available via the MCF, while the content to be shown and the duration of the demo were unknown to the developers. Content creators were to work in parallel while the application was being developed. Despite having only 10 days until the exhibition opened, the team held off on development to look back over prior MCF applications for code reuse possibilities.

By design, many features of the framework were able to be reused without attributing any additional effort to leverage them. If the developing team had to implement means of loading, unloading, displaying, transforming, and manipulating the wide range of media types, the completion would not have come until well beyond the tight deadline. Instead, the vast majority of development time was spent on the user interface that would control content and its transition logic.

First and foremost, the display size was unknown at the time of development. This was less of an issue than we first imagined, since content would have to be configured on a per wall basis. The individual needs for each wall, along with the parallel design of content and development, led to the apparent need for a scripting interface. Unfortunately, the only previously written application that involved any form of script only supported media that were shown deterministically, but always centered on the wall and sequentially ordered. The requirements for the Exodus conference added a spatial component where more than one media asset could be displayed at once, each with their own independent length of time. Required from the script per media item was the configurability of: the media type, location of the data file, a normalized position to display the data on the wall, a scale to modify the data's size, a starting time and the duration of time the data should be shown. Other requirements, such as the audio localization cues, needed further investigating. Different sound cues needed to be played depending on the position of the audience members. Though the MCF was already able to open and play sound files, there was not an integrated feature to trigger sounds depending on user positions.

The Exodus Storytelling Application employed a similar approach to the looping behavior seen in previous scripted applications; however, the implementation had to accommodate for the new requirements for positioning and starting each asset at their appropriate start time, as opposed to showing media items sequentially that were centered and scaled to perfectly fill up the tiled-display wall. Afterwards, development focus shifted to handling the audio cues generated by users' movements in front of the exhibits. To accomplish this, the application leveraged work accomplished by the Sonic Arts R & D audio team at the Qualcomm Institute (Qi), the University of California, San Diego (UCSD) branch of the California Institute for Telecommunications and Information Technology (Calit2). Audio cues were described within the script and sent off to a networked audio server at the appropriate time. The audio server, using data on the presence and location of audience members sent by custom depth camera tracking software, rendered spatialized sound to enhance the narrative presented by the visual displays. These audio cues were beneficial in involving the audience on a more personal and participatory basis, creating a greater sense of overall immersion within the media spaces. Examples of this included multilingual audio streams rendered via highly directional audio beamforming technology, and audio content that only began once users were standing in front of the visualization walls, as discussed in its respective chapter within this volume (Chap. 11).

The development of the Exodus Storytelling Application was completed within the allotted time. By being able to focus development time on nonexistent features and reusing many of the core modules present in the MCF, our development team successfully created a generic, flexible, and portable storytelling application.

Application of MCF to Archaeological Visualization and Storytelling

Archaeology strives to recreate accurate "stories" of past peoples by piecing together their material remains. The past is studied by examining sites and the objects within; this intersection of time and space stipulates that our interpretations are dependent upon the assessment of complex spatial relationships. As traditional "dirt" archaeology is a destructive science, extensive site documentation for visualization purposes—such as cartography, sketches, and photographs—is essential to preserve the spatial relationships destroyed by excavation. The emerging field of cyber-archaeology employs diagnostic imaging tools to gather enormous amounts of data, enabling 3D site visualization and virtual cultural heritage preservation, increasing the objectivity and integrity of our data sets, and facilitating public dissemination and interdisciplinary collaboration. These digital data sets pose new challenges to archaeologists, namely, how to streamline data integration for visualization techniques that prove most effective for analytics (Levy et al. 2010: 135–153). Standard computer and television displays lack the capacity to comprehensively view full data sets (without having to switch between dozens of windows), which previously limited the analytical value of digital documentation. The MCF, in contrast, offers the necessary analytical coherence and context by providing a holistic view of our data in a vast spectrum of formats, while the scalable environment allows us to investigate and compare the smallest details.

The MCF was heavily featured in the digital museum exhibit *EX3: Exodus, Cyber-archaeology, and the Future*. The aims of this exhibit were to demonstrate (1) the demand for transdisciplinary research to interpret the past and (2) the efficacy of public dissemination through digital museums. By incorporating transdisciplinary hypotheses to supplement contentious archaeological, historical, and textual evidence, EX3 provided new ways to rethink the Exodus and affirmed that there still remain many approaches to this historical problem that have yet to be pursued. EX3 also served as a model for a museum of the future firmly rooted in scientific storytelling through advanced visualization. MCF was used on three tiled display walls that each described a different story. The first detailed the cyber-archaeology process and its analytical value; the second relayed a narrative about mortuary sites that may evidence migratory "Exodus" events, and the third presented transdisciplinary approaches to investigate the "parting of the sea" narrative. The MCF storytelling platform permitted visitors to engage with both the archaeological content and the scientific methods that produce it, rendering science universally accessible and bridging the gap between researchers and the public.

Story 1: The Cyber-archaeology Process

The UC San Diego Edom Lowlands Regional Archaeology Project strives to integrate traditional fieldwork with advanced technologies to rapidly image and map archaeological sites. A 4 × 2 OptiPortable presented the story of how cyber-archaeology research is conducted by highlighting its three essential processes: data acquisition, data curation and analytics, and research dissemination. The three corresponding sections were spatially separated by different colored backgrounds on the display and matching colored lights on the floor so that the visitor could easily understand the dense visuals and see where to stand to hear the accompanying narrations (see Chap. 11). The left section contained pictures and videos of data acquisition techniques in the field, showcasing members of the ELRAP project and the advanced survey and imaging equipment currently in use, including laser scanning, GPS, aerial photography, ground penetrating radar, and more. The middle section highlighted data curation and analytics, showing pictures and videos of laboratory analysis and the data infrastructures and repositories that integrated data. The right section demonstrated efforts to disseminate cyber-archaeology research visually to the public and featured video clips of fly throughs of our 3D site models and pictures of outreach events.

Wrapping this all into one package was achieved using the scripting feature of the Exodus Storytelling Application. The individual sections could be fine tuned, timings adjusted on videos or images, as well as triggering the narration whenever necessary. The timing for each digital asset was scripted in an XML file, indicating the start and end times for the entire script. The positioning was done on a relative axis, with each asset being scaled to fit within its designated area. Furthermore, each asset could be assigned a "*z*" value, allowing it to be placed over or under other assets. This was necessary for the Cyber-archaeology layout, as a colored background filled the wall, and each asset appeared above it. This background was the basis for the coordination with the audio and visual dissemination of the presentation.

All three narratives were displayed simultaneously, uniquely paralleling how these processes are constantly ongoing and data sets ever-expanding. Thus, cyber-archaeology practices offer techniques to objectively record large amounts of data which researchers can interpret endlessly, minimizing bias and increasing scientific integrity. Within the controversial field of Biblical archaeology, cyber-archaeology proffers a novel approach to scientifically investigate the past.

Story 2: Mortuary Archaeology

A 2 × 2 OptiPortable highlighted mortuary archaeology excavations conducted by ELRAP in Wadi Fidan, located within the boundaries of the Biblical kingdom of Edom. Excavations at the enormous cemetery Wadi Fidan 40 and the nearby multi-individual tomb Wadi Fidan 61 provide a glimpse into the lives of the ancient inhabitants of tenth-century BCE Iron Age Edom (Levy et al. 1997: 293–308). Based on grave architecture, artifact assemblages, and an absence of nearby permanent settlements, the community is thought to have been nomadic and may have belonged to one of the Shasu tribes—the Shasu of Edom—referenced in ancient Egyptian texts (Levy et al. 2004: 63–89). A number of Egyptian artifacts were found in the graves as well, including a sixteenth century BCE Hyksos scarab dating to the approximate time of the Exodus, demonstrating a long-term connection between the nomadic populations of the Levant and the Nile Delta (Levy et al. 1997: 293–308). Many scholars suggest the early Israelites were one six nomadic Shasu tribes that made their way to Egypt as early as the sixteenth century BCE and later emerged in the Exodus narrative; the mortuary excavations in the Wadi Fidan provide evidence for such migrations (Levy et al. 2004: 63–89).

The mortuary archaeology OptiPortable looped one 4-min narration that relayed the story of these nomadic people by demonstrating how our researchers piece together archaeological and textual evidence. Content displayed on this OptiPortable fell into three categories: high-resolution photography, video clips, and flythroughs of 3D models. Photographs of the sites, grave architecture, artifact assemblage, skeletal remains, and landscape were paired with video clips demonstrating excavation methods, analytical laboratory procedures, and diagnostic imaging techniques. Flythroughs of 3D models of the Wadi Fidan 61 site and tomb provided visitors with a sense of the sites' context within the landscape that cannot be fully grasped with two-dimensional photographs alone. A 3D model of the standing stone displayed at the entrance of the exhibit exemplified the results of our efforts to virtually preserve cultural heritage with diagnostic imaging technologies.

By manipulating spatial and temporal elements within the MCF, a single piece of high-resolution content could be displayed across the full expanse of the OptiPortable or instead scale and position several multimedia sources to produce a coherent view of complex archaeological evidence, exhibiting detailed photographs, contextual information, and methodological techniques within one frame. For example, one segment displayed high-resolution photographs of the hieroglyphics of an Egyptian scarab, detail shots of preserved pomegranates, photographs of the artifacts *in situ* to show their spatial and contextual relationships within the burial, and finally a video clip that illustrated how radiocarbon samples from the pomegranates were processed and interpreted to date the grave. This degree of holistic visualization more closely reflects the process of generating archaeological interpretations and allows museum patrons to easily follow along and piece together the evidence for themselves, making our scientific methods more engaging and accessible to the general public.

To tell our mortuary archaeology story, we first created a "playlist" that listed the file names, formats, start and end times, position and scale on the display for each element. With this information, creating the script for the 4 min narration was quickly produced due to the framework's ease of manipulating spatial and temporal components. Once the script was written and the demo finalized, beginning and ending the playback loop was simple enough for undergraduates without any computer science background to master. From a content creator's perspective, the Exodus Storytelling Application used significantly less time than the creation of a standard HD video. Producing a video requires complicated professional editing software for precise temporal and spatial alignment with transitions between clips; moreover, it would be impossible to preview any such video during editing at the native high resolution of our tiled display walls. Comprehensive views of the complex relationships within a burial site are necessary to effectively relay a narrative about the identities and experiences of past peoples—a traditional video could not have provided such an encompassing view without extensive editing and dedicated man-hours. MCF presented a time-effective, flexible, and user-friendly alternative better suited for the visualization of a dynamic, ever-expanding data set.

The display's popularity indicated that a visual presentation of contextualized mortuary remains serves as a meaningful and intuitive method of communicating research to the public. Mortuary archaeology serves as an ideal springboard for storytelling within digital museums because it is inherently both narrative-based and transdisciplinary. This provides a unique opportunity for curators to unify scientific methods from disparate fields into a cohesive story about past livelihoods that the public can easily identify with. By utilizing the MCF, public engagement with the content becomes twofold: not only can we recount the story of an ancient population, we can simultaneously relate the story of scientific inquiry in archaeology.

Story 3: Transdisciplinary Approaches to the Exodus Narrative

Our largest display wall used 8 × 4 tiled HD screens (more than 64 million pixels) to present a transdisciplinary approach towards addressing the "parting of the Sea" in the story of the Exodus. Many researchers in archaeology, theology, and geophysics have investigated the possibility that the Thera (Santorini) volcanic eruption of the Late Bronze Age induced a tsunami reaching the Egyptian coast (Booysen 2013). However, these three fields have different perspectives on how this experience may have influenced the story. The event could have been experienced by the escaping Hebrews directly, the event may have been experienced by people hundreds of years prior to the Exodus and subsequently incorporated to the legend (Bruins et al. 2008; Halpern 2003: 50–57), or the eruption may have had no influence on the story of the Exodus at all. If it were the latter, then what other environmental phenomena could produce wave displacement in the Mediterranean? We used still images, text, animated fluid dynamics simulations, audio narration, and sound effects to frame this contentious story.

The MCF provided the spatiotemporal platform for integrating data from these three disciplines (geology, archaeology, and theology) into one research question. A series of still images were presented first, accompanied by audio narration to orient the visitor to the story, providing background context for why we must investigate this idea. The still images included photos of Middle Eastern excavation sites, active drilling of sediment cores to reveal preserved volcanic ash from the Theran eruption, current and historic maps, and photos of local geography highlighting characteristics of the environmental setting for the Exodus. Directional audio allowed for playing the narration simultaneously in English, Hebrew, and Arabic within three zones in front of the tiled display. Visitors could choose which language in which to listen based on where they stood in front of the tiled display.

Following this introduction, three possible geophysical scenarios, modeled by geophysicist Steven Ward of UC Santa Cruz, were presented that could explain the creation of a tsunami in the eastern Mediterranean Sea: a volcanic eruption at Santorini Island (left), a submarine Nile Delta landslide (middle), or a regional wind storm (right) (Fig. 13.4) (See also Salamon et al., this volume, Chap. 9).

Fig. 13.4 Three possible geophysical scenarios of Parted Sea in the Exodus shown in high-resolution computer simulations displayed at EX3 exhibition (left to right): Tsunami in the Eastern Mediterranean Sea from volcanic eruption at Santorini / Thera island *(left)*; Submarine Nile Delta landslide (middle); Regional wind storm *(right)*

The spatiotemporal relationship of these environmental processes to the Nile Delta is not intuitive to a general audience. The MCF provided a means to clearly describe these phenomena by grouping three animations per scenario, showing the physical processes on the Sea surface, in the subaqueous (underwater), and the subsequent local effect at the Delta. Without the large display and automation capacity to synchronize these scenarios to play simultaneously using the MCF, they would not have been easily contrasted and analyzed. Using the same sound-bending technology as for the multilingual narrations, ambient sound effects of an explosive volcano, a landslide, and a storm were used to further contextualize and enhance the contrast among the simulations, and the directional audio confined each sound effect to the correlative scenario.

Not only did the MC provide us with the unique ability to simultaneously present these three hypothetical scenarios to explain this part of the story of the Exodus, but it also told the story of archaeological and geological data collection and research. When viewed, the visitor could understand concisely why there is debate over when or how a tsunami could have been part of the Exodus, and why this question is being investigated with leading edge science and technology. These disparate data sets could not have been combined to convey the same story without the structure of the flexibility and power of the MCF.

The MCF provided EX3 content contributors with a fully automated platform that enabled contributors to efficiently build visual narratives containing a wide range of high-resolution data types. The ability to display multiple media items at once with precise control of their spatiotemporal components made it possible to present comparative and contextual information while also detailing how we acquired and analyzed our data. The flexibility of the MCF permits customization of audio and visual elements to effectively convey different stories and content can easily be added—features to be much desired by digital museum curators. Employing varying degrees of audio and visual complexity across the three tiled display walls allowed us to appeal to a wider audience and to better pinpoint visitor preferences for the design of future exhibits.

Future Work

Several features of the two MCF applications explored above can be unintuitive for those who are comfortable working with GUI applications as opposed to text-based scripts. Due to development time constraints, the Exodus Storytelling Application lacked some intuitive user interfaces and functions which would have made the framework more accessible to archaeologists, who then could have assumed a greater role in script creation. Incorporating a drag-and-drop timeline function into the menu structure will facilitate a more diverse set of users to dynamically build scripts *in situ*.

Memory management was also a concern with the first iteration of development for the Exodus Storytelling Application. Primarily, the decision was to load content when it was needed and unload it once its duration was over, which caused stuttering and delays when loading large data files such as 4K videos. Preloaded all content ensured that the requested media would load at the specified time, and this solution was viable for the exhibit as the totality of the data could fit in memory. It is, however, unscalable and would require some sort of prefetching and caching mechanisms for exhibitions whose data scales beyond the available hardware.

Finally, the playback of high resolution (4K+) video within the MCF depends upon another research product which introduces stricter video specification requirements than some commercial video players require. The extension to broaden these requirements would assist content producers and ultimately minimize the time they spend altering videos to smoothen playback.

In the future, we plan to capitalize on the features of the MCF to conduct extensive visual analytics of mortuary archaeology sites, commence international collaboration and data set sharing, and curate additional digital museum exhibitions.

Conclusion

The Exodus World Building Application being developed within such short guidelines show that the MediaCommons Framework successfully supported the display of numerous data types and formats on tiled display walls. As expected, the only required development for the applications was the user interface controls: application specific scripts which control the automated display of media content. Despite having written such applications with specific presentations in mind, the configurability via the scripting interface allows for the same storytelling applications to be used for future demonstrations as well.

The EX3 exhibit and its use of the MCF serve as an excellent model for museums of the future in which physical artifacts can be complemented with advanced visualization platforms adapted for scientific storytelling. Storytelling can be utilized to bridge the gap between public perception and the actual processes of scientific inquiry. The study and preservation of cultural heritage in the digital age will not be secure until the public can establish a personal interest in the field; effective and engaging presentations are thus vital and archaeologists must embrace transdisciplinary collaboration. To create a comprehensive narrative of the past, archaeologists cannot rely on specialized analysis alone; researchers must synthesize disparate data sets and diverse perspectives from many academic fields while continually advancing an objective archaeology rooted in the scientific method. Just as the utility of cyber-archaeology is dependent upon the available avenues for visualization, the movement towards digital museums can only be realized with the widespread adoption of cyber-archaeological practices. Through the MCF, we can integrate data into a coherent whole so that we may ensure that our work, in contributing to a collective memory, is the most complete and accurate representation of the past.

Acknowledgements This work was supported by the National Science Foundation under IGERT Award #DGE-0966375, "Training, Research and Education in Engineering for Cultural Heritage Diagnostics" at: UC San Diego Department of Anthropology, UCSD Levantine Archaeology Lab, UCSD Levantine Osteology

Lab, Department of Antiquities of Jordan. Qualcomm Institute, University of California, San Diego (UCSD).

References

Booysen, Riaan. 2013. *Thera and the Exodus: The Exodus Explained in Terms of Natural Phenomena and the Human Response to It*. Alresford, Hants: O-Books.

Bruins, Hendrik, Alexander MacGillivray, Costas Synolakis, Chaim Benjamini, Jörg Keller, Hanan Kisch, Andreas Klügel, and Johannes Van Der Plicht. 2008. Geoarcheological tsunami deposits at Palaikastro (Crete) and the Late Minoan IA eruption of Santorini. *Journal of Archeological Science* 35: 191–212.

Burns, Don, and Robert Osfield. 2004. Open Scene Graph. In *Proceedings of the IEEE Virtual Reality*, 265. Washington, DC: IEEE.

DeFanti, T., J. Leigh, L. Renambot, B. Jeong, A. Verlo, L. Long, M. Brown, D. Sandin, V. Vishwanath, Q. Liu, M. Katz, P. Papdopoulos, J. Keefe, G. Hidley, G. Dawe, I. Kaufman, B. Glogowski, K. Doerr, R. Singh, J. Girado, J. Schulze, F. Kuester, and L. Smarr. 2009. The OptiPortal, a Scalable Visualization, Storage, and Computing Interface Device for the OptiPuter. *Future Generation Computer Systems* 25: 114–123.

Doerr, Kai-Uwe, and Falko Kuester. 2011. CGLX: A Scalable, High-Performance Visualization Framework for Networked Display Environments. *IEEE Transactions on Visualization and Computer Graphics* 17: 320–332.

Halpern, B. 2003. Eyewitness testimony: Parts of exodus written within living memory of the event. *The Biblical Archeology Review* 29: 50–57.

Levy, T., R. Adams, and R. Shafiq. 1997. The Jabal Hamrat Fidan Project: Excavations at the Wadi Fidan 40 Cemetery, Jordan. *Levant* 31: 293–308.

Levy, T., R. Adams, and A. Muniz. 2004. Archaeology and the Shasu Nomads – Recent Excavations in the Jabal Hamrat, Fidan, Jordan. In *Le-David Maskil: A Birthday Tribute for David Noel Freedman*, vol. 9, ed. W.H. Propp and R.E. Friedman, 63–89. Winona Lake IN: Eisenbrauns.

Levy, T., V. Petrovic, T. Wypych, A. Gidding, K. Knabb, D. Hernandez, N. Smith, J. Schulz, S. Savage, F. Kuester, and T. DeFanti. 2010. On-site digital archeology 3.0 and cyber-archaeology: into the future of the past-new developments, delivery and the creation of a data avalanche. *Introduction to cyber-archaeology*, 135–153. Oxford: Archaeopress.

Renambot, L., A. Rao, R. Singh, B. Jeong, N. Krishnaprasad, V. Vishwanath, V. Chandrasekhar, N. Schwarz, A. Spale, C. Zhang, G. Goldman, J. Leigh, and A. Johnson. 2004. SAGE: the Scalable Adaptive Graphics Environment. *Proceedings of WACE* 9: 2004–2009.

Yamaoka, So., Kai-Uwe Doerr, and Falko Kuester. 2011. Visualization of High_resolution Image Collections on Large Tiled Display Walls. *Future Generation Computer Systems* 27: 498–505.

Part IV

The Exodus Narrative in its Egyptian and Near Eastern Context

Mythic Dimensions of the Exodus Tradition

14

Bernard F. Batto

Abstract

The received exodus narrative is an exilic or post-exilic Priestly redaction in which earlier JE traditions were deliberately rewritten so as to elevate the exodus "event" to mythic status, thereby glorifying both Yahweh and Israel. P portrayed the exodus as a continuation of the primordial battle between the creator and primeval Sea, a principal form of the ancient Near Eastern chaos monster. Pharaoh (Egypt) is depicted as the chaos monster—"the great dragon," to borrow the language of Ezekiel regarding Pharaoh. P manipulated the wilderness itinerary so as to have Pharaoh perish in *yam sûp*, geographically identifiable with our Red Sea but understood literally as "the Sea of End," a boundless expanse of ocean fraught with mythical overtones of non-creation, or chaos. Pharaoh is thus merged with primeval Sea and is defeated along with it. P employed Combat Myth motifs also to imply that the exodus was a second phase of creation whereby Israel, Yahweh's newest creation, emerged through the deity's splitting—or defeat—of the Sea.

This paper assumes a complex history of composition for the Pentateuch, as discussed by Richard Freedman, William Propp, Konrad Schmid, and others. It is unique among conference presentations for its focus within the biblical exodus tradition on mythological imagery derived from the Combat Myth. My thesis accords well with the positions of those who view the exodus narrative as a traditional story (Susan Hollis) or as cultural or collective memory (e.g., Jan Assmann, Ron Hendel, Aren Maeir). I agree with Christopher Brenner that the narrative reveals little about historical events. My thesis is not incompatible, however, with mediating positions (e.g., Brad C. Sparks) which posit that a historical event has been greatly embellished through poetic license or a religio-nationalistic agenda.

Introduction

For the sake of brevity I will here assume rather than argue that the exodus narrative found in the Book of Exodus is a traditional story (Propp 1999: 32–34; Sasson 1981: 84; Forsyth 1987:

B.F. Batto (✉)
DePauw University, 636 E. Seminary Street, Greencastle, IN 46135, USA
e-mail: bbatto@depauw.edu

T.E. Levy et al. (eds.), *Israel's Exodus in Transdisciplinary Perspective*, Quantitative Methods in the Humanities and Social Sciences, DOI 10.1007/978-3-319-04768-3_14,

3–17). The issue that I wish to address is not the historicity of the exodus[1] but rather the degree to which the exodus narrative has been adapted to the mythic traditions of the ancient Near East. The present form evidently was achieved at the hand of a Priestly redactor who sought to give the Mosaic Torah mythic status. Prima facie evidence comes from the manner in which Moses is elevated to semi-divine stature when his face "horned" (Exod 34:29–35), and how in death Moses is buried by God in mysterious secrecy (Deuteronomy 34; see Batto 1992: 123–126). P deliberately cloaked the exodus "event" in the mythic language, specifically in images drawn from the common Semitic Combat Myth. Moreover, Pharaoh and Egypt together are depicted as the incarnation of the Chaos Monster (Propp 1999: 560–561).

Exodus as an Adaptation of the Combat Myth

Israelite tradition has long associated the exodus with a divine deliverance at a sea, identified variously as "the sea" (*hayyām*) or by name as *yam sûp*. Whatever the identity of the sea, it quickly attracted to itself mythological imagery associated with "Sea" in ancient Near Eastern culture and especially in the common Semitic Combat Myth.

There is no doubt that *yam sûp* was the biblical name for our Red Sea, including its two northern extensions into the Gulf of Suez and the Gulf of Elath as evidenced by 1 Kings 9:26: "King Solomon built a fleet of ships at Ezion-geber, which is near Eloth on the shore of *yam sûp*, in the land of Edom."[2] Since rabbinic times there have been questions, however, about whether there existed a second, lesser body of water somewhere in the region that also bore the name *yam sûp*, to be translated literally as "Sea of (Papyrus) Reeds," or "Reed Sea." The principal stay of this argument is that *sûp* is an Egyptian loanword, derived from Egyptian *ṯwf(y)* and this loanword is attested in the Bible (Exod 2:3, 5 and Isa 19:6). Nevertheless, the Reed Sea hypothesis is flawed and, as I have argued elsewhere (Batto 1983: 27–35), must be given up. I follow Montgomery (1938: 131–132) and Snaith (1965: 395–98) in positing that *sûp* in the name *yam sûp* specifically is not an Egyptian loanword; rather as the LXX translators recognized, one may read *sôp* instead of *sûp*, or, better, regard *sûp* as a bi-form of Semitic *sôp* "end," "extremity."[3] *Yam sûp* should thus be translated as the "sea of end," that is, it is the last sea, the sea situated at the edge of creation—the created world being identical with *terra firma,* the solid earth, at the center of which stands the sacred mountain from where the Divine Sovereign rules (Batto 2004, Batto 2013b).

[1] See Propp (2006: 735–756) on the complex issues concerning historicity. S. C. Russell 2009 argues the attempt to find a single historical core for the exodus is misguided; close analysis of biblical traditions concerning Egypt and the exodus reveals considerable regional variation: Cisjordan-Israelite, Transjordan-Israelite, or Judahite. Oblath (2004: 194–195) concludes that the exodus narrative describes" a movement of people within and out of the region between the Negeb and the Gulf of Elath. . . . Egypt, no matter that it is a 'character' in the play, played no role in the events themselves."

[2] Propp (2006: 752) observes, "there is little doubt that in Exod 23:31; Num 14:25; 21:4; Deut 1:40; 2:1; Judg 11:16 [?]; 1 Kgs 9:26; Jer 49:21, the Suph Sea is the Gulf of Aqaba. In Exod 10:19; 13:18; Num 33:10–11, however, the Suph Sea appears to be the Gulf of Suez. In short, the Suph Sea is the Red Sea and its two northern arms."

[3] In the LXX *yam sûp* is consistently rendered as *Erythra Thalassa* except in two places: Judg 11:16 (*Thasassa Siph*, evidently [mis]reading *sîp* instead of *sûp*) and 1 Kgs 9:26 ("king Solomon built a ship in Gasion Gaber near Elath on the shore of the last sea [*tēs eschatēs thalassēs*] in the land of Edom." The latter was intended evidently as a literal rendering of *yam sûp/sôp*. Oblath (2004: 58–61) argues that the LXX translator being situated in Egypt, was intent on distinguishing the more distant (from Egypt) Gulf of Elath, which other Greek sources call the Aelanite Gulf, from the nearer and better known Red Sea/Gulf of Suez; but if such were the intention of the translator, why would other obvious references to the same Elath/Aelanite Gulf (Exod 23:31; Num 14:25; 21:4; Deut 1:40; 2:1; Jer 49:21) be rendered inconsistently in the LXX as *Erythra Thalassa*?

Patently, in various texts the name *yam sûp* had reference to the vast oceanic expanse that later the Greeks called *Erythra Thalassa* "Red Sea," subsequently translated into Latin as *Mare Rubrum* and into Aramaic as *yammā' śimmōqā'*, which included not only the Red Sea (with the Gulf of Elath and the Gulf of Suez) but also the whole expanse of oceans surrounding the Arabian peninsula including the Persian Gulf and the Indian Ocean—and presumably everything beyond.[4] As far as these ancients knew, *yam sûp* actually was located at the physical edge of *terra firma*, the boundary between creation and non-creation. It would be very appropriate to the ancient mythic mind that this sea be labeled as the "sea of end," that is, where creation ceases and non-existence takes over.[5]

The depiction of non-creation as untamed or chaotic waters was ubiquitous in the ancient Near East. In Genesis 1 God divided *tĕhôm*, using the *rāqîa'* to separate the waters above from those below to hallow out a living space for humankind and other creatures. In Egypt this trope was expressed as the primordial ocean Nun, whose chaotic waters were kept at bay by a barrier consisting of the sky god Nut and the earth god Geb. The taming or controlling of the chaotic waters of Nun was graphically depicted as the god Ra seated in his bark and riding across the sky, in the waters above Nut (Batto 2013b: 11–19, with illustrations). This was all positive from the Egyptian perspective, of course, but Israelite theologians would recast such mythemes as polemics to be used against the Egyptians!

P's Agenda of Mythopoeism

Turning now to the exodus narrative itself, I attend primarily to the Priestly tradition, as it is in P that the Combat Myth is most deeply entrenched. The Priestly Writer's agenda is most obvious when the P stratum of the Pentateuch is separated out from the other strata.

The Exodus Event in J+E

Recent biblical scholars dispute whether J and E once existed as discreet sources, or constituted but a single source that incorporated older fragments of tradition. Dozeman (2009), for example, posits two basic sources in the Book of Exodus, a "Non-P History" (matching JE of older source criticism) and a "P history." Whether originally one or two sources, most scholars hold that the JE elements constitute the most ancient layers of the exodus narrative. J and E are very difficult to disentangle, so I will treat them together, especially because in large measure they share a common narrative regarding what transpired in the deliverance at the sea.

According to JE, the Israelites in leaving Egypt traveled into the wilderness in the direction of *yam sûp*, i.e., the Red Sea. The text does not state that the Red Sea was their destination, only that the Israelites set forth in that direction, that is, in a southeastern direction. At some point Pharaoh mustered his chariot army to go after the Israelites. Pharaoh's motive is unclear, perhaps because J and E diverge in their presentations, e.g., flight versus deception versus Pharaoh changing his mind.

Pharaoh's forces caught up to the Israelites at "the sea" (*hayyām*). Since the wilderness itinerary given in Numbers 33 distinguishes a

[4] J. Fitzmyer (1971: 153–154): "According to M. Copisarow (1962) the term "Red Sea" originated with mariners of ancient Greece, independently of Egyptian or Hebrew influence; it designated the sea between Asia and Africa and was gradually extended from the Gulf of Suez to the Persian Gulf including the Indian Ocean."

[5] In discussion, James Hoffmeir noted that without vowel points *sôp* and *sûp* look the same, making possible in late texts a beautiful word play that enabled the mythic overtones posited herein. Aren Maeir questioned my contention that ancient Israelites attributed mythological attributes to *yam sûp*/Red Sea, noting that Iron Age and Persian period sailors navigated the Indian Ocean for spice trade with Arabia and India and so would not have considered it mythical. This ignores the fact that ancient mariners usually navigated close to known shorelines and that the Indian Ocean extends southward far beyond India (and Africa); and also that to the ordinary ancient Near Eastern mind oceans remained frightening entities fraught with mythic overtones, as attested in numerous texts.

camping station at the Red Sea (*yam sûp*, 33:10) some 3 days journey [three camping stations] from an earlier camping station at the sea of crossing (33:8), it is unclear whether JE intended to say that the Israelites found themselves trapped between the Egyptian army and *yam sûp* (already mentioned in Exod 13:18) or another, unidentified sea.[6] The Israelites find their progress blocked and believe their cause lost. Moses prays for help, and the deity responds with a call for faith, telling the Israelites to stand by and watch the deity's salvific actions unfold. The divine messenger/cloud then moves between the Israelites and the Egyptian army, with neither group moving during the night. During the night the deity causes the sea to recede, either by his direct breath or in the form of a strong east wind—possibly these are one and the same phenomenon. At the morning light the deity from his heavenly vantage panics the Egyptian army, which flees headlong into the temporarily dried sea bed, whereupon the deity allows the sea to flood back into its usual place, drowning the entire Egyptian force in the process. Apparently the Israelites never moved! (Noth 1962: 118; Childs 1974: 221).[7]

In my reading, JE placed the Israelite deliverance at an unnamed sea (in keeping with Numbers 33 which posits camping stations at two distinct seas). Moreover, only the Egyptians enter the Sea and are destroyed in it, leaving the Israelites free to continue unhindered on their journey to the mountain of God (Sinai/Horeb). The Israelites are mere spectators of the deity's awesome deeds wrought on their behalf. Having personally witnessed God's salvific acts, the Israelites gain renewed faith in God and his agent Moses.

P's Recasting of Exodus as an Act of Creation

If it is unclear in the JE tradition that the Israelites crossed "the sea" or that they ever reached *yam sûp* = the Red Sea, there can be no such doubt when it comes to the P tradition. P patently has the Israelites crossing on dry ground right through a "split" (*bqʿ*) Red Sea, with walls of water on their right and on their left (Exod 14:15–16, 21).

"The Sea" that P has in mind here is *yam sûp* = the Red Sea, for immediately following the deliverance at the Sea, P inserted in 15:22 the notice, "And Moses made Israel set forth from *yam sûp* and they went out into the Shur Wilderness" and they came to Marah where they encountered bitter, non-potable water. To achieve this version of Israelite journeying, P has redacted the JE exodus narrative to make the Israelites' trek through the wilderness conform to an old list of camping stations now found in Numbers 33, according to which the Israelites moved by distinct stages from Egypt to the promised land.[8] P made one major change to this itinerary, however. He contracted into a single station the two camping stations at the sea of crossing (Num 33:8) and at the Red Sea (Num 33:10). To accomplish this, P had to suppress the second station and transfer its name, *yam sûp* = the Red Sea, to the first sea, the sea that the Israelites were said to have crossed some three camping stations earlier. Manifestly, this contraction was a deliberate stratagem by which to reinterpret the exodus as an extension of Yahweh's combat against the Sea, which in turn was a continuation of the old Semitic Combat Myth.

[6] In previous publications I assumed that the latter was the case but now consider either case possible. I also opt for the integrity of Numbers 33 as an older and independent textual witness to the wilderness wanderings of the Israelites.

[7] Kloos (1986: 205) argues that the theme of the Israelites crossing a dried up sea was present already in the Song of the Sea.

[8] P is derived from Number 33 and not the other way round, as Numbers 33 would have no reason to add a second camping station at *yam sûp*, in addition to the one at the sea of crossing, had both seas not been already present in the tradition inherited by Numbers 33. The same cannot be said of P, against Propp (2006: 749–753).

Yahweh's Combat against the Sea is a topic that has been addressed many times in the past, by both supporters and detractors. I cannot delay over that debate here, except to note that I have recently reexamined the evidence for the presence of Combat Myth motifs in the HB, and again conclude that Combat Myth motifs are authentically present in numerous biblical passages (Batto 2013c). Hab 3:8–10, 15 is a prime example.

> Is not your wrath, O Yahweh, against River,
> Your anger against River,
> Your ire against Sea,
> When you drive your horses,
> Your chariot of salvation?
> You unsheathe your bow...
> The mountains see you and quake;
> Abyss cries aloud...
> With your horses you trample Sea,
> The raging of Mighty Water! (Hab 3:8–10, 15)

Despite textual corruptions and grammatical problems, it is patent that Habakkuk 3 has in view both the deliverance at the Sea and the Jordan crossing, and that both "events" have been homogenized as an extension of Yahweh's cosmic battle against chaotic Sea (Batto 1992: 146–147). The boundary between "history" and "myth" has been almost totally obscured.

Similarly in Psalm 74, composed in the shadow of the Babylonian destruction of the Jerusalem temple, the Psalmist attempts to rouse God's aid by contrasting his present apparent inaction in history with his awesome acts in primordial times in defeating the chaos monster.

> Why do you restrain your hand?
> your right hand remain idle inside your cloak?
> O God, my king from primeval times,
> who works salvation in the middle of the earth,
> It was you who broke apart the Sea by your might,
> who smashed the heads of the dragon on the water.
> It was you who crushed the heads of Leviathan,
> who gave him to the desert folk as food.
> It was you who opened springs and brooks;
> you who turned primordial rivers into dry land (Ps 74:11–15)

Implicit in the Psalmist's lament is a belief that in defeating the primordial chaos monster, Yahweh proved himself the creator of a world in which the forces of evil and non-creation are kept at bay (Batto 1987). Implicit here is an assumption that should "history" get out of sync with the myth, something is terribly amiss.

The image of the sea as the enemy of both Creator and creation/creature left its mark also on the exodus tradition. The author of Psalm 77, meditating upon God's primordial deeds (vs. 6, 12–13 [E.T.: 5, 11–12], portrays the exodus as an extension of the deity's battle against the sea:

> The waters saw you, O God,
> the waters saw you and shuddered;
> yea, the Abyss trembled.
> The clouds poured down water;
> the darkened skies rumbled;
> your arrows sped about.
> Your thunder resounded in the whirlwind;
> your lightning lit up the world;
> the earth quivered and quaked.
> Upon the Sea was your foot/dominion,
> upon Mighty Waters your treading,
> though no one saw your footprints.
> You led your people like a flock
> by the hand of Moses and Aaron (Ps 77:17–21 [E.T.: 16–20])

The last two cola make it clear that the psalmist has the exodus in mind. A similar mentality is at work in the Priestly Writer's reinterpretation of the exodus narrative.[9]

It is out of fashion these days to posit that the Priestly Writer composed his Genesis creation account against the backdrop of the Combat Myth, particularly the Babylonian myth *Enuma elish*. Granted, P consciously "demythologized" specific mythemes of the Combat Myth by eliminating a battle between the Creator and primeval Sea, and by reducing the chaos monster to mere creatures (*hattannînīm haggĕdōlîm*, Gen 1:21). But it is false to say that P completely eliminated entirely such mythemes from his creation account, as these form the backdrop for a new Priestly creation myth. As I have argued elsewhere, rather than eliminating Combat Myth motifs—e.g., the deity retiring his war bow by placing it in the sky where it can be seen by all (Gen 9:12–17)—P transformed them in such a

[9] For additional biblical examples of Combat Myth themes applied to historical events, see Batto (1992: 146–150).

way as to emphasize that Yahweh was the universal Divine Sovereign, and as such was also the creator of all and that everything in the heavens and on earth and in the underworld is completely under the domination of this great God, these being characteristics of the Divine Sovereign (Batto 2004, 2013b, 2013c).

Similarly, P skillfully recast the exodus as a continuation of creation—Act 2 of creation, as it were (Batto 1992: 118–123).[10] Just as in Genesis God divided the waters of *tĕhôm* to expose dry land as the first step in fashioning a realm where humankind and other creatures might flourish, so during the exodus God split the Sea in the process of creating a people for himself; the Israelites emerge from the sea on dry land to continue on the road to Sinai where they will complete the process of becoming God's covenanted people.

Combat Myth motifs in the Priestly exodus story deserve closer scrutiny, beginning with P's identification of the Sea of crossing as *yam sûp*, which as already stated, P understands as the "Sea of End."[11] *Yam sûp* carries connotations of non-creation or non-existence. Within the mythic world of the ancient Near East, *yam sûp* is but another name for the chaotic Sea against which the Divine Sovereign battled in primordial times, elsewhere named Tiamat, Yamm, Litan (Leviathan), or Rahab, among other labels.[12] Unlike *Enuma elish,* no HB text states that the deity actually "split" (*bqʿ*) primordial Sea (Propp 1999: 559), but this mytheme is implied. By whatever name, primordial Sea is pierced, split, slain or otherwise tamed by the Divine Sovereign (Isa 27:1; 51:9–11; Nahum 1:4; Ps 74:13–14; 89:9–10; 93:3–4; 104:6–9; Job 7:12; 9:8; 26:12; 38:7–11; 40:25–41:3 [ET 41:1–11]; 38:8–11; Prov 3:20; cf. Ps 18:16; 77:17–20; Rev 21:1) and in the process a sustainable "life bubble" or *Lebensraum* is established in which both celestial and terrestrial beings may live an orderly, secure existence (what the Greek called "cosmos"). Reminiscent of those primordial acts, P twice uses the verb *bqʿ* "to split" to describe Yahweh's action against the sea: God commanded Moses, "Raise your rod and extend your arm over the Sea and *split* it" (14:16; cf. 14:21; see Isa 63:11–13; Ps 78:13; Neh 9:9–11).[13] The sea of the exodus is suffused with mythic overtones elsewhere in the HB as well (Hab 3:8–10; Ps 114:1–8; 77:17–20). Within the P exodus narrative, the Song of the Sea functions to reinforce this viewpoint.

The Song of the Sea as Continuing Creation per the Combat Myth

Most scholars consider the Song of the Sea originally to have been an independent composition. Some think P was responsible for placing the Song in its present context; others argue this was done earlier, by a JE redactor. However it became joined to the exodus narrative, P found the Song very appropriate to his purpose; it enhanced his agenda of transforming the exodus into a second act of creation, following the pattern of the Combat Myth.

It has been observed that within the Song itself, the Sea is as much Yahweh's opponent as

[10] Conrad Schmid (oral comment) agreed that P makes creation and exodus parallel. Propp (1999: 560–561) finds that the Combat Myth pervades Exodus 14–15, in effect turning "the entire Torah" into "a Creation Myth"; but strangely, Propp does not find the Combat Myth "in Genesis 1–3, the Creation story proper."

[11] With some inconsistency Propp (1999: 34 and 560–561) similarly finds the Canaanite myth of the storm god Baʿlu as a prototype of the exodus tradition, but claims that the Sea is no longer the cosmic ocean but a specific body of water.

[12] Behemoth perhaps should be included here (Day 1985: 75–87; Batto 1995). In ancient Near Eastern literature and iconography the chaos figure(s) was (were) depicted in various forms: aquatic, bird-like, serpentine, draconic, seven-headed, or fully anthropomorphic; moreover, the monster may even exhibit multiformism, appearing in more than one shape within the same context (Batto 2013a: 244, Pitard 2007: 82–83, Wiggermann 1997: 37–39).

[13] The verb *bqʿ* is used in the Ugaritic Baʿlu Cycle for Anat's slaying of Mot "Death," another chaos figure (*KTU* 1.6 ii 32).

is Pharaoh and Egypt, if not more so.[14] At a blast from the deity's nostrils "the waters piled up, the floods stood up in a heap; the deeps congealed in the heart of the sea" (15:8). Compare the Priestly statement in Genesis 1 that the breath (or wind) of God in Genesis 1 churned the Abyss (*tĕhôm*) as God began to create and thus presumably was instrumental in causing dry land to appear, just as in Genesis 8:1 (also P?) it was a breath (or wind) from the deity that dried up the flood waters and allowed dry land to appear again. Most famous is the cleaving in twain of Ti'amat in *Enuma elish*.

Significant is the controverted phrase *ʿam-zû qānîtā* (15:16), which may be translated either as "the people whom you acquired" or as "the people whom you created." Based upon usage both in Canaanite literature and elsewhere in the HB, a strong case can be made for a meaning of "to create" for the verb *qānâ*, in addition to the more common meaning of "to acquire" (Batto 1992: 113 with n. 24). In that case, the exodus P narrative has been transformed into a second creation account, as not only the conceits of the Combat Myth but also the vocabulary of creation are present: the Israelites emerge from the split sea as the people whom God has "created" for himself.

There is one final Combat mytheme here. The Song concludes with the Israelites firmly planted on God's holy mount, in the shadow of the eternal throne of Divine Sovereign (15:17–18). The Combat Myth also climaxed in the building of a mountain sanctuary/palace for the Divine Sovereign, from which he then rules and maintains order over all (Levenson 1988: 78–99; Hurowitz 1992). Whatever its original identity—Sinai/Horeb, Gilgal, Shiloh, Jerusalem, or elsewhere (B. Russell 2007: 80–96)—in the Priestly redaction this sanctuary turns out to be Sinai temporarily, where God entered into covenant with his people, and proleptically, Jerusalem, the sanctuary that Yahweh eventually chose as his eternal "resting place" (Ps 132:13–14, cf. v. 8).[15]

Pharaoh/Egypt as the Incarnation of the Chaos Monster

Despite all the evils Assyria and Babylon inflicted upon Israel and Judah, they could be forgiven by Israelite prophets and theologians because they were God's instruments for punishing the many sins of God's people. Not so for Egypt. Ezekiel especially blames Egypt for corrupting Israel. Egypt was responsible for the "original sin" of the Israelites, for there Israel first learned idolatry (Ezekiel 20:7–8). In Egypt Israel acquired an addiction for apostasy, an addiction that neither Israel nor Judah could ever kick. As the prophet lamented in the allegory of the sisters Oholah and Oholibah, in their youth both Israel and Judah "played the harlot in Egypt" and enjoyed having their virgin breasts fondled (Ezek 23: 3, 19).

The prophet considered Egypt's corrupting influence on Israel and Judah so pervasive that he decried Egypt-Pharaoh as the chaos monster, a charge laid out in the first of eight oracles against Egypt: "Behold, I am against you, Pharaoh king of Egypt, the great dragon that lies in the midst of his Niles, that says, "My Nile is my own. I made myself!" (Ezek 29:3) (Batto 1992: 164). The "great dragon" (*hattannîm haggādôl*[16]) is, of

[14] Kloos (1986: 212) argues "that the Reed Sea story is a transformation of the [Canaanite] myth of the battle with the Sea"; also pp. 149–152. Dozeman (1996: 408–411) also recognizes that in Exod 15:5 the primary function of *yam sûp* is mythological, to reinforce "the power of Yahweh over sea," though elsewhere the phrase has greater geographical specificity. Cross (1973: 131–132) argued that the sea is only a passive instrument in Yahweh's control for defeating the Egyptian force, a historical enemy; similarly Forsyth 1987: 93–98. S. Russell (2009: 127–158) finds historical connections tenuous at best; the Song reflects a Judahite provenance, according to which the first half depicts victory over the Egyptians and the second half describes in language rooted in the mythological tradition of ancient Near Eastern kingship a victory tour by God's people.

[15] On the relation between the Divine Sovereign's "resting place" and creation, see Batto (1987: 148–155, 1992: 78–79).

[16] Perhaps read with some MSS *hattannîn haggādôl*; so BHS.

course, the primeval watery chaos monster of the Combat Myth.[17]

Labeling Pharaoh the chaos monster was the prophet's none too subtle attempt to counter Egyptian claims that Pharaoh was the incarnation of Egypt's chief god, frequently regarded as the creator. Pharaoh's statement in 29:3, "I made myself," likely alludes to Pharaoh's claim to being the creator, just as the words, "The Nile is my own" may imply not only ownership but also creation of the Nile. The prophet turned Egyptian hubris on its head: Pharaoh is not the creator but the anti-creator; what he creates is "his Niles," or to use the equivalent language from the parallel passage of 32:2 "your rivers"—this in parallelism to "the seas," in which this dragon dwells. The allusion to Yamm and Nahar ("Prince Sea" and "Judge River"), the opponents of Baʿlu in Canaanite myth, could hardly be more obvious. Egypt-Pharaoh is in league with the forces of chaos, not creation.[18]

In Ezekiel's sixth oracle against Egypt, the theme of anti-creation versus creation is continued. Borrowing perhaps from the imagery of Marduk battling Ti'amat, Yahweh promises to capture the dragon in his net and slay it, severing its arteries and drenching the land with its blood (32:6; cf. Ps 74:12–16). This action will cause the land of Egypt appropriately to revert to a precreation condition of darkness, in keeping with Egypt's true nature (32:7–8). With Pharaoh-Egypt and its nihilistic power eliminated, Yahweh will then cause paradisiacal conditions of plenty to prevail throughout the land as Egypt's streams overflow with oil (v. 14),[19] thus proving that Yahweh, not Pharaoh-Egypt, is the authentic creator.

It was not uncommon in the HB to dub Pharaoh-Egypt as the chaos monster (Day 1985: 88–101). Isa 30:7 underscores the futility of looking to Egypt for aid, likening Egypt to the defeated primordial foe of God: "Rahab the quelled" (reading *rahab hammōšbat* for MT *rahab hēm šābet*; cf. Isa 27:1; 51:9–11; Job 41). Egypt evidently is mentioned under the cipher of Rahab also in Ps 87:4. The contrast between "the Holy One of Israel" and Egypt as "human and not God" is emphasized in Isa 31:1–3. These biblical writers, like Ezekiel, likely wrote from an exilic context. P seemingly composed his work at about the same time and under similar circumstances. One should not be surprised, then, that P similarly excoriated Pharaoh-Egypt.

Already in the JE narrative there was a swipe at the supposed divinity of Pharaoh. In Exod 5:2, when Moses and Aaron go before Pharaoh with the message, "Thus says Yahweh, the God of Israel, 'Let my people go,'" Pharaoh retorts, "Who is Yahweh, that I should heed him and let Israel go: I do not know Yahweh and I will not let Israel go." Pharaoh correctly interprets Yahweh's demand as a challenge to Pharaoh's claim to being the supreme deity. Implicit in Pharaoh's hubris is a counterclaim that the Israelite god cannot hold a candle to divine Pharaoh.

P appropriated and augmented this divine contest. In Exod 7:8–11:10 the two opponents face off in a series of contests to see who is more powerful, with each working through human agents: Moses and Aaron for Yahweh, the Egyptian magicians for Pharaoh. In JE Pharaoh refused to concede because he "hardened his (own) heart." According to P, however, Pharaoh does not control even his own decisions; Yahweh "hardened Pharaoh's heart" so all Egypt will recognize Yahweh as superior, thereby gaining greater glory for Yahweh (14:17–18).

Indeed, Pharaoh is neither god nor creator, but rather the incarnation of the chaos monster. It is most appropriate in P's recasting of the exodus, therefore, that in the very act of attempting to prevent the Israelites from gaining their freedom and becoming God's new creation, Pharaoh and his army plunge headlong into *yam sûp*, the Sea of End/Extinction. Pharaoh and his forces are submerged into—better, merged with—the

[17] Gary Rendsburg (orally) identified Ezekiel's "great dragon" with the crocodile; while appropriate for Egypt specifically, Ezekiel's purview likely included a larger ancient Near Eastern mythic tradition; see Lewis (1996: 28–47).

[18] See also Hab 3:8–10, 15, above.

[19] This motif of paradisiacal plenty is attested in both biblical and Ugaritic literature (e.g., Job 29:6; *KTU* 1.6 iii 6–7).

"split" (defeated) Sea and perish along with it at the hand of Yahweh. In P's retelling of the exodus, Yahweh could gain no greater glory.

References

Batto, Bernard F. 1983. The Reed Sea: Requiescat in Pace. *Journal of Biblical Literature* 102: 27–35. Reprinted, with a postscript, in Batto 2013a: 158–174.

Batto, Bernard F. 1987. The Sleeping God: An Ancient Near Eastern Motif of Divine Sovereignty. *Biblica* 68: 153–157. Reprinted in Batto 2013a: 139–157.

Batto, Bernard F. 1992. *Slaying the Dragon: Mythmaking in the Biblical tradition*. Louisville, KY: Westminster John Knox.

Batto, Bernard F. 1995. Behemoth. Cols. 316–322. In *Dictionary of Deities and Demons in the Bible [DDD]*, 2nd ed, ed. Karel van der Toorn, Bob Becking, and Peter Horst. Leiden: Brill.

Batto, Bernard F. 2004. The Divine Sovereign: The Image of God in the Priestly Creation Account. In *David and Zion: Biblical Studies in Honor of J. J. M. Roberts*, 2nd ed, ed. Bernard F. Batto, Kathryn L. Roberts, and Peter Horst, 143–186. Winona Lake, IN: Eisenbrauns. Reprinted in Batto 2013a: 96–138.

Batto, Bernard F. 2013a. *In the Beginning: Essays on Creation Motifs in the Ancient Near East and the Bible*. Winona Lake, IN: Eisenbrauns.

Batto, Bernard F. 2013b. The Ancient Near Eastern context of the Hebrew ideas of creation, 7–53. In Batto 2013a.

Batto, Bernard F. 2013c. The Combat Myth in Israelite Tradition Revisited. In *Creation and Chaos: A Reconsideration of Herman Gunkel's* Chaoskampf *Hypothesis*, ed. JoAnn Scurlock and Richard H. Beal, [illegible]. Winona Lake, IN: Eisenbrauns.

Childs, Brevard S. 1974. *The Book of Exodus: A critical, Theological Commentary*. Old Testament Library. Philadelphia, PA: Westminster.

Copisarow, Maurice. 1962. The Ancient Egyptian, Greek and Hebrew Concept of the Red Sea. *Vetus Testamentum* 12: 1–13.

Cross Jr., Frank Moore. 1973. *Canaanite Myth and Hebrew Epic*. Cambridge, MA: Harvard University.

Day, John. 1985. *God's Conflict with the Dragon and the Sea: Echoes of a Canaanite Myth in the Old Testament*. Cambridge: Cambridge University. University of Cambridge Oriental Publications 35.

Dozeman, Thomas B. 1996. The *yam-sûp* in the Exodus and the Crossing of the Jordan River. *The Catholic Biblical Quarterly* 58: 407–416.

Dozeman, Thomas B. 2009. *Exodus. Eerdmans Critical Commentary*. Grand Rapids, MI: Eerdmans.

Fitzmyer, Joseph A. 1971. *The Genesis Apocryphon of Qumran Cave I: A Commentary*, 2nd ed. Rome: Biblical Institute Press. Biblica et Orientalia 18A.

Forsyth, Neil. 1987. *The Old Enemy: Satan and the Combat Myth*. Princeton, NJ: Princeton University.

Hurowitz, Victor (Avigdor). 1992. *I have Built You an Exalted House: Temple Building in the Bible in Light of Mesopotamian and Northwest Semitic Writings*. Sheffield: Sheffield Academic Press. JSOTSup 115; JSOT/ASOR Monographs 5.

Kloos, Corola. 1986. *Yhwh's Combat with the Sea: A Canaanite Tradition in the Religion of Ancient Israel*. Amsterdam/Leiden: van Oorschot/Brill.

Levenson, Jon D. 1988. *Creation and the Persistence of Evil: The Jewish Drama of Divine Omnipotence*. San Francisco: Harper & Row.

Lewis, Theodore J. 1996. CT 13.33-34 and Ezekiel 32: Lion Dragon Myths. *Journal of the American Oriental Society* 116: 28–47.

Montgomery, James A. 1938. Hebrew (2) *yam sûp* ('The Red Sea') = Ultimum Mare? *Journal of the American Oriental Society* 58: 132–133.

Noth, Martin. 1962. *Exodus: A Commentary*. Old Testament Library. Philadelphia: Westminster. [Trans. J. S. Bowden, from German *Das zweite Buch Mose, Exodus* [ATD 5], Göttingen: Vandenhoeck & Ruprecht, 1959.]

Oblath, Michael D. 2004. *The Exodus Itinerary Sites*. New York, NY: Lang.

Pitard, Wayne T. 2007. Just How Many Monsters Did Anat Fight (*KTU* 1.3 III 38-47)? In *Ugarit at Seventy-Five*, ed. K. Lawson Younger Jr., 75–88. Winona Lake, IN: Eisenbrauns.

Propp, William H.C. 1999. *Exodus 1-18*, Anchor Bible 2.1. New York, NY: Doubleday.

Propp, William H.C. 2006. *Exodus 19-40*, Anchor Bible 2.2. New York, NY: Doubleday.

Russell, Brian D. 2007. *The Song of the Sea: The Date of Composition and Influence of Exodus 15:1-21*. New York, NY: Lang.

Russell, Stephen C. 2009. *Images of Egypt in Early Biblical Literature*. Berlin: de Gruyter. Beihefte zur Zeitschrift für die altestamentliche Wissenschaft 403.

Sasson, Jack M. 1981. Literary Criticism, Folklore Scholarship, and Ugaritic Literature. In *Ugarit in Retrospect*, ed. G.D. Young, 81–98. Winona Lake, IN: Eisenbrauns.

Snaith, Norman. 1965. ים־סוף: The Sea of Reeds: The Red Sea. *Vetus Testamentum* 15: 395–398.

Stolz, Fritz. 1995. Sea. Cols. 1390–1401. In *Dictionary of Deities and Demons in the Bible [DDD]*, 2nd ed, ed. Karel van der Toorn, Bob Becking, and Peter Horst. Leiden: Brill.

Wiggermann, Frans A.M. 1997. Transtigridian Snake Gods. In *Sumerian Gods and Their Representations*, ed. Irving L. Finkel and Markham J. Geller, 33–55. Groningen: Styx. Cuneiform Monographs 7.

Egyptologists and the Israelite Exodus from Egypt

15

James K. Hoffmeier

Abstract

Early Egyptologists were steeped in interest in biblical history and in particular the Hebrew exodus story. Edouard Naville and W.M.F. Petrie were among the early pioneers. Of interest to early Egyptologists was the geography of the exodus and the route of the Hebrew departure from Egypt. By the mid-twentieth century, Egyptology's love affair with Old Testament matters had soured, but this allowed the discipline to develop as its own science.

Over the past decades, Biblical scholars have largely been swept into the current of historical minimalism, leaving Israel's origin story on the dust heap of history. This development serves as a pressing call for Egyptologists to return to the debate to bring data from Egypt to bear on historical and geographical matters. Indeed some have responded in constructive ways.

This chapter examines interaction between Egyptology and the exodus narratives and then reviews some of the newer archaeological, toponymical, and geological data from Northeastern frontier of Egypt that shed new light on the biblical narratives.

The Egyptian Origins of Israel: Recent Developments in Historiography

The Bible's portrayal of the children of Israel entering Egypt during a time of famine in Canaan, followed by a period of enslavement, and then a glorious exodus from Egypt under the leadership of Moses was largely viewed as reflecting historical reality in the field of biblical scholarship through much of the twentieth century. While biblical scholars debated the particular written sources behind the tradition and their reliability, the general picture was accepted as accurate.

In North America, the influence of W.F. Albright and his students, especially G. Ernest Wright and John Bright, contributed to this consensus. Not only did these scholars affirm the historicity of the sojourn–exodus tradition, but they also were convinced that the footsteps

J.K. Hoffmeier (✉)
Divinity School, Trinity International University, Deerfield, IL, USA
e-mail: jhoffmei@tiu.edu

T.E. Levy et al. (eds.), *Israel's Exodus in Transdisciplinary Perspective*, Quantitative Methods in the Humanities and Social Sciences, DOI 10.1007/978-3-319-04768-3_15,

of Joshua and the conquering Israelites could be traced through numerous archaeological sites in Israel. From the 1930s to 1970s Albright–Wright synthesis dominated the English-speaking academy. This solid superstructure began to experience fissures when Thomas Thompson (1974) and John Van Seters (1975) authored influential studies that weakened the scholarly foundations for historicity of the Genesis Patriarchal narratives established by the Albright school. Thompson and Van Seters were dismissive of the parallels drawn between Near Eastern social and legal texts and the Genesis narratives. And so began the slide down the slippery slope towards historical minimalism with the redefining of historiography that continued in the following decades until it reached the court of David and Solomon.[1] The 1980s saw the rise of skepticism towards the Israelite conquest of Canaan and a dismissive rejection of the Torah's stance that the Hebrews came from Egypt. In place of the traditional view, new models began to appear that explained Israel's origins as an indigenous development in the land. Some of the chief proponents of these views are Niels Peter Lemche (1985), Gösta Ahlström (1986), Giovanni Garbini (1988: 127–132), Israel Finkelstein (1988, 2006: 41–65), and William Dever (2003: 1–74).[2] If there is no evidence of a new people who conquered the land coming from outside of Canaan, they reasoned, then it seems unlikely that Israel originated in Egypt as the Pentateuch would have us believe. Consequently, Robert Coote (1990: 3) declared concerning the exodus and conquest, "these periods never existed." For Lemche, the lack of evidence of the Israelites in Egypt was enough reason for him to jettison the biblical tradition. He has opined that "the silence in the Egyptian sources as to the presence of Israel in the country" is "an obstacle to the notion of Israel's 400 year sojourn" (Lemche 1988: 31). It is fair to say that these statements of biblical scholars reflect a general skepticism of the last 25 years towards the Israelite origins as a people in Egypt.

The purpose of this chapter is not to further review the recent scholarly trends in the field of Old Testament studies, but rather to examine how Egyptologists have regarded the sojourn and exodus tradition. From the outset allow me to observe that I have not found the same level of skepticism among present-day Egyptologists towards the Egyptian origin traditions of the Bible as there is among Old Testament scholars and Syro-Palestinian/biblical archaeologists. Before examining the current situation, let us review the history of Egyptology and its relationship to Old Testament studies, particularly, the question of the historicity of the Israelite sojourn–exodus narratives.

Early Egyptology and the Hebrew Sojourn/Exodus Tradition

In the eighteenth and nineteenth centuries, Egyptology was considered by many to be the handmaiden of biblical studies, especially as it related to the stories in the Pentateuch. Abraham's encounter with a pharaoh in Genesis 12, Joseph's service in the court of pharaoh, and the narratives about Moses and the exodus were subjects of scholarly interest. It was certainly the hope of many that excavations in Egypt might provide direct or background information on the times, locations, events, and historical figures involved in the biblical narratives.

The Egypt Exploration Fund (now Society), that publishes the *Journal of Egyptian Archaeology*, was founded in 1882. In the Memorandum of Association, the EEF actually stated that one of its purposes was "to make surveys, explorations … for the purpose of elucidating or illustrating the Old Testament narrative, or any part thereof, insofar as the same is in any way connected with Egypt."[3]

[1] Regarding the skepticism of the historicity of the biblical accounts of David and Solomon, see Miller (1987, 1991: 28–31) and Garbini (1988: 21–32).

[2] For a critique of these various positions see Hoffmeier (1997: chapters 1 and 2).

[3] I am grateful to the secretary of the EES, Dr. Patricia Spencer, for providing me a copy of the original founding charter.

Because of the prominence of the Delta in the Pentateuchal stories, it was the focus of some of the early surveys and excavations. Two of the early Egyptologists to excavate under the auspices of the EEF were Sir William Matthew Flinders Petrie and Edouard Naville. The lesser known Naville was Swiss and a professor at the University of Geneva (Lesko 1997: 113). The interest of these scholars in the biblical history is well reflected in the titles of some of their early excavation report. Two of Naville's earliest ones were *The Store-City of Pithom and the Route of the Exodus* (1885) which dealt with his excavations at Tell el-Maskhuta and *The Shrine of Saft El Henneh and the Land of Goshen* (1887). The latter excavations were undertaken in 1885 at several sites: Saft el-Henneh, Khataanah-Kantir, and Tell el-Retabeh, all located in the eastern Delta and the Wadi Tumilat (Naville 1887). Meanwhile, Petrie's early work took him to San el-Hagar (biblical Tanis), Tell el-Yehudiah, and Tell el-Retabeh (20 years after Naville) (Petrie 1888). His interests in biblical history are also seen in the title of one of his publications, *Hyksos and Israelite Cities* (Petrie 1906a). The same year Petrie published a major monograph on his explorations in Sinai called *Researches in Sinai* and devoted chapter XIV to "Conditions of the Exodus" (Petrie 1906b). In the same volume, two chapters (XVI and XVII) were written by C.T. Currelly of the Royal Ontario Museum in which some possible locations for Mt. Sinai are examined. Five years later Petrie (1911) published his *Egypt and Israel*, which devoted two chapters to the sojourn and exodus.

Also during the first two decades of the twentieth century, the French Egyptologist Jean Clédat conducted surveys and some excavations in North Sinai and along the Isthmus of Suez during the second decade of the twentieth century (Clédat 1919: 210–228; Clédat 1920: 203–215). While his surveys and excavations were primarily Egyptological in nature, his publications reflected his interest in the exodus story. A section of one article is called on "le passage de la mer rouge" where he attempted to identify the toponyms of Exodus 14:2 (Clédat 1919: 201–228). Clearly many of the early Egyptologists were interested in the problem of the Israelite sojourn and exodus as reflected in the title of an article published in the *Irish Church Quarterly* in 1908 by L.E. Steele, viz., "The Exodus and the Egyptologist."

In 1922, Sir Alan Gardiner, the renowned Oxford Egyptologist, wrote a very sharply worded critique of Naville and others that he thought were naïvely using the Bible to find the Delta sites associated with the exodus story (Gardiner 1922: 203–215), followed by another in *Journal of Egyptian Archaeology* (Gardiner 1924: 87–96) that was a response to an article by Naville in the same volume of *JEA* (Naville 1924: 18–39). T.E. Peet, who was Brunner Professor of Egyptology at University of Liverpool, likewise rebuked Egyptologists whom he accused of being unduly influenced by the Bible in their Egyptological research or who were overly zealous to prove the historicity of the Old Testament narratives based upon questionable evidence (Peet 1922: 5–7). Naville took umbrage at Peet's charge that his suggestion that Tell el-Maskhuta was Pithom was the result of "guesses of early explorers, bent on finding biblical sites at any cost" (Naville 1924: 18). Additionally Naville demurred with Gardiner's classification of the Exodus narratives being "no less mythical than the details of creation recorded in Genesis. At all events our first task must be to attempt to interpret those details on the supposition that they are a legend" (Gardiner 1922: 205). The comparison of Genesis 1–3 with the Exodus narratives, Naville thought, was an invalid one and that they were not of the same literary type. In his rejoinder to Naville, Gardiner pointed out that he was misquoted, noting that he never used the phrase "all the story of the Exodus" (Gardiner 1924: 87). As it turns out this phrase had been accidentally or intentionally added by Naville. Thus Gardiner was not claiming that the entire Exodus narratives should be written off as legendary, although some elements in them appeared fanciful to him. Naville further took exception to Gardiner claiming that religious conservatism was compromising scholarly research (Gardiner 1922: 203–204). Naville charged that it was "Gardiner who introduces the religious element, which should be entirely left aside. He is strongly biased, not by religious conservatism, but by the opposite tendency and its conclusions" (Naville 1924: 18).

With the work of these pioneer Egyptologists, the search was on for the Biblical cities associated with the exodus. Unfortunately, Gardiner's strong condemnation of those whom we might call "Biblical Egyptologists," I believe, cast a pall for decades over serious investigation of biblical history by Egyptologists. Since the 1930s there have been only a few Egyptologists in Great Britain who actually integrated their work with biblical studies in general and in particular with the exodus tradition. One notable exception to this trend was a small book written by Alfred Lucas in 1938 called *The Route of the Exodus of the Israelites from Egypt*. Best known for his classic book, *Ancient Egyptian Materials and Industries* (1926), Lucas's book is still available in print in the fourth edition, revised by J.R. Harris (1962/1989). Although Lucas spent most of his career analyzing artifacts and materials from which they were made, he believed that his 40 years in Egypt gave him a basis to offer some insights into the biblical exodus story. Quite aware of the harsh tone of the debate about the location of the cities of the Exodus, Lucas pledged to follow the dictum of the chemist Robert Boyle who said: "a man may be a champion for truth without being an enemy of civility: and may confute an opinion without railing at them that hold it" (Lucas 1938: 8).

It seems that the heated debate of the 1920s has been ignited once again in the past few decades as more recent archaeological discoveries are being scrutinized along with even greater skepticism towards the Bible. The sage advice followed by Lucas is certainly appropriate today in the polarizing debate between historical minimalists and maximalists over the origins of Israel. In this dispute, Egyptologists, whom one would expect to have something to say on the subject, for the most part have been strangely silent. In fact, it was the intent of the "Egyptology and Ancient Israel" section of Society of Biblical Literature, which I established in the early 1990s, to provide a forum where Egyptologists and biblical scholars could meet to discuss matters of mutual interest.

One important question that emerges from this discussion is why have Egyptologists had so little to say on the subject of Israel's origin as presented in the Bible? In a sense, the rather harsh debate of Gardiner and Peet with Naville, I believe, had a chilling effect on scholarly integration of Egyptology and biblical studies for several generations of Egyptologists. Gardiner cast a giant shadow over the field of Egyptology for more than five decades. Who then would dare enter the arena of Hebrew-Egyptological investigation for fear of being criticized by him or to be accused of having a religious agenda by other scholars.

There is, of course, a positive side to this debate from the 1920s and that is that Egyptology was able to emerge as a discipline of its own, independent of the limited interests of the biblical historian. A perusal of major Egyptological journals shows that articles are scarcely found that deal with the Bible in general or the sojourn–exodus in particular.

In a sense, Gardiner and Peet did for Egyptology what William Dever did for Syro-Palestinian archaeology in the 1970s and 1980s by establishing it as a discipline in its own right apart from the interests and limitations of biblical archaeology (Dever 1992: 354–367). Dever's more recent proposal that there be a dialogue between Syro-Palestinian Archaeology and biblical studies to produce a new biblical archaeology is a model that would work well, I believe, for the disciplines of Egyptology and biblical studies.[4] As such, Old Testament scholarship could utilize Egyptian material where appropriate without hijacking the discipline. Unfortunately, there has been too little dialogue between the two disciplines for reasons that will be explored below. If those who are trained in the various specializations of Egyptology are not a part of the conversation,

[4] He speaks of a "dialogue" between Bible and archaeological data (Dever 1992: 358–359) and elaborates it in more detail recently (Dever 2001: chapter 3).

then, regrettably, biblical scholars who are not qualified to use Egyptian data will do so (as sometimes happens) in a way that will not do justice to Egyptian sources, resulting in poor integration with the Bible.

In Germany the kind of acrimony witnessed in Briton among Egyptology and biblical studies in the early twentieth century did not occur. There were some Old Testament scholars who worked Egyptological materials. Albrecht Alt (1883–1956) is a leading example. His doctoral dissertation from 1909 bore the title "Israel und Ägypten" (Fritz 1997: 79). Although he is generally recognized as a Hebrew Bible scholar, his work on historical geography demanded that he work with Egyptian toponym lists and other Egyptian sources. In his article on the "Die Deltaresidenz der Ramessiden" (Alt 1954: 1–13) he was one of the first continental scholars to recognize Qantir as the location of ancient Pi-Ramesses (see Chap. 8). He wrote a monograph on the *Hyksos Die Herkunft in neuer Sicht* in 1954.

Egyptologists and the Exodus: 1930s to the Present

To be sure, there were Egyptologists who occasionally wrote on the problem of the Israelite sojourn and exodus from the 1930s to the end of the twentieth century. Alfred Lucas has already been mentioned.

In Germany Wolfgang Helck produced a major work on the interconnections between Egypt and the Near East in the third and second millennium (Helck 1971), but this does not deal directly with the exodus. He did write an important article that argued that Hebrew writing Rameses corresponds to the Egyptian (Pi-) Ramesses (Helck 1965: 35–48) in response to the linguistic problems raised by Donald Redford on the correlation (Redford 1963: 408–418).

Siegfried Herrmann who authored *Israel in Egypt* (1973), a small monograph dealing with the sojourn and exodus, is another German scholar of our times who has used Egyptian sources to explicate the narratives in Exodus. These European scholars, however, were primarily Hebrew Bible scholars who had some training in Egyptology and used the materials in a responsible way, but they would not identify themselves as Egyptologists.

The late Manfred Görg (d. 9/2012) who was trained in Hebrew exegesis and Egyptology was the most prolific German scholar of our time to deal with Egyptian sources and the Old Testament. He has penned scores of articles dealing with Hebrew words that might be of Egyptian etymology and connecting Egyptian toponyms to biblical place names. Many of these articles are published in *Biblische Notizen* and in the monograph series, *Ägypten und Altes Testament*, both of which Görg edited. This series has been one venue where biblical issues and Egyptology have been discussed. In neighboring Switzerland, Othmar Keel has successfully employed Egyptian (along with other Near Eastern) iconography in his studies of Hebrew symbolism, but his studies tend to avoid historical questions (Keel 1978; Keel and Uelinger 1998).

Pierre Montet, the French excavator of San el-Hagar from 1928 to 1956, believed that he had discovered both Zoan/Tanis and Rameses of Exodus 1:11 and the Ramesside capital, Pi-Ramesses. He authored a book entitled, *Egypt and the Bible*, in which he presented his understanding of Egyptian data bearing on the Hebrew Scriptures and devotes a chapter to Moses and the exodus (Montet 1968: 16–34).

Bernard Couroyer and Henri Cazelles were also French biblical scholars who were well versed in Egyptology. They wrote on the exodus traditions as well as some other topics where Egyptian sources were brought to bear on the Bible. In addition to teaching Hebrew and Old Testament, Couroyer taught Coptic and Egyptian for more than 30 years (Puech 1997: 10),[5] and he wrote a number of articles dealing with the book of Exodus including "La résidence ramesside du Delta et la Ramsès biblique" (Couroyer 1946:

[5] For a complete bibliography of Couroyer, see Marcel Sigrist (1997: 20–28).

75–98), "Quelques égyptianismes dan l'Exode" (Couroyer 1956: 209–219), and "Un égyptianisme biblique": Depuis la fondation de l'Égypt (Exode, IX, 18) (Couroyer 1960: 42–48). An acknowledgement of his contributions in Egyptian-Hebrew studies, a memorial volume in his honor, was published under the title *Études Égyptologiques et Bibliques* (Sigrist 1997).

Cazelles worked competently with Egyptian materials. His most important contribution to the exodus tradition is his article "Les Localizations De L'exode et La Critique Litteraire" (Cazelles 1955: 346–358). It offers an excellent analysis of the toponym cluster in Exodus 14:2 and the Egyptian geographical names found in Pap. Anatasi III that describes the marshy areas on the eastern frontier.

Also in recent decades, my own mentors from the University of Toronto, the late Ronald Williams and Donald Redford, have included in their publication dossiers a number of important articles, dictionary entries, and books on Egypt and the Bible, some of which deal with the Exodus. One of William's seminal studies is "'A People Come Out of Egypt': An Egyptologist Looks at the Old Testament" (Williams 1975: 231–252). Williams wrote a number of articles in the *Hastings Dictionary of the Bible* including entries touching subjects in Genesis and Exodus, i.e., "Asenath," "Nile," "On," "Pharaoh, "Plagues of Egypt," and "Raamses, Rameses" (for his complete bibliography, see V. Williams 1983: 127).

Redford has been even more prolific over the past 50 years on integrating Egyptian data with matters related to Israel's origins. His monograph on the Joseph story remains a standard work on Genesis 39–50 after 40 years (Redford 1970), and *Egypt, Canaan, and Israel in Ancient Times* (1992) includes a major treatment of Israel's origins. His essays related to the Israelite sojourn and exodus have been very influential (Redford 1963, 1987, 1997, 2009), especially among minimalist leaning biblical scholars as he has argued that the geographical terms in the Exodus narratives point to the Saite Period (seventh century) and that the exodus story may be an adaptation of the Hyksos expulsion story that was applied to the Israelites. The linguistic, textual, and archaeological questions he has raised have not been ignored by Egyptologists, but have been thoughtfully answered (Helck 1965: 35–48; Kitchen 1998: 65–131; Hoffmeier 1997, 2005). The works of Williams and Redford have left their impact on the field of Hebrew Bible and Israelite history, and their careful and critical use of Egyptian materials has influenced me greatly.

Another distinguished Egyptologist who has written on matters related to the Israelite sojourn and exodus over the past 50 years is Kenneth Kitchen, a lonely voice among British Egyptologists. He is known in Egyptological circles as the leading Ramesside Period expert due to his seven-volume compilation of *Ramesside Inscriptions* (Blackwell), along with the *Ramesside Inscriptions Translated and Annotated: Notes and Comments* (Blackwell) that is still in progress. Drawing decades of work with Ramesside Period materials, Kitchen has written extensively on the Hebrew sojourn and exodus in his books (Kitchen 1966: 57–72, 1977: 75–91, 2003: 241–312) and countless articles dealing with questions related to background, chronology, and authenticity. He sees the setting of the exodus narratives as being the Ramesside Era. Although too many to cite, several of his articles stand out and are worthy of mention, such as "the Exodus" in the *Anchor Bible Dictionary* 2 (1992), "The Tabernacle—a Bronze Age Artifact" (Kitchen 1993: 119–129), and "Ancient Near Eastern Studies: Egypt," in *The Oxford Handbook of Biblical Studies* (2006). In collaboration with Paul Lawrence, Kitchen has recently produced a magisterial three-volume magnum opus, *Treaty, Law and Covenant in the Ancient Near East* (Kitchen and Lawrence 2012). Exceeding 1,000 pages in length, Volume 1 offers transcriptions and translations of every known law code and treaty text from the third through the first millenniums B.C., be they Sumerian, Eblaite, Akkadian, Hittite, Egyptian, Hebrew, or Aramean. The texts serve as the database for comparative study of biblical law and treaty texts (Volumes 2 and 3). Setting aside theories about the biblical text such as sources and their dating, Kitchen rather compares the ANE data directly to the biblical forms and concludes that the thirteenth century form of treaty texts best compares with the legal materials of Exodus and Deuteronomy.

My study, *Israel in Egypt: The Evidence for the Authenticity of the Exodus Tradition* (New York: Oxford University Press, 1996), was written from an Egyptological perspective to address the origins of Israel debate of the 1980s and 1990s since the historical minimalists who were setting the agenda were biblical scholars, biblical historians, and Syro-Palestinian archaeologists. Egyptologists, for the most part, have not been heard from in the 1980s and 1990s. There are a few exceptions. Unfortunately, by avoiding this interdisciplinary discussion, Egyptologists have allowed biblical scholars who are not trained to work with Egyptian texts to do so, often resulting in unwarranted conclusions. For example, Gösta Ahlström (1986: 40),[6] the biblical scholar, proposed that the hieroglyphic writing of "Israel" in the Merneptah Stela—despite the use of the people determinative—should refer to a geographical entity, not an ethnic group. Subsequently, Lemche (1998: 37) has expressed that "it is remarkable that other scholars have not taken up Ahltröm's interpretation." There is good reason why this interpretation was not embraced by scholars familiar with Egyptian orthography. No Egyptologists would ever read the signs of a foreign ethnic entity (𓌙𓀀𓁐𓏥) as indicating a foreign land, but a people group. Ahlström is forced to propose an error in the text and then emend it to fit his theory!

One critic of Ahltröm's theory, the late Anson Rainey (1991: 93), pointed out that this "simply demonstrated that Biblical scholars untrained in Egyptian epigraphy should not make amateurish attempts at interpretation." Rainey, though primarily known as a Semitist, an expert in the Amarna Letters, and a specialist in historical geography was trained in Egyptology by H. J. Polotsky. Rainey periodically entered the debate about the origins of early Israel, which included his analysis of the Merneptah reliefs at Karnak that presents a pictorial counterpart to military campaigns recorded on the "Israel" Stela (Rainey 2001: 57–75). Rainey's recent Bible atlas, already a classic, contains excellent treatment of the New Kingdom, in particular the Amarna Age and Ramesside Periods, as background to the origins of Israel, and his transcriptions, transliterations, and translations of Egyptian execration texts and toponymic lists are foundational to the understanding of the geopolitics of the Late Bronze and Iron Age when the Israelites first appear in Canaan (Rainey and Notley 2006: 58–121). He also investigates the toponyms of Egypt and Sinai from the Bible, which he maintained were based on "collective memories and legendary elaborations" . . . but that "such a powerful folk memory with so many ramifications can hardly be a strictly pure invention" (Rainey and Notley 2006: 118). He concluded by remarking that the geographical data in the exodus and wilderness traditions "does embody considerable geographic information" (Rainey and Notley 2006: 118).

While it is my contention in this chapter that Egyptologists have been for the most part silent during the debate of the 1980s and 1990s, some have engaged in the discussion in several multi-authored books which have included some Egyptological perspective on the exodus (some discussed already above). A seminar was held at Brown University in 1992, and its proceedings were published as *Exodus: The Egyptian Evidence* in 1997. This 112-page book contained the papers of the six participants, only three of whom were Egyptologists: Frank Yurco, James Weinstein, and Donald Redford. Yurco was the only one to present a positive case for using Egyptian evidence in understanding the origins of Israel, even proposing that Amun-her-khepsh-ef, the eldest son of Ramesses II and Queen

[6] This suggestion was first proposed by Ahlström a year earlier in an article which was coauthored by another Hebrew Bible scholar, Diana Edleman (Ahlström and Edelman 1985: 59–61).

Nefertari, might have been the crown prince who died in connection with tenth plague (Yurco 1997: 57–75). He believes that this prince's death can be fixed to the period between 1259 and 1249 B.C., a date for the exodus within the thirteenth century B.C. as proposed by Kitchen.

In the new *Oxford History of the Biblical World* (1998), Carol Redmount, an Egyptologist from University of California at Berkeley, wrote the chapter on the sojourn and exodus. Unfortunately, she uncritically accepts the views of historical minimalists about the nature and dating of the biblical materials to the late monarchy or exilic and postexilic periods (Redmount 1998: 79–121). Contrary to the views of many scholars who have examined the Egyptian backgrounds of the Exodus 1–14, Redmount claims "What is immediately striking about the earlier portions of the Exodus saga is the lack of distinctively Egyptian content and flavor, despite the Egyptian" (Redmount 1998: 87). Had Redmount consulted the works of Williams, Kitchen, and Hermann cited here, not to mention my *Israel in Egypt*, she would have been introduced to a wealth of Egyptian background data.[7]

1998 also saw the appearance of a volume from the Irene Levi-Sala Annual Research Seminar held the preceding year. It was called *The Origin of Early Israel—Current Debate* and was organized by Eliezer Oren who also edited the volume. The ten participants included eight biblical scholars and Syro-Palestinian Archaeologists, a classicist, and a lone Egyptologists, Kenneth Kitchen. It is curious that this seminar, usually held at Ben Gurion University (Beer Sheva), did not attract any of the fine Israeli Egyptologists.[8]

Finally, a collection of essays called *Ancient Israel* was published in 1998 by Biblical Archaeology Society and edited by Hershel Shanks. This book was revised and updated by different authors in 1999. It has a chapter on the sojourn and exodus that was originally written by Nahum Sarna, an excellent commentator on the book of Exodus (but not an Egyptologist), and was updated by Shanks.[9] These recent studies illustrate how Egyptologists, in my judgment, have not been sufficiently engaged in the origins of Israel debate even though Egypt does play a crucial role according to the biblical tradition.

What Do Egyptologists Really Think About the Exodus?

Despite these studies, the reality is that Egyptologists seem to show little interest in integrating their materials with biblical studies in general or with the exodus narratives in particular. In response to a rather negative seminar paper of Redford's in 1986, Manfred Bietak, the Austrian excavator of Tell el-Dab'a (Egypt), made a remarkable and telling rejoinder. He said, "Being an Egyptologist I feel somehow embarrassed to comment on problems surrounding the theme of 'the Exodus'" and then he proceeded to say, "I do not necessarily share Professor Redford's pessimism" (Bietak 1987: 163). So, what is behind this embarrassment?[10]

[7] (Hoffmeier 1997: 138–140). Interestingly, my book is listed in a "Select Bibliography" at the end of the chapter, and it offers the following annotation: "A detailed examination of the biblical account of the Exodus incorporating recent textual, historical, and archaeological scholarship, which concludes that the main points of the narratives are plausible" (120). It is not clear whether this is the conclusion of the author or the editor. Regardless, nowhere in Redmount's chapter is there evidence that *Israel in Egypt* was considered in drawing her minimalist conclusions.

[8] Surprisingly, Israeli Egyptologists have had little to say about the sojourn–exodus traditions. In a search of the *Egyptological Bibliography (1822–1997)*, I found only a few articles by Israeli Egyptologists that dealt with the sojourn–exodus narratives. One important contribution is by Sarah Israelit-Groll, "The Historical Background to the Exodus: Papyrus Anastasi VIII (Groll 1997: 109–115).

[9] For some reason, all the other chapters in this book are updated by leading scholars in their respective fields, while the Exodus chapter is revised by the editor!

[10] I point the readers to Bietak's paper in this volume. It is evident from his presentation at this conference on May 31, 2013, that his work in the NE Delta and particularly at

No doubt the fact that the Hebrew Bible remains the Scriptures of Jews and Christians alike automatically casts a pall of suspicion over it as a source for historical research. My own curiosity about this matter motivated me to investigate this question. In order to get some fresh data, I conducted a small, unscientific survey among members of the International Association of Egyptologists. Working from the IAE directory, I randomly selected 125 scholars from which to conduct a survey to gauge current attitudes among Egyptologists. I received 25 responses, a 20 % return. The only criteria I used in the selection process were not to include scholars whose views I already knew through personal communication or from their writings, and secondly, I attempted to cast my net wide so as to include scholars from a wide range of countries. Although I intentionally did not send the survey to Egyptian Egyptologists as the lines too often blur between academic study of ancient Israel and modern politics, making Egyptians reluctant to discuss the Bible,[11] I received responses from Egyptologists in the following countries: the United States (12), Great Britain (4), Germany (2), Belgium (2), and one each from Australia, France, Canada, Holland, Russia, Latvia, and Uruguay.

Four questions were posed:

1. Have you published any studies that deal with the Israelite/biblical sojourn and exodus story? Followed by "if not why not"? Five indicated that they had addressed the question in some manner, either in an article, section of a book, or book review, though none had engaged in a major project. Twenty answered NO, and I was able to place their reasons in four different categories:
   (a) No expertise in biblical studies or Hebrew: 8 1/2
   (b) No interest in the subject 4 1/2
   (c) Specialization in Egyptology too narrow to venture into another field: 6
   (d) To avoid the intensity of the debate about the Bible: 1
2. Do you think the early Israelites lived in Egypt and that there was some sort of exodus? Nineteen answered YES. None said NO, but four indicated that it possibly happened or that they were unsure. Only one who described himself as unsure had some 30 years before written positively about the exodus, but had grown skeptical in the intervening years. The strongest negative statement was that it was "unlikely." Interestingly, that opinion came from Oxford, Gardiner's old stomping grounds. And one chose not to answer the question.

Some who affirmed the historicity of an exodus from Egypt added interesting comments like "I don't think there is any doubt about it." Or "I see no reason why the Israelite sojourn in Egypt should have been fabricated. By the same token, I see no fundamental reason why an eventual exodus of the Israelites people could not have occurred."

I must admit to being surprised by the largely positive response to the question of the historicity of the sojourn–exodus story, but most gave no evidence of any knowledge of the debates of the past 30 years among Old Testament scholars and biblical archaeologists on the origins of Israel. The two final questions allowed opportunities for the respondents to give their reasons for their positions and to share any ideas or theories they had. These answers were not possible to quantify, and many left these questions blank. Those who offered additional thoughts indicated that given the regularity of Asiatics, to use the Egyptian term, entering Egypt during the days of famine or draught in the Levant it was likely that the biblical Hebrews were one such group. Several sought to associate the expulsion of the Hyksos with the Israelite exodus.

Tell el-Dab'a have provided extremely valuable information about the Semitic-speaking population (including the Hyksos) living in the Delta, which could well have included the Hebrews among them.

[11] The late Habachi Labib (2001: 119–127) wrote at some length on the sojourn–exodus in his publication of materials from his excavations in Qantir (Pi-Ramesses) in the 1950s, but his work only appeared in 2001, over 15 years after his death in 1984. Clearly in this chapter he demonstrates a rare interest among Egyptian Egyptologists in biblical history and the sojourn–exodus tradition, but that may be due to the fact that he was a Coptic Christian.

Another theme that came up with some frequency was the recognition that Egypt may never be able to produce positive archaeological evidence for the Hebrews in Egypt because there were large numbers of Semites in Egypt at various times during the second millennium B.C. and it would be impossible to distinguish one group from another.

I have already acknowledged that this was not a scientific survey; however, I think that it does offer some interesting insights which are offered here.

1. 80 % were either not interested in matters of biblical history or felt that they lacked the expertise to offer anything concrete to the origins of Israel debate. Those who had written on the subject have produced very little.
2. There was an important undercurrent I picked up from some of the respondents. Despite the fact that most felt that the exodus was a historical event, there was a feeling that this debate has such heavy religious implications that, as one Egyptologist admitted, "I have found it difficult to have unbiased discussions," and then he/she said, "I believe religion to be a private matter." Another scholar said, "Like most Egyptologists I suspect, I don't regard the whole Exodus thing as really relevant to us in a historical sense; I think it says more about the beliefs of those who are interested in it today than in ancient times." My survey, however, suggests otherwise. In fact there seems to be the attitude that the exodus is a religious matter, not one for **real** Egyptologists to investigate. This disposition came through very clearly in a statement by another scholar who protested: "The absence of Egyptologists from the exodus debate is indeed a conundrum. I am detecting almost an aversion in some circles to even discussing the exodus as a serious historical event, as if to discuss it seriously somehow leads people to question your credibility as a scholar." These two quotes perhaps offer the best testimony of what might be behind Bietak's reference to it being an embarrassment for an Egyptologist to discuss the exodus.

Thus I see a kind of disconnect. Egyptologists, on the one hand, seem to accept the historicity of the biblical sojourn and exodus narratives, but on the other hand either have no interest in investigating it using their discipline, or feel that it is a subject to be investigated by people with a religious agenda. This sounds like we remain stuck in the quagmire of the debates of the 1920s of Gardiner and Peet against Naville. But contrary to Gardiner (1922, 1924), who did write on the sojourn–exodus traditions, and Peet who authored *Egypt and the Old Testament* (1922), more recent Egyptologists have avoided the topic altogether.

Gardiner and Peet, it seems to me, were concerned with a critical approach to the use of Egyptology in studying the Bible rather than a simplistic and literalistic hermeneutic. It may well be that due to acrimonious feuds of the 1920s and the antireligious bias that pervades the western academy, and a long a history of anti-Semitism in the Middle East, for the present and the near future only a few Egyptologists will intentionally design research and excavation projects in an effort to answer questions of biblical history. This is regrettable since Egyptology is a cognate field to Hebrew studies and has much to contribute, offering both background and contextual data, and it can serve an important check and balance against the excesses of biblical scholarship that uncritically uses Egyptian sources.

It is my hope that Egyptologists will take a greater interest in bringing their expertise to the dialogue with Old Testament studies and that Hebrew Bible scholars will engage in a careful study of Egyptian history and archaeology before articulating rash conclusions about biblical history.

References

Ahlström, Gösta. 1986. *Who Were the Israelites?* Winona Lake, IN: Eisenbrauns.

Ahlström, Gösta, and Diana Edelman. 1985. Merneptah's Israel. *Journal of Near Eastern Studies* 44(1985): 59–61.

Alt, Albrecht. 1954. Die Deltaresidenz der Ramessiden. *Festschrift für Friedrich zum 70 Geburstage*, 1–13. Berlin: Akademie.

Beitak, Manfred. 1987. Comments on the "Exodus". In *Egypt, Israel, Sinai: Archaeological and Historical Relationships in the Bible Period*, ed. A.F. Rainey, 163–171. Tel Aviv: Tel Aviv University Press.

Cazelles, Henri. 1955. Les localizations De L'exode et La Critique Litteraire. *Revue Biblique* 62: 346–358.

Clédat, Jean. 1919. Notes Sur L'isthme De Suez. *Bulletin de l'Institut Français d''Archéologie Orientale* 16: 201–228.

———. 1920. Notes Sur L'isthme De Suez. *Bulletin de l'Institut Français d''Archéologie Orientale* 18: 203–215.

Coote, Robert. 1990. *Ancient Israel: A new horizon*. Minneapolis, MN: Fortress.

Couroyer, Bernard. 1946. La résidence ramesside du Delta et la Ramsès biblique. *Revue Biblique* 53: 75–98.

———. 1956. Quelques égyptianismes dan l'Exode. *Revue Biblique* 63: 209–219.

———. 1960. Un égyptianisme biblique 'Depuis la fondation de l'Égypt' (Exode, IX, 18). *Revue Biblique* 67: 42–48.

Dever, William. 1992. Archaeology, Syro-Palestinian and Biblical. In *Anchor Bible Dictionary 1*, ed. D.N. Freedman, 354–367. New York, NY: Double Day.

Dever, William. 2001. *What did the Biblical writers know & when did they know it? What archaeology can tell us about the reality of ancient Israel*. Grand Rapids, MI: William B. Eerdmans Publishing Company.

———. 2003. *Who Were the Early Israelites and Where Did They Come From?* Grand Rapids, MI: Eerdmans.

Finkelstein, Israel. 1988. *The Archaeology of the Israelite Settlement*. Jerusalem: Israel Exploration Society.

Finkelstein, Israel, and Neil Silberman. 2006. *David and Solomon: In search of the Bible's Sacred Kings and the Roots of the Western Tradition*. New York, NY: Free Press.

Fritz, Volkmar. 1997. Albrecht Alt. In *The Oxford Encyclopedia of Archaeology of the Ancient Near East 1*, ed. Eric Meyers, 79. New York/Oxford: Oxford University Press.

Garbini, Giovanni. 1988. *History & Ideology in Ancient Israel*. New York, NY: Crossroad.

Gardiner, Alan H. 1922. The Geography of the Exodus. In *Recueil D'etudes Egyptologiques Dediees a La Memoire De Jean-Francois Champollion a L'occasion Du Centenaire De La Lettre a M. Dacier*, ed. E. Champion, 203–215. Paris: Bibliotheque de L'Ecole Des Hautes Etudes.

———. 1924. The Geography of the Exodus: An Answer to Professor Naville and Others. *Journal of Egyptian Archaeology* 10: 87–96.

Groll, Sarah. 1997. The Historical Background to the Exodus: Papyrus Anastasi VIII. In *Études égyptologiquies et Bibliques à la mémoire dy Père B. Couroyer*, ed. Marcel Sigrist, 109–114. Paris: Cahiers de la Revue Biblique 36.

Habachi, Labib. 2001. *Tell El-Dab'a I: Tell El-Daba and Qantir the site and its connections with Avaris and Piramesse*. In Manfred Bietak & Ernst Czerny editors. Vienna: Verlag der Österreichischen Akademie Der Wissenschaften.

Helck, Wolfgang. 1965. *Tkw* und Ramses-Stadt. *Vetus Testamentum* 15: 35–48.

———. 1971. *Die Beziehungen Ägyptens zu Vorderasien im 3. und 2. Jahrtausend v. Chr., 2*, verbesserteth ed. Wiesbaden: Otto Harrassowitz.

Hoffmeier, James. 1997. *Israel in Egypt*. New York, NY: Oxford University Press.

———. 2005. *Ancient Israel in Sinai*. New York, NY: Oxford University Press.

Keel, Othmar. 1978. *The Symbolism of the Biblical World: Ancient Near Eastern Iconography and the Book of Psalms*. Trans. T.J. Hallett. New York, NY: Seabury.

Keel, Othmar, and Christoph Uelinger. 1998. *Gods, Goddesses and Images of God in Ancient Israel*. Trans. T.H. Trapp. Minneapolis, MN: Fortress.

Kitchen, Kenneth. 1966. *Ancient Orient and Old Testament*. Downers Grove, IL: Intervarsity.

———. 1977. *The Bible in Its world: Archaeology and the Bible Today*. Exeter: Paternoster.

———. 1993. The Tabernacle—a Bronze Age Artifact. *Eretz-Israel Avraham Malamat* 24: 119–129.

———. 1998. Egyptians and Hebrews, from Ra'amses to Jericho. In *The Origin of Early Israel—Current Debate: Biblical, Historical and Archaeological Perspectives*, ed. Shmuel A'ituv and Eliezer D. Oren, 65–131. London/Beer-Sheva: Ben-Gurion University of the Negev Press.

———. 2003. *On the Reliability of the Old Testament*. Grand Rapids, MI: Eerdmans.

Kitchen, Kenneth, and Paul Lawrence. 2012. *Treaty, Law and Covenant in the Ancient Near East, 3 volumes*. Wiesbaden: Harrassowitz.

Lemche, Niels Peter. 1985. *Early Israel: Anthropological and Historical Studies on the Israelite Society before the Monarchy*. Leiden: Brill.

———. 1988. *Ancient Israel: A New History of Israelite Society*. Sheffield: JSOT.

———. 1998. *The Israelites in History and Tradition*. Louisville, KY: Westminster/John Knox.

Lesko, Barbara S. 1997. Edouard Naville. In *The Oxford Encyclopedia of Archaeology in the Near East*, vol. 4, ed. Eric Meyers, 113. Oxford/New York: Oxford University Press.

Lucas, Alfred. 1938. *The Route of the Exodus of the Israelites from Egypt*. London: E. Arnold.

Lucas, Alfred. 1962. *Ancient Egyptian materials and industries*. 4th edition: Harris J.R. 1989. London: Histories & Mysteries of Man.

Miller, J. Max. 1987. Old Testament History and Archaeology. *Biblical Archaeologist* 50(1): 301–312.

———. 1991. Solomon: International Potentate or Local King? *Palestine Exploration Quarterly* 123: 28–31.

Montet, Pierre. 1968. *Egypt and the Bible*. Philadelphia, PA: Fortress.

Naville, Edouard. 1885. *The Store City of Pithom and the Route of the Exodus*. London: EEF.

———. 1887. *The Shrine of the Saft El-Henneh and the Land of Goshen*. London: Egypt Exploration Fund.

———. 1924. The Geography of the Exodus. *Journal of Egyptian Archaeology* 10: 18–39.

Peet, T.E. 1922. *Egypt and the Old Testament*. Liverpool, NY: University of Liverpool.

Petrie, W.M.F. 1888. *Tanis, Part II: Nebesheh and Defenneh (Tahpanhes)*. London: Egypt Exploration Fund.

———. 1906a. *Hyksos and Israelite Cities*. London: British School of Archaeology in Egypt.

———. 1906b. *Researches in Sinai*. New York, NY: E.P. Dutton.

———. 1911. *Egypt and Israel*. London: Society for Promoting Christian Knowledge.

Puech, Émile. 1997. Biographie Intellectuelle du R. P. B. Couroyer. In *Études égyptologiquies et Bibliques à la mémoire dy Père B. Couroyer*, ed. Marcel Sigrist, 9–19. Paris: Cahiers de la Revue Biblique 36.

Rainey, Anson F. 1991. Rainey's Challenge. *Biblical Archaeology Review* 16(6): 93.

———. 2001. Israel in Merenptah's Inscription and Reliefs. *Israel Exploration Journal* 51: 57–75.

Rainey, Anson F., and R.S. Notley. 2006. *The Sacred Bridge: Carta's Atlas of the Biblical World*. Jerusalem: Carta.

Redford, Donald. 1963. Exodus I 11. *Vetus Testamentum* 13: 408–418.

———. 1970. *A Study of the Biblical Story of Joseph*. Leiden: Brill.

———. 1987. An Egyptological Perspective on the Exodus Narrative. In *Egypt, Israel, Sinai: Archaeological and Historical Relationships in the Biblical Period*, ed. A.F. Rainey, 137–161. Tel Aviv: Tel Aviv University Press.

———. 1997. Observations on the Sojourn of the Bene-Israel. In *Exodus, the Egyptian Evidence*, ed. E.S. Frerichs and Lesko Leonard, 56–66. Winona Lake, IN: Eisenbrauns.

———. 2009. The Land of Ramesses. In *Causing His Name to Live: Studies in Egyptian Epigraphy and History in Memory of William J. Murnane*, ed. Peter Brand and Louise Cooper. Brill: Leiden.

Redmount, Carol. 1998. Bitter Lives: Israel in and out of Egypt. In *The Oxford History of the Biblical World*, ed. Michael D. Coogan, 79–121. New York/Oxford: Oxford University Press.

Sigrist, Marcel (ed.). 1997. *Études égyptologiques et bibliques à la mémoire du Père B. Couroyer*. Paris: J. Gabalda et Cie Éditeurs.

Thompson, Thomas. 1974. *The Historicity of the Patriarchal Narratives; the Quest for the Historical Abraham*. Berlin: de Gruyter.

Van Seters, John. 1975. *Abraham in History and Tradition*. New Haven, CT: Yale University Press.

Williams, Ronald. 1963. *Dictionary of the Bible*, ed. James Hastings, F. Grant, and H.H. Rowley.

———. 1975. 'A People Come Out of Egypt': An Egyptologist Looks at the Old Testament. *Vetus Testamentum Supplements* 28: 231–252.

Williams, Vivien. 1983. The Works of R. J. Williams. In: *Egyptological Miscellanies: A Tribute to Professor R. J. Williams*, ed. J.K. Hoffmeier, and E.S. Meltzer, 126–128. Ancient History 6, Ares Publishers.

Yurco, Frank. 1997. Merneptah's Canaanite Campaign and Israel's Origins. In *Exodus: The Egyptian Evidence*, ed. E.S. Frerichs and Lesko Leonard, 27–55. Winona Lake, IN: Eisenbrauns.

Out of Egypt: Did Israel's Exodus Include Tales?

16

Susan Tower Hollis

Abstract

Known as the home of the short story ancient Egyptian narratives have entertained many people for millennia, and, not surprisingly, a number of these narratives resonate with various biblical narratives. Indeed the Bible states that as the Israelites left Egypt in the Exodus, they took borrowed objects with them (Exod 12:35) which logically would have included intangible items like loanwords and concepts gained from living in the Egyptian intellectual and moral environment. Perhaps the most notable of these, and possibly the best known to biblical scholars, is the 19th dynasty "Tale of Two Brothers" from the Papyrus d'Orbiney (BM 10183) due to its similarities with the narrative of Joseph and Potiphar's wife in Genesis 39. Other tales also appear to be reflected in various biblical materials. Given, however, the worldwide appearances of similar tales and tale motifs to those of these two cultures, one must ask if the similarities between the Egyptian and biblical really represent direct borrowing from or influence of Egypt. Whatever the answer may be, the significance of the Joseph story lies in its incorporation of a memory of the Israelites' descent into Egypt.

The biblical narrative relating a vestigial memory of the Israelites' departure from Egypt reports that they left with borrowed objects (Exod 12:35) and, though the Bible does not say so, some Egyptian influences and/or borrowed ideas as well. Some commentators have considered that because of the similarity of various biblical narratives to some tales from ancient Egypt, mostly notably that of Joseph and Potiphar's wife in Genesis 39 to the late thirteenth century BC Egyptian "Tale of Two Brothers," the Israelites also borrowed some stories or parts of stories. Certainly the similarity of the respective heroes' responses to the attempted seduction by older, more powerful women in the two narratives mentioned makes the idea of borrowing a tempting one. The presence, however, of earlier narratives with powerful females attempting to seduce young males found in other parts of the ancient Middle

S.T. Hollis (✉)
SUNY Empire State College, State University, New York, New York, NY, USA
e-mail: Susan.Hollis@esc.edu

T.E. Levy et al. (eds.), *Israel's Exodus in Transdisciplinary Perspective*,
Quantitative Methods in the Humanities and Social Sciences, DOI 10.1007/978-3-319-04768-3_16,

East, along with the known mutability of tales and tale motifs commonly occurring within the pattern of a heroic journey, calls such a conclusion into question.

Out of Egypt: Did Israel's Exodus Include Tales?

Ancient Egypt, known since T. Eric Peet's 1929 Schweich Lectures as the "home of the short story" (Peet 1931: 27), has entertained people with its plethora of short narratives for millennia. Not surprisingly, some of these tales and their motifs resonate with various biblical narratives. Perhaps the most notable of these, and arguably the best known to biblical scholars, is the first section of the 19th dynasty "Tale of Two Brothers" from the Papyrus d'Orbiney (d'Orb.) (BM 10183) (Hollis 2008: 1–9; Wente 2003b)[1] due to its similarities to the narrative of Joseph and Potiphar's wife found in Genesis 39:1–20, each part of a larger, more complex narrative. Motifs from other tales such as the Middle Kingdom "Tale of Sinuhe" (P. Berlin 10499 and 3022) (Simpson 2003b: 54–66); the Middle Kingdom cycle of tales known as "King Cheops and the Magicians" (Simpson 2003a: 13–23),[2] commonly referred to as the Papyrus Westcar (P. Berlin 3033) (Parkinson 2001a; Simpson 1982)[3]; and the late New Kingdom narrative "The Tale of the Doomed Prince" (P. Harris 500, verso) (Lichtheim 1976: 200–203; Wente 2003a) also appear to be reflected in various biblical materials. Given the worldwide appearances of similar tales and their component motifs, especially in the ancient Middle East, however, one wonders whether or not the appearances of these tales and/or motifs in both ancient Egyptian and biblical narratives represent a direct borrowing of ancient Egyptian material or at least an Egyptian influence even as they appear as part of the Israelites' memory of their entrance into Egypt.

In general, current scholarship understands the Exodus, the foundational episode of ancient Israelite history, to include at least a kernel of historical truth even though modern commentators have been unable to definitively document it as a specific event. On the other hand, the constant movement of Semitic-speaking peoples in and out of Egypt has been documented as occurring throughout Egyptian history, particularly through trade, and, of course, some of these people actually settled in Egypt, beginning especially in the later Middle Kingdom and after, most notably in the northeast delta (Mumford 2001). The resulting interaction in the centuries prior to the late thirteenth or early twelfth century BC suggested date for the Exodus surely engendered considerable cultural exchange and influence. Indeed clear linguistic evidence exists for loanwords (Albright 1934; Cruz-Uribe 2001; Lambdin 1953). Furthermore the study of wisdom literature in its broadest sense has long suggested some very strong influence, even likely borrowing, particularly notable in the case of the relationship of the late New Kingdom Instruction of Amenemope and biblical Proverbs (Clifford 1998; Redford 1992: 389–394; Shupak 1993: 1–11; Washington 1994). Thus it is easy to understand that other types of exchanges took place between the two cultures. In fact, various commentators have suggested that Egypt exerted a significant influence on ancient Israelite culture as expressed, for example, by Niels Peter Lemche's statement that "Egypt became the cradle of the Israelite people" (Lemche 1998: 86). When it comes to narratives, however, at least some of what one finds in ancient Egyptian and biblical tales also appears in the literatures of other contemporaneous and earlier cultures (Gelander 2013: 181), thus calling into question any direct borrowing or influence.

[1] The hieroglyphic transcription of the original hieratic in which the story was written appears in Alan Gardiner's *Late Egyptian Stories* (Gardiner 1981), as does the transcription for the Doomed Prince mentioned below.

[2] Also known as the Tale of King Cheop's Court (Parkinson 1997: 102–127).

[3] The hieroglyphic copy for Sinuhe appears in Alyward Blackman's *Middle Egyptian Stories* (Blackman 1972: 1–41a), while that for P. Westcar appears in Adrian de Buck's *Egyptian Readingbook* (Buck 1977: 79–88).

In brief, the ancient Egyptian Tale of Two Brothers presents two brothers born of the same parents, Anubis, the older and married, and Bata, the younger, who lives with Anubis.[4] The older brother's wife has found her younger brother-in-law so attractive that she proposes sleeping with him and rewarding him with all kinds of fine things. Angry as an Upper Egyptian leopard, Bata refuses her, noting that she is like a mother to him, asking "what is this great wrong you said to me" (d'Orb. 4,1),[5] and stating he will say nothing further about this incident. She apparently does not trust or believe him and presents herself as assaulted by him when her husband Anubis returns that night. Anubis then becomes like an Upper Egyptian leopard and sets out to kill his younger brother[6] who, warned by his cattle as he was returning with them to the barn, turns tail and flees. Eventually Bata prays for help to the sun god Pre-Harakhty who assists him by separating the two men with a river populated by crocodiles. The following morning, Bata reveals the reality to his brother, and in an enigmatic move, perhaps to affirm the truth of his words, he severs his phallus, throwing it into the river where it is swallowed by a catfish.[7] Bata then tells his brother that he will go to live in the Valley of the Pine, a location in present-day Lebanon (Hollis 2008: 126–131 and index), where he will place his heart on a tree. Finally he makes a pact with Anubis, telling him that when his beer foams, he, Anubis, will know some evil has happened to his brother. He should then seek his brother's heart and, on finding it, place it in a bowl of cold water and then have him swallow it to revive him.

Once in the Valley of the Pine, Bata sets up housekeeping, but the gods feel that he is lonely and so they create a wife for him. In time, as Bata had predicted,[8] his divinely created wife, who had been carried off to Egypt, betrays him to the king who then arranges his death to avoid adultery. At the consequent receipt of the sign, Anubis carries out the necessary acts to revive his brother, and the two brothers, Bata now in the form of a beautiful bull, return to Egypt to a celebratory welcome. There the Bata-bull reveals himself to his former wife, who has him killed, but he is reborn as a pair of *Persea* trees from which he again reveals himself to his former wife, who has them cut down for Bata's third death.[9] As they are cut down, however, she swallows a splinter from one of them and becomes impregnated, eventually giving birth to Bata as the king's heir. In time the king dies and Bata assumes the throne, at which point the wife is judged and killed.

The similarity of the first part of this narrative to that of Joseph and Potiphar's wife struck commentators very early on, as comments by the Rev. Charles E. Moldenke, Ph.D., in his 1892 translation and publication of the Papyrus d'Orbiney illustrate[10]:

> We do not claim that Moses, who certainly knew and had studied the papyrus while a student at the University of Heliopolis, simply copied the situation, for only a slight portion of both accounts is similar, but he may have had the wording of our papyrus in mind while writing his story. (Moldenke 1892)[11]

[4] Significantly and commonly overlooked in discussions of this tale is that both brothers are identifiable ancient Egyptian deities (Hollis 2008: 47–87). In fact in the later part of the tale, Anubis acts in his role as mortuary deity (Hollis 2008: 74–85, 168). In addition, it is commonly thought that the nameless wife is also a deity, perhaps Hathor in her more dangerous modality (Hollis 2008: 98, 152–156, 162, 192).

[5] All translations related to Two Brothers are mine.

[6] Death was a possible consequence of adultery as the tale of Webaoner, the second tale from the Westcar Papyrus, shows in its report of the death of both Webaoner's adulterous wife and her paramour (Simpson 2003a: 14–16).

[7] This action relates Bata to Osiris whose phallus was lost when Seth dismembered him (Hollis 2008: 109).

[8] The fulfillment of this type of prediction occurs very commonly in myths and tales to the extent that the knowledgeable person can anticipate the outcome.

[9] Three actions to gain a goal commonly occur in traditional narratives worldwide.

[10] It had been published in facsimile form in Samuel Birch's *Select Papyri in Hieratic Character from the Collections of the British Museum, Part II* (Birch 1860).

[11] Moldenke reiterated these comments in 1895 (Moldenke 1895).

While there was never any evidence then, nor is there now, of any university at ancient Heliopolis in Egypt, Moldenke's comment underscores the understanding even in the late nineteenth century that the Israelites' sojourn in Egypt affected them in various ways, and this tale, announced first in 1852 (de Rougé 1852), certainly provided a major example of such Egyptian influence.

The tale's initial episode, formally known as Tale Type AaTh 318, the Faithless Wife, and even more commonly as the Potiphar's Wife motif K2111 in folkloristic scholarship,[12] exists worldwide although interestingly, despite the title, the motif does not require a wife as Stith Thompson's description illustrates: "woman makes vain overtures to a man and then accuses him of attempting to force her" (Thompson 1955–58: 4, 474).[13] Indeed looking at this motif and downplaying the "wife" in its title, one finds the motif present at least as early as the ancient Sumerian narrative of "Gilgamesh and the Bull of Heaven" (Frayne 2001: 120–127; George 1999: 166–175),[14] a narrative much better known later from the sixth tablet of the Epic of Gilgamesh, in both of which the female is a deity who attempts to seduce a human.[15] The fourteenth to thirteenth century BC Ugaritic myth of Aqhat also presents this motif, leading Delbert Hillers in his study of this myth to develop the outline of the motif's actions (Hillers 1973: 75–77). The similarity of the myth's motif to the same motif in Two Brothers is undeniable, though the Egyptian narrative has been little studied as a myth per se, but rather, as Donald B. Redford has noted, it is apparently the earliest documented tale with this motif (Redford 1970: 93). Considering Hiller's work and similar studies of Two Brothers (Hollis 1989: 30–31, 2008: 105–111), a pair of patterns emerges for its actions, one focusing on humans with little divine intervention save support and the other having a deity who is one of the two main protagonists:

1. A male, usually young and without exception virile, coexists with an older female who is in a position of authority.
2. The female, attracted by the male, attempts to seduce him or offer him marriage.
3. The male refuses for reasons appropriate to his culture and place.
4. The female falsely accuses him of attempting to seduce her.
5. Severe punishment is inflicted, commonly administered by another male in an authority position.[16]

[12] To make comparisons of traditional tales and their constituent motifs both within and across cultures, scholars use appropriate collections of indices of tale types and motifs. The oldest and best known of these comes from the work of Stith Thompson, who translated and enlarged the original work on tale types by Antti Aarne when he published *The Types of the Folktale* (Thompson 1964) as well as developed the well-known *Motif-Index of Folk Literature* (Thompson 1955–58). He defines a tale type as "a traditional tale that has an independent existence," a complete narrative (Thompson 1977: 415), while a motif is "the smallest element of a tale have a power to persist in tradition" (Thompson 1977: 415). Interestingly, in both publications he used the Egyptian Tale of Two Brothers as the defining example, clearly taking it from his 1946 volume *The Folktale* in which he provided a complete set of tale types and motifs present in Two Brothers (Thompson 1977: 275–276 and 482 [AaTh 315B* = later AaTh 318]), also outlined in my work *The Ancient Egyptian "Tale of Two Brothers"* (Hollis 2008: 22, n. 70).

[13] Significantly, like the motif K2111, Thompson's description of AaTh 318: Faithless Wife: "Plots with paramour against the life of her husband" (Thompson 1964: 112) is not strictly accurate for Two Brothers.

[14] The Sumerian tablets with this narrative are known from the eighteenth century BC, but are likely of earlier origin (George 1999: 140–141, 166).

[15] For an in-depth discussion of Ishtar's proposal, see Tzvi Abusch's article on Tablet Six (Abusch 1986).

[16] Potiphar exemplifies this action. As Hillers notes, several narratives including the Egyptian Two Brothers involve self-castration or castration, for example, the tales of Kombabos/Astronoe and Eshmun/Stratonice (Hillers 1973: 75–76). For a recent discussion of this motif, related to motif T418: The Chaste Youth and his Lascivious Stepmother, focusing on the trans-Asiatic areas, see William Propp's paper on the Eunuch Steward (Propp 2013). In addition, there is at least one African example in which the hero's phallus is severed by the

In myths, this pattern takes a slightly different form:

1. A human male, usually young and without exception virile, is found attractive or has something a female deity desires.
2. The female deity attempts to seduce him, provide a substitute for the desired object, or offer him marriage.
3. The human male refuses for reasons appropriate to his culture and place.
4. The female deity uses her power against the human male.
5. Severe punishment is inflicted, not uncommonly death, although not necessarily of the hero.

Clearly the female may be human or divine, but she is always in a power position relative to the male, and in every case, the male refuses the female's approaches and experiences punishment as a result. As noted earlier, the motif itself is virtually always set in the context of a larger narrative framework particular to its culture, and it often functions as a turning or a transition point which pushes the hero towards a new role, thus helping to identify why the motif occurs where it does, which is definitely the case for the biblical narrative of Joseph.

The development of the biblical variant of the motif begins as Joseph, who was living with his family in Canaan (Gen 37:1), ends up in Egypt due to a complicated set of factors resulting from his brothers' jealousy and resentment and his own rather arrogant behavior as his father's favorite that surely exacerbated his brothers' reactions. Sold as a slave to Potiphar, his abilities lead to his elevation in Potiphar's household to the role of household administrator through which he comes to the attention of Potiphar's wife. She, like the wife of Anubis, seeks to seduce Joseph, who like Bata, refuses her, stating "How . . . could I do this most wicked thing, and sin before God?" (Gen 39:9).[17] And like the Egyptian wife, Potiphar's wife wrongly accuses Joseph of attempting to seduce her. As a result, Joseph is thrown into jail rather than being executed, the ancient Israelite punishment for adultery (Lev 20:10; Deut 22:22) and also a customary practice in ancient Egypt.[18] While there, he accurately interprets the dreams of two fellow prisoners, which later leads to his interpretation of some of the king's dreams, resulting in his eventual elevation to become the king's administrator. Through this office Joseph is eventually able to provide food to his family in time of famine, resulting in their descent into Egypt. Thus the narrative embodies a memory of how the ancient Israelites arrived in Egypt, the departure from which is the Exodus narrative. Clearly the Joseph story as a whole functions as a bridge between the world of the patriarchs and the Israelites in Egypt, the exilic situation. Equally clear is that the central turning point in the narrative focuses on Joseph's encounter with Potiphar's wife, Mrs. Potiphar (Hollis 1989, 2013).

Given, however, the many examples of the Potiphar's wife motif, does its presence in the Bible truly reflect Egyptian influence or borrowing? The answer is equivocal, given the several unquestionably older narratives from the ancient eastern Mediterranean which exhibit this motif in its broadest sense. On the other hand, several examples where the descriptive language and/or actions appear similar might suggest at least an influence if not a borrowing. For example, Bata is described as a "beautiful young man, there being none of his form in the whole land; indeed the strength (*pḥty*) of a god was in him" (d'Orb. 1:3–4). This strength, both physical and sexual, for *pḥty* can mean both, attracts the attention of his brother's wife. Turning to the biblical analog, one finds a similar

aggrieved male (Paulme 1963: 11–20). William F. Hansen has documented 20 occurrences of this tale type in classical literature *Ariadne's Thread* (Hansen 2002: 332–352).

[17] All biblical quotations come from the Jewish Publication Society's *The Jewish Study Bible* (Berlin and Brettler 1999).

[18] See first complete tale in the Westcar Papyrus, the tale of prince Khafre (Parkinson 1997: 108–109, lines 3.13–4.10; Simpson 2003a: 14–16).

description for Joseph: "Now Joseph was well built and handsome" (Gen 39:6). While appearing to lack the sexual connotation of the Egyptian, the result was the same: the wife of Joseph's Egyptian master not only desired him but also attempted to seduce him. Nevertheless, when one knows of the similar description of Gilgamesh in the Mesopotamian sources, the putative origin of the biblical patriarchs, one can hardly assert the uniqueness of the Egyptian–biblical relationship of this phrase.

To argue the point, however, one might observe that while this comparison represents but one descriptive parallel between the Egyptian tale and Genesis 39, other parallels exist specific to Bata and Joseph. For example Bata's place in his brother's household was essentially that of a servant: he tended the cattle, plowed the fields, and brought in plants, milk, and wood, placing them before his brother and sister-in-law (d'Orb. 1,1–1,6). Here one can clearly see a comparison with Joseph who was not only a servant but also described as a slave (Gen 39:1; 39:17; 39:19). Initially both young men were relatively high-born, Bata of the same background as his brother, a landholder of some kind, and Joseph as the beloved son of a tribal leader, and through circumstance, each came to occupy a subservient role. Furthermore both young men engendered prosperity and fruitfulness in their work, that is, whatever each man did prospered. Most explicitly "the Lord blessed (Potiphar's) household through Joseph" (Gen 39:5), and in the Egyptian tale, Bata's cattle spoke to him, telling him where the good food was to be found, places to which he naturally took them, with the result that "the cattle which were before him became very beautiful, and they multiplied their offspring many times over" (d'Orb. 1,9–2,2). This latter image also resonates very strongly with that of Jacob and the multiplication of the herds he tended for his father-in-law Laban, who acknowledged to Jacob: ". . . Yahweh has blessed me on your account" (Gen 30:28). In addition, when Jacob sought compensation from Laban, asking for it only in flocks, the animals for which Jacob asked also multiplied greatly through the actions he took at their mating (Gen 30:37–40). Obvious, of course, is that both men began as shepherds.

Similar too is the shape of the narrative, that of the heroic journey, seen only when one views each narrative as a whole: notably the hero's departure from home—in both the Egyptian and biblical, under a cloud—leads to tests and trials with ultimate success and finally a return home and/or reconciliation. In the Egyptian case, the end of the narrative involves Bata's reconciliation and return, while for Joseph it involves the descent of his family to Egypt and their reconciliation. Overall this pattern, a common one in narrative, also moves the protagonist from one role in life to another, involving the hero's departure from his usual or initial location to a very different place and a liminal existence, that is, on the edges of society at best and equivalent to an exile or even a symbolic death, followed by his return in a different, commonly higher, status.

A number of scholars have discussed this pattern with its many variations in relation to the heroes and protagonists in tales and epics, some of which begin at or even before birth, including Otto Rank (Rank 1990), Lord Raglan (Raglan 1990), Joseph Campbell (Campbell 1968), and Albert Bates Lord.[19] In fact, this narrative sequence speaks to more than simply parallel motifs and descriptors in narratives, as J. Robin King has pointed out in his discussion of "The Joseph Story and Divine Politics" (King 1987). In this article, he builds on Vladimir Propp's *Morphology of the Folktale* (Propp 1968), a seminal work analyzing 100 Russian wonder tales which undergirds much of the modern analysis of traditional tales,[20] as he examines several specific ancient second and first millennia BC documents relating to different

[19] Lectures in Harvard University Course, Humanities 9a, Early and Oral Literature, Spring 1980.

[20] Propp's analysis shows the existence of 31 functions, that is, actions, and 7 dramatis personae who perform them (Propp 1968). In *International Folkloristics*, Alan Dundes provides a brief discussion and simple summary of Propp's functions along with a short discussion of the principles involved (Dundes 1999: 119–130).

individuals' rise to kingship. In doing so, King relates Propp's functions and dramatis personae or actors to the basic journey each individual had taken, using an analysis of the Egyptian Tale of Sinuhe as his foundation.

The Tale of Sinuhe itself is actually written within the framework of a tomb biography (Baines 1982: 33–34; Parkinson 1997: 21–22; 2001b: 149) which details the life of Sinuhe, initially a courtier of King Amenemhet I's queen, after he overhears a report of the king's death while on maneuvers with the king's son and heir Sesostris. At the news, Sinuhe panics and flees to the Levant as if in a dream, eventually coming to rest among the Retenu in Syria in the land of Yaa, a land flowing with honey and olives and other good things.[21] There he is welcomed and settled, being granted land and raising a family, even fighting on behalf of the tribal chief, leading his troop. In time he is approached and challenged by a strong man of the Retenu, a "peerless champion" (Sinuhe B 111) to single combat, a fight Sinuhe wins against apparent odds. Despite his success and real acceptance by the Retenu following his victory against the "peerless champion," Sinuhe wishes to return to Egypt, for to be away from Egypt for any ancient Egyptian is to be in exile (Hollis 1998), a desire that eventually comes to the attention of King Sesostris I. Sesostris then sends Sinuhe a letter inviting him back to Egypt so that he might have a proper ending to his life, an invitation Sinuhe accepts with alacrity. On his return, the exile is warmly welcomed and gifted with all he needs for a proper burial and life in the next world.

As King compared Sinuhe to his chosen ancient Middle Eastern accounts of Idrimi of Alalakh, Hattusilis of Anatolia, Esarhaddon of Assyria, and Nabonidus of Babylon as each struggles to gain the kingship (King 1987: 584–587), he analyzed the basic sequence of these narratives, with some variations, into ten steps: (1) an initial situation; (2) a threat and (3) its realization; (4) an exile with (5) success; (6) an exilic *agon* with (7) subsequent victory; (8) the threat overcome; (9) a return and reconciliation; and (10) finally an epilogue, though not all these actions appear in any given narrative. The actors in these narratives consist of the hero who is virtually always lowly born or a younger son, his patron, the threat or the threatener, a supporter in exile, the antagonist, and a powerful helper figure, again not all of whom appear in every narrative.[22] King also compared basic sequences and actors with a number of biblical narratives: those of Jacob, Joseph, Moses, and David. In this analysis King determines that the most significant characters in each historical narrative resolve into hero and deity (King 1987: 586–88), as Table 16.1 illustrates, with the whole serving to emphasize the power of the deity while highlighting the human hero.

More significant, however, is the larger pattern, that of the hero and the actions that make him a hero. As noted, each of the protagonists in the narratives discussed leaves his home, that is, his initial situation; overcomes some kind of threat; and returns home and/or achieves a reconciliation, not uncommonly at a higher status, although King does not note this last point. Sinuhe certainly fits this pattern with return and reconciliation, though at no higher status. Joseph, on the other hand, does not return home but does achieve reconciliation and a higher status. And of course Bata, initially a pastoralist assisting his agriculturalist brother as a virtual servant, ends up as king,[23] while the others detailed illustrate clear status change: Moses to the people's leader, David eventually to king, as Table 16.2 illustrates, and King's various historical figures achieve the kingship in their respective lands.

---

[21] This description reminds one of the biblical concept that the promised land was to be a land flowing with milk and honey (e.g., Exod 3:8; 33:3).

[22] King identifies the helper figure as divine as is the case for the specific political narratives with which he deals (King 1987: 584), but he overlooks the reality that there are helper figures in many tales who are not divine.

[23] Both Jan Assmann and I discuss Propp in relation to Two Brothers (Assmann 1977; Hollis 2008: 42).

Table 16.1 Narrative structure for ancient Middle Eastern narratives: Sinuhe and historical narratives

V. Propp	Sinuhe (King 1987: 580–582)	Idrimi (King 1987: 581; Oppenheim 1969c: 558)	Hattusilis III—1 (Goetze 1969: 393; King 1987: 582)	Hattusilis III—2 (Gurney 1954: 175–176; King 1987: 582)	Esarhaddon (King 1987: 582–583; Oppenheim 1969a: 289–290)	Nabonidus (Beaulieu 2000; King 1987: 583; Oppenheim 1969b: 560–562)
Initial situation	With king	Youngest son of king	Youngest son of king; dedicated to Ishtar; made governor of “Upper Country”	King without heir; Hattusilis set up new king; return to old territory	Youngest son of Sennacherib and Crown Prince; support from all Assyrian gods, especially Shamash	King of Babylon; favorite of Sin; mother promised he would build Sin’s temple, Ehulhul
Threat	Death of king	Evil deed	Threat from man he replaced	New king jealous of Ishtar support; harried Hattusilus	Jealous brothers	Evil actions of people
Threat realized	Panic and flees	Family fled evil deed	Lost king’s affection	Stripped Hattusilus of his lands	Plot against him	Citizens refuse to help build temple Ehulhul for Sin
Exile	Canaan/ Upper Retenu	Family in exile but Idrimi left for Canaan	Exile	Exile (support of disloyal king	Exile: went into hiding	Leaves city for 10 years
Success in exile	Prospers, commands	Success in exile gained supporters				Travels to reconcile with neighbors
Exilic agon	Retenu Hero challenge					
Exilic victory	Wins					
Threat overcome	Letter welcoming	Adad says to return home with army to Syria	Ishtar aided to regain affection	Successful revolt with Ishtar’s help	Defeat of brothers	People ready to receive him back; return to build Ehulhul
Return and reconciliation	Return with welcome and celebration	Welcome and reconciliation	Reconciliation	Return; gained kingship; gained kingship	Return to grateful nation	Completes temple; reconciliation
Epilogue	Gets tomb	Became king in Alalakh—30 years		Loyal to Ishtar	Consolidation of power; great and legitimate king	Rules according to Sin’s dictates

Table 16.2 Narrative structure for ancient Middle Eastern narratives: Bata and biblical narratives

V. Propp	Bata (d'Orb)	Joseph 1 (Gen 37)	Joseph 2 (Gen 39–47:27)	Jacob (Gen 25: 19–33:18)	Moses (Exod 2:1–12:51; 3:17–14:30)	David 1 (I Sam 16:1–18:5)	David 2 (I Sam 18:6–31:13)
Initial situation	Servant in Anubis's household	Home, younger son	Administers Potiphar's house	Home, younger son		Younger son; Samuel anoints as the next agent of Yahweh	Marriage to king's daughter Michal; at court
Threat	Attempted seduction	Jealous brothers, favorite and dreams	Mrs. Potiphar attempts to seduce Joseph	Esau loses birthright	Levite but to royal court	Threatens king	Jealousy at success vs. Philistines
Threat realized	False accusation	Sold to slavery	Mrs. Potiphar accuses Joseph of attempted seduction	Flees to Haran	Trouble; kills Egyptian		Saul seeks David's life
Exile	Valley of the Pine with heart on tree	Egypt in Potiphar's house	Joseph imprisoned	With Laban in Haran	Flees to Sinai	Leaves home to go to Saul's court	David flees
Success in exile	House, wife, hunting	Blessed by the Lord; runs house	Interprets dreams	Blessed by the Lord; success with flocks	Exile		
Exilic agon	Tree with heart cut down		Called to interpret Pharaoh's dreams	Compensation from Laban		Goliath	Contests vs. Saul
Exilic victory	Anubis restores heart		Appointed administrator		Runs off shepherds	Wins (threat to Saul)	Success with Jonathan's help
Threat overcome	Transforms to bull		Family to Egypt for food	Success in compensation	with Lord's support, challenge to pharaoh		Saul is killed
Epilogue	Becomes king		Israelites in Egypt	Brothers separate amicably	Organizes people		Yahweh's purpose established dynasty

This pattern also reflects the initiatory life cycle rite of passage originally outlined in 1909 by Arnold van Gennep (Gennep 1960) and later discussed and elucidated by Victor Turner (Turner 1967, 1969), which basically involves three stages: separation, transition (the liminal period), and (re)incorporation.[24] Each of the narratives mentioned here illustrates this pattern,[25] even Sinuhe, who hardly seems to have changed his status in Egypt on his return but who clearly altered his status within the narrative's central part during his stay in the Levant area.

Returning now to Moldenke's 1895 suggestion that Moses learned the story of Bata in Egypt and used it as he formulated the Joseph story, one surmises an underlying assumption that narratives like these traveled or were learned elsewhere rather than being individually invented or created, thus almost anticipating Hermann Gunkel's understanding, expressed in *Folktales in the Old Testament* (1917), that folktales travel and modify, and "since they always exist in a relatively indefinite form ... [they] are thus always capable of moulding themselves to fit the distinctive features of the people whom they reach and of adopting new shapes" (Gunkel 1987: 32). Gunkel noted furthermore that unless a tale moves as a whole story with the original motifs in the same order, the putative origin of a tale remains in question, since motifs similarly have the ability to travel (Gunkel 1987: 32).

Thus it is not surprising to find motifs in addition to the examples of the Potiphar's wife motif, K2111, in biblical and other traditional narratives. For example, several birth-related motifs from Egypt recall specific biblical motifs, as in The Tale of the Doomed Prince which opens with a childless king's prayer to his god requesting a son and heir, clearly reminding biblical scholars of the barren wives of the patriarchs and of Hannah in 1 Sam 10:20–21, as well as the king's prayer in the Ugaritic Tale of Aqhat (Coogan 1978: 27–47; Coogan and Smith 2012: 27–55). Other Egyptian narratives suggestive of divine intervention in procreation and/or birth include the narratives of the Birth of the King found in temples related to the 18th and 19th dynasty kings Hatshepsut, Amenhotep III, and Ramesses II (Brunner 1964: 1–9) and the last tale from the Papyrus Westcar in which the wife of a priest of the sun god gives birth with the aid of goddesses (Simpson 2003a: 21–24).[26] Of course, Sinuhe's combat with the strong man from Retenu has been compared to David's battle with Goliath (2 Sam 17),[27] since both men battle successfully against tremendous odds in their respective narratives,[28] and in the third tale from the Westcar Papyrus, a magician separates waters to retrieve a lost hairpiece, reminding one of God's reported actions at the Reed Sea (Exod 14:21–22; 26–29). Finally the biblical pattern of promoting the younger brother or relative over the older brother(s) reflects not only the Egyptian Tale of Two Brothers but also the Egyptian narrative of Horus and Seth in which Horus, the nephew and younger, succeeds to the throne

[24] Dundes presents a succinct discussion of van Gennep's ideas and an extract of his work in *International Folkloristics* (Dundes 1999: 99–108).

[25] For discussion of Two Brothers as representing initiation, one of the five life cycle rites of passage, see discussions by Assmann (1977) and Hollis (2008: 189–190).

[26] The birth of the divine kings from the wife of a sun priest may possibly be read as divine procreation.

[27] Among the various commentaries on the parallels, see discussions by Harry Hoffner (Hoffner 1968), Philip J. King (King 2007: 351–352), and Azzan Yadin (Yadin 2004: 380). While one might analyze the two fights as each reflecting Motif G512, Ogre killed, a major difference between them centers around the context. David's challenge is part of a battle of his people against the Philistines, termed a "contest of champions" by Yadin (2004: 379), which occurs in a physical space between the two armies in place of whole armies fighting, while Sinuhe's combat is unrelated to any army action, for the Retenu champion approaches Sinuhe in his tent, simply challenging him to fight (Sinuhe B 110).

[28] For this motif, see http://www.ruthenia.ru/folklore/thompson/index.htm, though neither tale is included as part of the Thompson index. Curiously much from the ancient world as well as from the Bible appears to be absent in Thompson's indices, so his use of the Egyptian Tale of Two Brothers seems to constitute an exception.

rather than his uncle Seth as well as the political narratives described earlier.[29] Clearly the traditional motifs and narrative patterns one finds in the Hebrew Bible reflect strongly those of other contemporaneous and older cultures, and numerous biblical commentators, most notably those who bring a folkloristic approach to biblical studies, continue to elaborate on the topic of international commonalities of tale types and motifs. For example, Susan Niditch has stated, "Cross-cultural comparisons reveal what is unique and what is not" (Niditch 1993: 11) asserting that "[t]he sharing of patterns and motifs found in various cultures is itself a trait of folklore—a trait evidenced by the Bible" (Niditch 1993: 11).

Not surprisingly, however, scholars continue to question both the origins of the motifs and tale types and the mode of their transmission.[30] For example, various attempts to locate the origins of Two Brothers elsewhere have occurred (Hollis 2008: 22, 25), though not related to Genesis 39.[31] In fact, Anthony Spalinger discussed it (Spalinger 2007: 141–143), specifically building on Carl Wilhelm von Sydow's earlier discussion of Oikotypes (von Sydow 1948).[32] These discussions, however, lack dated ancient concrete examples or any suggestions about how transmission from the noted areas, specifically East European and Slavic, might have occurred aside from a general mention of oral transmission and oral narrators. And curiously an approach to origins of motifs as reflecting human experience appears to be lacking in any discussions. Yet the motifs commonly present actual human issues, as for example the desire of a childless couple to have a child, an heir, or of a mature female's attraction to a much younger male. These kinds of issues form part of the common human experience and would seem to provide basic material for narratives. At some level, then, it seems that one must consider the human experience as a valid source for motifs.

Ultimately then, Jon Levinson's words regarding the Potiphar's wife episode accurately sum up its situation in relation to the Egyptian narrative: "Although some dependence is likely, the biblical narrative adapts its prototype to the characteristic Israelite theological and ethical convictions" (Levinson 2004: 78). Thus while one may easily find tale types, motifs, and patterns from Egyptian narratives reflected in the biblical materials, their ubiquitous nature worldwide and the focus and purpose of the materials within the Bible itself hardly support more than a minimal dependence on Egyptian materials. The limitation here lies in the words "tale," "tale type," "motif," and "patterns," for the focus of these materials, as in folklore in general as expressed by William Bascom, lies in their value as entertainment, validation of the culture, education, and maintenance of conformity to the culture's values (Bascom 1954: 342–346), which each of these tales does for its particular culture: the Egyptian narrative of Two Brothers perhaps representing the foundation for a dynastic change (Hollis 2008: 192–193; Wettengel 2003; Spalinger 2007: 138, 143–137) and the biblical narrative recalling the vestigial memory of the people's descent to Egypt from which the Exodus then occurred. In the end, while certainly various Egyptian tales were known to some, if not many, of the ancient Israelites who left Egypt in the Exodus, whatever direct influence existed was adapted to their needs. Thus the answer to the question posed in the title of this discussion suggests that Egyptian influence may be possible, but if so, it has been adapted and used for the ancient Israelites' own purposes and bears that stamp.

[29] The younger brother's rise over the elder brother(s) appears ubiquitously in traditional tales.

[30] For an excellent discussion of origins, albeit in relation to ballads but nevertheless pertinent, see Alan Bold's discussion on ballads (Bold 1979: Ch. 1).

[31] Such a discussion of the biblical origins along these lines might be very interesting. Might it be possible to separate Genesis 39 from its Egyptian analog, perhaps considering Tablet VI of the Epic of Gilgamesh in light of the putative Mesopotamian origin of the Abraham?

[32] Alan Dundes provides a discussion of von Sydow's work along with an extract of it in his *International Folkloristics* (Dundes 1999: 137–151) with a specific mention of the Egyptian Tale of Two Brothers (Dundes 1999: 150).

References

Abusch, Tzvi. 1986. Ishtar's Proposal and Gilgamesh's Refusal: An Interpretation of *The Gilgamesh Epic*, Tablet 6, Lines 1-79. *History of Religions* 26: 143–187.

Albright, William F. 1934. *The vocalization of the Egyptian Syllabic Orthography*. New Haven, CT: American Oriental Society. Kraus Reprint, Millwood, NY, 1974.

Assmann, Jan. 1977. Das ägyptische Zweibrüdermärchen. *Zeitschrift für Ägyptische Sprache und Alterstumkunde* 104: 1–24.

Baines, John. 1982. Interpreting *Sinuhe*. *Journal of Egyptian Archaeology* 68: 31–44.

Bascom, William R. 1954. Four functions of Folklore. *Journal of American Folklore* 67: 333–349.

Berlin, Adele, and Marc Z. Brettler (eds.). 1999. *The Jewish Study Bible*. New York, NY: Oxford University.

Beaulieu, Paul-Alain. 2000. Cylinder of Nabonidus. *Legal and Administrative Texts from the Reign of Nabonidus*. Yale Oriental Series, Babylonian Texts 19, New Haven, CT: Yale University. Available at http://en.wikipedia.org/wiki/Cylinder_of_Nabonidus#Translation

Birch, Samuel. 1860. *Select Papyri in Hieratic Character from the Collections of the British Museum, Part II*. London: Woodfall and Kinder.

Blackman, Alyward M. (ed.). 1972. *Middle-Egyptian Stories*. Bibliotheca Aegyptiaca II, Bruxelles: Édition de la Fondation Égyptologique Reine Élisabeth.

Bold, Alan. 1979. *The Ballad*. New York, NY: Methuen.

Brunner, Helmut. 1964. *Die Geburt des Gottkönigs: Studien zur Überlieferung eines altägyptischen Mythos*. Ägyptologische Abhandlungen 10, eds. W. Helck and E. Otto, Otto Harrassowitz: Wiesbaden.

Buck, A. de. 1977. *Egyptian Readingbook, vol. I: Exercises and Middle Egyptian Texts*. Chicago, IL: Ares Publishers. Reprint of 1948 edition.

Campbell, Joseph. 1968. *The hero with a Thousand Faces*. Bollingen series XCII, 2nd ed. Princeton, NJ: Princeton University.

Clifford, Richard J. 1998. The Wisdom Literature. *Interpreting Biblical Texts*, ed. G. M. Tucker. Nashville, TN: Abingdon.

Coogan, Michael (ed.). 1978. *Stories from Ancient Canaan*. Philadelphia, PA: Westminster.

Coogan, Michael D., and Mark S. Smith (eds.). 2012. *Stories from Ancient Canaan*, 2nd ed. Louisville, KY: Westminster John Knox.

Cruz-Uribe, Eugene. 2001. Vocabulary. In *The Oxford Encyclopedia of Ancient Egypt, vol. 3*, ed. D.B. Redford, 482–483. New York, NY: Oxford University.

de Rougé, Emmanuel. 1852. Notice sur un manuscrit égyptian en écriture hiératique écrit sous le règne de Merienphthah, fils du grand Ramses vers le XVe siècle chrétienne. *Revue Archéologique Series* I, VIII: 1–15.

Dundes, Alan (ed.). 1999. *International Folkloristics. Classic Contributions by the Founders of Folklore*. Lanham, MD: Rowman & Littlefield.

Frayne, Douglas. 2001. The Sumerian Gilgamesh Poems. In *The Epic of Gilgamesh. A Norton Critical Edition*, ed. B. R. Foster, 99–155. New York, NY: W. W. Norton.

Gardiner, Alan H (ed.). 1981. *Late-Egyptian Stories. Biblioteca Aegyptiaca I*, Bruxelles: Édition de la Fondation Égyptologique Reine Élisabeth.

Gelander, Shamai. 2013. Folk Narratives in the Bible. In *Encyclopedia of Jewish Folklore and Traditions*, vol. 1, ed. F.E. Raphael Patai and H. Bar-Itzhak, 181–183. Armonk, NY: M. E. Sharpe.

Gennep, Arnold v. 1960. *The Rites of Passage*. (Trans. M. B. Vizedom, and G.L. Caffee) Chicago, IL: University of Chicago.

George, Andrew. 1999. *The Epic of Gilgamesh. The Babylonian Epic Poem and Other Texts in Akkadian and Sumerian*. New York, NY: Penguin.

Goetze, Albrecht. 1969. Prayer of Pudu-hepas to the Sun goddess of Arinna and her Circle. A. A prayer to the Sun-Goddess herself. In *Ancient Near Eastern Texts Relating to the Old Testament, 3rd ed. with Supplement*, ed. J. B. Pritchard, 393. Princeton, NJ: Princeton University.

Gunkel, Hermann. 1987. *The Folktale in the Old Testament*. trans. M. D. Rutter from the German (1917). Historic Texts and Interpreters in Biblical Scholarship, ed. J. W. Rogerson, Sheffield: Almond.

Gurney, Oliver R. 1954. *The Hittites*. Baltimore, MD: Penguin.

Hansen, William F. 2002. *Ariadne's Thread. A Guide to International Tales Found in Classical Literature*. In Myth and Poetics, ed. G. Nagy, Ithaca, NY: Cornell University.

Hillers, Delbert. 1973. The bow of Aqhat: The Meaning of a Mythological Theme. In *Orient and Occident: Essays Presented to Cyrus H. Gordon on the Occasion of his Sixty-fifth Birthday. Alter Orient und Altes Testament 22*, ed. H.A. Hoffner, 71–80. Neukirchner: Neukirchen-Vluyn.

Hoffner, Harry. 1968. A Hittite Analogue to the David and Goliath Contest of Champions. *Catholic Biblical Quarterly* 30: 220–225.

Hollis, Susan T. 1989. The Woman in Ancient Examples of the Potiphar's Wife Motif, K2111. In *Gender and Difference in Ancient Israel*, ed. P.L. Day, 28–42. Minneapolis, MN: Fortress.

———. 1998. Ancient Israel as the Land of Exile and the 'Otherworld' in Ancient Egyptian Folktales and Narratives. In *Boundaries of the Ancient Near Eastern World: A Tribute to Cyrus H. Gordon. Journal for the Study of the Old Testament Supplement Series 273*, ed. M. Lubetski, C. Gottlieb, and S. Keller, 320–337. Sheffield: Sheffield Academic.

———. 2008. *The Ancient Egyptian "Tale of Two Brothers." A Mythological, Religious, and Historico-Political Study*, 2nd ed. Oakville, CT: Bannerstone.

———. 2013. Joseph. In *Encyclopedia of Jewish Folklore and Traditions*, vol. 1, ed. F.E. Raphael

Patai and E. Haya Bar-Iszhak, 284–286. Armonk, NY: M. E. Sharpe.

King, J. Robin. 1987. The Joseph Story and Divine Politics: A Comparative Study of a Biographic Formula from the Ancient Near East. *Journal of Biblical Literature* 106: 577–594.

King, Philip J. 2007. David Defeats Goliath. In *"Up to the Gates of Ekron": Essays on the archaeology and history of the eastern Mediterranean in honor of Seymour Gitin*, ed. S.W. Crawford and A. Ben-Tor, 350–357. Jerusalem: W. F. Albright Institute.

Lambdin, Thomas O. 1953. Egyptian Loan Words in the Old Testament. *Journal of the American Oriental Society* 73: 145–155.

Lemche, Niels P. 1998. *The Israelites in History and Tradition*. Library of Ancient Israel, ed. D. A. Knight, Louisville, KY: Westminster John Knox.

Levinson, Jon. 2004. Genesis (introduction and annotations). In *The Jewish Study Bible Featuring the Jewish Publication Society Tanakh Translation*, ed. A. Berlin and M.Z. Brettler, 8–101. New York, NY: Oxford University.

Lichtheim, Miriam. 1976. *Ancient Egyptian Literature: A Book of Readings. Volume II: The New Kingdom.* Berkeley, CA: University of California.

Moldenke, Charles E. 1892. *The Oldest Fairy Tale Translated from the Papyrus D'Orbiney, with Notes.* Watchung, NJ: The Elsinore Press. https://books.google.com/books?id=BoYKAAAAIAAJ Accessed 6 May 2013.

———. 1895. The Oldest Fairy Tale. *Transactions of the Meriden (Conn.) Scientific Association* 7: 32–81.

Mumford, G.D. 2001. Syria-Palestine. In *The Oxford Encyclopedia of Ancient Egypt, vol. 3*, ed. D.B. Redford, 335–343. New York, NY: Oxford University.

Niditch, S. 1993. *Folklore and the Hebrew Bible*. Guides to Biblical Scholarship: Old Testament Series, ed. G. M. Tucker, Minneapolis, MN: Fortress.

Oppenheim, A. Leo. 1969a. Historical Documents. 9 Esarhadden (a) Fight for the Throne. In *Ancient Near Eastern Texts Relating to the Old Testament, 3rd Edition with Supplement*, ed. J. B. Pritchard, 289–290. Princeton, NJ: Princeton University.

———. 1969b. Nabonidus and His God. In *Ancient Near Eastern Texts Relating to the Old Testament, 3rd Edition with Supplement*, ed. J. B. Pritchard, 562–563. Princeton, NJ: Princeton University.

———. 1969c. The Story of Idrimi, King of Alalakh. In *Ancient Near Eastern Texts Relating to the Old Testament, 3rd Edition with Supplement*, ed. J. B. Pritchard. 557–558. Princeton, NJ: Princeton University.

Parkinson, Richard B. 1997. *The Tale of Sinuhe and Other Ancient Egyptian Poems 1940–1640 BC*. Oxford: Oxford University.

———. 2001a. Papyrus Westcar. In *Oxford Encyclopedia of Ancient Egypt*, vol. 3, ed. D.B. Redford, 24–25. New York, NY: Oxford University.

———. 2001b. Sinuhe. In *Oxford Encyclopedia of Ancient Egypt*, vol. 3, ed. D.B. Redford, 292. New York, NY: Oxford University.

Paulme, Denise. 1963. Un conte de fées africain: Le garçon travesti ou Joseph en Afrique. *L'Homme. Revue française d'anthropologie* 3: 5–21.

Peet, Thomas E. 1931. *A Comparative Study of the Literatures of Egypt, Palestine and Mesopotamia. Schweich Lectures of the British Academy; 1929.* London: Pub. for the British Academy by H. Milford, Oxford University.

Propp, Vladimir. 1968. *Morphology of the Folktale*, ed. L. Wagner (Trans. L. Scott) 2nd ed. Austin, TX: University of Texas.

Propp, William H.C. 2013. The *Eunuch Steward*: A Trans-Asian Tale of Seduction and Castration. *International Journal of Applied Psychoanalytic Studies* 10. wileyonlinelibrary.com

Raglan, Lord. 1990. The Hero: A study in Tradition Myth and Drama part II. In *In Quest of the Hero. Mythos*, vol. 3, ed. R.A. Segal, 89–175. Princeton, NJ: Princeton University.

Rank, Otto. 1990. The Myth of the Birth of the Hero. In *In Quest of the Hero*. Mythos, vol. 3, ed. R.A. Segal, 3–86. Princeton, NJ: Princeton University.

Redford, Donald B. 1970. *A Study of the Biblical Story of Joseph (Genesis 37-50), Supplements to Vetus Testamentum XX*. Leiden: E. J. Brill.

———. 1992. *Egypt, Canaan, and Israel in Ancient Times*. Princeton, NJ: Princeton University.

Shupak, Nili. 1993. Where can Wisdom be found? The Sage's Language in the Bible and in Ancient Egyptian Literature. *Orbis Biblicus et Orientalis* 130, ed. O. Keel, Freiburg, Schweiz: Universitätsverlag & Göttingen: Vandenhoeck & Ruprecht.

Simpson, William K. 1982. Pap. Westcar (Berlin 3033) Cols. 744–746. In *Lexikon der Ägyptologie*, vol. 4, ed. W. Helck and E. Otto. Harrassowitz: Wiesbaden.

———. 2003a. King Cheops and the Magicians. In *The Literature of Ancient Egypt. An Anthology of Stories Instructions, Stelae, Autobiographies, and Poetry*, ed. W.K. Simpson, 13–24. New Haven, CT: Yale University.

———. 2003b. The story of Sinuhe. In *The Literature of Ancient Egypt. An Anthology of Stories Instructions, Stelae, Autobiographies, and Poetry*, ed. W.K. Simpson, 54–66. New Haven, CT: Yale University.

Spalinger, Anthony J. 2007. Transformations in Egyptian Folktales: The Royal Influence. *Revue d'Égyptologie* 58: 137–156.

Thompson, Stith (ed.). 1955–58. *Motif-Index of Folk Literature, revised and enlarged*, vol. 4. Bloomington, IN: Indiana University.

——— (ed.). 1964. *The Types of the Folktale: A Classification and Bibliography*. Helsinki: Soumalainen Tiedeakateimia Scientiarum Fennica (Trans. and enlarg. of Antti Aarne, Verzeichnis der Märchentypen, Folklore Fellows Communications 3, Folklore Fellows Communications 183).

———. 1977. *The Folktale*. Berkeley, CA: University of California Press. Reprint of 1946, Holt, Reinhart, & Winston.

Turner, Victor. 1967. Betwixt and Between: The Liminal Period in *Rites de Passage*. In *The Forest of Symbols: Aspects of Ndembu Ritual*, ed. V. Turner, 93–111. Ithaca, NY: Cornell University.

———. 1969. *The Ritual Process: Structure and Anti-Structure*. Ithaca, NY: Cornell University.

von Sydow, Carl W. 1948. On the Spread of Tradition. In *Selected Papers on Folklore on the Occasion of his 70th birthday*, ed. L. Bødker, 11–43. Copenhagen, Denmark: Rosenkilde and Bagger.

Washington, Harold C. 1994. *Wealth and Poverty in the Instruction of Amenemope and the Hebrew Proverbs*. Society of Biblical Literature: Dissertation Series 142, eds. D. L. Petersen and P. Perkins, Atlanta, GA: Scholars.

Wente, Jr., and F. Edward. 2003a. The Tale of the Doomed Prince. In *The Literature of Ancient Egypt. An Anthology of Stories Instructions, Stelae, Autobiographies, and Poetry*, ed. W.K. Simpson, 75–79. New Haven, CT: Yale University.

———. 2003b. The Tale of Two Brothers. In *The Literature of Ancient Egypt. An Anthology of Stories Instructions, Stelae, Autobiographies, and Poetry*, ed. W.K. Simpson, 80–90. New Haven, CT: Yale University.

Wettengal, Wolfgang. 2003. *Die Erzählung von den beiden Brüdern. Der Papyrus d'Orbiney und die Königsideologie der Ramessiden*. Orbis Biblicus et Orientalis 195, Freiburg, Schweiz, Vandenhoeck & Ruprecht: Universitätsverlag & Göttingen.

Yadin, Azzan. 2004. Goliath's Armor and Israelite Collective Memory. *Vetus Testamentum* 54: 373–395.

17 The Egyptian Origin of the Ark of the Covenant

Scott B. Noegel

Abstract

The best non-Israelite parallel to the Ark of the Covenant comes not from Mesopotamia or Arabia, but from Egypt. The sacred bark was a ritual object deeply embedded in the Egyptian ritual and mythological landscapes. It was carried aloft in processions or pulled in a sledge or a wagon; its purpose was to transport a god or a mummy and sometimes to dispense oracles. The Israelite conception of the Ark probably originated under Egyptian influence in the Late Bronze Age.

The Ark of the Covenant holds a prominent place in the biblical narratives surrounding the Israelites' exodus from Egypt. Its central role as a vehicle for communicating with Yahweh and as a portable priestly reliquary distinguishes it from all other aspects of the early cult. In varying detail, biblical texts ascribe to the Ark a number of functions and powers, which have led scholars to see the Bible's portrayal of the Ark as the result of historical development and theological reinterpretation.[1] While some have looked to Mesopotamia and premodern Bedouin societies for parallels to the Ark, the parallels have remained unconvincing and have contributed to the general view that the Ark was uniquely Israelite. Today I propose that we can gain greater insight into the Israelite Ark and the narratives in which it appears by looking to a hitherto overlooked parallel: the Egyptian sacred bark.

Ark of the Covenant

Biblical texts describe the Ark of the Covenant as a sacred object containing five major features. The first is a wooden box (Heb. *ʾarōn*), roughly 4 ft. × 2.5 × 2.5, and overlaid with gold.[2] The second is a lid (Heb. *kappōreth*), made entirely of gold, not plated like the box,[3] which contains a molding running along its top edge. Its third component is a pair of gold *kerubîm*, i.e.,

[1] See Dietrich (2007: 250–252).

S.B. Noegel (✉)
Near Eastern Languages and Civilization (NELC), University of Washington, Seattle, WA, USA
e-mail: snoegel@uw.edu

[2] The word *šittîm* "acacia" is a loan from Egyptian. See Muchiki (1999: 256). There are a number of species of acacia that grow in Egypt, the Sinai peninsula, the Judean desert, and the Negev.

[3] The lid also is translated "mercy seat," based on an etymological association. However, the word *kappōreth* simply means "covering."

T.E. Levy et al. (eds.), *Israel's Exodus in Transdisciplinary Perspective*, Quantitative Methods in the Humanities and Social Sciences, DOI 10.1007/978-3-319-04768-3_17,

"sphinxes,"[4] that rest on top of the lid and face each other with their wings touching. Note that the lid was understood as God's "throne," whereas the box was viewed as his "footstool" (e.g., 1 Chron 28:2, 2 Chron 9:18, Ps 99:5, 132:7). The Ark's fourth feature was its wooden poles, which were inserted through four gold rings and never removed.[5] Only the priestly tribe of Levi was permitted to carry the Ark, and even then, only after they had veiled it (Exod 40:3, 40:21).[6] No one of non-priestly descent was allowed to touch it. The Ark's fifth feature was its contents: the tablets of the law (Deut 10:1–5, *ʾarōn hab-berîth*; Exod 25:22, *ʾarōn ha-edūth*), a jar of manna (Exod 16:33–34), and possibly the rod of Aaron (Heb 9:4, cf. Num 17:10).[7]

In addition to serving as a reliquary, texts attribute two other functions to the Ark. Most prominently, it served as the symbolic presence of Yahweh. In times of war, Yahweh led as the Lord of Hosts, seated upon the *kerubîm*, surrounded by standard bearers preceding him. Each standard was topped with a banner representing an Israelite tribe or family line (Num 2:1–34, 10:35, Ps 132:8).[8]

As the symbolic presence of Yahweh, the Ark was connected to miracles and oracles. Thus, when the priests carried the Ark into the Jordan River the waters parted (Josh 3:8–17), and Moses, Phinehas, Samuel, Saul,[9] and David each received divine direction from the Ark (Exod 25:22, 30:6, Num 7:89, Judg 20:27–28, 1 Sam 3:3, 1 Sam 14:18, cf. 2 Sam 2:1, 5:19, 11:11, 15:24).[10]

Before the temple was built, the Ark stayed at a number of sanctuaries including Gilgal (Josh 7:6), Shechem (Josh 8:33), Bokhim (Judg 2:1–5), Bethel (Judg 20:27), Shiloh (1 Sam 3:3), Kiriath-Jearim (1 Sam 7:1–2), and Gibeon (1 Kgs 3:4, 1 Chron 16:37–42, 21:29, 2 Chron 1:3–4). During the visits Yahweh would accept sacrifices and bless his sanctuaries. Finally, the Ark acquired a ritual function. On Yom Kippur the high priest would sprinkle bull's blood onto and in front of the Ark's lid (Lev 16:14).

[4] On the Egyptian origin of this creature, see already Albright (1938) and now Mettinger (1999). Attestations of the Assyrian cognate *ᵈkurību* do not permit a precise or a consistent description of the creature. Thus, some appear to have animal heads while others have human heads. Nevertheless, the *ᵈkurību* commonly are described as fashioned images that either stand at entrances to portals or face each other. The use of the cuneiform DINGIR sign marks them as divine. See *CAD* K, 559, s. v. *kurību*. Even in the Bible, there is some variation concerning this creature. Thus, the *kerubîm* on the Ark have two wings (Exod 25:20, 37:9), but four wings in Ezekiel's vision (Ezek 1:6). The closest parallels are the sarcophagus of Ahiram, king of Byblos, and an ivory found at Megiddo. Both objects are highly Egyptianized and depict a king seated on a *kerubîm*-flanked throne. The latter item also features a winged solar disk and lotus offering. See Kyrieleis (1969: 41–81). Many objects found at Megiddo dating to this period evince Egyptian influence, if not also a presence. See Novacek (2011). On other possible parallels, including a stone throne from Lebanon and a divine statue from Cyprus, see Zwickel (1999: 101–105). On archaeological evidence for the Israelite cult, see Zwickel (1994).

[5] On two occasions oxen pulled the Ark on a "newly constructed wagon" (1 Sam 6:7, 2 Sam 6:3), though this was not ordinary practice.

[6] Moreover, the priests were forbidden from looking at the *kappōreth* "lid." Hence, it was veiled. Only the high priest could look at the *kappōreth* on Yom Kippur, provided he has undertaken a special rite and has changed his garments (Lev 16:4). On the veil and the lid, see Bordreuil (2006).

[7] According to Josephus, *Jewish Wars* 5.219, the innermost sanctum was empty.

[8] In Exod 17:15, Moses built an altar to Yahweh after his battle against the Amalekites and named it יְהוָה נִסִּי "Yahweh is my banner." The identification of Yahweh with a banner is reminiscent of the Egyptian hieroglyphic representation of *nṯr* "god" with a banner (i.e., 𓊹).

[9] 1 Chron 13:3 suggests that people did not seek oracles from the Ark during Saul's reign.

[10] The LXX of 1 Sam 14:18 reads "ephod." The instrument of divination in 2 Samuel is less clear, but Van der Toorn and Houtman (1994) argue that "ephod" here stands for "Ark" and that the Ark functioned for divination. They also opine that there were multiple Arks in the region whose existence was blurred by later Deuteronomist editing. If the authors are correct in arguing that the Ark that David brought to Jerusalem was not a national symbol, but a Saulide cult object, then perhaps we should look to the tribe of Benjamin as the original locus for the object.

Fig. 17.1 Divine palanquins, relief of Tiglath Pileser III

Previously Proposed Parallels to the Ark

Scholars have cited two objects as possible parallels for the Ark. The first is a divine palanquin as seen notably in the Assyrian reliefs of Tilglath Pileser III (744–727 B.C.E.).[11] The panel shows the king's seizure of foreign gods from their temples (Fig. 17.1).[12]

While some of the gods sit on thrones and might have served as a source of oracles, a number of differences remain. No poles were used to transport them, and there are no boxes and no lids. They were not covered in gold, nor do any of them contain relics *or kerubîm*. There is no evidence that the statues were carried into battle. Finally, there appear to have been no restrictions on who could touch them.

A second object previously compared to the Ark is the Bedouin *ʿuṭfa* (also called a *maḥmal*, *abu-dhur*, *markab*, and *qubba* [Fig. 17.2]).[13] It accompanied tribes into battle and signalled the presence of the divine. However, the Bedouin transported them on horses or camels. It contained no box, no lid, and no poles. Some were inscribed with spells and Quranic verses, but they never served as reliquaries or as the throne and footstool of God. They were not overlaid in gold, and they contained no *kerubîm*. There also were no restrictions on who could touch them.

While the palanquin and Bedouin objects offer some parallels, the dissimilarities limit their usefulness as analogues. Indeed, Menaham Haran long ago observed that the Ark's origins must be sought not in nomadic life, but in a

[11] Zwickel (1999: 106) also suggests a parallel with Egyptian divine palanquins, but he appears to reject it, because the Bible refers to the *ʾărōn* as a footstool. He does not consider a connection to the barks. See also Zwickel (1994).

[12] The relief, which is on display in the British Museum, was photographed by the author.

[13] The image of the *ʿuṭfa* appears in Musil (1928: 573). The *ʿuṭfa*, *maḥmal*, *abu-dhûr*, *markab*, and *qubba* have been treated rather loosely as a collective by earlier biblicists who proposed them as parallels to the Ark (e.g., Morgenstern 1942; de Vaux 1965: 9, 296–297), and since that time they have been adopted somewhat uncritically into the scholarly literature. Nevertheless, the items are rather distinct in appearance and function, and each has its own history. The *ʿuṭfa* generally refers to the hooded camel saddle used by married women of Sudan, Arabia, Tripoli, etc. It cannot be traced to pre-Islamic times. See Robinson (1931b). Tradition places the origin of the *maḥmal* in Mamluk Cairo in the thirteenth century CE. See Robinson (1931a). The *merkab* and *abu-dhûr* appear to be synonyms for the ostrich-feather litter that sits upon camels. They are recorded in premodern Bedouin society, but not pre-Islamic society. See Musil (1928: 571–574). The Egyptian *merkab* cannot be dated before the eleventh century CE, when the Persian traveler Nasir-i Khusrau described its use in conjunction with a Nile inundation ceremony, see Sanders (1994: 103). Only the *qubba* dates to pre-Islamic times, as it is represented on the temple of Bel at Palmyra (first century CE). Nevertheless, all of these litters are tent-like structures, and thus, they are more fruitfully compared to the tabernacle. See Homan (2002: 90–94). Homan does not discuss the *abu-dhur*. The Hebrew cognate *qubbāh* in Num 25:8 also refers to a tent.

Fig. 17.2 Bedouin *utfa*

sedentary community, since the Israelite priests carried it on foot (Haran 1985: 270). Moreover, as Michael Homan has shown (Homan 2002: 113–114), the strongest parallels for the tabernacle in which the Ark was placed are ancient Egyptian military and funerary tents including the tent-like coverings for funerary barks.[14] This suggests even greater propriety in looking to Egypt for an analogue.[15]

Egyptian Sacred Barks

With this in mind, I should like to propose that the Egyptian sacred bark offers a more compelling and complete parallel for the Ark. Of course, the bark was not merely a boat, but a sacred ritual object deeply imbedded in the ritual and mythological landscapes of the Egyptians. Though they resembled boats, they rarely, if ever, were set in water. Even when they needed to cross the Nile, they were loaded onto barges. Usually, they were carried by hand or in some cases dragged on a sledge or placed on a wagon (Fig. 17.3).[16]

The bark's most basic function was to transport gods and mummies. When transporting gods, the bark was fitted with a gold-plated naos containing a divine image seated on a *ḥwt-*

[14] Curiously, Homan (2002: 113) does not discuss a possible parallel between the Ark and the Egyptian bark, but instead he notes that Ramesses' golden throne appears in the Qadesh record as "flanked by falcon wings, just as the Ark is flanked by winged cherubim." Moreover, Homan (2002: 145–147) notes that the construction of the tabernacle's frame employs the term *qerāšîm* "(thin) boards," a word of nautical importance that elsewhere (i.e., Ezek 27:6) refers to the main cabin on a boat. See also Kitchen (1993: 119–129).

[15] We may add to this the fact that biblical tales set in Egypt often show a close knowledge of Egyptian practices and beliefs and, in some cases, draw upon Egyptian literary traditions. See, e.g., Sarna (1986) and the brief discussion by Currid (1997: 23–32) and his bibliography.

[16] Photograph of sacred barks at Medinet Habu by the author.

Fig. 17.3 Barks on stands with carrying poles, Medinet Habu

block throne,[17] which was veiled with a thin canopy of wood or cloth (Fig. 17.4).[18]

When transporting the dead, it carried the sarcophagus within a covered gold-plated catafalque (Fig. 17.5).[19] There is no one type of sacred bark, but rather many variations on a theme, each with its own set of accouchement.[20]

Many barks were decorated with protective *kerubîm*, such as the naos of the bark of Amun

[17] On the *ḥwt*-block throne, *srḫ*-block throne, and the "lion-throne," see Kuhlmann (2008). For a comparative work on thrones, see Metzger (1985).

[18] The veiled bark of Amun here comes from a relief at Karnak, photographed by the author.

[19] The photograph of the bark transporting the catafalque in the tomb of Userhat (TT 56) was taken by the author.

[20] See Göttlicher (1992: 13–75), who divides the cultic barks into four basic types: those belonging to districts, states, gods, or of non-locale or unspecified nature, with each category containing many variations. Most of the barks are given epithet-like names, though the general term for bark appears to have been *wi3*, perhaps related to the verb *wi3* "to be separated, secluded, segregated." See *WÄS* 1982: 272.

Fig. 17.4 Veiled bark of Amun, Karnak

found on Seti I's mortuary temple in Qurna (Fig. 17.6, left) and the bark of Horus in the temple at Edfu (Fig. 17.6, right).[21]

Like the Ark of the Covenant, sacred barks were carried on poles by priests, the so-called pure ones (Egyptian: *wꜥbw*), who had performed purification rituals in order to hoist the bark. Though most Egyptian rituals were never witnessed by the public, the procession of the sacred bark was an important exception. It was the focus of an intense series of festivals throughout the year, as many as five to ten per month, which involved loud music and dancing.[22]

During the celebrations, priests carried the bark from one shrine to another, and made stops along the way, during which they dramatized mythological scenes. The route and length of the processions varied depending on the gods they carried and their mythologies.[23]

The bark also gave oracles. While resting at one of the stations it could be consulted by written oracles, and while en route during the procession, it could be asked a question to which it would respond yes or no by bowing fore or aft. Some priests marched before the bark wafting incense and others alongside and behind. Some bore standards representing nomes, much like the tribal procession of the Ark of the Covenant (Barta 1965–66).

While I know of no sacred bark whose footstool contained relics, the placing of oaths beneath the feet of statues is attested. Thus, in a letter from Ramesses II to the Hittite king Hattusilis III, we find the following reference: "The writing of the covenant that [I made] to the Great King, and which the King of Hattu has made with me, lies beneath the feet of [the god Ra]. The great gods are witnesses [to it]."[24] Scholars have long likened this practice to the

[21] Photographs by the author.

[22] Stadler (2008). On Theban barks, see Bell (1985: 251–294).

[23] See Sauneron (1960: 93); Teeter (2011: 56–75).

[24] A copy of the letter also was placed at the feet of the Hittite god Teshub. On the correspondence between these kings see Edel (1994: 1/16–29, 2/27–29). For the Egyptian texts of the treaty, see Kitchen (1971: 225–232); Edel (1983: 135–153). Note that Beckman (1996: 125) treats the god in the broken portion of the letter as the Hittite storm god.

Fig. 17.5 Bark on catafalque, tomb of Userhat

placing of the covenantal tablets in the Ark's footstool.[25]

In addition, from the 18th Dynasty well into the Roman period, Egyptians fashioned statues of the god Ptah-Sokar-Osiris standing upright on their own coffins (Fig. 17.7).[26]

Of interest here is that the coffins often housed copies of the *Book of the Dead* or small corn mummies. While the Ramesside letter and statue are not exact parallels to the Ark, they share the concept of texts placed beneath the feet of a god.

Like their divine counterparts, funerary barks functioned as a means of transport and mythological invocation. However, rather than transport images, they ferried the deceased to their tombs. As in the festivals, loud music accompanied burial processions. These processions too were public, though the number of attendees naturally varied.[27]

The bark's trip to the tomb invoked the journey of the sun as it sailed to the land of the west. Like Ra in his solar bark,[28] the deceased hoped to sail on a cycle of renewal and emerge with him at dawn.

Even from this cursory treatment, it should be clear that the Ark and the bark share much in common in both design and function, and each, in its own way, was connected to a historicized

[25] See already Herrmann (1908). The platforms on which Arks were placed also sometimes stored texts. Thus, spell 64 of the *Book of Going Forth by Day* (lines 25–26) concludes by noting that the spell was discovered by a master-worker in a plinth belonging to the god of the *Hennu*-bark (i.e., Sokar or Horus). P. London BM EA 10477 (P. Nu), Tb 064 Kf (line 25), P. Cairo CG 51189 (P. Juya), Tb 064 (line 284). Moreover, in the 18th Dynasty the term *s.t wr.t* "great seat," which usually referred to the throne of a king or a god, came to be used for the pedestal on which one rested a divine bark or the bark shrine itself. Eventually, it became a metonym for the temple. See McClain (2007: 88–89). Herrmann (1908: 299–300) also draws attention to the parallel. In 1 Samuel 10:25, Samuel also places a scroll containing the duties of kingship before the Ark.

[26] The Late Period exemplar shown here is courtesy of the British Museum (E9742).

[27] Teeter (2011: 57) remarks: "Festivals also illustrated how little separation there was between the concepts of funerary and nonfunerary practices. For example, festivals of Osiris, the god of the afterlife, were celebrated in the Karnak Temple and recorded in detail at the Temple of Hathor at Dendara, structures that are not usually associated with mortuary cults."

[28] The solar god rode one boat (*mʿnḏ.t*) during the day and another (i.e., *mskt.t*) at night. On the orientation of these boats, see Thomas (1956: 56–79).

Fig. 17.6 Naoi containing *kerubîm*

mythology of return. Of course, I am not suggesting that the Ark of the Covenant *was in fact* a bark; only that the bark served as a model, which the Israelites adapted for their own needs. Thus, the Israelites conceived of the Ark not as an Egyptian boat with a prow and stern and oars,[29] but as a rectangular object, more akin to the riverine boat that informs the shape of Noah's Ark (6:14–16).[30] Nevertheless, some of the bark's other aspects remained meaningful in Israelite priestly culture. It still represented a throne and a footstool and so it still served as a symbol of the divine presence. It continued to be a sacred object that one could consult for oracles, and its maintenance continued to be the exclusive privilege of the priests.

Moreover, there is evidence that it retained the chthonic import of its Egyptian prototype. In part this comes from the very name that the Israelites gave the object, an *ʾarōn*, which also, and perhaps primarily, means "coffin."[31] As such it appears in the narratives concerning the deaths of the patriarch Jacob (Gen 50:1–14) and his son Joseph (Gen 50:26), both of whom were embalmed according to Egyptian practice and placed in an *ʾarōn*.[32]

[29] Of course, the Israelites dispensed with the Egyptian practice of placing an image of the God's head on the prow and stern.

[30] The term for Noah's Ark is *tēbāh* (Gen 6:14). It is also used for the small chest into which the infant Moses was floated to safety (Exod 2:3, 2:5). The word *tēbāh* is a loan from the Egyptian *ḏb3(t)* "naos, casket, sockel for a throne." Interestingly, like the Hebrew word *kissēʾ* "throne," the Israelites did not use the term *tēbāh* for the Ark of the Covenant, even though it was available to them. It is plausible that the Israelites used the term *ʾarōn* instead of *tēbāh* (or *kissēʾ* "throne"), because it distinguished the object from a boat while retaining its chthonic associations. On the Hebrew and Egyptian lexemes, see *HALOT*, p. 1678, s.v. תֵּבָה; *WÄS* 5: 555–562, and Hannig (1995: 1003), s.v. *ḏb3(t)*. The meaning "coffin" is spelled *ḏb3(t)*. On the word as a loan into Hebrew, see Muchiki (1999: 258). On the LXX's rendering of both *ʾarōn* and *tēbāh* as κιβωτός, see Loewe (2001).

[31] The word *ʾarōn* appears in 2 Kgs 12:9–16 (=2 Chron 24:8–12), where it is often translated "(money) chest." However, the passage carefully states that the priest Jehoida took an *ʾarōn* and bored a hole into its lid (i.e., *delet*, lit. door). This clarifies that the coffin was repurposed as a coffer. The Akkadian cognate *arānu* similarly means coffin and cashbox, *CAD* A 2, p. 231, s.v. *arānu*. Note also that the Phoenician cognate *ʾarōn* appears on a number of royal memorial inscriptions in reference to heavily Egyptianized Phoenician sarcophagi. See *KAI*, nos. 1, 9, 13, 13, 29. If the wood used to build the Ark (i.e., *šittîm* "acacia") is to be identified with *spina aegyptiaca*, then it is noteworthy that the Egyptians also used it to construct coffins.

[32] Gen 50:2 states that Joseph ordered his servants and physicians to do the embalming, but they are not identified as Egyptians.

Fig. 17.7 Ptah-Sokar-Osiris figure standing on coffin

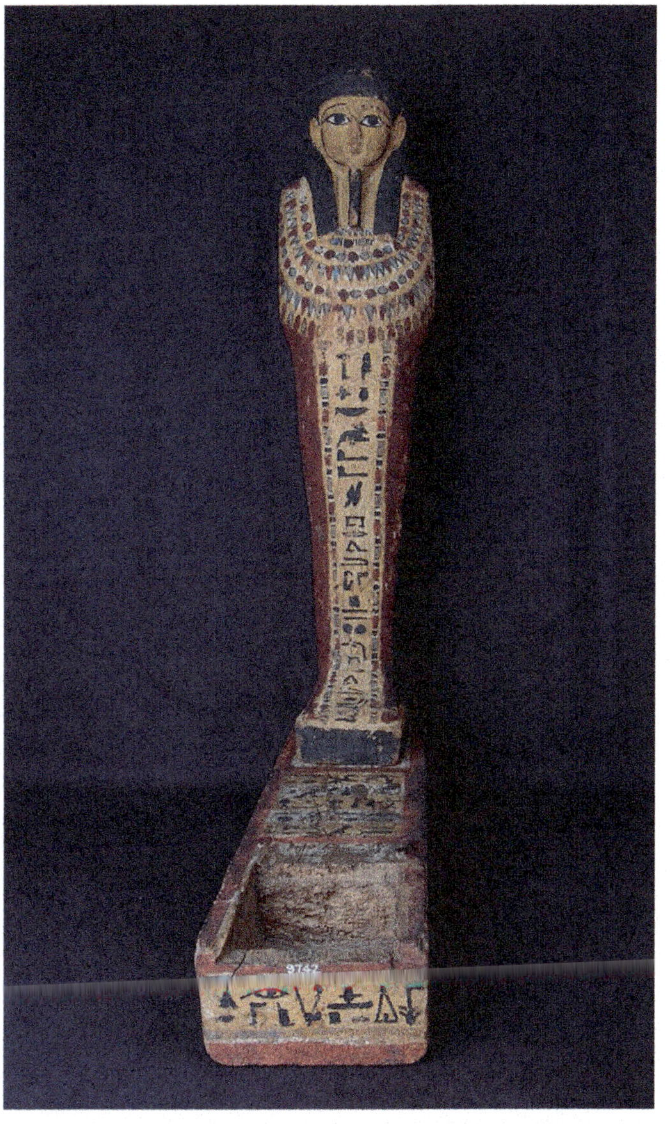

Underscoring the chthonic nature of the *ʾarōn* is its frequent association with threshing floors. See, for example, the account of Joseph's return to Canaan:

> When they reached the threshing floor of the bramble, near the Jordan, they lamented loudly and bitterly; and there Joseph observed a seven-day period of mourning for his father. When the Canaanites who lived there saw the mourning at the threshing floor of the bramble, they said, "The Egyptians are holding a solemn ceremony of mourning." That is why that place near the Jordan is called *Ābēl-Miṣrayîm* (lit. the "Mourning of the Egyptians," Gen 50:10–11).

The narrator does not say why the procession stopped here, but readers *are* forced to wonder, because Jacob was to be buried at Machpelah (Gen 49:30).[33] Also unclear is what the Canaanites saw that suggested an *Egyptian*

[33] See, for example, Sarna (1989: 348), who asks "Why does the procession stop at just this place?," and suggests that the region might have had Egyptian connections.

mourning practice.[34] While the text mentions the presence of Egyptian officials, they were far outnumbered by the elders of Israel, the household of Joseph and his brothers, and all the members of their father's household. Indeed, Gen 50:9 states that the group constituted a *ham-maḥaneh kābēd me'ōd*, "an exceedingly large camp." We are told nothing of professional wailing women nor of people dancing nor of an Opening of the Mouth ceremony. Even the length of the event, 7 days, suggests an Israelite mourning practice.[35] Instead, the narrator twice states that the rite took place at a threshing floor. We must consider this as more than a passing reference, for throughout the Near East threshing floors were regarded as numinous places rich with chthonic and fertility associations, and thus, they were loci for cultic activity.[36] However, since the Canaanites identified the mourning ritual as an Egyptian practice, we must ask more specifically what cultic significance the threshing floor had in Egypt.

In Egypt, the threshing floor was most widely associated with Osiris and his cult. I need not dwell here on the complex origins and nature of Osiris.[37] Suffice it to say that he was connected *inter alia* to the resurrection of the dead[38]; and though the etymology of his name is disputed, it is clear already in the Pyramid Texts that the Egyptians identified him with a divine throne, perhaps as the "Seat of Creation" or the "Throne of the Eye (i.e., Sun)."[39] The identification of Osiris with new grain is attested abundantly in the mythological corpora as well as in ritual practices, such as the making of corn mummies and Osiris beds,[40] the rites found in the *Dramatic Ramesside Papyrus*,[41] and the "Driving of the Calves" (*ḥwỉ bḥsw*) ritual.[42] The latter rite was enacted at a number of public festivals,[43] during which the threshing of grain was interpreted as the dismemberment of Osiris.[44] After mourning

[34] Cf. the mourning over the men whom Yahweh slew for looking into the Ark in 1 Sam 6:18–19. On the peculiarities of this passage and proposed connection to Ark narratives, see Tur-Sinai (1951: 275–286).

[35] When Jacob died, the narrator noted that the Egyptians bewailed him for 70 days (Gen 50:3). Herodotus relates that the body was placed in niter for 70 days (*Histories* 2.86). Diodorus Siculus states that the preparation of the body took 30 days and the wailing another 72 days (*Histories* 1.91). However, Job and his friends mourn for 7 days (Job 2:13). Cf. 1 Chron 10:12.

[36] Aranov (1977) supplies a wealth of comparative data on the subject, though his approach is rather Frazerian in orientation. For the cultic use of the threshing floor in Mesopotamia, see Jacobsen (1975: 65–97). At Ugarit, threshing floors also were tied to mourning and fertility rites and used as sites for divination (*CAT* 1.141–145, 1.155) and summoning the dead (*CAT* 1.20–22). Similar cultic activity took place in the Aegean world (*Homeric Hymn to Demeter*, 185–189). The threshing floor shared a number of these associations in ancient Israel as well. Thus, Gideon sought an oracle by means of divination at a threshing floor (Judg 6:11–20). Prophecy and royal judgment also took place there (1 Kgs 22:10–11), the latter, even during the period of the Sanhedrin (Aranov 1977: 161–176). The association of the threshing floor with fertility is suggested also in the book of Ruth, in which Ruth and Boaz have sex at a threshing floor (Ruth 3). See also Hos 9:1. That some sexual activity took place in or near the Israelite temple is clear by legal and prophetic pronouncements against such acts (see, e.g., Deut 23:18–19, Hos 4:14, 1 Kgs 14:24, 15:12, 22:38–47, 2 Kgs 23:7, Jer 2:20, 5:7, Ezek 16:31, Mic 1:7). See also Littauer et al. (1990: 15–23).

[37] On the complex history of Osiris and the use of corn mummies, see Griffiths (1980).

[38] Though neither Osiris nor the deceased whom he judged ever returned to the land of the living. Instead, they were resurrected in the afterlife.

[39] *Pyr.* 2054. See Griffiths (1980: 87–99). On the etymology of his name, see Kuhlmann (1975: 135–138) and Westendorf (1977: 95–113).

[40] Griffiths (1980: 167–168); Tooley (1996: 167–179).

[41] See Sethe (1928); Gardiner (1955); Quack (2006: 72–89); and Geisen (2012).

[42] The verb *ḥwỉ* means "beating, threshing." See Egberts (1995) for a comprehensive study of this ritual. Though details mainly come from temples of the Graeco-Roman period, the original contexts for the ritual belong to Theban festival processions for Osiris in the Ramesside period, which themselves derive in part from festivals at Memphis (Egberts 1995: 182–183).

[43] Including the Sokar festival, Osiris Mystery, Min festival, festival of Behdet, Opet festival, and perhaps also the festival of the first month of summer. See Egberts (1995: 412).

[44] The ritual also involved the royal consecration of four *mr.t*-chests, reliquaries that contained four differently colored linen bandages for Osiris' mummy. Some texts appear to refer to garments worn by a divine statue, but their use as bandages for the mummification of Osiris is

over Osiris, his members were reunited, reinvigorated, and concealed beneath the threshing floor.[45]

Since the mourning rites for Osiris took place at a threshing floor, we perhaps can understand why the Canaanites perceived the Israelites' mourning for Jacob as an Egyptian event, and while the narrator does not offer more than the twofold mention of the threshing floor by way of explanation, the references to the embalming of the patriarchs and their interments in an *ʾarōn* naturally evoke an Egyptian, if not Osirian, subtext.[46]

Additional support comes from a number of talmudic and midrashic traditions, which Rivkah Ulmer has shown[47] to draw heavily upon Osiris mythology when discussing the burial of Joseph in an *ʾarōn* and the bringing of his bones back to Canaan.[48] Like Osiris, Joseph is said to be buried among the kings and also in the Nile. According to midrash, the Egyptians placed Joseph in a metal coffin and buried him in the Nile in order to bless the river with fertility. Later Moses summons the coffin from the water by using language and paraphernalia suggestive of Egyptian magic.[49] When the coffin surfaces, it is then compared to a sprouting stalk of reed. Moses then carries the coffin away on his shoulders, much like a sacred bark, with all of Israel in procession.

Moreover, like the Egyptian *ʾarōn* in which the patriarchs were buried, the Ark of the Covenant was associated with new grain and the threshing floor. Thus, we find that the Ark's miraculous crossing of the Jordan took place during harvest time (Josh 3:15).[50] Later, when the Philistines captured the Ark they placed it in the temple of Dagon (1 Sam 1:5). Like Osiris, Dagon was associated with new grain and fertility[51] and possessed chthonic aspects,[52] with titles linking him to rites for the dead.[53] Clearly, the chthonic aspect

their primary function. The boxes were consecrated by dragging them and beating them with scepters, as one would do to grain. Since the *mr.t*-chests represented the cardinal points, the rite enacted the king's dominion over Egypt and his leading of Egypt to the gods. Another rite involved the carrying of two sticks, one topped with a serpent's head. According to one text, each stick represented one half of a severed worm. This ritual was interpreted as driving out the enemy, like a worm, which is both a grain eater and corpse eater. The ritual of the *mr.t*-chests preceded that of the driving of the calves, the former rite standing for the mummification of Osiris and the latter for the protection of his tomb after burial. During the Osiris Mystery, these rites were performed at the necropolis over an underground structure in which Osiris effigies were interred (Egberts 1995: 185, 388, 438–439).

[45] On rituals for assembling Osiris' body, see Egberts (1995: 200).

[46] In Egyptian mythology and ritual, the living king Horus (in the form of pharaoh) performs the mourning rites for his father and deceased king Osiris. Interestingly, in Gen 50:10–11, Joseph mourns for his father, the deceased patriarch Jacob.

[47] See Ulmer (2009: 107–142), for the texts paraphrased here (i.e., Exod. Rab. 20:19, Deut. Rab. 11:7, b. Soṭah 13a, Mek. de Rabbi Ishmael, Pesiq. Rab. Kah. 11, Vayehi Beshallaḥ) and additional evidence.

[48] There is some discrepancy concerning when Joseph's body was taken from Egypt. Exod 13:19 states that Moses took the bones of Joseph with him. Jub 46:9 claims that the Israelites took all the bones of Jacob's sons from Egypt, except those of Joseph. Josephus, *Antiquities*, 2.195–200, places the retrieval of Joseph's bones at a much later time.

[49] That Moses was learned in Egyptian magic appears also in Acts 7:22, Pliny, *Mos.* 1.6, 1.21, 1.24.

[50] Here the spring barley harvest is meant, since Josh 5:10 mentions the celebration of the passover.

[51] The Hebrew word for grain is *dāgān*. It appears in conjunction with the threshing floor in Num 18:27. The prophet Hosea too punfully identifies Dagon with the threshing floor (Hos 9:1). On Dagon, see Singer (1992: 431–450); Healey (1999: 216–219); and Felie (2003: 279–280). On the identification of Dagon as a god of storms, see Green (2003: 63–72). However, see Schwemer (2001), for a more exhaustive treatment of weather gods, which does not include Dagon. If Dagon was in any way identified with Osiris, then the story of the dismemberment of Dagon's statue before the Ark of the Covenant would echo the mythology of Osiris' dismemberment. However, I have found no evidence for the connection. For the view that the dismemberment represents the brutality of warfare as seen in Ugaritic texts, see Wiggins (1993).

[52] See already Roberts (1972: 18–19), who argues that this occurs chiefly through his identification with Enlil and the types of sacrifices offered to him, which are identified as sacrifices for the dead.

[53] At Terqa, his temple was called the "temple of the funerary ritual" (Akkadian: *bīt-kispi*). See Felie (2003: 96). At Mari, he was given the epithet *bēl pagrē*, i.e.,

of the Israelite Ark and its association with grain were not lost on the Philistines.[54] Moreover, when the Philistines found their god dismembered before the Ark, they sent it back on a newly constructed wagon. When it reached Beth-Shemesh, the villagers were harvesting grain (1 Sam 6:15).

Later, in David's time, the Ark was heading to the threshing floor of Nakhon,[55] when Yahweh killed a layperson for touching it (2 Sam 6:6).[56] Even when the Ark accompanied the Israelites on the battlefield, it was housed in a *sukkāh*, a temporary "booth" made from foliage (2 Sam 11:11).[57] This term, of course, lies behind the name for the Festival of Sukkoth, which commemorates the Israelites' exodus from Egypt (Lev 23:42, Deut 16:13–16, Zech 14:16–19, Ezra 3:4, Neh 8:14–17).[58] When David conquered Jerusalem, he purchased the city's main threshing floor after encountering an angel there (2 Sam 24:16–17). He then built an altar on the spot in order to avert a plague, which the Septuagint places during the wheat harvest (2 Sam 24:15–25, 1 Chron 21:16).[59] In a lengthy procession amidst music, shouting, frenetic dancing, and burnt offerings, David later would don a priestly linen ephod, lead the Ark to the threshing floor, and place it in a tent.[60] David marked the event as a fertility rite by giving the people gifts of bread loaves and cakes of dates and raisins (2 Sam 6:19).[61] The same threshing floor became the site on which Solomon built the temple (1 Kgs 6:19, 8:1–9, 2 Chron 3:1). Moreover, Solomon moved the Ark into the temple during the Festival of Sukkoth (1 Kgs 8:2),[62] which necessitated processions, dancing, and sacrifices.[63]

"Lord of the Dead." See G. Dossin, *ARM* 10 63:15–16, C. -F. Jean, *ARM* 2 90; 137: 43–44, J.-R. Kupper, *ARM* 3 40. Cf. Ezek 43:7. At Ugarit too he received sacrifices to the deceased. See Neiman (1948) and Dussaud (1935).

[54] Note also the narrator's statement in 1 Sam 23:1 that the Philistines were fighting at Qeilah and plundering its threshing floors. This act led David to seek Yahweh's oracle twice, presumably by way of the Ark, as to whether to battle the Philistines (1 Sam 23:2–4).

[55] The word *nākōn* might also be read as an adjective meaning "prepared, right." 1 Chron 13:9 reads *kîdōn* instead of *nākōn*. Tur-Sinai (1951: 282–285) argues that *nākōn* and *kîdōn* refer to "pestilence" and "affliction."

[56] The Ark also was stored in private homes such as that of Abinadab, whose house, which the narrator twice emphasizes, was located on "the hill" (2 Sam 6:3–4), and of Obed-Edom (2 Sam 6:10–11), whose household prospered on account of the Ark.

[57] According to Lev 23:40–41, the foliage included "good fruits," "palm branches," "boughs of leafy trees," and "willows of the brook." This differs slightly in Neh 8:15, which calls for "olive branches," "branches of wild olive," "myrtle branches," "palm branches," and "branches of thick trees." The Mishnah clarifies the fruit as a "citron" (Sukkah I iii 8; I iii 12). The Mishnah's prohibition against using any plants from an Asherah (Sukkah I iii 1–3, 5) implies that at one time some people did obtain foliage from an Asherah, thus again attesting to the festival's early fertility associations.

[58] Note that Neh 8:14 historicizes the festival of Sukkoth as an institution created to remember the Israelite's departure from Egypt. Nevertheless, as 8:17 clarifies: "... the Israelites had not done so from the days of Joshua son of Nun to that day." I take this gloss to refer not to the festival itself, but to the erection of booths within the courtyards of the temple (8:16), i.e., an innovation that required historical justification.

[59] See the *Kaige* recension of the LXX for 2 Sam 24:15. Araunah (also called Ornan) was a Jebusite and, thus, an inhabitant of Jebus (i.e., Jerusalem) before David conquered it. Araunah appears to be a Hittite name or title. See Sayce (1921) and Rosén (1955). Wyatt (1985) argues that "the Araunah" (the name contains the definite article in 2 Sam 24:16) was the last Jebusite king (cf. Ezek 16:1).

[60] 1 Chron 15:27 adds that David was wearing a robe of fine linen as were also the Levites, singers, and Chenaniah, the music master. The passage lists the instruments as including a shofar, trumpets, cymbals, loud harps, and lyres.

[61] The event was identified as a fertility rite by Smelik (1992: 52–53), though I disagree with his dating of the narrative to the post-exilic period. On raisin cakes as a fertility food, see also Hos 3:1 and Song 2:5. The biblical writer has inverted the theme of fertility by informing the reader at story's end that David's wife Michal died childless (2 Sam 6:23).

[62] The passage refers to the month by its Canaanite name Ethanim, rather than Tishri, the name used after the introduction of the Babylonian calendar. 2 Macc 10:6–8 also informs us that the temple's renovation closely followed the pattern of the Festival of Ingathering.

[63] Sukkoth here is referred to simply as "the Festival" (*he-ḥāg*). See similarly in 1 Kgs 8:65, Ezek 45:23, Neh 8:14, and 2 Chron 7:8. Ezra 3:1–7 links the dedication of the altar with Sukkoth. On Sukkoth, see Haran 1985: 298–300. A similar annual harvest festival existed at

Such references suggest a correlation between the Ark's movement and the harvest and again demonstrate that the Hebrew *ʾarōn*, whether understood as an Egyptian coffin or the Ark of the Covenant, was intimately connected to threshing floors and their fructifying and chthonic associations.[64]

Possible Context for the Integration and Adaptation of the Bark into the Israelite Cult

To this point, I have argued that Egyptian sacred barks served as models for the Israelite Ark of the Covenant and that consequently the two objects

Shiloh in the period of the Judges (see Judg 21:19–23). Perhaps this explains the Ark's trip to Shiloh in 1 Sam 3:1–31. The Mishnah elucidates the passage in Judges by connecting it to courtship rites: "And the daughters of Jerusalem went forth in the vineyards. And what did they say? 'Young man, lift up your eyes and see who you would choose for yourself (as a wife). Set not your eyes on beauty, but set your eyes on family" (Taanith iv 8). Note also the mention in Judg 9:27 of a vintage feast at Shechem before Yahweh was worshiped there. See the insightful query and response concerning celebration during Sukkoth found in Sukkah iv 4: "How was the rite of the palm branch fulfilled [on the Sabbath]? If the first festival day of the Feast fell on a Sabbath, they brought their palm branches to the Temple Mount and their ministers took them and set them in order on the roof of the portico, but the elders set theirs in a [special] chamber. The people were taught to say, 'Whoever gets my palm branch, let it be his as a gift.' The next day they came early and the ministers threw the palm branches down before them and the people snatched at them and beat each other." The beatings mentioned here clearly mimic the act of threshing. Compare, e.g., the similar report of R. Jonathan b. Baroka who noted: "They use to bring palm branches and beat them on the ground at the sides of the Altar, and that day was called: 'The day of the Branch threshing'" (Sukkah iv 6).

[64] Moreover, each of the sanctuaries that the Ark visited has chthonic associations and connections to fertility. In 1 Sam 7:1, the men of Kiriath-jearim move the Ark to the house of a man named Aminadab, which is said to be on a hill. The gloss concerning the hilltop suggests that it was an open space, much like those on which threshing floors were situated. On the connection between threshing floors and high places in ancient Israel, see Aranov (1977: 51–52). Note also that we are told that the Ark rested there for 20 years, during which the people "mourned" (i. e., וַיִּנָּהוּ in 1 Sam 7:2). The verb used here usually appears in reference to mourning for the dead. Though the account of the Ark at the fall of Jericho contains no reference to grain, the circumambulation of the city seven times is suggestive of pilgrimage dances and the round shape of the threshing floor (note the pleonastic etymological connection between Hebrew חגג "Festival" and חוג "circle, vault of heaven"). It is of note that the Mishnah describes the rituals that took place on the seventh day of Sukkoth as including the blowing of the shofar and a sevenfold circumambulation of the altar with willow branches (Sukkah iv 5). Moreover, each of the sanctuaries that housed the Ark in the pre-monarchic period has numinous, chthonic, and fertility associations. At Gilgal, Joshua erected 12 stones to commemorate the Ark's miraculous crossing of the Jordan River (Josh 4:19–24). Joshua reinstituted the rite of circumcision at Gilgal (Josh 5:2–6), and there Yahweh gave him an oracle (Josh 10:8). Judg 3:19 mentions Gilgal as a place known for its "carved idols" (*pəsilîm*). Later prophets associate the site with idolatry and temple prostitution (e.g., Hos 4:15, 9:15, 12:11, Amos 4:4, 5:5). The sanctuary site of Shechem is connected to fertility in that it was set up near an oak tree (Gen 12:6–7, Josh 24:26, Judg 9:6). Judg 8:33, 9:44 associates Shechem with Baal-Berith. Moreover, when Joshua renewed the covenant there, he commemorated the occasion by erecting a stone (Josh 24:26). Bokhim (lit. "weepers") is given an aetiology that connects it to a divine encounter during which the people acknowledged following other gods and wept (Judg 2:1–5). This pericope suggests that mourning rites were performed there. Bethel was the site of a divine encounter with the patriarch Jacob after which he erected a *maṣṣēbāh* and anointed it with oil (Gen 28:10–19). Hos 12:4 adds that this event was accompanied by weeping. Bethel was also the site of a mourning rite during which the Israelites wept (Judg 20:26, 21:2–4). Though dating from a later period (fourth c. BCE), Papyrus Amherts 63 attests to bovine imagery at Bethel and associates it with a fusion between Yahweh and Horus. See Nims and Steiner (1983: 261–274). At Shiloh, Joshua cast lots before Yahweh in order to divine which lands belonged to which tribes (Josh 18:1). Shiloh also is the site of a harvest festival during which young men selected spouses from among the young women who danced in the vineyards (Judg 21:19–24). The event was precipitated by the aforementioned mourning of the tribe of Benjamin at Bethel. When the Ark stayed at Kiriath-Jearim it was placed on a hilltop (1 Sam 7:1–2). Initially, it was not an Israelite settlement, but rather a Hivite one (Josh 9:7–17), and at one time it was home to the cult of Baal, as its other name Kiriath-Baal attests (Josh 15:9, 15:60, 18:4, 1 Chron 13:6). On Kiriath-Jearim and its relationship to Gibeon, see Blenkinsopp (1969). Gibeon was the seat of an ancient sanctuary called "the great high place" (1 Kgs 3:4). This site too is characterized as numinous. Thus, during the battle at Gibeon, Yahweh halted the motion of the sun (Josh 10:12–13), and Joshua allotted it to the Levites, making it a priestly town (18:25, 21:17). Later God appeared and spoke to Solomon at Gibeon (1 Kgs 3:4, 1 Kgs 9:2, 1 Chron 1:2–7).

share much in common in design, function, and cosmic import. While no particular bark can be singled out as a prototype for the Ark's design, the object's associations with death and fertility, its close relationship with the threshing floor, and the mention of the Israelites' practice of Egyptian embalming in conjunction with the *ʾărōn* are suggestive of the cult of Osiris.[65]

Exactly how and when the object became an appurtenance of the Israelite cult is difficult to say since biblical texts mythologize the Ark's creation. Nevertheless, archaeological evidence suggests that the Late Bronze Age provided the best opportunity,[66] because it saw an increased Egyptian presence in the Levant (Giveon 1978; Wimmer 1990: 1065–1106; Nakhai 2001). The Egyptians built garrisons, administrative offices, and Egypto-Canaanite temples to facilitate the collection of grain.[67] The Canaanite elites, meanwhile, often sought to emulate Egyptian customs (Higginbotham 1996: 154–169). Of course, they did not adopt Egyptian practices wholesale, but adaptated them to fit their own religions, as Beth Nakhai observes:

> These temples blended Egyptian structural or decorative elements with Canaanite architectural forms. Egyptian ritual and other objects were found alongside Canaanite cultic paraphernalia. Iconography often exhibited qualities of synthesis rather than exclusivity, but in general the gods and goddesses of Canaan prevailed (Nakhai 2001: 154–155).[68]

The Late Bronze Age also saw movements of Canaanites and Egyptians in both directions, and it is at this time that the Merneptah stele records the earliest written evidence for Israelites in Canaan. This was a formative and flexible period in the history of Israelite religion as it also saw the gradual fusion of the Canaanite god El with Yahweh.[69] However, of particular relevance here

[65] Barkay (1994) has shown that Egyptian embalming practices were employed by some within Israel well into the Iron Age. See also Zevit (2001: 247, n. 198), who, citing Barkay's study on embalmment in Israel, observes: "... perhaps some Nilotic mythology accompanied the science, if learned from the Egyptians."

[66] It is not my intention to enter the debate concerning the historicity of the Exodus. The subject has been covered amply by biblical scholars, Egyptologists, and archaeologists and little by way of consensus has emerged (as can be seen by many of the essays in this volume). Some suggest that it was a plural phenomenon that took place in stages over time. Others view it as a single but much smaller event. Some scholars indentify the nascent Israelites with the marauding ʿApiru, though in more recent years, the Shasu have become the comparative group of choice. See Greenberg (1955); Giveon (1974: 267–271); Redford, (1992: 269–280); and Rainey (2008: 51–55). Dating the Exodus has proved even more difficult, though most positivist views place it sometime between the fourteenth and twelfth centuries B.C.E. In my view, while some of the proto-Israelites might have lived in settlement communities in Egypt, such as the House of Joseph, the Levites, and perhaps elements of the Benjaminite and Judahite tribes, and others might have been among the Shasu, the overwhelming archaeological evidence suggests an indigenous Late Bronze Age Canaanite origin for most of the Israelites. In general, I concur with Weinstein (1997: 98), who remarks: "If there was an historical exodus, it probably consisted of a small number of Semites migrating out of Egypt in the late thirteenth or early twelfth century B.C., ultimately settling in southwestern Canaan, where their Egyptian heritage would allow them to melt into the local populace ...."

[67] Temples include those at Apheq, Ashdod, Ashqelon, Beth Shean, Gaza, Jaffa, Lachish, Megiddo, and Tell Abu Hawam. Note in particular the remark of Nakhai (2001: 151) that "Egypt dominated LB II Ashdod and Ashod's sacred site should be considered Egypto-Canaanite." Over a century ago, the remains of an Egypto-Canaanite temple were discovered north of the Damascus Gate in Jerusalem. The finds, which date to the Ramesside period, included an Egyptian stele dedicated to Osiris. Also discovered in 1975 was a serpent statue. See Wimmer (1990: 1073).

[68] A similar process of integration and adaptation in which the foreign elements did not compromise the identity and worship of the local gods obtains in the Iron Age, when Egyptian interests and influence shifted to the Phoenician world, as Giveon (1978: 31) remarks: "In spite of the strong influence of Egyptian culture on the Canaanites in general and Byblos in particular, the fundamental religious concepts of the Egyptians were not copied or even adopted by the peoples of Western Asia: it is only the iconography of the Egyptians which was used as a means of expressing the religious beliefs of the Canaanites. In the process of transfer Egyptian pictorial concepts were changed in varying degrees, the changes being due sometimes to a lack of understanding of their real meaning and sometimes to the need to use similar pictographs to express different ideas."

[69] In some circles, this also led to identifying Asherah as Yahweh's consort. See Olyan (1988: 38–61). On the syncretism of El and Yahweh, see Smith (2003), who

are a great many seals, scarabs, and other objects found in Canaan, which Othmar Keel has shown, attest to the Canaanites' fascination with Memphis and its god Ptah,[70] also known by his syncretistic name, Ptah-Sokar-Osiris.[71] Ptah-Sokar-Osiris was at once the creator, the sun, and judge of the underworld, and he was the patron god of craftsmen. Like Yahweh, he created the world by fiat and was a god of justice who rewarded the righteous and punished sinners with death.[72]

It is in this context of Egyptian-Canaanite exchange, Israelite religious syncretism, and Levantine interest in Ptah-Sokar-Osiris that I envision the Israelites' adaptation of the bark. Those features, functions, and associations that the Ark shared with its prototype represented facets of a shared taxonomy, aspects that made sense in both Egyptian and Israelite religious contexts before the object was integrated.[73] Thus, the Ark retained its significance as a throne and footstool and as the symbol of God's presence, and it continued to be a source of oracles. By calling the object an *ʾarōn*, the Israelites retained the Ark's chthonic associations that resonated with the priestly conception of El Yahweh as a creator god to whom the first fruits are offered (Lev 23:9–14).[74]

On the other hand, those features that did not resonate with the priestly conception of God were refashioned or reconceptualized. Thus, the object's connection to a boat was obscured by referring to it as a "throne and footstool" and by naming it an *ʾarōn*, which suggested its use as a coffin.[75] These labels served a dual purpose. As a

argues that El was first identified as the god of the Exodus and that El was identified with Yahweh in the premonarchic period.

[70] According to Keel (2006: 248), this process began already at the end of the Middle Bronze Age IIB: "The scarabs previously discussed here thus testify to the Middle Bronze Age Canaanites' fervor and enthusiasm for Egyptian culture in general and for the god of Memphis in particular. The predilection for this god can be explained by trade connections with Memphis and by the fact that many of the Canaanites coming to Egypt during the Thirteenth and Fifteenth Dynasties were craftsmen." See also the twelfth century B.C.E. ivory plaques from Megiddo inscribed in honor of a Canaanite temple musician named Kurkur, who had been trained at Memphis and was serving the court of Ashqelon (Wimmer 1990: 1091–1093; Lippke 2011). Florian Lippke has informed me by personal communication (April 15, 2013) of a number of additional scarabs and seals found in the southern Levant that feature Ptah iconography, including those from Abu Hawam, Achsib, a good portion from Tell Agul and Akko, Aschkelon, at least 5 from Beth Shean as well as Der Balah, Dotan, Ekron, En Samije, Tell Eschtori, 13 from Tel Fara, 5 from Tel Gamma, 1 from Gath, 1 from Gerisa, 3 from Gezer, and 1 from Tell Hesi. In Jordan there is one from Amman, one from Tell Deir Alla, and one cylinder seal from there as well. See Zecchi (1996) for the spread of the cult of Osiris after the eleventh century B.C.E.

[71] On the fusion of the three deities already during the Old Kingdom, see Gaballa and Kitchen (1969). Note that unlike some of the other Egyptian festivals, the Festival of the Mystery of Osiris was celebrated throughout the country (Teeter 2011: 59).

[72] Keel (2006: 258–259) discusses four scarabs that feature Ptah standing before Re and Osiris in the form of two birds. He argues that the items demonstrate a knowledge of the concepts represented in the *Memphite Theology*. See Koch (1965), cited by Keel (2006: 261, n. 103). With Schlögl (1980), Keel dates the *Memphite Theology* to the reign of Ramesses II.

[73] Note that Yahweh also was connected with horses in his capacity as a solar deity (e.g., 2 Kgs 23:11, Hab 3:8, Mal 4:2, Ps 19:5–7, 84:12). See Ahlströhm (1984: 22–23) and Stähli (1985).

[74] Interestingly, Exodus 25 portrays Yahweh as a god of craftsman. After giving Moses detailed instructions for building the Ark, he personally selects the craftsmen and fills them with his spirit and wisdom (Exod 35:30–36:7). Note too that the word "firmament" (Heb. *rāqiʿa*), which God created to support the heavens in Gen 1:6, derives from a root whose basic meaning is to "beat out metal."

[75] Other cults apparently were more receptive to the notion of a god who can die and be resurrected. Such might account for shared aspects of the Levantine cults of Osiris, Baʿal, and ʾAdonis, of which scholars have long been aware. Strange (2004: 350) has argued that the Osiris cult was refracted in or synchronized with the cults of Baʿal and ʾAdonis. Redford (1992: 43–44) similarly argues that the comparative study of Osiris, Baʿal, and ʾAdonis has fallen into disfavor largely because scholars have tried to distance themselves from the earlier methodological pitfalls that beset the works of Sir James Frazer. According to Redford, scholars have erred too much on the side of caution and have "thrown out the baby with the bath water." For a more recent critical treatment of the topic, see Mettinger (2001). The spread

throne without a statue, the Ark visually conveyed the aniconic nature of Yahweh (see Mettinger 1995). As a casket without a body, it engrained the notion of Yahweh as a god who cannot die. Concomitantly, the deposition of the tablets into the *ʾarōn* replaced any suggestion of a divine judge of the underworld with objects that represented Yahweh's role as divine judge and lawgiver.[76] Thus, Yahweh became the ruler of the entire cosmos from the heavens to the underworld,[77] and "threshing" became an idiom for the judgment and dismemberment of *Yahweh's* enemies.[78]

of the cult of Osiris throughout the Mediterranean world is certainly in evidence during the Graeco-Roman period. See, in particular, the host of later myths involving Osiris and/or accounts involving floating chests observed by Plutarch (*De Iside et Osiride*, 15–16) and others and discussed by Holley (1949: 39–47); Griffiths (1980: 28–34). Of note is the depiction of the cult of Osiris on a frescoe at Pompeii, which shows a temple devoted to Isis. Within the temple is a naos, which is constructed from the disassembled parts of an Osiris coffin. Inside the naos is painted a bark of Osiris. The frescoe shows that the bark of Osiris could be reimagined in other ways outside of Egypt. For an image of the frescoe, see Merkelbach (2001: 505).

[76] According to the Hebrew Bible, Aaron's rod stood near the Ark (Num 17:1), though a tradition found in the New Testament (Heb 9:4) places the rod inside the Ark. Nevertheless, much like the Ark itself, the wonders ascribed to it, including its transformation into a serpent (Exod 7:10) and its blossoming and production of almonds (Num 17:8), convey a sense of chthonic power and fertility. Its transformation into a serpent has parallels in Egyptian magical praxis. See Noegel (1997). The parallels to Egyptian magic and the comparative evidence gathered in this essay suggest that the Levites possessed at least some knowledge of Egyptian religious practices. It long has been observed that several individuals connected to the early Israelite priesthood possessed Egyptian names, including Assir, Hofni, Pinehas, Hor, Merari, and of course Moses. See already Noth (1928: 63–64). Some have opined that the name Aaron too is Egyptian. See Spencer (1992) and Muchiki (1999). If one concedes that the Levites possessed knowledge of Egyptian religion, might it also be that the solar bark's encounter with the great serpent Apep informs the Israelite conception of the great serpent Leviathan (also called *tannîn*)? Not only is this creature identified with the Pharaoh (Ezek 29:3, 32:2), but its dismemberment by Yahweh is credited with enabling the Israelites to cross the Reed Sea (Isa 51:9–10). In Exod 7:10, Aaron's staff becomes a *tannîn* before devouring the serpent-staffs of the Egyptian magicians. Moreover, Ritner (1993: 165–167) has discussed how Egyptian priests employed rituals to ensure the safe nightly passage of the solar bark. These rites involved paralyzing Apep by severing, burning, or otherwise destroying an effigy of the serpent. No such rite is attested in the Hebrew Bible. However, the many references to the great serpent as an enemy of Yahweh and the role that priests played in maintaining cosmic order through ritual suggest that perhaps we should read more into the etymological connection between the priestly tribe of "Levi" and the "Leviathan" (the etymological connection was first suggested to me by the late Cyrus H. Gordon). As many have pointed out, Ps 74:13–14 refers to the beast as having multiple heads and, thus, more akin to the multi-headed creature *ltn* in Ugaritic texts (e.g., *CAT* 1.5 I 1–8) and artistic representations from Arslan Tash and Diyala. Nevertheless, some representations of chaos serpents in Egypt also have multiple heads. See, e.g., the panel in the tomb of Tuthmosis III (KV 34, chamber J, right wall) that depicts the falcon-headed Sokar overpowering a many-headed chaos serpent in a cave during the fifth hour of the Amduat. Moreover, the aforementioned passage in Psalm 74 is followed rather fittingly with chaotic water and solar imagery in vv. 15–16 with: "You broke the springs and the torrents, you dried ever-flowing streams. To you belongs day and also night, you have prepared the light and the sun." I also note that Isaiah 27 opens by referencing the punishment of Leviathan (v. 1) and concludes with Yahweh threshing his enemies (v. 12). Additional evidence for the Levitical association with serpents appears in Num 21:8–9, in which Moses cures the people of snake bites by fashioning a bronze serpent, and in the later mention that the priests had allowed the item to be venerated until Hezekiah destroyed it (2 Kgs 18:4).

[77] In Isa 66:1 Yahweh declares: "Heaven is my throne, and the earth (*ʾereṣ*) is my footstool. Where is the house you will build for me? Where will my resting place be?" If the word *ʾereṣ* here means "underworld," as it does elsewhere in the Bible (e.g., Job 10:21–22, Ps 139:15, Isa 44:23, Jonah 2:6) and in Ugaritic and Akkadian texts, then the passage also connects the *kappōreth* "lid/throne" with the heavens and the *ʾarōn* "Ark/footstool" with the underworld. Since heaven and underworld constitute a more apt merism and a better case of cosmic symmetry than heaven and earth, one naturally might question whether *ʾereṣ* means "underworld" in Gen 1:1: "In the beginning God created the heavens and the earth (*ʾereṣ*)." That *ʾereṣ* here might mean underworld is suggested later in Gen 7:11, which describes the cosmic collapse in a way that reverses the processes of creation. Thus, Yahweh rains down the reservoir of heavenly water and ushers up the fountains of the deep.

[78] See, e.g., 2 Kgs 13:7, Isa 21:10, 27:12, Jer 51:33, Hos 13:3, Amos 1:3, Mic 4:12, Hab 3:12. A full discussion appears in Aranov (1977: 177–181).

By the time of the early Israelite monarchy, the Ark, like the god Yahweh, was perceived as entirely Israelite, though memory of its origins likely remained and required negotiation.[79] Hence, it was integrated retroactively into the national epic of the Exodus and intimately tied to Egypt and the first harvest festival.[80] This gave the object an aetiology that distinguished it from Egyptian religious practice,[81] one that served the needs of the priesthood, the royal house, and the national epic. As the centerpiece of the Israelite sanctuary, the Ark of the Covenant stood as the symbolic presence of Yahweh and his legitimation of the Levitical priesthood. As the Lord of Hosts who rides upon the *kerubîm*, the Ark legitimated the royal house and its wars. As an Egyptian object transformed and détourned, it offered visual and literary validation of the Exodus.

[79] The processes of innovation and negotiation find an apt parallel in the establishment of temples at Dan and Bethel under Jeroboam I. To decentralize the religious and political authorities in Jerusalem, Jeroboam set up golden calves in the temples and told the people, "Behold O Israel, your gods who brought you up from the land of Egypt" (1 Kgs 12:28). Jeroboam had strong support from Egypt and even lived there for a time (1 Kgs 11:40, 12:2). On the complex relationship between the account of Jeroboam's calves and Exodus 32, see Knoppers (1995) and Russell (2009: 41–43). Note that 2 Chron 13:8 appears to understand the calves as a war standard and thus mobile like the Ark of the Covenant. As Russell (2009: 27–30) points out, Jeroboam I's use of bovine imagery might reflect Egyptian, Syrian, or Mesopotamian influence or the renovation of older imagery associated with El to which Yahweh was fused. In my view, the power of icons lies in their ability to communicate on multiple levels to multiple audiences. A bovine image could have served this purpose.

[80] The Exodus stories appear to have originated in the northern kingdom of Israel and to have been integrated into Judahite historiographical tradition at a later time (see Hoffman 1989; Na'aman 2011). The construction of the Ark with acacia suggests that the Ark's origins were south of the kingdom of Israel.

[81] On the theory that the Ark narratives serve as an aetiology for its use in the Jerusalem cult, see already Rost (1926), though Rost does not discuss parallels to Egyptian barks.

References

Ahlströhm, G.W. 1984. An Archaeological Picture of Iron Age Religions in Ancient Palestine. *Studia Orientalia* 55: 117–145.

Albright, W.F. 1938. What Were the Cherbim? *Biblical Archaeologist* 1: 1–3.

Aranov, M.M. 1977. The Biblical Threshing-Floor in the Light of the Ancient Near Eastern Evidence: Evolution of an Institution. Ph.D. Dissertation, New York University.

Barkay, G. 1994. Tombs and Burial Practices in Judah in the Biblical Period. In *Graves and Burial Practices in the Land of Israel in the Ancient Period*, ed. I. Singer, 96–164. Jerusalem: Yad Yitzhaq Ben Zvi (Hebrew).

Barta, Winfried. 1965–66. Zum Ritual der Götterbarke im Neuen Reich. *Jaarbericht van het Vooraziatisch egyptisch Genootschap "Ex Oriente Lux"* 19: 462–463.

Beckman, G. 1996. *Hittite Diplomatic Texts*. Atlanta, GA: Scholars Press.

Bell, L. 1985. Luxor temple and the cult of the Royal *Ka*. *Journal of Near Eastern Studies* 44: 251–294.

Blenkinsopp, J. 1969. Kiriath-Jearim and the Ark. *Journal of Biblical Literature* 88: 143–156.

Bordreuil, P. 2006. pårōket et kappōret. À propops du saint des saints en Canaan et en Judée. In *Les espaces syro-mésopotamiens: Dimensions de l'expérience humaine au proche-orient ancien*, Subartu, vol. 17, ed. P. Butterlin et al., 161–168. Brussels: Brepols.

Currid, J.D. 1997. *Ancient Egypt and the Old Testament*. Grand Rapids, MI: Baker Books.

Dietrich, W. 2007. *The Early Monarchy in Israel. The Tenth Century B.C.E.* Trans. J. Vette from German. Society of Biblical Literature Biblical Encyclopedia Series 3. Leiden: Brill.

Dussaud, R. 1935. Deux stèles de Ras Shamra portant une dédicace au dieu Dagon. *Syria* 16: 177–180.

Edel, E. 1983. Der ägyptisch-hethitische Friedensvertag zwischen Ramses II. und Hattusilis III. In *Texte aus der Umwelt des Alten Testaments*, ed. R. Borger et al., 135–153. Gütersloh: Gütersloher Verlagshaus Gerd Mohn. B1, L2.

———. 1994. *Die ägyptisch-hethitische Korrespondenz aus Boghazköi in babylonischer und hethitischer Sprache*, Rheinisch-Westfälische Akademie der Wissenschaften 77, vol. 1–2. Opladen: Westdeutscher Verlag.

Egberts, A. 1995. *In Quest for Meaning: A Study of the Ancient Egyptian Rotes of Consecrating the Meret-Chests and Driving the Calves*, Egyptologische Uitgaven 8, vol. 1–2. Leiden: Nederlands Instituut voor het Nabije Oosten.

Felie, L. 2003. *The God Dagan in Bronze Age Syria*. Leiden: Brill.

Gaballa, G.A., and K.A. Kitchen. 1969. The Festival of Sokar. *Orientalia* 38: 1–76.

Gardiner, A.H. 1955. *The Ramesseum Papyri*. Oxford: Oxford University Press.

Geisen, C. 2012. The Ramesseum Dramatic Papyrus: A New Edition, Translation, and Interpretation. Ph.D. Dissertation, University of Toronto.

Giveon, R. 1974. *Les bédouins Shosou des documents égyptiens*. Leiden: Brill.

———. 1978. *The Impact of Egypt on Canaan*, Orbis biblicus et orientalis, vol. 2. Göttingen: Vandenhoeck and Ruprecht.

Göttlicher, A. 1992. *Kultschiffe und Schiffskulte im Altertum*. Berlin: Gebr. Mann Verlag.

Green, A.R.W. 2003. *The Storm God in the Ancient Near East*. Winona Lake, IN: Eisenbrauns.

Greenberg, M. 1955. *The Hab/piru*. New Haven, CT: American Oriental Society.

Griffiths, J.G. 1980. *The Origins of Osiris and His Cult*, Studies in the History of Religions, vol. 40. Leiden: Brill.

Hannig, R. 1995. *Großes Handwörterbuch Ägyptisch-Deutsch*, Kulturgeschichte der antiken Welt, vol. 64. Mainz: Verlag Philipp von Zabern.

Haran, M. 1985. *Temples and Temple Service in Ancient Israel*. Winona Lake, IN: Eisenbrauns.

Healey, J.F. 1999. Dagon דגון. In *Dictionary of Deities and Demons in the Bible*, ed. Karel van der Toorn et al., 216–219. Leiden: Brill.

Herrmann, J. 1908. Ägyptische Analogien zum Funde des Deuteronomiums. *Zeitschrift für die Alttestamentliche Wissenschaft* 28: 291–302.

Holley, N.M. 1949. The Floating Chest. *Journal of Hellenic Studies* 69: 39–47.

Homan, M.M. 2002. *To Your Tents, O Israel! The Terminology, Function, Form, and Symbolism of Tents in the Hebrew Bible and the Ancient Near East*, Culture and History of the Ancient Near East, vol. 12. Leiden: Brill.

Higginbotham, C. 1996. Elite Emulation and Egyptian Governance in Ramesside Canaan. *Tel Aviv* 23: 154–169.

Hoffman, Y. 1989. A North Israelite Typological Myth and a Judaean Historical Tradition: The Exodus in Hosea and Amos. *Vetus Testamentum* 39: 169–182.

Jacobsen, Th. 1975. Religious Drama in Ancient Mesopotamia. In *Unity and Diversity: Essays in the History, Literature, and Religion of the Ancient Near East*, ed. Hans Goedike and J.J.M. Roberts, 65–97. Baltimore, MD: Johns Hopkins Press.

Keel, O. 2006. Reflections of Ptah and Memphite Theology from the Soil of Palestine: Iconographic and Epigraphic Evidence. In *Text, Artifact, and Image: Revealing Ancient Israelite Religion*, Brown Judaic Series, vol. 346, ed. Gary M. Beckman and Theodore J. Lewis, 239–272. Providence, RI: Brown Judaic Studies.

Kitchen, K.A. 1971. *Ramesside Inscriptions, Historical and Biographical*, vol. 2. Oxford: B. H. Blackwell Ltd.

———. 1993. The tabernacle—A Bronze Age artifact. *Erets Israel* 24: 119–129.

Knoppers, G.A. 1995. Aaron's Calf and Jeroboam's Calves. In *Fortunate the Eyes that See: Essays in Honor of David Noel Freedman in Celebration of His Seventieth Birthday*, ed. Astrid B. Beck et al., 92–104. Grand Rapids, MI: Eerdmans.

Koch, K. 1965. Wort und Einheit des Schöpfergottes in Memphis und in Jerusalem. *Zeitschrift für Theologie und Kirche* 62: 251–293.

Kuhlmann, K.P. 2008. Thrones. In *UCLA Encyclopedia of Egyptology*, ed. Willeke Wendrich, 1–13. Los Angeles, CA: University of California.

———. 1975. Zur Etymologie des Göttesnamens Osiris. *Studien zur Altägyptischen Kultur* 2: 135–138.

Kyrieleis, H. 1969. *Throne und Klinen*, Jahrbuch des Deutschen Archäologischen Instituts, vol. 24. Berlin: Walter de Gruyter.

Lippke, F. 2011. The Southern Levant in context: A Brief Sketch of Important Features Related to the Religious Symbol System in the Bronze Ages. In *Egypt and the Near East—The crossroads: Proceedings of an International Conference on the Relations of Egypt and the Near East in the Bronze Age, Prague, September 1–3, 2010*, ed. Jana Mynářová, 221–233. Prague: Czech Institute of Egyptology, Charles University in Prague.

Littauer, M.A., J.H. Crouwel, and P. Steinkeller. 1990. Ceremonial Threshing in the Ancient Near East. *Syria* 52: 15–23.

Loewe, R. 2001. Ark, archaism and misappropriation. In *Biblical Hebrew, Biblical Texts: Essays in Memory of Michael P. Weitzman*, Journal for the Study of the Old Testament Supplement Series, vol. 333, ed. Ada Rapoport-Albert and Gillian Greenberg, 113–145. London: Sheffield Academic Press.

McClain, J.B. 2007. The Terminology of Sacred Space in Ptolemaic Inscriptions from Thebes. In *Sacred Space and Sacred Function in Ancient Thebes*, Studies in Ancient Oriental Civilization, vol. 61, ed. Peter Dorman and Besty M. Bryant, 85–95. Chicago, IL: Oriental Institute of the University of Chicago.

Merkelbach, R. 2001. *Isis Regina-Zeus Sarapis: Die griechisch agyptische Religion nach den Quellen Dargestellt*. Berlin: Walter de Gruyter.

Mettinger, T.N.D. 1999. Cherubim. In *Dictionary of Deities and Demons in the Bible*, ed. Karel van der Toorn et al., 189–192. Leiden: Brill.

———. 2001. *Riddle of Resurrection: "Dying and Rising Gods" in the Ancient Near East*, Coniectanea Biblica, Old Tesament, vol. 50. Stockholm: Almqvist and Wiksell.

———. 1995. *No Graven Image? Israelite Aniconism in Its Ancient Near Eastern Context*, Coniectanea Biblica, Old Testament Series, vol. 42. Stockholm: Almqvust and Wiksell.

Metzger, M. 1985. *Königsthron und Gottesthron; Throndarstellungen in Ägypten und im Vorderen Orient, im dritten und zweiten Jahrtausend vor Christus, und Bedeutung für das Verständnis Aussagen über den Thron im Alten Testament*, Alter Orient und Altes Testament Bd. 15, 1-2. I: *Text*; 11: *Katalog und Bildtafeln*, vol. 1–2. Kevelaer: Verlag Butzon und Bercker: Neukirchen-Vluyn: Neukirchener Verlag.

Morgenstern, J. 1942. The Ark, the Ephod, and the 'Tent of Meeting'. *Hebrew Union College Annual* 17: 153–266.

Muchiki, Y. 1999. *Egyptian Proper Names and Loanwords in North-West Semitic*, SBL Dissertation Series, vol. 178. Atlanta, GA: Society for Biblical Literature.

Musil, Alois. 1928. *The Manners and Customs of the Rwala Bedouins*, American Geographical Society Oriental Explortations and Studies, vol. 6. New York, NY: Czech Academy of Sciences and Arts.

Na'aman, N. 2011. The Exodus Story: Between Historical Memory and Historiographical Composition. *Journal of Near Eastern Religions* 11: 39–69.

Nakhai, B. 2001. *Archaeology and the Religions of Canaan and Israel*. Boston, MA: American Schools of Oriental Research.

Neiman, D. 1948. PGR: A Canaanite Cult Object in the Old Testament. *Journal of Biblical Literature* 67: 55–60.

Nims, C.F., and R.C. Steiner. 1983. A Paganized Version of Psalm 20:2-6 from the Aramaic text in Demotic script. *Journal of the American Oriental Society* 103: 261–274.

Noegel, S.B. 1997. Moses and Magic: Notes on the Book of Exodus. *Journal of the Ancient Near Eastern Society* 24: 45–49.

Noth, M. 1928. *Die israelitischen Personennamen*. Stuttgart: Kohlhammer.

Novacek, G.V. 2011. *Israel: Highlights from the Collections of the Oriental Institute of Chicago, Featuring Objects from the Haas and Schwartz Megiddo Gallery*. Chicago, IL: Oriental Institute.

Olyan, S.M. 1988. *Asherah and the Cult of Yahweh*, Society of Biblical Literature Monograph Series, vol. 34. Atlanta, GA: Scholars Press.

Quack, J. 2006. Zur Lesung und Deutung des Dramatischen Ramesseumpapyrus. *Zeitschrift für Ägyptische Sprache und Altertumskunde* 133: 72–89.

Rainey, A. 2008. Shasu or Hapiru: Who Were the Early Israelites? *Biblical Archaeology Review* 34: 51–55.

Redford, D.B. 1992. *Egypt, Canaan, and Israel in Ancient Times*. Princeton, NJ: Princeton University Press.

Ritner, R.K. 1993. *The Mechanics of Ancient Egyptian Magical Practice*, Studies in Ancient Oriental Civilization, vol. 54. Chicago, IL: Oriental Institute of the University of Chicago.

Roberts, J.J.M. 1972. *The Earliest Semitic Pantheon: A Study of the Semitic Deities Attested in Mesopotamia before Ur III*. Baltimore, MD: Johns Hopkins University Press.

Robinson, A.R. 1931. The Mahmal of Moselm pilgrimage. *Journal of the Royal Asiatic Society of Great Britain and Ireland* 1: 117–127.

———. 1931. The "Utfa" or camel-litter of the Arabs. *Journal of the Royal African Society* 30: 69–78.

Rosén, H.B. 1955. Arawna-nom Hittite? *Vetus Testamentum* 5: 318–320.

Rost, L. 1926. *Die Überlieferung von der Thronnachfolge Davids*, Beiträge zur Wissenschaft vom Alten und Neuen Testament, vol. 42. Stuttgart: Kohlhammer.

Russell, S.C. 2009. *Images of Egypt in Early Biblical Literature: Cisjordan-Israelite, Transjordan- Israelite, and Judahite Portrayals*. Berlin: Walter de Gruyter.

Sanders, P. 1994. *Ritual, Politics, and the City in Fatimid Cairo*. Albany, NY: State University of New York Press.

Sarna, N.M. 1986. *Exploring Exodus: The Heritage of Biblical Israel*. New York, NY: Schocken.

———. 1989. *The JPS Torah Commentary: Genesis*. Philadelphia, PA: Jewish Publications Society.

Sauneron, S. 1960. *The Priests of Ancient Egypt*. London: Evergreen Books.

Sayce, A.H. 1921. The Hittite Name Araunah. *Jewish Theological Studies* 22: 267–268.

Schlögl, H.A. 1980. *Der Gott Tatenen: Nach Texten und Bildern des Neuen Reiches*, Orbis biblicus et orientalis, vol. 29. Freiburg: Universtätsverlag.

Schwemer, D. 2001. *Wettergottgestalten. Die Wettergottgestalten Mesopotamiens und Nordsyriens im Zeitalter der Keilschriftkulturen: Materialien und Studien nach den schriftlichen Quellen*. Wiesbaden: Harrassowitz.

Sethe, K. 1928. *Dramatische Texte zu altägyptischen Mysterienspielen*, Untersuchungen zur Geschichte und Altertumskunde Ägyptens, vol. 10. Leipzig: Hinrichs.

Singer, I. 1992. Towards the Image of Dagon the God of the Philistines. *Syria* 69: 431–450.

Smelik, K.A.D. 1992. Hidden in the Ark Narrative: An Analysis of I Samual iv-vi and II Samuel vi. In *Converting the Past: Studies in Ancient Israelite and Moabite Historiography*, ed. Klaas A.D. Smelik, 35–58. Leiden: Brill.

Smith, M.S. 2003. Astral Religion and the Representation of Divinity: The Cases of Ugarit and Judah. In *Prayer, Magic, and the Stars in the Ancient and Late Antique World*, ed. Scott B. Noegel et al., 186–206. University Park, PA: Pennsylvania State University Press.

Spencer, J.R. 1992. Aaron. In *Anchor Bible Dictionary*, vol. 1, ed. D.N. Freedman, 1–6. New York, NY: Doubleday.

Stadler, M. 2008. Procession. In *UCLA Encyclopedia of Egyptology*, ed. Willeke Wendrich, 1–12. Los Angeles, CA: University of California-Los Angeles.

Stähli, H.-P. 1985. *Solare Elemente in Jahweglauben des Alten Testaments*, Orbis biblicus et orientalis, vol. 66. Freiburg: Universitätsverlag.

Strange, J. 2004. Some Notes on Biblical and Egyptian Theology. In *Egypt, Israel, and the Ancient Mediterranean World: Studies in Honor of Donald B. Redford*, Probleme der Ägyptologie, vol. 20, ed. Gary K. Knoppers and Antione Hirsch, 345–360. Leiden: Brill.

Teeter, E. 2011. *Religion and Ritual in Ancient Egypt*. Cambridge: Cambridge University Press.

Thomas, E. 1956. Solar Barks from Prow to Prow. *Journal of Egyptian Archaeology* 42: 56–79.

Tooley, A.M.J. 1996. Osiris Bricks. *Journal of Egyptian Archaeology* 82: 167–179.

van der Toorn, K., and C. Houtman. 1994. David and the Ark. *Journal of Biblical Literature* 113: 209–231.

Tur-Sinai, N.H. 1951. The Ark of God at Beit Shemesh (1 Sam. VI) and Peres 'Uzza (2 Sam. 6; 1 Chr. XIII). *Vetus Testamentum* 1: 275–286.

Ulmer, R. 2009. *Egyptian Cultural Icons in Midrash*, Studia Judaica, vol. 52. Berlin: Walter de Gruyter.

de Vaux, R. 1965. *Ancient Israel*, vol. 2. London: Darton, Longman, and Todd Ltd.

Weinstein, J. 1997. Exodus and Archaeological Reality. In *Exodus: The Egyptian Evidence*, ed. E.S. Frerichs and L.H. Lesko, 87–103. Winona Lake, IN: Eisenbrauns.

Wiggins, S.A. 1993. Old Testament Dagan in the Light of Ugaritic. *Vetus Testamentum* 43: 268–274.

Wimmer, D. 1990. Egyptian temples in Canaan and Sinai. In *Studies in Egyptology Presented to Miriam Lichtheim*, vol. 2, ed. S. Israelit–Groll, 1065–1106. Jerusalem: Magnes Press, Hebrew University of Jerusalem.

Westendorf, W. 1977. Zu frühformen von Osiris und Isis. *Göttinger Miszellen* 25: 95–113.

Wyatt, N. 1985. 'Araunah the Jebusite' and the Throne of David. *Studia Theologica* 39: 39–53.

Zecchi, M. 1996. *A Study of the Egyptian God Osiris Hemag*, Archeologia e Storia della civiltà egiziana e del vicino oriente antico, vol. 1. Imola: Editrice la mandragorta.

Zevit, Z. 2001. *The religions of Ancient Israel: A synthesis of Parallactic Approaches*. London: Continuum.

Zwickel, W. 1999. *Der salomonische Temple*, Kulturgeschichte der Antiken Welt, vol. 83. Mainz am Rhein: Verlag Philipp von Zabern.

———. 1994. *Der Tempelkult in Kanaan und Israel. Ein Beitrag zur Kultgeschichte Palästinas von der Mittelbronzezeit bis zum Untergang Judas*, Forschungen zum Alten Testament, vol. 10. Tübingen: Mohr Siebeck.

Moses the Magician

18

Gary A. Rendsburg

Abstract

Exodus 1–15 repeatedly shows familiarity with Egyptian traditions: the biblical motifs of the hidden divine name, turning an inanimate object into a reptile, the conversion of water to blood, a spell of 3 days of darkness, the death of the firstborn, the parting of waters, and death by drowning are all paralleled in Egyptian texts, and, for the most part, nowhere else.

Before reaching the substantive portion of this article, I begin with a few introductory notes. (a) The present article is the natural follow-up to my earlier essay, "Moses as Equal to Pharaoh" (Rendsburg 2006), which dealt with aspects of Exodus 1–15 that evoke Egyptian motifs relevant to Horus, the god of kingship, and/or to the Pharaoh, as royal figure. In this second article, we will focus upon reverberations of Egyptian magical and (additional) literary motifs within the Exodus account. (b) I ask the reader's forbearance regarding the title, "Moses the Magician," which should be understood loosely, since not every feature to be discussed herein relates to the role of the Egyptian lector-priest or magician-priest per se. Nevertheless, as we shall see, such matters dominate the biblical story—and besides, the alliterative title resonates felicitously. (c) Many of the items to be presented in this article have been treated by others before me (see especially Zevit 1990; Noegel 1996; Currid 1997: 83–120), though hopefully this essay still will present some new perspectives, including, I believe, one particular item (see below §8) hitherto not commented upon, at least to the best of my knowledge. (d) Finally, my method in this article, as was also the case in the first article, is to consider Exodus 1–15 as a narrative whole,[1] and to proceed through the biblical text in its canonical order.

§1. We start with Exod 3:13–15, which constitutes the second of Moses's four objections to God, along with God's response, in the scene atop Mt Horeb (see Exod 3:1; identified with Mt Sinai in later Jewish tradition). Moses anticipates that the people of Israel will ask him for God's name: וְאָמְרוּ־לִי מַה־שְּׁמוֹ מָה אֹמַר אֲלֵהֶם "and they shall say to me, 'What is his

G.A. Rendsburg (✉)
Rutgers University, New Brunswick, NJ, USA
e-mail: grends@rci.rutgers.edu

[1] Though clearly the Song of the Sea in the final chapter is an earlier poetic version (indeed most likely the oldest piece of literature in the entire Bible), predating the prose account in Exodus 14.

T.E. Levy et al. (eds.), *Israel's Exodus in Transdisciplinary Perspective*,
Quantitative Methods in the Humanities and Social Sciences, DOI 10.1007/978-3-319-04768-3_18,

name?'—what shall I say to them?"[2] God responds not with any of the standard divine names used in the Bible—יהוה "YHWH," אֱלֹהִים "Elohim" ("God"), שַׁדַּי "Shaddai," etc.—but rather proclaims אֶהְיֶה אֲשֶׁר אֶהְיֶה *ʾɛhyɛ ʾašɛr ʾɛhyɛ* "I am that I am" (v. 14), a name used nowhere else in the Bible, shortened later in the verse to the simple אֶהְיֶה *ʾɛhyɛ* "I am".[3] This unique divine name is to be understood as "the unknown name" of Yahweh, a parallel to "the unknown name" of Ra, as narrated in Pap. Turin 1993 (c. 1300 B.C.E.).[4] In this Egyptian myth, Isis seeks to learn the secret name of Ra, though the great god refuses, with the comment, "My father and mother told me my name. I have hidden it in my body since birth, so as to prevent the power of a male magician or a female magician from coming into existence against me" (lines cxxxii.11–12).[5] Isis, in turn, produces a venomous snake, which bites Ra, thereby forcing him—with no alternative, as he suffered tremendously from the burning poison—to ultimately disclose his name: "The great god announced his name to Isis, the Great One of Magic" (line cxxxiii.14).

The parallel between the two stories is clear: in both cases the great god has a secret name. But the key differences are illuminating:

(a) In the biblical account, God is not fearful of disclosing his name: Moses asks, and God divulges—because in the biblical conception of the single deity, Yahweh has no fear of falling under the influence of magical praxes (see, for example, Num 23:23). In the Egyptian story, by contrast, Ra explicitly declares his dread of the powerful magicians, and hence he does everything "humanly" possible not to divulge his name.[6]

(b) In the biblical story, not only does God reveal "the unknown name" to Moses, but the reader of the story learns the name as well—again, because there can be no concern with magical abuse or misuse of this appellation. In the Egyptian tale, by contrast, the reader does not learn the special name of Ra; he/she learns only that Isis learned the name.

§2. The Exodus narrative continues with Moses's third objection to God atop Mt Horeb, namely, that the people will not believe him when he returns to Egypt: וַיַּעַן מֹשֶׁה וַיֹּאמֶר וְהֵן לֹא־יַאֲמִינוּ לִי וְלֹא יִשְׁמְעוּ בְּקֹלִי כִּי יֹאמְרוּ לֹא־נִרְאָה אֵלֶיךָ יְהוָה׃ "And Moses answered, and he said, 'but behold, they will not believe me, and they will not listen to my voice, for they will say, "YHWH did not appear unto you"'" (Exod 4:1). So, how best to impress people in Egypt with one's power, to instill belief in them? The answer: to empower the hero with the ability and capacity associated with the lector-priest. Accordingly, God instructs Moses at this point to cast down his shepherd's staff, which turns into a snake (נָחָשׁ), and then, upon Moses's grasping the snake by the tail, it reverts to a staff (vv. 2–4). A very similar maneuver is

[2] Translations of the Hebrew herein are my own. I also have translated the shorter Egyptian passages quoted, though for longer Egyptian texts I have relied on the translations of others, as indicated.

[3] This short form may be alluded to in one other passage, Hos 1:9.

[4] For the four other sources, all Ramesside, all from Deir el-Medina, see http://www.digitalegypt.ucl.ac.uk/literature/isisandra.html (along with transliteration of Pap. Turin 1993).

[5] Adapted from the translation of Robert K. Ritner in *COS*, 1.33–34. For other versions, see Wilson (1969: 12–14) and McDowell (1999: 118–120). For the standard edition, see Pleyte and Rossi (1869–1876: 1.173–180) (with Plates CXXXI-CXXXIII, LXXVII+XXXI in vol. 2).

[6] On the fear expressed by the Egyptian gods concerning the threat of magic directed against them, see Ritner (1993: 21–22).

employed by the chief lector-priest (*ḫry-tb*) Webaoner and the caretaker of his gardens in "The Wax Crocodile" story, the second of the tales appearing in Papyrus Westcar (Pap. Berlin 3033, c. 1600 B.C.E., though the composition is several centuries earlier), as a means to avenge the indiscretions of the local townsman[7]:

> When day broke, and the second day came, the caretaker informed Webaoner of the matter .... Then he lit a fire and said, 'Bring me my chest of ebony and electrum', and he opened it and made a crocodile of wax seven fingers long. He read out his magic words saying ... 'If anyone comes to bathe in my lake ... the townsman'. Then he gave it to the caretaker, and he said to him: 'After the townsman goes down to the pool, as is his daily fashion, you shall cast the crocodile after him'. The caretaker went forth, and he took the crocodile of wax with him.
>
> ...
>
> After night fell, the townsman returned as was his daily fashion, and the caretaker threw the crocodile of wax behind him into the water. At once it grew into a crocodile of seven cubits, and it took hold of the townsman.
>
> Webaoner tarried with His Majesty the King of Upper and Lower Egypt, Nebka, the vindicated, for seven days, all the while the townsman was in the lake without breathing. After the seventh day came, His Majesty the King of Upper and Lower Egypt, Nebka, the vindicated, came forth, and the chief lector-priest Webaoner placed himself in his presence and said to him, 'May Your Majesty come and see the marvel which has taken place in Your Majesty's time'. His Majesty went with Webaoner. He called out to the crocodile and said, 'Bring back the townsman'. The crocodile came out of the water. Then the chief lector-priest Webaoner said, 'Open up!' And he opened up. Then he placed...
>
> Said His Majesty the King of Upper and Lower Egypt, Nebka, the vindicated, 'This crocodile is indeed fearful!' But Webaoner bent down, and he caught it and it became a crocodile of wax in his hand.

To be sure, "The Wax Crocodile" tale does not mention Webaoner's grasping the crocodile by the tail, and of course Moses's action atop Mt Horeb involves a snake. But there is additional data to flesh out the relationship between the two stories. On the biblical side, we note that when Moses and Aaron appear before Pharaoh and produce this wonder, the word נָחָשׁ "snake" is not used, but rather the term תַּנִּין "crocodile" occurs[8]: וַיָּבֹא מֹשֶׁה וְאַהֲרֹן אֶל־פַּרְעֹה וַיַּעֲשׂוּ כֵן כַּאֲשֶׁר צִוָּה יְהוָה וַיַּשְׁלֵךְ אַהֲרֹן אֶת־מַטֵּהוּ לִפְנֵי פַרְעֹה וְלִפְנֵי עֲבָדָיו וַיְהִי לְתַנִּין׃ "and Moses and Aaron came unto Pharaoh, and they did such, as YHWH had commanded; and Aaron cast-down his staff before Pharaoh and before his servants, and it became a crocodile" (Exod 7:10).

The difference in word usage is to be explained on two grounds: (a) in the wilderness setting of Mt Horeb, the snake is appropriate; while in the Pharaoh's court, which could never be too far from the Nile, the crocodile is more suitable (see already Cassuto 1967: 94); and (b) a holistic approach to the narrative suggests that an "upgrade" is operative here, for while the snake trick is clearly something special, how much more impressive is the transformation of the staff into a crocodile.[9] On the Egyptological side,

---

[7] Standard editions of the entire Papyrus Westcar are Blackman-Davies (1988) and (Lepper 2008). The translation here, excerpted from the second tale, is adapted from that of William Kelly Simpson in Simpson (2003: 13–16) (in particular pp. 15–16). The reader should be aware that the manuscript is much more fragmentary than my clean prose would suggest (especially in this selection, col. 2, line 15, through col. 4, line 3), though the narrative thread is clear nonetheless. For other English renderings, see Parkinson (1997: 102–109) (esp. pp. 107–108); and Quirke (2004: 77–81) (esp. pp. 78–80) (with transliteration). See also the German translation by Lepper (2008: 29–35) (esp. pp. 31–34) (with transliteration).

[8] There is a debate amongst scholars concerning the meaning of the word תַּנִּין, though in my opinion "crocodile" is the only possible option here and in other Egyptian contexts (e.g., Ezek 29:3, 32:2). For those in agreement with this conclusion, see the references in Noegel (1996: 47, n. 12). For full treatment, notwithstanding a contrary view, see Cohen (1991).

[9] The source-critical approach assigns Exodus 4:1–16 to the Yahwist account, which uses נָחָשׁ "snake" both here and in its creation account in Genesis 2–3, and Exodus 7:1–13 to the Priestly source, which uses תַּנִּין "crocodile, sea-monster" both here and in the first creation account in Genesis 1, and which elevates Aaron to greater prominence. For convenient orientation, see Friedman (2003: 130 n. *).

we may note a series of seals portraying an individual (most likely a magician-priest) holding crocodiles by the tail, one in each hand; see Fig. 18.1 for two examples[10]—in addition to the famous Horus stelae of the god Horus holding snakes, scorpions, etc., in similar fashion (see Rendsburg 2006: 213–215). A chart may be useful here to summarize the evidence:

Action	Biblical account	Egyptian evidence
Snake from inanimate to animate	Moses atop Mt Horeb	
Holding snake by the tail	Moses atop Mt Horeb	Horus stelae
Crocodile from inanimate to animate	Moses and Aaron before the Pharaoh	Webaoner in "The Wax Crocodile"
Holding crocodile by the tail		Egyptian seals (from Egypt and Canaan)

Finally, we note the response by the Egyptians in the court: וַיִּקְרָא גַּם־פַּרְעֹה לַחֲכָמִים וְלַמְכַשְּׁפִים וַיַּעֲשׂוּ גַם־הֵם חַרְטֻמֵּי מִצְרַיִם בְּלַהֲטֵיהֶם כֵּן׃ "and Pharaoh also called the wise-men and the sorcerers; and they did likewise, the lector-priests of Egypt, with their spells thus" (Exod 7:11). Which is to say, naturally the Egyptian lector-priests are able to reproduce the wonder produced by Moses and Aaron, since, as we learn from "The Wax Crocodile" story, the ancient Egyptians themselves believed that such individuals could transform an inanimate object into a crocodile.[11]

§3. We turn now to the extended narrative concerning the Ten Plagues, several of which evoke Egyptian tropes. The first of these is Plague One, turning the Nile into blood: וַיַּעֲשׂוּ־כֵן מֹשֶׁה וְאַהֲרֹן כַּאֲשֶׁר ׀ צִוָּה יְהֹוָה וַיָּרֶם בַּמַּטֶּה וַיַּךְ אֶת־הַמַּיִם אֲשֶׁר בַּיְאֹר לְעֵינֵי פַרְעֹה וּלְעֵינֵי עֲבָדָיו וַיֵּהָפְכוּ כָּל־הַמַּיִם אֲשֶׁר־בַּיְאֹר לְדָם׃ "And Moses and Aaron did thus, as YHWH commanded, and he raised the staff, and he struck the water that is in the Nile, before the eyes of Pharaoh and before the eyes of his servants; and all the water that is in the Nile was turned into blood" (Exod 7:20). The Egyptian parallel is well known, from the "Admonitions of Ipuwer," the sage who described the chaotic state of the land, with reference to either the First Intermediate Period or the Second Intermediate Period, though the sole surviving manuscript, Pap. Leiden 344, dates to the 19th Dynasty, centuries later.[12] The key passage reads as follows: *ỉw ms ỉtrw m snf swrỉ.tw ỉm.f nyw.tw m rmṯ ỉn.tw mw* "Indeed, the river is blood, yet one drinks from it; one turns away from people, yet one thirsts for water" (col. 2, line 10) … *[iw]ms dšrt ḫt tꜣ ḫꜣswt ḫbꜣ […] psḏt rwty iy.ti n kmt* "[In]deed, the desert is throughout the land, the nomes are ravaged; foreign-tribes (lit. "bowmen") have come into Egypt" (col. 3, line 1). Note not only the parallel to the first plague, but also the equally important fact that the upheaval is associated with the presence of foreigners in the land.[13]

[10] For examples from Egypt, see Quibell (1898: pl. 30, no. 26) (from the Ramesseum); Petrie (1906: pl. 11, no. 222) (from Tell el-Yehudiyeh); and Petrie (1909: pl. 34, no. 92) (from Memphis). For seals of this sort from the land of Israel, see Keel (1997): Achsib, no. 115; Tell el-ʿAǧul, nos. 200 and 996; and Akko, no. 115; and Keel (2010): Beth-Shan, no. 87; Beth-Shemesh, no. 10; and Dor, no. 26. For a general survey, including reproductions of some of the above seals, see Münger (2003: 69), fig. 2, nos. 11–15. I am extremely grateful to Dr Münger for directing my attention to these seals and for this wealth of bibliography.

[11] For additional aspects of this episode, see Rendsburg (2006: 209–210).

[12] For discussion of these matters, see Enmarch (2005, 2008), now the standard treatments of this important composition. English translations include Lichtheim (1973: 149–163), Parkinson (1997: 166–199), Nili Shupak in *COS*, 1.93–98; Vincent A. Tobin in Simpson (2003: 188–210); and Quirke (2004: 140–150) (with transliteration of the text).

[13] For another reference to water turning to blood in an ancient Egyptian text, namely Setne II (on which see below, §5), see Lichtheim (1973–1980: 3.148); and Robert K. Ritner in Simpson (2003: 485). The parallel is less apt, though, since it is the Nubian magician's mother's water which will turn to blood, should he be defeated whilst performing sorcery in Egypt.

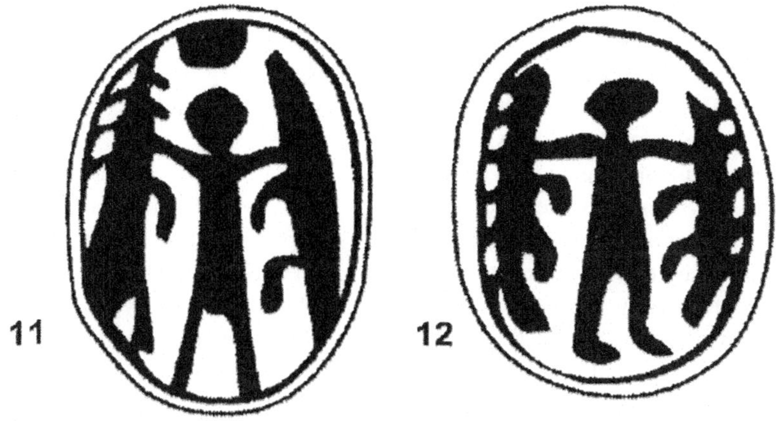

Fig. 18.1 Seals depicting an individual (most likely, a magician-priest) holding two crocodiles by the tail, one in each hand. Original publications: on the left, Petrie 1909: pl. 34, no. 92 (from Memphis); on the right, Petrie 1906: pl. 11, no. 222 (from Tell el-Yehudiyeh). The whereabouts of the former is now unknown. The latter is housed in the Petrie Museum of Egyptian Archaeology, catalogue no. UC61221, and may be viewed via online search at: http://petriecat.museums.ucl.ac.uk/search.aspx. My gratitude to Dr Alice Stevenson for this information and to Prof Stephen Quirke for additional assistance

§4. Plagues Three and Four both involve insects, כִּנִּים "lice" and עָרֹב "gnats, flies," respectively.[14] In these two instances, the supporting Egyptological evidence comes not from Egyptian sources directly, but rather from Herodotus's detailed description of the land of Egypt in *The Histories*, Book Two. In section 37, he reports as follows: "Their priests shave the whole body every other day, that no lice or aught else that is foul may infest them in their service of the gods"[15]; while in section 95, he states very succinctly: "Gnats are abundant,"[16] after which follows a long section detailing how the Egyptians protect themselves from this pest. In light of these statements, the presence of these two annoyances within the plagues account in the book of Exodus is rather appropriate. Also noteworthy is Herodotus's observation regarding the priests who shave their bodies regularly, so as not to become impure, and hence be disqualified from temple service; this will explain the presence of the *ḥartumim* = Eg., *ḥry-tp* "lector-priests" in the short accounts of Plague Three and Plague Six.

The key passage in the former pericope is the following: וַיַּעֲשׂוּ־כֵן הַחַרְטֻמִּים בְּלָטֵיהֶם לְהוֹצִיא אֶת־הַכִּנִּים וְלֹא יָכֹלוּ וַתְּהִי הַכִּנָּם בָּאָדָם וּבַבְּהֵמָה. "And the lector-priests did thus, by their spells, to bring-out the lice, but they were not able; and the lice were upon human and beast" (Exod 8:14). Ironically, had the *ḥartumim* succeeded in their attempt to duplicate Moses's and Aaron's action (see vv. 12–13), the lice would have been even more present. Though even without such action, the lice that Moses and Aaron brought forth already had infested "human and beast," which would have included the priests presumably.

The second relevant verse occurs later, within the Plague Six pericope: וְלֹא־יָכְלוּ הַחַרְטֻמִּים לַעֲמֹד לִפְנֵי מֹשֶׁה מִפְּנֵי הַשְּׁחִין כִּי־הָיָה הַשְּׁחִין בַּחַרְטֻמִּם וּבְכָל־מִצְרָיִם׃ "And the lector-priests were not able to stand before Moses, on account of the boils; for the boils

[14] For the identification of עָרֹב as "gnats, flies," see Rendsburg (2003).

[15] Translation of Godley (1921–1924: 1.319). For discussion, see Lloyd (1976–1988: 1.165–166).

[16] Translation of Godley (1921–1924: 1.381). For discussion, see Lloyd (1976–1988: 1.382).

were upon the lector-priests and upon all of Egypt" (Exod 9:11). While not lice, the boils constitute a different skin affliction, which also would have rendered the Egyptian priests unable to serve the gods. To my mind, it is not a coincidence that the contest between the *ḥartumim*, on the one hand, and Moses and Aaron, on the other, ends with the third plague of lice, and that the *ḥartumim* reappear only once (in a cameo appearance, as it were), during the telling of the sixth plague of boils. The attack on the lector-priests by extension represents an assault on the heart of Egyptian religion, for without the priestly service in the temples, the cults are inoperative, the deities are ineffective, and all of Egypt descends into turmoil.

§5. Our next topic is Plagues Eight and Nine, the plagues of locusts and darkness, respectively. The two are clearly distinct, but they share the motif of darkness descending upon the land:

וַיְכַס אֶת־עֵין כָּל־הָאָרֶץ וַתֶּחְשַׁךְ הָאָרֶץ וַיֹּאכַל אֶת־כָּל־עֵשֶׂב
הָאָרֶץ וְאֵת כָּל־פְּרִי הָעֵץ אֲשֶׁר הוֹתִיר הַבָּרָד וְלֹא־נוֹתַר
כָּל־יֶרֶק בָּעֵץ וּבְעֵשֶׂב הַשָּׂדֶה בְּכָל־אֶרֶץ מִצְרָיִם׃

And it [sc. the locust swarm] covered the eye of the whole earth, and the earth was darkened, and it ate all the vegetation of the land and all the fruit of the trees, which the hail left over; and not a single green was left on the trees or on the vegetation of the field, in all the land of Egypt. (Exod 10:15)

וַיֵּט מֹשֶׁה אֶת־יָדוֹ עַל־הַשָּׁמָיִם וַיְהִי חֹשֶׁךְ־אֲפֵלָה בְּכָל־אֶרֶץ
מִצְרַיִם שְׁלֹשֶׁת יָמִים׃
לֹא־רָאוּ אִישׁ אֶת־אָחִיו וְלֹא־קָמוּ אִישׁ מִתַּחְתָּיו שְׁלֹשֶׁת יָמִים
וּלְכָל־בְּנֵי יִשְׂרָאֵל הָיָה אוֹר בְּמוֹשְׁבֹתָם׃

And Moses extended his hand towards the heaven, and there was a deep darkness in all the land of Egypt (for) 3 days. One could not see his fellow, and no one rose from his place (for) 3 days; and for all the children of Israel there was light in their dwellings. (Exod 10:22–23)

In the first verse above, note that "the eye of the whole earth" is a Hebrew quasi-calque on the Egyptian phrase *ỉr.t rʿ* "the eye of the sun," a metaphor for Ra, and by extension, the land of Egypt (Erman-Grapow 1926–1931: 1.107; Yahuda 1933: 62–63; Rendsburg 1988: 7). Indeed, Targum Onqelos renders the Hebrew phrase עֵין שִׁמְשָׁא דְכֹל אַרְעָא "the eye of the sun of the whole earth," inserting the word שִׁמְשָׁא "sun" in the middle of this phrase, thereby departing from its typical word-for-word rendering (Rendsburg 1990).

The relevant Egyptian texts here are the "Prophecy of Neferti" and "Setne Khamwas and Si-Osire" (=Setne II). The former is a Middle-Egyptian composition, though it is known to us only from New-Kingdom copies, to wit, the complete 18th-Dynasty Pap. St Petersburg 1116B (whose lines are cited below), along with many fragmentary copies from the 19th and 20th Dynasties.[17] The pertinent passage reads as follows: *ỉtn ḥbs nn psd.f mȝȝ rḫyt nn ʿnḫ.tw ḥbsw šnʿ* "the sun-disc is covered, it does not shine for people to see; no one can live, when the clouds cover" (line 25) ... *styw ḫtyw tȝ ỉw ḫrw ḫpr ḥr ỉȝbtt ỉw ʿȝmw hȝt r kmt* "Syrians (*styw*) are throughout the land, enemies emerge in the east, Asiatics (*ʿȝmw*) have come-down into Egypt" (lines 32–33). Once more we note how the disorder, characterized by the concealment of the sun, is connected to the arrival of Syrians/Asiatics (i.e., Semites).

The second parallel appears in the late Demotic text "Setne Khamwas and Si-Osire" (=Setne II) (Pap. British Museum 604 verso). This composition relates a series of stories concerning Setne Khamwas, son of Rameses II, high priest of Memphis, and renowned magician. While the text is late (indeed, the manuscript dates to the first century C.E., based on the information provided on the recto), and indeed the

[17] The standard editions are Goedicke (1977) and Helck (1992). For English translations see Lichtheim (1980: 138–151), Parkinson (1997: 131–143), Nili Shupak in *COS*, 1.106–110, Vincent A. Tobin in Simpson (2003: 214–220), and Quirke (2004: 135–139) (with transliteration of the text).

traditions concerning Setne Khamwas grew over the course of time,[18] a number of the motifs hark back to earlier Egyptian tropes.[19] The relevant episode within the narrative actually concerns not Setne Khamwas, but rather his son Si-Osire, who in fact surpasses his father in wisdom and magic. In the course of this story, an unnamed Nubian magician states the following: "Were it not that Amun would find fault with me, and that the lord of Egypt might [punish me], I would cast my sorceries upon Egypt and would make the people of Egypt spend three days and three nights seeing no light, only darkness."[20] In short, we have here another Egyptian tale centered on darkness, in fact, specifically 3 days of darkness, in accordance with the biblical description of the ninth plague.

§6. The culminating plague, as is well known, is the death of the firstborn, narrated in Exod 11:1–10, 12:29–30.

וַיְהִי ׀ בַּחֲצִי הַלַּיְלָה וַיהוָה הִכָּה כָל־בְּכוֹר בְּאֶרֶץ מִצְרַיִם מִבְּכֹר פַּרְעֹה
הַיֹּשֵׁב עַל־כִּסְאוֹ עַד בְּכוֹר הַשְּׁבִי אֲשֶׁר בְּבֵית הַבּוֹר וְכֹל בְּכוֹר בְּהֵמָה׃

And it was, in the middle of the night, and YHWH struck all the firstborn of the land of Egypt, from the firstborn of Pharaoh who sits on his throne, unto the firstborn of the captives who are in the house of the pit, and all the firstborn of the animals. (Exod 12:29)

The same motif occurs in Egyptian funerary literature, both the Pyramid Texts of the Old Kingdom period and the Coffin Texts of the Middle Kingdom.[21] The relevant passages from these collections were brought to the attention of scholars by Mordechai Gilula (1977):

(a) Pyramid Texts, par. 399a–b (within the "Cannibal Hymn") from pyramids of Unas, c. 2350 B.C.E., and Teti, c. 2320 B.C.E.

(wnỉs)/(ttỉ) pỉ wḏʿ mdw ḥnʿ ỉmn rn-f / hrw pw n rḫs smsw

(Sethe 1908–1922: 1.208; see also Faulkner 1969: 81)[22] (see Fig. 18.2)

It is the king who will be judged with Him-whose-name-is-hidden on this day of the slaying of the first-born (*smsw*). (Unas 508//Teti 322)

(b) Coffin Texts, §178p (Spell 573) (c. 2000 B.C.E.)
- from Asyut [Siut] coffin, S1C, inner coffin of *msḥt* = Cairo 28118
- parallel: Asyut [Siut] coffin S2C, outer coffin of *msḥt* = Cairo 28119

ỉnk wḏʿ mdw ḥnʿ ỉmn rn-f / grḥ pw n r‹ḫ›s wrw

(de Buck 1935–1961: 6.178; see also Faulkner 1973–1978: 2.176) (see Fig. 18.3)

I am he who will be judged with Him-whose-name-is-hidden on this night of the slaying of the first-born (*wrw*).

(c) Coffin Texts, §163b–c (=Spell 136) (c. 2000 B.C.E.)
- from Saqqara coffin Sq3Sq, coffin of *snny* in Saqqara storeroom with three parallels:
  - B2L, outer coffin of *gwꜣ-tp*, from el-Barsha = BM 38039
  - B2P, inner coffin of *spỉ*, from el-Barsha, now in the Louvre[23]
  - Sq4C, coffin of *ḥnw*, from Saqqara = Cairo J39052

[18] This includes the transformation of the title *stm* "priest" (>Setne) into part of the protagonist's name.

[19] In addition to which, one of the tales is known from an earlier Aramaic version found at Elephantine; see Robert K. Ritner in Simpson (2003: 471), with n. 1.

[20] Translation of Lichtheim (1980: 138–151) (in particular p. 144). For another rendering, see Robert K. Ritner in Simpson (2003: 471–489) (in particular p. 480). The standard edition remains, Griffith (1900), with translation and transliteration on pp. 142–207.

[21] For basic orientation into these two genres, see Hornung (1999: 1–6, 7–12), respectively.

[22] For a different approach on how to render this passage, see Allen (2005: 51, 91). Eyre (2002: 85) includes nary a comment about this line.

[23] The sigla for these coffins (S1C, B2L, etc.) are those employed by de Buck. In this particular case, for coffin B2P, de Buck did not supply a museum accession number, beyond indicating its current location in the Louvre.

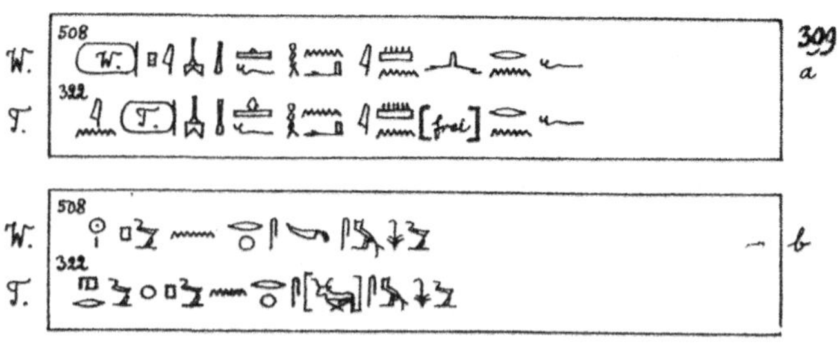

Fig. 18.2 Pyramid Text §399a–b, from pyramids of Unas [W] and Teti [T], from Sethe (1908–1922: 1.208). Used with kind permission of Georg Olms Verlag

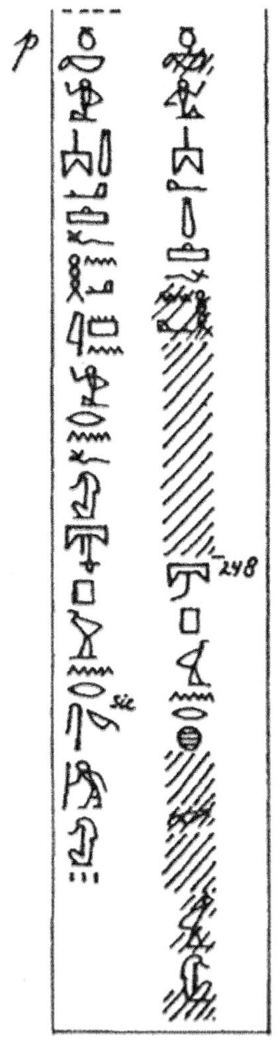

Fig. 18.3 Coffin Text §178p, from de Buck (1935–1961: 6.178). Used with kind permission of the Oriental Institute of the University of Chicago

. . . *grḥ pw n rḥs wrw* / *hrw pw n rḥs wrw*

(de Buck 1935–1961: 2.163; see also Faulkner 1973–1978: 1.117) (see Fig. 18.4)

. . .this night of the slaying of the first-born (*wrw*),

this day of the slaying of the first-born (*wrw*).

Now, to be sure, we know very little about "this day of the slaying of the first-born" (in the first text above), or with even greater relevance "this night of the slaying of the first-born" (in the second text), or the composite version with both "night" and "day" (in the third text). In addition to the change from *smsw* in the Pyramid Texts to *wrw* in the Coffin Texts, one also should note that the latter is followed by the "deity" determinative, in all copies and in all instances (see Figs. 18.3 and 18.4). The same holds for the expression *ỉmn rn-f* "He-whose-name-is-hidden" in those instances where the phrase is extant (§178p [again, see Figs. 18.3 and 18.4] and §163a [not included above]).

Regardless of how this echo of a myth is to be understood, one will agree with Gilula (1977: 95): "These passages are strong evidence that a mythological tale once circulated in which some or all of the first-born in Egypt—whether gods, mortals or animals—were slain on a certain day or night. Such a myth may very likely lie in the background of the biblical account."

§7. While the ten plagues suffice to demonstrate Yahweh's might and salvific power (that is, from the perspective of the biblical narrator and his audience), the story line allows for one ultimate act which eclipses all the previous ones. I refer, naturally, to the splitting of the Reed Sea and the subsequent drowning of the Egyptians, recounted both in prose (Exodus 14) and in poetry (Exodus 15). Our treatment

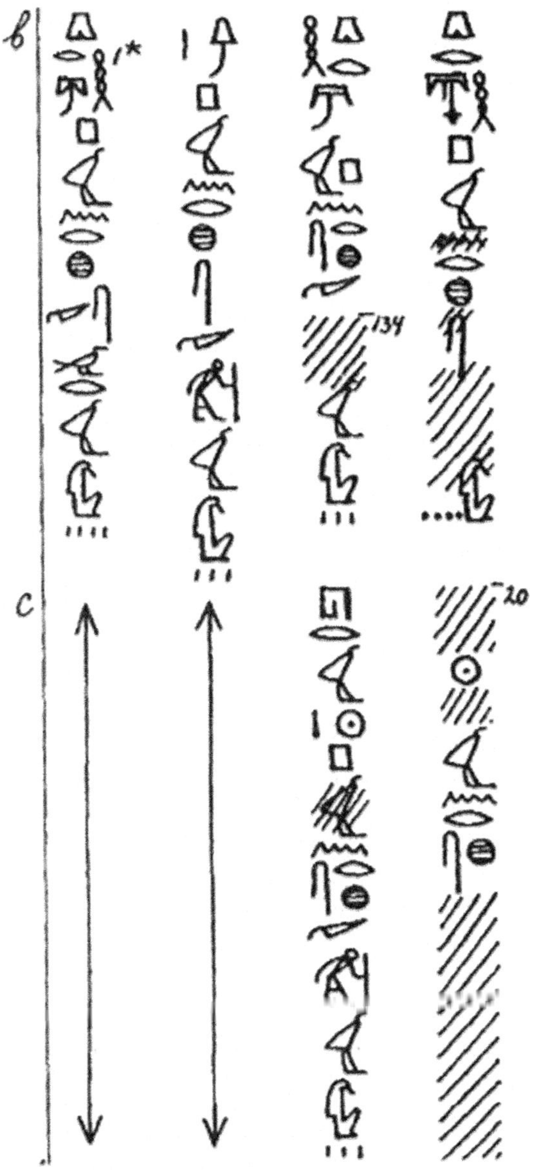

Fig. 18.4 Coffin Text §163b–c, from de Buck (1935–1961: 2.163). Used with kind permission of the Oriental Institute of the University of Chicago

divides this episode into its two component parts, beginning with the splitting of the sea, for which there are two germane texts from ancient Egypt. We begin with the less famous and indeed slightly less comparable story, namely, an episode which appears in the Demotic tale "Setne Khamwas and Na-nefer-ka-ptaḥ," known as "Setne I" for short, dated to the Ptolemaic period (Pap. Cairo 30646).[24] (For general orientation regarding the figure of Setne Khamwas, see above, §5) In the relevant snippet, the magician-prince Na-nefer-ka-ptaḥ seeks to locate and obtain the "book" written by Thoth himself, now housed inside a box (actually, within a series of boxes) in the middle of the Nile at Coptos. Na-nefer-ka-ptaḥ creates a model boat, replete with rowers and sailors, then brings it all to life via the recitation of a spell. At which point, the story reads:

> He said to the rowers, "Row me to the place where that book is!" [They rowed him by night] as by day. In three days he reached it. He cast sand before him, and a gap formed in the river. He found six miles of serpents, scorpions, and all kinds of reptiles around [the place where the book was]. He found an eternal serpent around this same box. He recited a spell to the six miles of serpents, scorpions, and all kinds of reptiles that were around the box, and did not let them come up.[25] (page 3, lines 29–32)

The narrative continues with Na-nefer-ka-ptaḥ's opening of the series of nested boxes and his successful acquisition of the "book." In short, we see in this tale the belief that an Egyptian wise man could part the waters (viz., "a gap formed in the river") in order to reach the desired object, in this case, the "book" written by Thoth himself.

The second pertinent text returns us to Papyrus Westcar (see above, §2), though this time our focus is on the third tale in the collection, "The Boating Party" story. In this tale, the royal family is enjoying a day of leisure on the lake, when a pendant of one of the princesses falls into the water. The king (Seneferu [4th Dynasty, c. 2600 B.C.E.]) commands that the chief lector-priest

[24] As with Setne II discussed above, the standard treatment remains, Griffith (1900), with translation and transliteration on pp. 82–141.

[25] Translation of Lichtheim (1980: 127–138) (esp. p. 130). For another translation, see Ritner in Simpson (2003: 454–469).

(*ḫry-tb ḥry-tp*) Djadja-em-ankh be brought, at which point we pick up the story as follows (col. 5, line 25, through col. 6, line 15):

> Said his majesty, "Djadja-em-ankh, my brother, I did as you had said. My majesty's heart was refreshed seeing them row. Then a pendant of new turquoise of one of the leaders fell into the water. She stopped rowing and thereby spoiled her side. I said to her, "Why have you stopped rowing?" She said to me, "Because the pendant of new turquoise fell into the water." I said to her, "Row! I shall replace it for you!" She said to me, "I prefer my thing to one like it."
>
> Then the chief lector-priest Djadja-em-ankh said his say of magic. He placed one side of the lake's water upon the other; and he found the pendant lying on a shard. He brought it and gave it to its owner. Now the water that had been twelve cubits deep across (lit. 'on its back') had become twenty-four cubits when it was turned back. Then he said his say of magic and returned the waters of the lake to their place.
>
> His majesty spent the day feasting with the entire palace. Then he rewarded the chief lector-priest Djadja-em-ankh with all good things.[26]

In other words, the marvel produced by Moses at the Sea of Reeds is of a piece with the marvel produced by the famous lector-priest Djadja-em-ankh in "The Boating Party" tale. Though whereas the dividing of the waters saves the day in the Egyptian tale, in the biblical narrative this action serves to outwit the Egyptians, allowing the Israelites to flee safely to the other side of the body of water. It is as if the biblical writer wishes to state the following (both here and in several instances above): if you Egyptians believe that magician-priests are capable of such praxes, we will use those very same actions to bring about your ruin and defeat. At the same time, for these tropes to be meaningful to an Israelite audience, who after all were the consumers of the biblical literature, one must assume a considerable knowledge amongst the Israelites of ancient Egyptian beliefs and practices, perhaps even specific literary motifs.

§8. The final item to be discussed here is the climactic act in the extended narrative of Exodus 1–15, to wit, the drowning of the Egyptians. We begin not with evidence from Egypt directly, but rather with the testimony of Herodotus, *The Histories*, Book Two, section 90:

> When anyone, be he Egyptian or stranger, is known to have been carried off by a crocodile or drowned by the river itself, such an one must by all means be embalmed and tended as fairly as may be and buried in a sacred coffin by the townsmen of the place where he is cast up; nor may any of his kinsfolk or his friends touch him, but his body is deemed something more than human, and is handled and buried by the priests of the Nile themselves.[27]

In short, according to Herodotus, death by drowning in ancient Egypt was a noble death.[28]

Corroboration for this statement is forthcoming from native Egyptian sources, especially the two interrelated New Kingdom funerary texts, Amduat, 10th hour, and Book of Gates, 9th gate,[29] both of which portray the drowned ones afloat in the river in their respective registers. See below, line drawings in Figs. 18.5 and 18.6 and photo in Fig. 18.7, from KV-9, the tomb of Rameses VI.[30] A representative passage from the accompanying hieroglyphic text in the former runs as follows: "You are those who are within Nun, the drowned who are in his following. May life belong to your *bas*!"[31] Another excellent exemplar of the Book of Gates, 9th gate, is visible in KV-14, the tomb shared by Tausert (pharaoh-queen,

[26] Translation of Lichtheim (1973: 216–217) (esp. p. 217). For other renderings see Parkinson (1997: 109–112) (esp p. 111), Simpson (2003: 16–18) (esp. pp. 17–18), and Quirke (2004: 81–83) (esp. pp. 82–83) (with transliteration). See also the German translation by Lepper (2008: 36–40) (esp. pp. 38–39) (with transliteration).

[27] Translation of Godley (1921–1924: 1.375).

[28] For further discussion, see Griffith (1909), Rowe (1940: 3–30), and Lloyd (1976–1988: 1.366–367).

[29] For orientation, see Hornung (1999: 27–53, 55–77), respectively.

[30] All to be seen in the unsurpassed magisterial edition of Piankoff and Rambova (1954).

[31] Translation of Manassa (2007: 350).

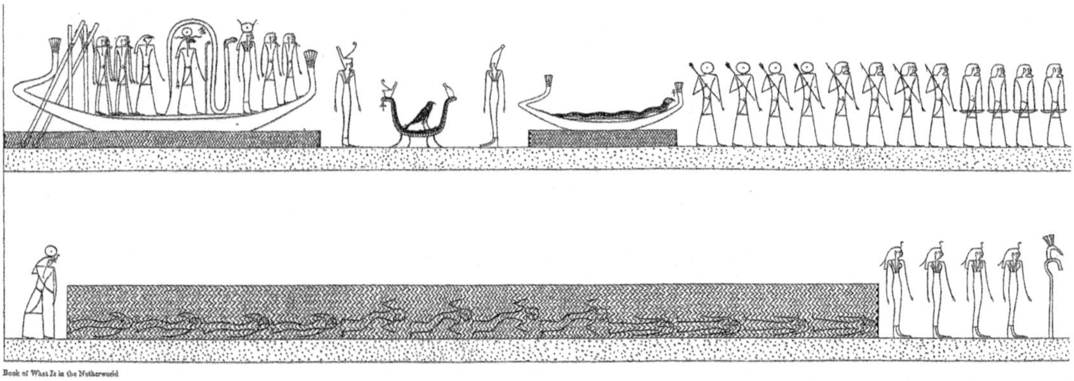

Fig. 18.5 KV-9, tomb of Rameses VI, Right Wall of Corridor G, Amduat 10th hour (Piankoff and Rambova 1954: fig. 85 [double foldout facing p. 299], with text on p. 304). Reprinted by permission of Princeton University Press. This image is also available online at http://www.egiptologia.org/textos/amduat/10/; see also Hornung (1999: 51), fig. 23

last monarch of Dynasty 19) and Setnakht (first pharaoh of Dynasty 20).[32]

To relate this concept to a narrative already discussed herein, note that in Setne I (see above, §7), three different characters, including the great magician Na-nefer-ka-ptaḥ himself, drown in the Nile and become "the praised one" of Ra—in Ritner's words, "an expression for the deified 'drowned'" (Ritner in Simpson 2003: 460, n. 19). Lichtheim (1973–1980: 3.131) goes so far to render the expression as simply "drowned," though with a note "Lit., 'He became one praised of Re'" (Lichtheim 1973–1980: 3.138, n. 12).

To return now to the biblical account of Exodus 14–15: it is as if the biblical author is stating: okay, Egyptians, if you believe that drowning is such an honorable death, then fine, we will arrange your demise in just such manner. Which is to say, the Torah's narrative turns the death of honor into the death of dishonor.[33]

Time and again we have seen how the Exodus narrative evokes Egyptian tropes and turns them on their head. The imagery of baseball may be helpful here, since in this sport the ground rules of the home team and its ballpark are in effect during the game. Since the Egyptians are the "home team" in the Exodus story, the biblical author plays by their rules, whereby inanimate objects may be transformed into crocodiles, the death of the first born is an important theme, waters can be divided in order to restore joy to the royal family, death by drowning is honorific, and more. The biblical author subverts all of these notions as he leads his readers through the sustained narrative. To repeat what I stated at the end of the previous section: in order for this technique to be meaningful, one must assume a considerable knowledge amongst the Israelites of ancient Egyptian beliefs and practices, perhaps even specific literary motifs.

§9. In the above sections, we have seen repeatedly that Moses (or at times Aaron) performs the same actions as those achieved by magicians, lector-priests, and the like in

[32] For superb color images, go to: http://www.thebanmappingproject.com/database/image.asp?ID=14638, http://www.thebanmappingproject.com/database/image.asp?ID=14641, http://www.thebanmappingproject.com/database/image.asp?ID=14642, especially the middle one.

[33] Given the centrality of the drowning motif in New Kingdom texts of the afterlife, I would argue that the presence of this trope in Exodus 14–15 is essential to the biblical tradition, and does not constitute a later development. For discussion, see Loewenstamm (1972: 101–120) (with English summary on pp. viii–ix).

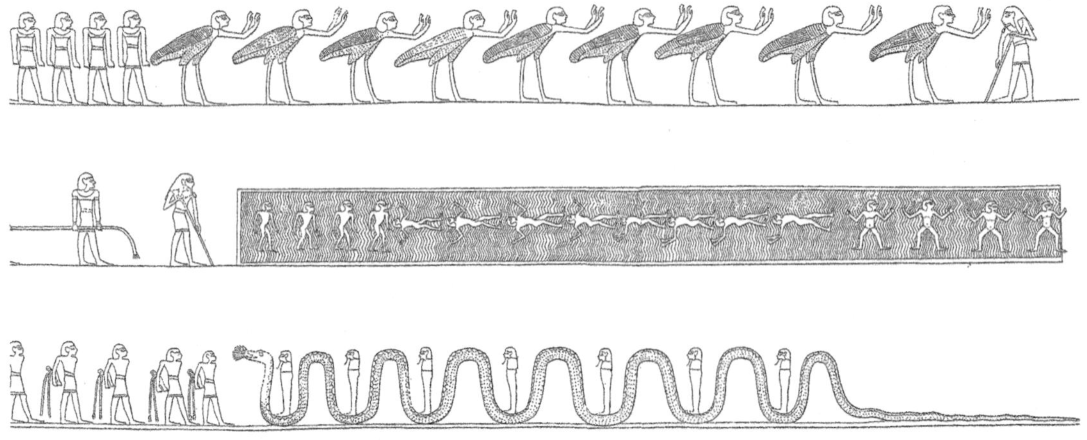

Fig. 18.6 KV-9, tomb of Rameses VI, Left Wall of Corridor, Book of Gates, 9th Gate, (Piankoff and Rambova 1954: fig. 54 [double foldout facing p. 190], with text on pp. 193–194). Reprinted by permission of Princeton University Press. This image is also available online at http://www.egiptologia.org/textos/puertas/08/; see also Hornung (1999: 74), fig. 38

ancient Egyptian texts. In the biblical presentation of the narrative, however, one detects an important distinction. When the Egyptian magicians execute their magic, they do so via the recitation of magical spells, as indicated in the following verses:

Exodus 7:11
וַיִּקְרָא גַּם־פַּרְעֹה לַחֲכָמִים וְלַֽמְכַשְּׁפִים וַיַּעֲשׂוּ גַם־הֵם חַרְטֻמֵּי מִצְרַיִם בְּלַהֲטֵיהֶם כֵּן׃
And Pharaoh called the wise-men and the sorcerers, and they also did, the magician-priests of Egypt, by their spells likewise.

Exodus 7:22 וַיַּעֲשׂוּ־כֵן חַרְטֻמֵּי מִצְרַיִם בְּלָטֵיהֶם
And the magician-priests of Egypt did likewise by their spells.

Exodus 8:3 וַיַּעֲשׂוּ־כֵן הַחַרְטֻמִּים בְּלָטֵיהֶם
And the magician-priests did likewise by their spells.

Exodus 8:14 וַיַּעֲשׂוּ־כֵן הַחַרְטֻמִּים בְּלָטֵיהֶם
And the magician-priests did likewise by their spells.

This presentation of the Egyptian *ḥartumim* accords with the narratives we have examined herein, in which Webaoner "read out his magic words," Djadja-em-ankh "said his say of magic," Na-nefer-ka-ptaḥ "recited a spell," and so on (quoting the translations utilized above). Magical praxes in ancient Egypt were almost always accomplished through the recitation of magical spells (Ritner 1993: 35–49; see Noegel 1996: *passim*)—a point clearly recognized by the biblical author. By contrast, when Moses (and/or Aaron) engage in such acts, the biblical text is mindful never to ascribe the results to the magical arts. The leaders of the Israelites are able to accomplish such tasks because God empowered them to do so, pure and simple. In the words of Nahum Sarna (1986: 59), "Moses knows no techniques, recites no spells, utters no incantations or magical formulae."[34] Let us recall here the famous passage in Num 23:23, uttered by Balaam (mentioned above [§1] in passing): כִּי לֹא־נַחַשׁ בְּיַעֲקֹב וְלֹא־קֶסֶם בְּיִשְׂרָאֵל "for there is no magic in Jacob, and no sorcery in Israel"—a point which holds throughout the Bible, including, as we have seen, the book of Exodus. In short, while the ends are the same, the means are profoundly different.

One additional point is worthy of mention here, as it once more speaks to the manner in which the Exodus narrative is thoroughly anchored in its Egyptian milieu. I refer here to the manner in which these actions

[34] See also Sarna (1991: 37).

Fig. 18.7 KV-9, tomb of Rameses VI, Left Wall of Corridor, Book of Gates, 9th Gate; note the drowned ones in the upper register; photo © Gary A. Rendsburg

(whether induced by magic or via divine empowerment) are considered perfectly natural and acceptable—one might even say, expected—within the general story line. Which is to say, of course these things can happen. As Ritner (1993: 8–9) observed, "No suggestion of trickery is ever implied in Egyptian terms for magic. Even where theatrical feats are described in literature, there is no indication that writer or audience disbelieved the *possibility* of such feats."[35] The biblical text enters the mindset of ancient Egypt so thoroughly that this is equally true of the book of Exodus—on both sides, whether the *ḥartumim* or whether Moses and Aaron are the initiators of these feats.

§10. Finally, we turn to the issue of the wide chronological range of the Egyptian parallels evoked in this article. As we have seen, the texts derive from the full range of Egyptian chronological periods: Old Kingdom (Pyramid Texts), First or Second Intermediate Period (Ipuwer), Middle Kingdom (Pap. Westcar, Neferti, Coffin Texts), New Kingdom (Pap. Turin 1993, Amduat, Book of Gates), Persian period (Herodotus), Ptolemaic rule (Setne I), and Roman period (Setne II). The warnings of parallelomania are clearly in mind here.

There are several controls, though. First we should keep in mind the tenacity of Egyptian religion and tradition, with beliefs and customs present already during the Old Kingdom still reverberating in the late period. One need only peruse Alan Lloyd's (1976) commentary on Herodotus, *The Histories*, Book Two, to realize the number

[35] To which Ritner added a footnote, "Compare the miraculous events narrated in Pap. Westcar, Setna I, and Setna II" (p. 9, n. 24), all of which have been discussed above. Ritner further commented that this Egyptian attitude to magic stands in contrast to the Greco-Roman world, in which magic was judged with skepticism and distrust (p. 9).

of points raised by the fifth-century Greek historian with antecedents stretching back one or two millennia. To provide just one illustration from a passage cited earlier, note that the depilation practiced by the Egyptian priests (see above, §4) is depicted throughout the earlier epochs as well, even if it did not become standard until the 19th Dynasty apparently (te Velde 1995: 1733; Green 2001: 73; Filer 2001: 135). Or to use an example not forthcoming from Herodotus, note that the motif of darkness occurs both in the "Prophecy of Neferti," composed during the Middle Kingdom, with textual witnesses from the New Kingdom, and in "Setne Khamwas and Si-Osire" (=Setne II), dated to the first century C.E.—a span of approximately two millennia.

Accordingly, regardless of how we envision the production of the book of Exodus (multiple sources, single unified text, etc.), and no matter to when we date these sources and/or the final product,[36] to my mind, and as I hope to have demonstrated by now, an educated Israelite writer and his well-informed Israelite audience would have been familiar with the Egyptian cultural context which motivated a good portion of the dramatic narrative of Exodus 1–15.

The second control is that almost without exception the parallels to the Exodus narrative are known only from Egypt, and not from other sources, such as Canaan (especially Ugarit) and Mesopotamia. Motifs such as the hidden name of the deity, turning an inanimate object into a snake or a crocodile, the casting of darkness, the death of the firstborn, the splitting of the waters, and the drowning theme find a home in Egyptian culture only, without an echo (to the best of my knowledge) in other ancient Near Eastern societies. The one exception is the first plague, turning the water into blood, which occurs in two Sumerian compositions, both involving Inanna, "Exaltation of Inanna" and "Inanna and the Gardener" (for references, see Propp 1998: 349). This concession, however, should not serve to overturn the larger picture presented here.

In sum, the narrative that encompasses Exodus 1–15 evokes the Egyptian setting at every turn. I, for one, like to imagine an ancient Israelite audience enjoying the recitation, with complete understanding of the nature of the composition, which both subverts Egyptian religious notions and simultaneously expresses Israel's national heritage in exquisite literary fashion.

[36] As intimated above on several occasions, I prefer a holistic approach to the narrative. As to its date, while a full treatment is not possible here, I would place the composition at the time of the early monarchy. For general background, albeit with a focus on the book of Genesis, see Rendsburg (2005).

References

Allen, James P. 2005. *The Ancient Egyptian Pyramid Texts* (Writings of the Ancient World), vol. 23. Atlanta, GA: Society of Biblical Literature.

Blackman, A.M., and W.V. Davies. 1988. *The Story of King Kheops and the Magicians: Transcribed from Papyrus Westcar (Berlin Papyrus 3033)*. Reading, MA: J.V. Books.

de Buck, Adriaan. 1935–1961. *The Egyptian Coffin Texts*. 7 vols. Oriental Institute Publications 34, 49, 64, 67, 73, 81, 87. Chicago, IL: Oriental Institute. http://oi.uchicago.edu/research/pubs/catalog/oip/

Cassuto, Umberto. 1967. *A Commentary on the Book of Exodus*. (trans. Israel Abrahams). Jerusalem: Magnes.

Cohen, Chaim. 1991. The Other Meaning of Biblical *tannin*: Snake or crocodile? (in Hebrew). In *Sefer Prof. H. M. Y. Gevaryahu: Meḥqarim ba-Miqra' u-ve-Maḥševet Yiśra'el*, 2 vols., ed. Ben-Zion Luria, 75–81. Jerusalem: Ha-Ḥevra le-Ḥeqer ha-Miqra' be-Yiśra'el.

COS = Hallo, William W., and Younger, K. Lawson. 1997–2002. *The Context of Scripture*. 3 vols. Leiden: Brill.

Currid, John D. 1997. *Ancient Egypt and the Old Testament*. Grand Rapids, MI: Baker.

Enmarch, Roland. 2005. *The Dialogue of Ipuwer and the Lord of All*. Oxford: Griffith Institute.

———. 2008. *A World Upturned: Commentary on and Analysis of The Dialogue of Ipuwer and the Lord of All*. Oxford: British Academy / Oxford University Press.

Erman, Adolf, and Grapow, Hermann. 1926–1931. *Wörterbuch der ägyptischen Sprache*, 5 vols. Berlin:

Akademie-Verlag. http://www.egyptology.ru/lang.htm#Woerterbuch

Eyre, Christopher. 2002. *The Cannibal Hymn: A Cultural and Literary Study*. Liverpool: Liverpool University Press.

Filer, Joyce M. 2001. Hygiene. In *The Oxford Encyclopedia of Ancient Egypt*, vol. 2, ed. Donald B. Redford, 133–136. New York, NY: Oxford University Press.

Faulkner, Raymond O. 1969. *The Ancient Egyptian Pyramid Texts*. Oxford: Clarendon.

———. 1973–1978. *The Ancient Egyptian Coffin Texts*, 3 vols. Oxford: Clarendon. Reprinted in one volume, retaining original pagination. Oxford: Aris & Phillips, 2004.

Friedman, Richard E. 2003. *The Bible with Sources Revealed*. San Francisco, CA: HarperCollins.

Gilula, Mordechai. 1977. The Smiting of the First-Born: An Egyptian Myth? *Tel-Aviv* 4: 94–95.

Godley, A.D. 1921–1924. *Herodotus: The Persian Wars*, 4 vols. LCL, vols. 117–120. London//New York: W. Heinemann/G.P. Putnam's Sons.

Goedicke, Hans. 1977. *The protocol of Neferyt: The prophecy of Neferti*. Baltimore, MD: Johns Hopkins University Press.

Green, Lyn. 2001. Hairstyles. In *The Oxford Encyclopedia of Ancient Egypt*, vol. 2, ed. Donald B. Redford, 73–76. New York, NY: Oxford University Press.

Griffith, F.Ll. 1900. *Stories of the High Priests of Memphis: The Sethon of Herodotus and the Demotic Tales of Khamuas*. Oxford: Clarendon.

———. 1909. Herodotus II. 90. Apotheosis by drowning. *Zeitschrift für ägyptische Sprache und Altertumskunde* 46: 132–134.

Helck, Wolfgang. 1992. *Die Prophezeihung des Nfr.tj*, 2nd ed. Wiesbaden: Otto Harrassowitz.

Hornung, Erik. 1999. *The Ancient Egyptian Books of the Afterlife*. (trans. David Lorton). Ithaca, NY: Cornell University Press.

Keel, Othmar. 1997. Corpus der Stempelsiegel-Amulette aus Palästina/Israel, Von den Anfängen bis zur Perserzeit, Band I: *Von Tell Abu Farağ bis ʿAtlit*. Orbis Biblicus et Orientalis, Series Archaeologica 13. Freiburg/Göttingen: Universitätsverlag Freiburg/Vandenhoeck & Ruprecht.

———. 2010. *Corpus der Stempelsiegel-Amulette aus Palästina/Israel, Von den Anfängen bis zur Perserzeit*, Band II: *Von Bahan bis Tel Eton*. Orbis Biblicus et Orientalis, Series Archaeologica 29. Freiburg/Göttingen: Universitätsverlag Freiburg/Vandenhoeck & Ruprecht.

Lepper, Verena M. 2008. *Untersuchungen zu pWestcar. Eine philologische und literaturwissenschaftliche (Neu-)Analyse, Band* 70. Wiesbaden: Otto Harrassowitz. Ägyptologische Abhandlungen.

Lichtheim, Miriam. 1973–1980. *Ancient Egyptian Literature*, 3 vols. Berkeley, CA: University of California Press.

Lloyd, Alan B. 1976–1988. *Herodotus, Book II*, 2 vols. Etudes préliminaires aux religions orientales dans l'Empire romain, t. 43. Leiden: E. J. Brill.

Loewenstamm, Samuel E. 1972. *The Tradition of the Exodus in Its Development* (in Hebrew). Jerusalem: Magnes.

Manassa, Colleen. 2007. *The Late Egyptian Underworld: Sarcophagi and Related Texts from the Nectanebid Period*. Wiesbaden: Otto Harrassowitz.

McDowell, A.G. 1999. *Village Life in Ancient Egypt: Laundry Lists and Love Songs*. Oxford: Oxford University Press.

Münger, Stefan. 2003. Egyptian Stamp-Seal Amulets and Their Implications for the Chronology of the Early Iron Age. *Tel-Aviv* 30: 66–82.

Noegel, Scott B. 1996. Moses and Magic: Notes on the Book of Exodus. *Journal of the Ancient Near Eastern Society* 24: 45–59.

Parkinson, R.B. 1997. *The Tale of Sinuhe and Other Ancient Egyptian Poems 1940-1640 BC*. Oxford: Oxford University Press.

Petrie, W. M. Flinders. 1906. *Hyksos and Israelite Cities*. British School of Archaeology in Egypt, 12. London: School of Archaeology in Egypt.

———. 1909. *Memphis I*. British School of Archaeology in Egypt, 14. London: School of Archaeology in Egypt.

Piankoff, Alexandre, and Natacha Rambova. 1954. *The Tomb of Ramesses VI*. Bollingen series. New York, NY: Pantheon Books.

Pleyte, Willem, and Rossi, Francesco. 1869–1876. *Papyrus de Turin*, 2 vols. Museo egizio di Turino. Leiden: E. J. Brill. Reprint: Wiesbaden: LTR-Verlag, 1981 (in a single volume).

Propp, William H.C. 1998. *Exodus 1-18*. Anchor Bible 2. New York, NY: Doubleday.

Quibell, J.E. 1898. *The Ramesseum*. Publications of the Egyptian Research Account, vol. 2. London: B. Quaritch.

Quirke, Stephen. 2004. *Egyptian Literature 1800 BC: Questions and Readings*. London: Golden House.

Rendsburg, Gary A. 1988. The Egyptian Sun-God Ra in the Pentateuch. *Henoch* 10: 3–15.

———. 1990. Targum Onqelos to Exod 10:5, 10:15, Numb 22:5, 22:11. *Henoch* 12: 15–17.

———. 2003. Beasts or Bugs? Solving the Problem of the Fourth Plague. *Bible Review* 19/2: 18–23.

———. 2005. The Genesis of the Bible. In *The Blanche and Irving Laurie Chair in Jewish History*, 11–30. Separatum published by the Allen and Joan Bildner Center for the Study of Jewish Life, Rutgers, The State University of New Jersey. http://jewishstudies.rutgers.edu/component/docman/doc_view/117-the-genesis-of-the-bible?Itemid=158

———. 2006. Moses as Equal to Pharaoh. In *Text, Artifact, and Image: Revealing Ancient Israelite Religion*, ed. Gary M. Beckman and Theodore J. Lewis, 201–219. *Brown Judaic Studies* 346. Providence, RI: Brown Judaic Studies.

Ritner, Robert K. 1993. *The Mechanics of Ancient Egyptian Magical Practice*. Chicago, IL: Oriental Institute of the University of Chicago. *Studies in Ancient Oriental Civilization*, no. 54.

Rowe, Alan. 1940. Newly-identified Monuments in the Egyptian Museum Showing the Deification of the

Dead. *Annales du Service de Antiquities de l'Egypte* 40: 1–50.

Sarna, Nahum M. 1986. *Exploring Exodus*. New York, NY: Schocken.

———. 1991. *The JPS Torah Commentary: Exodus*. Philadelphia, PA: Jewish Publication Society.

Sethe, Kurt H. 1908–1922. *Die altägyptischen Pyramidentexte*. 4 vols. Leipzig: J. C. Hinrichs. Reprint: Hildesheim: Georg Olms, 1969. http://www.lib.uchicago.edu/cgi-bin/eos/eos_title.pl?callnum=PJ1553.A1_1908_cop3

Simpson, William Kelly (ed.). 2003. *The Literature of Ancient Egypt*. New Haven, CT: Yale University Press.

te Velde, Herman. 1995. Theology, Priests, and Worship in Ancient Egypt. In *Civlizations of the Ancient Near East*, vol. 3, ed. Jack M. Sasson, 1731–1749. New York, NY: Charles Scribner's Sons.

Wilson, John A. 1969. Egyptian Myths, Tales, and Mortuary Texts. In *Ancient Near Eastern Texts Relating to the Old Testament*, 3rd ed, ed. James B. Pritchard, 3–36. Princeton, NJ: Princeton University Press.

Yahuda, A.S. 1933. *The Language of the Pentateuch in its Relation to Egyptian*. London: Oxford University Press.

Zevit, Ziony. 1990. Three Ways to Look at the Ten Plagues. *Bible Review* 6/3: 16–23. 42.

Egyptian Texts relating to the Exodus: Discussions of Exodus Parallels in the Egyptology Literature

19

Brad C. Sparks

Abstract

Some 30 ancient Egyptian texts with Exodus "parallels" or Exodus-like content have been identified by 56 Egyptologists, archaeologists, and Semiticists from 1844 to date in the professional literature. Additional texts are identified in the present study for a total of more than 90 Egyptian texts containing Exodus parallels. The Egyptian texts are mainly of politico-literary and religious genres, most ranging in date from the New Kingdom to the Greco-Roman era and appearing in hieroglyphic-hieratic-Demotic forms (a few in Greek and Coptic and one trilingual in hieroglyphic-Demotic-Greek, the Rosetta Stone). The Egyptological publications and Egyptian texts were recently found in the course of research on the history of the long-standing Exodus problem, in effect answering the call of Julius Wellhausen for a much-needed investigation of Egyptian traditions underlying the Exodus.

Based on the new work presented at the UCSD Exodus conference (see this volume) Jan Assmann has ventured beyond his pioneering concept of cultural "mnemohistories" to comment that consensus views of the Exodus are "now highly contested" because there has been "Perhaps too much unanimity as to the non-historicity of the Exodus"; the "old certainties" of Exodus as pure myth are "gone." Thomas Schneider (this volume) finds a potentially significant parallel to the Exodus 12 Plague on the Firstborn in several late Egyptian texts. Gary Rendsburg (this volume) finds several Egyptian texts with parallels to the Exodus including the drowned men

Early versions of this chapter were read at the Egyptology session of American Schools of Oriental Research (ASOR) annual meeting, on November 17, 2007, and as updated for the UCSD Exodus Conference, June 1, 2013. My special thanks to Thomas Schneider, John Bloom, Scott Noegel, and David Frankfurter for helpful editorial comments, discussions, and critiques of various drafts of this chapter as well as to J. Assmann, S. Hollis, W. Dever, L. Geraty, B. Brandstater, B. Waltke, E. Merrill, R. Younker, P. Machinist, W. Hallo, W. Schniedewind, H. Goedicke, and W. Wendrich, and for very early drafts my thanks to the late David Noel Freedman, A. Maeir, the late Anson Rainey, E. Hornung, A. Loprieno, J. Hoffmeier, J. Currid, and the late David Lorton, but the present author is, of course, responsible for the contents and any errors herein.

B.C. Sparks (✉)
Managing Editor (this volume)
Principal Organizer, UCSD Exodus Conference,
Archaeological Research Group
Los Angeles-San Diego, CA, USA
e-mail: exodus@eng.ucsd.edu

T.E. Levy et al. (eds.), *Israel's Exodus in Transdisciplinary Perspective*,
Quantitative Methods in the Humanities and Social Sciences, DOI 10.1007/978-3-319-04768-3_19,

(soldiers) in the Amduat and Book of Gates paintings in royal tombs ca. 1500/1400 BC. The present author independently has found similarly dramatic Egyptian imagery of the Walls of Water in a Parted Sea (the Flaming Red Sea, Eg. *Yam Nesret*) in the same wall-painted Amduat and Gates hieroglyphic books.

Introduction

"Striking" and "astounding" are common words used in the Egyptology literature citing Egyptian text parallels to the Exodus:

- Erik Hornung (Hebrew Divine Name of Exod. 3:14 "Parallele" in an Egyptian text); John Gwyn Griffiths noted it is a "startling" parallel (review in *Journal of Egyptian Archaeology*).[1]

  Hornung and Gerhard Fecht noted the importance of the "parallel drawn to the Old Testament passage Exodus 3:14" (Moses at the burning bush) in the Destruction of Mankind text of royal pharaonic tombs ca. 1300 BC (also known as the Book of the Celestial Cow):

  > . . . *jw.j-jm.j* evidently means "I am I" or "I am that I am." Since in the given context it must mean: ". . . as whom I have proven to be" . . ., the phrase indeed recalls something Old Testament-like: see Exodus 3, 14 "I am that I am". . ..
  >
  > What is here of interest is of course the early theology surrounding God's name YHWH, but not its origin and actual etymology [i.e., whether ultimately Egyptian or Hebrew].[2] (Hornung 1997: 125 n.aa, citation omitted; cf. 63 n.121; transl.)

- Jan Assmann ("astounding parallels" to the Golden Calf incident of the Exodus in an Egyptian text, plus other Exodus parallels in other texts).[3]
- Donald Redford ("The striking resemblance between this catastrophic storm" in the Rhind Mathematical Papyrus, "and some of the traditional 'plagues' seems more than fortuitous").[4]
- Hans Goedicke (nonreligious Egyptian pharaonic narrative's similarity to the Exodus "so striking that it is difficult to overlook it").[5]
- Roland Enmarch ("striking parallel" between the Ipuwer Admonitions, and a later text, and "the description of the Nile flowing with blood in the Old Testament").[6]
- Susan Hollis ("striking" "close parallel" to the Joseph seduction story in the ca. 1200 BC Egyptian papyrus text of Tale of Two Brothers).[7]
- James Allen (Biblical "foreshadowing" of Story of Sinuhe "resembles" Moses' story).[8]
- Nili Shupak ("quite extraordinary" and "astonishing" Egyptian parallels to the Joseph story in the Sehel 7-Year Famine Stela).[9]

These parallels to the Exodus are not said by the scholars to be coincidental, but those in

[1] Hornung (1982/1997: 63 n.121), Griffiths (1988: 276).

[2] In a 1991 addendum printed without change in the 1997 edition, Hornung defends his and Fecht's "ich bin, der ich bin" translation but offers an intriguing but speculative redactional prehistory of the text, which however has no textual-variant support or other manuscript evidence. Hornung suggests the possibility ("möglich") that the original text read *jw.j-jm* and thus could be read as "I am there (in the sky [implied])," and later the particle *.j* was added, rendering the final form "I am (that) I am." Hornung had noticed a text (Book of the Dead, Spell 1) where in manuscript variants the addition of *.j* converted the *m* of locality ("there") into an *m* of identity ("I am"). But this hypothetical redaction sequence in the Destruction does not avoid the Exodus, since the proposed textual alteration resulting in the only actual known manuscript would likely have been motivated in the first place by wordplay to imitate the Hebrew meaning of "I am that I am," thus subversively implying the divinity of Pharaoh as one and the same as the Hebrew God named the "I am that I am." Moreover, the implication of "I am there" in the suggested *ur*-text as the sun god being "there in the sky" can also be interpreted as implying "I am there in the (celestial) ocean" (=primordial sea where Ra was born)—or once again, an allusion to the terminal pharaonic death scenario in the Combat at the (Red) Sea or Yam Suph, Sea of Annihilation (Batto, this volume; Batto 1984, 1992, 2013a, 2013b), and thus once again the Exodus.

[3] Assmann (2002: 406).

[4] Redford (1992: 420).

[5] Goedicke (2004: 102).

[6] Enmarch (2009: 28).

[7] Hollis (2008: 12).

[8] Allen (2010: 285).

[9] Shupak (2006: 126, 138).

which the scholar asserts or implies that the Exodus tradition had a literary or a historical dependency on Egypt and cites specific Egyptian texts.

The focus of this work is on genuine parallels involving mythological narratives and historical traditions of the Exodus and their composition, transmission, and borrowing, rather than accidental or coincidental ancient assemblages of Biblical-like themes that only appear Biblical when viewed by scholars through their modern interpretive framework. Diagnostic criteria for identifying Egyptian Exodus parallels are discussed.

This chapter collects in one place the known publications of Egyptian Exodus parallels up to 2014 and highlights some of the more recent work, calling attention to their potential to guide future research. Space limitations here preclude in-depth treatment of each of more than 90 Egyptian text parallels (see Sparks, forthcoming, for more detailed consideration).

By investigating the genre categories[10] and potentially relevant iconography for Egyptian works of "similar ... content" (Hornung 1999: 27) and by following other clues left in the Egyptology literature, nearly 60 additional Egyptian texts have been added to the previous 30 or so in the literature. Other conference presenters have added several more texts for a total of more than 90 Egyptian documents containing Exodus parallels at present. Thus there are much greater numbers of Egyptian parallels to the Exodus than have been previously suspected. In fact, it is usually asserted that no such parallels exist.

The "extraordinary" Egyptian parallels to the Exodus, noted below, are generally reported in isolated fashion during the course of work whose focus is on other subjects and have never been brought together in one place and studied as a whole until now.

As ASOR's Chester McCown noted long ago, such Egyptian Exodus parallels have largely "escaped notice" and "not been collected in any one place" (McCown 1925: 359, 362–363).

This was despite the urging of the father of modern Biblical criticism, Julius Wellhausen, who in 1885 called for a much-needed investigation of any Egyptian textual "traditions" behind the Exodus (Wellhausen 1885: 440). Wellhausen noted that the Egyptian priest-historian Manetho (ca. 260 BC) had referred to a number of stories about the Exodus that the latter identified with the Exodus (Hyksos expulsion, leper expulsion) mostly "malicious inventions," but whether "any genuine tradition underlies them at all is a point much needing to be investigated" (ibid.). Unfortunately no such investigation of Egyptian textual traditions underlying the Exodus has ever been carried out until now; only isolated examples of Egyptian parallels to the Exodus have been pointed out plus endless reworkings of the confusing and enigmatic references to the Hyksos and leper expulsion stories.

Certain genres of Egyptian literature are particularly fruitful for searches for Exodus parallels and demonstrate the potential for current scholarship to be enriched and enhanced with our expanding knowledge. In his book, *Moses the Egyptian,* Jan Assmann alludes to the Exodus connections to late texts in the genre of Egyptian apocalypse and national distress (cf. Rothöhler 2003: 392/ fn.9; Enmarch 2009: 28):

> Both oracles [of the Potter and the Lamb] seem to belong to the same kind of nationalistic [Egyptian] tradition which spread under the experience of foreign domination and which formed the context of the Egyptian 'Exodus' stories. (Assmann 1997: 225 n.29, cf. 34)

In his recent critical edition of the famed Admonitions of Ipuwer, British Egyptologist Roland Enmarch called attention to the "striking parallel" between Egyptian literature and the Plague of Blood in the Exodus. Enmarch cited the "parallels" given in the work of Benedikt Rothöhler, though skeptical of the latter's "historicist" views of Egyptian apocalyptic texts.

Rothöhler commented on a late Egyptian Coptic text referring to the Blood Nile (Enmarch 2009: 39, cp. 28, citing Exodus 8:17 [7:17] and Rothöhler 2003: 392). Rothöhler describes this Blood Nile as evidence of "national" or "physical" catastrophe in Egypt, which he says should not be dismissed as "meant purely symbolically"

[10] No technical definition of genre is used here in view of Di Biase-Dyson's comments (2013: 50): "Defining genre has long been a problem in Egyptian literary studies, as in other scholarly fields."

but may have a "real descriptive narrative level" tying disastrous later events in Egypt to allusions to what occurred in "the past" (cp. Rothöhler 2003: 391.3, 392, transl.). Hence, Rothöhler points us to the ancient Egyptian literature of pessimism and lamentation, *Chaosbeschreibungen* ("chaos descriptions": Assmann 1983: 349, 360; "national disaster" Williams 1992: 2:397b). (See discussion of David Frankfurter's work in the "Methodology" section, below.)

Much of this work has been published in the past two decades (see Table 19.2, below, for a comprehensive catalog). Past Egyptologists pointing to such Exodus parallels include John Wilson (Director, Oriental Institute, Chicago), Wolfgang Helck, Georges Posener, and Jozef Vergote.

Table 19.1 Major Egyptian texts presenting Exodus parallels

Destruction of Mankind
Admonitions of Ipuwer
Hatshepsut's Speos Artimedos text
Tale of Two Brothers
Story of Sinuhe
Sehel 7-Year Famine Stela
Rosetta Stone

Historical Survey

Since 1844 tentative parallels to the Exodus have been found in dozens of Egyptian texts and published in the literature by professional Egyptologists, archaeologists, and Biblical scholars, some of whom were presenters at the UCSD Exodus Conference in May–June 2013 and are contributors to this volume. The Egyptian texts include the Destruction of Mankind, Admonitions of Ipuwer, the Tale of Two Brothers, the Sehel 7-Year Famine Stela, and even the Rosetta Stone, plus many others. The texts and publications are briefly reviewed here—due to space limitations—for light shed on the intellectual history and development of the Exodus problem and how they may assist further scholarly investigation. Some of the Egyptian texts will be presented followed by a discussion of methodology and the diagnostic criteria used in identifying the Egyptian Exodus parallels (Table 19.3).

In the early days of Egyptology and modern Biblical studies, Exodus parallels were noted in such newly translated texts as the famed Rosetta Stone (Browne 1844: 607–610, 613) and in classics of ancient Egyptian literature such as the Tale of Two Brothers (pap. d'Orbiney) first translated in 1852, the Story of Sinuhe first translated in 1865, and the Admonitions of Ipuwer (Exodus parallels first noted by Alfred Jeremias 1913: 220 n.5). The first translation of the Destruction of Mankind (or Book of the Celestial Cow) by Swiss Egyptologist Edouard Naville in 1874 (Naville 1875) led to further studies and discoveries of related literature and late Egyptian synopses that described what became known among specialists as the primordial rebellion of mankind.

The story of how Joseph married into the priesthood of Heliopolis, followed by the migration of the rest of Joseph's family, the children of Israel, set the stage for the ultimate Exodus departure. The Joseph seduction story is told in the famed Egyptian "Tale of Two Brothers," which is "one of the best known and most discussed Egyptian literary texts, and attested in the ca. 1200 BC Papyrus d'Orbiney in the British Museum (Schneider 2008: 315). As the late UCLA Egyptologist Miriam Lichtheim wrote: "The episode of Bata and his brother's wife has a remarkable similarity with the tale of Joseph and Potiphar's wife, a similarity that has often been commented on" (Lichtheim 2006: 2:203). John Romer directly connects the Egyptian and Hebrew stories: "... the biblical story of Joseph and Potiphar's wife was also a well-known Egyptian tale that first appears in the ancient literature about 1200 BC" (Romer

Table 19.2 Published tentative Egyptian parallels to the Exodus in Egyptological literature since 1980

Egyptologists, Semiticists, Archaeologists publishing Exodus parallels[a]	Year of first publication[b]	Egyptian texts with Exodus parallels or Exodus data cited [insufficiently Exodus-related in brackets]	Scholar's position on the Exodus or the texts: myth and/or event	Exodus parallels cited
1. Hans Goedicke	1981	El Arish Stela, Speos Artemidos, Hearst Medical Papyrus, London Medical Papyrus, Apophis and Sekenen-re, Rhind Math Papyrus	Event + Myth	"Canaanite Illness" Plague, Ten Plagues Walls of Water (Thera tsunami) Drowning of Egyptian Army
2. Erik Hornung	1982	Destruction of Mankind, Setna II (Ipuwer, Meri-ka-re, Book of the Dead Spell 175, Book of Gates, El Arish Stela, Esna, Kom Ombo, Edfu, Book of Fayum, Insinger Papyrus)	Myth + Event?	Hebrew Divine Name of Exodus 3:14, Moses' Contest with Pharaoh's Magicians (Primeval Revolt)
3. Gerhard Fecht	1982	Destruction of Mankind	Myth + Event?	Hebrew Divine Name of Exodus 3:14
4. Susan Tower Hollis	1982	Tale of Two Brothers	Myth + Event	Joseph Seduction
5. Rudolph Cohen	1983	Ipuwer	Event	Ten Plagues
6. Nahum Sarna	1986	Ipuwer, Setna II	Event + Myth	Ten Plagues Plague of Blood
7. Emmanuel Anati	1986	Ipuwer, Neferti, Meri-ka-re, Story of Sinuhe	Event + Myth	Ten Plagues Moses' Flight from Egypt
8. John Romer	1988	Tale of Two Brothers	Myth + Event	Joseph Seduction
9. William W. Hallo	1990	Sehel 7-Year Famine Stela	Event + Myth?	Joseph's 7-Year Famine
10. Leonard Lesko	1991	Meri-ka-re	Myth + Event?	Parting of Waters?
11. Donald Redford	1992 (1970)	Rhind Mathematical Papyrus [Tempest Stela] (El Arish Stela) (Tale of Two Brothers)	Myth + Event	Ten Plagues "tremendous catastrophe" (Pi-[ha]-Khiroth at[Red] Sea) (Joseph Seduction)
12. Manfred Görg	1992	Metternich Stela	Event + Myth	Moses' Mother Jochebed
13. George Coats	1992	Tale of Two Brothers, Story of Sinuhe	Myth + Event	Joseph story incl. Seduction
14. David Frankfurter	1993	Ipuwer, Asclepius Apocalypse, Elijah Apocalypse (Neferti)	Myth + Event?	Plague of Blood (Darkness, Mass Death, Disintegration of Political + Social Order)
15. Jan Assmann	1996	Demotic Chronicle, Papyrus CPJ 520, Oracle of Potter [Oracle of Lamb]	Myth + Event	Golden Calf Plague of Blood, "Egyptian 'Exodus'" National Distress
16. Scott Noegel	1996	Ipuwer, Setna II	Myth + Event?	Plague of Blood
17. William Propp	1999	Setna II (Neferti?, Ipuwer?, Tempest Stela?)	Myth + Event	3-Day Plague of Darkness (Plague of Blood?)

(continued)

Table 19.2 (continued)

Egyptologists, Semiticists, Archaeologists publishing Exodus parallels[a]	Year of first publication[b]	Egyptian texts with Exodus parallels or Exodus data cited [insufficiently Exodus-related in brackets]	Scholar's position on the Exodus or the texts: myth and/ or event	Exodus parallels cited
18. David Alon	1999	Ipuwer, Sehel 7-Year Famine Stela	Event	Ten Plagues Joseph's 7-Year Famine
19. Sarah Israelit-Groll	1999	Neferti [Papyrus Anastasi VIII]	Event + Myth?	Drought [Semites in E Delta]
20. Thomas Schneider	2000	Setna II, Pap. Cairo 58027	Myth + Event?	Moses' Contest with Pharaoh's Magicians, Passover-like Protection of Pharaoh
21. Nili Shupak	2003	Ipuwer, Sehel 7-Year Famine Stela	Event + Myth	Plague of Blood Joseph's 7-Year Famine
22. Thomas E. Levy	2003	Story of Sinuhe	Myth + Event	Moses' flight from Egypt
23. Roland Enmarch	2004	Ipuwer, Asclepius Apocalypse [Oracle of Lamb] (Oracle of Potter)	Myth + Event	Plague of Blood
24. Christiane Zivie-Coche	2004	Setna [I] + II [Westcar Papyrus]	Myth + Event	Moses' Contest with Pharaoh's Magicians [Parting of (Red) Sea]
25. Edmund Meltzer	2004	Story of Sinuhe	Myth + Event	Joseph story
26. Caterina Moro	2005	(Tale of Two Brothers), Early SIP Egyptian source (via Artapanus, et al.) (Oracle of Lamb)	Myth + Event	(Joseph), Asiatic Exodus from Egypt
27. Benjamin Scolnic	2005	Pap. Brooklyn 35.1446	Event + Myth	Joseph and Israelite slaves ruling houses
28. Gary Rendsburg	2006	Setna I & II, Ipuwer, Neferti, Amduat, Book of Gates, Coffin Text Spells 136 + 573	Event + Myth	Waters Split; Plague of Darkness, Plague of Blood, Drowned (Soldiers?), Asiatics, Death of Firstborn (?)
29. Pieter van der Horst	2006	PGM IV.3007-3086	Event + Myth?	Red Sea Crossing, Plagues, Pillar of Cloud
30. Victor Matthews	2006	Story of Sinuhe, Tale of Two Brothers	Myth + Event?	Moses' flight from Egypt, Joseph seduction
31. Don Benjamin	2006	Story of Sinuhe, Tale of Two Brothers	Myth + Event?	Moses' flight from Egypt, Joseph seduction
32. Bernhard Lang	2009	Sehel 7-Year Famine Stela	Myth + Event?	Joseph's 7-Year Famine
33. Richard Freund	2009	Ipuwer [Tempest Stela]	Event + Myth	Ten Plagues
34. James Allen	2010	Story of Sinuhe	Myth + Event?	Moses' exile from Egypt
35. Kerry Muhlestein	2011	Pap. Brooklyn 35.1446	Event?	Joseph and Asiatic slavery

(continued)

Table 19.2 (continued)

Egyptologists, Semiticists, Archaeologists publishing Exodus parallels[a]	Year of first publication[b]	Egyptian texts with Exodus parallels or Exodus data cited [insufficiently Exodus-related in brackets]	Scholar's position on the Exodus or the texts: myth and/ or event	Exodus parallels cited
36. Victor P. Hamilton	2012	Tale of Two Brothers	Event	Joseph seduction
37. Mary Leith	2012	Isis and Horus in Swamp [=Metternich Stela]	Myth + Event	Moses' Birth story

Where in certain cases it is difficult to determine the scholar's position on the significance of his/her specific Egyptian parallels and their effect on the literary dependence and/or possible historicity of the Exodus tradition, the scholar's overall position on the Exodus has been estimated instead

Assertedly Exodus-related documents are listed by the authors publishing the connection, but documents that fail to satisfy the present author's diagnostic criteria for Exodus connection (see main text) or are insufficiently documented by the listed authors are set off within square brackets [ ] along with the corresponding claimed Exodus thematic elements in brackets [ ] (last column of table). Documents and parallels set off in parentheses ( ) are those only indirectly indicated by the cited scholar as Exodus-related

The scholars in the fields of Egyptology (including papyrology), archaeology, and Semitics that are most directly relevant to the study and recovery of ancient Egyptian texts are selected for inclusion

In the Scholar's Position column, "Event" signifies the position that at least some aspects of the Exodus represent a historical event and some may be distorted in the course of transmission of oral tradition or may be mythological, in varying degrees depending on the scholar. "Myth + Event" means that the Exodus is deemed primarily mythological and only partially historical. "Event + Myth" means roughly the reverse, partially historical, partially mythological

[a]Some of these Egyptian Exodus parallels have been published by Egyptologists and Semiticists as far back as 1844. There are at least 19 such scholars pre-1980 which are listed in this footnote rather than in the table above, which focuses on recent work (see Browne 1844: 607–610, 613; Brugsch 1864; Lange 1874; Moldenke 1895: 33; Meyer 1906: 151; Jeremias 1913: 220 n.5, 1916: 360, 362; McCown 1925: 374, 382–386; Yahuda 1934: 84; Oldfather 1935: 206; Vergote 1959: 23, 35, 47, 66; cf. 1985: 289–290; Posener 1960: 90, 1971: 220–256; Wolf 1962: 400–401; Helck 1965: 48, on Pap. Jumilhac parallel to Moses' birth legend; Gray 1965: 82; Greenberg 1969: 40, 157, 198–200, 202, 204; Wilson ANET 1969: 441c, 683c; De Vries 1975: 953b; Lichtheim 1976: 2:209, 2006: 2:209; Purdy 1977: 122)

[b]Goedicke 1984, 1987, 1988: 173; 2004: 102–104; Hornung 1982/1997: 63n.121, 90–95, 130; 2002: 55; Fecht 1982/1997: 125 n.aa; Hollis 2008: 12; Cohen 1983: 24b–25a; Sarna 1986: 69–70, 1992: 2:698a; Anati 1986: 287, 291; 1997: 246–247; 2001: 158–159; Romer 1988: 52–53; Hallo 1990/2010: 681; Lesko 1991: 103; Redford 1992; cf. Redford 1970: 88, 93; 1992: 420, 2006: 10:08:39; Görg 1992: 13; Frankfurter 1993: 203–204, cf. 171, 184/fn.102; Assmann 1997: 225 n.29, 2002: 398–399, 406; Noegel 1996: 51; Propp 1999: 318–319, 351–352; Aling 1999: 15; Israelit-Groll 1999: 161, Schneider 2008; Schneider, this volume Chap. 43; Shupak COS 2003: 1:94; 2006: 126, 138; Levy 2003: 5; Enmarch 2004: 2:366–367, 379; 2009: 28, 39; Zivie-Coche 2004: 124–125; Meltzer 2004: 79–80; 2007; Moro 2005: 79–83, 2008: 299–307, 2010; Rendsburg 2006: 211/fn.33; this volume, Chap. 18; van der Horst 2006: 281; Matthews-Benjamin 2006: 65, 137; Lang 2009: 285–286; Freund 2009: 80–82; Allen 2010: 285; Muhlestein 2011: 206b–207a; Hamilton 2012: 42b; Leith 2012: 34:01–35:00

1988: 52). There may well be further Exodus parallels too.[11]

SUNY Egyptologist Susan Tower Hollis sums up decades of analysis of the Tale of

[11] Following the Joseph-like events in the papyrus narrative, a number of additional episodes are mentioned that are interesting and Exodus-like (Purdy 1977: 122 notices that "two stories are involved," only one like that of Joseph): The Egyptian-named figure Anubis is apparently the "acting king" (Schneider 2008: 321b), is armed, and chases the possibly Semitic figure Bata (or Semitic *Bet* representing a household servant: Schneider 2008: 321b–322a), intending to kill him. Divine intervention causes a "great body of water" to separate the two (Pap. d'Orb. 6:7–9; Lichtheim 1976/2006: 2:206). The body of water has "sides," using the same Egyptian word *ru-'i* for "sides" used elsewhere in the papyrus to refer to the palace doorway or gate wall (cp. d'Orb. 6:7, 9; 16:10; cf. Hollis 2008: 50). The Semitic figure then departs to the east to dwell in the Semitic Levant (d'Orb. 7:1–8:1; cf. *'š*-trees of Retenu or Syria-Palestine: Hollis 2008: 128–129) after phallic self-mutilation (extreme "circumcision" ?: Hollis 2008: 126; d'Orb. 7:10). The pharaoh puts drops of sacrificial blood "*beside* the two door posts" of the great palace gate (though blood is not directly *on* the gate), which grow into strong, protective persea trees (d'Orb. 16:10, emphasis added).

Two Brothers in her discussion of its Biblical parallels:

> While there is no question that Semitic peoples, including Israelites, lived in Egypt during the second millennium BCE and that occasionally famines occurred, some of them lengthy, the actual historicity of the episode can not be proved or disproved. Most significantly, there is no attestation of Joseph or Potiphar in Egyptian sources, although the name *P3-di-p3-R* is known from the late period .... Nevertheless, such an episode could have occurred, and the Israelites understood that it did, its features most likely being transmitted orally before being included within the Genesis narrative and written down....
>
> It is possible that the Egyptian narrative contains reflexes of an actual historical situation ... over who succeeded Merneptah in the last years of the nineteenth dynasty .... (Hollis 2008: 108, citations omitted)

The Story of Sinuhe (ca. 1800–1700 BC) seems to combine both Joseph-like and Moses-like plot elements. The "striking and unmistakable ... mirror–image relationship between major plot elements of Sinuhe and of the Biblical Joseph narrative" has been noticed (Meltzer 2007; cf. Meltzer 2004: 79–80; Coats 1992: 2: 979b). Egyptologist James Allen notes the Biblical "foreshadowing" in "the story of Sinuhe's long exile [that] resembles that of Moses in the story of the Exodus" (Allen 2010: 285). Archaeologist Thomas Levy of UC-San Diego suggests that Sinuhe was a recurring theme in ANE literature that also appeared in the Exodus (Levy 2003: 5). Egyptologist Richard Parkinson considers these "coincidental resemblances" (2010: 150).

There was a general understanding among scholars of the time, in the wake of the 1872 discovery of the Babylonian flood story in cuneiform tablets, that if all of humanity was intended in the divine punishment recounted in the Destruction of Mankind narrative then it seemed to be a major Egyptian parallel to the Bible, i.e., to the divine destruction of mankind in the flood story in the Bible (Gen. 6–9). In 1881 French Egyptologist Gaston Maspéro noted that the parallel was a "dry deluge" and did not seem to discount it despite the apparent discrepancy (Maspéro quoted in King 1918: 48). The Destruction of Mankind is commonly cited as a Biblical Flood parallel (Lenormant 1882: 451–452; King 1918: 48; Gordon and Rendsburg 1997: 48 fn.32; Leeming 2004: 112a), but such correlations have always labored under the fundamental difficulty that there is in fact in the text no flood, no ark, no warning to mankind, and no preservation of animal life. The traditional Egyptian punishment for sin was by fire not water (Hornung 1997: 94). Later, parallels to the Exodus were found in the Destruction of Mankind (more below).

By the turn of the century Egyptology shook off the vestiges of Biblical concerns, thus accounting for a drop in interest in Biblical parallels (Hollis 2008: 12/fn.9, 17; Trigger 1995: 29–30). A public clash between Swiss Egyptologist Edouard Naville and Britain's Alan Gardiner in 1922–1924 over arguments about the Route of the Exodus evidently had a chilling effect on subsequent interest in pursuit of Biblical parallels in Egyptology (Hoffmeier, this volume, Chap. 15). Interest has been increasing in recent decades nonetheless (see Table 19.2, above, for Egyptian Exodus parallels published since 1980).

Egyptian "Primeval Revolt" as the Exodus: An Integrated Narrative Not a Coincidental Assemblage of Exodus-Like Motifs

British Biblical scholar Henry Browne first found an allusion to what later became known as the primeval rebellion of mankind in the Rosetta Stone, and he identified it with the events of the Biblical revolt of the Israelites in the Exodus (Browne 1844: 607–610, 613). The Rosetta Stone was inscribed by Ptolemy V Epiphanes ca. 196 BC as he was grappling with internal rebellion and foreign threat. Ptolemy called upon the memory of primeval revolt in the monumental text of the Rosetta Stone as an event he would not let happen again.

Within the primary text of the Destruction of Mankind,[12] Egyptologists eventually identified a singular event they term the "primeval revolt" or the "rebellion of mankind" in the Heliopolis/Eastern Nile Delta area of Lower Egypt (see Sauneron 1962: 5:298, 322–327, 339; Yoyotte 1972, 2013: 346–352, transl.), which is a theme then identified in a dozen or more Egyptian texts by in-depth analysis. It is an early religious or mythological event in ancient Egyptian literature,[13] which resembles the Exodus. This singular event has Exodus-like parallels that are functionally integral to the narrative, rather than a text merely aggregating disparate Exodus-like motifs. The central theme of the revolt is the challenge to sun god Ra as Pharaoh of Egypt posed by evidently non-Egyptian people in northern Egypt, with their escape and ensuing armed pursuit. The rebels become "refugees" (Yoyotte 1972: 164, 2013: 348, transl.) who merely wish to leave Egypt, not overthrow the pharaoh or the government.

The "primeval revolt" proceeds through a series of Exodus-like events that parallel the sequence of events in the Book of Exodus, in the same general order presented in the Biblical text, thus making it difficult to dismiss as an accidental assemblage of unrelated, merely illusory Exodus-like motifs. The general course of these texts in composite is as follows: the Blood Plague, a skin plague that nearly kills pharaoh, an abnormal darkness that traps the army with the pharaoh in the royal palace, armed pursuit of escaping foreign population of the Heliopolis area (Eastern Delta) headed east to return to their enemy god Apophis in the mountains of sunrise (Sinai), army failure to slaughter the escapees, and the implied death of the firstborn and the army (in the celestial ocean) and the pharaoh (by water serpent owing to negligence of Nun, the god of the ocean). Some interpretations may clash with this scenario, for example one focused on Herakleopolis (Backes 2010), but such a scenario had already been previously criticized and rejected by Spalinger (2000: 265–266).

By 1982, in a major new translation, some "primeval revolt" scholars explicitly linked the main "revolt" hieroglyphic text to the Exodus (Hornung and Fecht 1982/1997), referring to the Destruction of Mankind and the Divine Name revelation in Exodus 3:14.

Hornung further associates the revolt in the Destruction of Mankind with references in earlier texts, those of Ipuwer, the Instructions for Merika-re, Book of the Dead Spell 175, and the Book of Gates, and in several later texts: El Arish shrine, temple texts of Esna, Kom Ombo, and Edfu, the papyrus Book of the Fayum and the Insinger Papyrus (Hornung 1997: 90–95).

Other Egyptologists have pointed out what appear to be clear Exodus parallels (blood plague and Egyptian armed pursuit of unarmed foreign population fleeing Egypt) in this main Egyptian "revolt" narrative, the Destruction of Mankind, but without explicitly calling attention to a

[12] The Destruction of Mankind is often regarded as a myth of a "golden age" ruined by human rebellion against the gods, an etiological myth explaining how the present world emerged, or as a solar cycle myth. But these ideas have been heavily criticized if not refuted and have even been declared to be modern-day "Egyptological myth" (Baines 1991: 92) and yet continue to circulate (cf. Baines ibid., 1996: 364; Zivie-Coche 2004: 38, 53; Spalinger 2000: 276; Guilhou 1989: 144 n.62; Naville 1875: 16.). It is also widely recognized that in the Destruction as in other texts generally, Ra is "identical to Pharaoh" (Spalinger 2000: 261, emphasis added, cf. 258, 269; cf. Yoyotte 1972: 163, 2013: 347). In the Destruction (verses 75 and 83), Ra is named in cartouches and titled as "King of Upper and Lower Egypt."

[13] Though usually regarded as a mythological text, the Destruction of Mankind is structured as a pharaoh's historical annal (Redford 1986: 94) conforming precisely to the standard nine-element structure for military annals of the Dominion Record type with rebels resisting not attacking pharaoh (see Lundh 2002: 24–27, 238, for the standard annalistic structure, and Sparks, forthcoming, for full analysis). Spalinger has identified the Destruction as a transitional literary form derived from the stock "literary device of the speech in a war setting where king addresses his army," the *Königsnovelle*, comparable to the historical war speeches of pharaohs of the New Kingdom (Spalinger 2000: 280–281). The Destruction of Mankind is a "legend ... which also has political resonances" with history (Grimal 1994: 172). Richard Parkinson allows "specific historical events" in the Destruction of Mankind narrative (Parkinson 1997: 233 n.46). The most extensive commentator, French Egyptologist Nadine Guilhou suggests that the Destruction is perhaps a "mythical reflection of the historical reality" of the turbulent First Intermediate Period (Guilhou 1989: 138, transl.).

Biblical connection (Grimal 1994: 44; Mojsov 2005: 84; Spalinger 2000: 266; Naville 1875: 13, 18; Yoyotte 1972: 164).[14]

For example, French Egyptologist Nicolas Grimal has noted that the Destruction text describes elaborate magical efforts to duplicate the Plague of Blood, using red pigment, making "the ***waters of the Nile*** . . . tak[e] on the appearance of ***blood***" (Grimal 1994: 44, emphasis added). "For many nights the waters of the ***Nile ran red with the blood***," writes Bojana Mojsov of the earliest plague-like scene depicted in the Destruction of Mankind text before the attempted duplication (Mojsov 2005: 84, emphasis added). Christopher Eyre suggests that the "wading in blood" might have occurred on ground where the Nile's extent is widened in the annual flood and thus is not actually ground anymore at least temporarily: "The image of the goddess ***wading in blood*** might potentially be an image of ***wading in the [Nile] inundation water***" (Eyre 2002: 162 fn.46).

"Destruction of Mankind" is a misnomer recognized even by the scholar who coined the term in 1874, Edouard Naville, who realized that it "does not extend to all of humanity" but to those not of the "Egyptian race" (Naville 1875: 18, transl.). There are "two groups" involved in the Destruction text: (1) ***Egyptians*** and (2) the ***non-Egyptian*** "enemies of Re" (Spalinger 2000: 278; Naville 1875: 18, 1876: 108–109, 1885: 415; Clagett 1989: 1:2:544 n.11). Or two groups "divided into good and evil" (Brugsch 1881: 37, transl., cf. 37–39).[15] And Naville's 1874–1885 translations have long since been superseded by modern renderings based on a more complete set of texts (e.g., Hornung 1997 presents an all-source collated hieroglyphic text, transliteration, critical translation, and commentary) (See Figs. 19.1–19.3).

Needless to say there was no "destruction of mankind" in the Destruction of Mankind. Nor was there any destruction of the Revolt, as sun god Ra (Pharaoh) "tried in vain to destroy the people" but "did not succeed" (Naville 1875: 13, 18, transl.; Spalinger 2000: 266) and the Revolt still existed later (verses 183, 234–236). The possibility of partial massacres of rebels seems to conflict with the text's theme of a singular massacre event that was averted by Ra/Pharaoh's ruse with the red-colored beer and the Revolt's apparent success in escaping Egypt. Hornung expresses surprise that "The event itself, the destruction of the rebellion, is not even mentioned!" (Hornung's exclamation) (1997: 55 n.35, transl.). Hornung writes that the Destruction "text continues laconically, omitting the actual punishment of the rebels . . . in the desert" (1990: 109).

As mentioned above, Hornung and Fecht translated "I am that I am" ("Ich bin, der ich bin"), the full Hebrew Divine Name, in the Destruction of Mankind, where it appears as a two-word phrase in the Destruction, with the first word as the Egyptian *jw.i* (like the Hebrew root verb in *Yahweh* as a century of Egyptology has recognized with *jw*)[16] or perhaps **'eye* as

[14] Related Egyptian hieroglyphic "books" or documents in this genre of royal underworld books, though having structure and order that may differ, have "similar" content (Hornung 1999: 27). These books, which include the Amduat and Book of Gates, contain color pictures of what appear to be dramatic scenes of an Exodus-like event at a body of water (see Rendsburg, this volume, Chap. 18; Sparks, Nov 17, 2007, ASOR paper; see Figs. 19.4 and 19.5, below). This large body of potentially Exodus-related material (e.g., El Arish text cited by Hornung) provides new details not previously known, particularly as to apparent episodes in Egypt after the Israelites left, which though unreported in the Biblical narrative make natural logical sense within the narrative (e.g., the queen leading the search for the missing pharaoh at Pi-Khiroth, the Egyptians dividing up the land left behind by the departed slaves, an attempted palace coup, and the subsequent foreign invasion of a militarily defenseless and plague-devastated Egypt).

[15] Naville's title "Destruction of Mankind" may have been inspired by another paper before the same London society on a seemingly analogous text (a similarity not borne out by later study) that had recently described it as reciting a "destruction . . . of the race of man in the city of Heliopolis" and likened to the Greek Deucalion legend of the destruction of mankind (Goodwin 1873: 104, 107).

[16] Egyptologists have long recognized the Egyptian verb particle *jw* to be phonetically and semantically identical to the Hebrew verb forming the Divine Name Yahweh, the only argument being at a step further to decide whether they are also etymologically derived one from the other (Depuydt 1998: 29; Gardiner *Egyptian Grammar* 1957: 384, 551b; Erman-Grapow *Wörterbuch* 1926: 1:42; Lacau

Fig. 19.1 Exodus parallels are found in the Destruction of Mankind inscribed in the tomb of Seti I, ca. 1300 BC, tomb KV 17 in the Valley of the Kings, Thebes (Luxor), Egypt. This is a portion of one of the four walls inscribed in tomb chamber Je (courtesy of Theban Mapping Project, Cairo/Luxor)

phonologically reconstructed by Thomas Schneider (pers. comm. June 18, 2012) (arguments over the semantics of the phrase in Egyptian or in Hebrew are beyond the scope of this chapter).[17]

The Hornung–Fecht translation of the Egyptian text of the full Hebrew Divine Name, "I am that I am" (or "I am who I am"), has been accepted by subsequent scholars such as John Gwyn Griffiths, Edward Wente at the Oriental Institute/Chicago, and William Kelly Simpson at Yale (2003: 294), without commenting on the Exodus aspects. Griffiths states that the "I am I" translation "[illegible] the only one possible" (Griffiths 1988: 276).

Recent decades have seen an expansion of published references and some discussion of possible and probable Egyptian Exodus parallels (see Table 19.2).

Methodology

Diagnostic Criteria for Identifying Egyptian Exodus Parallels

Not every Egyptian document is an Exodus-related text, of course. Texts must contain sufficient elements to meet the objective criteria set forth below. Most Egyptian literature contains nothing apparently related to the Exodus or an Exodus-like event. Such works as, say, the Wisdom of Ptah-hotep, the war annals of Ramses II, the Satire on Trades, the Voyages of Wenamun, or the Pyramid Texts, are not apparently Exodus related.

Even some of the texts with parallels to the Exodus seen by Egyptologists do not automatically qualify as Exodus-related here in our research unless they satisfy the appropriate criteria mentioned herein. A late document such as the Demotic Chronicle, for example, with a single brief possible parallel with the Golden Calf does not seem sufficiently compelling to be included here as one of the identified Exodus-like texts, as we would need multiple Exodus elements and/or unique Exodus themes. The Tempest Stela has also been suggested as a parallel to the Exodus in some of the Ten Plagues, but a powerful and unusual storm that does not create pitch-black darkness does not

Recueil Philologie Égyptiennes et Assyriennes 1913: 35:63; Hornung/Fecht 1997: 125 n.aa, 130–131, in the context of the Destruction of Mankind text).

[17] Semantics of the phrase in Hebrew go back through centuries of disputes over whether it should be rendered "I am that I am," "I am the existing one," "I will be who I will be," "I-will-be," "The Existent One," etc., arguments over etymology, phonology, relationship between *ehyeh* and *yahweh*, etc. There is a vast literature on the subject which cannot be cited here (briefly, see Houtman 1993: 1:92–100, 368; Propp 1999: 1:203–226; Schniedewind 2009; Childs 1974: 60–89).

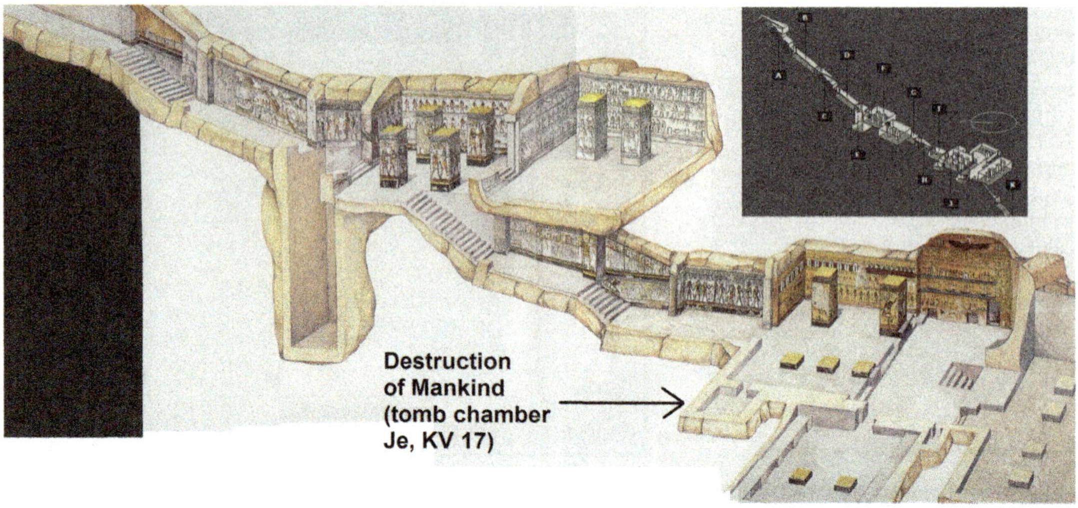

Fig. 19.2 Seti I tomb complex in the Valley of the Kings (KV 17), showing the location of room (chamber Je) containing the Exodus-like Destruction of Mankind text (after Weeks 2001)

Fig. 19.3 Edouard Naville, Swiss Egyptologist and archaeologist. First translator of the Destruction of Mankind, in a paper presented to the Society of Biblical Archaeology, London, Nov 3, 1874 (Naville 1875)

seem to satisfy the requirement for a parallel to the Plague of Darkness.[18] For another example, Batto (2013a: 173) notes that rather than a unique Exodus-like event, Papyrus Anastasi VIII merely documents "ordinary occurrences" such as a drought, not a drying of a sea to cross, and "Moreover, I do not find any of the events noted in P. Anastasi VIII to parallel closely the 'events' of the exodus as depicted in the Book of Exodus, even in their aggregate."

When the Egyptian texts are brief, laconic, or fragmentary, an element of interpretation is necessarily involved. However, when multiple lines of these evidentiary criteria converge in a single

[18] Gilula (1977) claimed that the Death of the Firstborn derived from an early Egyptian tradition, but this seems untenable as the tradition is the well-known "Cannibal Hymn" in the Pyramid and Coffin Texts, which concerns slaughtering of elder or (ambiguously) firstborn *gods* and *humans* for cooking in a cannibal pot (Pyr.Text §405b) with the eaten body parts enumerated in detail—heart, lungs, legs, bones, etc. (Eyre 2002: 8–10, passim). This is very unlike the Plague on the Firstborn which involves no cannibalism or cooking in pots. The "night" of the slaughter is linked to the "day" as well, forming a complete 24-hour period, not the Biblical "midnight" firstborn death (Exod 11:4, 12:29). "Eldest" (*smsw*) here merely means the *older half of the family* of 9 gods, the eldest 4 or 5, since the cannibal slaughter is of "gods" (plural *nṯrw*) of a single divine family not just the one firstborn (Pyr.Texts §§400a, 409c, 413b). Also excluded is treatment of the Egyptian Leper Expulsion tradition whose Exodus connection hinges on the isolated assertion of the Osarsiph = Moses equation in a single passage at the end of long narrative, appearing to be a scribal gloss, with otherwise very little narrative or thematic resemblance to the Exodus (literature too extensive to cite, but see Gruen 1998: 57–72). Even Wellhausen rejected the Leper Expulsion = Exodus as a "malicious invention" (1885: 440).

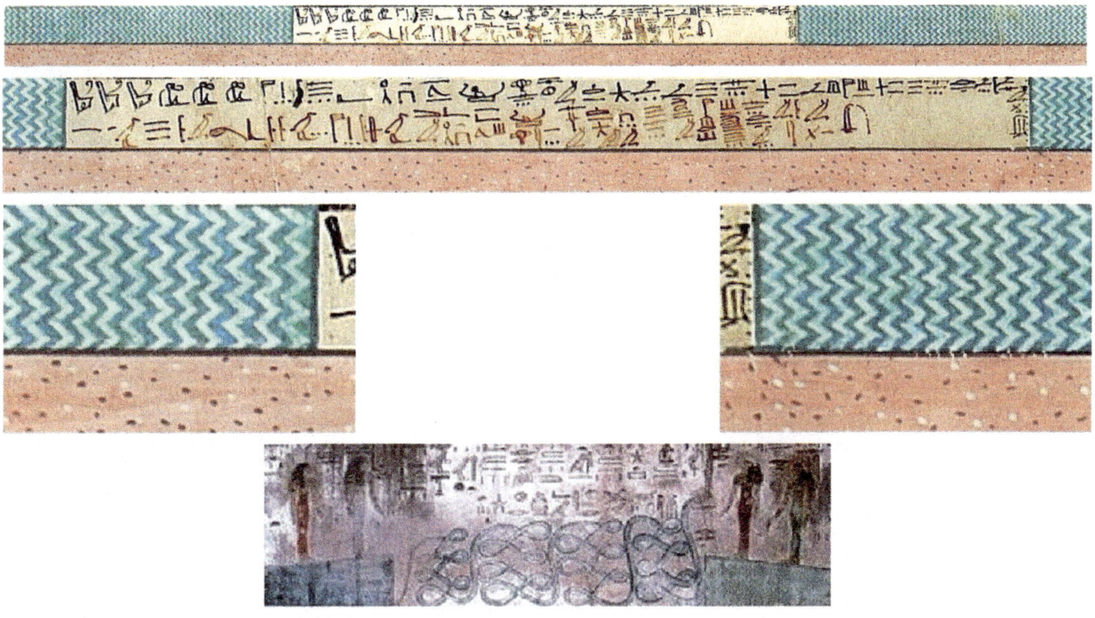

Fig. 19.4 (*Upper image, with 2 enlargements*): Walls of Water in a Parted Sea, the Flaming (Red) Sea (Egyptian *Yam Nesret*, El Arish Stela), scene in the Amduat book, Hour 5, a royal pharaonic underworld book with "content ... similar" to the Destruction of Mankind (Hornung 1999: 27). Hieroglyphic text caption between walls states that water had once been present and will return in deadly fashion (lines H-453-455, W-K 1453-1457); see subsequent connected Hour 10 scene of Drowned Soldiers in the same waterway or body of water (Fig. 19.5, below) 5 hours later (Thutmose III tomb KV 34, ca. 1500 BC, image courtesy of Theban Mapping Project). (**b**) (*Lower image*): This second image is the Amduat-parallel scene in the Book of Gates, Hour 4, with Hour Goddesses standing upright on top of the walls of divided waters, showing that the Walls of Water in the corresponding Amduat Hour 5 scene are vertical and not an overhead image or a map of pools or lakes. *Coiled serpent* represents enemy entity Apophis clouding the view, not a picture of blessed dead (image courtesy of Theban Mapping Project)

document of considerable length and/or with [illegible] imagery, we can have greater confidence in the identification of the Exodus parallel. And when the criteria are absolutely unique to the Exodus (such as assertion of the drowning of the Pharaoh and his army) we should have the greatest confidence of all. A recurring natural phenomenon would not be unique, so we would have less confidence in such a report unless corroborated by additional Exodus-like elements. A Plague of Blood, however, is absolutely unique as it is not a recurring natural phenomenon. It cannot be caused by red algae in the Nile or by red Nile silt as is often claimed without scientific foundation. No photographs have ever shown the "naturally occurring" Nile red in color, and no red algae have ever been found by biologists in the Nile or anywhere in East Africa even microscopically. A Red Nile does not naturally occur (Sparks 2003, 2004).

At least one and preferably two (or more) unique or distinctive thematic elements must occur in a document in order to constitute an Exodus parallel and may be classified by whether the motif is unique or distinctive. If distinctive, it is non-unique but must still be rare or unusual to qualify here.

These criteria are based on Assmann's prominent criteria of Egyptian "army pursuit" of the "mass release" of fleeing people for identifying "explicit parallels to the biblical Exodus story" in Egyptian documents (Assmann 2002: 402).

Other obvious signs of the Exodus include the Blood Nile, plagues attacking animals and humans, destroying crops, death of firstborn, three or more days of intense darkness, and plundering of Egypt by female slaves. Collectively these destructive events describe the mythological or the historical collapse of the nation of Egypt (geoscientists advocating the Thera

Fig. 19.5 Drowned Soldiers scene in the royal Amduat underworld book of ca. 1500 BC, 10th Hour, lower register, with enlargement. "Soldiers of Ra" (of Pharaoh) per Naville 1909: 97 (Amenhotep II tomb KV 35, ca. 1450 BC copy, images courtesy of Theban Mapping Project)

Table 19.3 Comparison of Egyptian chaos literature with the Exodus

Main thematic elements of Egyptian chaos literature (selected from table in Frankfurter 1993: 183–185)	*Exodus thematic elements*
"1. Chaos in Society: Disintegration of the Social Order	
"a. ... ['mass death']	Mass Death from Ten Plagues
"b. internal social strife and rebellions	Slave Rebellion
"c. abandonment of villages and cities	Mass Exodus from Cities
"2. Chaos in Earth: ... and the Nile	
"b(1) famine	Famine from Ten Plagues
"c(1) Nile running with blood (instead of water)	Plague of Blood
"3. Collapse of Borders	
"a(2) invasion of Asiatics	Invasion Threat from the Asiatics (Exod. 1:10)
"4. Chaos in the Heavens	
"a. disappearance (or darkening) of sun ...."	Plague of Darkness
Additional major Egyptian thematic elements (from Frankfurter 1993: 202, 204)	*Exodus thematic elements*
Egyptians "dug in the rivers ... and found no water"	Plague of Blood
"farm animals ... die in a catastrophe"	Plagues on Cattle
"plants die"	Seventh + Eighth Plagues
"Birds ... fall on the ground dead"	Dead Quail for Israelites in the Sinai Wilderness

volcanic tsunami theory of the Exodus might prefer the historical). These "scientific diagnostic signs" (cf. Machinist 1996: 402) are basic common sense assessments.

As new literature is reviewed these criteria may be refined further and augmented. Refinements may sometimes be necessary to sharpen a line of demarcation between a valid Exodus parallel and an indeterminate or an invalid parallel. To date, Exodus or Exodus-like parallels identified by 56+ scholars are *tentative* and subject to discussion and debate.

David Frankfurter has made a special study of Egyptian apocalyptic literature and its earliest forms found in the *Chaosbeschreibung*, the literature of national disaster, which he links to the formation of the Exodus tradition. Frankfurter infers an indirect literary dependence on words and themes, especially the Plague of Blood topos, as seen in several Egyptian texts of a wide range of dates and in the Book of Exodus, including the Ipuwer Admonitions, the Asclepius Perfect Discourse, and the Coptic Apocalypse of Elijah (1993: 203–204).

Frankfurter has identified several diagnostic motifs in this classical Egyptian disaster tradition which remained relatively fixed and "constant" from the Middle Kingdom to the Roman era, as it became idealized as archaic topos of chaos resulting from the loss of a powerful god-ruler pharaoh (Frankfurter 1993: 183; Schipper 2002: 282–286 has followed Frankfurter's diagnostic motif analysis). Each of his four main categories is readily identifiable with Exodus themes, besides the obvious Blood Nile which Frankfurter himself points out (ibid.: 183–185, footnotes omitted; cf. 171, 203). Certain additional related themes he has noted also can be related to Exodus themes (Table 19.3).

Discussion

Discussion of the significance of these Egyptian parallels to the Exodus inevitably revolves around questions of historicity and methodology. But most of the discussion will have to await extensive treatment in book form to do adequate justice to the sheer volume of material (at least 90 Egyptian texts) as well as the complexity of the issues involved. Some points below will be listed merely as outline points for reasons of space.

1. Mythology and Historicity of the Exodus

An evenhanded summary of the present range of thought about the historicity of the Exodus has been given by Nadav Na'aman (2011: 39–40; also see Na'aman, this volume, Chap. 42):

> The theme of Egyptian subjugation and the Israelite Exodus is the most frequently mentioned historical event in biblical literature . . .. The effort to establish the historical nucleus of the narrative is disputed among scholars.
>
> The range of opinions stretches from those who suggest that the nucleus of the story is basically authentic and the episode reflects an important event in the early history of Israel, on the one hand, to those who entirely dismiss the historicity of the episode, emphasize that the story was written at a later time and suggest that it mainly reflects the time of its composition, on the other hand. According to the latter view, the Exodus story is essentially myth that was formulated in late time and does not reflect the reality of the early history of Israel.
>
> Between the two extremes lie scholars who accept the historicity of a few details in the story and suggest that the story includes a nucleus—albeit small—of historical events that took place on Egyptian soil.

The vexing problem of historicity of the Exodus should be reassessed in light of all of the Egyptian Exodus parallel texts. It has been recognized in the field of Mesopotamian Biblical parallels, which include such famous examples as the Babylonian flood account, Epic of Gilgamesh Tablet 11, that "it is important for comparative studies to go as far as possible to ***include all relevant and available sources*** for comparison" (Chen 2010, emphasis added; cf. Maeir, this volume, Chap. 31).

The naive approach would be to deem these texts as purely historical documents, when in fact most (but not all) of the texts are literary or religious in nature though containing historically useful material. The opposite end of the spectrum would see these materials as simply containing random Exodus-like themes assembled accidentally and having no bearing on any historical Exodus-like event. But that minimalistic "coincidence" approach also must be rejected as it would prove too much because it would also require us to deny the development of unified Egyptian mythological, literary, or folklore narratives on *any subject* as mere "coincidence." So our approach must be in the middle.

With the Egyptian Exodus parallels we have an unprecedented opportunity to trace the movement and development of a literary tradition—whether historical, mythological, or folkloristic—in time and space across millennia and continents. This is because Exodus parallels can be found in more than 90 Egyptian texts spread across 2,000 or more years and approximately 300 more can be found in other ancient literatures of neighboring nations and more distant locales in the Eastern Mediterranean (see next).

2. Exodus Themes Are Not Common or Universal Motifs

It has been suggested that these Exodus-like themes are merely universal human thought patterns and experiences common to every culture and time period, as represented in the Aarne–Thompson–Uther (ATU) Motif Index, though the vast majority of sources for which are modern not ancient (the Motif Index is only occasionally cited in Egyptology, e.g., Assmann 1977: 2; also see Hollis, this volume, Chap. 16). There is a certain amount of circularity inherent in the argument because it assumes what it seeks to prove; namely, it assumes that the appearance of Exodus-like folklore themes in various cultures was more or less random and independent of each other, arising *sui generis* from the collective subconscious (whether Jungian psychological or social-cultural), and therefore any cohesive, integral Exodus-like narrative must consist of merely these random themes.

Thus, the argument goes, the integral narrative is deemed illusory, even when an Egyptian text follows the approximate order of events in the Book of Exodus, and a common Egyptian origin by mythological borrowing and/or historical transmission of folklore narratives is dismissed as sheer accident. In fact, the spread of the Exodus story, and its genetic dependencies and linkages, can be traced through time and space century by century, through neighboring geographic regions, and can be mapped in a stemma (Sparks, forthcoming). Thus, the appearance of Exodus themes in neighboring literatures is evidence of diffusion not evidence of independent invention via a common subconscious.

The universality of Exodus-like motifs is not at all apparent, unlike obvious universal or pervasive experiences such as the seasons, war, pestilence, crime, and childbirth. The Blood Nile, Death of the Firstborn, and the Parting of the Red Sea are hardly universal experiences that reside in a cultural subconscious.

Motifs involving bloody rivers, mass death of firstborn, drowning of armies or kings, and the splitting of seas ***do not appear*** in the worldwide ATU *Types of International Folktales* (check online indices; see Uther 2004; Thompson 1955–58) or in El-Shamy's classification of 12,600 motifs from the Arab-Islamic world in his two-volume, 1,000-page *Folk Traditions of the Arab World: A Guide to Motif Classification* (El-Shamy 1995). Ordinary river crossings by victorious armies like Alexander's or Napoleon's are hardly the splitting of a sea enabling escape of a pursued slave population, and the successful armies do not drown.

3. **Scholars must have Biblical traditions and motifs in mind in order to find Egyptian parallels, so it cannot be a discrediting bias**
4. **Mainstream Scholars with No Religious Motivation Find the Egyptian Exodus Parallels**
5. **Archaeology of Large Numbers in the Exodus and "Archaeological Invisibility"**

Long-standing arguments over the seeming lack of archaeological traces of the Israelites in the Sinai have been revisited in this conference without resolution (see various contributions to this volume of Dever, Halpern, Maeir, Propp, et al.; Finkelstein and Silberman 2001: 62).

Richard Friedman pointed out at the conference that the failure to find archaeological artifacts in Egypt or the Sinai would not prove the Exodus did not happen at all, because the Sinai has not been fully explored archaeologically even to date.

Some Sinai archaeologists have reported archaeological evidence of the Exodus in the Sinai, Negev, and Canaan, in the EB-MB transition (IBA or EB IV) that they correlate with Egyptian text parallels to the Exodus (Cohen 1983; Anati 1986, 1997, 2001, this volume, Chap. 35; Alon 1999; cf. refs. in Geraty, this volume, Chap. 4).

Halpern (this volume) notes "land armies did traverse that terrain, without leaving detected archaeological traces."

Seminomadic peoples and continual caravan crossings do not leave special, identifiable ruts in the hard gravel or the soft sand—one rut looks like another. Nor do they usually leave inscriptions with labels identifying the travelers or the herders and the dates of their migration.

Ernst Knauf quips that pottery does not give "passport information" and "Almost never is it possible to identify the nationality of a cooking pot" (Knauf 2013: 66, 68). The "archaeological invisibility of tent-dwelling nomads" as well as transient caravans and migrants, and the lower classes in sedentary populations, make it difficult though not impossible to find some scattered traces (Martin and Finkelstein 2013: 39 fn.39; Fantalkin 2008: 21; Zorn 1994: 45a n.5; Finkelstein 1995: 23–30, 79–85, 97–101, 122, 155–156; Finkelstein and Perevolotsky 1990: 75). Again, finding some "traces" still is not necessarily the finding of the ethnic identity that left those traces.

Referring now to the most commonly cited Exodus number as the ***extreme*** case—"two million" people have the same wear and tear and leave the same amount of trace evidence in 40 years of wandering or crossing the Sinai as 20,000 Bedouins and caravaners wandering or crossing the Sinai over 4,000 years. They are indistinguishable. The math is simple:

2 million × 40 years = 80 million person-years
20,000 × 4,000 years = 80 million person-years

The present author is not asserting that there were two million or asserting any particular number here, simply using the extreme to make the most forceful rhetorical point. Order-of-magnitude numbers like these can easily be in error by a factor of 10× without affecting the overall conclusions. If the Exodus wandering involved far fewer than two million (as is often asserted too) then even less trace evidence would be left behind and hence less of a problem of non-identification. Censuses of the population of Bedouin tribes in the Sinai in the early twentieth century counted about 40,000, and the population appears to be static (cf. Har-el 1983: 114). "Trace evidence" includes evidence of the dead. The granite shield of the southern Sinai would make burials in hard gravel very difficult; hence the Israelites would likely take the discarnated bones for secondary burial (as with Joseph's bones: Exod 13:19; Josh 24:32) when reaching better soil. The greater question is where are the bodies or the burials of hundreds of thousands to *millions* of Bedouins in some 100–200 generations since the Bronze Age? No one suggests that the Bedouins never existed from ancient times to date just because millions of remains are not found.

6. **Lateness of Text Does Not Nullify Historicity**

   Manetho is a classic example of an Egyptian historical source universally used by every Egyptologist as the foundation of Egyptian history and chronology, despite the problems of textual transmission, inclusion of mythological and folkloristic elements, errors and confusion of rulers, dynasties, and capitals. Yet Manetho wrote his 30-dynasty history ca. 260 BC, thousands of years after the earliest dynasties he recounts.

7. **Genre Does Not Automatically Exclude Historicity—Myth and History Coexist**

   As Jan Assmann puts it (this volume): "Myths may very well be based on historical experiences."

Conclusions

Long ago ASOR's Chester McCown attempted to draw the Exodus parallels between Hebrew and Egyptian apocalyptic literature and prominently sounded the alarm in the *Harvard Theological Review* about the effects of neglecting this vital and necessary work (McCown 1925: 359):

> So far as I know, the available material [comparing Egyptian and Hebrew apocalyptic] has not been collected in any one place, and its weight, therefore, has not been carefully ascertained; nor have the critical and historical conclusions which follow been fully drawn.

Such a study is long overdue.

References

Allen, James P. 2010. *Middle Egyptian*, 2nd ed. Cambridge, UK: Cambridge University Press.

Alon, David. 1999. *New Light on the Exodus*. Jerusalem: Ariel [Hebrew].

Anati, Emmanuel. 1986. *The Mountain of God*. New York, NY: Rizzoli (Eng. transl.) [=*Montagna di Dio*. Milan, Italy: Editoriale Jaca Book, 1986].

———. 1997. *Esodo tra mito e storia*. Valcamonica: Edizioni del Centro.

———. 2001. *The Riddle of Mount Sinai: Archaeological Discoveries at Har Karkom* [Studi Camuno, vol. 21]. Valcamonica, Italy: Edizioni del Centro.

Assmann, Jan. 1977. Das ägyptische Zweibrüdermachen (Papyrus d'Orbiney): eine Textanalise auf drei Ebenen am Leirfanen der Einheits frage. *Zeitschrift für Ägyptische Sprache und Altertumskunde* 104: 1–25.

———. 1983. Königsdogma und Heilserwartung: Politische und kultische Chaosbeschreibungen in ägyptischen Texten. In *Apocalypticism in the Mediterranean World and the Near East*, ed. David Hellholm, 345–377. Tübingen, Germany: J.C.B. Mohr. Proceedings of the International Colloquium on Apocalypticism, Uppsala, 12–17 Aug 1979.

———. 1997. *Moses the Egyptian* (transl. Andrew Jenkins). Cambridge, MA/London: Cambridge Univ. Press.

———. 2002. *The Mind of Egypt* (transl. Andrew Jenkins). New York, NY: Metropolitan Books/Henry Holt & Co. (Eng. transl.) [=*Ägypten: Eine Sinngeschichte*. Munich: Carl Hanser Verlag, 1996].

Backes, Burkhard. 2010. Das Massaker von Herakleopolis. In *Honi soit qui mal y pense. Studien zum pharaonischen, griechisch-römischen und spätantiken Ägypten zu Ehren von Heinz Josef Thissen*, ed. Hermann Knuf, Christian Leitz, and Daniel von Recklinghausen, 15–23 [Orientalia Lovaniensia Analecta vol. 194]. Leuven: Peeters.

Baines, John R. 1991. Egyptian Myth and Discourse: Myth, Gods, and the Early Written and Iconographic Record. *Journal of Near Eastern Studies* 50(April 1991): 81–105.

———. 1996. Contextualizing Egyptian Representations of Society and Ethnicity. In *The Study of the Ancient Near East in the Twenty-First Century: The William Foxwell Albright Centennial Conference*, ed. Jerrold S. Cooper and Glenn M. Schwartz, 339–384. Winona Lake, IN: Eisenbrauns.

Batto, Bernard F. 1984. Red Sea or Reed Sea? *Biblical Archaeology Review* 10(4): 56–63.

———. 1992. *Slaying the dragon: Mythmaking in the Biblical tradition*. Louisville, KY: Westminster/John Knox.

———. 2013a. *In the beginning: Essays on creation motifs in the Bible and the Ancient Near East*. Louisville, KY: Eisenbrauns.

———. 2013b. The Combat Myth in Israelite tradition revisited. In *Creation and chaos: A reconsideration of Hermann Gunkel's Chaoskampf hypothesis*, ed. JoAnn Scurlock, and Richard Henry Beal. Winona Lake, IN: Eisenbrauns.

Biase-Dyson (see Di Biase-Dyson).

Browne, Henry. 1844. *Ordo saeclorum*. London: John W. Parker.

Brugsch, Heinrich. 1864. Ueber die Gruppe für "Schwimmen" in den hieroglyphischen Texten.

Zeitschrift für Ägyptische Sprache und Altertumskunde 2: 65–67.

Brugsch, Heinrich. 1881. *Die neue Weltordnung nach Vernichtung des sündigen Menschengeschlechtes*. Berlin: Verlag S. Calvary.

Chen, Yi Samuel. 2010. *Comparative Studies of the Mesopotamian and Biblical Flood Stories and the Comparative Method*. abstract of paper delivered at the Society of Biblical Literature (SBL), Atlanta, Georgia, 20 Nov 2010. http://www.sbl-site.org/meetings

Childs, Brevard S. 1974. *The Book of Exodus: A Critical, Theological Commentary [The Old Testament Library]*. Philadelphia, PA: Westminster.

Clagett, Marshall. 1989. *Ancient Egyptian Science: A Source Book*, vol. 1 (tomes 1 and 2) [APS Memoirs, vol. 214]. Philadelphia, PA: American Philosophical Society.

Coats, George W. 1992. Joseph, Son of Jacob. In: David Noel Freedman (ed.) *Anchor Bible Dictionary*. New York, NY: Doubleday. *Anchor Yale Bible Dictionary*, vol. 3, 976b–981b. New Haven, CT: Yale Univ. Press (2007 reprint).

Cohen, Rudolph (Rudolf). July/Aug 1983. The Mysterious MB I people: Does the Exodus Tradition in the Bible Preserve the Memory of their Entry into Canaan? *Biblical Archaeology Review* 9(4): 16–29.

Depuydt, Leo. 1998. The Meaning of Old and Middle Egyptian 𓇋𓅱 *Jw* in light of the Distinction between Narration and Discussion. In *Jerusalem Studies in Egyptology*. Egyptology, Ägypten und Altes Testament 40, ed. Irene Shirun-Grumach, 19–36. Harrassowitz: Wiesbaden.

De Vries, Carl E. 1975. Plague. In *Wycliffe Bible Encyclopedia*, vol. 2, ed. Charles F. Pfeiffer et al., 1349–1352. Chicago, IL: Moody Press.

Di Biase-Dyson, Camilla. 2013. *Foreigners and Egyptians in the Late Egyptian Stories*. Leiden: Brill [Probleme Der Ägyptologie, vol. 32].

El-Shamy, Hasan M. 1995. *Folk Traditions of the Arab World: A Guide to Motif Classification*. Bloomington, IN: Indiana Univ. Press. 2 vols.

Enmarch, Roland. July 2004. *The Dialogue of Ipuwer and the Lord of All*, 2 vols. Doctoral dissertation. Oxford, England: Oxford Univ., Worcester College.

———. May 2009. *A world upturned: Commentary on and Analysis of The Dialogue of Ipuwer and the Lord of All* [British Academy Postdoctoral Fellowship Monographs]. Oxford/NY: British Academy/Oxford University Press. [Revised edition of 2004 doctoral dissertation].

Erman, Adolf, and Hermann Grapow. 1926, 1928, 1929, 1930, 1931, 1950, 1963. *Wörterbuch der Aegyptischen Sprache*, 7 vols. Leipzig: Hinrich's.

Eyre, Christopher J. 2002. *The Cannibal Hymn: A Cultural and Literary Study*. Liverpool: Liverpool Univ. Press.

Fantalkin, Alexander. 2008. The Appearance of Rock-Cut Bench Tombs in Iron Age Judah as a Reflection of State Formation. In *Bene Israel: Studies in the Archaeology of Israel and the Levant during the Bronze and Iron Ages in Honour of Israel Finkelstein*, ed. Alexander Fantalkin and Assaf Yasur-Landau, 17–44. Leiden/Boston: Brill [Culture and History of the Ancient Near East, 31].

Fecht, Gerhard. 1982/1991/1997. Metrische Umschreibung mit Anmerkungen zum Verständnis von Metrik und Aussage. In *Der Ägyptische Mythos von der Himmelskuh: Eine Ätiologie des Unvollkommenen [OBO vol. 46]*, ed. Erik Hornung, 109–127. Freiburg/Göttingen: Universitätsverlag/Vandenhoeck & Ruprecht (1982 orig. ed.; 2nd ed. rev. 1991; 1997 reprint).

Finkelstein, Israel. 1995. *Living on the Fringe*. Sheffield: Sheffield Academic.

Finkelstein, Israel, and Avi Perevolotsky. 1990. Processes of Sedentarization and Nomadization in the History of Sinai and the Negev. *Bulletin of the American Schools of Oriental Research* 279: 67–88.

Finkelstein, Israel, and Neil Asher Silberman. 2001. *The Bible Unearthed*. New York, NY: Free Press.

Frankfurter, David. 1993. *Elijah in Upper Egypt: The Apocalypse of Elijah and Early Egyptian Christianity*. Minneapolis, MN: Augsburg Fortress.

Freund, Richard A. 2009. *Digging Through the Bible: Understanding Biblical People, Places, and Controversies through Archaeology*. Lanham, MD/New York, NY: Rowman & Littlefield.

Gardiner, Alan H. 1957. *Egyptian Grammar*, rev. 3rd ed. London: Griffith Institute/Oxford Univ. Press.

Gilula, Mordechai. 1977. The Smiting of the first-born: An Egyptian Myth? *Tel Aviv* 4: 94–95.

Goedicke, Hans. 1984. The Canaanite Illness. *Studien zur Altägyptischen Kultur* 11: 91–105.

———. 1987. *Exodus: The Egyptian Evidence. Johns Hopkins University*, 23 April 1987, 21 pp. Paper presented to "Who Was the Pharaoh of the Exodus?" Near Eastern Archaeology Society (NEAS) Symposium, Memphis, TN, 23–25 April 1987.

———. 1988. The Northeastern Delta and the Mediterranean. In *The Archaeology of the Nile Delta, Egypt: Problems and Priorities*, ed. Edwin C.M. van den Brink, 165–175. Amsterdam: Netherlands Foundation for Archaeological Research in Egypt [Proceedings of Cairo seminar, 19–22 Oct 1986, for 15th anniversary of Neth. Inst. of Archaeology and Arabic Studies in Cairo].

———. 2004. *The Speos Artemidos Inscription of Hatshepsut and Related Discussions*. Oakville, CT: HALGO, Inc.

Goodwin, Charles W. Sep–Oct 1873. On chap. 115 of the Book of the Dead. *Zeitschrift für Ägyptische Sprache und Altertumskunde*, vol. 11, 104–107.

Gordon, Cyrus H., and Gary A. Rendsburg. 1997. *The Bible and the Ancient Near East*. Norton, New York, NY: W. W. June 1997 rev.

Görg, Manfred. 1992. Jochebed und Isis. *Biblische Notizen* 61: 10–14.

Gray, John. 1965. *Archaeology and the Old Testament World*. New York, NY: Harper Torchbook. Reprint [Thomas Nelson, London, 1962].

Greenberg, Moshe. 1969. *Understanding Exodus* [Heritage of Biblical Israel, vol. 2], Part 1. New York, NY: Behrman House.

Griffiths, John Gwyn. 1988. Review of *Der ägyptische Mythos von der Himmelskuh.* Eine Ätiologie des Unvollkommenen by Erik Hornung et al. *Journal of Egyptian Archaeology* 74: 275–277.

Grimal, Nicolas. 1994. *A History of Ancient Egypt.* London: Blackwell. Eng. transl. [Barnes & Noble, New York, 1997 reprint]) = [*Histoire de l'Egypte ancienne* (Librarie Arthème Fayard, 1988)]

Groll, Sarah, see Israelit-Groll, Sarah.

Gruen, Erich S. 1998. *Heritage and Hellenism: The Reinvention of Jewish Tradition.* Berkeley/Los Angeles, CA: Univ. of California Press.

Guilhou, Nadine. 1989. *La Vieillesse de Dieux.* France: Univ. of Montpellier.

Hallo, William W. 1990. Compare and Contrast: The Contextual Approach to Biblical Literature. In *The Bible in the Light of Cuneiform Literature* [Scripture in Context III] [Ancient Near Eastern Texts and Studies, vol. 8], ed. William W. Hallo, Bruce William Jones, and Gerald L. Mattingly, 1–30. Lewiston, NY: Edwin Mellen [reprinted in Hallo 2010: 677-697].

———. 2010. *The World's Oldest Literature: Studies in Sumerian Belles-Lettres.* Leiden: Brill.

Hamilton, Victor P. 2012. Genesis. In *Baker Illustrated Bible Commentary*, ed. Gary M. Burge and Andrew E. Hill. Grand Rapids, MI: Baker.

Har-El (Har'el), Menashe. 1983. *The Sinai Journeys: The Route of the Exodus.* San Diego, CA: Ridgefield Co. (rev. Eng. transl.). [*Mase Sinai* (Tel Aviv: Am-Oved Publ. Ltd. 1968)].

Helck, Wolfgang. Jan 1965. *Ṯkw* und die Ramses-stadt. *Vetus Testamentum* 15(1): 35–48.

Hollis, Susan Tower. 1982. The New Egyptian 'Tale of Two Brothers': A Mythological, Religious, Literary, and Historico-Political Study of the Papyrus d'Orbiney. Ph.D. dissertation. Univ. Microfilms, Ann Arbor, MI.

———. 1990. *The Ancient Egyptian "Tale of Two Brothers"* [Oklahoma Series in Classical Culture 12]. Oklahoma City, OK: Univ. of Oklahoma Press.

———. 2008. *The Ancient Egyptian "Tale of Two Brothers": A Mythological, Religious, Literary, and Historico-Political Study*, rev. 2nd ed. Oakville, CT: Bannerstone Press/David Brown Book.

Hornung, Erik. 1982. *Der Ägyptische Mythos von der Himmelskuh: Eine Ätiologie des Unvollkommenen* [OBO vol. 46]. Freiburg/Göttingen: Universitätsverlag/Vandenhoeck & Ruprecht. Orig. ed.

———. 1990. (transl. David Warburton). *The Valley of the Kings: Horizon of Eternity.* NY: Timken. (Engl. rev. transl.) [= *Tal der Königer*, Zurich, Munich: Artemis Verlag, 1982 Ger. orig.]).

———. 1997. *Der Ägyptische Mythos von der Himmelskuh: Eine Ätiologie des Unvollkommenen [OBO vol. 46].* Freiburg/Göttingen: Universitätsverlag/Vandenhoeck & Ruprecht. 2nd ed. rev. 1991; 1997 reprint.

———. 1999. (transl. David Lorton). The Ancient Egyptian Books of the Afterlife. Ithaca, NY/London: Cornell Univ. Press.

———. 2002. *The secret Lore of Egypt: Its Impact on the West.* Ithaca, NY: Cornell University Press (Transl. David Lorton).

Houtman, Cornelis. 1993. (transls. Johan Rebel and Sierd Woudstra), *Exodus* [*Historical Commentary on the Old Testament*] vol. 1. Kampen, Neth: Kok.

Hornung-Fecht (Hornung and Fecht). 1982/1997. (=Hornung 1982/1997)

Israelit-Groll, Sarah. 1999. Historical Background to the Exodus: Papyrus Anastasi VIII. In *Gold of Praise: Studies on Ancient Egypt in Honor of Edward F. Wente* [Studies in Ancient Oriental Civilization 58], ed. Emily Teeter and John A. Larson, 159–162. Chicago, IL: Oriental Institute of the University of Chicago.

Jeremias, Alfred. 1913. *Handbuch der altorientalischen Geisteskultur.* Leipzig: J. C. Hinrichs.

———. 1916. *Das Alte Testament im Lichte des Alten Orients*, 3rd ed. Leipzig: J. C. Hinrichs [2nd ed. 1913].

King, Leonard W. 1918. *Legends of Babylon and Egypt in Relation to Hebrew Tradition* [Schweich Lectures 1916]. Oxford: Oxford Univ. Press/British Academy.

Knauf, Ernst Axel. 2013. (with Konrad Schmid, Hermann Michael Niemann, and Silvia Schroer, eds.). *Data and Debates: Essays in the History and Culture of Israel and its Neighbors in Antiquity* [*Alter Orient und Altes Testament* (*AOAT*) vol. 407.] Munster: Ugarit-Verlag.

Lacau, Pierre. 1913. Notes de Grammaire: A propos de la grammaire égyptienne de M. Erman. *Recueil de Travaux Relatifs à la Philologie et à l'Archéologie Égyptiennes et Assyriennes*, vol. 35, 59–82. Paris: Librairie Honoré Champion.

Lang, Bernhard. 2009. *Joseph in Egypt.* New Haven, CT: Yale Univ. Press.

Lange, Johann. 1874. (John Peter) (transl. C. M. Mead; ed. Philip Schaff). *Commentary on the Holy Scriptures: Exodus.* Grand Rapids, MI: Zondervan (1876, 1949 reprint). [English translation of 1874 German original].

Leeming, David. 2004. *The Oxford Companion to World Mythology.* Oxford: Oxford Univ. Press.

Leith, Mary Joan Winn. 2012. *Shock and Awe: The Exodus Narrative.* Biblical Archaeology Society Lecture Series, DVD Item 9HLP7. Washington, DC: BAS, April.

Lenormant, François. 1882. (transl. Francis Brown). *The Beginnings of History* (Engl. transl. of French 2nd ed.). New York: Scribner's Sons.

Lesko, Leonard H. 1991. Ancient Egyptian Cosmogonies and Cosmology. In *Religion in Ancient Egypt*, ed. Byron E. Shafer, 88–122. Ithaca, NY/London: Cornell Univ. Press.

Levy, Thomas E. 2003. Foreword. In Miroslav Bárta (transl. Renata Landgráfová). *Sinuhe, the Bible, and the Patriarchs*, 5–7. Prague: Set Out.

Lichtheim, Miriam. 1973/1973/1980/2006. *Ancient Egyptian Literature.* Berkeley/Los Angeles/London: Univ. of Calif. Press. (1973/1976/1980, 2nd ed. 2006) vol. 1 (1973), vol. 2 (1976), vol. 3 (1980).

Lundh, Patrik. 2002. *Actor and Event: Military Activity in Ancient Egyptian Narrative Texts from Tuthmosis II to Merenptah* [Uppsala Studies in Egyptology, vol. 2]. Sweden: Uppsala University.

Machinist, Peter. 2009. The Road Not Taken: Wellhausen and Assyriology. In *Homeland and Exile: Biblical and Ancient Near Eastern Studies in Honour of Bustenay Oded*, ed. Gershon Galil, Mark Geller, and Alan Millard, 469–531. Leiden/Boston: Brill [Vetus Testamentum Supplements, vol. 130].

Martin, Mario A.S., and Israel Finkelstein. 2013. Iron IIA Pottery from the Negev Highlands: Petrographic Investigation and Historical Implications. *Tel Aviv* 40: 6–45.

Matthews, Victor H., and Don C. Benjamin. 2006. *Old Testament Parallels: Laws and Stories from the Ancient Near East*, 3rd rev. ed. Mahwah, NJ: Paulist.

McCown, Chester C. 1925. Hebrew and Egyptian Apocalyptic Literature. *Harvard Theological Review* 18: 357–411.

Meltzer, Edmund S. 2004. Sinuhe, Jonah and Joseph: Ancient 'Far Travellers' and the Power of God. In *God's Word for our World: Volume II—Theological and Cultural Studies in Honor of Simon John De Vries*, ed. J. Harold Ellens et al., 77–81. London/NY: T & T Clark/Continuum [JSOT Supp. 389].

———. 2007. *In search of Sinuhe: 'What's in a Name?'* Paper presented at American Research Center in Egypt (ARCE) annual meeting, Toledo, Ohio, 20 April 2007. http://www.ceae.unlugar.com/meltzer.htm

Meyer, Eduard, and Bernhard Luther. 1906. *Die Israeliten und ihre Nachbarstämme*. Germany: Max Niemeyer Halle.

Mojsov, Bojana. 2005. *Osiris: Death and Afterlife of a God*. Malden, MA/Oxford: Blackwell.

Moldenke, Charles E. 1895. The Oldest Fairy Tale. *Transactions of the M[illegible]ciation 7*: 32–81.

Moro, Caterina. 2005. L'identificazione tra Ebrei e Hyksos nelle fonti alessandrine. In *Definirsi e definire: percezione, rappresentazione e ricostruzione dell'identità*, ed. Massimo Gargiulo, Chiara Peri, and Giuseppe Regalzi, 71–88. Rome: Associazione Orientalisti. [Proceedings of the 3rd Conference of Orientalists, Rome, 23–25 Feb 2004]. http://www.orientalisti.net/wp-content/uploads/2009/06/atti2004.pdf

———. Dec 2008. Lo storico Artapano e il passato multietnico. In *Traduzione di tradizioni e tradizioni di traduzione*. ed. Benedetta Bellucci, Elio Jucci, Alfredo Rizza, and Bianca Maria Tomassini Pieri, 295–315. Milan: Qu.A.S.A.R. [Pavia, Italy, 19–21 April 2007] http://www.orientalisti.net/wp-content/uploads/2009/06/atti2007.pdf

———. Feb 2010. L'historien Artapan et le passé multiethnique. In *Interprétations de Moïse: Égypte, Judée, Grèce et Rome*, ed. Philippe Borgeaud, Thomas Römer, and Youri Volokhine with Daniel Barbu, 43–82. Leiden/Boston: Brill. [Jerusalem Studies in Religion and Culture, vol. 10].

Muhlestein, Kerry. June 2011. Levantine Thinking in Egypt. In *Egypt, Canaan and Israel: History, Imperialism, Ideology and Literature*, ed. Shay Bar, Dan'el Kahn, and J.J. Shirley, 190–235. Leiden/New York, NY: Brill. [Proceedings of Conference at Univ. of Haifa, 3–7 May 2009] [Culture and History of the Ancient Near East, vol. 52].

Na'aman, Nadav. 2011. The Exodus Story: Between Historical Memory and Historiographical Composition. *Journal of Ancient Near Eastern Religions* 11: 39–69.

Naville, Edouard. June 1875. La Destruction des Hommes par les Dieux. *Transactions of the Society of Biblical Archaeology* 4 (SBA, London): 1–19.

———. 1876. The Legend of the Destruction of Mankind. *Records of the Past*, vol. 6. London: SBA 103–112

———. 1885. L'inscription de la destruction des hommes dans le tombeau de Ramses III. *Transactions of the Society of Biblical Archaeology* 8 (SBA, London): 412–420.

———. 1909. (transl. Colin Campbell). *The Old Egyptian Faith*. [=Engl. transl. of Naville, *La Religion des Anciens Égyptiens*, 1907]. London + Putnam, NY: Williams & Norgate.

Noegel, Scott B. 1996. Moses and Magic: Notes on the Book of Exodus. *Journal of Ancient Near Eastern Studies* 24: 45–59.

Oldfather, Charles H. (transl.). 1933/1935. *Diodorus of Sicily*, vols. 1 and 2. [Loeb Classical Library]. Cambridge, MA: Harvard Univ. Press.

Parkinson, Richard B. 1997. *The Tale of Sinuhe and Other Ancient Egyptian Poems 1940–1640 BC*. Oxford: Oxford Univ. Press.

———. Dec 2010. *Poetry and Culture in Middle Kingdom Egypt* [Studies in Egypt[illegible] Near East]. London/Oakville, CT: Continuum/Equinox (ed. with addenda).

Posener, George. 1960. *De la Divinite du pharaon*. Paris: Imprimerie Nationale.

———. 1971. Literature. In *The Legacy of Egypt*, rev. ed, ed. J.R. Harris. Oxford: Oxford University Press.

Propp, William H.C. Sept 1999. *Exodus 1-18* [Anchor Bible Commentary series]. New York, NY/London: Doubleday.

Purdy, Strother. 1977. Sinuhe and the Question of Literary types. *Zeitschrift für Ägyptische Sprache und Altertumskunde* 104: 112–127.

Redford, Donald B. 1970. *A Study of the Biblical Story of Joseph*. Leiden: Brill. 290 pp.

———. 1986. *Pharaonic King-Lists, Annals, and Day-Books: A Contribution to the Study of the Egyptian Sense of History*. Mississauga, Ontario: Benben [SSEA Publication 4].

———. (Redford ABD 1992). Pi-Hahiroth. In *The Anchor Bible Dictionary*, 6 vols, vol. 5, 371a, ed. David Noel Freedman. New York, NY: Doubleday

[*Anchor Yale Bible Dictionary*. New Haven, CT: Yale Univ. Press, 2007 reprint].

———. 1992. *Egypt, Canaan, and Israel in Ancient Times*. Princeton, NJ: Princeton Univ. Press.

———. 2006. Video interview in Associated Producers. *The Exodus Decoded* documentary, broadcast 23 April 2006. "The Exodus Decoded Transcript." 18 Aug 2006. http://theexodusdecoded.com/simcha.jsp. accessed 2 Sept 2006

Rendsburg, Gary A. 2006. Moses as Equal to Pharaoh. In *Text, Artifact, and Image: Revealing Ancient Israelite Religion*, ed. Gary M. Beckman and Theodore J. Lewis, 201–219. Providence, RI: Brown Judaic Studies.

Romer, John. 1988. *Testament: The Bible and History*. New York: Holt.

Rothöhler, Benedikt. 2003. Hermes und Plotin: Mögliche Berührungen zwischen dem Hermetischen Asklepiustraktat und Plotins Schrift 'Gegen die Gnostiker' (Enneaden II, 9). In *Egypt, Temple of the Whole World: Studies in Honour of Jan Assmann*, ed. Sibylle Meyer, 389–407. Leiden: Brill [Studies in the History of Religions vol. 97].

Sarna, Nahum M. 1986. *Exploring Exodus*. New York, NY: Schocken Books.

———. 1992. (Sarna ABD 1992) Exodus, Book of. In *The Anchor Bible Dictionary*, 6 vols., 2:689b–700a, ed. David Noel Freedman. New York, NY: Doubleday. [*Anchor Yale Bible Dictionary*. New Haven, CT: Yale Univ. Press, 2007 reprint].

Sauneron, Serge. 1962. *Le Temple d'Esna* [Esna V]. Cairo: IFAO.

Schipper, Bernd U. 2002. Apokalyptik und Ägypten?—Erkenntnisse und Perspektiven: 2) Die Texte und die ägyptische Tradition. In *Apokalyptik und Agypten: Eine Kritische Analyse der Relevanten Texte aus dem Griechisch-Römischen Ägypten*, ed. Andreas Blasius and Bernd U. Schipper, 282–294. Leuven/Paris/Sterling, VA: Peeters [Orientalia Lovaniensia Analecta, vol. 107].

Schneider, Thomas. Sept 2000. Reviews of Manfred Görg, *Die Beziehungen zwischen dem Alten Israel und Ägypten. Von den Anfängen bis zum Exil*, Darmstadt 1997, and John D. Currid, *Ancient Egypt and the Old Testament*, Grand Rapids 1997, *Theologische Literaturzeitung*, vol. 125, 882–885.

———. 2008. Innovation in Literature on Behalf of Politics: The Tale of the Two Brothers, Ugarit, and 19th Dynasty History. *Ägypten und Levante* 18: 315–326.

Schniedewind, William M. 2009. Calling God names: an inner-biblical approach to the Tetragrammaton. In *Scriptural Exegesis: The Shapes of Culture and the Religious Imagination. Essays in Honour of Michael Fishbane*, ed. Deborah A. Green and Laura S. Lieber, 74–86. Oxford: Oxford University Press.

Shupak, Nili. 2003. (Shupak COS 2003) The Admonitions of an Egyptian Sage: The Admonitions of Ipuwer (1.42). In *The Context of Scripture: Canonical Compositions, Monumental Inscriptions, and Archival Documents from the Biblical World*, 1:93a–98b, ed. William W. Hallo. Leiden/Boston: Brill.

———. 2006. A Fresh Look at the Dreams of the Officials and of Pharaoh in the Story of Joseph (Genesis 40–41) in the Light of Egyptian Dreams. *Journal of Ancient Near Eastern Studies* 30: 103–138.

Simpson, William Kelly (ed.). 2003. *The Literature of Ancient Egypt*, 3rd rev. ed. New Haven, CT: Yale Univ. Press.

Spalinger, Anthony John. 2000. The Destruction of Mankind: A transitional literary text. *Studien zur Altägyptischen Kultur* 28: 257–282.

Sparks, Brad C. 2003/2004. Red Algae Theories of the Ten Plagues [Part 1]. *Bible and Spade*, vol. 16, no. 3(Summer), 66–77; [Part 2] vol. 17, no. 1 (Winter), 17–27; [Part 3] vol. 17, no. 3(Summer), 71–82.

Thompson, Stith. 1955–58. *Motif-Index of Folk Literature*. Bloomington, IN: Indiana Univ. Press. 6 vols.

Trigger, Bruce G. 1995. Egyptology, Ancient Egypt, and the American Imagination. In *The American Discovery of Ancient Egypt*, ed. Nancy Thomas, 21–35. New York, NY: Los Angeles County Museum of Art/Harry NAbrams.

Uther, Hans-Jörg. 2004. *The Types of International Folktales: A Classification and Bibliography*. Helsinki: Academia Scientiarum Fennica. 3 vols.

van der Horst, Pieter W. 2006. *Jews and Christians in their Graeco-Roman context: Selected Essays on Early Judaism, Samaritanism, Hellenism, and Christianity*. Tübingen: Mohr Siebeck [Wissenschaftliche Untersuchungen zum Neuen Testament, vol. 196].

Vergote, Jozef (Joseph). 1959. *Joseph en Égypte: Genèse chap. 37-50 à la lumière des ètudes égyptologiques*. Louvain, Belgium: Publications Universitaires [Orientalia et Biblica Lovaniensia, vol. 3].

———. 1985. 'Joseph En Egypte': 25 Ans Apres. In *Pharaonic Egypt: The Bible and Christianity*, ed. Sarah Israelit-Groll, 289–306. Jerusalem: Magnes Press/Hebrew University.

Weeks, Kent R. (ed.). 2001. *Valley of the Kings*. Vercelli, Italy/New York, NY: White Star/Michael Friedman.

Wellhausen, Julius. 1885. (transls. J. Sutherland Black and Allan Menzies). *Prolegomena to the History of Israel: With a Reprint of the Article* Israel *in "Encyclopædia Britannica"*. Edinburgh: Adam & Charles Black. [=*Prolegomena zur Geschichte Israels* (G. Reimer, Berlin, 1883, 2nd ed. German) + *Britannica* article, "Israel," 1881].

Wente, Edward F. 2003. (Wente-Simpson 2003) Destruction of Mankind. In *The Literature of Ancient Egypt*, 3rd rev, 289–298, 560, ed. William Kelly Simpson. New Haven, CT: Yale Univ. Press.

Williams, Ronald J. 1992. Egyptian Wisdom Literature. In *The Anchor Bible dictionary*, 6 vols., 2:395b-399b, ed. David Noel Freedman. New York, NY: Doubleday. [=*Anchor Yale Bible Dictionary*. New Haven, Ct: Yale Univ. Press, 2007 reprint].

Wilson, John A. 1969. (Wilson ANET 1969) Egyptian Oracles and Prophecies: The Admonitions of Ipu-wer. In *Ancient Near Eastern Texts Relating to the Old Testament*, 3rd rev., 441–449c, ed. James B. Pritchard. Princeton, NJ: Princeton Univ. Press.

Wolf, Walther. 1962. *Kulturgeschichte des alten Ägypten*. Stuttgart: Kroner.

Yahuda, Abraham S. 1934. *The Accuracy of the Bible*. London: Heinemann [New York, NY: E. P. Dutton, 1935].

Yoyotte, Jean. 1972. *Religion de l'Égypte ancienne: Annuaire 1971–1972*, 162–167. Paris: Paris: École Pratique des Hautes Études, 5th Section—Sciences Religieuses.

Yoyotte, Jean. 2013. Histoire, géographie et religion de l'Égypte ancienne: Opera selecta par Jean Yoyotte. *Orientalia Lovaniensia Analecta* 224, ed. Ivan Guermeur. Leuven: Peeters

Zivie-Coche, Christiane. 2004. Book I. Pharaonic Egypt. In *Gods and Men in Egypt: 3000 BCE to 395 BCE*, 1–191, ed. Françoise Dunand and Christiane Zivie-Coche. (transl. David Lorton). Ithaca, NY: Cornell Univ. Press.

Zorn, Jeffrey R. 1994 Estimating the Population Size of Ancient Settlements: Methods, Problems, Solutions, and a Case Study. *Bulletin of the American Schools of Oriental Research* 295: 31–48.

Part V

The Exodus Narrative as Text

The Exodus Narrative Between History and Literary Fiction: The Portrayal of the Egyptian Burden as a Test Case

20

Christoph Berner

Abstract

It is beyond dispute that the Exodus Narrative (Exod 1–15) provides a detailed account of the Egyptian bondage and the circumstances of the Israelites' liberation. At the same time, it is obvious that these details do not form a coherent picture, but often stand in tension with, if not in flat contradiction to each other. The paper argues that the complexity of the account is best understood as the result of a continuous process of literary expansions (*Fortschreibungen*) by which the Exodus Narrative gradually took shape. Taking the references to the Israelites' forced labor as a test case, the paper suggests that substantial parts of the narrative did not exist prior to the earliest priestly layer (P 1), but belong to an extensive post-priestly stage of development. As a result, the Exodus Narrative reveals only little about the historical circumstances of the Exodus events, yet all the more about how different generations of postexilic scribes imagined these circumstances and took their share in the literary development of the biblical account.

This paper deals with the literary history of the biblical account on the Exodus and thus contributes a further perspective on the issues discussed by Garrett Galvin, Richard Friedman, Thomas Römer, Stephen Russell, and Konrad Schmid. It shows that the biblical portrayal of the Israelites' situation in Egypt is the result of a complex literary development which took place to a large extent during the postexilic period. As a result, I will argue that the biblical account may not be equated with a historical source providing firsthand information on the Israelites' sojourn in Egypt. With this critical appraisal of the biblical account, the paper also contributes to the discussion on which basis and to what extent it is possible to reconstruct the history of the Exodus events (see esp. the contributions of Manfred Bietak, Israel Finkelstein, Lawrence Geraty, James K. Hoffmeier, and Donald B. Redford).

The Biblical Account as a Historical Source?

Despite the scarcity of external evidence, it cannot be denied that the key points of the biblical account of the Exodus can be related to certain

C. Berner (✉)
Theologische Fakultät, Universität Göttingen,
Gottingen 37073, Germany
e-mail: email@christophberner.de; cberner@gwdg.de

T.E. Levy et al. (eds.), *Israel's Exodus in Transdisciplinary Perspective*,
Quantitative Methods in the Humanities and Social Sciences, DOI 10.1007/978-3-319-04768-3_20,

historical phenomena. Egyptian sources testify to the fact that nomadic groups from Palestine would migrate to Egypt during times of famine. Moreover, there is clear evidence that people of Semitic origin were enslaved by the Egyptians and forced to work at the construction sites; and finally, it can be assumed that occasionally a group of slaves tried to escape and was pursued by a company of Egyptian chariots. In the light of these and other points of contact between the Egyptian sources[1] and the biblical account in Gen 37–Exod 15, it is certainly imaginable that the latter reflects a historical reality. Yet, this alone is definitely no sufficient proof of the historicity of the Exodus events. The decisive question that needs to be answered before any historical reconstruction may be undertaken concerns the literary character of the Biblical account: Is it a reliable source based on ancient traditions about the Exodus or rather a literary fiction which develops an idealized picture of Israel's origins from a much later point of view?

Half a century ago, probably most biblical scholars would have argued in favor of the first option. At that time, the Documentary Hypothesis, according to which the Pentateuch consists of several old sources based in turn on reliable oral traditions, was still widely accepted. In recent years, however, the shortcomings of the Documentary Hypothesis have become more and more apparent.[2] The alleged sources are often fragmentary and do, moreover, display a considerable number of explicit literary interconnections which make it very unlikely that they should represent originally separate documents. In consequence, a number of scholars did not only abandon the paradigm of the Documentary Hypothesis, but at the same time decided to dispense with any kind of critical research in the literary development of the biblical text. The tensions and doublets which were previously explained by the assumption of different sources are either ignored by these scholars[3] or they are interpreted as a specific of Israelite literature, for instance by Thomas Thompson according to whom "the traditional complex-chain narrative is not an editorial or redaction structure, but a type of literature in its own right: one of the ways in which ancient Israel told long stories" (Thompson 1987: 157).

In my opinion, simplistic models of this kind are highly questionable. They not only ignore more than two centuries of critical scholarship, but they also equate the present biblical text with ancient Israelite literature without adducing any evidence to support this claim. To avoid any misunderstanding, I wish to stress that a strictly synchronic approach to the Hebrew Bible is certainly a legitimate option, as long as one is only interested in a reception-aesthetic or a similar perspective. Yet, if the biblical text is used as a historical source, its own literary history cannot be ignored.

The traces of this literary history are so evident that it must be regarded as a fact which cannot seriously be called into question. For instance, the basic distinction between the priestly and the non-priestly parts of the Pentateuch is a well-established result of biblical criticism which remains a cornerstone for any Pentateuchal theory.[4] Yet, despite the evident conceptual and theological differences between the priestly and the non-priestly material, it is at the same time apparent that these two literary domains have not developed independently from each other. Rather, it seems that there is an ongoing dialogue between the different literary voices, with the priestly text commenting on and revising parts of the non-priestly narrative, while other non-priestly passages already seem to react to priestly texts.[5] This phenomenon of

[1] On the Egyptian evidence see, e.g., Hoffmeier (1996), Frerichs and Lesko (1997), Otto (2000).

[2] The present state of debate in Pentateuchal Criticism is reflected in Dozeman et al. (2011).

[3] See, e.g., Fischer and Markl (2009). Similarly Houtman (1993–2002).

[4] See, e.g., Dozeman (2009) whose recent commentary on the Book of Exodus is entirely based on this basic distinction.

[5] The *locus classicus* for this literary interaction is represented by the two revelation scenes in Exod 3–4 (non-P) and Exod 6–7 (P) which are discussed in the contribution by Thomas Römer. One might equally point to the plague narrative (Exod 7:8–11:10). Here, a

literary interaction calls for an interpretive model different from the one provided by the Documentary Hypothesis. It is best explained by the Supplementary Hypothesis which describes the literary development of the Pentateuch as a sequence of successive expansions and/or redactions.

Especially among European scholars, this redaction historical approach has found an increasing number of supporters.[6] While some still maintain the distinction between a pre-priestly and an independent priestly narrative and claim that the text has considerably been expanded in the course of and/or after their redactional combination (Levin 1993; Otto 1996; Achenbach 2003; Kratz 2005), others have abandoned even this last relic of the Documentary Hypothesis and assume that the Pentateuch has developed gradually in a sequence of several compositional or redactional layers (Blum 1990; Berner 2010; Albertz 2012; Utzschneider and Oswald 2013). Despite these divergences, the different manifestations of the redaction historical approach more or less agree in their general perspective on the development of the Pentateuch. They suggest that considerable parts of the non-priestly text do not represent elements of one or more ancient sources, but rather expansions of the priestly narrative that date from the postexilic period. At the same time, it becomes apparent that the hermeneutics and editorial techniques employed by the scribes who created the respective expansions are strikingly similar to what we know from the early reception history of the "biblical" text at Qumran. With a certain right, one might, therefore, speak of a constant process of interpretation which commences with the redaction history of the "biblical" text itself and continues in its different versions rewritings and early commentaries.[7]

It goes without saying that this critical approach to the literary history of the Pentateuch does considerably reduce its value as a source for reconstructing the history of ancient Israel. It does, however, provide new insight into the intellectual history of those postexilic circles who shaped the biblical picture of Israel's origins and thus created a collective memory which should determine the identity of early Judaism (Kratz 2013).

Naturally, the literary development of the Exodus Narrative is governed by the same hermeneutic principles and redactional dynamics, which can be traced throughout the Pentateuch. The present text of Exod 1–15 is the result of a complex history of literary expansions which have contributed to the gradual development of the narrative elements as well as their theological interpretation.[8] To illustrate this model, one can

basic, pre-priestly layer consisting of three plagues (Exod 7:14–8:28*) has been changed by the priestly writer into an account of a contest between Aaron, Moses and the Egyptian magicians who compete as miracle workers (Exod 7:8–9:12*). This notion then continues in the post-priestly plagues (Exod 9:13–10:27) which combine the aspects of divine sanctions and miraculous demonstrations of YHWH's power. For a detailed analysis of the development of the plague narrative see Berner (2010: 168–266).

[6] In contrast, the paradigm of the Documentary Hypothesis is still much more common among scholars from Israel and the United States. See, e.g., the conference comments of Richard Friedman or the commentary of William H. Propp (Propp 1999). Moreover, in recent years one can notice a growing influence of the school of the so-called Neo-Documentarians, most prominently represented by Baruch J. Schwartz and Joel S. Baden, who in contrast to and despite the developments in Europe promote a highly mechanical interpretation of the Documentary Hypothesis, which leaves no room for redactional activity at all.

[7] Until today, the development of the "biblical" text and its early Jewish reception history are mostly treated as independent issues, while attempts to establish a comprehensive history of postexilic literature are exceptions rather than the rule. See however Zahn (2011), Dimant and Kratz (2013).

[8] Strictly speaking, it is possible to distinguish four major stages in the development of the Exodus Narrative which each comprise a variety of literary layers sharing a similar narrative profile and theological agenda (Berner 2010). The pre-priestly text represents the first stage in this development and was already composed as a reaction to the downfall of the Northern Kingdom in 722 BCE. In contrast, the earliest priestly layer (P 1 = stage 2) already seems to date from the early postexilic period (around 500 BCE). Later in postexilic times, this priestly adaption of the pre-priestly narrative has been subject to an extensive reworking by late

freely choose any thematic aspect or literary unit within the Exodus Narrative. In the following, I will focus on the different descriptions of the Egyptian bondage,[9] as this motif does not only represent a key element of the narrative, but proves to be crucial for any historical reconstruction of the Exodus events as well.

The Different Descriptions of the Egyptian Bondage and Their Literary Development

The tension between slavery and liberation dominates the entire plot of the Exodus Narrative and may thus be labeled as the central *Leitmotiv*. Surprisingly, however, only few passages provide details concerning the exact nature of the Israelites' forced labor. Apart from the fairly elaborate account of Pharaoh increasing the Israelites' burden in Exod 5, there is only a set of brief notes in Exod 1:11–14 stressing different aspects of the tasks the Israelites were involved in. Over the last decades, there has been a highly controversial discussion concerning the interpretation of the respective verses, especially of the place names Pithom and Ramesses mentioned in Exod 1:11 and their historical implications (e.g., Redford 1963; Schmidt 1990; Hoffmeier 1996). Yet, while the passage has been extensively treated from a philological and archaeological point of view, the crucial issue of its literary history has not always been paid due attention to. This is problematic for two reasons: Firstly, Exod 1:11–14 are not the work of a single author, but consist of a non-priestly (Exod 1:11–12) and a priestly part (Exod 1:13–14) (e.g., Dozeman 2009: 69). Secondly, the passages providing details on the Israelites' forced labor are only loosely connected to their literary context, and it is therefore questionable, whether they represent an original part of the text or a later addition.

Taking a closer look at the non-priestly text in Exod 1:11–12, it becomes clear at once that the reference to the building of the store cities in 1:11b stands out from its immediate context and interrupts the argument in 1:11a, 12:

Exod 1:11 וַיָּשִׂימוּ עָלָיו שָׂרֵי מִסִּים לְמַעַן עַנֹּתוֹ בְּסִבְלֹתָם
וַיִּבֶן עָרֵי מִסְכְּנוֹת לְפַרְעֹה אֶת־פִּתֹם וְאֶת־רַעַמְסֵס׃
12 וְכַאֲשֶׁר יְעַנּוּ אֹתוֹ כֵּן יִרְבֶּה וְכֵן יִפְרֹץ וַיָּקֻצוּ מִפְּנֵי בְּנֵי יִשְׂרָאֵל׃

Exod 1:11a And they set work masters over them to oppress them with forced labor.
11b And they built store cities for Pharaoh, Pithom and Ramesses.
12 But the more they oppressed them, the more they multiplied and spread, so that they came to dread the Israelites.

In the light of the literary evidence it is reasonable to follow the suggestion made by Donald B. Redford and to regard Exod 1:11b as a later addition which aims at providing a more specific picture of the forced labor of the Israelites (Redford 1963: 414–415).[10] While Exod 1:11a, 12 are probably part of the pre-priestly narrative (Gertz 2000: 380–388; Berner 2010: 31–33), it proves difficult to determine the precise place of 1:11b within the literary history of the Exodus Narrative. At the earliest, it might represent an addition from the Neo-Assyrian Period, which should be regarded as the *terminus post quem* for internal reasons, such as the reference to the city Pithom and the use of the Akkadian loanword ערי מסכנות ("store cities") (Lemche 1996: 63–65). It must, however, not be overlooked that most biblical references to "store cities" belong to the Chronistic literature.[11] A postexilic origin

dtr circles. Considerable portions of the non-priestly text are part of this late dtr stage of development (stage 3). Finally, a fourth stage may be detected which again adjusts the narrative to priestly concerns (P 2).

[9] For a more detailed analysis of the bondage motif see Berner (2011: 211–240).

[10] In fact, the case is even more complex, as there is a certain tension within Exod 1:11b as well. From a syntactical point of view, the names of the two cities given in Exod 1:11bβ do not connect smoothly to the preceding introduction of the building of store cities (ערי מסכנות) which lacks the determination (ערי המסכנות) one would expect if the sentence had been conceived by a single author. Thus, it is possible that the explication of the store cities' names in Exod 1:11bβ represents an even later addition.

[11] Apart from Exod 1:11b and 1 Kgs 9:19 the term occurs only in 2 Chr 8:4, 6; 16:4; 17:12; 32:28.

of Exod 1:11b, therefore, remains an option that should not be dismissed too quickly.

Regardless of the precise origin of Exod 1:11b, it is evident that the verse betrays a certain knowledge of text-external realities, as the places Pithom and Ramesses were not invented by an Israelite scribe. At the same time, it is clear that a scribe introduced these place names into the Exodus Narrative in order to highlight that the Israelites were involved in the building of the two cities. Similar attempts of establishing etiological connections between certain places and the (pre-)history of Israel are attested throughout the Hexateuch and beyond.[12] As a rule, they may prove instructive for our understanding of the historical situation in which the text was written, but they provide no reliable data to reconstruct the historical circumstances of the events narrated in the text. Or, to use the terminology employed by Konrad Schmid in his contribution, they facilitate a certain insight into the world of the author, yet not into the world of the narrative.

Similar methodological caution should be exercised with regard to the details of the Israelites' forced labor provided by Exod 1:13–14, although in this case the situation is slightly different, as there is no decisive evidence for a text-external reference point. According to Exod 1:14, the Israelite slaves were assigned to the production of bricks as well as different kinds of labor in the fields. It is often overlooked and thus all the more important that both aspects are first attested in the context of the earliest priestly layer of the Exodus Narrative (P 1). In consequence, they should first and foremost be addressed as parts of the priestly concept of the Egyptian bondage and not as firsthand historical information.

However, the case is even more complex, as it is doubtful whether the specific details provided in Exod 1:14 were originally part of the priestly text at all. In Exod 6:9, the priestly writer only refers to the "hard service" (עבודה קשה), while its details are no longer an issue. Moreover, at least the reference to the Israelites' service in the fields in Exod 1:14aβ shows clear traces of a secondary addition. It is only loosely attached to the previous sentence, and its connection to the preceding introduction of the forced labor is primarily established by the resumptive clause in 1:14b. Exod 1:14b rephrases the wording from Exod 1:13 and thus reflects a typical editorial technique employed to integrate later additions into a given literary context. Without the secondary appendix in Exod 1:14aβb, the two verses show a perfect parallel structure framed by the chiastic use of the term עבד resp. עבדה that serves as the Leitmotiv of the short priestly passage:

Exod 1:13 ויעבדו מצרים את־בני ישראל בפרך׃ 14 וימררו את־חייהם בעבדה קשה בחמר ובלבנים

ובכל־עבדה בשדה

את כל־עבדתם אשר־עבדו בהם בפרך׃

Exod 1:13 And the Egyptians made the Israelites serve with rigor 14 and they made their life bitter with hard service in mortar and brick.

And with all kinds of service in the field.

With all of their service which they made them serve with rigor.

It can be asked if the reference to mortar and brick in Exod 1:14aα could not represent another later addition. For instance, one might think of a secondary explication of the Israelites' forced labor in the light of Gen 11:3 where the same building materials, mortar and brick, are mentioned (Berner 2010: 36–37). Yet in the end there is no decisive evidence to support this claim, and it is thus the easiest explanation that it was the priestly writer who first expressed the idea that the Israelites used to work in mortar and brick.

In contrast, I have argued above that the introduction of the field labor in Exod 1:14aβb should be regarded as a post-priestly addition. This raises the question as to what was the trigger of this addition. That the Israelites were forced to work in the fields is a motif that is only this once referred to in the Exodus Narrative and beyond. There is, however, one parallel that could prove relevant. Interestingly, the semantic field introduced in Exod 1:14aβ recurs in the context of the hail plague (Exod 9:19–21) where Pharaoh's servants (עבדים) in the field (בשדה) are mentioned.

[12] Cf., e.g., Gen 4:17; 11:9; 12:8; 13:4; 21:31; 22:14; 32:31; Exod 17:7, 15; Num 11:3, 34; 13:24; 21:3; 32:38, 42; Dtn 3:14; Josh 5:9; 7:26; 19:47.

Exod 9:19 וְעַתָּה שְׁלַח הָעֵז אֶת־מִקְנְךָ וְאֵת כָּל־אֲשֶׁר לְךָ בַּשָּׂדֶה כָּל־הָאָדָם
וְהַבְּהֵמָה אֲשֶׁר־יִמָּצֵא בַשָּׂדֶה וְלֹא יֵאָסֵף הַבַּיְתָה וְיָרַד עֲלֵהֶם הַבָּרָד
וָמֵתוּ׃ 20 הַיָּרֵא אֶת־דְּבַר יְהוָה מֵעַבְדֵי פַּרְעֹה הֵנִיס אֶת־עֲבָדָיו
וְאֶת־מִקְנֵהוּ אֶל־הַבָּתִּים׃ 21 וַאֲשֶׁר לֹא־שָׂם לִבּוֹ
אֶל־דְּבַר יְהוָה וַיַּעֲזֹב אֶת־עֲבָדָיו וְאֶת־מִקְנֵהוּ בַּשָּׂדֶה׃

Exod 9:19 Send, therefore, and have your livestock and everything that you have in the open field brought to a secure place; every human or animal that is in the open field and is not brought under shelter will die when the hail comes down upon them.'" 20 Those officials of Pharaoh who feared the word of YHWH hurried their slaves and livestock off to a secure place. 21 Those who did not regard the word of YHWH left their slaves and livestock in the open field.

Although the author of this passage had certainly been thinking of Egyptian servants, the correspondence with the priestly term for the Egyptian bondage (עבדה) allows for a different reading as well. Read against the background of the priestly terminology, it suggests itself that the servants mentioned in Exod 9 were in fact Israelite slaves. In my opinion, it is precisely this interpretation which has triggered the post-priestly expansion in Exod 1:14aβb. The idea that the Israelites were forced to work in the fields thus proves to be the result of inner-biblical exegesis (Berner 2010: 37).

So far, we have dealt with three considerably different types of literary characterizations of the bondage motif in Exod 1:11–14. While the notion of the Israelites working in mortar and brick (Exod 1:14aα) was most likely introduced by the priestly writer and reflects his conception of the Egyptian bondage, the reference to the building of the store cities in Exod 1:11b represents a secondary addition to the pre-priestly text motivated by an etiological interest. In contrast, the idea that the Israelites used to work in the fields (Exod 1:14aβb) shows no text-external reference point at all, but proves to be the result of the exegetical reflections of a postexilic scribe.

Of the three different characterizations of the Egyptian bondage established in Exod 1:11–14, only one has had a discernible impact on the literary development of the Exodus Narrative. The priestly motif of forced labor in mortar and brick recurs in Exod 5 where Pharaoh orders that the Israelites' burden should henceforth be increased:

Exod 5:6 וַיְצַו פַּרְעֹה בַּיּוֹם הַהוּא אֶת־הַנֹּגְשִׂים בָּעָם וְאֶת־שֹׁטְרָיו
לֵאמֹר׃ 7 לֹא תֹאסִפוּן לָתֵת תֶּבֶן לָעָם לִלְבֹּן הַלְּבֵנִים כִּתְמוֹל שִׁלְשֹׁם
הֵם יֵלְכוּ וְקֹשְׁשׁוּ לָהֶם תֶּבֶן׃ 8 וְאֶת־מַתְכֹּנֶת הַלְּבֵנִים אֲשֶׁר הֵם עֹשִׂים
תְּמוֹל שִׁלְשֹׁם תָּשִׂימוּ עֲלֵיהֶם לֹא תִגְרְעוּ מִמֶּנּוּ כִּי־נִרְפִּים הֵם עַל־כֵּן
הֵם צֹעֲקִים לֵאמֹר נֵלְכָה נִזְבְּחָה לֵאלֹהֵינוּ׃ 9 תִּכְבַּד הָעֲבֹדָה
עַל־הָאֲנָשִׁים וְיַעֲשׂוּ־בָהּ וְאַל־יִשְׁעוּ בְּדִבְרֵי־שָׁקֶר׃

Exod 5:6 That same day Pharaoh commanded the taskmasters of the people, as well as their supervisors, 7 "You shall no longer give the people straw to make bricks, as before; let them go and gather straw for themselves. 8 But you shall require of them the same quantity of bricks as they have made previously; do not diminish it, for they are lazy; that is why they cry, 'Let us go and offer sacrifice to our God.' 9 Let heavier work be laid on them; then they will labor at it and pay no attention to deceptive words."

A close reading of the passage reveals that the author makes use of two key terms which are firmly rooted in the priestly layer. The author firstly refers to the Israelites' labor as עבדה and furthermore elaborates on the motif that this labor was connected to the making of bricks (לבנים). In short, Exod 5 evidently presupposes the priestly text in Ex 1:13, 14aα. This clearly speaks against the still widespread belief that Exod 5 represents an early element of the Exodus Narrative which according to some even provides firsthand information on the historical situation of the Israelites in Egypt (e.g., Weimar and Zenger 1975: 26–36; Albertz 2012: 102–106). Quite the contrary, the text belongs to a post-priestly stage in the literary development of the Exodus Narrative as has been convincingly demonstrated by Jan Christian Gertz (Gertz 2000: 335–345).

If Exod 5 represents a late addition, this consequently raises the question as to the main purpose of the story. Although the text contributes significantly to fleshing out the scarce description of the Israelites' situation given in Exod 1, it is certainly more than a mere aggadic embellishment. Rather, the author of Exod 5 was first and foremost interested in solving a tension that had emerged in the course of the literary development of the Exodus Narrative. The tension results from the post-priestly, late dtr account in Exod 4 which aims at showing that Moses made the people believe in his message of deliverance by working miraculous signs.

[Exod 4:31]וַיַּאֲמֵן הָעָם וַיִּשְׁמְעוּ כִּי־פָקַד יְהוָה אֶת־בְּנֵי יִשְׂרָאֵל וְכִי
רָאָה אֶת־עָנְיָם וַיִּקְּדוּ וַיִּשְׁתַּחֲווּ׃

[Exod 4:31] The people believed; and when they heard that YHWH had given heed to the Israelites and that he had seen their misery, they bowed down and worshiped.

However, this faith of the people referred to in Exod 4:31 stands in stark contrast to the disobedience, with which they react to Moses' message according to the priestly text in Exod 6:

[Exod 6:9]וַיְדַבֵּר מֹשֶׁה כֵּן אֶל־בְּנֵי יִשְׂרָאֵל וְלֹא שָׁמְעוּ אֶל־מֹשֶׁה
מִקֹּצֶר רוּחַ וּמֵעֲבֹדָה קָשָׁה׃

[Exod 6:9] Moses told this to the Israelites; but they would not listen to Moses, because of their broken spirit and their cruel slavery.

In order to explain this sudden change of behavior of the formerly faithful people the author of Exod 5* creates a new situation by making Pharaoh increase the burden. The faith of the people is thus heavily unsettled, and it becomes understandable why the Israelites do no longer want to listen to Moses (Gertz 2000: 344; Berner 2010: 144).

As a result, the motif of Pharaoh increasing the Israelites' burden should not be isolated from the purpose of Exod 5 within the Exodus Narrative. The motif does not provide insight into the historical situation of the Israelites prior to the Exodus, but rather represents the literary means by which a connection between the non-priestly and the priestly material is established. Thus, Exod 5 is a crucial text for our understanding of the literary history of the Exodus Narrative,[13] while it fails to qualify as a source of the history of Ancient Israel.

Conclusions

To sum up, I would like to start with the basic observation that the biblical picture of the Egyptian bondage is a composite one. It is composed of different literary layers stressing distinct aspects which do not add up to a coherent whole. Considering this literary evidence, it is not likely that these different layers should reflect echoes from the late second millennium which can provide reliable information on the Israelites' sojourn in Egypt. Rather, it can be assumed that they represent different literary explications of the bondage motif which have no immediate value as historical sources.

The brief analysis of Exod 1:11–14 and Exod 5 has shown that the literary explications of the bondage motif each betray a distinct background and purpose. The isolated mention of the store cities in Exod 1:11b was motivated by an etiological interest and reflects a text-external reference point. In contrast, the priestly and to a large extent post-priestly explications of the bondage motif in Exod 1:13–14 and Exod 5 developed obviously within the literary confines of the Exodus Narrative. They can be seen as part of a continuous process of innerbiblical interpretation. This process has been the driving force in the development of the biblical text and thus had a crucial impact on the portrayal of Israel's early history.

However, this is not to say that the Egyptian bondage or the Exodus in general should be regarded entirely as scribal inventions. On the contrary, it is reasonable to assume that the Exodus tradition had not developed out of thin air, but was based on a historical event of some sort, even if it was a marginal one. Nevertheless, it is crucial to maintain a sharp distinction between this event and its literary portrayal in the Exodus Narrative. The biblical picture of the Egyptian bondage is the work of learned scribes and has emerged mainly in the postexilic period. As a result, the biblical text reveals little to nothing about the historical circumstances of the Exodus events. Rather, its exegetical and theological value lies in shedding light on how different generations of postexilic scribes imagined these circumstances and thus contributed to the literary development

[13] In this context, it should be pointed out that the story of the increasing of the burden in Exod 5 has had some impact on the development of Exod 3 as well. The reference to the Israelites' taskmasters in Exod 3:7b represents obviously a secondary addition inspired by Exod 5 where the taskmasters are frequently referred to (Exod 5:6, 10, 13, 14); see Berner (2010: 105).

of the biblical account. As such, the Exodus Narrative remains an important source, yet not for uncovering the origins of Ancient Israel, but for reconstructing the ideological and theological concepts which were dominant at the dawn of Early Judaism.

References

Achenbach, Reinhard. 2003. *Die Vollendung der Tora: Studien zur Redaktionsgeschichte des Numeribuches im Kontext von Hexateuch und Pentateuch. Beihefte zur Zeitschrift für altorientalische und biblische Rechtsgeschichte [BZAR] 3*. Harrassowitz: Wiebaden.

Albertz, Rainer. 2012. *Exodus, vol. 1: Ex 1–18. Zürcher Bibelkommentare AT [ZBK.AT] 2.1*. Zürich: Theologischer.

Berner, Christoph. 2010. *Die Exoduserzählung: Das literarische Werden einer Ursprungslegende Israels. Forschungen zum Alten Testament [FAT] 73*. Tübingen: Mohr Siebeck.

———. 2011. The Egyptian Bondage and Solomon's Forced Labor: Literary Connections between Exod 1–15 and 1 Kgs 1–12? In *Pentateuch, Hexateuch, or Enneateuch: Identifying Literary Works in Genesis through Kings. SBL Ancient Israel and its Literature 8*, ed. T.B. Dozeman et al., 211–240. Atlanta, GA: Society of Biblical Literature.

Blum, Erhard. 1990. *Studien zur Komposition des Pentateuch. Beihefte zur Zeitschrift für die alttestamentliche Wissenschaft [BZAW] 189*. Berlin/New York: de Gruyter.

Dimant, Devorah, and Reinhard G. Kratz (eds.). 2013. *Rewriting and Interpreting the Hebrew Bible: The Biblical Patriarchs in the Light of the Dead Sea Scrolls. Beihefte zur Zeitschrift für die alttestamentliche Wissenschaft [BZAW] 439*. Berlin/New York: de Gruyter.

Dozeman, Thomas B. 2009. *Commentary on Exodus. Eerdmans Critical Commentary [ECC]*. Grand Rapids, MI/Cambridge, UK: Eerdmans.

Dozeman, Thomas B., Konrad Schmid, and Baruch J. Schwartz (eds.). 2011. *The Pentateuch: International Perspectives on Current Research. Forschungen zum Alten Testament [FAT] 78*. Tübingen: Mohr Siebeck.

Fischer, Georg, and Dominik Markl. 2009. *Das Buch Exodus. Neuer Stuttgarter Kommentar Altes Testament [NSK.AT] 2*. Stuttgart: Verlag Katholisches Bibelwerk.

Frerichs, Ernest S., and Leonard H. Lesko. 1997. *Exodus: The Egyptian evidence*. Winona Lake, IN: Eisenbrauns.

Gertz, Jan Christian. 2000. *Tradition und Redaktion in der Exoduserzählung: Untersuchungen zur Endredaktion des Pentateuch. Forschungen zur Religion und Literatur des Alten und Neuen Testaments [FRLANT] 186*. Göttingen: Vandenhoeck & Ruprecht.

Hoffmeier, James K. 1996. *Israel in Egypt: The evidence for the Authenticity of the Exodus Tradition*. New York: Oxford University Press.

Houtman, Cornelis. 1993–2002. *Exodus, 4 vols., Historical Commentary on the Old Testament [HCOT]*. Kampen/Leuven: Kok Publishing House/Peeters.

Kratz, Reinhard G. 2005. *The Composition of the Narrative Books of the Old Testament*. London: T&T Clark.

———. 2013. *Historisches und biblisches Israel: Drei Üerblicke zum Alten Testament*. Tübingen: Mohr Siebeck.

Lemche, Niels Peter. 1996. *Die Vorgeschichte Israels: Von den Anfängen bis zum Ausgang des 13. Jh. v. Chr., Biblische Enzyklopädie [BE] 1*. Stuttgart: Kohlhammer.

Levin, Christoph. 1993. *Der Jahwist Forschungen zur Religion und Literatur des Alten und Neuen Testaments [FRLANT] 157*. Göttingen: Vandenhoeck & Ruprecht.

Otto, Eckart. 1996. Die nachpriesterschriftliche Pentateuchredaktion im Buch Exodus. In *Studies in the Book of Exodus: Redaction—Reception—Interpretation. Bibliotheca Ephemeridum Theologicarum Lovaniensium [BETL] 126*, ed. M. Vervenne, 60–111. Peeters: Leuven.

Otto, Eckart (ed.). 2000. *Mose: Ägypten und das Alte Testament. Stuttgarter Bibelstudien [SBS] 189*. Stuttgart: Verlag Katholisches Bibelwerk.

Propp, William H.C. 1999. *Exodus 1–18: A New Translation with Commentary and Introduction. The Anchor Bible [AB] 2*. New York, NY: Doubleday.

Redford, Donald B. 1963. Exodus I 11. *Vetus Testamentum* 13: 401–418.

Schmidt, Werner H. 1990. *Exodus, Sinai und Mose: Erwgungen zu Ex 1–19 und 24. Erträge der Forschung [EdF] 191*. Darmstadt: Wissenschaftliche Buchgesellschaft.

Thompson, Thomas L. 1987. *The Origin Tradition of Ancient Israel: I. The Literary Formation of Genesis and Exodus 1–23. Journal for the Study of the Old Testament: Supplement Series [JSOTSup] 55*. Sheffield: JSOT.

Utzschneider, Helmut, and Wolfgang Oswald. 2013. *Exodus 1–15. Internationaler Exegetischer Kommentar zum Alten Testament [IEKAT]*. Stuttgart: Kohlhammer.

Weimar, Peter, and Erich Zenger. 1975. *Exodus: Geschichten und Geschichte der Befreiung Israels, Stuttgarter Bibelstudien [SBS] 75*. Stuttgart: Verlag Katholisches Bibelwerk.

Zahn, Molly M. 2011. *Rethinking Rewritten Scripture: Composition and Exegesis in the 4QReworked Pentateuch Manuscripts. Studies in the Texts from the Desert of Judah [STDJ] 95*. Leiden: Brill.

Fracturing the Exodus, as Told by Edward Everett Horton

21

Baruch Halpern

Abstract

The Exodus was a fable inspired by possible events of Israel's past although its historical genesis will be as irretrievable to us as will be its original narrators. It is important to understand that the text's modern discussants to not wield the tools necessary to confront the epistemological challenges that we face. The true question is: "What do we need to know in order to know what we want to know?" Faced with storytellers and their audiences who contributed historical detritus while adding artistic value to the story, the subject's sole value is to recover the story's magic: to understand Israel's modes of social thought over time and the culture that immortalized the Exodus.

The Exodus as a historical issue is a faith question.

Does God Exist?

We ask that about what? Ghosts, alien abductors—the modern incubi. "They told Marconi," runs the Cole Porter song, "Wireless was a phony." Is the Exodus a steamship or cold fusion?

Evidence for the Exodus starts with poetry. Its marriage with science bids: How far does knowledge propel imagination, and the reverse?

Some think the Exodus a fable, others a datable event as described. But most scholars believe the Exodus happened to a particular crew. Many compare the Exodus to the Pilgrims in the American Revolution, their foundation myths taken at face value, who somehow assimilated others. Ecumenists imagine an ancient DAR. Others think Team Moses consisted only from Mayflower Brahmins.[1]

A master of such emended claims, Ephraim Kishon, recalled Amos 3:1: a "whole family [Yhwh] raised from Egypt." A family, Kishon

B. Halpern (✉)
University of Georgia,
Athens, GA, USA
e-mail: b.halpern@gmail.com

[1] A later Brahmin, Emerson (1836: 1) reports: "Our age is retrospective. It builds the sepulchres of the fathers.... The foregoing generations beheld God and nature face to face; we, through their eyes. Why should we not also enjoy an original relation to the universe? Why should we not have a poetry and philosophy of insight and not of tradition, and a religion by revelation to us, and not the history of theirs? . . . Let us demand our own works and laws and worship." Newton's Book of Nature is a child of the paleolithic, and an adult in the seventh century BCE.

T.E. Levy et al. (eds.), *Israel's Exodus in Transdisciplinary Perspective*,
Quantitative Methods in the Humanities and Social Sciences, DOI 10.1007/978-3-319-04768-3_21,

reasoned, included grandparents, parents, and two children, six people in all.

A contemporary biographer describes St. Cuthbert's last days as solitary, except for one healing. 25 years later, Bede reports elaborate communications to Cuthbert's abbot on doctrinal issues. Jenkins (2013) describes the discrepancy as his introduction to historical criticism.

In a 1937 study, a witness is challenged to sharpen his senses. A horse and cart becomes a zebra and chariot, then on to giraffes, police, airplanes.

Finally, poor Marco, sheepishly admits he saw:

> "Nothing," I said, growing red as a beet,
> "But a plain horse and wagon on Mulberry Street."

The author (Geisel 1937) entitled this account, "A Tale That No One Can Beat." Just so, several Biblical texts portray the Exodus. Assmann describes it likewise. Theologians call this incomparability.[2]

So which is the Exodus? Cuthbert, or Geisel?

We distinguish details—the "destruction of the world" story (Sparks) has an Egyptian orientation, omitting resettlement of the escapees. But Semites always swarmed in the Delta (so anthropometry). A report that "Shasu of Edom" passed the Wadi Tumeilat (Pap. Anastasi VI 4.11–5.5) or that two slaves escaped past Migdol (Pap. Anastasi V 19.2–20.6; both ANET 259) may be routine police blotter material. Or, the Exodus.

Egypt dominated Canaan from EB I through the Ottoman. From 1520–1150, it exerted supreme influence. It recruited elites. Dever correctly asks whether the Exodus story stemmed from an elite.

The *story* was typologically true. It engages Egypt, conflates pertinent stories. It identifies Joseph with the Hyksos.[3] So, one desert rat claimed to have escaped Egypt. Another knew of sorcery, as channeling supernatural powers. Yet another experienced the twilight onslaught of Delta insects. And this little piggy squealed, "Egyptians sank."

Of special moment is the wandering.[4] In Near Eastern culture, a bandit background was romantic: David, Idrimi, even Sinuhe. To have braved the uncultivated land is a badge of courage, though mainly reflecting tax evasion, draft-dodging, and smuggling.[5] In no bureaucratic milieu does allegiance attach to apparatchiks. Participation in the wandering was an ideal, because of some primal sense of allochthony, at play in the formation of Israel's identity.[6]

Storytellers formulated an Israelite national myth. They festooned it with spectacle. They mocked obtuse Egyptians, like the Nile, running in the wrong direction. "And the sea returned . . . to its tide as they fled toward it." Storytellers mystified, and edified. They explained Pesah, circumcision. Even before our versions were codified, they acted the story. We have scores, or scripts, from JEPD and elsewhere, with a message to succeeding generations.

Scripts are not performances. The score was *performed* in individual households, even according to P.[7] Most such households were

[2] 2 Sam 7:23 = 1 Chr 17:21; Deut 4:7, 34–39, for example. The motif of incomparability in conjunction with the Exodus begins with Exod 15:11. Cf. also Deut 33:29; Ps 147:19–20; contrast, 2 Kgs 18:35 = Isa 36:30 and 2 Chr 32:14. The specificity of the comparison is absent in post-exilic texts.

[3] Halpern 1993. To the treatment of Joseph's Egyptian name there as a Semitic-Egyptian portmanteau should be added a possible tie to the god, Baal Zaphon, on whose prominence particularly in the New Kingdom see Bietak.

[4] Based on interviews conducted at Bedouin encampments and settlements in the company of Emmanuel Marx, 1984.

[5] After 1967, Israeli police found over 70,000 stolen automobiles buried in the Sinai for resale in Egypt.

[6] In 1983, Yigael Yadin described (in conversation) his first meeting with the cabinet of Jordan's King Abdullah. The king introduced his ministers by name and role, and in some cases added the remark, "He's an Arab." Asked what distinguished "Arabs" from the others, Abdullah responded that they had epic poetic traditions. Yadin asked, "Like this?" and recited the few lines of Bedouin epic he recalled from his student days. The king embraced him: "Ya Yadin! You're an Arab!" Cf. Talmon 1966; Dozeman 2000.

[7] Exod 12:1–27, 43–49; 13:1–16. Friedman (2003) assigns 12:1–20, 40–49 to P, and 13:1–16 to E. Neither envisions the sacrifices of Pesah/Mazzot as public, but rather as particular to households or compounds.

elite (with young ewes to slaughter).[8] But in traditional cultures, elites mobilize biddable dependents. Imagine that 5,000 households rehearsed the story of the Exodus. That means 5,000 performances annually. The divergences must have been impressive.

This unmanageably mixed multitude begot the scripts our authors produced.

Can we even approach this Exodus? Scholars once guessed what underlies the Eden story. The Flood, popularly, remains a live subject thanks to folklore and archeologists' recovering Noah's Ark or settlements beneath the Black Sea (cf. Atlantis). The Tower of Babel reflects the community and architecture of southern Mesopotamia. And *scholarly* discussion of the patriarchs outlived the 1970s.

As Dever (2003) says, the Conquest of Joshua 4–13 is a dead issue. Nineteenth-century scholars dismissed it, based on Judges.[9] Its mainstream defenders persisted into the 1980s: they cited "correspondences" to archeological and other sources.

We frame issues as the "Exodus," or the "Conquest," or the "Davidic Empire." Biblical presentations damn us, to think in categories, to argue accuracy and even authenticity. We anchor doctrine in lapidary history. We might believe what Ezekiel saw.

We *sell* faith, on say-so, or Kipling's "Just So," or Pogo's "Go-Go Stories." Like these, our Exodus texts begin—explicitly, as children's stories. Like Homer, whom Augustine's culture took literally. Like our literature, Homer underwent a metamorphosis in genre.

Children's literature doesn't get our historical dander up until it gets into the court system. Then it becomes *homoousios* versus *homoiousios*.

My instinct is, there was an exodus. But my instinct is also, it bore no resemblance to ancient or modern accounts. Stories, like Marco's, grow in reimagining. Demote our pharaoh: make him a nomarch. No, a mayor, or a brewer, or just the owner of a field. Then the story becomes pedestrian. So do I believe there was an Exodus?

What's the historical kernel of Little Red Riding Hood? A child goes through the woods, and into danger. But did Hansel and Gretel really lose signal on their primitive GPS? Do our autumn gingerbread houses celebrate liberation from child sacrifice? What is the background of Puss-in-Boots? Such tales have *some* inspiration—Daube's suggestion (1956) for Humpty-Dumpty remains controversial (Opie 2004: 176). Still, magic beans are a children's market.

Addressee Versus Audience

We intuit that the question of a historical Exodus is charged; it is church polemics. Cicero's argues that belief in gods' activity inspires the piety that maintains social relations (*de natura deorum* 1.4): is this the case here?

We dismiss a three-million-man march through the Sinai, even allowing they recycled garbage.[10] To whom is an argument against a mass Exodus addressed? It is something Tom Paine might have written (1961). Addressing the Bible, however, is not for a scholarly audience. It is a part of that distancing from naive literalism that earlier characterized Greek myth, Roman myth and lore, even ancient annals.

To locate the Exodus, to rescue its kernel, respondents modify scale, date, even the events. They cite terrain, goods, language, names, customs, and monuments: the stage sets coincide, imaginatively, with realia. In *this* metanarrative,

[8] Hendel's intimations lead to the inference, it was town elites who eventually shed Egypt's longstanding sovereign demands for logistical support. The model of the Peloponnesian War or of Byblos in the Amarna archive suggests a rise of competing parties in each town. Initially, the ruling elite and their village client elites depend on the Egyptian network; their counterparts ally with mobile elements, who disrupt the crops and the caravans of the rulers. More difficult to account for is the role of Philistia in this equation.

[9] Stade 1881; Halpern 1988: 182: Joshua distinguishes between "conquest" and "supplanting."

[10] Of course, land armies did traverse that terrain, without leaving detected archaeological traces. As Friedman observes, the census numbers are from P.

the addressee is not the Bible, but our first group, who discredit detail to discredit the story. In unison, we shake our heads at the sophomoric dismissal of a cultural heritage. At the least, we say, we can trace the history of the tradition. Which writer or redactor added what detail?

Unfortunately, redaction-criticism is too imprecise for the task. We have 5+ sources. Even the document whose identity is agreed, P, is dated from ca. 700–400. P ratified older practices, and stories, too. Some of its usage anticipates, or programs, later times; some accrued later, as it did with the gospels. To pinpoint its date is a tall order.

The most misunderstood principle in Biblical Studies is that a text testifies only about the time of its writing. The idea holds water. But to look at the text's past is to perform (Wimsatt and Beardsley 1946) "an act of criticism." Every artifact has a prehistory, an imprint of its past. Analysis based on cherry-picked details from a narrative is implicitly a history of the text and of its reproduction.

Bluntly, most of our assumptions about things like post-P additions (Schmid; Römer) fail the historical test of sociological imagination, to invoke C. Wright Mills. Especially in the post-exilic era, the likelihood that lone scribes made quirky amendments (versus errors) seems small. These required communal approval. Normally, they involved doctrinal freighting (the Jubilee, for example, affirms a policy ethic). Where such changes are documented, as through Diatesseronic readings in the New Testament, they are generally of this sort. A late addition, thus, of the names Ramesses and Pithom, would be gratuitous, neither important nor accidental. This lack of sociological vision explains unflagging inquietude in the history of tradition.[11]

To discuss history, whether of a text's insuscitated past or of its present,[12] we need to understand the intentions *behind* the particulars. We must experience the author's worldview, so finely as this is possible. Categories are too totalizing to explain particulars: E is northern, or P is Aaronide, or Jeremiah is a partisan marginalized after Josiah's death.[13]

We need to know who is telling us the story (R. E. Friedman). I add, we need to know the teller intimately, in his bureaucratic, political and economic life, and even his personality. But here, an adjustment to any one assumption requires adjustment to the next, like leveling a set of hangings in a house that's out of plumb. This means you never really finish adjusting, but just reach a point where you're willing to live with what you've got. Euphemistically put, it's the "Oh, good enough for now" moment. Which is the uniform maximal claim of all scientific thought.

Further, if every reading is individual (Wimsatt and Beardsley 1949, anticipating "reader response"), a scholar rarely persuades others about hypothesized formative stages. The very fragmentation of visions licenses ongoing

[11] That intentional amendment is communal is a lesson that should have been learned already when the Samaritan Pentatuech was introduced into text-critical discourse in the eighteenth century.

[12] For one trajectory of this concept from Collingwood to Gadamer, see Lucas 1997: 111–112.

[13] The last is a warning: it is a terrible misreading, with a bearing on the evolution of the book itself. Note the survival of DtrH as the definitive account of Israelite history, or of Deuteronomy, or of other Josianic partisans (Zephaniah and Ezekiel) and the canon all these appealed to (Amos, Hosea, Micah, Isaiah): our informants left a very specific discussion, within a narrow tradition, first to themselves and ultimately to us, as David Vanderhooft has pointed out (in conversation); all this material, like summations of pre-Socratic fragments, was party literature. Likewise, the survival of the ideology of the reform in 2 Isaiah, 3 Isaiah and the whole establishment of the post-exilic era. Churches have a penchant for portraying themselves as marginal, particularly in their first centuries. The zeal of reform abates only long after a party licenses its intellectual terrorists to conduct witch-hunts. At first, parties plead for pluralism, diversity. Power diminishes their tolerance. It also bears recollection that Jeremiah was appointed one of the top (three) officials of the Babylonian provincial administration. His story is reminiscent of that of Yohanan ben-Zakkai's relationship with Vespasian (T.B. Gittin 56a-b), and probably a model for it. This is an established authority (Sommer 1998; Halpern 1998), the founding figure, the truest oracle, of a post-exilic party. Even from Egypt, even under the Persians, his position and faction emerge from the Exile with the upper hand. Pseudepigraphic recourse to Isaiah is meant to undermine his influence.

discussion. Why are medieval metaphysics a thing of the past? Because epistemology, beginning with Hume, like the novel in Joyce, formally acknowledged its limitations. My fear is, the Exodus event belongs with medieval metaphysics. If so, its discussants are very late to acknowledge their limitations.

Third, if dating P is complicated, think of this: P is only an inferred entity, although I think it real enough, like electrons in cloud chambers. We turn to the Exodus. This is merely something *IN* P, just like the Jubilee year. Scholars *know* they're taking intellectual shortcuts when they pronounce on it. Or, on the fact that P makes Sinai the goal of the Exodus, and so interprets Exodus 15 (Halpern 1983; R.E. Friedman; Maeir, this volume, Chap. 31). In other sources, the Conquest is the matter of moment: no land, no covenant.

Amaleq, What Is It Good for?

Even on workaday subjects, responsible scholars can hold odd opinions—that is the history of our field. But why don't we hear about comets, volcanoes and red tide in connection with Josiah's reform? Some say it never happened; some compare David to King Arthur. But we have no claim that Huldah translated Josiah to Avalon, or that the Witch of En-Dor summoned up an alien.

Which does suggest a certain poverty of imagination in the field. Troy, too, has its Mudville Moments: the mighty alien, divine intervention, invisibility and invulnerability, have struck out there, as well. The Classics tell us that Caesar rose three days after his death, or that Nero disappeared to the east, to India. They even give us Atlantis.[14] And all that, of course, flows from a long, uncodified intermixing of local cultures.

Worse yet, unfortunately, is the absence of imagination in the Plague cycle. Even a historian with an imagination stunted by professional training might add earthquakes or extraordinary flooding. Why is there no attack by leopards, asps, mythical beasts, mythical armies (*a là* Genesis 14), Sodomite brimstone rather than hail? The Egyptians could be just as obtuse with more spectacular plagues.

Possibly, until the final plague or two, the prosaic progression explains Egypt's reluctance to acknowledge Yhwh's signs. If the glacial escalation of spectacle is intentional, then the authors' touch is remarkable, a form of Stan Laurel restraint. It just takes a while to dawn on the Egyptians that these plagues are serious. (Contrast the rapid series of disasters that befall Job.)

It seems reasonable to speculate: one does not turn hell into chaos; hell is the region that creation does not turn into abiding order. It is fleeting. Egypt is the underworld, and the Israelite battle, fought by Yhwh, is not against Sea (surely not before P), but against Mot. The Exodus is the *maqlû*, the day of the meeting of heavens, the biosphere, and the underworld, all the divisions of our cosmos: its festival is All Souls' Day. The Hosts[15] of Yhwh depart the netherworld, to enter the biosphere, later to gain eternal continuity in community. The promise is that they will scale the heavens of land and God, Eden's Hill. If the object may culminate in a day of darkness and not light, in the designee's death (cf. Am 5:25 ff.), perhaps the build-up, albeit relatively pedestrian, could be eerie.

This sort of speculation about sensibilities seems purely literary. It invites no historical or archeological test (despite ample evidence in each case) of an ideological bent. Yet to dispense with it is to miss the depth of the story.

Beyond this, others discuss the full narrative cycle (as Rendsburg). Exodus 15:1–18 is our

[14] Another need: a serious discussion of the claim that archaeologists are licensed to choose just any interpretation of text, after having rejected text, because they cannot distinguish between those who date texts rationally and those who do not. A tit-for-tat response would be, archaeologists find Noah's Ark, Atlantis, alien colonists, and "Malibu Barbie" (Rowe 1994), on occasion. Hard to distinguish between those archaeologists and the others?

[15] P's conception intentionally demands the plural.

oldest extensive composition linking an Egyptian disaster with the Conquest. J already incorporates that song to sustain his reconstruction of Israel's victory; he has no truck with divided waters.[16]

Later, the division of the Jordan becomes the sea dividing, paralleling creation, in P.[17] Also in P, Sinai becomes the goal of the Exodus. In other Pentateuchal sources, it is the Conquest that is of moment: no land, no covenant. One wonders how P's idea of successive covenants correlates to the Greek "ages of man," or even to the successive cosmoi of some pre-Socratics.

Tantalizing, and instructive with regard to our own limitations, are Amaleq and Midian. Judges lumps the two together with easterners.[18] Pentateuchal sources associate them with the Exodus.

J's Amaleq, in the Negev (Num 13:29), attacked Israel between Exodus and Sinai (Exodus 17). D's Amaleq attacked stragglers (Deut 25:17). Both commemorate erasing Amaleq's memory—a self-working contradiction, like Max Beerbohm's (1919) "Enoch Soames."

In J, a Midianite, Reuel, furnishes Moses's wife; E calls him Jethro. E associates Midian with Balaam and Moab, motivating P's choices, as we shall see.[19] J has Edom attack Midian (Gen 36:35) nonetheless, perhaps meaning some part of it.

P makes Amaleq part of Edom (Gen 36:12, 16). He also associates the name of Reuel with Edom (Gen 36:10, 13, 17).[20] But he makes

[16] That earlier commemorations of the action at the Sea of Reeds do not describe the cleavage was observed by Cross 1973. Instead, they use terms such as "turned dry" (= was fordable?). But the agreement is with the Jordan River division in Joshua: as in J, and evidently in E, the Sea is withheld, and then restored to its flow. That J incorporated Exodus 15 is certain, since the E doublet has Miriam sing the incipit: R_{JE} relates both, rather than combining them; J's account thus ended with Exod 15:1–18. Antique poems are used by all Pentateuchal sources as warrants, evidence for, their narratives. This is not a strategy one finds, however, in the layers of the Deuteronomistic History from the late seventh century forward. A question arises, then: what event is in mind in Psalm 68? Zebulun and Naphtali are singled out as in SDeb; Benjamin is youngest; Yhwh's in Jerusalem. We have flooding and kings fleeing, as in SDeb; there are echoes of Deut 33, probably Gen 49, and a J reflex of its processional use in Num 10:35 f. (Halpern 1981: ch. 3; cf. Knohl 2010). Booty and tribute eventuate. This song requires fresh attention, as do a number of lesser early compositions.

[17] P has the wind as the first principle in Gen 1. In J's Exodus 14, wind is motive, driving water back; in P, Moses divides water without wind. In Exodus 14, P reenacts the flood, the divided water combined again, reversing the cosmology of Genesis 1. Versus J, where the change back to dry land as in Gen 2:2 ff. is more natural.

[18] Judg 6:3, 33; 7:12; cf. 3:13; 10:12. There is also a mixture of Moab and Midian in the Balaam story.

[19] Num 22:4, 7. Friedman (2003) 280 argues that Midian enters the Balaam text secondarily, as a reaction to the R_{JEP} combination of J's Moab narrative in Num 25:1–5 with P's Peor dalliance in 25:6–19. This division, to begin with, is problematic. There is no JE conclusion in Num 25, a genuine rarity, as Friedman has shown. Further, there are no close parallels in R's technique: the gratuitous intrusion of Midian into Moab seems sudden, not subtle. I understand Midian here as a representative of the mystical east, as though their elders are magi. They are thus appropriate porters of magical tokens to Balaam—tokens used in divination (hair, hems, figures, arrows). Balaq in Num 22:15 sends weightier ambassadors than those in 22:7, indicating that Midian's elders were recruited to impress the seer. In the continuation, it is the god who yields to their import. (The mockery of Balaam for not divining his ass's message perhaps engages the purposeful straying of asses in 1 Sam 9.) God's acquiescence to the journey is a doublet: 22:20–21//22:22–35. Balaam's foot being crushed as a result of the angel's action (22:25) also relates to Jacob's lamed thigh in Gen 32:26, 32–33, with the pun on *ngd* (C) in the same passage. That Israel "covers the eye of the land" (22:5, 11) relates to the locust plague (Exod 10:5, 15, E; and note Deut 33:28, which relates, too, to Num 23:9).

[20] P tying the same name to Gad (Num 2:14). Josh 13:21–22 follows P (Num 31). In Num 10:29–32, J also assigns Moses's in-laws to Midian. J makes the assignment in Exod 2:16–22, E in Exod 3:1; 4:18. That Judges 4 follows J rather than E probably has to do with Judahite antecedent exegeses of the Song of Deborah. Note further that P agrees on the name of the mountain (Sinai *v* Horeb) with J against E. However, the more freighted revelation of the divine name has PE agreement against J (along with the Midianite execration and the absence of imprecation against Amaleq; and, plague narratives). Since so important a concurrence suggests other such choices are charged, P's relegation of Reuel to an Edomite lineage is probably a comment on J's naming of him as Moses's

Midian Abraham's offspring, the target of a refined *ḥerem* in Num 31.[21]

Judg 4:11 traces Jael's Heber, the Qenite, to J's Midianite scout Hobab, Moses's brother in-law. Moab, Edom, Amaleq and Qen appear in sequence in Balaam's oracles (Num 24:17–22). Balaam calls Amaleq "the first of nations," possibly suggesting antiquity, and contemplates its end.

Still, only Saul (1 Sam 14:48) and David (1 Sam 30) defeat Amaleq, and an Amaleqite kills Saul (2 Sam 1). Midian is legendary for its defeat by Gideon, commemorated in Isaiah (9:3; 10:26, with puns on Oreb). Neither Amaleq nor Midian appears as a political entity thereafter. However, David's Edomite prince flees to Egypt from (a geographical) Midian (1 Kgs 11:18), perhaps explaining (versus J in Gen 36:35) P's association of Reuel with Edom (Gen 36:10–17).

So why attach Amaleq or Midian to the Exodus? Is the Exodus, after all, a catch-all for any folkloristic background? Is it a repository to acclaim David, to justify esnecy, to introduce new laws about the *ḥerem*, to justify judicial structure? *Was* there a hostile Amaleq in Sinai?

As a community of scholars, we almost all hope to identify, or despair of identifying, one thing:

a plain horse and wagon on Mulberry Street.

That weary progress, the Israelite carnival. This is a historiographic question, not an archeological one, poetic, not scientific. We apply the methods of history, to be sure. And so, elements in our texts that we judge unknowable to Israelite storytellers in the Iron Age are the only things that require explanation and contextualization—changing geomorphology (as Mosier), or the identification of Avaris (Bietak). If the Israelite scribes knew the 400-year-stela, we can only call on it as evidence of a later portrayal of Israel as Hyksos (Halpern in Shanks, et al. 1992; 1993). It's what they could not be expected to reconstruct, if anything, that deserves our attention.

Professor Dever and I discussed these issues back in the 1980s and again at a Hershel Shanks Smithsonian conference (Shanks et al. 1992). We also discussed the conquest. My, fallible, recollection is that we agreed what the evidence was, what the possibilities were for interpretation. And so, we left medieval metaphysics behind, though, as his paper shows, each of us retains a view (and Dever 1997). So perhaps we just carry the metaphysics more personally than publicly.

"What You Mean, 'We?'"

I have few thoughts about the Exodus, nor a synthesis, that readers don't anticipate. So, I am left with a simple thesis.

Our subject's sole value is *Kulturgeschichte*, understanding Israel's modes of social thought over time.

Our storytellers, and their audiences, contributed historical detail while adding artistic value to the story. We can all develop reasoned, albeit creative, reconstructions. None of us can write—*EXODUS*: *THE EVENT*. We *can* know what the next answers will look like.

No synthesis will unify every detail in the narrative. When archeological inference coincides with textual, alternatives are almost always possible.

We are addressing questions more basic than historical ones, such as Omride foreign policy. We are not testing for the God Particle. This is string theory, in 19 dimensions, and dark energy at 24 % of the universe. We do not wield the tools necessary to confront the issues.

The question is always:
What do we need to know,
In order to know,
What we want to know?

father-in-law, like the separation of Qen from Amaleq in 1 Sam 15. In fact, E refers to one of Moses's wives as a Cushite (Num 12:1). Some identify her with Cushan, parallel in poetry to Midian in Hab 3:7.

[21] The text may comment on Saul's handling of Amaleq in 1 Samuel 15 and simultaneously on David's distribution of spoil in 1 Sam 30:24–25. It also establishes the rules of Holy War in P. The term, *zikkārôn* in 31:54 obliquely recalls the Amaleqite issue. A Midianite woman violates the tent in Nu 25:6, 14; Midian are children of Qeturah in Gen 25.

If we can't answer that compound question, we have no business asking the question from which we started. If we don't know what we need to know in order to know what we want to know, our means are deficient.

Only a rock-solid history of traditions could yield plausible scenarios by which the Exodus story developed. The Song of the Sea dates with reasonable security to ca. 1,150–850. In it, an Israelite authority connects Egyptian defeat and the Conquest. Other archaic poetry, such as Balaam's ("El who brought him out from Egypt has bison horns": shades of the calf), also alludes to the Exodus, and anticipates the Conquest.[22] The Exodus appears in the Covenant Code (Exod 23:9), yet another source of E (like the Song of the Sea in Exod 15:21). Even the idea of a forty-year detour is already enshrined in Amos (2:10, before the felling of Amorite titans; 5:25).

But historically, we do not look for split seas.[23] Nor do we seek out suites of disasters: these furnish material for stories without locating historically-connected events (Hendel 2001). The chain-of-evidence principle demands that we make choices, even in imagining the tradition's growth.[24]

And the worst of it: what archeological reflex do you expect for your version of the story? Hypothetical: we have an ocean of DNA evidence from bones in the Delta and tombs in Israel in Iron II. Make the match perfect: we can filiate forty Egyptian cemeteries with 200 Israelite tombs. Can we say we've found the community of the Exodus, at whatever time, or do we need still more data? Winston Churchill once responded in question time to an inquiry on rearmament. He said, My friend puts me in mind of the gentleman cabled for instructions about his mother-in-law's death in Brazil. He wired, "Embalm, cremate, bury at sea. Take no chances." When will it be enough? When we are taking no chances?

[22] Linguistic features of Exodus 15 precede standard Hebrew. Semantically, it deserves and has not received thorough review, but "driver" for *rkb* seems to fall out of the lexicon, as do *nd* (but for Isa 17:11) and transitive *r'ṣ* (otherwise, paired with *rṣṣ*, in phonetic play (*r'd* may be related), both used in quotations; note *ṣll* with the meaning, "plumb" and transitive *nwy* (cf. Hab 2:5; and, the pun in Exod 15:13). The absence of light cavalry, consonant with Egyptian practice, no longer characterized Israel by the late ninth century. The poem is not twelfth-century, as it presents Philistia as a regional designation (though not as a polity). Its political geography, including Edom and Moab as regions (without naming Ammon, possibly for poetic reasons), is more plausibly Iron IIA than Iron IA. Another archaic poem, Balaam's "El, who brought ...," denies wizardry in/against Israel (23:23; 24:1) and may have inspired P's contest with Egyptian magicians. The use of the verb *yṣ'*, C, of god in these passages perhaps programs its use in E and P, principally, in connection with the Exodus. The other verb commonly used for the Exodus, *'lh*, C, often takes Moses as a subject, negatively in P especially (with exceptions). Like this, other poetry with Exodus or Conquest associations, including Balaam on Amaleq and Qen, deserves detailed treatment.

[23] That seventh-century concept is absent from Isaiah (11:15–16; 19:5), with his seven branches of the "Sea of Egypt" and Yhwh's five towns, altar and stela in Egypt, and repetition of Egypt's humiliation (19:17–25). The idea of a sea's division comports with his imagery, as with the language of hydrological engineering, but enters the Reed Sea tradition afterward. (Sennacherib, in 2 Kgs 19:24; Isa 37:25, boasts of drying up the Niles.) Drying also appears in Isa 42:15; in a closer echo of Exodus in 44:27; and with passage across waters, 50:2; 51:10; Ps 106:9. Jer 51:36 connects it with Babylon; Hos 13:15 has east wind drying, but not connected with Exodus. Am 7:4's dried Tehom is cosmostrophic, as Nah 1:4. *bq'* as a verb here occurs in Exod 14; Isa 63:12–14; Ps 78:15; Neh 9:11; it is cosmogonic. Ps 74:15. *ybš* C appears in Josh 2:10; 4:23; 5:1, referring, respectively, to the Reed Sea, the Jordan and Reed Sea, and the Jordan; the deverbal noun appears in Exod 14; 15:19; Ps 66:6. Note the relative paucity of reference to the event at the Sea in any form, especially in earlier materials.

[24] Here, the test of historical intellect is most acute: what can be confirmed or refuted, will, for a time. First, fallacious arguments and, then, dead assumptions are by-products of inadequate tools. So some historians dwell among contemporary horizons of assumption; others imagine developments beyond that 360° horizon. Yet scholarly contributions about even a minute issue have a butterfly effect. Anticipating, not trends in emplotment or rhetoric but in areas where evidence will be sought, is more like writing Verne-like science fiction: futurology more than divination.

Identity Hell

A sign on the path: Taboo. In American culture, it's Nazis, Vegans, Creation, or the Boston Red Sox. Each marks identity, always permeable and thin. How permeable is the membrane between programmatic cultural history and our own?

Why do classicists discuss a Trojan War? It is their guild's national myth. What fascination origin myths hold! Identity. Everyone in the Mediterranean world is archaizing in the Neo-Assyrian era and after, and Levantines throw national history into the Late Bronze Age in order to exaggerate their antiquity. Babylonians and Egyptians extend historical romance farther, with omens running to Sargon, forgeries to Manishtushu (Rubio 2009). We do have a signet ring of Ahab from the first century CE.[25] You need not forge the cosmos, the tabernacle, Gudea's temple, to a divine blueprint, just a saleable souvenir—an object that replaces a journey. Were there indeed multiple "arks" (an idea reviewed here by Noegel)? Representations sold like hotcakes. What else are "model shrines"?

Some souvenirs, such as Deuteronomy, inspire the renewal, rejection, or renegotiation of traditional identity: Jeremiah says, Yhwh will no longer be known as the god who brought Israel out of Egypt, but as the god who gathered them from the lands to which they were scattered by Mesopotamian powers (16:14; 23:7); Deutero-Isaiah expands on the idea. The Exodus as a national myth is supplanted by a return from exile; or, the appeal of the Exodus as a national myth is the regular presence of such a need.

Just that reformulation explains what the Exodus is good for. What's so great about leaving Egypt, and its sensuous fleshpots? Yhwh doesn't evict you, but installs you. The Exodus is the prequel to the possession of Canaan.

The patriarchal narratives don't front the Exodus as an incentive. Genesis 15 mentions it to explain the delay between the promise of the land and its fulfilment—like the *katechon* in 2 Thessalonians, or even like the leisurely cruise from Exodus 8 to the end of Deuteronomy. The Exodus features quirky stories about Egyptian dupes. It's raining frogs and the Egyptians hardly notice. The Conquest, by contrast, attracts short, dull shrift—the Jordan splits, a wall falls, the Gibeonites dress up for a party, and, finally, the sun stands still. The rest is a yawn.

And when *do* Israel celebrate Mazzot? J already places it with the Exodus. Again, American scholars routinely compare the Exodus to the Pilgrims' Thanksgiving, adopted by the state only in 1863. But even Puritans, with their blarney about persecution in England, can't name the day the Mayflower was built, or set sail, or exactly from where, or specify which of the various ships named Mayflower it was.

Thanksgiving sanctifies the settlement. It celebrates colony foundation, as ancient Greeks did. Where is the Israelite festival celebrating the Conquest? On an optimistic reading, is it Pesach itself, as in Joshua, or is it Sukkot? Sukkot is now the festival of lawgiving, just as Sinai has supplanted the Conquest as P's focus of fulfilment. In discussion after Assmann's paper, embarkation was treated as utopian. This is far from being at the story's heart, but fits with the legislation of an ideal society. Advertising, after all, is always utopian.

Who celebrates departure, Byron shaking the dust of England from his feet? Is Israel Nora in Ibsen's *Doll's House*? The story has become background, like the patriarchal narratives, for normative elements in the culture. Is the myth of the 10th plague merely a substitute for the Aqeda (or the reverse), as the etiology of a sacrificial debt?[26] Is the Passover, as Cooper and Goldstein suggest, a nationalization of an earlier ancestor feast? Are we looking, foremost, at Ishtar's descent, at the harrowing of hell, at Odysseus in Hades or on Aeaea, in fact, at a dyshemerized resurrection?

[25] Like Deuteronomy in de Wette's estimation, not in ancient script; Uehlinger 1997; Harvey and Halpern 2008.

[26] Remembering always that the sacrifice is for a kinship unit, and thus, the *yāḥîd* is the representative not of a nuclear family, but of a community, QR *yaḥad*.

Another thought experiment: consider, the Exodus is never severed from the Conquest, because no one was grateful to leave cultivated land. But setting a story in Lotus-Land, storytellers had more latitude to indulge their imaginations, and cover to sneak coded messages into their work. So, P, naming names, gets downright mystical at Sinai, and in the construction of a cosmological sanctuary that actually was mobile. Antiquarian-like, he plumbs detail.

The Exodus story is like ancient history recovered late in Babylon, through the archeology of Nabopolassar, Nebuchadnezzar, especially Nabonidus, but also of their predecessors.[27] It is, *faut de mieux*, itself a tapestry, not artistically unlike that more sophisticated one of Bayeux, detailing currents in ancient culture.

To return to Exodus as a story: J imagines magic boils. This is Judahite humor, as in Job, *Schadenfreude*. All the sources mock Egypt's pretense to knowledge and know-how (perhaps the best is the inability of Egypt's sorcerers to appear, in P, once stricken with boils). They paint the Egyptians as butts because Yhwh has become the trickster here. R.E. Friedman (1986) had the distinction of identifying how important to our narrative a cycle of trickery is in Genesis—it is a variant of Delphic prophecy, or self-working ancient prophecy.[28] "Deception for Deception", as he called it, continues into Exodus. First with tricky midwives, then with tricky interpretations of what "throwing into the River" means, and, finally, in the form of a trickster *god*.[29] In this material, Yhwh plays the role of the eldest of The Billy Goats Gruff.

Egyptians built wonders, husbanded wealth and power. Yhwh befuddled them. Joseph could read dreams and Egyptians couldn't. Yhwh stiffens the pharaoh's resolve, possibly through a measured sequence of plagues.[30] Israelite humor here deserves attention. From it, one might produce an empathetic retelling: The Bible, a Sitcom.

Who Killed the Exodus?

Nothing better exemplifies our selective identification with the Exodus than the degree to which we neglect P's skills, in shaping it, augmenting it, making it the inescapable center of creation.

So a last thought experiment. Consider that P, too, was writing children's literature. But he mistook his audience. Forget the mind-numbing detail about the tabernacle, the laws of purification and the like. Think in terms of his narratives: "The earth was void and empty, and darkness...," or the purchase of Machpelah. Or, even take the spies story. Imagine his recitation:

> OK, Muffy, so they decided to check out their land. So, Moses sent them from Paran, at Yhwh's command. Got that, Muffy? No, at Yhwh's command, never mind what Deuteronomy says. All of them chiefs, one from each tribe of Israel. Nope, no girls. Anyway, from Reuben, they picked, Shammua son of Zakkur. Remember that one. And from Simeon, there was a guy named Shaphet son of Hury, and from Judah, Rudolph son of ... Now, who have I named so far?

Children would be climbing walls.

P can only be compared to a contemporary gamer geek designing Dungeons and Dragons, tackling questions about time travel, insisting on numbers and names of tribal leaders to be immortalized as characters only, his own Enoch

[27] Starting at least with the case of Tiglath-Pileser I, who erected a stela in the Lebanon in sight of the one that he encountered of Anum-hirbi.

[28] For self-working prophecy and double causation in texts from 2 Samuel and the Joseph story down to Josiah's death and Oedipus, see Halpern 2001.

[29] One might see in Balaq another ruler whose own naiveté renders him the object of divine trickery as well, and question how the Gibeonite deception fits the pattern. The reversal of divine permission for Balaam's journey is probably a result of JE source combination (see above on the doublet); the divine attempt on Moses may, oddly, be more ironic. It certainly turns on Zipporah's having outwitted god by fulfilling the condition that Moses be "a bridegroom of bloodguilt," perhaps that he kill for her, but with respect to circumcision. For further divine trickery, see Jer 4:10; 1 Kgs 18:37; 22:19–23; Ezekiel 20.

[30] I prescind from the division of J from E in this narrative chapter, but insist that within JE's plague sequence diachronic development is easily discerned.

Soameses. He even enters into the kinds of details that in modern technology would be incorporated into a video game (with a build-your-own-tabernacle module). A century ago, P would have been a member of the Baker Street Irregulars. He would inhabit the world of Sherlock Holmes, loading the stories up with pseudoscientific explanations. In good Mesopotamian manner, he would compile menus and manuals and samples of clay. He would construct a fantasy-land, a theme park, and demand mastery of its every complexity. P would introduce into Disneyland a halakhic world of priorities, taboos and tariffs, capped with obscure wisdom about chickpea reproduction and watering times. P takes the narrative of an Exodus from children's literature, like Homer, to the point of being children's games, with their elaborate and superstitious rules.

If you don't share his passion, P is Bede on Cuthbert on steroids. He drives the wooden stake into the heart of the Exodus. He starts with a fairy tale. Like many such tales, it evolved with the mores of its audience: it was traditional, then bowdlerized, and finally recast as politically correct, as doctrinal. Before P, one did not even begin to think the miracles literal. And P makes the miracles so much less entertaining than they would have been had they been concocted by Ridley Scott or the authors of Final Destination 10. P is the reason that we have *chad gadya*', "I don't know why she swallowed a fly," "that lived in the house that Jack built" interpretations of the plagues as following one from the next.[31]

Fairy tales start out as entertainment, undergo recasting, and finally go out of print or circulation or public consciousness (Job, narrative Megillot, and deutero-canonical literature). Sometimes, however, the fairy tale is evidence in some Kafkaesque court. That is what P does to the Exodus: the miracles are literal, the issues now doctrinal, *homoousios* versus *homoiousios*. P delivers the Axial blow to what was once poetry: he turns it literal, into pseudoscientific history.

The Exodus event, of which there was, I am confident, at least one, is not to be reached through our story's details. Our most valuable mission is to recover, instead, the story's magic, to understand the culture that immortalized it. It is perhaps the calling of the next generation to restore the Exodus's standing as a fairy tale.

But as historians, not just of society and science but of culture and religion, we should bear it in mind that P levers children's stories into apocalyptic, even Qabbala, as well as canon law. P divides the world of Homer from the speculations of the pre-Socratics, and he will divide the latter, too, from neo-Platonism and further precursors to science. P, in a blue star-field-decorated dunce cap, divides heliocentrism from relativity, ultimately, and yet is building on a foundation of imagination. Marco's imagination, the urge to tell the tale that cannot be beat.

That is the way of religion. That is the way of science. Axial moments indeed.

[31] So, blood kills fish, and frogs therefore mature in numbers, but their death leads to a profusion of insects, etc. See for example the National Geographic "documentary," *The Ten Plagues of the Bible*, which aired 4 April 2010 on its channel.

References

Beerbohm, M. 1919. *Seven Men*. London: Heinemann.

Daube, D. 1956. Nursery Rhymes and History. *The Oxford Magazine* 74: 230–232. 272–274, 310–312.

Dever, W.G. 1997. Is There Any Archaeological Evidence for the Exodus? In *Exodus: The Egyptian Evidence*, ed. E. Frerichs and L. Lesko, 67–86. Winona Lake: Eisenbrauns.

———. 2003. *Who Were the Early Israelites and Where Did They Come From?* Grand Rapids: Eerdmans.

Dozeman, T.B. 2000. Hosea and the Wilderness Wandering Tradition. In *Rethinking the Foundations*, ed. S. McKenzie, T. Römer, and H.H. Schmid. Fs.J. van Seters, 55–70. Beiträge zur Zeitschrift für die Alttestamentliche Wissenschaft, 294. Berlin: de Gruyter.

Emerson, R.W. 1836. *Nature. Addresses and Lectures*. Boston: James Munroe and Company.

Friedman, R.E. 1986. Deception for Deception. Who Breaks the Cycle? *Bible Review* 68:22–31. (spring).

———. 2003. *The Bible with Sources Revealed*. San Francisco: HarperSanFrancisco.

Geisel, Theodor S. 1937. *To Think That I Saw It on Mulberry Street*. New York: Vanguard.

Halpern, B. 1981. *The Constitution of the Monarchy in Israel*, Harvard Semitic Monographs, vol. 25. Chico, CA: Scholars Press.

Halpern, B. 1983. *The Emergence of Israel in Canaan*. SBLMS, 29; Chico: SBL.

———. 1988. *The First Historians: The Hebrew Bible and History*. San Francisco: Harper & Row.

———. 1993. The Exodus and the Israelite Historians. *Eretz Israel* 23: 89*–96*.

———. 1998. Why Manasseh is Blamed for the Babylonian Exile: the evolution of a biblical tradition. *Vetus Testamentum* 48: 473–514.

———. 2001. *David's Secret Demons*. Grand Rapids: Eerdmans.

Harvey, P., and B. Halpern. 2008. W. M. L. De Wette's *Dissertatio critica-exegetica* ... Context and Translation. *Zeitschrift für die alttestamentliche Rechtsgeschichte* 14: 47–85.

Hendel, R. 2001. The Exodus in Biblical Memory. *Journal of Biblical Literature* 120: 601–622.

Jenkins, J.P. 2013. http://www.patheos.com/blogs/anxiousbench/2013/03/dying-quietly/.

Knohl, I. 2010. *Pharaoh's War with the Israelites: The Untold Story*. Jerusalem: Shalem.

Lucas, G.R. 1997. Recollection, Forgetting and the Hermeneutics of History: Meditations on a Theme from Hegel. In *Hegel, History and Interpretation*, ed. S. Gallagher, 97–117. SUNY: Albany.

Mills, C.W. 1959. *The Sociological Imagination*. New York: Oxford University.

Opie, I. 2004. Playground rhymes and the oral tradition. In *International Companion Encyclopedia of Children's Literature*, ed. P. Hunt and S.G. Bannister Ray, 173–186. London: Routledge.

Paine, T. 1961 [1776]. Common Sense; Addressed to the Inhabitants of America, on the Following Interesting Subjects. In *The Life and Major Writings of Thomas Paine I*, ed. P.S. Foner. New York: Citadel.

Rowe, Hartvey. 1994. Smithsonian Barbie Letter. Unpublished ms. http://www.snopes.com/humor/letters/smithsonian.asp

Rubio, G. 2009. Secrets and Antiquarian Nostalgia: Writing and Scholarship in Ancient Mesopotamia. In *Reconstructing a Distant Past*. Fs. Silva, ed. D. Barreyra and G. Del Olmo, 153–180. Aula Orientalis Suppl. 25. Barcelona: Ausa.

Shanks, H., W.G. Dever, B. Halpern, and P.K. McCarter. 1992. *The Rise of Ancient Israel*. Washington, DC: Biblical Archaeological Society.

Sommer, B. 1998. *A Prophet Reads Scripture: Allusion in Isaiah 40–66*. Stanford: Stanford University.

Stade, B. 1881. Zur Entstehungsgeschichte des vordeuteronomischen Richterbuches. *Zeitschrift für die Alttestamentliche Wissenschaft* 1: 339–343.

Talmon, S. 1966. The "Desert Motif" in the Bible and Qumran literature. In *Biblical motifs: Origins and Transformations*, ed. A. Altmann, 31–63. Waltham: Philip W. Lown Institute of Advanced Judaic Studies, Brandeis University.

Uehlinger, C. 1997. The Seal of Ahab, King of Israel. *Michmanim* 11: 39–52.

Wimsatt, W.K. 1954. History and Criticism: A Problematical Relationship. In *The Verbal Icon: Studies in the Meaning of Poetry*, ed. W.K. Wimsatt, 252–265. University of Kentucky: Lexington.

Wimsatt, W.K., and M.C. Beardsley. (1946) 1954. The Intentional Fallacy. In *The Verbal Icon: Studies in the Meaning of Poetry*, ed. W. K. Wimsatt, 3–18. Lexington: University of Kentucky. Reissued from *Sewanee Review* 54: 468–488.

———. 1949. The Affective Fallacy. *Sewanee Review* 57: 31–55.

22 The Revelation of the Divine Name to Moses and the Construction of a Memory About the Origins of the Encounter Between Yhwh and Israel

Thomas Römer

Abstract

This chapter deals mainly with three questions: (1) The literary formation of the book of Genesis: In this regard the chapter interacts with the conference comments and/or contributions of Christoph Berner, Richard Friedman, and Konrad Schmid. (2) It also asks about the origins of the Exodus tradition(s) in the biblical texts by using the texts from Kuntillet Ajrud and comes to similar conclusions as Israel Finkelstein (Chap. 3). (3) It then addresses the question of how much "cultural memory" (see in Chap. 1 and also Chaps. 5 and 31) is contained in the two accounts of Moses' call in regard to the origins of the deity Yhwh and its veneration by seminomadic groups, a question also dealt with in contributions of Thomas Levy and Manfred Bietak.

The non-priestly and priestly stories in Exodus 3:1–4:18 and 6:2–8 agree on the idea that the name of the God of Israel was not revealed to the Hebrews before the time of Moses. In the context of the construction of the Pentateuch both texts underline somewhat differently Moses' role as mediator, who after the fall of Jerusalem becomes a substitute for royal mediation (although the royal image of Moses was already invented in the seventh century BCE). Both texts are not older than the sixth century, but they may preserve the historical memory that Yhwh had not always been the god of "Israel." This older memory can be traced back through texts such as Hosea 12 and the inscription of Kuntillet Ajrud in the monarchic period and perhaps even earlier.

Introduction: The Exodus, Yhwh, and Moses

There is no doubt that the Exodus tradition is at the very center of the "historical memory" of the Hebrew Bible. For instance, it begins the Decalogue with Yhwh presenting himself as the god who has brought Israel out of Egypt:

T. Römer (✉)
Collège de France, University of Lausanne, Lausanne, Switzerland
e-mail: thomas.romer1312@gmail.com

T.E. Levy et al. (eds.), *Israel's Exodus in Transdisciplinary Perspective*, Quantitative Methods in the Humanities and Social Sciences, DOI 10.1007/978-3-319-04768-3_22,

אָנֹכִי יְהוָה אֱלֹהֶיךָ אֲשֶׁר הוֹצֵאתִיךָ מֵאֶרֶץ מִצְרַיִם מִבֵּית עֲבָדִים (Exod 20:2; Deut 5.6). Interestingly here Yhwh appears as the only protagonist of the Exodus without any mention of Moses. This is also the case in other allusions to the Exodus tradition. The so-called historical credo in Deuteronomy 26:5–9 also presents Yhwh as the author of the exodus. This is further the case in texts like Amos 2:10 and especially in the Psalms. The case of the Psalms is particularly interesting.

In contrast to the traditions of the patriarchs, the *exodus tradition* is at the very heart of the "historical retrospectives" in the Psalms.[1] Interestingly Moses is only mentioned in a few late Psalms: Psalm 77:21 and 105:26. In Psalm 99:6 Moses and Aaron appear as priests: in Psalm 103:7 as mediator of Yhwh's will, and in 106:16 (together with Aaron) and 23, they are mentioned in the context of the revolt of the people in the wilderness. The other Psalms that evoke the Exodus do not mention Moses, even those that allude to the plagues, like Psalm 78:43–51; 111:4(?); 135:8–9 (especially the destruction of Egypt's firstborn); and 136:10 (similar to 135:8–9), and allude to the miracles at the Sea of Reeds (especially the annihilation of the Egyptian army: Ps 76:7; 78:13; and 136:15; the repelling of the Sea: Ps 114:3–6; the partition of the Sea: Ps 77:10; 78:12; 136:13–14).

Moses is also missing in general allusions to the exodus in Psalm 80:9–10. Psalm 135 links the exodus tradition with the conquest of the Transjordanian territory without mentioning Moses or the conquest of Canaan. In the allegorical Psalm 80, the exodus is linked to Israel's implantation into the land (and also to the loss of that land); in the same way Psalm 111:4–6 combines Yhwh's miracles in Egypt with the evocation of his "eternal" covenant (with the patriarchs? or at Sinai?) and the conquest of the land.

The very few mentions of Moses in the allusion to the exodus are confirmed by his sparse appearances outside the Deuteronomistic History (DtrH) (and the book of Chronicles and Ezra-Nehemiah): he appears linked with the exodus only in Isaiah 63:11–12 and Micah 6:4 (the other few mentions in Jer 15:1, Mal 3:22, and Dan 9:11, 13 relate to the law or his function as an intercessor). This observation may indicate that there perhaps existed an exodus tradition without Moses.

The story of 1 Kings 12 may also support this idea. According to this narrative Jeroboam I builds two sanctuaries in Bethel and Dan where he places bull statues: "So the king took counsel, and made two calves of gold. He said to the people, 'You have gone up to Jerusalem long enough. Here are your gods, O Israel, who brought you up out of the land of Egypt.' He set one in Bethel, and the other he put in Dan" (1 Kgs 12:28–29):

הִנֵּה אֱלֹהֶיךָ יִשְׂרָאֵל אֲשֶׁר הֶעֱלוּךָ מֵאֶרֶץ מִצְרָיִם

The plural, which appears also in the story of the golden calf in Exodus 32:4, is intriguing. Even if the text speaks of two sanctuaries, it is clear that the bulls or the calves are not representing different deities, but the national god. Should one understand then the plural as alluding to the national god and his consort, Ashera, as suggested by E. Axel Knauf (Knauf 1998: cols. 1375–1376)? However, there are no clear hints elsewhere that Ashera might have been associated with the exodus; thus this idea remains very speculative.

Or should the plural allude to Yhwh in his different manifestations: the Yhwh from Bethel and the Yhwh from Dan? The easiest solution might be to understand the plural as polemical, as a transformation of an original cultic exclamation. A comparison of 1 Kings 12:28 with the opening of the Decalogue shows that both exclamations are very similar. If there was an original singular behind 1 Kings 12:28 the similarity would be even more important.

The Judean redactors of 1 Kings 12 apparently wanted to convince their audience that the Northern cult in Bethel and Dan (and elsewhere) was a "polytheistic" one.

The mention of Dan in 1 Kings 12 is also intriguing. According to Eran Arie, Dan became part of Israel only in the eighth century (Arie 2008:

[1] For more details on this question see Römer (2011).

34–38). In this case it is possible that 1 Kings 12 is a retroprojection from the time of Jeroboam II. One may even consider whether the figure of Jeroboam I as a whole is a creation based on the figure of king Jeroboam from the eighth century. But this speculation is beyond the topic of our chapter.

Coming back to Yhwh and the exodus it clearly appears that, at least since the eighth century, Yhwh was venerated in Israel (probably not yet in Judah) as a deity who brought his people out of Egypt. But in biblical texts, which can be confidently dated to the monarchic times, there is no mention of Moses; for now, we leave aside the Pentateuchal texts, whose dates are conspicuously complicated.

The construction of the exodus as the "real" national memory can be traced in chapter 12 of the book of Hosea. This chapter may reflect, if not the voice of the prophet himself, the situation in the North of the second half of the eighth century, although much later dates have also been suggested.[2] As A. de Pury (1992 and 2006) has shown, this text opposes the Jacob and the exodus traditions. Jacob is depicted very negatively in this text: he supplanted his brother and has become a "Canaanite," a merchant with false scales who likes to oppress (Hos 12:4 and 8). Even his battle with God is, in contrast to Genesis 32, related in a different and negative manner (Hos 12:4–5). Already at the very beginning of this poem, it becomes obvious that "Jacob" will be judged by Yhwh (12:3). Whereas Jacob is related to a deity that is called "elohim" or "el,"[3] Yhwh presents himself as the God from the land of Egypt: וְאָנֹכִי יְהוָה אֱלֹהֶיךָ מֵאֶרֶץ מִצְרָיִם (Hos 12:10). Again this statement is reminiscent of the Decalogue, although there is no verb for "bringing out"; Yhwh himself is described as a deity whose origins are related to Egypt.

In Hosea 12:13–14, Jacob's flight to Aram and his "slavery" on behalf of a woman are opposed to Yhwh's prophet who is leading Israel out of Egypt and who guards it:

וַיִּבְרַח יַעֲקֹב שְׂדֵה אֲרָם וַיַּעֲבֹד יִשְׂרָאֵל
בְּאִשָּׁה וּבְאִשָּׁה שָׁמָר

וּבְנָבִיא הֶעֱלָה יְהוָה אֶת־יִשְׂרָאֵל
מִמִּצְרָיִם וּבְנָבִיא נִשְׁמָר

The mention of the prophet is prepared in v. 11, which claims that Yhwh reveals his will through his prophets. It is usually assumed that the prophet mentioned in v. 14 is Moses, but interestingly he is not named. Why is that so? Probably because of the prophetic group behind Hosea 12, a group which seeks legitimacy by claiming that there was already a prophetic mediation at the time of the exodus.

Summing up: Hosea 12 can be understood as a polemical text against the Jacob tradition. Against the attempt to establish the Jacob tradition as the official national origin myth in the North (Hos 12:4–5 hints at the change of Jacob's name into Israel), the author of Hosea 12 claims that Yhwh is related to Egypt and not to the Patriarch. This also means that the relation between Israel and his god is not "hereditary" or mediated by a Patriarch; it is the result of an encounter, and the mediator of this relation is a prophet. Hosea 12 is perhaps one of the first attempts to emphasize Moses' role in the exodus tradition. Interestingly the Pentateuchal narrative of the exodus also highlights the idea that Yhwh was known by the Hebrews only in relation with the Exodus and also constructs Moses as a prophet.

Exodus 3–4 and 6 and Their Functions in the Non-priestly and Priestly Exodus Narratives: Divergences and Convergences

In the current debate about the formation of the Torah, the traditional consensus built upon the documentary hypothesis has faded away. In Europe, most scholars have given up the Wellhausen paradigm, whereas in North

[2] For an eight century date see de Pury (1992, 2006) and Blum (2009). For an exilic or a postexilic date: Whitt (1991) and Pfeiffer (1999).

[3] מַלְאָךְ in 12:5 is probably a gloss that transformed the original "El" into a gloss, see, e. g., Gertner (1960: 277, 281).

America the traditional documentary hypothesis is still popular, but the so-called Neo Documentarians have developed a variant that has not much to do with the traditional model.[4] In this difficult situation the distinction between P and non-P is apparently one of the few results of scientific analysis of the Pentateuchal text with which most scholars would agree. However the question whether P was originally an independent document or conceived as the redaction of an older narrative remains disputed. And many "non-P" texts (traditionally J/E) are now considered to be post-priestly.

This is also the case for the two variants of Yhwh's encounter with Moses (in Exod 3–4 and 6) in which Yhwh appoints him as the one who should lead the Hebrews out of Egypt into the land of Canaan.

Recently K. Schmid and others have argued that Exodus 3–4 should be understood as a unified text written by one author who already knew the priestly text of the revelation of the divine name to Moses and who wanted to question the notion that the divine revelation of the divine name happened first in Egypt by transferring it to the "mountain of God" (Schmid 1999: 186–208; Otto 1996; Kegler 2003). According to Schmid, Yhwh's statement according to which he has heard the cry of the Hebrews in 3:7 is taken over from the priestly passage in Exodus 6:2–8. Indeed the cry of the Israelites occurs in Exodus 3:7–9 (צעקה) as well as in Exodus 2:23 (ויזעקו). But this does not prove that the author of Exodus 3–4 already knew the priestly account in Exodus 6. The idea that the Israelites cried out because of their oppression also occurs in the Dtr historical credo in Deuteronomy 26:7 (וַנִּצְעַק), with the same orthography as in Exodus 3 (צעק instead of זעק in Exod 2:23). Thus, it is not necessary to postulate a literary dependency of Exodus 3–4 on the P-texts in Exodus 2:23–35 and 6.

It also seems difficult to maintain the idea that Exodus 3–4 is basically a uniform text written by one author. The literary analysis will demonstrate that the non-P version of Moses' call was revised and broadened several times. Therefore I prefer the traditional option that dates the first edition of Exodus 3–4 somewhat earlier than the P variant in Exodus 6:2–8 (see also Gertz 1999: 254–326). In the context of this chapter I cannot deal with all the literary questions of these chapters of Exodus. I will focus on the question of the revelation of the divine name and the construction of the figure of Moses in both texts.

Exod 3: Moses, the Prophet, and Yhwh, the Unknown God

In the present form of the book of Exodus, Exodus 3:1–4:18 is clearly a unit; it is framed by the mention of Jethro, Moses' father-in-law, in 3:1 and 4:18, and by Moses' arrival at the mountain of God and his return from there.

There is a strange repetition between 4:18 and 4:19. In 4:18, Moses tells Jethro that he must return to Egypt to see his brethren and Jethro tells him to go in peace. In 4:19 however we read: "Yhwh said to Moses in Midian, 'Go back to Egypt; for all those who were seeking your life are dead'." This verse does not make much sense after 4:18, since Moses had already informed Jethro about his return to Egypt. Verse 4:19 however fits very well after the beginning of Exodus 2:23: "After a long time the king of Egypt died" (2:23aßb–25 are commonly considered to be part of P). If we can read 2:23aα together with 4:19, one may conclude, as already envisaged by Julius Wellhausen (1963: 71), that there was an older story, into which Moses' call narrative had later been inserted. Apparently then the original account of Exodus 3:1–4:18 did not belong to the oldest Moses story.

As William Propp states in his commentary, "Exodus 3-4 is a key passage for the documentary analysis of the Torah" (1999: 190). However it has always been very difficult to reconstruct two parallel narratives in this text (see on this Römer (2006)). In regard to the use of the divine names

[4] See on this question also the contributions of Schmid (Chap. 24) and Berner (Chap. 20) as well as the evaluation of the current situation in Pentateuchal scholarship in Römer (2013).

yhwh and *elohim*, it should be noted that there is a third variant: *ha-elohim* (3:6,11–13). The expression "ha-elohim" often contains the idea of a "mysterious" or an "unknown" God, and its concentrated use in Exodus 3 is probably related to the revelation of the divine name. The expression appears until Exodus 3:13, where Yhwh tells Moses about his identity. In an absolute form this relatively rare term, compared to *elohim*, no longer appears in the exodus narrative until Exodus 18, the story of Jethro's visit to Moses and his sacrifice for Yhwh. This already indicates a relationship between Exodus 3 and 18.

There is some redundancy in verses 7–10, but this redundancy underlines the importance of the divine speech, which, as sometimes observed, is chiastically structured: ABCB'A'. To Yhwh seeing his people's (עַמִּי) oppression in Egypt (בְּמִצְרָיִם) in v. 7a corresponds Moses' mission to bring Yhwh's people (עַמִּי) out of Egypt (מִמִּצְרָיִם), v. 10. Yhwh hearing (שָׁמַעְתִּי) the people's cry (צַעֲקָתָם) in v. 7 is taken up in v. 9 by the statement that Israel's cry (צַעֲקַת בְּנֵי־יִשְׂרָאֵל) has come up (בָּאָה) to Yhwh. In the middle, in v. 8, we find the promise of Israel's transfer from "this land" into a good and spacious land, flowing with milk and honey. It is difficult to imagine that such a clear structure would only be an accidental result of the pasting together of two different documents.

V. 10 is linked to the following because it inaugurates Moses' appointment, which is constructed as a prophetic call. The closest parallel to Exodus 3:10–12 is Jeremiah's call in Jeremiah 1:4–10 (Köckert 2000 and Grätz 2007). Both passages contain the following elements:

	Exodus 3	Jeremiah 1
Sending	v. 10: Go, I will send you (לְכָה וְאֶשְׁלָחֲךָ)	v. 7: You shall go where I will send you (עַל־כָּל־אֲשֶׁר אֶשְׁלָחֲךָ תֵּלֵךְ)
Objection	v. 11: Who am I that I should go	v. 6: I do not know how to speak, for I am a boy
Promise of assistance	v. 12: I will be with you (אֶהְיֶה עִמָּךְ)	v. 8: I am with you (אִתְּךָ אֲנִי)
Sign	v. 13: this shall be the sign for you that it is I who sent you: ... you shall worship God on this mountain	v. 9: then Yhwh ... put out his hand and touched my mouth

Moses appears as the prophet by whom Yhwh will lead his people out of Egypt in agreement with Hosea 12. Exodus 3:10–13 expresses the same idea as Deuteronomy 18:15–20, a passage which by the way also displays parallels with Jeremiah 1:4–10. Therefore, with E. Blum and others, it seems appropriate to label the original narrative of Exodus 3 a "D-composition" (Blum 1990: 17–43) and to date it in the sixth century BCE.

After Moses' (prophetic) call, the narrative turns to the question of the identity of the deity that is about to appoint him. This question is already brought up in v. 6 in the scene of the burning bush where the divine self-presentation, "I am the God of your father," is followed by the apposition "the god of Abraham, the God of Isaac, and the God of Jacob." This is grammatically awkward and already emended in the Samaritan Pentateuch and some Greek manuscripts. The apposition appears as a later attempt to create a link with the Patriarchal traditions (Weimar 1980: 38, 341).

One may recall here an observation made by Rendtorff, according to which the land that God promises to the Israelites is introduced in Exodus 3 as if it were a completely unknown land (Rendtorff 1990: 85). Interestingly it is not said to have been promised to the Patriarchs as is the case in the priestly story of Moses' call in Exodus 6:8. This may indicate that in the original story of Exodus 3 there was no mention of the Patriarchs at all.

After Moses makes a second objection—he does not know the name of the ancestral god in the name of whom he should speak to the Israelites—and Yhwh reveals himself (or not) through the abundantly commented expression אֶהְיֶה אֲשֶׁר אֶהְיֶה, the following verse identifies Yhwh again as the "God of Abraham, Isaac and Jacob." There are strong arguments that this verse did not belong to the original narrative. First of all, this speech is introduced after v. 14 by "*wayyomer 'od elohim*." As in Genesis 22:15, *'od* ("further") already indicates that the following is an addition. This addition, which can be compared to Psalm 135:13, may be understood as an attempt to create a parallel with the priestly idea expressed in Exodus 6:2 that even if Yhwh did not appear to the Patriarchs under his real name, he is of course the god of the Patriarchs.

If one considers v. 15 as an addition then the transition from v. 14 to v. 16 is even smoother. There is only one divine speech which starts with the word play on the tetragrammaton, finally revealed in two steps (*'ehyeh*, then *Yahweh*).

If it is true that the names of the Patriarchs in v. 6 and 15 belong to a reworking of the original text, one may wonder whether this is also the case in v. 16. (Weimar 1980: 332–333, 341). This I admit may be suspected of circular reasoning; the decision is of course dependent on an overall theory about the composition of this text. One may however note that the mention of the patriarchal names is separated from *'elohe 'aboteka* by the verb and comes somewhat late. This could indicate that these words were first written on the margins of the scroll, before a later copyist integrated them into the text.

Time and space do not allow to demonstrate that the original account ended in 3:17 followed by 4:18. It contained approximately 3.1–2* (without the *mal'ak*?), 3–4, 6aαb, 7–14, 16aαb, 17, 4.18. (See for a similar reconstruction Gertz (1999: 394).) The intention of the original story is twofold: It legitimates Moses' status as Israel's proto-prophet, and it recognizes that knowledge of the divine name is connected to the exodus. The story, as observed by Michaeli, Berge, Schmid,[5] and others, shares with Exodus 6 the idea that the revelation of the divine name Yhwh is something new. In the original text, the deity presented itself as Moses' patriarchal God (v. 6) and Moses identifies this god with the ancestral deity of the Israelites. The fact that ancestral gods do not bear personal names is attested by texts from Ugarit that often mention an *'ilu 'ibi* ("god of the father") (Van der Toorn 1993). The author of Exodus 3* wants to emphasize that this unknown god is in fact the deity Yhwh. A similar procedure can be observed in Exodus 6.

Before turning to this text, it should be mentioned again that Exodus 3 was not initially part of the oldest exodus-Moses story in which it has been inserted. The oldest story started with a brief description of the difficult situation of the Hebrews in Egypt and the story of Moses' birth and his "adoption" by the daughter of Pharaoh. The story of his birth and exposure displays literary dependence on the birth legend of Sargon, the legendary founder of the Assyrian Empire, as has often been observed (Cohen 1972; Ardiñach 1993).

Sargon and Moses are both exposed by their mothers, both of whom are in some ways related to the priesthood. Sargon's mother is a priestess, and Moses' mother is the daughter of Levi, the ancestor of Israel's priestly tribe. Their fathers do not intervene. They are set adrift on a river in a basket, to be found and adopted. In both cases, the adoption is presented as royal adoption: Sargon is "loved" by Ishtar, and Moses becomes the son of Pharaoh's daughter.

Even though the Sargon story is about the third-millennium Assyrian king, it was written under Sargon II, his namesake, at the end of the eighth century. It contains Neo-Assyrian orthographic forms and idiomatic expressions attested only in this period (Lewis 1980: 98–110). Therefore the story of Moses, modeled on it, cannot be dated prior to the seventh century BCE. Exodus 2 presupposes no knowledge of Moses, his origins, or his name; everything needs to be explained. It is tempting, then, to understand the first written story about Moses (which cannot be reconstructed in detail) to be a reaction to Neo-Assyrian royal ideology, elaborated at Josiah's court. The Assyrian background is also present in the mention of the "store cities" (עָרֵי מִסְכְּנוֹת) in Exodus 1:11, which uses a loanword

[5] Michaeli (1974:65); Berge (1997: 116): "Moses, already knowing the identity of the speaking God, now asks for his name because he does not know it"; Schmid (1999: 206).

from the Assyrian *maškanu*.[6] If a seventh-century setting of the oldest Moses narrative is plausible, one may speculate that the insertion of the figure of Moses into the narrative and its construction as a royal figure are linked to the Judean rewriting of an older Northern exodus tradition. But let us now turn to Exodus 6:2–8.

Exodus 6:2–8: The Unknown Name of Yhwh and the Theory of the Divine Revelation

The priestly account of the revelation of the divine name in Exodus 6:2–8 displays a clear structure (for a similar proposal see Magonet 1983):

v.2 אני יהוה
v.3 וארא אל אברהם אל יצחק ואל יעקב
v.4 הקמתי את בריתי אתם לתת להם את ארץ כנען
v.5 נאקת בני ישראל אשר מצרים מעבדים אתם
ואזכר את בריתי
v.6 אני יהוה
והוצאתי אתכם מתחת סבלת מצרים

וגאלתי אתכם
v.7 ולקחתי אתכם לי לעם
והייתי לכם לאלהים
אני יהוה
המוציא אתכם מתחת סבלת מצרים
v. 8 והבאתי אתכם אל הארץ אשר
נשאתי את ידי לתת אתה לאברהם ליצחק וליעקב ונתתי אתה לכם מורשה
אני יהוה

This structure reveals the importance of the divine presentation, since the statement "I am Yhwh" appears four times. The self-presentations in v. 2 and 8 frame the divine speech, whereas the formula in v. 6 and v. 7 is in both cases followed by the almost identical statement: "who will bring you out from the burdens of Egypt." Here, as in Exodus 3, Yhwh characterizes himself as the god that brings out of Egypt. In contrast to the original version of Exodus 3, Exodus 6 insists on the strong continuity between the patriarchs and the exodus. The exodus and the conquest of the land are presented in the divine speech as the results of the divine covenant and promises to the Patriarchs.

This relation is theorized in v. 3, where P constructs a theology of the divine revelation: "I appeared to Abraham, Isaac and Jacob as El Shadday but by my name Yhwh I did not make myself known to them." This is a clear reference to Genesis 17:1 ("… Yhwh[7] appeared to Abraham and said to him: I am El Shadday …"), which allows the priestly author to construct a history of the divine revelation in three stages:

In the primeval history, God is known to all humans as "elohim." For Abraham and his descendant he is "El Shadday"; and only Moses

[6] See Knauf (1988: 104); the rare description of the "bracketing of the bricks" in Exod 5:7 (לבן לבנים, cf. also in Exod 1:14 בַּעֲבֹדָה קָשָׁה בְּחֹמֶר וּבִלְבֵנִים) can be related to the Accadian expression *libnate labanu*, which is for instance used in a building inscription of Esarhaddon (Uehlinger 1990: 361, cf. 250–251); these texts are however often attributed to priestly redactors.

[7] Some commentators have thought that the name Yhwh in Gen 17:1 does not fit with P's theory of the divine revelation. But this is not true: The tetragrammaton is used by the narrator in order to inform the reader about the identity of El Shadday. In the narrative, Abraham does not get any information on this.

and Israel in Egypt are instructed about God's personal name, Yhwh. This means that Israel's singular privilege is the knowledge of the divine name and through this privilege Israel becomes the only nation capable of worshipping God by means of an adequate sacrificial cult. On the other hand, however, P advocates—contrary to the Deuteronomists—an inclusive monotheism: all people of the earth venerate the same god, irrespective of whether they address him as *elohim*, *El*, or *El shadday*. This idea works better if P was the author of an independent document and not a redactor of older non-P narratives.

Intriguingly, God's revelation to Moses happens in Egypt, as opposed to Exodus 3, where the divine name is revealed to Moses at the "mountain of God." The idea of a divine revelation places Exodus 6 in parallel with Ezekiel 20 (v. 5: "I made myself known to them (*yd'*, Nif) in the land of Egypt," cf. Exod 2:25 and 6:3 where the same root occurs). According to the priestly tradition God disclosed his true name in Egypt. For the authors of P and of Ezekiel 20, the story of the exodus also and above all remains the story of the revelation of the divine name. The divine speech to Moses is, according to P, the last step in the history of God's revelation in which Israel, through Moses' intermediary, is informed of his real name.

Exodus 3 and Exodus 6: A Brief Comparison

Although Exodus 3 locates the divine revelation on the mountain of God (by using three expressions, חֹרֵב, הַר הָאֱלֹהִים and the rare expression הַסְּנֶה which only occurs in Exod 3:2–4; Deut 33:17 is probably also an allusion to the Sinai) and Exodus 6 places it in Egypt, both texts basically agree that Yhwh was not always Israel's God but he revealed himself to the people by the intermediary of Moses. And, even if Exodus 3 is constructed as an anticipation of the Sinai theophany, it is inserted in a narrative context in which Moses is sojourning in Madian, in the "south."

In the context of the Pentateuchal narrative this presentation emphasizes the central role of the exodus tradition (by transforming the patriarchal narratives into a prologue of a sorts) and also legitimates the figure of Moses as the exclusive mediator and Israel's first prophet.

Both texts are not older than the sixth century BCE, but they may preserve the historical memory that Yhwh has not always been the god of "Israel." For sure, neither Exodus 3 nor Exodus 6 are historical texts. But they may preserve a "longue durée," a long-term memory of the "adoption" of the deity Yhwh in relation to Egyptian or southern traditions.[8]

Some Historical Speculations About the Origins of Yhwh and His Adoption by "Israel"

We may start with a very basic observation about the name of Israel, which is attested outside the Bible at the end of the thirteenth century BCE in the stele of Pharaoh Merneptah and perhaps even somewhat earlier on a Statue Pedestal from the time of Ramses II (but this interpretation remains very speculative and, according to an oral communication of Thomas Schneider, is unconvincing).[9] The name "Israel" contains the theophoric element "El" and not Yhwh or Yhw. Even as Nadav Na'aman rightly emphasizes that the location of the entity "Israel" in the Merneptah stele "cannot be established with certainty and all attempts to locate it in the central highlands ... rest on a pre-conceived idea of its place" (Na'aman 2011: 47), it is clear that the stele refers to a group located in the Levant whose patron deity is apparently El, or Ilu, like in Ugarit.

[8] For the construction of a "cultural memory" see also the contributions of Jan Assmann (Chap. 1) and Aren Maeir (Chap. 31).

[9] Van der Veen et al. (2010: 15–25). The authors suggest to read "Ia-cha-ri" or "Ia-cha-l," which is quite different from the "Isrial" of the Merneptah stele. In the pedestal the toponym is written in an enclosure that indicates the name of a land or a city.

On the other hand, there are five biblical texts that locate Yhwh in the South and that describe an encounter between him and Israel.[10]

In Judges 5:4–5 and his "elohistic" parallel Psalm 68:8–9, Yhwh seems to be identified with the Sinai, and he is coming from Seir, according to Judges 5:4.

> Yhwh, when you went out from Seir, when you marched from the region of Edom, the earth trembled, and the heavens poured, the clouds indeed poured water. The mountains quaked before Yhwh, the Sinai, before Yhwh, the God of Israel.

A similar statement is found in Deuteronomy 33:2:

> Yhwh came from Sinai, and dawned from Seir upon them; he shone forth from Mount Paran. He arrived from Meribat Qadesh; at his right, to the slopes, for them.[11]

Here again Yhwh comes from Seir, which is located in parallel to Mount Paran, whose location cannot be established.[12]

And finally, Habakkuk 3:3:

> God (*Eloah*) comes from Teman, the Holy One from Mount Paran. *Selah*. His glory covers the heavens, and the earth is full of his praise.

In this verse the name Yhwh is replaced by Eloah, but v. 2 and 8 suggest an identification of Eloah with Yhwh. Interestingly, this psalm locates Yhwh's origin in Teman, a name that appears in Gen 36 as the name of a clan in Edom. The Edomite connection for Teman is also clear in other Biblical texts (Jer 47:7,20; Ezek 25:13; Amos 1:11–12, Obad 8–9). An Edomite location of Teman would also fit with the above-quoted texts mentioning Seir.

Of course it is also possible that Teman is a more general term for the South, but then the South evidently also includes the Edomite territory.

In regard to Teman, the inscriptions of Kuntillet Ajrud are of major interest. Even after the recently published editio princeps (Meshel et al. 2012), several questions remain debated, especially the function of the site. Was it a resting place or even a sanctuary in which Asherah played a central role as recently suggested by Nurid Lissovsky and Nadav Na'aman (2008)? According to I. Finkelstein and E. Piasetzky, "Kuntillet 'Ajrud functioned between ca. 795 and 730/20 BCE" (Finkelstein and Piasetzky 2008: 178). Two inscriptions mention a "Yhwh from Teman" (Meshel et al. 2012: 95, Inscription 3:6, and 98, Inscription 3:9 with article), associated with Asherah. This is an indication that still in the eighth century Yhwh was venerated as a deity from the South. On the other hand, another inscription invokes a "Yhwh from Shomron" with his Ashera (Meshel et al. 2012: 87, Inscription 3.1). If the site was used by travelers or worshippers from Israel/Samaria, it is interesting that they still acknowledge the existence of a "southern Yhwh."

The existence of a deity Yhwh from Teman in the eighth century BCE may then tentatively be related to the famous Shasu-nomads which in some Egyptian inscriptions, especially from the time of Amenophis III and Ramses II, appear as *tꜣšꜣśw yhwꜣ*. The expression *yhwꜣ* seems to be a toponym, which may also designate a deity (cf. the identification of Yhwh and Sinai in Judges 5). In the list from Amara, the different Shasu groups are listed under *tꜣšꜣśwś'rr* (the Shasu land Seir), which according to Manfred Weippert could be a kind of title indicating the location of the different Shasu tribes (Weippert 1974: 270–271). An Edomite location for those Shasu groups was made plausible by the excavations of Thomas Levy and his team in the Jabal Hamrat Fidan, who states that in the case of Wadi Fidan "the archeological record supports the biblical and historical evidence" (Levy et al. 2004: 89).

[10] Of course, there is also a discussion about the age of these texts. According to Pfeiffer (2005: 268) the idea of Yhwh's original location in the Sinai is a late invention from the time of the exile after the destruction of the temple. This seems to me a kind of an "allegorical exegesis." For the possibility that those texts conserve old memories see also Leuenberger (2010).

[11] The second half of the verse raises a number of text-critical problems, see, e.g., Pfeiffer (2005: 182–183).

[12] Mount Paran which is only attested here and in Hab 3 in the HB, contrary to the wilderness of Paran, may be already a learned speculation; its identification with Qadesh, *'En el-qederat*, allows us to date the text between the tenth and sixth century BCE.

It might therefore be plausible that the veneration of Yhwh as a god who defeats the Egyptians was brought to Israel by a Shasu group. As N. Na'aman observed, "the biblical description of Egypt as a 'house of bondage' reflects very well the Egyptian reality of the New kingdom" (2011: 49). It is therefore a plausible speculation that Yhwh was brought to Israel by a group that worshipped an "Edomite" or a "Southern" Yhwh. Maybe there was also a narrative tradition about a figure like Moses, since his Midianite connections can hardly be explained as an invention, as well as the kernel of Exodus 18, where his Midianite father-in-law offers a sacrifice to Yhwh (Blenkinsopp 2008). Of course, any precise reconstruction is impossible. The biblical texts of the divine revelation of Yhwh's name however retain "traces of memories"—to use an expression of Jan Assmann—of a non-autochthonous origin of Yhwh.

Conclusion

The biblical exodus narrative was written down for the first time in Judah. Moses appears here as a prototype for Josiah, and the situation of Egyptian oppression seems to reflect the Assyrian situation. The exodus tradition is of course older and came to Judah from Israel after 722 BCE. The literary contours of this tradition cannot be reconstructed. Hosea 12 shows however how Yhwh, the God of the exodus, is opposed to the Jacob tradition. This may reflect the attempt to make the exodus the "official" foundation myth of Israel. The two accounts of Yhwh's revelation to Moses, although written in the sixth century, still keep the memory that Yhwh was not an autochthonous deity but was "imported" from the South. This theory gains support from the inscriptions of Kuntillet Ajrud but also from the evidence about the *Shasu* groups, since some of them apparently worshipped a deity called *Yahu*. Even if this brings us to the last centuries of the second millennium BCE, the biblical texts have preserved a *long-term* memory about the exodic origins of Yhwh.

References

Ardiñach, P.R. 1993. La leyenda acádia de Saegón. *Rivista Bíblica* 50: 103–114.

Arie, Eran. 2008. Reconsidering the Iron Age II Strata at Tel Dan: Archaeological and Historical Implications. *Tel Aviv* 35: 6–64.

Berge, Kåre. 1997. *Reading Sources in a Text. Coherence and Literary Criticism in the Call of Moses*, Arbeiten zu Text und Syntax des Alten Testaments, vol. 54. St. Ottilien: EOS Verlag.

Blenkinsopp, Joseph. 2008. The Midianite-Kenite Hypothesis Revisited and the Origins of Judah. *Journal for the Study of the Old Testament* 33: 131–153.

Blum, Erhard. 1990. *Studien zur Komposition des Pentateuch*, Beihefte zur Zeitschrift für die alttestamentliche Wissenschaft, vol. 189. Berlin: de Gruyter.

———. 2009. Hosea 12 und die Pentateuchüberlieferungen. In *Die Erzväter in der biblischen Tradition. Festschrift für Matthias Köckert*, Beihefte zur Zeitschrift für die alttestamentliche Wissenschaft, vol. 400, ed. A.C. Hagedorn and H. Pfeiffer, 291–321. Berlin: W. de Gruyter.

Cohen, C. 1972. The Legend of Sargon and the Birth of Moses. *Journal of Ancient Near Eastern Studies* 4: 46–51.

de Pury, Albert. 1992. Osée 12 et ses implications pour le débat actuel sur le Pentateuque. In *Le Pentateuque. Débats et recherches*, ed. P. Haudebert, 175–207. Lectio Divino 151. Paris: Editions du Cerf.

———. 2006. The Jacob Story and the Beginning of the Formation of the Pentateuch. In *A Farewell to the Yahwist? The Composition of the Pentateuch in Recent European Interpretation*, Society of Biblical Literature. Symposium series, vol. 34, ed. T.B. Dozeman and K. Schmid, 51–72. Atlanta, GA: Society of Biblical Literature.

Finkelstein, Israel, and E. Piasetzky. 2008. The Date of Kuntillet 'Ajrud: The ^{14}C perspective. *Tel Aviv* 35: 135–185.

Gertner, M. 1960. The Masorah and the Levites. An Essay in the History of a Concept. *Vetus Testamentum* 10: 241–284.

Gertz, Jan Christian. 1999. *Tradition und Redaktion in der Exoduserzählung. Untersuchungen zur Endredaktion des Pentateuch*, Forschungen zu Religion und Literatur des Alten und Neuen Testaments, vol. 186. Göttingen: Vandenhoeck & Ruprecht.

Grätz, Sebastian. 2007. "Einen Propheten wie mich wird dir der Herr, dein Gott, erwecken". Der Berufungsbericht Jeremias und seine Rückbindung an das Amt des Mose. In *Moses in biblical and extra-biblical traditions*, Beihefte zur Zeitschrift für die alttestamentliche Wissenschaft, vol. 372, ed. A. Graupner and M. Wolter, 61–77. Berlin: de Gruyter.

Hendel, Ron. 2001. Exodus in Biblical Memory. *Journal of Biblical Literature* 120: 601–622.

Kegler, Jürgen. 2003. Die Berufung des Mose als Befreier Israels. Zur Einheitlichkeit des Berufungsberichts in Exodus 3–4. In *Freiheit und Recht. Festschrift für Frank Crüsemann zum 65. Geburtstag*, ed. C. Hardmeier, R. Kessler, and A. Ruwe, 162–188. Gütersloh: Chr.Kaiser.

Köckert, Matthias. 2000. Zum literargeschichtlichen Ort des Prophetengesetzes Dtn 18 zwischen dem Jeremiabuch und Dtn 13. In *Liebe und Gebot. Studien zum Deuteronomium. Festschrift zum 70. Geburtstag von Lothar Perlitt*, ed. R.G. Kratz and H. Spieckermann, 80–100. Göttingen: Vandenhoeck & Ruprecht.

Leuenberger, Martin. 2010. Jhwhs Herkunft aus dem Süden. Archäologische Befunde—biblische Überlieferungen—historische Korrelationen. *Zeitschrift für die alttestamentliche Wissenschaft* 122: 1–19.

Levy, Thomas E., Russel B. Adams, and Adolfo Muniz. 2004. Archaeolgy and the Shasu Nomads: Recent Excavations in the Jabal Hamrat Fidan, Jordan. In *Le-David maskil: A Birthday Tribute for David Noel Freedman*, ed. R.E. Friedman and W.H.C. Propp, 63–89. Winona Lake, IN: Eisenbrauns.

Knauf, Ernst Axel. 1988. *Midian. Untersuchungen zur Geschichte Palästinas und Nordarabiens am Ende des 2*, Abhandlungen des deutschen Palästina-Vereins. Harrassowitz: Jahrtausends v.Chr. Wiesbaden.

———. 1998. Bethel. In *Die Religion in Geschichte und Gegenwart*, 4th ed. H.D. Betz, D.S. Browning, B. Janowski, E. Jüngel. Tübingen: Mohr Siebeck vol. 1, cols. 1375–1376.

Lewis, Brian. 1980. *The Sargon Legend. A Study of the Akkadian Text of the Tale and the Tale of the Hero who was Exposed at Birth*, ASOR Dissertation Series, vol. 4. Cambridge, MA: American Schools of Oriental Research.

Magonet, Jonathan. 1983. The Rhetoric of God: Exodus 6.2-8. *Journal for the Study of the Old Testament* 27: 56–67

Meshel, Zeev et al. 2012. *Kuntillet ʻAjrud (Ḥorvat Teman): An Iron Age II Religious Site on the Judah-Sinai border*. Jerusalem: Israel Exploration Society.

Michaeli, Frank. 1974. *Le livre de l'Exode*, Commentaire de l'Ancien Testament, vol. 2. Neuchâtel: Delauchaux et Niestlé.

Na'aman, Nadav. 2011. The Exodus story: Between Historical Memory and Historiographical Composition. *Journal of Ancient Near Eastern Religions* 11: 39–69.

Na'aman, Nadav, and Nurit Lissovsky. 2008. Kuntillet 'Ajrud, Sacred Trees and the Asherah. *Tel Aviv* 35: 186–203.

Otto, Eckart. 1996. Die nachpriesterliche Pentateuchredaktion im Buch Exodus. In *Studies in the Book of Exodus. Redaction—Reception—Interpretation*, Bibliotheca Ephemeridum theologicarum Lovaniensium, vol. 126, ed. M. Vervenne, 61–111. Leuven: University Press - Peeters.

Pfeiffer, Henrik. 1999. *Das Heiligtum von Bethel im Spiegel des Hoseabuches*, Forschungen zu Religion und Literatur des Alten und Neuen Testaments, vol. 183. Göttingen: Vandenhoeck & Ruprecht.

———. 2005. *Jahwes Kommen von Süden: Jdc 5, Hab 3, Dtn 33 und Ps 68 in ihrem literatur- und theologiegeschichtlichen Umfeld*, Forschungen zu Religion und Literatur des Alten und Neuen Testaments, vol. 211. Göttingen: Vandenhoeck & Ruprecht.

Propp, William H.C. 1999. *Exodus 1-18*, Anchor Bible, vol. 2. New York, NY: Doubleday.

Rendtorff, Rolf. 1990. *The Problem of the Process of Transmission in the Pentateuch* J.J. Scullion, transl. Journal for the Study of the Old Testament. Supplement, vol. 89. Sheffield: JSOT Press. German original: *Das überlieferungsgeschichtliche Problem des Pentateuch*. Beihefte zur Zeitschrift für die alttesttamentliche Wissenschaft, vol. 147. Berlin: de Gruyter (1976).

Römer, Thomas. 2006. Exodus 3-4 und die aktuelle Pentateuchdiskussion. In *The Interpretation of Exodus. Studies in Honour of Cornelis Houtman*, Contributions to Biblical Exegesis and Theology, vol. 44, ed. R. Roukema, 65–79. Leuven: Peeters.

———. 2011. Extra-Pentateuchal Biblical Evidence for the Existence of a Pentateuch? The Case of the "Historical Summaries," Especially in the Psalms. In *The Pentateuch. International Perspectives on Current Research*, Forschungen zum Alten Testament, vol. 78, ed. T.B. Dozeman, K. Schmid, and B.J. Schwartz, 471–488. Tübingen: Mohr Siebeck.

———. 2013. Zwischen Urkunden, Fragmenten und Ergänzungen: Zum Stand der Pentateuchforschung. *Zeitschrift für die alttestamentliche Wissenschaft* 125: 2–24.

Schmid, Konrad. 1999. *Erzväter und Exodus. Untersuchungen zur doppelten Begründung der Ursprünge Israels innerhalb der Geschichtsbücher des Alten Testaments*, Wissenschaftliche Monographien zum Alten und Neuen Testament, vol. 81. Neukirchener Verlag: Neukirchen-Vluyn.

Uehlinger, Christoph. 1990. *Weltreich und »eine Rede«. Eine neue Deutung der sogenannten Turmbauerzählung (Gen 11,1-9)*, Orbis Biblicus et Orientalis, vol. 101. Freiburg (CH): Universitätsverlag.

van der Toorn, Karel. 1993. Ilib and the 'God of the Father'. *Ugarit Forschungen* 25: 379–387.

van der Veen, Peter, et al. 2010. Israel in Canaan (Long) before Pharaoh Merenptah? A Fresh Look at Berlin Statue Pedestal Relief 21687. *Journal of Ancient Egyptian Interconnections* 2: 15–25.

Weimar, Peter. 1980. *Die Berufung des Mose: Literaturwissenschaftliche Analyse von Exodus 2,23-5,5*, Orbis Biblicus et Orientalis, vol. 32. Freiburg (CH): Universitätsverlag.

Weippert, Manfred. 1974. Semitische Nomaden des zweiten Jahrtausends. Über die Š3św der ägyptischen Quellen. *Biblica* 55: 265–280.

Wellhausen, Julius. 1963. *Die Composition des Hexateuchs und der historischen Bücher des Alten Testaments*. Reprint of the 3rd edition 1899. Berlin: de Gruyter.

Whitt, William D. 1991. The Jacob traditions in Hosea and their relation to genesis. *Zeitschrift für die alttestamentliche Wissenschaft* 103: 18–43.

The Structure of Legal Administration in the Moses Story 23

Stephen C. Russell

Abstract

Although Exodus 18:13–26 is set in the period of desert wandering, scholars have generally understood the text as reflecting the social world of the monarchic period.[1] Some locate it more specifically in the time of Jehoshaphat, who, according to 2 Chronicles 19:4–11, appointed local judges and established a high court in Jerusalem. According to this view, Exodus 18:13–26 was composed as an etiology for the system of royal judges attested in 2 Chronicles 19:4–11. I propose that the structure of the legal world envisioned by Exodus 18:13–26 is much more closely paralleled by that assumed in Ezra 7:12–26, where the Persian king Artaxerxes instructs Ezra to appoint judges who know the Mosaic law. As such, and in light of literary-historical considerations, Exodus 18:13–26 is best understood as a postexilic expansion of Exodus 18. The expanded Chapter 18 now serves as a major bridge in the book of Exodus by summarizing the deliverance from Egypt and anticipating the revelation at Sinai.

According to Exodus 18:13–26, on the advice of his Midianite father-in-law, Moses appointed officials to judge legal disputes.[2] They shared his juridical authority by deciding minor cases but referred hard ones to him. Through this system direct access to the highest court was restricted so that the burden of Moses' case load was made more manageable without compromising his unique position. A parallel account of the establishment of the judiciary is contained in Deut 1:9–18 and the texts are generally regarded as having a literary relationship to one another, with

[1] Most influentially, Knierim (1961). And see: Childs (1974: 325), Reviv (1982), Schäfer-Lichtenberger (1985), Crüsemann (1996: 83–90), Cook (1999), Meyers (2005: 138–140).

[2] I am grateful to Mark S. Smith for his very helpful comments on a written draft of this paper and to Daniel E. Fleming, Jacob L. Wright, and Christoph Berner, who made useful suggestions after hearing the paper as it was presented orally.

S.C. Russell (✉)
Department of History, John Jay College, CUNY, New York, NY, USA

2209 Mount Vernon St, Philadelphia, PA 19130, USA
e-mail: stephen.christopher.russell@gmail.com

T.E. Levy et al. (eds.), *Israel's Exodus in Transdisciplinary Perspective*,
Quantitative Methods in the Humanities and Social Sciences, DOI 10.1007/978-3-319-04768-3_23,

connections also to Numbers 11:10–30, according to which Yahweh shared Moses' spirit with 70 of Israel's elders.[3] In their chapters, Thomas Römer, Konrad Schmid, Christoph Berner, and Daniel Fleming conceive of the book of Exodus as being edited over a long period of time.[4] I share this conviction. My chapter argues that Exodus 18 was expanded in the Persian period by the addition of vv. 13–26. Ronald Hendel argued in his chapter that the Exodus story was a living tradition that was intimately connected to the contexts in which it was told. Jan Assmann and Israel Finkelstein emphasized the early, northern prophetic context of the story, while Baruch Halpern emphasized its familial setting. For Hendel and Assmann, the story changed according to the milieu in which it was told. My chapter demonstrates such a change to Exodus 18 in the postexilic period. Verses 13–26, I propose, were added to the chapter in the postexilic period and reflect a postexilic understanding of the structure of the legal world.[5]

[3] On the literary relationship between the texts, see Rose (1981: 226–257), Van Seters (1985, 1994: 208–219), Blum (1990: 153–163), Crüsemann (1996: 87–88), Johnstone (1998: 257–259), Cook (1999), Schwartz (2009), Schmid (2010: 235), Berner (2010: 424–425, 428–429), Carr (2011: 267), Baden (2012: 141–142), Albertz (2012: 312–313).

[4] Dozeman expresses caution about pinpointing the precise social setting of the narrative (2009: 409–410). Houtman likewise cites the opinion of several scholars but he argues that it is not possible to reconstruct a precise historical figure behind the figure of Moses in the story (1996: 397–399).

[5] At the conference where this paper was originally presented, Thomas Römer mentioned to me his student Daniele Garrone, who is writing a dissertation on Exodus 18 and who takes up the question of a possible connection to Ezra 7. Authors writing outside of the historical-critical tradition have also occasionally observed a connection between Exodus 18 and Ezra 7. J.H. Sailhamer posits that Jethro's actions "foreshadow" those of Cyrus (2009: 376). R.H. Isaacs suggests that the judicial system established by Moses was "reinstituted" by Ezra (2000: 65). Oswald (2011) compares the extended prophetic role of Moses in Exodus 18:13–27 to that of Jeremiah in Jer 42:1–6 and suggests that both texts refer to the reorganization of Judah in the sixth century BCE.

Legal Administration in Exodus 18 and 2 Chronicles 19

A half century ago, Rolf Knierim (1961) put the traditio-historical study of Exodus 18:13–26 on new footing. While an older generation had imagined various premonarchic traditions lying behind the narrative (Albright 1963; Gressmann 1913: 161–180; Noth 1972: 136–141), Knierim located this etiology of the judicial system firmly in the monarchic period.[6] Jethro concludes his advice to Moses with the assurance that all the people will return "to their place," in peace (v. 23b).[7] The reference to each Israelite having their own מָקוֹם suggests a social setting after the settlement in the land. Knierim further narrowed the social setting of the text. Exodus 18:13–26 betrays no anxiety over Moses' own authority, which is simply assumed. Rather, the text is concerned with transferring Moses' authority to other judges. In Knierim's view, neither clan elders nor priests of local shrines would have needed such an etiology, but only a newly appointed category of judge. He therefore sought the etiology's background in the changes to the judicial system attributed to Jehoshaphat in 2 Chron 19:4–11. Jehoshaphat, according to the text, reorganized the traditional juridical system by establishing a high court in Jerusalem and appointing local judges throughout the land. Knierim argued that Exodus 18:13–26 was composed in order to lend Mosaic authority to these new, royally appointed judges.[8]

[6] A useful summary of the main lines of Knierim's argument can be found in Benjamin (1983: 151–154).

[7] Cf. Noth (1962: 150). To return "in peace," as J. Gerald Janzen notes, means to return "reconciled with one another," i.e., with the legal dispute settled (1997: 130).

[8] The thesis has been enormously influential. Even where commentators have expressed doubts about a specific connection to Jehoshaphat, they have tended to understand Exodus 18:13–26 as reflecting changes to the judicial system in the era of the monarchy. For example, in dialogue with Knierim's work, Reviv (1982) and Schäfer-Lichtenberger (1985) locate the narrative in the time of King David.

The theory, however, does not adequately explain several features of Exod 18:13–26 and 2 Chon 19:4–11.[9] I wish to highlight two shortcomings of this hypothesis that will form a backdrop to the alternative I propose.[10] First, in my assessment, Knierim his overestimated the tension in 2 Chron 19:4–11 between centrally appointed royal judges and other forms of legal authority. The chapter portrays a judicial system that is composed of several overlapping and complementary power structures. Rather than a neat hierarchy of judges all having the same type of authority, there are interconnected roles for judges drawn from different sectors of society. By using the preposition מן, v. 8 portrays Jehoshaphat as appointing to the central court some priests and Levites and some of "the heads of the ancestors of Israel" (v. 8), apparently a term for a traditionally recognized leadership structure based on the language of kinship (cf. Ezra 4:3; 2 Chron 23:2). Jehoshaphat recognized their traditional authority. Furthermore, the judges that Jehoshaphat appointed in the cities of Judah would defer difficult cases to the multipartite central court (v. 10). In other words, the narrative does not understand the royally appointed judges as operating independently of traditional forms of distributed authority held by clan leaders, priests, and Levites. Royal and traditional authority are portrayed as operating coherently within a single system. In Exod 18:13–26, as I argue further below, juridical authority is portrayed as coming only from the top down with no connection to traditional forms of authority.[11] In arguing that only royal judges required an etiology like Exod 18:13–26, Knierim has, in my assessment, overestimated the tension in 2 Chron 19:4–11 between different forms of legal authority and has not fully come to terms with the exclusively top-down approach to authority in Exod 18:13–26.

A second difficulty with the hypothesis is its failure to account for the prominent role Moses' non-Israelite father-in-law plays in Exod 18:13–26. In 2 Chron 19:4–11, the judicial reforms are initiated by Jehoshaphat, son of Asa, of the line of David. There is no hint in 2 Chron 19:4–11 that the appointment of local judges or the creation of a Jerusalem high court came as a result of foreign influence. Exodus 18:13–26, on the other hand, credits a foreigner with initiating the appointment of local judges. The father-in-law's foreignness, as I argue further below, is central to Exodus 18:13–26 and is not merely the result of the larger narrative frame. Whether or not one regards as historically plausible the theses of Gressmann, Albright, and Noth, who saw behind the text one form of premonarchic encounter or another, a great strength of their analyses was their recognition of the startling nature of this foreign attribution. In understanding Exod 18:13–26 as an etiology

[9] Since Knierim's hypothesis is based on a purportedly shared historical setting assumed by the texts rather than on any literary dependency between them, the historical reliability of 2 Chron 19:4–11 must also be established. The description of Jehoshaphat's judicial reforms in 2 Chon 19:4–11 has no parallel in the Book of Kings. The main activity attributed to Jehoshaphat here, his establishment of a system of judges, reads suspiciously like an extended wordplay on his name, as Julius Welhausen pointed out long ago (Wellhausen 1961: 191). While wordplay is a well-known feature of the Chronicler's work, it is difficult to find another indisputable instance of such an extended wordplay. Pamela Barmash posits the existence of another extended pun in 2 Chron 16:12, this one bilingual. Asa, whose name relates to an Aramaic root meaning to "heal," reportedly consults healers instead of Yahweh (cf. Barmash 2005: 34). More to the point, Gary Knoppers and Steven McKenzie have shown the pervasiveness of the Chronicler's hand in 2 Chron 19:4–11 (Knoppers 1994; McKenzie 2007: 309). The pericope is rich with language, themes, and content characteristic of the Chronicler. As such, it was probably composed by him rather than copied from a source. Albright has also highlighted the evidence for the Chronicler's hand in the text, but reached the opposite conclusion about its historical reliability (1950). Crüsemann argues that the Chronicler here draws on a source (1996: 90–98). See also Japhet (1993: 770–779), Klein (1995, 2012: 271–278).

[10] For a thorough critique of Knierim's hypothesis, see Cook (1999), Graupner (1999). Cook asserts that Exodus 18 belongs to the same stream of tradition as Deuteronomism and the Book of Deuteronomy (Cook 1999). The structure of the legal system envisioned in Deuteronomy 1:9–18, however, is quite different from that envisioned in Exodus 18:13–26.

[11] On the nature of delegated authority in the text see also Janzen (1997: 131), Harkam (1999). Cf. Hoftijzer (2001).

ascribing Mosaic authority to royal judges, Knierim has, in my view, overestimated the role of Moses in the text and underestimated that of Moses' father-in-law.

Legal Administration in Exodus 18 and Ezra 7

I propose that in understanding judicial authority as coming only from a central leader and in attributing the judicial system to a foreigner, Exod 18:13–26 compares better with the structure of the legal world envisioned by Ezra 7:12–26. According to Exod 18:13–26, a single leader, Moses, stands at the top of a single judicial hierarchy. Lower judges are appointed solely on the basis of Moses' authority. He chooses "men of valor, who fear god, men of truth who despise ill-gotten gain," as though the choice depended solely on individual character. There is no hint in the language used to describe these officials that they have any relationship to traditional structures of governance based on assumed kinship or to priestly or Levitical status.[12] Rather, the judges are described in bureaucratic and military language as "men of valor" and as "commanders"—the latter term, in my view, being taken over from Deuteronomy 1.[13] Nor is there any room in the system for the existence of other judges who happen to go unmentioned here. According to v. 14, all the people go to Moses for judgment, and according to v. 22 every dispute is to be settled within the system as it is outlined in the chapter. A similar view of legal authority is reflected in Ezra 7:12–26. According to v. 25, Ezra is instructed to appoint judges and magistrates who "know the laws of God." As with Exodus 18:13–26, the language used for these judges betrays no connection to traditional forms of authority. While priests and Levites are mentioned in vv. 13, 16, and a variety of temple officials in v. 24, the text does not assign them any judicial role. The exclusive legal authority of judges in Ezra 7:12–26 is highlighted by a comparison to Ezra 10, where elders and centrally appointed judges jointly investigate the marriages of certain Israelites to foreigners. This picture of an ongoing role for traditional leadership based on the language of kinship is also found in Deut 1:15, where Moses confirms the authority of tribal heads (ראשי שבטיכם) and in Num 11:24, where Yahweh places some of Moses spirit on 70 of the elders of Israel (שבעים איש מזקני ישראל).[14] In contrast, both Ezra 7:12–26 and Exod 18:13–26 contain an idealized view of judicial power, which is imagined as emanating from a central authority without regard to forms of traditional authority held by elders, priests, or Levites.[15]

Both texts also credit a foreigner with the idea for a system of judges. In Ezra 7:12–26, the command to establish a judicial system is given

[12] Cook, following Knierim, argues that the phrase "those who hate ill-gotten gain" implies those who are already in a position to exercise judicial authority (Cook 1999: 296; Knierim 1961: 149–150). Here, I emphasize that the language used is not that of traditional kinship—elders, fathers, ancestral houses, clans, tribes, and so on—nor that of priestly or Levitical status.

[13] As several commentators point out, the seventh-century BCE Yabneh Yam Ostracon attests the judicial function of the official known as a שׂר (Sarna 1991: 100; Van Seters 1994: 213 n. 18; Weinfeld 1991: 138). A number of biblical texts likewise suggest the title was used for royal functionaries in the monarchic period (1 Sam 8:12, 13; 1 Sam 22:7; 2 Sam 18:1; 2 Kgs 1:9, 11, 13; Isa 3:3; 1 Chron 12:35; 2 Chron 17:14; but cf. Priestly use of the term in Num 31:14, 48, 52, 54). Their role in the judicial system is discussed in Frymer-Kenski (2003). On the term שׂר, see also Fox (2000: 158–163). She observes similarities and differences between the function of the Hebrew שׂר and the Egyptian *śr* and cites Egyptian texts showing the judicial function of these officials.

[14] Note that Exodus 18:12 recognizes a leadership role for the elders of Israel (זקני ישׂראל), though there is no indication that they perform any judicial function. In contrast, compare Exodus 24:14. According to this late gloss, Aaron, Hur, and the elders of Israel take on juridical functions. While MT reads, "all Israel's elders" in Exod 18:12, the Samaritan Pentateuch reads "some of Israel's elders," with partitive *min* (Propp 1999: 625).

[15] In his analysis of Exodus 18, LeFebvre distinguishes between law that derives from custom and law that derives from divine oracle (2006: 40–47). My analysis here focuses instead on the authority of the individuals who offer judgment.

by Artaxerxes to Ezra in the form of an Aramaic letter. The identification of Artaxerxes as King of Persia in the narrative frame (v. 1) and the fact that the letter is written in Aramaic rather than Hebrew highlight the foreign character of the patriarchal figure who suggests the system of judges. Likewise, in Exod 18:13–26, the suggestion for a system of judges comes from a foreigner, Moses' father-in-law. The peculiarity of this arrangement is brought out by a comparison with the alternative tradition of the establishment of the judiciary in Deut 1:9–18. There, Moses himself was responsible for devising the system of judges. He proposed it to the people, who actively participated in the decisions about whom should serve as judges. That text acknowledges the existence of foreigners and grants appointed judges some legal authority over resident aliens (v. 16), but the system is designed and established exclusively by Israelites. In Exod 18:13–26, on the other hand, the narrative credits a foreigner with the idea for the system of judges.

Furthermore, the foreignness of Moses' father-in-law is not only the result of the larger narrative frame in which the story occurs. Exodus 18:13–26 also highlights the fact that the father-in-law is a non-Israelite by attributing to him idiosyncratic speech patterns.[16] A careful examination of the father-in-law's speech shows that it contains a density of unusual morphological, syntactical, and lexical features. Mordechay Mishor points to several of these (Mishor 2006; cf. Jacob 1992: 507; Greenstein 1999: 160).[17] The narrative frame uses the expression מן הבקר in v. 13, with the preservation of the *nun* before a definite noun.[18] Such usage is quite normal in standard Biblical Hebrew. The father-in-law, however, uses מן בקר in v. 14, with the preservation of the *nun* before an indefinite noun. The verb נבל (v. 18) is used 18 times in biblical poetry but this is its only use in a prose text. In v. 18 the father-in-law uses the form עֲשֹׂהוּ where one might better expect עֲשׂתוֹ.[19] The preposition מול, itself occurring only 26 times in the Bible, is used in connection with a deity only in Exod 18:19. In v. 20 the father-in-law uses the exceptionally rare אֶתְהֶם instead of the common form אֹתָם.[20] In v. 20 he uses an asyndetic relative clause (הַדֶּרֶךְ יֵלְכוּ בָהּ), which is a rare construction in prose except in the book of Chronicles.[21] In v. 21 the father-in-law also uses the verb חזה with the meaning "to choose," which is otherwise unattested in Biblical Hebrew. In v. 23 he uses the verb בוא with the preposition על instead of the much more common אל, which is used by Moses in v. 15.[22] Individually, each of these features might be dismissed as holding no particular significance, but taken together they suggest a deliberate attempt by the narrator to characterize the father-in-law's speech as stilted, unusual, and foreign.

Exodus 18:13–26 and Ezra 7:12–26 thus share two key perspectives that are not shared by any of the other texts usually considered in relation to the establishment of the judiciary: Deuteronomy 1, Numbers 11, and 2 Chronicles 19. At the same time, the texts do not have particularly strong linguistic connections at the level of shared

[16] On the use of dialogue as a means of characterization in biblical narrative, see especially Alter (1981: 79–110). On the use of unusual speech patterns to characterize foreigners in biblical narrative, see Kaufman (1988: 54–55), Rendsburg (1995).

[17] In addition to the features noted below, Mishor argues that the Hiphil of זהר, used in v. 20, may be an Aramaic loanword (2006: 228). He points to וְזָהִירִ֖ין in Ezra 4:22. The form, however, is not widely recognized as an Aramaic loanword.

[18] In v. 13, the definite article is missing from "morning" and apparently also from "evening" in 4QpaleoExod[m] (Sanderson 1986: 333).

[19] The latter form is found in Gen 41:32; Exod 12:48; Deut 30:14; Jer 1:12; 23:20; 30:24; Job 23:9; 28:26; 2 Chron 30:3.

[20] The form אֶתְהֶם occurs elsewhere only in Gen 32:1; Num 21:3; Ezek 34:12; 1 Chron 6:50. The form אֹתָם occurs some 285 times.

[21] On the rarity of this syntactic construction see GKC §155d. 4QpaleoExod[m] includes the relative pronoun [ר]אש, but MT is to be preferred as *lectio difficilior* (Sanderson 1986: 334).

[22] The Samaritan Pentateuch has substituted אל for MT על (Sanderson 1986: 334).

phraseology and are written in different languages. They do not come from the same scribal hand or school. Rather, they both reflect the same social milieu, in which it was possible to imagine the judicial system as being organized exclusively from the top down and as being initiated by a foreigner. Both texts, in my assessment, date from the postexilic period, whether or not the system they envision was ever implemented.[23] In the monarchic period, official, bureaucratic structures of governance associated with the royal court had always shared juridical and other forms of power with priestly groups and with traditional forms of leadership based on kinship, especially town elders.[24] They all functioned as part of a single system. It is only in the postmonarchic period, under the influence of the great empires and after the disruption of traditional modes of life, that biblical scribes came to imagine an exclusively top-down approach to judicial governance like that presented in Exod 18:13–26 and Ezra 7:12–26.[25] It is in this period also that at least some circles sought to lend legitimacy to Israelite systems of law by associating them with foreign sanction.[26]

Exodus 18 and Literary-Critical Considerations

A postexilic date for Exod 18:13–26 does not contradict linguistic evidence and is supported by literary-critical and traditio-historical considerations. A detailed discussion of the linguistic evidence for dating the text lies well beyond my aims here. In this short pericope, the majority of which is dialogue and some of which may be copied from Deuteronomy 1, there is insufficient linguistic data to characterize the text as being written in Standard Biblical Hebrew or Late Biblical Hebrew, which are only subtly different from one another.[27] With

[23] On the difficulty of dating the current form of Ezra 7:12–26, see Pakkala (2004: 49–53).

[24] A comprehensive reconstruction of the history of judicial administration in ancient Israel and Judah lies well beyond my aims here. Of interest to the legal historian are several biblical narratives. In Jeremiah 26:8–24, prophets, priests, and royal officials play complimentary roles in Jeremiah's trial. Town elders adjudicate between Boaz and his relative in Ruth 4:1–12. 1 Chron 26:29–32 describes David's judicial system and distinguishes between matters of the King and matters of Yahweh. According to 1 Samuel 15, Absalom sought to undermine David's legitimacy by raising doubts about his administration of justice. Mic 3:9–12 ascribes judicial functions to priests and officials. In a forthcoming article, I emphasize the importance of town elders in the system of legal administration (Russell 2014). I agree with Barmash, who argues, "a great deal of legal authority remained in the local community throughout the First Temple period" (2005: 35). On the history of Israelite and Judahite legal administration, see also Macholz (1972), Wilson (1983a, b), Niehr (1987), Westbrook and Wells (2009: 35–52), Wells (2010).

[25] In fact, to judge by Ezra 10:8, 14, where elders have judicial functions, if the idealized system of Ezra 7:12–26 was ever implemented, it must have been implemented very late indeed. Ezra 7:12–26 itself may reflect multiple stages of editorial activity. Pakkala argues that the original text within the unit consisted only of Ezra 7:11a, 12–15, 16b, 19–22 (2004: 32–40). For Pakkala, the system of legal administration in Ezra 7:25 postdates Ezra 10. On Ezra 7:12–26, see also Grätz (2004). He regards this edict of Artaxerxes as a fictional composition from the Hellenistic period.

[26] My thesis here does not depend on a narrow theory of imperial authorization of the Torah as it has been advocated by P. Frei (1984). For an overview of the issues involved, see Schmid (2007).

[27] The linguistic evidence for a date for Exod 18:13–26 is equivocal. Propp (1999: 627) considers יִשְׁפּוּטוּ in v. 26 a late spelling (cf. תַעֲבוּרִי in Ruth 2:8 and תִּשְׁמוּרֵם in Prov 14:3). GKC § 47g and Joüon-Muraoka § 44c, on the other hand, attribute the forms to their pausal or prepausal position. Polzin has shown that non-assimilation of *nun* before an indefinite noun is characteristic of Late Biblical Hebrew (Polzin 1976: 66; cf. Rendsburg 1980: 72), but this feature occurs in Moses' father-in-law's speech (v. 14) and may thus reflect a deliberate attempt by the narrator to portray the father-in-law in a particular light. According to Hurvitz, the verb עמד acquired an expanded semantic range in the postexilic period (1982: 95–96). The use of the verb in v. 23 may reflect this expanded range. Even if this could be shown definitively from the context, however, the occurrence of the word is again in speech attributed to Moses' father-in-law. The noun phrase "men of truth" (v. 21) is used elsewhere only in Neh 7:2, in the singular. There the noun phrase is used in close proximity to "fearer of god" and "commander," both also used in Exod 18:21. As such, one text may be deliberately alluding to the other and this noun phrase in itself cannot therefore constitute an argument about linguistic dating. Rose had pointed out that Exodus 18 speaks of *torot*, in the plural, not Torah, in the singular (1981: 229).

regard to literary-critical and traditio-historical considerations, I limit myself to brief observations on the relationship of Exod 18:13–26 to Deut 1:9–18 and Num 11:12–30, the relationship of Exod 18:13–26 to Exodus 18 as a whole, and the relationship of Exodus 18 to the structure of the book of Exodus.

Exodus 18:13–26 is widely regarded as containing thematic and linguistic similarities to Deut 1:9–18 and Num 11:12–30. In my view, the connections to Num 11:12–30 are not particularly strong. Although Exod 18:13–26 and Num 11:16–17, 24b–30 are both interested in the nature of Moses' leadership, there is in Exod 18:13–26 no sense that appointed leaders will share in Moses' spirit. Nor is there any sense that Numbers 11 has legal administration as a concern of leadership. Rather, Num 11:16–17, 24b–30 shares much more in common with Deut 1:9–18 than it does with Exodus 18:13–26.[28] At most, Exodus 18 and Numbers 11 share the use of the root כבד to describe the essential problem being addressed: the responsibility is too heavy for Moses. But the root is too common in Biblical Hebrew to serve as evidence of direct literary borrowing in one direction or another (Carr 2011: 268 n. 33). In sum, Num 11:12–30 and Exod 18:13–26 do not share particularly strong linguistic or thematic links.[29] There is little reason to posit that one text is literarily dependent on the other.

The connections between Exod 18:13–26 and Deut 1:9–18 are considerably stronger. They share the same essential problem: the people are too many. And they share the same solution: Moses appoints a tiered structure of officials to carry out judicial functions. Furthermore, the literary connections between the texts are palpable. Both texts contain the phrase "officials of thousands, officials of hundreds, officials of fifties, and officials of tens."[30] Between them, the two chapters share the only three occurrences of this extended noun phrase in the Hebrew Bible. Yet, there is no syntactic reason for considering one occurrence more fitting to its context than the other. There is also close resemblance between Exod 18:26, "the hard matter they would take to Moses" and Deut 1:17, "the matter which is too hard for you, you shall bring near to me." But again, there is no grammatical reason to posit one particular direction of dependence over the other.[31]

But the plural, which is relatively rare throughout the Bible, is used also in Dan 9:10; Neh 9:13. In sum, there is insufficient data to make a strong linguistic case for a date for the text.

[28] David Carr concisely summarizes scholarly treatment of Numbers 11 (Carr 2011: 267). The chapter has long been viewed as having two principal layers. The main narrative concerns the people's murmuring about the lack of food and Yahweh's provision of quail. A second layer concerns the distribution of Moses spirit to 70 of the elders of Israel in vv. 16–17, 24b–30 and perhaps also includes Moses' complaint in vv. 11–12, 14–15. Carr cites Blum (1990: 82–84) and Baden (2009: 108–109) as regarding all of Moses' complaint in vv. 11–12, 14–15 as belonging to the same layer as the elders story in vv. 16–17, 24b–30 and Sommer (1999: 611–612) as regarding none of it as belonging to the same layer. Both Numbers 11 and Deuteronomy 1 have a common setting at the departure from Horeb/Sinai and both contain the theme of Moses' ability to bear the burden of the people. Carr, drawing on the work of Martin Rose (1981: 226–257), notes that Num 11:16–17, 24b–30 conforms to the murmuring story structure of the quail episode and as such should be viewed as a harmonizing expansion to Numbers 11 that drew on Deut 1:9–18. While some case can be made for the dependence of Num 11:16–17, 24b–30 on Deut 1:9–18, however, there is no basis for establishing a direction of dependence, if there is any, between Num 11:16–17, 24b–30 and Exod 18:13–26. On Moses' role as intercessor in relation to these texts, see Aurelius (1988: 180–183).

[29] Contra Childs (1974: 324–325) and Seebass (2002: 43).

[30] Although officials were evidently responsible for varying numbers of individuals, they are not envisioned in the narrative as reporting to one another up a chain of command. Rather, to judge by vv. 22, 26, the system imagines only two tiers of judges: Moses and officials. Any case that is too difficult for the officials is passed on to Moses directly.

[31] Although Edward Greenstein (1999) has argued that the root דבר is a *Leitmotif* in Exodus 18, the root is very common in Biblical Hebrew and is no more at home in Exodus 18 than in Deuteronomy 1, where it occurs 18 times, five of which are in vv. 9–18. The root is used in both the Covenant Code (e.g., Exod 22:8) and in the Deuteronomic Code (e.g., Deut 15:2) with the technical meaning "legal case." The word thus offers little reason for suggesting one direction of dependence or another between Exodus 18 and Deuteronomy 1.

Thematic evidence, though not definitive, suggests that Deut 1:9–18 may have been the original text. John Van Seters notes that Deut 1:9–18 is explicable entirely on the basis of the Deuteronomic Code.[32] In Deut 16:18–19, the people are commanded to appoint tribal judges and officials in every city gate and are charged with executing justice without partiality. According to Deut 17:8–13, a legal case which is too baffling to judge must be brought to a central court consisting of Levitical priests and a judge. Deuteronomy 1:9–18 can thus be explained as a Deuteronomic reflection on the themes of the Deuteronomic Code that retrojects legal structures and procedures from the Code back into the time of Moses himself. In contrast, Exod 18:13–26 is isolated thematically from its context. Outside of Exod 18:13–26 and a late gloss about the temporary delegation of legal authority to Aaron and Hur in Exod 24:14, the themes of legal administration that I have been discussing are not taken up directly in the book of Exodus nor in the non-Priestly material in the Pentateuch.[33] In light of these considerations, it seems more likely that Exod 18:13–26 is literarily dependent on Deut 1:9–18, which is, in turn, based on the themes of the Deuteronomic Code, rather than the other way around. At the same time, this line argument should be regarded as suggestive rather than conclusive.[34]

Let us turn, then, to the relationship of Exod 18:13–26 to Exodus 18 as a whole. Form critics of Exodus 18 have viewed the narrative's structure as key to understanding its traditio-historical background.[35] The story's main events occur over two days, with quite distinct activities on each. On day one, Jethro meets Moses in the desert, listens to all that Yahweh had done for Israel, and celebrates a feast to God. On day two, Moses' father-in-law observes Moses administering justice to all the people and recommends a new system of judicial administration, which is adopted. In addition to the thematic contrast between the two halves there are important distinctions in terminology. In vv. 1–12, Jethro's name is used seven times and the title father-in-law is used three times, somewhat interchangeably. In vv. 13–26, however, the character is referred to only as Moses' father-in-law and the name Jethro does not appear.[36] Likewise, the noun אלהים and the divine name יהוה are both used in vv. 1–12, while only אלהים

[32] In Deuteronomy 1, the problem is the number of the people, a theme also found in 1 Kgs 3:8–9.

[33] Propp notes that Exod 24:14 is marked as an editorial insertion by the resumptive repetition, "Moses ascended the mountain" in vv. 13, 15 (2006: 299). Note also that in Numbers 25:5, traditionally considered non-Priestly, the "judges of Israel" execute Moses' sentence of death upon their men who attached themselves to Baal-Peor. The theme of fair judgment is taken up in the Holiness Code in Lev 19:15. In the Priestly text Num 35:24, the whole community is involved in judgment.

[34] M. Rose proposed that Number 11 is dependent on Deuteronomy 1, which is in turn dependent on Exodus 18 (1981: 226–257). For J. Baden, Deut 1 is dependent on both Exod 18 and Num 11 (2012: 141). It seems implausible to him that Num 11 and Exod 18, which describe separate events in Israel's past, should both be derived from a single episode in Deut 1. I do not share his skepticism in this regard. Crüsemann argues that a stronger case can be made for literary stratification in Deuteronomy 1:9–18 than in Exodus 18:13–26 (1996: 87–88). However, Deut 1:9–18 reads sensibly as a unit.

[35] For Hugo Gressman, the events now recounted as part of day one were originally an etiology of the Yahweh cult at Kadesh, to which, he believes, the narratives were originally attached (1913: 161–180). On the other hand, the events now recounted as part of day 2 were originally an etiology of a dual system of justice based on lay judges and priestly oracles. For Gressman, Jethro's role as instructor brought the two originally independent tales together. Along related lines, Noth believed that the figure of Moses' father-in-law had brought the two traditions together and that the cultic tradition was more central than the legal one (1972: 136–141). According to Noth, behind Exodus 18:1–12 lay a tradition of pilgrimage to a mountain held sacred by both Midianites and some early Israelites. Childs also notes the tenuous connection of the tradition about legal administration to Midian (1974: 326). See also Albright (1963, 1968: 39–42, 1970). On Midianites-Kenites as possible bearers of early Yahwistic traditions, see also Blenkinsopp (2008).

[36] Cf. Schäfer-Lichtenberger (1985: 61), Albertz (2012: 299). In v. 14, LXX reads "Jethro," where MT, 4QpaleoExod^{m}, and the Samaritan tradition all read "father-in-law of Moses" (Sanderson 1986: 333; Propp 1999: 626).

appears in vv. 13–26.[37] Given these linguistic and thematic differences between vv. 1–12 and vv. 13–26, I agree with an older generation of form critic that understood the two halves of the chapter as having different traditio-historical backgrounds. At the same time, the chapter reads sensibly as a whole. It opens with Jethro hearing all that God had done for Moses and for Israel and it closes with him departing for his home. Furthermore, vv. 13–26 offer no explanation of how Moses' father-in-law reenters the narrative after such a long absence. As such, vv. 13–26 seem to assume the existence of vv. 1–12. These apparently contradictory observations are easily reconciled on the hypothesis that vv. 13–26 constitute an expansion of vv. 1–12, 27 (so also Berner 2010: 406–429).[38] In my view, the expansion was partly based on material from Deut 1:9–18 but it reflected a fundamentally different view of the structure of legal administration than that text.

Finally, I offer some brief observations on the relationship of Exodus 18 to the book as a whole. E. Carpenter has shown that the chapter sits at a transitional point in the book (Carpenter 1997).[39] The first half of the chapter recapitulates the deliverance of the people from Egypt narrated in the preceding chapters while the second half anticipates the revelation of God's law from Sinai as reported in the chapters that follow. The relationship of Exodus 18:1–12, 27 to the material that precedes can be refined further. Konrad Schmid, drawing on the work of Erhard Blum, has noted that the chapter has particularly close affinities with Exodus 3–4: the name Jethro, the term "mountain of God," the description of the exodus as an act of Yahweh's goodness (טובה) and as his deliverance (Hiphil of נצל), and the use of the definite article with God (האלהים) (Schmid 2010: 235; Blum 1990: 360–365). For Schmid, Exodus 18 belongs to a post-Priestly redactional layer that included Exodus 3–4, a layer that he dates to the early fifth century BCE. Setting aside the use of the definite article with God, which is too common in the Pentateuch to be diagnostic in and of itself, I would point out that the connections to Exodus 3–4 occur only in vv. 1–12, 27.[40] Exodus 18:13–26 is linguistically and thematically quite different, as I have noted above. If vv. 13–26 is indeed an expansion to vv. 1–12 + 27, then it would be later than Schmid's post-Priestly redactional layer.

Exodus 18:13–26 looks forward to the revelation at Sinai.[41] Edward L. Greenstein (1999) has shown how the narrative has artfully deployed the leitwort דבר, generally, "thing, word, matter," but used at times in Exodus 18 with the more narrow meaning "legal case." In his view, it serves to introduce the motif of the "words of

---

[37] There is some variation in the manuscript traditions in the language used for the deity. In v. 1, MT reads "God," while a Genizah manuscript, LXX, and the Targumim read "Yahweh" (Propp 1999: 624).

[38] The opening words of v. 13, וַיְהִי מִמָּחֳרָת, may have belonged to the original narrative. Verses 3–4, which contain the names of Moses' sons, contain awkward grammatical shifts in subject. They may also contain later material. On the redaction history of the chapter, see Berner (2010: 406–429). Berner posits some ten redactional layers in the chapter. I am more cautious than he is about the ability of scholarship to reconstruct the many layers of the text. Schäfer-Lichtenberger argues that a basic narrative in Exodus 18 was supplemented by three additions: vv. 15b, 20b; vv. 16b, 20a; and vv. 21b, 25b (1985).

[39] Carpenter writes, "Exodus 18 is perhaps the major transitional chapter in the book of Exodus, summarizing the past events (Exod. 1–17) and preparing for the coming revelations at Sinai (Exod. 19–40). To be sure, there are other transitional passages (e.g., 1.1–7; 15.22–24). But ch. 18 seems to be the major hinge in the structure of the total composition, serving both as a prologue and an epilogue" (1997: 91–92). In my view, Carpenter has not adequately accounted for the several chapters devoted to the description of the tabernacle (Exod 25–40). On the bridging function of the chapter, see also Smith (1997: 228–231), Propp (1999: 633–634), Meyers (2005: 136), Schmid (2010: 235 n. 447). On Exodus 18 as a prologue to Exodus 19–24, see Cassuto (1967: 211). See also Frevel (2003), Römer (2009).

[40] Compare Frevel (2003: 5–20).

[41] In this regard, Erhard Blum is correct in asserting that Exodus 18 was inserted only after the Priestly editors had finished the beginning of the Sinai pericope in Exodus 19 (Blum 1991: 54–56, 2006: 94 n. 18).

Yahweh," in chapters 19–33.[42] To my mind, as a prologue to the revelation at Sinai, Exodus 18:13–26 performs at least three functions. First, it addresses one major shortcoming of the Covenant Code, namely the lack of a mechanism for the implementation of the laws it contains. Second, it reminds the audience that although the revelation at Sinai in 34:16 includes an injunction against marrying foreign women, even Moses himself, at the very genesis of the judicial system, had a foreign wife. As such, the narrative in its current form can be interpreted as advocating a different response to foreign marriage than Ezra 7–10.[43] Third, although Aaron and the elders of Israel play important leadership roles in Exodus 24, the prologue in Exod 18:13–26 establishes the priority of another kind of judicial system. That system, like the one described in Ezra 7:12–26, had an exclusively centralized structure rooted in administrative terminology and was implemented on the advice of a foreigner.

References

Albertz, R. 2012. *Exodus, Band I: Ex 1–18*, Zürcher Bibelkommentare AT 2.1. Zurich: Theologischer Verlag Zürich.

Albright, W.F. 1950. The Judicial Reform of Johoshaphat. In *Alexander Marx: Jubilee Volume on the Occasion of His Seventieth Birthday*, ed. S. Lieberman, 61–82. New York: Jewish Theological Seminary of America.

———. 1963. Jethro, Hobab and Reuel in Early Hebrew Tradition. *Catholic Biblical Quarterly* 25: 1–11.

———. 1968. *Yahweh and the Gods of Canaan: A Historical Analysis of Two Contrasting Faiths*, Jordan lectures in comparative religion, vol. 7. London: Athlone Press.

———. 1970. Midianite Donkey Caravaneers. In *Translating & Understanding the Old Testament: Essays in Honor of Herbert Gordon May*, ed. H.T. Frank and W.L. Reed, 197–205. Nashville, TN: Abingdon.

Alter, R. 1981. *The Art of Biblical Narrative*. New York: Basic Books.

Aurelius, E. 1988. *Der Fürbitter Israels: Eine Studie zum Mosebild im Alten Testament*, Coniectanea Biblical Old Testament Series, vol. 27. Stockholm: Almqvist & Wiksell International.

Baden, J.S. 2009. *J, E, and the Redaction of the Pentateuch*, Forschungen zum Alten Testament, vol. 68. Tübingen: Mohr Siebeck.

———. 2012. *The Composition of the Pentateuch: Renewing the Documentary Hypothesis*. New Haven, CT: Yale University Press.

Barmash, P. 2005. *Homicide in the Biblical World*. New York: Cambridge University Press.

Benjamin, D.C. 1983. *Deuteronomy and City Life: A Form Criticism of Texts with the Word City (ʿîr) in Deuteronomy 4:41–26:19*. Lanham, MD: University Press of America.

Ber, V. 2008. Moses and Jethro: Harmony and Conflict in the Interpretation of Exodus 18. Communio viatorum 50: 147–170.

Berner, Christoph. 2010. *Die Exoduserzählung: Das literarische Werden einer Ursprungslegende Israels*, Forschungen zum Alten Testament, vol. 73. Tübingen: Mohr Siebeck.

Blenkinsopp, J. 2008. The Midianite-Kenite Hypothesis Revisited and the Origins of Judah. *Journal for the Study of the Old Testament* 33: 131–153.

Blum, E. 1990. *Studien zur Komposition des Pentateuch*, Beiheft zur Zeitschrift für die alttestamentliche Wissenschaft, vol. 189. Berlin: Walter de Gruyter.

———. 1991. Gibt ist die Endgestalt des Pentateuch. In *Congress Volume: Leuven, 1989*, Supplements to Vetus Testamentum, vol. 43, ed. J.A. Emerton, 46–57. Leiden: Brill.

———. 2006. The Literary Connection between the Books of Genesis and Exodus and the End of the Book of Joshua. In *A Farewell to the Yahwist? The Composition of the Pentateuch in Recent European Interpretation*, ed. Thomas B. Dozeman and Konrad Schmid, 89–106. Atlanta, GA: Society of Biblical Literature.

Carpenter, E. 1997. Exodus 18: Its Structure, Style, Motifs and Function in the Book of Exodus. In *A Biblical Itinerary: In Search of Method, Form and Content: Essays in Honor of George W. Coats*, ed. E. Carpenter, 91–108. Sheffield, England: Sheffield Academic Press.

Carr, D.M. 2011. *The Formation of the Hebrew Bible: A New Construction*. Oxford: Oxford University Press.

Cassuto, U. 1967. *A Commentary on the Book of Exodus*. Trans. I. Abrahams. Jerusalem: Magnes Press.

Childs, B.S. 1974. *The book of Exodus: A Critical, Theological, Commentary*, Old Testament Library. Louisville, KY: Westminster John Knox Press.

[42] On the theme of God's mighty acts in the chapter, see also Houtman (1996: 394). Cf. Dozeman (2009: 408), Cassuto (1967: 222).

[43] Whatever meaning the emphatic לו in v. 27 may have carried in the original narrative, before the addition of vv. 13–26, in its current context it reads as though Moses' father-in-law leaves by himself, without Moses' wife and children, who apparently remain with Moses. Contrary to the procedures narrated in Ezra 10, even Moses kept his foreign wife with him. 4QpaleoExod^{m} and LXX omit לו in v. 27 (Sanderson 1986: 334). On הלך with ל, cf. 1 Sam 26:12 Jer 5:5; Song 2:11. On varying biblical attitudes towards outsiders, see Leveen (2010).

Cook, S.L. 1999. The Tradition of Mosaic Judges: Past Approaches and New Directions. In *On the Way to Nineveh: Studies in Honor of George M. Landes*, ed. S.L. Cook and S.C. Winter, 286–315. Atlanta, GA: Scholars Press.

Crüsemann, F. 1996. *The Torah: Theology and Social History of Old Testament Law*. Trans. A.W. Mahnke. Minneapolis, MN: Fortress Press.

Dozeman, T.B. 2009. *Exodus*, Eerdmans Critical Commentary. Grand Rapids, MI: Eerdmans.

Fox, N.S. 2000. *In the Service of the King: Officialdom in Ancient Israel and Judah*, Monographs of the Hebrew Union College, vol. 23. Cincinnati, OH: Hebrew Union College Press.

Frei, P. 1984. Zentralgewalt und Lokalautonomie im Achämenidenreich. In *Reichsidee und Reichsorganisation im Perserreich*, Orbis biblicus et orientalis, vol. 55, ed. P. Frei and K. Koch, 8–131. Fribourg: Univesritätsverlag.

Frevel, C. 2003. 'Jetzt habe ich erkannt, dass YHWH grösser ist als alle Götter': Ex 18 und seine kompositionsgeschichtliche Stellung im Pentateuch. *Biblische Zeitschrift* 47: 3–22.

Frymer-Kenski, T. 2003. Israel. In *A History of Ancient Near Eastern Law*, vol. 2, ed. Raymond Westbrook, 986–987. Leiden: Brill.

Grätz, S. 2004. *Das Edikt des Artaxerxes: Eine Untersuchung zum religionspolitischen und historischen Umfeld von Esra 7, 12–26*, Beiheft zur Zeitschrift für die alttestamentliche Wissenschaft, vol. 337. Berlin: Walter de Gruyter.

Graupner, A. 1999. Exodus 18, 13–27: Ätiologie einer Justizreform in Israel? In *Recht und Ethos im Alten Testament: Gestalt und Wirkung; Festschrift für Horst Seebass zum 65. Geburtstag*, ed. S. Beyerle, G. Mayer, and H. Strauss, 11–26. Neukirchen-Vluyn: Neukirchener.

Greenstein, E.L. 1999. Jethro's Wit: An Interpretation of Wordplay in Exodus 18. In *On the Way to Nineveh: Studies in Honor of George M. Landes*, ed. S.L. Cook and S.C. Winter, 155–171. Atlanta, GA: Scholars Press.

Gressmann, H. 1913. *Mose und seine Zeit: Ein Kommentar zu den Mose-Sagen*, Forschungen zur Religion und Literatur des Alten und Neuen Testaments, vol. 2/1. Göttingen: Vandenhoeck & Ruprecht.

Harkam, G. 1999. 'Die Sache ist dir zu schwer…': Das Entlastungsmotiv in Ex 18:13–27 und das Modell von 'management by concentration and delegation'. In *Vielseitigkeit des Alten Testaments: Festschrift für Georg Sauer zum 70. Geburtstag*, ed. J.A. Loader and H.V. Kieweler, 49–61. New York: Peter Lang.

Hoftijzer, J. 2001. Description or Categorical Rule? Some Remarks on Ex. 18:16. In *Veenhof Anniversary Volume: Studies Presented to Klaas R. Veenhof on the Occasion of His Sixty-Fifth Birthday*, ed. W.H. van Soldt, J.G. Dercksen, N.J.C. Kouwenberg, and Th.J.H. Krispijn, 193–196. Leiden: Nederlands Instituut voor het Nabije Oosten.

Houtman, C. 1996. *Exodus*. Vol. 2. Trans. Sierd Woudstra. Historical Commentary on the Old Testmanet. Kampen: Kok.

Hurvitz, A. 1982. *A Linguistic Study of the Relationship between the Priestly Source and the Book of Ezekiel: A New Approach to an Old Problem*, Cahiers de la Revue biblique, vol. 20. Paris: J. Gabalda.

Isaacs, R.H. 2000. *Every Person's Guide to Jewish Law*. Northvale, NJ: Jason Aronson.

Jacob, B. 1992. *The Second Book of the Bible*. Trans. W. Jacob. Hoboken, NJ: Ktav.

Janzen, J.G. 1997. *Exodus*, Westminster Bible Companion. Louisville, KY: Westminster John Knox Press.

Japhet, S. 1993. *I & II Chronicles*, Old Testament library. Louisville, KY: Westminster John Knox Press.

Johnstone, W. 1998. *1 and 2 Chronicles. Volume 1: 1 Chronicles 1–2 Chronicles 9; Israel's Place Among Nations*, Library Hebrew Bible/Old Testament Studies, vol. 253. London: T&T Clark.

Kaufman, S.A. 1988. The Classification of North West Semitic Dialects of the Biblical Period and Some Implications Thereof. In *Proceedings of the Ninth World Congress of Jewish Studies (Panel Sessions: Hebrew and Aramaic Languages)*, 41–57. Jerusalem: World Union of Jewish Studies.

Klein, R.W. 1995. Reflections on Historiography in the Account of Jehoshaphat. In *Pomegranates and Golden Bells: Studies in Biblical, Jewish, and Near Eastern Ritual, Law, and Literature in Honor of Jacob Milgrom*, ed. D.P. Wright, D.N. Freedman, and A. Hurvitz, 643–657. Winona Lake, IN: Eisenbrauns.

———. 2012. *2 Chronicles*, Hermeneia. Minneapolis, MN: Fortress Press.

Knierim, R. 1961. Exodus 18 und die Neuordnung der mosäischen Gerichtsbarkeit. *Zeitschrift für die Alttestamentliche Wissenschaft* 73: 146–171.

Knoppers, G.N. 1994. Jehoshaphat's Judiciary and the 'Scroll of YHWH's Torah'. *Journal of Biblical Literature* 113: 59–80.

LeFebvre, M. 2006. *Collections, Codes, and Torah: The Re-characterization of Israel's Written Law*, Library of Hebrew Bible/Old Testament Studies, vol. 451. New York: T & T Clark.

Leveen, A.B. 2010. Inside Out: Jethro, the Midianites and a Biblical Construction of the Outsider. *Journal for the Study of the Old Testament* 34: 395–417.

Macholz, G.Z. 1972. Zur Geschichte der Justizorganisation in Juda. *Zeitschrift für die alttestamentliche Wissenschaft* 84: 314–340.

Mayes, A.D.H. 1979. *Deuteronomy*, New Century Bible Commentary. Grand Rapids, MI: Eerdmans.

McKenzie, S.L. 2007. The Trouble with King Jehoshaphat. In *Reflection and Refraction: Studies in Biblical Historiography in Honour of A. Graeme Auld*, Supplements to Vetus Testamentum, vol. 113, ed. R. Rezetko, T.H. Lim, and W.B. Aucker, 299–314. Leiden: Brill.

Meyers, C. 2005. *Exodus*, New Cambridge Bible Commentary. Cambridge: Cambridge University Press.

Mishor, M. 2006. On the Language and Text of Exodus 18. In *Biblical Hebrew in Its Northwest Semitic Setting: Typological and Historical Perspectives*, ed. S.E. Fassberg and A. Hurvitz, 225–229. Winona Lake, IN: Eisenbrauns.

Niehr, H. 1987. *Rechtsprechung in Israel: Unterschungen zur Geschichte der Gerichtsorganisation im Alten Testament*, Stuttgarter Bibelstudien, vol. 130. Stuttgart: Katholishches Bibelwerk.

Noth, M. 1962. *Exodus: A Commentary*. Trans. J.S. Bowden. Old Testament Library. Philadelphia, PA: Westminster Press.

———. 1972. *A History of Pentateuchal Traditions*. Trans. B.W. Anderson. Englewood Cliffs, NJ: Prentice-Hall.

Oswald, W. 2011. Jeremiah and Moses, A Comparison of Their Public Offices in Exod. 18:13–27 and Jer. 42:1–6. In *"My Spirit at Rest in the North Country" (Zechariah 6.8): Collected Communications to the XXth Congress of the International Organization for the Study of the Old Testament, Helsinki 2010*, Beiträge zur Erforschung des Alten Testaments und des antiken Judentums, vol. 57, ed. H.M. Niemann and M. Augustin, 265–272. Frankfurt: Peter Lang.

Pakkala, J. 2004. *Ezra the Scribe: The Development of Ezra 7–10 and Nehemia 8*, Beiheft zur Zeitschrift für die alttestamentliche Wissenschaft, vol. 347. Berlin: Walter de Gruyter.

Polzin, R. 1976. *Late Biblical Hebrew: Toward an Historical Typology of Biblical Hebrew Prose*, Harvard Semitic Monographs, vol. 12. Missoula, MT: Scholars Press.

Propp, W.H.C. 1999. *Exodus 1–18: A New Translation with Introduction and Commentary*, Anchor Bible, vol. 2. New York: Doubleday.

———. 2006. *Exodus 19–40: A new translation with introduction and commentary*, Anchor bible, vol. 2A. New York: Doubleday.

Rendsburg, G.A. 1980. Late Biblical Hebrew and the Date of "P". *Journal of Ancient Near Eastern Studies* 12: 65–80.

Reviv, H. 1982. The Traditions Concerning the Inception of the Legal System in Israel: Significance and Dating. *Zeitschrift für die Alttestamentliche Wissenschaft* 94: 566–575.

Reviv, H. 1995. Linguistic Variation and the 'foreign' Factor in the Hebrew Bible. *Israel Oriental Studies* 15: 177–190.

Römer, T. 2009. Provisorische Überlegungen zur Entstehung von Exodus 18–24. In *"Gerechtigkeit und Recht zu üben" (Gen 18,19): Studien zur altorientalischen und biblischen Rechtsgeschichte, zur Religionsgeschichte Israels und zur Religionssoziologie; Festschrift für Eckart Otto zum 65. Geburtstag*, Beihefte zur Zeitschrift für altorientalische und biblische Rechtsgeschichte, vol. 13, ed. R. Achenbach and M. Arneth, 128–154. Wiesbaden: Harrassowitz.

Rose, M. 1981. *Deuteronomist und Jahwist: Untersuchungen zu den Berührungspunkten beider Literaturwerke*, Abhandlungen zur Theologie des Alten und Neuen Testaments, vol. 67. Zürich: Theologischer Verlag.

Russell, S.C. Gate and Town in 2 Sam 15:1–6: Spatial and Corporate Dimensions of Absalom's Political Strategy. *Journal of Ancient History*, in press.

Sailhamer, J. 2009. *The Meaning of the Pentateuch: Revelation, Composition, and Interpretation*. Downers Grove, IL: IVP Academic.

Sanderson, J.E. 1986. *An Exodus Scroll from Qumran: 4QpaleoExod^{m} and the Samaritan Tradition*, Harvard Semitic Studies, vol. 30. Atlanta, GA: Scholars Press.

Sarna, N. 1991. *The JPS Torah Commentary: Exodus*. Philadelphia, PA: Jewish Publication Society.

Schäfer-Lichtenberger, C. 1985. Exodus 18: Zur Begründung der königlichen Gerichtsbarkeit in Israel-Juda. *Dielheimer Blätter zum Alten Testament und seiner Rezeption in der Alten Kirche* 21: 61–85.

Schmid, K. 2007. The Persian Imperial Authorization as Historical Problem and as Biblical Construct: A Plea for Differentiations in the Current Debate. In *The Pentateuch as Torah: New Models for Understanding Its Promulgation and Acceptance*, ed. G.N. Knoppers and B.M. Levinson, 22–38. Winona Lake, IN: Eisenbrauns.

———. 2010. *Genesis and the Moses Story*. Winona Lake, IN: Eisenbrauns.

Schwartz, B.J. 2009. The Visit of Jethro: A Case of Chronological Displacement? The Source-Critical Solution. In *Mishneh Todah: Studies in Deuteronomy and Its Cultural Environment in Honor of Jeffrey H. Tigay*, ed. N.S. Fox, D.A. Glatt-Gilad, and M.J. Williams, 29–48. Winona Lake, IN: Eisenbrauns.

Seebass, H. 2002. *Numeri 10:11–22:1*, Bibel und Kirche, vol. 4. Neukirchen-Vlyun: Nerkirchener Verlag.

Smith, M.S. 1997. *The Pilgrimage Pattern in Exodus*. With contributions from E.M. Bloch-Smith. Journal for the Study of the Old Testament Supplement Series, vol. 239. Sheffield: Sheffield Academic Press.

Sommer, B. 1999. Reflecting on Moses: The Redaction of Numbers 11. *Journal of Biblical Literature* 118: 601–624.

Van Seters, J. 1985. Etiology in the Moses Tradition: The Case of Exodus 18. *Hebrew Annual Review* 9: 355–361.

———. 1994. *The Life of Moses: The Yahwist as Historian in Exodus-Numbers*. Louisville, KY: Westminster John Knox Press.

Weinfeld, M. 1991. *Deuteronomy 1–11: A New Translation with Introduction and Commentary*, Anchor Bible, vol. 5. New York: Doubleday.

Wellhausen, J. 1961. *Prolegomena to the history of ancient Israel*. Cleveland: Meridian.

Wells, B. 2010. Competing or Complementary? Judges and Elders in Biblical and Neo-Babylonian Law. *Zeitschrift für Altorientalische und Biblische Rechtsgeschichte* 16: 77–104.

Westbrook, R., and B. Wells. 2009. *Everyday Law in biblical Israel: An Introduction*. Louisville, KY: Westminster John Knox Press.

Wilson, R.R. 1983a. Enforcing the Covenant: The Mechanisms of Judicial Authority in Early Israel. In *The Quest for the Kingdom of God: Studies in Honor of George E. Mendenhall*, ed. H.B. Huffmon, F.A. Spina, and A.R.W. Green, 33–69. Winona Lake, IN: Eisenbrauns.

———. 1983b. Israel's Judicial System in the Preexilic Period. *Jewish Quarterly Review* 74: 229–248.

Distinguishing the World of the Exodus Narrative from the World of Its Narrators: The Question of the Priestly Exodus Account in Its Historical Setting

24

Konrad Schmid

Abstract
When interpreting biblical texts historically, it is crucial to acknowledge the difference between the world of the narratives and the world of the narrators and to account for this difference in a methodologically controlled manner. The biblical account of Israel's exodus from Egypt narrated in Exodus 1–15 is a highly debated topic in Hebrew Bible studies, in terms of both its compositional and also its historical evaluation. Nonetheless, in the global discussion on the Pentateuch, there is broad consensus about the Priestly texts in Exodus 1–15 (Exod 1:7, 13–14; 2:23*–25; 6:2–12; 7–11*; 14–15*). They provide an apt starting point for inquiry into the different historical settings of the biblical exodus account. The present chapter will discuss several narrative peculiarities in the Priestly exodus account that might be explained by their authorial setting in the early Persian period.

Divergences in Current Pentateuchal Scholarship

It is a more or less obvious observation that Pentateuchal research in Israel, North America, and Europe differs widely in terms of presuppositions, methods, and results.[1]

Actually, there are very few points on which scholars agree, and these basically pertain to four specific elements: (1) The Pentateuch is a composite text. (2) The Pentateuch is, as a literary entity, a product of the first millennium BCE.[2] (3) The composition of the Pentateuch cannot be sufficiently explained without the assumption of sources, and by "source" I mean a self-contained literary piece that once existed independently. And (4), among the alleged sources, there is one textual layer that is less controversial than

[1] See Dozeman et al. (2011); cf. Schmid (2010); Baden (2012a); idem (2012b); idem (2012c); Schwartz (2012a). For overviews of scholarship see, e.g., Fischer (2003); Römer (2004a, b); Otto (2009); Schmid (2008: 37–41).

[2] Regarding its different final shapes, see Blum (1991).

K. Schmid (✉)
University of Zurich
Zurich, Switzerland
e-mail: konrad.schmid@access.uzh.ch

T.E. Levy et al. (eds.), *Israel's Exodus in Transdisciplinary Perspective*,
Quantitative Methods in the Humanities and Social Sciences, DOI 10.1007/978-3-319-04768-3_24,

others, which is P—the so-called Priestly Document.[3]

Regarding the theory that there are earlier sources included in the Pentateuch, scholars often identify a great divide between "documentarians" and "supplementarians," referring thereby to the proponents of the Documentary Hypothesis located mainly in North America and Israel on the one hand, and the—primarily—European scholars who envision a more complex genesis of the Pentateuch on the other hand.

However, the nomenclature of this distinction is not very precise. Why? The difference between the Documentary Hypothesis and more recent European approaches arising in the wake of Rendtorff and Blum [4] does not lie in the question of whether or not one reckons with independent documents that have been incorporated in the Pentateuch.[5] The difference between the two positions involves *how many and which kinds* of documents are assumed and how the process of their compilation and redactional expansion is best reconstructed. The documentarians limit themselves often to three or, including D, four sources, and one or more generally mechanical redactors who basically compiled these texts.[6] The so-called non-documentarians reckon with more sources (for instance, P, D, a Primeval History, the Abraham cycle, the Jacob cycle, the Joseph story, the exodus story).[7] This approach also assigns more text to redactional layers in the Pentateuch that compile and update the different source texts.

One very basic and important commonality among the so-called non-documentarian approaches regarding the exodus story is that this story was neither from the beginning nor from the early periods of biblical literature merely one *episode* within a much larger story spanning from the patriarchs or even creation to the death of Moses or the conquest of the land. It enjoyed a significant existence as a literary entity unto itself. Only for P do we have clear evidence that the exodus is merely the second act in Israel's foundational history.[8] Whether or not this is to be evaluated in terms of the notion of Genesis and Exodus as two formerly independent traditions of origins for Israel prior to P has remained contested up to present.[9]

It is, however, important to see that the divergences in global Pentateuchal scholarship are not just the result of different assessments of individual texts within an otherwise shared approach to the Pentateuch. The differences arise from disparate understandings about how to do historical exegesis. For instance, how does one assess the degree of dependence of a biblical author on his historical and cultural contexts?[10] How is one to evaluate literary discontinuities and continuities in ancient narrative texts? Or, how should one conceive of ancient authors' imagination and creativity? Of course, all these questions cannot be decided *more geometrico*. Nevertheless, there is still a great need for a

[3] See the standard text assignments by Elliger (1952); repr. 1966; Lohfink (1978/1988); Otto (1997). There is new debate, especially among European scholars, regarding the original end of P, especially in the wake of Perlitt (1988/1994). Compare the general thematic agreement, but variability with regard to the literary end at either Exodus 29 (Otto 1997), Exodus 40 (Pola 1995; Kratz 2000: 102–117; Bauks 2000), Leviticus 9 (Zenger 1997; idem 2004: 156–175), Leviticus 16 (Köckert 2004: 105; Nihan 2007: 20–68), or Numbers 27 (Ska 2008). A staggering of endings within the priestly document between Exodus 40 and Leviticus 26 is suggested by Gertz (2007: 236). Frevel (2000) supports the traditional conclusion in Deuteronomy 34 (cf. Schmidt 1993: 271; Weimar 2008: 17). Blenkinsopp (1976); Lohfink (1978/1988); Knauf (2000b); Guillaume (2009), see the conclusion of P^g in Joshua.

[4] See, e.g., Rendtorff (1977a); idem (1975/1977b); Blum (1984); Kratz (2005); Otto (2007); Schmid (2010); idem (2012b); Berner (2010).

[5] For a more detailed treatment of these processes, see Schmid (2010: 7–16, 334–347); idem (2011a).

[6] See Schwartz (2012b).

[7] See, e.g., the charts in Kratz (2000: 331), Otto (2003: 1099), and Gertz (2007: 216).

[8] See Schmid (2010), building *inter alia* on Römer (1990) and de Pury (1991).

[9] See, e.g., Carr (2001); Dozeman (2006); Van Seters (2006); Schmitt (2009a); Davies (2010).

[10] See, e.g., Sommer (2011), who is albeit overstating his case.

discussion of such problems in a historically informed way.

In order for progress to be made in Pentateuchal research in global terms, it is necessary to be both cautious and transparent in terms of methodology and basic presuppositions. And therefore, the following paper will mainly address some foundational methodological considerations and their relevance for a specific test case, the alleged P layer in Exodus.

Determining the Relationship Between the World of the Narrative and the World of the Narrator

When interpreting the Bible historically, and this is especially true for the Pentateuch because its stories are at least history-*like*, it is crucial to acknowledge the difference between the world of the narratives in the Bible and the world of the narrators, and to account for this difference in a methodologically controlled manner.

To illustrate this point with a nonbiblical example: In 1804, the German writer Friedrich Schiller published a play by the name "Wilhelm Tell," who is seen as one of the founding heroes of Switzerland in the thirteenth century.[11] The world of this narrative is, accordingly, the early Middle Ages—the thirteenth century. The world of its author is the early nineteenth century. Of course, it is not impossible that this play about "Wilhelm Tell" reworks and includes some historically accurate memories from the thirteenth century—one could perhaps argue this on the basis of a number of historical observations, but it is much more likely, and likewise much easier to demonstrate, that this play reflects foremost the time of Schiller himself, the nineteenth century.

To identify historical elements in such a history-like story pertaining to the world of its authors is a complicated and demanding task. It also depends considerably upon the specific nature and the genre of the narrative itself. The best results are attained when certain elements appear in the story that cannot be explained fully within its narrative plot, seeming instead to resonate with the author's own historical and cultural contexts.

If we approach the exodus story from such a perspective, then the starting point must account for this difference: The narrative of the exodus story plays out in the second half of the second millennium BCE—to be precise: in the year 2666 *anno mundi*, according to the chronology of the biblical text, which is 480 years before the dedication of the Solomonic temple (cf. 1 Kgs 6:1).[12] It is, however, important to keep in mind that there are also elements in the exodus story that seem to blur a specific historical location of the events—"Pharaoh" remains constantly nameless, and the 430 years of Israel's oppression in Egypt seem likely to serve as a counterpart to the 430 years of the monarchy in (Israel and) Judah rather than as a historical statement.[13] A possible reason for this blurred perspective might be the "mythical" quality of the events depicted.[14]

The world of the narrators of the exodus story is, as virtually all scholars agree, not identical with the world of the narrative. Some scholars, especially in the wake of Frank Moore Cross, view the Song of the Sea in Exodus 15 as a very ancient piece of literature,[15] but this is probably untenable, given the textual links in Exodus 15 to the preceding narrative in Exodus 14 (including its Priestly portions).[16]

At any rate, it is safe to say that a matter of dispute is just how distant and how different the world of the narrators is from the world of the narrative. In addition, it is probably also safe to say that there are *several* distances between the

[11] Schiller (1804) (1996), see the commentary on 735–850.

[12] For the details of the chronology, also regarding the different textual versions, see Hughes (1990).

[13] Schmid (2010: 19).

[14] See below n. 23.

[15] See the discussion in Russell (2007); Dozeman (2009: 336–337); Utzschneider/Oswald (2013: 339–341).

[16] See, e.g., Berner (2010: 389–400), especially 395; Klein (2012), see also the discussion in Albertz (2012: 253); Utzschneider/Oswald (2013: 341).

different worlds of narrators from the world of the narrative, which vice versa also makes the notion of the world of the narrative more complicated.[17]

In terms of methodology, it is very important to avoid relating the world of the narrative to the worlds of the narrators in a simplistic allegorical way, e.g., the motive of the oppression and the corvée labor in Egypt is *only* a projection of Neo-Assyrian practices, or the exodus from Egypt is *only* a camouflage for the exodus from Babylonia. To be sure, the assertion of these historical backgrounds is possible and to a certain extent even probable, but there are at least two additional considerations that need to be taken into account. First and foremost, each literary text develops its own fictive universe, which itself includes various indispensable elements. Second, a foundational myth such as the exodus story is likely to have reworked a very complex set of memories and traditions.[18] Therefore, the biblical exodus account is not just a text to be decoded through the reconstruction of experiences from another, later time. It should also be evaluated in terms of a literary text that has been shaped—I make this assertion with caution—*both* by earlier historical memories (maybe even from the world of the narrative itself) and contemporaneous influences (from the worlds of the narrators).

Given the current state of scholarship, the best initial approach to gaining an adequate historical-critical interpretation of the exodus story is to discuss and evaluate the literary layer that is the least contested, the Priestly version of the exodus story. This layer has received considerable attention in past and present scholarship. A few of the many investigations include the monograph by Peter Weimar and a more recent essay by Thomas Römer.[19] A new approach has recently been presented by Christoph Berner in his *Die Exoduserzählung: Das literarische Werden einer Ursprungserzählung Israels*.[20] He argues that the P texts in Exodus are not part of a source but are of a redactional nature. Jakob Wöhrle's study *Fremdlinge im eigenen Land: Zur Entstehung und Intention der priesterlichen Passagen der Vätergeschichte* seems to have a more convincing point for a comparable approach to Gen 12–50, but I will not go into that discussion here.[21] For my argument, the theological perspective and the alleged historical context of P in Exodus are more important than their literary nature.

Narrative and Authorial Aspects of the Priestly Exodus Story

The Priestly version of the Exodus story is usually, with minor variations in detail, considered to be made up of the following verses in Exodus: 1:7, 13–14; 2:23*–25; 6:2–12; 7:1–2, 4–7, 8–10a, 11–13, 19–20*, 21b, 22; 8:1–3, 11*, 12–14a, 15; 9:8–12; 11:10; 12:1, 3–8*, 18–20; 12:40–41; 14:1–4*, 8a, 10*, 15, 16–18a*, 21–23*, 26–29*.[22] It includes the basic elements of the oppression of the Israelites in Egypt, the commissioning of Moses, the contest with the Egyptian magicians, the setting up of the Pessach, Israel's departure from Egypt, and the death of Pharaoh and his army in the sea.

While this narrative is about the early history of Israel, as Norbert Lohfink and Ernst Axel Knauf in particular have pointed out,[23] it is less helpful to approach the Priestly Document, including its exodus story, as a historiographical work. Rather, the Priestly Document intends to

[17] For the composite nature of the exodus account see, e.g., Gertz (2000); Dozeman (2009), but also Berner (2010); Albertz (2012).

[18] See Schmid (2012a: 83–84), see also Hendel (2001); Bishop Moore and Kelle (2011: 77–95).

[19] Weimar (1973); Römer (2009), see also Utzschneider/Oswald (2013: 50–52).

[20] Berner (2010) (cf. my review in Schmid 2011c), see also Van Seters (1995: 574); Albertz (2012: 10–26).

[21] Wöhrle (2012).

[22] Following basically the delineations proposed by Gertz (2000: 394–396), cf. also Lohfink (1978/1988: 222–223 n. 29).

[23] Cf. Lohfink (1978/1988: 227–242) (227: "Die Rückverwandlung der Geschichte in Mythus"); Knauf (2000a).

present Israel's beginnings not in terms of history, but in terms of a foundational myth. It is easier to describe this difference in German terms: P as a whole does not write *Geschichte*, but *Urgeschichte*. The importance of this differentiation will become clear in the next section.

The following discussion will address four textual elements of P's narrative world that might open up a window into the specific historical world of its narrators.

1. The first one is the decidedly anti-Egyptian stance of the narrative and its emphasis on divine violence against the Egyptians. It appears most poignantly in Exodus 14:28a, one of the final statements of the description of the crossing of the sea[24]:

וַיָּשֻׁבוּ הַמַּיִם	The waters returned
וַיְכַסּוּ אֶת־הָרֶכֶב	and covered the chariots
וְאֶת־הַפָּרָשִׁים	and the chariot drivers,
לְכֹל חֵיל פַּרְעֹה	the entire army of Pharaoh
הַבָּאִים אַחֲרֵיהֶם בַּיָּם	that had followed them into the sea.

The Egyptians are buried in the sea after the Israelites have safely reached the opposite shore. The specific description of this act reveals its significance for P: The wording of Exodus 14 shows clearly that the salvation of Israel and the destruction of Egypt in P are not based on an arbitrary act of God; both elements are divine *creational activities*. This may be demonstrated first by Exodus 14:22:

וַיָּבֹאוּ בְנֵי־יִשְׂרָאֵל בְּתוֹךְ הַיָּם בַּיַּבָּשָׁה	The Israelites went into the sea *on dry*
וְהַמַּיִם לָהֶם חֹמָה	*ground*, the waters forming a wall for them
מִימִינָם וּמִשְּׂמֹאלָם	on their right and on their left.

In the crossing of the sea, the Israelites went on *dry ground*, in Hebrew: ביבשה.. The term יבשה only appears once in the Priestly Document before Exodus 14:22. This is the statement in Gen 1:9, in the Priestly account of the creation:

וַיֹּאמֶר אֱלֹהִים	And God said,
יִקָּווּ הַמַּיִם מִתַּחַת הַשָּׁמַיִם	"Let the waters under the sky be gathered
אֶל־מָקוֹם אֶחָד	together into one place,
וְתֵרָאֶה הַיַּבָּשָׁה	and let *the dry ground* appear."
וַיְהִי־כֵן	And it was so.

In the miracle at the Sea of Reeds something similar to the third day of creation happens: the *dry ground* can be seen. The Priestly Document apparently intends to shape the presentation of this miracle in the same mold as the creational activity of God during the very first days of creation.[25]

The wording of Exodus 14:28a also shows a similar affiliation with God's creational activity reported at the very beginning of world history:

וַיָּשֻׁבוּ הַמַּיִם	*The waters* returned
וַיְכַסּוּ אֶת־הָרֶכֶב	and *covered* the chariots
וְאֶת־הַפָּרָשִׁים	and the chariot drivers,
לְכֹל חֵיל פַּרְעֹה	the entire army of Pharaoh
הַבָּאִים אַחֲרֵיהֶם בַּיָּם	that had followed them into the sea.

Within the Priestly Narrative, this statement is quite similar in literary terms to the covering of the earth by the waters of the flood in Gen 7:19–20:

וְהַמַּיִם גָּבְרוּ מְאֹד מְאֹד	The waters swelled mightily
עַל־הָאָרֶץ	on the earth
וַיְכֻסּוּ כָּל־הֶהָרִים הַגְּבֹהִים	that all the high mountains under the
אֲשֶׁר־תַּחַת כָּל־הַשָּׁמָיִם׃	whole heaven were *covered*;
חֲמֵשׁ עֶשְׂרֵה אַמָּה מִלְמַעְלָה	fifteen cubits deep
גָּבְרוּ הַמָּיִם	*the waters* swelled,
וַיְכֻסּוּ הֶהָרִים׃	and the mountains were *covered*.

The implicit theological argument underlying this thematic and terminological link can be described as follows: The destruction of the Egyptian army in the sea is tantamount to the

[24] On P in Ex 14 see Levin (2013: 104–111).

[25] See Schmid (2011b: 280).

eradication of the sinful creatures during the flood. Erasing the Egyptian army is thus another element of the establishment of God's creational world order. These observations show why P is more *Urgeschichte* than *Geschichte*.

Why is the destruction of Egypt's military power noteworthy? Outside of this episode, P projects a very peaceful view of the world. The flood story of Genesis 6–9 seems to be an obvious exception, but even here, P actually criticizes the notion of divine violence. In Genesis 6:11–13, P takes up the judgment prophecies of Am 8:2 and Ezekiel 7:2–6 and in fact argues that while there was once a divine proclamation concerning a divine destruction of the world, this event took place in the primordial age of world history and had been settled once and for all in Genesis 9.[26]
Therefore, within this overall peaceful world-view of P, the case of Egypt, especially the destruction of *Egypt's military power* at the crossing of the sea, is quite a striking exception.[27]

It even seems that P makes a distinction between Egypt's military and Egypt's "civilian population," as for instance also P's reinterpretation of the plagues against Egypt as a contest of magicians suggests.[28] It seems to have been influenced to some extent by P's notion of a peaceful world. P's "plague" account includes five miracles (I 7:8–13*: staff to snake, II 7:19, 20a*, 21b, 22*: nile water to blood, III 8:1–3, 11*: frogs, IV 8:12–15*: lice, V 9:8–12: boils) that are arranged in climactic fashion: the Egyptian magicians are able to repeat the first three miracles, they are unsuccessful regarding the fourth and have to acknowledge the power of the God, and finally, they are afflicted by the fifth miracle, the boils, and have to give in.[29] The damaging impact of the plagues of Egypt is very limited in P. A look at the blood episode in Exodus 7:19.20a*, 21b, 22* (P)[30] is instructive in this respect:

וַיֹּאמֶר יְהוָה אֶל־מֹשֶׁה	And YHWH said to Moses,
אֱמֹר אֶל־אַהֲרֹן	"Say to Aaron,
קַח מַטְּךָ וּנְטֵה־יָדְךָ	'Take your staff and stretch out your hand
עַל־מֵימֵי מִצְרַיִם עַל־נַהֲרֹתָם	over the waters of Egypt, over its rivers,
עַל־יְאֹרֵיהֶם	its canals,
וְעַל־אַגְמֵיהֶם	and its ponds,
וְעַל כָּל־מִקְוֵה מֵימֵיהֶם	and all its pools of water,
וְיִהְיוּ־דָם	so that they may become blood;
וְהָיָה דָם בְּכָל־אֶרֶץ	and there shall be blood throughout the whole
מִצְרַיִם	land of Egypt,
וּבָעֵצִים וּבָאֲבָנִים׃	even in wood and stones.'"
וַיַּעֲשׂוּ־כֵן מֹשֶׁה וְאַהֲרֹן	Moses and Aaron did just
כַּאֲשֶׁר צִוָּה יְהוָה	as YHWH commanded
וַיְהִי הַדָּם	and there was blood
בְּכָל־אֶרֶץ מִצְרָיִם׃	in all the land of Egypt.
וַיַּעֲשׂוּ־כֵן	But the magicians of Egypt did the same by
חַרְטֻמֵּי מִצְרַיִם בְּלָטֵיהֶם	their secret arts;
וַיֶּחֱזַק לֵב־פַּרְעֹה	so Pharaoh's heart remained hardened,
וְלֹא־שָׁמַע אֲלֵהֶם	and he would not listen to them;
כַּאֲשֶׁר דִּבֶּר יְהוָה׃	as YHWH had said.

In this "plague," unlike its non-Priestly counterpart, no one suffers.[31] All water in Egypt is turned into blood by Moses and Aaron, and there is an implicit assumption that after they had performed this miracle, the blood immediately turned back into water. Otherwise the Egyptian magicians would not have been able to repeat the miracle. Thus the event apparently lasted only for a very short time—and it was a miracle, not a plague.[32]

[26] Cf. Schmid (2012a: 166–167).

[27] Berner (2010: 375–382) assumes a complicated literary process for the depiction of the Egyptian army in Exodus 14.

[28] Cf. Van Seters (1995); Römer (2003). For the delimitation of P in the plague cycle see Gertz (2000: 79–97).

[29] Cf. Gertz (2000: 82).

[30] See n. 22.

[31] Such an interpretation implies, of course, the original independence of the P source, *contra*, e.g., Van Seters (1995: 574). Blum (1990: 250–252) acknowledges the self-contained character of the Priestly plagues within the framework of his contextual interpretation of the Priestly plague cycle and he assumes that P has reworked a preexisting tradition.

[32] Even the subsequent slaying of the firstborn is presented in a very reduced manner (as an announcement in two verses in Exodus 12:12–13, embedded in an

Be this as it may: P envisions wide-reaching political, cultural, and religious peace for the whole known ancient world, for everyone except for Egypt. Why?

It is difficult to see a sufficient basis for this motivation solely within the narrative world of P's exodus account. P is ultimately interested in the establishment of the sanctuary. The destruction of Egypt at the crossing of the sea is not really necessary for such a narrative development. Of course, it may have been given for P's authors from the exodus traditions they knew, but the specific interest in divine violence against Egypt remains noteworthy.

Albert de Pury suggested that the violence towards Egypt might have arisen in response to the specific constellation of the world of P's authors in the early Persian period.[33] Of course, there is considerable debate over the possible date of P. Scholars often argue for a Neo-Babylonian or, like de Pury, an early Persian origin, but especially in Israel it is common to view P as a preexilic text, a conclusion also shared by American scholars like Richard Friedman.[34] Others, like William Propp, seem to allow for some fluidity and interpret P as the result of a process which began in the preexilic period and extended into the Persian period.[35]

In my opinion, the basic arguments regarding the date of P by Julius Wellhausen still seem valid today: P presupposes the cult centralization of Deuteronomy, which can be dated to the Josianic period, and the classical prophets do not presuppose the legislation of P.[36] There is, however, also some need to point out a specific shortcoming in Wellhausen's understanding. His arguments were overly focused on the internal intellectual developments in ancient Israel and Judah. Especially in the last several decades studies have shown that ancient Israel and Judah's intellectual history is heavily influenced by the imperial ideologies of the Assyrians, the Babylonians, the Persians, and sometimes even the Greeks. P, for example, seems to respond to basic conceptions from the Persian worldview and political theology, chief among them being the peaceful, well-ordered organization of the world according to different nations, all of which dwell in their lands with their own language and culture. This is, for instance, reflected in P's share in the Table of Nations in Genesis 10[37]:

בְּנֵי יֶפֶת [...] בְּאַרְצֹתָם אִישׁ לִלְשֹׁנוֹ לְמִשְׁפְּחֹתָם בְּגוֹיֵהֶם׃

Gen 10:2,5: The sons of Japheth [. . .] in their lands, with their own language, by their families, by their nations.

אֵלֶּה בְנֵי־חָם לְמִשְׁפְּחֹתָם לִלְשֹׁנֹתָם בְּאַרְצֹתָם בְּגוֹיֵהֶם׃

Gen 10:20: These are the sons of Ham, by their families, by their languages, in their lands, and by their nations.

אֵלֶּה בְנֵי־שֵׁם לְמִשְׁפְּחֹתָם לִלְשֹׁנֹתָם בְּאַרְצֹתָם לְגוֹיֵהֶם׃

Gen 10:31: These are the sons of Shem, by their families, by their languages, in their lands, and by their nations.

It has long been recognized that one of the closest parallels to the basic idea of Genesis 10 is found in Persian imperial ideology, as attested, e.g., in the Behistun inscription, which was disseminated widely throughout the Persian Empire.[38]

According to its political ideology, the Persian Empire was structured according to the different nations. The imperial inscriptions declare that every nation belongs to their specific region and has their specific cultural identities. This structure is the result of the

elaborated Pessach account); the execution is not reported within P (Gertz 2000: 394–396).

[33] de Pury (2007/2010).

[34] Friedman (1987: 161–216), see also Hurvitz (1988: 88–100); idem (2000).

[35] Propp (1999: 730–732).

[36] Wellhausen (1886: 385–445).

[37] See Vink (1969: 61); Knauf (2000b: 104–105); Nihan (2007: 383), see also Vermeylen (1992). Levin (1993: 124) takes a different stance.

[38] Schmitt (1991); idem (2009b); Greenfield and Porten (1982).

will of the creator deity, as Klaus Koch has pointed out in his "Reichsidee und Reichsorganisation im Perserreich," where he identifies this structure as "Nationalitätenstaat als Schöpfungsgegebenheit."[39]

The picture of Egypt in P, as a nation needing to be tamed in an otherwise well-organized and disciplined world, might suggest that P predates 525 BCE, the date of the Persian conquest of Egypt by Cambyses.[40] P seems to reflect the peaceful world order of the Persian Empire at a point in time that it includes the whole ancient world—except for Egypt. This constellation in the world of P's authors might also explain why the divine violence against Egypt seems to be directed more towards its army than towards its population. This differentiation seems to play a role in Exodus 14:4: "I will harden Pharaoh's heart, and he will pursue them, so that I will gain glory (כבד ni.) for myself over Pharaoh and all his army; and the Egyptians shall know that I am YHWH." The "Egyptians" in this verse probably do not refer to "Pharaoh and all his army" because they are facing their imminent destruction. It is not them who need to know "that I am YHWH." Rather, the remaining Egyptians shall learn from the death of their king and the destruction their army "that I am YHWH."[41]

2. There is, secondly, another striking element in P that pertains to the depiction of Egyptian religion in P. In Exodus 12:12b, God tells Moses:

וּבְכָל־אֱלֹהֵי מִצְרַיִם	On all the gods of Egypt
אֶעֱשֶׂה שְׁפָטִים	I will execute judgments:
אֲנִי יְהוָה׃	I am YHWH.

This is the only instance in P where אלהים denotes a *plurality* of deities, and where deities other than YHWH himself are envisioned. P is a decidedly monotheistic text,[42] propagating a sophisticated version of inclusive monotheism that reflects the empirical diversity of different religions in the world that are, however, all transparent guideposts pointing to the one creator deity that is ultimately presented in the narrative flow of P as YHWH.

William Propp has drawn attention to the fact that Exodus 12:12 is formulated as *yiqtol*: "I will punish."[43] This precludes the possibility that Exodus 12:12 is referring to the earlier humiliation of the Egyptian gods in the plague cycle that has already taken place. Exodus 12:12 is apparently a narrative element that is not fully integrated into the world of the narrative, but again provides a window into the world of the narrator.[44]

3. The introduction of the theme of the כבד יהוה in P, thirdly, is remarkable. From Exodus 16 on, the כבד יהוה is the most prominent mode of God's revelation. The concept, however, does not seem to be properly introduced within the narrative. But Exodus 14:17–18 uses כבד ni. in order to describe the details of the destruction of the Egyptian army in the sea. Specifically highlighted are the chariots and the horsemen. God's victory over the Egyptians apparently establishes his כבוד in

[39] Frei and Koch (1996: 201).

[40] Von Beckerath (2002); Briant (2002: 50–55); Cruz-Uribe (2003).

[41] The redactional verse 14:25 (see Krüger 1996: 532, see also n. 43 below) then interprets the Egyptians as the Egyptian soldiers who recognize, just before their death, that it is YHWH himself who fights against them.

[42] See Schmid (2011b: 278–289).

[43] Propp (1999: 400).

[44] It might be possible to relate these "judgements" on the gods of Egypt to P's specific location of the miracle at the sea "in front of Ba'al Zaphon" (Exod 14:2). The place is probably the antecedent to the sanctuary of Zeus Casios mentioned by Herodotus (II, 6, 158; III, 5) and is to be identified with *Ras Qasrun* on the sandbar of the *Sabakhet (Sabkhat) el Bardawil*. Excavations show no evidence reaching back prior to the Persian conquest of Egypt, see Davies (1990), especially 162–164. It is noteworthy that, according to P, the Israelites are commanded to head back (*šwb*) to "Ba'al Zaphon" in order that the miracle can take place, see Krüger (1996: 521f). The miracle in P is mainly a demonstration of God's power, not necessary for the deliverance of the Israelites.

P's eyes.[45] This results in Exodus 16 being able to expect that the כבד יהוה is a concept that the audience understands.[46]

4. Finally, P's introduction to the figure of Moses in its exodus account also seems to have been influenced by the world of the author, especially when compared with the earlier tradition in Exodus 2.

Within the P narrative, Moses' appearance shows up somewhat surprisingly in the report of his commissioning in Egypt, in Exodus 6:2–8. He is more properly introduced by the presentation of his ancestry in Exodus 6:16–20, especially in Exodus 6:20:

וַיִּקַּח עַמְרָם	And Amram took
אֶת־יוֹכֶבֶד דֹּדָתוֹ	Jochebed, his aunt,
לוֹ לְאִשָּׁה	for himself as a wife
וַתֵּלֶד לוֹ אֶת־אַהֲרֹן וְאֶת־מֹשֶׁה	and she bore him Aaron and Moses,
וּשְׁנֵי חַיֵּי עַמְרָם	and the years of Amram's life
שֶׁבַע וּשְׁלֹשִׁים וּמְאַת שָׁנָה׃	were one hundred thirty-seven years.

Three elements are striking in this introduction. Firstly, we are told that Moses' father married his aunt. Secondly, Moses has an older brother, Aaron. Thirdly, we do not hear a word about Moses miraculous deliverance while a baby, which appears in the non-Priestly account in Exodus 2.

The third point is especially noteworthy because we can assume quite safely that Exodus 6:20 is acquainted with and reworks Exodus 2,[47] given the somewhat difficult relationship between Moses parents in terms of their kinship (which according to Lev 18:12; 20:19 is illegitimate). This can be explained best by taking into account the introduction of Moses' birth story in Exodus 2:1. There Moses' origins are depicted as follows:

וַיֵּלֶךְ אִישׁ מִבֵּית לֵוִי	And a man from the house of Levi went
וַיִּקַּח אֶת־בַּת־לֵוִי׃	and took the daughter of Levi.

This verse poses many problems that I cannot deal with here, but it seems clear that Exodus 6:20 interprets the description of Moses' father in Exodus 2 ("a man from the house of Levi") as someone who is at least a grandson of Levi, whereas Moses' mother ("the daughter of Levi") seems to be a direct daughter of Levi, therefore making Moses' parents nephew and aunt.

If we can reasonably assume that Exodus 6:20 is acquainted with and reworks Exodus 2:1, then it is possible to go a step further in comparing the texts. They are quite similar to each other, but Exodus 6:20 introduces several changes. The most important are the following: (1) Moses' parents bear names (Amram and Jochebed). (2) Their kinship is explicitly defined as nephew and aunt, (3) They are explicitly married (in Exodus 2 Moses' nameless father "took" the daughter of Levi, but Amram "took" Jochebed "for himself as a wife"). (4) Moses has an elder brother, Aaron. And (5), most importantly, Exodus 6 is completely silent about Moses miraculous deliverance.

Regarding a possible window into the world of the author, especially points (4) and (5) are remarkable.

The introduction of Aaron as *elder* brother in Exodus 6, a fact that deprives Moses of the claim to being the firstborn son as suggested in Exodus 2, is probably to be interpreted in line with P's general tendency to eliminate any connotations of Moses as an exceptional hero with extraordinary powers. P has a decidedly theocentric view of history, and Moses' task in the exodus is simply *to announce* what *God* will do for Israel, as Exodus 6:6 clearly states:

לָכֵן אֱמֹר לִבְנֵי־יִשְׂרָאֵל	Therefore say to the Israelites:
אֲנִי יְהוָה	I am YHWH,
וְהוֹצֵאתִי אֶתְכֶם	and I will lead you out
מִתַּחַת סִבְלֹת מִצְרַיִם	from the labor of the Egyptians
וְהִצַּלְתִּי אֶתְכֶם מֵעֲבֹדָתָם	and I will deliver you from slavery to them.
וְגָאַלְתִּי אֶתְכֶם בִּזְרוֹעַ נְטוּיָה	I will redeem you with an outstretched arm and
וּבִשְׁפָטִים גְּדֹלִים׃	with mighty acts of judgment.

It is not Moses, but God who leads Israel out of Egypt. Furthermore, the goal of the exodus is not the conquest of the land, but

[45] See Wagner (2012: 68–72). Utzschneider/Oswald (2013: 320) also highlight the use of כבדת in Exodus 14:25.

[46] See also Struppe (1988: 139–143).

[47] For Exodus 2 as the original beginning of the exodus story, see Otto (2000); Carr (2001); Schmid (2010).

the dwelling of God amidst his people, cf. Exodus 29:46:

וְיָדְעוּ	And they shall know
כִּי אֲנִי יְהוָה אֱלֹהֵיהֶם	that I am YHWH their God,
אֲשֶׁר הוֹצֵאתִי אֹתָם	who brought them out
מֵאֶרֶץ מִצְרַיִם	of the land of Egypt
לְשָׁכְנִי בְתוֹכָם	that I might dwell among them;
אֲנִי יְהוָה אֱלֹהֵיהֶם׃	I am YHWH their God.

P depicts Moses as merely an agent of God whose primary task is to establish the sanctuary. This is possibly also the way the Priestly authors perceived the Persian kings of their time: They too are commissioned by God, and their task is to rebuild the temple. The ideal king, according to P, is only portrayed in terms of what he is not: He is the *opposite* of the Pharaoh of the exodus, who does not listen to Israel's God and hardens his heart.

It is quite apparent that P's Pharaoh is shaped as something of an "Anti-Cyrus," in Isaiah 45:3[48]:

וְנָתַתִּי לְךָ	And I will give you
אוֹצְרוֹת חֹשֶׁךְ	the treasures of darkness
וּמַטְמֻנֵי מִסְתָּרִים	and riches hidden in secret places,
לְמַעַן תֵּדַע	so that *you may know*
כִּי־אֲנִי יְהוָה	that it is I, YHWH,
הַקּוֹרֵא בְשִׁמְךָ	who call you by your name,
אֱלֹהֵי יִשְׂרָאֵל׃	the God of Israel.

[48] Cf. Kratz (1991: 104 n. 388); see for Isaiah 45 also Leuenberger (2010). The most fitting counterpart for Isa 45:3 is Exodus 5:2:

וַיֹּאמֶר פַּרְעֹה	But Pharaoh said,
מִי יְהוָה	Who is YHWH,
אֲשֶׁר אֶשְׁמַע בְּקֹלוֹ	that I should listen to his voice
לְשַׁלַּח אֶת־יִשְׂרָאֵל	and let Israel go?
לֹא יָדַעְתִּי אֶת־יְהוָה	I do not know YHWH,
וְגַם אֶת־יִשְׂרָאֵל לֹא אֲשַׁלֵּחַ׃	and I will not let Israel go.

However, the literary-history location and affiliation of this verse are unclear (see, e.g., Gertz 2000: 335–339). At any rate, it does not seem to be part of P.

This apolitical function for Moses in P is further accentuated by the omission of his birth account.[49] It has often been noted that Exodus 2 is an adaptation of the Sargon legend which, as has especially been pointed out in more recent scholarship, is subversively reworked by Exodus 2. Not the Assyrian king, but Moses is the one chosen by God to be a mighty leader of his people. The Neo-Assyrian background and the anti-imperial stance are elided from P's account of the Moses story. Moses, according to P, is the voice of God, and his task is to inform Israel about God's actions and to establish the sanctuary. He even has an elder brother, Aaron, who is to be the ancestor of the sanctuary's priests.

Dating P in the early Persian period, however, is often contested by way of the argument about its linguistic character. Especially in Israel and North America, but less so in Europe, the fact that P is written in Classical Biblical Hebrew (CBH) is viewed as support for a preexilic date: CBH belongs, according to the epigraphical evidence of the basic referential corpus, to the eighth through sixth centuries BCE. Therefore, P is to be dated to this period as well.[50] From my perspective, the debate about the conclusiveness of this historical linguistic argument is only about to begin. There is no room to deal with this issue here, but I would like to mention my main reservations. Firstly, the fact that a text is written in CBH and not in Late Biblical Hebrew (LBH) informs us primarily about its theological perspective within the biblical tradition, and not, at least not directly, about its historical date. Secondly, there is a significant gap in the external, nonbiblical control corpora for Hebrew from the sixth to second centuries BCE: There are many inscriptions from that period, but they are in Aramaic rather than Hebrew. Whether that in itself is

[49] Rendsburg (2006).

[50] See above n. 34. A good overview on the overall debate is provided by the contributions in Young (2003); Miller-Naudé and Zevit (2012).

a telling fact is contested. At any rate, we are not able to define a clear *terminus ante quem* for CBH from the external evidence. Thirdly, there is a basic asymmetry between the methods used by linguists to date CBH and LBH texts. Biblical texts written in CBH belong to eighth to sixth century because the external evidence dates to that period. The external evidence for LBH is mainly found in the texts from the Dead Sea from the second and first century BCE, but the biblical texts and books written in LBH, like Chronicles, Ezra, Nehemiah, Daniel, and Esther, are dated much earlier because they are, at least in parts, obviously older than the second or first century. The arguments regarding LBH show at minimum that a multitude of arguments need to be taken in account when dating biblical texts, and the external evidence is but one of them. Neither is it a decisive one.

Conclusions

Whether P's exodus account is a Persian period text or not is and will probably remain a topic of debate. To be sure, I do not preclude the possibility that P has reworked earlier material, especially in its legal section—rather, this is a quite probable assumption.[51] But, at any rate, there is no other method than carefully distinguishing between the world of the narrative and the world of the author for evaluating biblical texts in historical terms. Only in this way can we take steps forward in global Pentateuchal research and link it with other historical disciplines.[52]

References

Albertz, R. 2012. *Exodus 1–18*, Zürcher Bibelkommentare 2.1. Zürich: Theologischer Verlag.

Baden, J.S. 2012a. The Continuity of the Non-Priestly Narrative from Genesis to Exodus. *Biblica* 93: 161–186.

———. 2012b. From Joseph to Moses: The Narratives of Exodus 1–2. *Vetus Testamentum* 62: 133–158.

———. 2012c. *The Composition of the Pentateuch: Renewing the Documentary Hypothesis*. New Haven, CT: Yale University Press.

Bauks, M. 2000. La signification de l'espace et du temps dans "l'historiographie sacerdotale". In *The Future of the Deuteronomistic History*, Bibliotheca Ephemeridum Theologicarum Lovaniensium, vol. 147, ed. T. Römer, 29–45. Leuven: Peeters.

Berner, C. 2010. *Die Exoduserzählung: Das literarische Werden einer Ursprungserzählung Israels*, Forschungen zum Alten Testament, vol. 73. Tübingen: Mohr.

Bishop Moore, M., and B.E. Kelle. 2011. *Biblical history and Israel's Past: The Changing Study of Bible and history*. Grand Rapids, MI: Eerdmans.

Blenkinsopp, J. 1976. The Structure of P. *Catholic Biblical Quarterly* 38: 275–292.

Blum, E. 1984. *Die Komposition der Vätergeschichte*, Wissenschaftliche Monographien zum Alten und Neuen Testament, vol. 57. Neukirchen-Vluyn: Neukirchener.

Blum, E. (1990) *Studien zur Komposition des Pentateuch, Beihefte zur Zeitschrift für alttestamentliche Wissenschaft*, vol. 189. Berlin: de Gruyter.

———. 1991. Gibt es die Endgestalt des Pentateuch? In *Congress Volume Leuven 1989*, Vetus Testamentum Supplements, vol. 43, ed. J.A. Emerton, 46–57. Leiden: Brill.

Briant, P. 2002. *From Cyrus to Alexander: A History of the Persian Empire*. Translated by Peter T. Daniels. Winona Lake, IN: Eisenbrauns.

Carr, D.M. 2001. Genesis in Relation to the Moses Story. Diachronic and Synchronic Perspectives. In *Studies in the Book of Genesis. Literature, Redaction and History*, Bibliotheca Ephemeridum Theologicarum Lovaniensium, vol. 155, ed. A. Wénin, 273–295. Leuven: Peeters.

Cruz-Uribe, E. 2003. The Invasion of Egypt by Cambyses. *Transeuphratène* 25: 9–60.

Davies, G.I. 1990. The Wilderness itineraries and recent archaeological research. In *Studies in the Pentateuch*, Vetus Testamentum Supplements, vol. 41, ed. J.A. Emerton, 161–175. Leiden: Brill.

———. 2010. The transition from Genesis to Exodus. In *Genesis, Isaiah and Psalms*, Vetus Testamentum Supplements, vol. 135, ed. K.J. Dell, G.I. Davies, and Y. Von Koh, 59–78. Leiden: Brill.

de Pury, A. 1991. Le cycle de Jacob comme légende autonome des origines d'Israël. In *Congress Volume Leuven 1989*, Vetus Testamentum Supplements, vol. 43, ed. J.A. Emerton, 78–96. Leiden: Brill.

———. 2007/2010. P^g as the Absolute Beginning. In *Les dernières rédactions du Pentateuque, de l'Hexateuque et de l'Ennéateuque*, ed. T. Römer, and K. Schmid. Bibliotheca Ephemeridum Theologicarum Lovaniensium 203. 99–128, Leuven: Peeters, repr. 2010. In *Die Patriarchen und die Priesterschrift: Les Patriarches et le document sacerdotal. Gesammelte Studien zu seinem 70. Geburtstag: Recueil d'articles,*

[51] Cf., e.g., Nihan (2007: 608–619).

[52] See also Bishop Moore and Kelle (2011).

à l'occasion de son 70e anniversaire, ed. J.-D. Macchi et al. Abhandlungen zur Theologie des Alten und Neuen Testaments 99. 13–42, Zürich: Theologischer Verlag Zürich.

Dozeman, T.B. 2006. The Commission of Moses and the Book of Genesis. In *A farewell to the Yahwist? The Composition of the Pentateuch in Recent European interpretation*, Society of Biblical Literature symposium series, vol. 34, ed. T.B. Dozeman and K. Schmid, 107–129. Atlanta, GA: Society of Biblical Literature.

———. 2009. *Commentary on Exodus*, The Eerdmans Critical Commentary. Grand Rapids, MI: Eerdmans.

Dozeman, T.B., K. Schmid, and B.J. Schwartz. 2011. *The Pentateuch: International Perspectives on Current Research*, Forschungen zum Alten Testament, vol. 78. Tübingen: Mohr.

Elliger, K. 1952. Sinn und Ursprung der priesterlichen Geschichtserzählung. *Zeitschrift für Theologie und Kirche* 49: 121–143, repr. 1966, In *Kleine Schriften zum Alten Testament*, ed. H. Gese, and O. Kaiser. Theologische Bücherei 32. 174–198, Munich: Kaiser.

Fischer, G. 2003. Zur Lage der Pentateuchforschung. *Zeitschrift für die alttestamentliche Wissenschaft* 115: 608–616.

Frei, P., and K. Koch. 1996. *Reichsidee und Reichsorganisation im Perserreich*, Orbis Biblicus et Orientalis, vol. 55. Fribourg: Editions Universitaires.

Frevel, C. 2000. *Mit Blick auf das Land die Schöpfung erinnern*, Herders Biblische Studien, vol. 23. Freiburg: Herder.

Friedman, R.E. 1987. *Who Wrote the Bible?* San Francisco, CA: Harper.

Gertz, J.C. 2000. *Tradition und Redaktion in der Exoduserzählung: Untersuchungen zur Endredaktion des Pentateuch*, Forschungen zur Religion und Literatur des Alten und Neuen Testaments, vol. 186. Göttingen: Vandenhoeck & Ruprecht.

———. 2007. *Grundinformation Altes Testament*, Uni-Taschenbücher, vol. 2745. Göttingen: Vandenhoeck & Ruprecht.

Greenfield, J.C., and B. Porten. 1982. *The Bisitun Inscription of Darius the Great. Corpus Inscriptionum Iranicarum, Part I. Inscriptions of Ancient Iran, Vol. V. The Aramaic Versions of the Achaemenian Inscriptions, Texts I.* Lund Humpries: Aramaic version. London.

Guillaume, P. 2009. *Land and Calendar: The Priestly Document from Genesis 1 to Joshua 18*, Library of Hebrew Bible, Old Testament Studies, vol. 391. New York, NY: T&T Clark.

Hendel, R. 2001. The Exodus in Biblical Memory. *Journal of Biblical Literature* 120(2001): 601–622.

Hughes, J. 1990. *Secrets of the Times: Myth and History in Biblical Chronology*, Journal for the Study of the Old Testament. Supplement series, vol. 66. Sheffield: Sheffield Academic Press.

Hurvitz, A. 1988. Dating the Priestly Source in Light of the Historical Study of Biblical Hebrew. A Century after Wellhausen. In *Lebendige Forschung im Alten Testament*, ed. O. Kaiser, 88–100. Berlin: Walter de Gruyter.

———. 2000. Once again: The Linguistic Profile of the Priestly Material in the Pentateuch and its Historical Age. A Response to J. Blenkinsopp. *Zeitschrift für die alttestamentliche Wissenschaft* 112(2): 180–191.

Klein, A. 2012. Hymn and History in Ex. 15: Observations on the Relationship between Temple Theology and Exodus Narrative in the Song of the Sea. *Zeitschrift für die alttestamentliche Wissenschaft* 124: 516–527.

Knauf, E.A. 2000a. Der Exodus zwischen Mythos und Geschichte: Zur priesterschriftlichen Rezeption der Schilfmeer-Geschichte in Ex 14. In *Schriftauslegung in der Schrift. Festschrift für Odil Hannes Steck zu seinem 65. Geburtstag*, Beihefte zur Zeitschrift für die alttestamentliche Wissenschaft, vol. 300, ed. R.G. Kratz et al., 73–84. Berlin: de Gruyter.

———. 2000b. Die Priesterschrift und die Geschichten der Deuteronomisten. In *The Future of the Deuteronomistic History*, Bibliotheca Ephemeridum Theologicarum Lovaniensium, vol. 147, ed. T. Römer, 101–118. Leuven: Peeters.

Köckert, M. 2004. *Leben in Gottes Gegenwart. Studien zum Verständnis des Gesetzes im Alten Testament*, Forschungen zum Alten Testament, vol. 43. Tübingen: Mohr.

Kratz, R.G. 1991. *Kyros im Deuterojesaja-Buch: Redaktionsgeschichtliche Untersuchungen zu Entstehung und Theologie von Jes 40–55*, Forschungen zum Alten Testament, vol. 1. Tübingen: Mohr.

———. 2000. *Die Komposition der erzählenden Bücher des Alten Testaments*, Uni-Taschenbücher, vol. 2157. Göttingen: Vandenhoeck & Ruprecht.

———. 2005. *The Composition of the Narrative Books of the Old Testament*. London: T&T Clark.

Krüger, T. 1996. Erwägungen zur Redaktion der Meerwundererzählung (Exodus 13,17-14,31). *Zeitschrift für die alttestamentliche Wissenschaft* 108: 519–533.

Leuenberger, M. 2010. *Ich bin Jhwh und keiner sonst (Jes 45,5f): Der exklusive Monotheismus des Kyros-Orakels Jes 45,1–7 in seinem religions- und theologiegeschichtlichen Kontext*, Stuttgarter Bibelstudien, vol. 224. Stuttgart: Katholisches Bibelwerk.

Levin, C. 1993. *Der Jahwist*, Forschungen zur Literatur und Geschichte des Alten und Neuen Testaments, vol. 157. Göttingen: Vandenhoeck & Ruprecht.

———. 2013. Source criticism: The Miracle at the Sea. In *Re-Reading the Scriptures: Essays on the Literary History of the Old Testament*, Forschungen zum Alten Testament, vol. 87, 95–114. Tübingen: Mohr.

Lohfink, N. 1978/1988. Die Priesterschrift und die Geschichte. In *Congress Volume Göttingen 1977*, ed. J.A. Emerton. Vetus Testamentum Supplements 29. 183–225, Leiden: Brill, repr. 1988, In *Studien zum Pentateuch.* Stuttgarter biblische Aufsatzbände 4. 213–253, Stuttgart: Katholisches Bibelwerk.

Miller-Naudé, C., and Z. Zevit. 2012. *Diachrony in Biblical Hebrew*, Linguistic Studies in Ancient West Semitic, vol. 8. Winona Lake, IN: Eisenbrauns.

Nihan, C. 2007. *From Priestly Torah to Pentateuch. A Study in the Composition of the Book of Leviticus*, Forschungen zum Alten Testament, vol. II/25. Tübingen: Mohr.

Otto, E. 1997. Forschungen zur Priesterschrift. *Theologische Rundschau* 62: 1–50.

———. 2000. Mose und das Gesetz: Die Mose-Figur als Gegenentwurf Politischer Theologie zur neuassyrischen Königsideologie im 7. Jh. v. Chr. In *Mose: Ägypten und das Alte Testament*, Stuttgarter Bibelstudien, vol. 189, 43–83. Stuttgart: Katholisches Bibelwerk.

———. 2003. *Pentateuch. Religion in Geschichte und Gegenwart*, vol. 6. Tübingen: Mohr.

———. 2007. *Das Gesetz des Mose*. Darmstadt: Wissenschaftliche Buchgesellschaft.

———. 2009. Kritik der Pentateuchkomposition: Eine Diskussion neuerer Entwürfe. In *Die Tora: Studien zum Pentateuch: Gesammelte Aufsätze*, Beihefte zur Zeitschrift für altorientalische und biblische Rechtsgeschichte, vol. 9, ed. O. Otto, 143–167. Wiesbaden: Harrassowitz.

Pola, T. 1995. *Die ursprüngliche Priesterschrift. Beobachtungen zur Literarkritik und Traditionsgeschichte von Pg*, Wissenschaftliche Monographien zum Alten und Neuen Testament, vol. 70. Neukirchen-Vluyn: Neukirchener.

Perlitt, L. 1988/1994. Priesterschrift im Deuteronomium? *Zeitschrift für die alttestamentliche Wissenschaft* 100 Supplements: 65–88, repr. 1994, In *Deuteronomium-Studien*, ed. L. Perlitt. Forschungen zum Alten Testament. 123–143, Tübingen: Mohr.

Propp, W.H. 1999. *Exodus 19–40: A New Translation with Introduction and Commentary*, Anchor Bible 2A, New York, NY: Doubleday.

Rendsburg, G.A. 2006. Moses Equal to Pharaoh. In *Text, Artifact, and Image. Revealing Ancient Israelite Religion*, Brown Judaic studies, vol. 346, ed. G.M. Beckman and T.J. Lewis, 201–219. Providence, RI: Brown University.

Rendtorff, R. 1977a. *Das überlieferungsgeschichtliche Problem des Pentateuch*, Beihefte zur Zeitschrift für die alttestamentliche Wissenschaft, vol. 147. Berlin: de Gruyter.

———. 1975/1977b. Der "Jahwist" als Theologe? Zum Dilemma der Pentateuchkritik. In *Congress Volume Edinburgh 1974*, ed. G.W. Anderson et al. Vetus Testamentum Supplements 28. 158–166, Leiden: Brill, trans. 1977 The "Yahwist" as Theologian? The Dilemma of Pentateuchal Criticism. *Journal for the Sutdy of the Old Testament* 3: 2–10.

Römer, T.C. 1990. *Israels Väter: Untersuchungen zur Väterthematik im Deuteronomium und in der deuteronomistischen Tradition*, Orbis Biblicus et Orientalis, vol. 99. Fribourg: Academic.

———. 2003. Competing magicians in Exodus 7-9: Interpreting magic in the priestly theology. In *Magic in the biblical world. From the rod of Aaron to the ring of Solomon*, Journal for the Study of the New Testament. Supplement series, vol. 245, ed. T.E. Klutz, 12–22. London: T & T Clark.

———. 2004a. Hauptprobleme der gegenwärtigen Pentateuchforschung. *Theologische Zeitschrift* 60: 289–307.

———. 2004b. La formation du Pentateuque: histoire de la recherche. In *Introduction à l'Ancien Testament*, Le Monde de la Bible, vol. 49, ed. T. Römer, J.-D. Macchi, and C. Nihan, 67–84. Geneva: Labor et Fides.

———. 2009. The Exodus Narrative according to the Priestly Document. In *The Strata of the Priestly Writings: Contemporary Debate and Future Directions*, Abhandlungen zur Theologie des Alten und Neuen Testaments, vol. 95, ed. S. Shectman and J.S. Baden, 157–174. Zürich: Theologischer Verlag Zürich.

Russell, B.D. 2007. *The Song of the Sea: The Date of Composition and Influence of Exodus 15:1–21*, Studies in Biblical Literature, vol. 101. New York, NY: Peter Lang.

Schiller, F. 1996. Tell: Schauspiel. In *Werke und Briefe, Dramen IV*, ed. F. Schiller, 385–505. Frankfurt am Main: Deutscher Klassiker Verlag.

Schmid, K. 2008. *Literaturgeschichte des Alten Testaments*. Darmstadt: Wissenschaftliche Buchgesellschaft, trans. 2012a, *The Old Testament: A Literary History*. Minneapolis, MN: Fortress.

———. 2010. *Genesis and the Moses story: Israel's dual origins in the Hebrew Bible*, Siphrut, vol. 3. Winona Lake, IN: Eisenbrauns.

———. 2011a. Has European Pentateuchal Scholarship Abandoned the Documentary Hypothesis? Some Reminders on Its History and Remarks on Its Current Status. In *The Pentateuch: International Perspectives on Current Research*, Forschungen zum Alten Testament, ed. T. Dozeman, K. Schmid, and B. Schwartz, 17–30. Tübingen: Mohr.

———. 2011b. The Quest for "God:" Monotheistic Arguments in the Priestly Texts of the Hebrew Bible. In *Reconsidering the Concept of Revolutionary Monotheism*, ed. B. Pongratz-Leisten, 271–289. Winona Lake, IN: Eisenbrauns.

———. 2011c. Rez. zu C. Berner. Die Exoduserzählung. *Zeitschrift für die alttestamentliche Wissenschaft* 123: 292–294.

———. 2012a. See Schmid K. 2008.

———. 2012b. Genesis in the Pentateuch. In *The Book of Genesis. Composition, Reception, and Interpretation*, Vetus Testamentum Supplements, vol. 152, ed. C.A. Evans, J. Lohr, and D. Petersen, 27–50. Leiden: Brill.

Schmidt, L. 1993. *Studien zur Priesterschrift*, Beihefte zur Zeitschrift für die alttestamentliche Wissenschaft, vol. 214. Berlin: de Gruyter.

Schmitt, H.-C. 2009a. Erzväter- und Exodusgeschichte als konkurrierende Ursprungslegenden Israels – ein Irrweg der Pentateuchforschung. In *Die Erzväter in der biblischen Tradition*, ed. A.C. Hagedorn, and H. Pfeiffer. Beihefte zur Zeitschrift für die alttestamentliche Wissenschaft 400. 241–266, Berlin: de Gruyter.

Schmitt, R. 1991. *The Bisitun Inscriptions of Darius the Great: Old Persian Texts*. Corpus inscriptionum Iranicarum. Part I: Inscriptions of Ancient Iran. Vol. I: The Old Persian Inscriptions. Texts I. London: School of Oriental and African Studies.

———. 2009. *Die altpersischen Inschriften der Achaimeniden: Editio minor mit deutscher Übersetzung*. Wiesbaden: Reichert.

Schwartz, B.J. 2012a. The Pentateuch as Scripture and the Challenge of Biblical Criticism: Responses among modern Jewish Thinkers and Scholars. In *Jewish Concepts of Scripture: A Comparative Introduction*, ed. B.D. Sommer, 203–228. New York: NYU Press.

———. 2012b. How the Compiler of the Pentateuch Worked: The Composition of Genesis 37. In *The Book of Genesis – Composition, Reception and Interpretation*, ed. C.A. Evans, J.N. Lohr, and D.L. Petersen, 263–278. Leiden: Brill.

Ska, J.-L. 2008. Le récit sacerdotal. Une "histoire sans fin"? In *The Books of Leviticus and Numbers*, Bibliotheca Ephemeridum Theologicarum Lovaniensium, vol. 215, ed. T. Römer, 631–653. Leuven: Peeters.

Sommer, B.D. 2011. Dating Pentateuchal Texts and the Perils of Pseudo-Historicism. In *The Pentateuch: International Perspectives on Current Research*, ed. T.B. Dozeman, K. Schmid, and B.J. Schwartz, 85–108. Tübingen: Mohr.

Struppe, U. 1988. *Die Herrlichkeit Jahwes in der Priesterschrift*, Österreichische Biblische Studien, vol. 9. Klosterneuburg: Österreichisches Katholisches Bibelwerk.

Utzschneider, H./Oswald, W. 2013. *Exodus 1–15*. Internationaler exegetischer Kommentar. Stuttgart: Kohlhammer.

Van Seters, J. 1995. A Contest of Magicians? The Plague Stories in P. In *Pomegranates and Golden Bells. Studies in Biblical, Jewish, and Near Eastern Ritual, Law, and Literature in Honor of Jacob Milgrom*, ed. D.P. Wright, D.N. Freedman, and A. Hurvitz, 569–580. Winona Lake, IN: Eisenbrauns.

———. 2006. The Patriarchs and the Exodus: Bridging the Gap between Two Origin Traditions. In *The Interpretation of Exodus*, Contributions to Biblical Exegesis and Theology, vol. 44, ed. R. Roukema, 1–15. Leuven: Peeters.

Vermeylen, J. 1992. La "table des nations" (Gn 10): Yaphet figure-t-il l'Empire perse? *Transeuphratène* 5: 113–132.

Vink, J.G. 1969. The Date and the Origin of the Priestly Code in the Old Testament. In *The Priestly Code and Seven Other Studies*, Oudtestamentische studiën, vol. 52, ed. J.G. Vink et al., 1–144. Leiden: Brill.

von Beckerath, J. 2002. Nochmals die Eroberung Ägyptens durch Kambyses. *Zeitschrift für ägyptische Sprache und Altertumskunde* 129: 1–5.

Wagner, T. 2012. *Gottes Herrlichkeit: Bedeutung und Verwendung des Begriffs kābôd im Alten Testament*, Supplements to Vetus Testamentum, vol. 151. Leiden: Brill.

Weimar, P. 1973. *Untersuchungen zur priesterschriftlichen Exodusgeschichte*, Forschung zur Bibel, vol. 9. Würzburg: Echter.

———. 2008. *Studien zur Priesterschrift*, Forschungen zum Alten Testament, vol. 56. Tübingen: Mohr.

Wellhausen, J. 1886. *Prolegomena zur Geschichte Israels*, 3rd ed. Berlin: Reimer.

Wöhrle, J. 2012. *Fremdlinge im eigenen Land: Zur Entstehung und Intention der priesterlichen Passagen der Vätergeschichte*, Forschungen zur Religion und Literatur des Alten und Neuen Testaments, vol. 246. Göttingen: Vandenhoeck & Ruprecht.

Young, I. 2003. *Biblical Hebrew. Studies in Chronology and Typology*, Journal for the Study of the Old Testament. Supplement series, vol. 369. London: T &T Clark.

Zenger, E. 1997. Priesterschrift. *Theologische Realenzyklopädie* 27: 435–446.

———. 2004. *Einleitung in das Alte Testament*, Studienbücher Theologie 1,1. Stuttgart: Kohlhammer.

Part VI

The Exodus in Later Reception and Perception

The Despoliation of Egypt: Origen and Augustine—From Stolen Treasures to Saved Texts

25

Joel S. Allen

Abstract

Few passages have evoked such embarrassment, both to Jewish and Christian expositors, as the despoliation of Egypt (Gen 15:14; Exod 3:21–22, 11:2–3, and 12:35–36). This chapter examines key allegorical "text therapies" by which Jews and Christians sought to heal the text of its improprieties. This allegory warranted the appropriation of their classical heritage into the realm of the spirit—thus despoiling Egypt's treasures again. Philo finds in plundered Egyptian vessels an emblem of the pedagogical provisions necessary for the soul to make its journey to its heavenly home. He thus transforms the Greek educational curriculum from a path for social advancement into a highway for the soul's transformation and elevation. Origen similarly saw in Egyptian plunder a reflection of spiritual treasures necessary for true knowledge of God. Beginning with Philo, this liberal posture was adopted first in the east by Origen and later in the west by Augustine. They sought to provide justification for those who wished to have the best of both the biblical and classical worlds and to balance the faith of scriptures with the reason of Greek philosophy.

This chapter examines the interpretive history (Philo, Origen, and Augustine) of one particular aspect of the exodus story: namely the despoliation of Egypt. As other contributions in this volume show (namely, Pieter van der Horst, in "From Liberation to Expulsion: The Exodus in the Earliest Jewish-Pagan Polemic"), the Jews of antiquity had detractors who sought to depict Moses and his people as devious and disingenuous. While many Jews argued that the plundering of Egypt amounted to a fair wage, this chapter examines allegorical interpretations which transform the plundering of Egypt into that which provides justification for spiritual appropriation of pagan texts. Since text survival is the unintended result of being copied for reading, this embrace of pagan education and

J.S. Allen (✉)
Dakota Wesleyan University
709 W. 4th Street, Mitchell, SD 57301, USA
e-mail: joallen@dwu.edu

T.E. Levy et al. (eds.), *Israel's Exodus in Transdisciplinary Perspective*,
Quantitative Methods in the Humanities and Social Sciences, DOI 10.1007/978-3-319-04768-3_25,

literature provided the ethos necessary for the survival of many which remain extant.

Few passages have evoked such embarrassment, both to Jewish and Christian expositors, as the despoliation of Egypt (Gen 15:14; Exod 3:21–22, 11:2–3, and 12:35–36). Since attacks on Israel's behavior necessarily involved an attack on the Hebrew Bible and upon the God it claims to reveal, Jews and Christians found themselves arguing along similar lines against similar opponents. This essay compares two influential allegorical "text therapies" by which Jews (or "Jew," that being Philo) and Christians sought to heal the text of its apparent impropriety while providing warrant for what they wanted to do anyway; appropriate the best of the classical tradition and baptize it into the realm of the spirit. The allegory was understood thusly: just as the children of Israel appropriated the treasures of Egypt ultimately for the construction of the wilderness tabernacle, we are duty-bound to use the best of our heritage in the classical world for spiritual purpose. This interpretive agenda sprang from a relatively humanistic ethos which encouraged the copy, study, and thus preservation of the literary output of the Greco-Roman world.

It comes as no surprise that Philo is the first of all ancient interpreters to allegorize the despoliation of Egypt (*Heir* 272–274). What comes as something of a surprise is the fact that Philo chose Gen. 15:14 as the text upon which to practice his allegorizing arts. The whole of the Christian exegetical tradition looks to the despoliation texts in Exodus. The reason for this distinction is quite simple: Gen. 15:14 works better for Philo's allegory and the Exodus traditions suit the interpretive goals of the Christian allegory. For the purposes of this chapter, I will only briefly summarize Philo and move on to Origen and Augustine.

As stated, Philo's text of choice is Gen 15:14, which in the LXX reads as follows:

τὸ δὲ ἔθνος ᾧ ἐὰν δουλεύσωσιν κρινῶ ἐγώ μετὰ δὲ ταῦτα ἐξελεύσονται ὧδε μετὰ ἀποσκευῆς πολλῆς. On the literal plane, Philo views μετὰ ἀποσκευῆς πολλῆς—which could be translated, "with a lot of luggage"—as travel provisions for the journey from Egypt to Canaan.[1] In the allegory, these provisions represent Greek encyclical education (the *paideia*) which nourish the soul—as a sort of spiritual provision—with virtue thus enabling the soul to ascend from earthly bondage to a heavenly homeland. This interpretation could not have been supported by the exodus versions of the despoliation since vessels of silver and gold and clothes can hardly be considered traveling provisions.

Philo chose the encyclia, the most mundane strata of the Greek educational system and incorporated it into his soteriology (Mendelson 1983: 81).[2] He sought to accommodate the fact that many elite Jewish youths of Alexandria would be enrolled in Greek educational institutions. Alexandrian Jews, he insists, should utilize the encyclia in their strivings toward divine knowledge instead of exploiting it to further social and/or political ambition. Philo sought to provide a spiritual purpose for knowledge which originated in secular education (Ibid., 82).

While Philo's interpretation stands alone in many regards, this allegory, beginning with Origen, had a strong impact on the theology and practice of late antiquity by providing a model for the integration of faith and reason. It also encouraged an intellectual yet spiritual openness to classical learning that allowed its texts to

[1] The ὧδε (or "here") refers to the promised land which for Philo represents the soul's true destination in God. Rene Bloch (in "Leaving Home: Jewish-Hellenistic Authors on the Exodus") has pointed out that in Philo's *Life of Moses*, the destination point for the exodus is always left uncertain. Bloch suggests that Philo wants us to view Moses as a cosmopolitan citizen of the world and thus he sought to de-emphasize his status as founder of the land of the Jews. I add that Philo's de-emphasis on the destination point also helps the biblical story to function as allegory. The real destination of Moses and his people who have escaped Egyptian bodily passions is the ultimate "here" of heaven; the soul's true home.

[2] Bloch (in *Leaving Home*) describes the fact that, for Philo, Egypt is not only the home of the bodily passions and a place from which the soul must escape, but also a place of cosmopolitan learning. It is this beneficial aspect of Alexandrian culture especially in the *paideia* which can instruct the soul in the sturdy virtues necessary for its heavenward ascent.

survive. We will return to this "saving texts" aspect in due course.

Origen's *Letter to Gregory*

Origen's influence as a biblical critic and exegete constitutes what is essentially the beginning of the golden era of Christian exegesis. The text which treats the despoliation of Egypt is now chapter 13 of the *Philocalia* ("Love of what is beautiful"), an anthology of Origen's writings compiled by Basil and Gregory of Nazianzus (Robinson 1893: xiv). We will follow the English translation of Joseph Trigg (Trigg 1998: 210–13).[3] Origen's letter has traditionally been known as his *Epistula ad Gregorium*.[4]

> I have desired that with all the power of your innate ability you would apply yourself, ultimately, to Christianity. I have, for this reason, prayed that you would accept effectively those things from the philosophy of the Greeks that can serve as a general education or introduction for Christianity (εἰς χριστιανισμὸν δυνάμενα γενέ σθαι ἐγκύκλια μαθήματα ἢ προπαιδεύματα) and those things from geometry and astronomy that are useful for the interpretation of the Holy Scriptures. For just as the servants of philosophers say concerning geometry, music, grammar, rhetoric and [astronomy][5] that they are adjuncts to philosophy, we say this very thing about philosophy itself with regard to Christianity (περὶ αὐτῆς φιλοσοφίας πρὸς χριστιανισμόν).

In paragraph 2, Origen introduces his allegory of the plundering of Egypt.

> 2. And indeed Scripture hints at this principle in Exodus (αἰνίσσεται τὸ ἐν Ἐξόδῳ γεγραμμέ νον), where, with God himself the person speaking, the children of Israel are told to ask their neighbors and cohabitants for vessels of silver and gold and for clothing (Exod 11:2 and 12:35). Having in this way despoiled the Egyptians, they may find material among the things they have received for the preparation of divine worship (εἰς τὴν πρὸς θεὸν λατρείαν). For, from the spoil taken from the Egyptians the children of Israel prepared the appurtenances of the holy of holies, the ark with its covering, the cherubim, the mercy-seat, and the golden vessel in which was stored the manna, the bread of angels. These articles appear to have been made from the finest of the Egyptian's gold. From gold of second quality they made wide solid gold candelabrum, close to the interior veil; and the golden table, on which was placed the bread of offering; and between them the golden incense altar. They used any gold of third or fourth quality, if available, for the construction of the holy vessels. The silver of the Egyptians became other things. For, while sojourning in Egypt, the children of Israel had an abundant quantity of such precious materials for the worship of God to show for their experience there.[6] From the clothing of the Egyptians it seems that they had such things as they needed for what Scripture describes as embroidered works. Assisted by God's wisdom (μετὰ σοφίας θεοῦ), tailors stitched the articles of clothing together to serve as veils and tapestries for the interior and exterior.
>
> 3. (abridged) I could speak from knowledge gained by experience that few are those who have taken anything useful from Egypt and have come out from there and have prepared objects for the worship of God, but many have followed the example of Hadad, the Idumaean.[7] These are those who have used some Greek ingenuity to beget heretical ideas and have so to speak prepared golden calves in Bethel, which means 'house of God.'

What is particularly striking about Origen's argument is what is left unsaid. While he clearly desires the allegory to provide biblical warrant for theological students to study these classical texts, yet the allegory is not explained as allegory. He does say that this principle is "hinted" in Exodus (αἰνίσσεται—a standard verb implying allegory), but he provides no explanation. This indicates the readers were already familiar with the allegory.

[3] Trigg bases his translation on the critical edition of the text found in Grégoire le Thaumaturge (Crouzel 1969: 186–94). This is the Greek text followed here.

[4] We assume that the recipient of Origen's advice was Gregory Thaumaturgus (Bishop Gregory of Neocaesarea). Nautin has recently argued that the recipient of Origen's letter was not the same Gregory as the Bishop of Neocaesarea (Nautin 1977: 161).

[5] Trigg, for some reason, has "geometry" repeated here. The SC text of Crouzel and PG have "astronomy."

[6] The translation of Crouzel is "car les fils d'Israël exilés en Égypte, ont gagné de leur séjour là-bas d'avoir en abondance quantité de matériaux précieux pour fabriquer les objets utiles au culte divin" (Crouzel 1969: 189).

[7] He is mistaken here and means Jeroboam I who build calves at Bethel and Dan.

Central to Origen's interpretation is the narrative *haggadah* around which he builds the allegory. There is no biblical claim that plundered Egyptian treasure was used to construct the wilderness tabernacle. Yet this is central to the allegory, that is, just as the Hebrews used the plundered treasures of Egypt to build the tabernacle, so we ought to plunder the knowledge of the Greeks for Christian purposes. This connection was certainly being made before Origen. We have a hint of it in Jubilees 48:18. It does not, however, play a central interpretive role in the extant literature until Origen.[8] Origen is not only dependent upon *haggadic* addition here, but may be passing on other elements of the tradition that have been lost (viz. the varying qualities of gold taken).

We can be relatively certain Origen was familiar with the Philo passage above, but he draws on a different biblical text which suits his own purposes. Exod 12:35–36 is necessary for Origen since he needs to associate the plundered riches of Egypt (their gold and silver particularly) to the riches needed for the construction of the wilderness tabernacle (gold and silver). The general μετὰ ἀποσκευῆς πολλῆς in LXX Gen 15:14 lacks specificity. Origen applies the allegory to both the encyclia and philosophy, not just to the encyclia as does Philo. Origen states that philosophy should be considered preparation for Christian theology just as encyclical education was adjunct to the study of philosophy.

Origen draws attention to the divine wisdom given to Bezalel and Oholiab for the construction of the tabernacle (Exod 31). A key element in Origen's interpretation is the necessity of divine wisdom to successfully transform Egyptian treasures for true divine worship. Just as divine wisdom was given for the construction of the tabernacle, so divine wisdom is necessary to transform Greek learning into that which affords true worship to the true God (the tabernacle). It is easy to misuse Greek philosophy by building a golden calf (paragraph 3), but with divine wisdom, Egyptian treasures can be claimed and transformed. Origen contrasts the wisdom given by God for construction of the tabernacle to the lack of wisdom displayed by the Egyptians who used their wealth for idols.

If Origen has been drawing upon an unknown narrative haggadic tradition of varying qualities of gold and silver taken from Egypt, he has done so uniquely and his purpose is to illustrate the need for the Christian to discriminate between varying levels of helpfulness of Greek learning for biblical interpretation. Perhaps we could even say that the gold and silver of lower quality relate to the encyclical learning which is helpful only with literal interpretation. These precious metals are usable only in the outer courts of literal exegesis. The gold and silver of finer qualities represent philosophical learning that inspires and relates to allegorical interpretation. These qualities of gold and silver are the very holy of holies of allegorical hermeneutic. In paragraph 4, Origen describes prayerful allegorical exegesis as a sort of participation in the godhead—a "holy of holies" kind of experience.

In paragraph 3, Origen expresses great concern about over-plundering Egypt (thus the need for divine wisdom). Many others have been overly accommodating and overly platonized; their heresy springs from reckless prayerless interpretation which lacks divine wisdom. Only few do this correctly; most—like Jeroboam whom Origen inadvertently calls Hadad—end up building a Golden Calf instead of a tabernacle. Origen is eager to discourage excessive accommodation and only allows for usage of philosophy as an adjunct to theology and for biblical hermeneutics.

Let us conclude with some brief observations. Origen is the first Christian writer to identify the treasures of Egypt with classical education and in doing so clearly follows Philo. Origen, like Philo, seeks to walk in balance between an overly accommodative attitude and an overly isolationist attitude toward Greek learning. He wants his students to learn the encyclical subjects since they are necessary for the literal explication of the Scriptures. He also sees philosophy as an important introduction to Christianity; it lays the groundwork for a philosophically sensitive

[8] On the Jubilees section, see chapter 2 of my *The Despoliation of Egypt* (2008).

theology. Philosophy inspires the right questions and prayerful allegorical biblical interpretation provides the right answers. The allegory of Hadad (properly Jeroboam I) in paragraph 3 outlines the danger; it is easier to build a calf than a tabernacle.

According to P. Beatrice, "On the whole, all these texts offer a sufficiently coherent picture of Origen's position: the liberal arts and pagan philosophy, which constitute the wisdom of this world, can be redeemed only if they enter the service of Christian theology and biblical exegesis" (Beatrice 2007: 172). While it is likely that Origen was familiar with the Philo passage, he draws on a different biblical text for different purposes. Philo's allegory seeks to spiritualize Greek education as the very provision needed for the soul's journey to home to God. Origen seeks to give academic substance to theological and biblical inquiry and thus craft a Christian faith acceptable to intellectual non-Christian world.

Augustine: *Doctr. Chr.* 2.40.60–61

Augustine's affinity for Platonic thought is well known and Platonism seemed particularly compatible with the Christian faith. He was under the influence of Plotinus whom he read in the Latin translation of Victorinus (Brown 1967: 92–93). His philosophical readings had a distinct role to play even in a world dominated by revelation. He held that "a mind trained on philosophical methods could think creatively within the traditional orthodoxy of the church" (Ibid., 113).

Our passage (*Doctr. chr.* 2.40.60–61) plays a central role in this process of cultural integration. It is here he provides biblical justification for the liberal posture toward pagan culture, an exegesis which generally but uniquely follows the lines laid down by Origen (Simonetti 1994: 118).[9]

60.a. Any statements by those who are called philosophers, especially the Platonists, which happen to be true and consistent with our faith should not cause alarm, but be claimed for our own use, as it were from owners who have no right to them. Like the treasures of the ancient Egyptians, who possessed not only idols and heavy burdens, which the people of Israel hated and shunned (*quae populus Israhel detestaretur et fugeret*) but also vessels and ornaments of silver and gold and clothes (*sed etiam uasa atque ornamenta de auro et argento et uestem*), which on leaving Egypt the people of Israel, in order to make better use of them, surreptitiously (*clanculo*) claimed for themselves (they did this not on their own authority but at God's command—and the Egyptians in their ignorance (*nescienter*) actually gave them the things of which they had made poor use) [Exod 3:21–2, 12:35–6]—(60.b) similarly all the branches of pagan learning contain not only false and superstitious fantasies and burdensome studies that involve unnecessary effort, which each one of us must loathe and avoid as under Christ's guidance we abandon the company of pagans, but also studies for liberated minds which are more appropriate to the service of the truth (*liberales disciplinas usui veritatis aptiores*), and some very useful moral instruction, as well as the various truths about monotheism to be found in their writers. These treasures—like the silver and gold, which they did not create but dug, as it were, from the mines of providence, which is everywhere (*quasi metallis divinae providentiae quae ubique infusa est*)—which were used wickedly and harmfully in the service of demons (*quo perverse atque iniuriose ad obsequia daemonum abutuntur*) must be removed by Christians, as they separate themselves in spirit from the wretched company of pagans, and applied to their true function, that of preaching the gospel. As for their clothing—which corresponds to human institutions, but those appropriate to human society, which in this life we cannot do without—this may be accepted and kept for conversion to Christian purposes.

61.a. This is exactly what many good and faithful Christians have done. We can see, can we not, the amount of gold, silver, and clothing with which Cyprian, that most attractive writer and most blessed martyr, was laden when he left Egypt; is not the same true of Lactantius, and Victorinus, of Optatus, and Hilary, to say nothing of people still alive, and countless Greek scholars? Isn't this what had been done earlier by Moses himself, that most faithful servant of God, of whom it is written that he was trained in all the wisdom of the Egyptians [Acts 7:22]? 61.b. Pagan society, riddled with superstition, would never have given to all these men the arts which it considers useful—least of all at a time when it was trying to shake off the yoke of Christ and persecuting Christians—if it had suspected that they would be adapted to the

[9] I follow the translation of Green (Green 1997: 124–6) to which I will add some Latin text and the subdivisions 60.a,b and 61.a,b.

> purpose of worshipping the one God, by whom the worship of idols would be eradicated. But they did give their gold and silver and clothing to God's people as it left Egypt, little knowing (*nescientes*) that the things they were giving away would be put back into the service of Christ. The event narrated in Exodus was certainly a figure, and this is what it foreshadowed. (I say this without prejudice to any other interpretation of equal or greater importance.)

In 60.a., he claims that the Hebrew people had to discriminate between good and bad treasures. They despised and rejected the bad Egyptian treasures (idols, etc.) and appropriated good treasures. The Egyptians had proven they were unfit for this wealth by turning good gold into bad idols. This is, of course, not in the Bible nor have we encountered it in the rabbinic traditions.[10] It is also different from what we read in Origen where the emphasis is not on discriminating between good and bad treasures but in knowing how to utilize Egypt's treasures once they are plundered; you can build either a calf or a tabernacle from the same gold.

Augustine adds this element to the story (or received it from an unknown *haggadic* tradition) on the literal level to fine-tune the allegory; the Christian theologian is also duty-bound to discriminate between good and bad ideas in philosophy. This theme he expands in 60.b.—the theologian is to adjudicate between the "false and superstitious fancies and heavy burdens of unnecessary labor" and the "fields of bountiful knowledge apt for the use of the truth, and useful precepts dealing with ethics." A Christian should consider Platonism and philosophical monotheism gold worth having. Augustine adds gold and silver *ornamenta* to the biblical *uasa* in light of the use of these treasures for the ornamentation of the tabernacle.

Another interesting element for his hermeneutic is the notion that the Egyptians were lending things unknowingly (*nescienter*) and that the Israelites were to take these treasures "surreptitiously" (*clanculo*). Augustine is highlighting the view that the Egyptians loaned these treasures without knowledge of the Israelite plan of escape. The secretive nature of this operation is the very groundwork of the anti-Jewish attack; deception smacks of fraud. Pompeius Trogus refers to the stealth by which the people of Moses plundered Egyptian temples.

> Becoming leader, accordingly, of the exiles, (Moses) carried off by stealth the sacred utensils of the Egyptians, who, trying to recover them by force of arms, were compelled by tempests to return home (Stern 1976: 1:337).

Our key phrase "He carried off *by stealth* the sacred utensils" is an indication that there was some broad-based knowledge of a secretive despoliation. The same seems to be operative when Lysimachus claims that Jerusalem was originally called "Hierosyla" ("temple robbery") because of Jewish "sacrilegious propensities" (*Ag. Ap.* 1.311). Artapanus also provides the reason for the Egyptian pursuit of the escaping Hebrews; they wanted to retrieve what had been stolen from them (Holladay 1984: 223). It is possible this was an accumulation of older traditions concerning Egypt's other enemies retrojected back upon the Jews.[11] But while these may be little more than bloated rumors several generations removed from the Bible, they provide evidence that there was knowledge of a secretive despoliation abroad in anti-Jewish writers.[12] It is therefore surprising that Augustine is willing to highlight such an unsavory feature.

[10] Mek (Piskha 13) has almost the exact opposite idea—Israel took idols from Egypt specifically to destroy them. "And they despoiled the Egyptians" shows that that their idols melted and ceased to exist (as idols) and returned to their initial state.

[11] Schürer et al. note that the common notion that Moses led a group of lepers out of Egypt may have been an older Egyptian tradition about the expulsion of a defiled people that was transferred to the Jews at a later stage (Schürer 1986: 3.1:601). Stern proposes that the temple-robbing tradition arose under the influence of the Egyptian's later experience with their Persian enemies (Stern 1976: 1:340). It is more likely, however, that we are here dealing with the Gentile misunderstanding of the later interpretive tradition that the vessels taken from Egypt were used to construct the wilderness tabernacle.

[12] Pieter van der Horst, in "From Liberation to Expulsion: The Exodus in the Earliest Jewish-Pagan Polemic," has described the charges made by anti-Jewish pagan

It is also likely that the biblical story of despoliation would have been seen as a particularly egregious flouting of propriety since manumission carried with it certain social obligations toward the previous master (Glancy 2002: 14, 53, 69, and 161, n. 84). In the face of these social obligations, the plundering becomes particularly noxious. By emancipation, master becomes patron and is to be treated like a parent. The Jews ought to have treated Pharaoh with great deference but instead they showed disrespect by robbing him through deceit on their way out the door.

The textual reason why Augustine admits to the secrecy of the event is the addition of the word κρυφῇ (LXX Exod 11:2) to the translation of the Hebrew דַּבֶּר־נָא בְּאָזְנֵי הָעָם. In other words, the Septuagint translation of Exodus 11:2 already had highlighted the secrecy of the event in a way not necessitated by the Hebrew text itself. An Augustinian rendering of Exod 11:2 in Old Latin reads: *loquere ergo* <u>*secreto*</u> *in aures populi.* (Sabatier 1743: 155).[13] The secrecy of the affair has importance for Augustine here since, as we noted, in the allegory, Christians could only secretly plunder the intellectual riches of the Greek world. Pagan society would never have willingly shared its useful arts with Christians, "if it had suspected that [this knowledge] would be adapted to the purpose of worshipping the one God, by whom the worship of idols would be eradicated." Augustine highlighted God's role in the deception noting that it occurred *non auctoritate propria sed praecepto Dei*.

In 60.b. we encounter the phrase, "These things they did not make themselves but they dug them up out of certain, as it were, mines of divine providence, which have been scattered about everywhere."[14] This has absolutely no connection to the biblical text or previous exegesis and is an interesting phrase because it provides a glimpse into Augustine's rationale for openness to pagan philosophy. Just as the gold used to create idols comes from God's earth and ultimately is owned by God, so the truths in pagan texts are God's truths, and were placed by his providence throughout the world to be recovered and put to holy use. This is not the only place in Augustine's writings where this notion appears; in *Conf.* (7.9.15) he compares the Platonic doctrine of the *Logos* with the better teachings of Christianity:

> And so I came to You [Lord] from the Gentiles, and I sought gold which You wanted Your people to take from Egypt since it was Yours wherever it was.

God intends the despoliation to occur since these treasures belong to God in the first place. This notion may be best understood against the background of the Stoic doctrine of the *spermatikos logos* which is known to have played a large role in Justin's belief that Christ was the embodiment of the logos of which every person partakes (Droge 1987: 313). This seminal reason has been planted by God in humans at birth and, with pedagogical cultivation, is able to blossom into virtue. One Greek fragment reads, "By nature, we are all born with the seeds of virtue... We must develop them with learning of virtue."[15]

Toward the end of 60.b., we read, "As for their clothing—which corresponds to human institutions, but those appropriate to human society, which in this life we cannot do without—this may be accepted and kept for conversion to

writers—possibly based on the word "plunder" in Exod 12:36—that the Jews had stealthily robbed the Egyptians during the exodus. Jewish responses in Artapanus, Ezekiel the Tragedian and the author of *Wisdom of Solomon* typically claim that what was taken from Egypt amounted to nothing more than overdue wages.

[13] The reading comes from *Quaestiones de Exodo 39*.

[14] P. Beatrice notes that in his first Cassiciacum dialogue, *Contra Acedemicos* (III.17.38), which dates to the Milanese period, Augustine claims that Arcesilaus hid the Platonic doctrine of the immortality of the soul in the ground to be found by posterity. "Augustine was thinking of Platonism precisely as the 'Egyptian' gold and silver the pagans dug from the mines of providence, as he was to write ten years later in *De doctrina christiana* II, 145. The image of the buried treasure was the same" (Beatrice 2007: 181).

[15] Quoted by M. Horowitz 1974: 12.

Christian purposes." This refers to the Christian participation in the trade guilds and other social institutions (the *collegia*) of late antiquity, a topic which takes up much of Augustine's attention in *Doctr. chr.* 2.38ff. While he personally withdrew from the world and its institutions as a monk and priest, Augustine recognized the world as a place where the Christian and non-Christian must interact, where God's word and kingdom must be extended. Augustine justifies this exegetical jump by the fact that participants in the *collegia* were known for their distinctive dress (*Doctr. chr.* 2.39–40).

By the 390s, it became urgent to him to understand the structures of secular society and their relationship to the possibilities of Christian life (Rist 1994: 216). The *collegia* of Roman culture were to be despoiled, not for social or financial gain, but for Christian purposes. Eusebius of Caesarea went further claiming that Rome itself and all its social institutions had become, in Constantine's empire, a divine vehicle to promote the spread of Christian culture (Ibid., 208). Marrou describes Augustine's process of accommodation of ancient culture: to isolate those unhelpful elements associated with paganism itself, and to purify, refashion, and utilize the rest (Marrou 1983: 393).

We see here a broadening of application of the despoliation motif. Philo focuses on the soul's ascent to God nourished by standard encyclical education. Origen adds philosophy to the mix and moves from personal spiritual progress to philosophically sensitive biblical exegesis. Augustine, by focusing on the presence of clothing with the plunder, emphasizes not only theological inquiry but full Christian participation in the social *collegia* of common life. He even mentions that philosophy trains the mind to think about ethics. Augustine was after more than the Christianization of literature; he sought to despoliate even the social institutions of Rome for divine purpose.

One very telling difference is evident between Origen and Augustine. For Augustine, there are many examples of Christians who have used pagan learning rightly. He says (61.a. above):

> For what else have our many good and faithful done? Do we not see with what great quantity of gold and silver and garments Cyprian, that most persuasive and blessed martyr, stuffed his pockets and made his exit from Egypt? How about Lactanius? Consider Victorinus, Optatus and Hilary. Their spoils were so great that I need say nothing of those who are still alive.

An innumerable number of Greek theologians have successfully plundered Egypt according to Augustine. Origen is of the opposite opinion; for him, hardly anyone can navigate this balancing act rightly. The tendency is to build a golden calf rather than a tabernacle. In paragraph 3 of his letter to Gregory, we read,

> I could speak from knowledge gained by experience that few are those who have taken anything useful from Egypt and have come out from there and have prepared objects for the worship of God, but many have followed the example of Hadad, the Idumaean. These are those who have used some Greek ingenuity to beget heretical ideas and have, so to speak, prepared golden calves in Bethel, which means 'house of God.'

Augustine is clearly more robust in his attitude toward Egyptian plundering. Augustine's enthusiasm results in part from the success of Origen's example. Augustine begins our passage with the bold claim that, not only should pagan learning not be feared (did he have Origen's reticence in mind?), it should be claimed for Christian usage. This increasingly positive attitude reflects the historical situation after the time of Constantine when it was possible for a Christian to accept that Roman culture could have a positive role in salvation history.

Significance of the Allegory

Even after Christianity was firmly in place as the leading religion of the empire, Christians found themselves in a dilemma. Outright rejection of the classical heritage was too great a price to pay for faith. Stoics had long been allegorizing the moral improbity out of poetry. Rhetoric could train persuasive preachers and philosophy had obvious value for the theologian. "In general it was recognized that pagan literature could be plundered with profit provided that due caution was observed and the end justified the means" (Reynolds and Wilson 1991: 38). Beginning with Philo, this humanitarian posture was adopted in the east by Origen and in the west by Augustine. It provided a handy justification for those who wished to have the best of both Classical and Christian worlds. The old Greco-Roman system of education continued right through late antiquity and no attempt was made by Christians to replace it.

If a less accommodating posture has been adopted, as the new religion became universal, it might have imposed a ban on pagan texts. Such a censorship, if it had persisted, would have doomed present day classical studies. Texts are not preserved intentionally; they are preserved as the accidental outcome of being studied and copied over long periods of time. It was this process that the despoliation allegory encouraged. It is a fact that the vast majority of texts we have from the classical period were written by Christian hands, or at least, written during the period in which Christianity was the universal religion. That some classical texts remain is due to the fact that Christian students and academics were plundering them for Christian purposes. Said differently, if it were not for this attitude of accommodation, reflected and justified in our allegory, and if the opposing position had been allowed to prevail, countless extant texts would have been lost.

The church, in spite of what some might suspect, never imposed a ban on classical literature. Church fathers of the highest repute happily encouraged the reading of pagan texts as an adjunct to theology and biblical interpretation. Only one attempt to develop a distinct Christian curriculum is known from antiquity. Appollinaris (c. 310–90), with the aid of his father, briefly attempted to develop a complete Christian curriculum by writing a Homeric-style poem about the antiquity of the Jews and a hexameter version of the Psalms. But this was a response to Julian's persecution of Christians in 362 which included a ban on Christian teachers in public schools. As soon as the ban was lifted, Christians and pagans went back to studying pagan texts side by side.

Julian's proscription against Christian teachers is instructive in itself; it was necessary because many leading scholars, rhetors, and philosophers were Christians. The irony of it all is quite remarkable; for Julian to turn Rome back to paganism he had to forbid Christians from teaching *pagan* texts in their academies! When he died shortly later, the ban was dropped and Egypt went back to being plundered. Christians had become completely immersed in the academic world and Christianity itself came to be considered a philosophical school. The influence of this allegory played no small role in this matter.

Other metaphors for Christian appropriation were also used (primarily that of the beautiful captured non-Israelite woman based on Deut. 21:10–4 — see Origen's *Homily on Leviticus* 7:6). Yet the metaphor of the plundering of Egypt played a much larger role right through the middle ages.[16] Tertullian protested, "What has Jerusalem to do with Athens?" He was a voice in the wilderness. If Origen and Augustine were to answer, they might have said, "Quite a lot, actually!" Most Christians viewed the classical tradition as a problematic yet necessary preparation for the gospel almost on equal grounds as the Hebrew Bible itself.

References

Allen, Joel. 2008. *The Despoliation of Egypt: In pre-Rabbinic, Rabbinic and Patristic Traditions. Supplements to Vigiliae Christianae*, vol. 92. Leiden: Brill.

Beatrice, P.F. 2007. A Chapter in the History of Patristic Exegesis and Late Antique Culture. *Studia Patristica* 39: 159–183.

Brown, Peter. 1967. *Augustine of Hippo*. Berkley, CA: University of California Press.

Crouzel, Henri. 1969. *Grégoire le Thaumaturge. Remerciement à Origène suivi de la lettre d' Origène á Grégoire. Sources chrétiennes (SC) 148*. Paris: Cerf.

Droge, Arthur. 1987. Justin Martyr and the Restoration of Philosophy. *Church History* 56: 303–319.

[16] Folliet collects these sources with a short commentary on each (Folliet 2002: 47). See also L. Frizzell 1994: 383–391.

Folliet, Georges. 2002. La Spoliatio Aegyptiorum (Exode 3:21–23; 11:2–3; 12:35–36) Les Interprétations de cette Image chez les Pères et Autres Écrivains Ecclésiastiques. *Tradition* 57: 1–48.

Frizzell, L. 1994. "Spoils from Egypt", between Jews and Gnostics. In *Hellenization Revisited*, ed. Wendy Helleman, 383–384. Chico, CA: University Press of America.

Glancy, Jennifer A. 2002. *Slavery in Early Christianity*. Oxford: Oxford University Press.

Green, R.P.H. 1997. *Augustine, De Doctrina Christiana*. Oxford: Clarendon Press.

Holladay, Carl. 1984. *Fragments from the Hellenistic Jewish Authors*. Missoula, MT: Scholars Press.

Horowitz, M.C. 1974. The Stoic Synthesis of the Idea of Natural Law in Man: Four themes. *Journal of the History of Ideas* 35: 3–16.

Marrou, H.I. 1983. *Saint Augustin et la fin de la culture antique*. Paris: E. de Boccard.

Mendelson, Alan. 1983. *Secular Education in Philo of Alexandria*. Cincinnati, OH: Hebrew Union College Press.

Nautin, P. 1977. *Origène, sa vie et son oeuvre*. Paris: Beauchesne.

Reynolds, L., and N. Wilson. 1991. *Scribes and Scholars*, 3rd ed. Oxford: Clarendon Press.

Rist, John M. 1994. *Augustine: Ancient thought baptized*. Cambridge: Cambridge University Press.

Robinson, J. Armitage. 1893. *The Philocalia of Origen*. Cambridge: Cambridge University Press.

Sabatier, P. 1743. *Bibliorum Sacrorum Latinae Versiones Antiquae*. Reginaldum Florentiane: Remis.

Schürer, E. 1986. *The History of the Jewish People in the Age of Jesus Christ*. vol. 3.1 of 4 vols. Revised and edited by G. Vermes, F. Millar, M. Goodman, and M. Black. Edinburg: T and T Clark.

Simonetti, M. 1994. *Biblical Interpretation in the Early Church*. Edinburgh: T & T Clark.

Stern, Menahem. 1976. *Greek and Latin Authors on Jews and Judaism*. vol. 1 of 3 vols. Jerusalem: The Israel Academy of Sciences and Humanities.

Trigg, Joseph. 1998. *Origen*. London: Routledge.

Leaving Home: Philo of Alexandria on the Exodus

26

René Bloch

Abstract

For Jewish-Hellenistic authors writing in Egypt, the Exodus story posed unique challenges. After all, to them Egypt was, as Philo of Alexandria states, their fatherland. How do these authors come to terms with the biblical story of liberation from Egyptian slavery and the longing for the promised land? In this chapter I am taking a close look at Philo's detailed discussion of the Exodus and locate it within the larger context of Jewish-Hellenistic literature (*Wisdom of Solomon*, Ezekiel's *Exagoge*).

In Philo's rewriting of the Exodus the destination of the journey is barely mentioned. Contrary to the biblical narrative, in the scene of the burning bush, as retold by Philo, God does not tell Moses where to go. Philo's main concern is what happens in Egypt: both in biblical times and in his own days. The Exodus is nevertheless important to Philo: He reads the story allegorically as a journey from the land of the body to the realms of the mind. Such a symbolic reading permitted him to control the meaning of the Exodus and to stay, literally and figuratively, in Egypt.

The myth of the Exodus as narrated in the Hebrew Bible is an unusually powerful story. But it did not end with the Bible. Jewish, pagan, Christian, and Muslim authors pondered the meaning of the Exodus, commented on it, or simply rewrote the story. This chapter on the dilemma of Jewish-Hellenistic authors such as Philo of Alexandria, writing on the Exodus in Egypt, is part of a group of contributions in this volume which ask questions about the intentions of new versions and interpretations of the Exodus from the Hellenistic period till Late Antiquity (see, e.g., the contributions by Caterina Moro on the Jewish-Hellenistic author Artapanus, not discussed in my chapter, and by Pieter van der Horst on the "dialogue" between pagan and Jewish authors on the Exodus). It is, however, important to keep in mind that already the

R. Bloch (✉)
Institute for Jewish Studies and Institute of Classical Studies, University of Bern, Switzerland
e-mail: rene.bloch@theol.unibe.ch

T.E. Levy et al. (eds.), *Israel's Exodus in Transdisciplinary Perspective*,
Quantitative Methods in the Humanities and Social Sciences, DOI 10.1007/978-3-319-04768-3_26,

original story as told in the Bible is very much a literary construct with its own agenda. This chapter thus asks similar questions of the Exodus myth as some of the Biblical scholars do (see, e.g., the contribution by Ron Hendel).

Moses at the burning bush in chapter three of the Exodus book is one of the most dramatic scenes in the Hebrew Bible. All necessary ingredients for a dramatic epiphany are at play: Moses, who is watching the flock of his father-in-law Jethro near the mountain Horeb, the "mountain of God," becomes aware of a paradoxical scenery of nature: there is a flame of fire coming out of a bush, but the bush is not consumed by the fire. Moses hears a divine voice calling him by name and telling him to keep away from the site. The scene of the burning bush is a classic example of how sacred space is created: "Remove the sandals from your feet, for the place on which you are standing is holy ground."[1] This is the site where Moses is informed by God about the land where the Israelites will dwell in the future. It is here, on sacred ground, that the territory of Israel's future home is described and announced to Moses. In this fiery and, one can imagine, smoky scene of Exodus 3, the destination of the journey becomes clear:

> and I have come down to deliver them from the Egyptians, and to bring them up out of that land to a good and broad land, a land flowing with milk and honey, to the country of the Canaanites, the Hittites, the Amorites, the Perizzites, the Hivites, and the Jebusites.[2]

This is the ticket. To leave no doubt where the journey is going the destination is repeated a little later: "I declare that I will bring you up out of the misery of Egypt, to the land of the Canaanites, the Hittites, the Amorites, the Perizzites, the Hivites, and the Jebusites, a land flowing with milk and honey."[3]

Among the several ancient Jewish authors who comment on and make use of this scene, the Jewish philosopher Philo of Alexandria is particularly interesting. In his vast oeuvre, Philo comes to talk most prominently about the Exodus in his *Life of Moses*, a fascinating Jewish-Hellenistic rewriting of the biblical Moses narrative, and in his *Questions and Answers on Exodus*. In the former tractate the biblical scene of the burning bush is treated in detail and with quite a few additions. Philo reads the burning bush which is not consumed by fire allegorically as a symbol for the resistance of the Jewish people:

> For the burning bramble was a symbol of those who suffered wrong, as the flaming fire of those who did it. Yet that which burned was not burnt up, and this was a sign that the sufferers would not be destroyed by their aggressors, who would find that the aggression was vain and profitless while the victims of malice escaped unharmed.[4]

Philo imagines that the biblical bush had thorns: it "is a very weakly plant, yet it is prickly and will wound if one do but touch it."[5] The miracle that the bush was not devoured by fire refers in Philo's reading to the nation's condition, and to Israel's strength in difficult times. God tells Moses: "Do not lose heart; your weakness is your strength, which can prick, and thousands will suffer from its wounds."[6] As in the biblical version God then strengthens Moses' limited self-confidence by showing him additional miracles and by assuring him that he would support Moses when it comes to rhetorics. To the Jewish people, Moses would not just be "an assistant to their liberation," but a "leader" of their "migration" (*apoikias*).[7]

In comparison with the biblical report there is something crucial missing in God's speech to Moses and in the lengthy passage on the biblical

[1] Exod 3:5. Translations of biblical passages follow *The New Oxford Annotated Bible*, Oxford 2010[4].

[2] Exod 3:8.

[3] Exod 3:17.

[4] Philo *De Vita Mosis* 1.67. All translations of Philo are, if not stated otherwise, from the English edition published in the *Loeb Classical Library*.

[5] Philo *De Vita Mosis* 1.68.

[6] Philo *De Vita Mosis* 1.69.

[7] Philo *De Vita Mosis* 1.71.

scene of the burning bush[8]: Where is Moses supposed to go? There is no mention of any destination. Philo's God tells Moses that he would soon become the leader of their migration from Egypt, but forgets to tell Moses where to take his people. All he is told is to "make a three-days's journey beyond the bounds of the country" and to sacrifice there.[9]

In this chapter I would like to ask what it meant for an author such as Philo of Alexandria who lived in Egypt to write on the Exodus, the biblical story of Israel's migration from Egypt to a better place. For Jewish-Hellenistic authors writing in Egypt the Exodus story posed unique challenges. After all, to them Egypt was, as Philo of Alexandria states, their fatherland (*patris*).[10] How does Philo come to terms with the biblical story of liberation from Egyptian slavery and the longing for the promised land? I will focus on Philo for most of the chapter. Towards the end of my chapter I will add two more examples from Jewish-Hellenistic literature, *Wisdom of Solomon* and Ezekiel's Exodus tragedy *Exagoge*.

Perhaps a few biographical remarks on Philo of Alexandria are in order. As a matter of fact, there is not much one can say for sure about his life: In spite of his impressively large oeuvre (37 tractates have survived, but Philo wrote quite a few more), very little can be said with certainty about his life and activities.[11] Contrary to the other major literary figures of Diaspora Judaism, Josephus and Paul, Philo very rarely speaks of himself, and there is no autobiography. Even the dates of his life are debated. Philo was probably born around 20 BC and died around AD 50. Even when Philo discusses the Jewish embassy to the Roman emperor Caligula (AD 39/40) of which he was most probably the leader, we learn very little about the author himself.[12] Josephus has a few lines on Philo and his family which indicate that Philo grew up in an upper class Jewish family in Alexandria.[13] The very existence of his large oeuvre implies that he must have enjoyed financial independence. Philo probably lived his whole life in Alexandria, which at the time was the intellectual center of Greek-speaking Judaism. He surely left the city for traveling: In the context of the biblical report on Abraham's departure from Ur, Philo makes the point that "the stay-at-home is to the travelled as the blind are to the keen-sighted." One reason which makes people travel is, "the chance of benefiting their country, when occasion offers, in its most vital and important interests."[14] Philo certainly lived up to this ideal, when he participated in the Jewish embassy to Rome when the Jews of Alexandria were under attack. We will come back to this later.

Philo was at least once in Jerusalem and visited the temple there.[15] In Philo's understanding Jerusalem is the *mētropolis* or "mother city" of all the Jews, while their "fatherland" (*patris*) is the place where they actually live. The *locus classicus* for this is *In Flaccum* 46:

> it is the holy city where the sacred temple of the Most High God stands, that they (the Jews) regard as their mother city, but the regions they obtained from their fathers, grandfathers, great-grandfathers, and even more remote ancestors, to live in, (they regard) as their fatherland where they were born and brought up.[16]

To Philo, Egypt, or at least the city of Alexandria, was his fatherland. He refers to the city as "our Alexandria."[17] Philo repeatedly, as in the scene of the burning bush, uses the Greek term *apoikia*, "migration," when writing about

---

[8] The passage comprises some 20 paragraphs: *De Vita Mosis* 1.65–84.

[9] Philo *De Vita Mosis* 1.73.

[10] Philo *In Flaccum* 49, see below.

[11] Cf. Schwartz (2009: 9–31).

[12] Philo *Legatio ad Gaium*; *In Flaccum*.

[13] Josephus *Antiquitates Iudaicae* 18.257–260.

[14] Philo *De Abrahamo* 65.

[15] Philo *De Providentia* 2.64.

[16] Philo *In Flaccum* 46 (transl. van der Horst 2005).

[17] Philo *Legatio ad Gaium* 150. Cf. Gruen (2002: 74).

the Exodus. *Apoikia* literally means "settlement far from home."[18]

Let us return to Philo's Moses: According to Philo, Moses who grew up in the royal family of the Pharaoh had the opportunity to make a political career in Egypt.[19] But being the prototype of the ideal wise man, Moses showed no interest in the superficial symbols of power. In Philo's reading, it is at the scene of the burning bush that Moses is rewarded for his wise behavior and becomes the leader of the Jewish people. But again: what is the destination? Instead of making Moses explicitly the leader who would bring the Israelites to Canaan, Philo states that God gave Moses "the whole world":

> And that is but natural, for he is a world citizen (*kosmopolitēs*), and therefore not on the roll of any city of men's habitation, rightly so because he has received no mere piece of land but the whole world as his portion.[20]

While the Biblical text makes it very clear that Moses would lead the Israelites out of Egypt to the land of Canaan, Philo's Moses, right after he has been made the leader and savior of the Jews, is presented as a cosmopolitan citizen. Philo uses the term *kosmopolitēs*, "citizen of the world," only eight times in his oeuvre, but his understanding of the term is of great importance for his concept of the world and the law. According to Philo, as he states at the beginning of his tractate on the creation of the world, "the world is in harmony with the Law, and the Law with the world" and "a man who observes the law is constituted thereby a loyal citizen of the world." For Philo there is no difference between the law of nature and the law of Moses.[21] Therefore, Moses must be a citizen of the world and not just of a specific city or land. The biblical narrative of Moses who would not live to arrive in the promised land suited Philo's understanding of Moses as a cosmopolitan citizen (and more generally of a "cosmic" Judaism). Moses reaches out beyond a limited territory. And it comes as no surprise when Philo at the end of his tractate on Moses does not stick to the biblical description of Moses' death. According to Philo, Moses is not buried in the valley in the land of Moab (as stated in Deut 34:6), but he migrates to heaven. Here, too, as in the passage on the burning bush, Philo uses the word *apoikia* for Moses' migration. But here it is clear where this migration would lead: "the migration from there . . . to heaven."[22] God resolved Moses' "twofold nature of soul and body into a single unity, transforming his whole being into mind, pure as the sunlight."[23] Moses becomes immortal and joins God. This is Moses' Exodus in Philo: It is a spiritual journey towards God.

It is not that the destination of the journey of the Israelites is not mentioned at all in Philo's *Life of Moses*. Eventually, more as an aside, Philo mentions the destination of the Exodus by name: Moses was to lead the Israelites to "Phoenicia and Coelesyria and Palestine, then called the land of the Canaanites."[24] But the narrative in Philo's *Life of Moses* does not focus on the promised land; rather, it focuses on the point of departure, it focuses on Egypt.

In Philo, Egypt is repeatedly described as a country to be avoided and to be left behind—not in a geographic sense, though, or with a specific destination in mind, but from a philosophical point of view: For Philo Egypt is the "symbol" of the body and of passion.[25] Egypt is a "bodily land" (*sōmatikē chōra*).[26] In her thorough study *The Land of the Body* Sarah Pearce has shown how for Philo Egypt consistently symbolizes the

[18] *LSJ*, s.v. On the other hand by calling Jerusalem *mētropolis* Philo uses a term which is regularly used for "*the mother-state,* in relation to colonies" (*LSJ*, s.v.). Jerusalem can thus be both the origin of the Jewish colony and its destination. Cf. on this Pearce (2004: 19–36).

[19] Philo *De Vita Mosis* 1.150–154. For a detailed study on Philo's Moses in comparison with the biblical narrative cf. Feldman (1997).

[20] Philo *De Vita Mosis* 1.157.

[21] Philo *De Opificio Mundi* 3. Cf. on this Martens (2003).

[22] Philo *De Vita Mosis* 2.288; the phrasing (*tēs enthende apoikias*) is identical with the one in 1.71 on the Exodus.

[23] Philo *De Vita Mosis* 2.288.

[24] Philo *De Vita Mosis* 1.163.

[25] Philo *De Migratione Abrahami* 77.

[26] Philo *De Migratione Abrahami* 151.154; *De Mutatione Nominum* 90.

material sphere, the land of the body.[27] In Philo's *Life of Moses* there are certainly also attempts to stress some sort of an Egyptian-Jewish "symbiosis." Thus Philo imagines for Moses an international education in Egypt: Wise teachers from different parts of the world come to the royal palace to assure a first class education, starting with the *enkyklios paideia* and leading to the most important field of study, philosophy. Among Moses' many teachers Egyptian scholars are mentioned with praise.[28] Nevertheless, Pearce is right in stressing the overall negative image of Egypt in Philo. Egypt stands for the body, and the Exodus stands for the migration of the soul from the body. When the Israelites departed from Egypt, they thus approached a higher stage of understanding. In his comment on Gen 26:2 ("The Lord appeared to Isaac, and said, 'Do not go down to Egypt'"), Philo explains the etymology of Egypt:

> Egypt is to be translated as "oppressing," for nothing else so constrains and oppresses the mind as do desire for sensual pleasures and grief and fear. But to the perfected man (Isaac), who by nature enjoys the happiness of virtue, the sacred and divine word recommends all perfection and not to go down into the passions but to accept impassivity with joy, bidding (the passions) a fond farewell.[29]

To avoid Egypt or to leave Egypt behind, as in the Exodus, is in Philo's reading thus an allegorical way of saying farewell to the body and its passions.

Philo's understanding of Egypt is ambivalent: On the one hand, he understands Egypt as his home. Philo clearly participated in the cultural life of Alexandria—as becomes obvious, e.g., when he describes his visits to the theater—,[30] and time and again he demonstrates a very thorough knowledge of Greco-Roman culture.[31] On the other hand, Egypt is negatively understood as a symbol for the body and passion. These are two different ways of understanding Egypt, but they are not mutually exclusive. The journey of the soul towards the world of God is a central theme in Philo and the Exodus was *the* obvious example for a symbolic reading of an actual journey. For Philo, it seems to me, such a symbolic (instead of a literal) reading of the Exodus also permitted him to keep the Exodus under control, so to speak, and to stay in Egypt.

God's curious silence on the destination of Moses' migration in the scene of the burning bush is no exception. Philo generally shows very little interest in Canaan/Palestine.[32] His discussion of the biblical narrative focuses on the books of Genesis (mainly) and Exodus, not on the books of Joshua and Judges. When Philo comes to talk about Gen 15:18, God's promise to Abram that his offspring would receive the land "from the river of Egypt to the great river, the river Euphrates," he prefers an allegorical over a geographical reading.[33] Only rarely is the conquest of Canaan a topic.[34] In his treatise *Questions and Answers on Exodus*, of which only a fraction survives (covering Exodus 12–28), Philo asks and answers 124 questions on the book of Exodus. But nowhere is the final aim of the Exodus mentioned. When he

[27] Pearce (2007: 89) on the phrase "bodily land".

[28] Philo *De Vita Mosis* 1.21–24. Cf. Bloch (2012: 80–83).

[29] Philo *Quaestiones in Genesim* 4.177. Philo seems to connect the Hebrew word for Egypt, *mizrajim*, with the Hebrew verb *zarar* (to bind, tie up; to show hostility). In this passage, Philo interprets the physical oppression, as related in the book of Exodus, allegorically as an oppression of the mind: On this etymology cf. Pearce (2007: 82–85).

[30] Philo *De Ebrietate* 177. Cf. Bloch (2009: 72–74).

[31] Cf. Bloch (2011: 173–189).

[32] On Philo and Palestine cf. Schaller (2001: 13–27) and de Vos (2012: 87–100). Feldman (1997:78) suspects an apologetic reason behind Philo's reluctance to mention the destination of the Exodus: He could thus avoid the charge of dual allegiance. But this seems unlikely. The apologetic agenda of *Life of Moses* has been overstated.

[33] Philo *De Somniis* 2.255: "Again God promises wise Abraham a portion of land "from the river of Egypt to the great river Euphrates", not mentioning a section of country, but rather the better part in ourselves. For our body and the passions engendered in it or by it are likened to the river of Egypt, but the soul and what the soul loves to the Euphrates." Cf. Pearce (2007: 104–105).

[34] Cf. Berthelot (2007: 39–56).

comments on Exod 23:20 ("I am going to send an angel in front of you, to guard you on the way and to bring you to the place that I have prepared"), Philo interprets the "place" once more allegorically:

> (...) the entry into the land, (that is) an entry into philosophy, (which is), as it were, a good land and fertile in the production of fruits, which the divine plants, the virtues, bear.[35]

This is where the philosopher Philo wants to go and this is the ideal destination of the Israelites.

The Exodus is important to Philo: in part because of the allegorical message which he sees behind the story, but also in part because of certain parallels between Egypt of biblical times and his own days. It seems to me that Philo's way of talking about Egypt is heavily colored by personal impressions. I have recently argued that Philo's *Life of Moses*, or at least parts of this tractate, can be read as a text in which Philo ponders on his own life, not the least his response to the anti-Jewish riot in Alexandria in AD 38. As a matter of fact, Philo's description of the suffering of the Israelites under the cruel Egyptian despots in *Life of Moses* is very similar to his report on the suppression of the Jews at the time of the riot in the tractate *Embassy to Caligula* (*Legatio ad Gaium*). Moreover, by going to Rome and pleading at the emperor's palace on behalf of the Jews, Philo slips into the role of Moses who did the same at Pharao's palace. Moses is obviously out of reach, but he is Philo's *paradeigma*, his constant point of orientation.[36] When Philo writes about Moses and the Exodus he is very much in Egypt. At times Moses and Philo even merge into one person. They enjoyed the same kind of education[37] and were both confronted with animosity in Egypt. Philo's main concern is Egypt.

How did other Jewish-Hellenistic authors conceptualize the Exodus? I will have to be very brief on this and will only hint at two other examples from Jewish-Hellenistic literature: *Wisdom of Solomon* and Ezekiel's tragic play *Exagoge* on the Exodus.

Wisdom of Solomon is a pseudepigraphic text which became part of the Septuagint. The author, most likely writing in Alexandria and possibly a contemporary of Philo, takes up the identity of King Solomon.[38] The book, written in a Greek which shows both Hebrew and pagan influences, belongs to Jewish Wisdom literature. It comprises three sections, of which the first compares the life of the righteous who pursues wisdom to that of the wicked who first ridicules the values of Jewish wisdom, but then realizes that in the end righteousness wins over injustice. The middle section of the book is an intense praise of wisdom in the name of King Solomon. The third and last section (Chaps. 10–19) contrasts, like the first, righteous and wicked approaches to life, but this time the author presents examples from Israel's "history": in the center of the discussion is the biblical story of the Exodus and the destiny of the Israelites and the Egyptians. The author is eager to show that the Egyptians are punished by the same means as the Israelites are delivered[39]: The Egyptians try to kill Moses in the Nile, but in the end their men drown in the water of the sea of reeds. God shows the Israelites the path to water in the desert, but makes the water of the Egyptians undrinkable. In the words of the author:

> For by those very things through which their enemies were punished, they in their want were benefited.[40]

A major critique in *Wisdom of Solomon* is, as in Philo, the Egyptian worship of animals. For

[35] Philo *Quaestiones in Exodum* 2.13.

[36] Bloch (2012: 69–84).

[37] Philo *De Vita Mosis* 1.21–24 (on Moses' education); Philo *De Congressu Eruditionis Gratia* 74–76 (on Philo's education).

[38] On content, structure, and authorship of *Wisdom of Solomon* cf. the concise introduction by Chesnutt (2010: 1242–1244).

[39] Chesnutt (2010: 1243).

[40] *Wisdom of Solomon* 11:5. I follow the translation by David Winston, *The Wisdom of Solomon*, New York 1979.

this, too, the Egyptians are being punished adequately by the plagues:

> In repayment for their wicked and witless reasoning, by which they were misled into worshiping brute reptiles and worthless beasts, you sent against them a swarm of creatures devoid of reasoning; that they might know that by those things through which a man sins, through them he is punished.[41]

Another parallel with Philo—and this brings us back to our earlier argument—is the remarkable fact that in *Wisdom of Solomon* after a detailed description of the suffering of the Israelites and the punishment of the Egyptians, the Exodus does not really happen anyway. The Israelites do not get further than to the Sea of Reeds. One may say that they remain in Egypt. *Wisdom of Solomon* ends somewhat abruptly with a statement about God's constant help for Israel "in every time and place."[42] Some scholars have wondered about the rather abrupt end of the book. Why does the author make no mention of the subsequent arrival in the promised land?[43] The reason for our author's silence seems to be the same as in the case of Philo: his main concern is Egypt.

Let me add a third example of such a reluctance to really leave Egypt: the drama *Exagoge* by the Jewish-Hellenistic poet Ezekiel. The play—probably also written in Alexandria, maybe in the second century BC—has only survived in fragments.[44] However, the extant 269 Greek verses allow for more than a brief insight. The play sets out with a typical prologue as we know it from many Greek tragedies: It is Moses himself who dares to enter the stage all alone (something the rhetorically challenged Moses of the Torah would never have done) and tells the audience of the beginnings of the Jewish people as well as his own origins. This is the beginning of the first fragment:

> When Jacob left Canaan
> he came to Egypt with seventy
> souls and fathered a great
> people that has suffered and been oppressed.
> Till this day we have been ill-treated
> by evil men and a powerful regime.[45]

We don't know whether this was the actual beginning of the play. It might very well be the case. In any case, Moses early on in the drama explains the Israelites' presence in Egypt: Jacob had left Canaan and migrated to Egypt and Moses is a descendent of Jacob. Moses then reports how he was saved by the Egyptian princess and how, after having killed an Egyptian, he had to flee, and how he met the daughters of Raguel (Jethro). The fragmentary nature of the transmission of the text leaves many questions open. But of Ezekiel's version of the biblical scene of the burning bush a substantial part has survived. Strikingly, here, too, God does not tell Moses where to go:

> Now go, and report in my words
> to all the Hebrews and
> then to the king my instructions to you,
> that you lead my people from the land.[46]

A little later, God speaks of the destination as "your own land" (*idion chōron*).[47] But the name of the land is not mentioned here or anywhere else in the surviving fragments. In the drama, there is certainly a movement out of Egypt and towards the end there is mention of reports by scouts who visited the land. But to Ezekiel what happened in Egypt appears to be of greater importance. In what seems to have been the fourth act of the play there is a long speech by an Egyptian soldier who returns home, apparently the only survivor of the Egyptian disaster at the Sea of Reeds. The Egyptian messenger tells the probably mostly Jewish audience—which must have enjoyed this scene—how the armed forces of the Pharaoh had

[41] *Wisdom of Solomon* 11:15–16.

[42] *Wisdom of Solomon* 19:22.

[43] Cf. the discussion in Reider (1957: 224–225).

[44] On Ezekiel cf. the excellent editions by Jacobson (1983) and Lanfranchi (2006).

[45] Ezekiel *Exagoge* 1–6. I follow the translation by Jacobson (1983).

[46] Ezekiel *Exagoge* 109–112.

[47] Ezekiel *Exagoge* 167.

no chance against the Israelites. Too late did the Egyptians realize "that God was helping them."[48] As in *Wisdom of Solomon*, this is a central message of the text: the wrongdoers are punished for their unjust behavior and the righteous prevail. And here too, as in Philo, the main interest is Egypt. Shortly after the Israelites had left Egypt, the Egyptian messenger returns (to Alexandria, one can assume) and continues the dialogue with the Jews in Egypt.

We have looked at three Jewish-Hellenistic forms of rewriting the Exodus: Philo's *Life of Moses*, and, very briefly, *Wisdom of Solomon* and Ezekiel's *Exagoge*. In many ways the three examples could not be more different from each other: Philo's *Life of Moses* is a philosophical biography, *Wisdom of Solomon* is a piece of Wisdom literature, and Ezekiel's drama is a Hellenistic tragedy. But the three authors, who most probably all wrote in Egypt, all have in common that in their rewriting of the Exodus they remain very much in Egypt and barely leave home.

References

Berthelot, Katell. 2007. Philo of Alexandria and the Conquest of Canaan. *Journal for the Study of Judaism* 38: 39–56.

Bloch, René. 2009. Von Szene zu Szene: Das jüdische Theater in der Antike. In *Juden in ihrer Umwelt. Akkulturation des Judentums in Antike und Mittelalter*, ed. R.C. Schwinges and M. Konradt, 57–86. Basel: Schwabe.

———. 2011. *Moses und der Mythos: Die Auseinandersetzung mit der griechischen Mythologie bei jüdisch-hellenistischen Autoren*. Leiden: Brill.

———. 2012. Alexandria in Pharaonic Egypt: Projections in *De Vita Mosis*. *The Studia Philonica Annual* 24: 73–88.

Chesnutt, Randall D. 2010. Solomon, Wisdom of. 1242–4. In *The Eerdmans Dictionary of Early Judaism*, ed. J. Collins and D. Harlow. Grand Rapids, MI: Eerdmans.

de Vos, Jacobus C. 2012. *Heiliges Land und Nähe Gottes. Wandlungen alttestamentlicher Landvorstellungen in frühjüdischen und neutestamentlichen Schriften*, Forschungen zur Religion und Literatur des Alten und Neuen Testaments, vol. 244. Göttingen: Vandenhoeck & Ruprecht.

Feldman, Louis H. 1997. *Philo's Portrayal of Moses in the Context of Ancient Judaism*. Notre Dame, IN: University of Notre Dame.

Gruen, Erich S. 2002. *Diaspora: Jews Amidst Greeks and Romans*. Cambridge, MA: Harvard University.

Jacobson, Howard. 1983. *The Exagoge of Ezekiel*. Cambridge: Cambridge University.

Lanfranchi, Pierluigi. 2006. *L'Exagoge d'Ezéchiel le Tragique. Introduction, texte, traduction et commentaire*, Studia in Veteris Testamenti pseudepigrapha, vol. 21. Boston, MA: Brill.

LSJ = Liddell, Henry G., Robert Scott, Henry S. Jones. 1996. *A Greek-English Lexicon*. 9th ed. Oxford: Clarendon.

Martens, John W. 2003. *One God, One Law: Philo of Alexandria on the Mosaic and Greco-Roman Law*. Boston, MA: Brill.

Pearce, Sarah. 2004. Jerusalem as 'Mother-City' in the Writings of Philo of Alexandria. In *Negotiating Diaspora. Jewish Strategies in the Roman Empire*, Library of Second Temple Studies, vol. 45, ed. J.M.G. Barclay, 19–37. New York, NY: T&T Clark.

———. 2007. *The Land of the Body: Studies in Philo's Representation of Egypt*, Wissenschaftliche Untersuchungen zum Neuen Testament, vol. 208. Tübingen: Mohr Siebeck.

Reider, Joseph. 1957. *The Book of Wisdom: An English Translation with Introduction and Commentary*. New York, NY: Harper & Brothers.

Schaller, Berndt. 2001. Philon von Alexandreia und das "Heilige Land". In *Fundamenta Judaica. Studien zum antiken Judentum und zum Neuen Testament*, Studien zur Umwelt des Neuen Testaments, vol. 25, ed. Berndt Schaller, 13–27. Göttingen: Vandenhoeck & Ruprecht.

Schwartz, Daniel R. 2009. Philo, His Family, and His Times. In *The Cambridge Companion to Philo*, ed. A. Kamesar, 9–31. Cambridge: Cambridge University.

van der Horst, Pieter W. 2005. *Philo's Flaccus: The First Pogrom: Introduction, Translation and Commentary*, Philo of Alexandria Commentary Series, vol. 2. Atlanta, GA: Society of Biblical Literature.

[48] Ezekiel *Exagoge* 236.

Hero and Villain: An Outline of the Exodus Pharaoh in Artapanus

27

Caterina Moro

Every villain is the hero of his own myth
(Christopher Vogler, *The Writer's Journey*)

Abstract

This chapter explores the possible identification of Chenephres, the first opponent of Moses in the story of Exodus as told by the Jewish-Hellenistic historian Artapanus, with Khaneferra, the royal name of Sobekhotep IV of the XIII dynasty. Sobekhotep was perhaps the last great "king of Upper and Lower Egypt" before foreign dynasties, the so-called Hyksos, took control of the whole country. The missing link between the historical king and Artapanus' character may have been a piece of fictionalized history that exalted Sobekhotep beyond his real merits, for example crediting to him the reconstruction of Amun's temple in Karnak (Diospolis; *Praep. Ev.* 9.27.11), that took place during the XII dynasty. After an examination of some sources from Sobekhotep's times, the chapter will consider the Egyptian historical and pseudo-historical literature, searching for a likely date for this biography of Khaneferra. Finally, the chapter will discuss the place and meaning of this hypothetical source in Artapanus' [illegible] response to Manetho's version of the Exodus from Egypt.

The possible parallel between an Egyptian historical figure and the Pharaoh of Artapanus, proposed in my first international publication (Moro 2010), drew the attention of Brad C. Sparks, principal organizer of this UCSD Exodus Conference, who is now involved in a project on the history of identification of "Exodus parallels" in Egyptian texts (Chap. 19). This nonbiblical account of Exodus by Artapanus is investigated as an example of ideological use of history in the context of Jewish-pagan polemics in Greco-Roman Egypt, a topic dealt with in this volume also by the articles of Pieter van der Horst (Chap. 29) and René Bloch (Chap. 26). This chapter is part of a research on the appropriation of Egyptian culture by Jewish tradition from the Persian Period to Late Antiquity, relating not only to Jewish-Hellenistic authors but also to the evolution of the "Moses myth" in the biblical text,[1] and in the midrash (Moro 2011: 156–175).

I am obliged to my friend and colleague R. Antonio Di Gesù who kindly improved the English of this chapter.

C. Moro (✉)
Sapienza Università di Roma, Rome, Italy
e-mail: moroclarab@libero.it

[1] Moro (2011: 40–104). Just like Christopher Berner (Chap. 20) I think that the Exodus narrative can be used

T.E. Levy et al. (eds.), *Israel's Exodus in Transdisciplinary Perspective*,
Quantitative Methods in the Humanities and Social Sciences, DOI 10.1007/978-3-319-04768-3_27,

Introduction

The Hellenistic historian Artapanus (quoted by Eusebius, *Praep. Ev.* 9.27) is the only Jewish[2] source that identifies Moses' first opponent by name. Artapanus wrote his story of Moses in Greek, most probably in Egypt and certainly before the first century BC, when Alexander Polyhistor included it in his *Peri Ioudaion*.[3] Even though he knew *Genesis* and *Exodus*, perhaps already in the translation by the LXX (Denis 1987: 64) he felt at liberty to retell in his own way the stories on biblical characters. It is evident that for Artapanus national tradition was still living and growing,[4] as it was, maybe to a lesser degree of liberty, for the authors of "rewritten Bibles" of Palestinian origin such as *Jubilees* or *Liber Antiquitatum Biblicarum*. While it is customary for scholars to use the Bible as a first source when discussing a phenomenon, I don't believe that there was a time or a place in which the Bible, as we have it today, was the *only* extant source of Jewish tradition, either religious or narrative.

Why Artapanus did choose to give a name to the anonymous Pharaoh of Exodus 1–2? The first motivation of Artapanus could have been to arouse interest in his tale. Just like a modern author of sensationalistic books or of TV programs on ancient mysteries, he knew the name of the culprit and the part of the story the Bible did not tell us. Moreover, as Christopher Vogler says, "a story is only as good as its villain" (Vogler 2007: 66): the stronger the personality of the *Shadow*, the opponent, the more intriguing the story. In the Bible this is especially true for Moses' second opponent, the Pharaoh of the plagues' narrative (Exod 7–13), who is treated by the author as a real character with his own personality and his tragic destiny—the "hardening of the heart" can be another name for Ate, the godsent blindness of classical tragedy.[5] As Niehoff remarks, by narrating Exodus from Egypt as a personal conflict between Moses and a villain Artapanus denies its interpretation as an ethnic conflict between the Hebrews and the native Egyptians (even in modern storytelling, focusing on the struggle between hero and villain understates the importance of political, economic, and social motivations) (Niehoff 2001: 73).

Generally speaking, one of Artapanus' strategies is to make more explicit themes that are already present in the Bible or, if we prefer, themes that were already part of that cluster of traditions which found a final, somehow condensed form in the biblical text of Exodus (Moro 2011: 9–12, 61–64). In my opinion, the most important and productive theme in all the narratives about Moses is the establishing of a new model of king/leader/judge who mediates between men and the Deity,[6] and who replaces the idea of sacred kingship common in preexilic Israel and Judah. Deuteronomistic history had discredited this idea but it was still at work in the story of Moses by Hecateus of Abdera (where Moses is a conquering king who founded a new city, like Omri, and built the dynastic temple like Jeroboam)[7] and, as we shall see, also in Artapanus. The story of Moses' adoption by an Egyptian princess, a reversal of the more common narrative motive of the adoption of the infant hero-to-be by a commoner,[8] is a sophisticated means to relate Moses' kingship to the

to reconstruct the intellectual history of Judaism, not historical events.

[2] Some scholars (e.g., Jacobson 2006) tried to demonstrate that Artapanus was not a Jew, but I do not find their arguments convincing.

[3] For an extensive bibliography on Artapanus, see Borgeaud et al. (2010).

[4] Unlike Gruen (2002: 201–212) I do not believe that Artapanus had humorous intentions.

[5] In the words of Propp's commentary (Propp 1998: 346): "it is clear that the cause of ruin is Pharaoh himself."

[6] See, e.g., Mathews (2012) on Moses as king in the Pentateuch (but I don't agree on the very high dates proposed by the author for the biblical texts in question).

[7] Diod. Sic. 1.40.3 (Stern 1974: 26–35; Gruen 1998: 55; Moro 2011: 40–59).

[8] On the tale type known as "the hero who was exposed at birth" see Lewis (1980); on Moses birth story see Zlotnick-Sivan (2004), Moro (2011: 60–104). Unlike Rendsburg (2006: 204–205) I do not think that the Egyptian connotations of the Moses' infancy story (including the allusions to the birth of Horus) are incompatible with the influence of this tale type, whose ultimate origins are Mesopotamian, and this influence goes beyond the story about the birth and affects other parts of Moses' career (e.g., Exod 2: 11–17; see Moro 2011: 96–99).

idealized version of Egyptian royal ideology that became so popular in the ancient world since the Persian conquest of the country.[9]

The book that conveys at best this utopian concept of Egyptian institutions was also the main literary source of Artapanus: we know it as the first book of Diodorus' *Bibliotheca Historica* (published after 36 BC) but he might have read it in a previous form, maybe as the original work on Egypt by Hecateus of Abdera (end of fourth century BC).[10] Artapanus uses this source to demonstrate that the best part of Egyptian, and then Greek culture, has Jewish origins. In the story by Artapanus young Moses was the teacher of Orpheus, the author of sacred poems who taught Egyptian religion to Hellenes, according to Diodorus (1.96.2–5), or the belief in one only God, according to the Jewish versions of the Orphic hymn (see Holladay 1996: 128–133, 196–218). Moses' victorious expedition in Ethiopia is similar to the war waged in far countries by Sesoosi (Sesostris I),[11] but also to the pacific expeditions of Osiris[12]: like the god of ancient times, Moses is loved by the people he conquered,[13] and instead of agricultural techniques he propagates the custom of circumcision.[14] We shall mention other parallels in the course of the chapter.

Moses as an Adoptive Son and a Royal Heir

The description of the first part of Moses' career has also the scope of showing his qualities as a honorable adoptive son and a worthy heir to the throne of Egypt. Young Moses helps Chenephres strengthen his kingdom in a time in which different dynasties share the power over Egypt:

> Now [Palmanothes] dealt meanly with the Jews. First he built *Tessan* [sic], then he set up the temple there. Later he built the sanctuary in Heliopolis. He fathered a daughter Merris, whom he betrothed to a certain Chenephres, a ruler of the regions above Memphis. At that time there were many rulers in Egypt (Eusebius *Praep. Ev.* 9.27.2–3).[15]

> When he reached manhood, he [Moses] bestowed on humanity many useful contributions, for he invented ships, machines for lifting stones, Egyptian weapons, devices for drawing water and fighting, and philosophy[16] (*Praep. Ev.* 9.27.4).

> He also divided the state into thirty-six [36] nomes, and to each of the nomes he assigned the god to be worshipped; in addition, he assigned the sacred writing to the priests. The god he assigned were cats, dogs and ibises. He did all these things for the sake of keeping the monarchy stable for Chenephres, for prior to this time the masses were disorganized and they would sometimes depose, sometimes install rulers, often the same person, but sometimes others (*Praep. Ev.* 9.27.5).

We see here an exact correspondence with the number of nomes in Sesostris' I times (Obsomer 1989: 43), as we read in Diodorus (in Roman Egypt there were 42), but also with the political motivations adduced by the historian:[17]

> He (Sesoosis) divided the whole country into six and thirty [36] districts, which the Egyptians call nomes. (Diod. Sic. 1.54.3)[18]

---

[9] See the description of Egyptian kingship in Diod. Sic. 1.70–72 and the [illegible] of Darius as an ideal king obeying divine law in Diod. Sic. 1.94.4–5 (Assmann 2002: 368); on Moses as an heir to the Egyptian throne see Moro, *Mosè erede al trono di Egitto nelle fonti giudeo-ellenistiche*, to be published in *Aegyptus*.

[10] On Hecateus of Abdera see Fraser (1972: 496–505), Burstein (1992), Zamagni (2010: 133–139). On the sources of Diodorus first book see Burton (1972: 1–34). For a list of parallels between Artapanus and Diodorus' first book see, e.g., Zellentin (2008: 35–46).

[11] Diod. Sic. 1.53–58. About the tradition on Sesostris and the historical figures behind it see, e.g., Burton (1972: 163–165), Lloyd (1982: 37–40), Obsomer (1989), Ladynin (2010). On Artapanus' Moses and Sesostris see Tiede (1972: 153–167).

[12] Diod. Sic. 1.17–20. On Osiris as a conquering king in the native tradition see the inscription from Dendera (first century BC) in Yoyotte (1977).

[13] Diod. Sic. 1.18.5–6 and 1.20.5.

[14] On the Egyptian origins of circumcision see Hdt. 2.104; Diod. Sic. 1.28.1–3; 1.55.5.

[15] English translations of Artapanus' fragments are from Holladay (1983).

[16] I wonder if we can infer here a particular meaning of "philosophy" (priestly expertise?), like in Philo *Vita Mosis* 1.23 (Moses was taught by Egyptian teachers on the "philosophy" of "sacred characters of hieroglyphics" and animal cult). See note 18.

[17] English translations of Diodorus' first book are from Murphy (1985).

[18] Grajetzki (2006: 42): "Senuseret seems to have reorganized or at least confirmed the frontiers of the

> In days of the ancient kings, when the people frequently conspired together and revolted against their sovereigns, a certain king who excelled in wisdom divided the country into a great number of districts and taught the inhabitants of each one to worship a certain animal, or not to eat a certain food; thus with each locality respecting its own fetishes but scorning those of the others, the people of Egypt as a whole would never be able to unite (Diod. Sic. 1.89.5).

> Thus, for these reasons Moses was loved by the masses, and being deemed worth of divine honor by the priest, he was called Hermes because of his ability to interpret [*hermeneia*] the sacred writing (*Praep. Ev.* 9.27.6).

The identity of Moses with Hermes (Thoth) that Artapanus attributes to Egyptian priests[19] parallels Diodorus' portrayal of this god:

> Above all others, Osiris esteemed Hermes, who was endowed with a superior natural faculty for inventing things of service to the human race. For this god was the first to bring language to perfection; he named many nameless things, invented the alphabet, and ordained ceremonies governing divine worship and sacrifices to the gods. (...) And he taught the Greeks eloquence [*hermeneia*], which is why he is called Hermes. In short, they told him to have been the sacred scribe of Osiris, the one to whom he confided all things[20] and on whose counsel he especially relied (Diod. Sic. 1.15–16).

The success of Moses causes a change of attitude in the king. Chenephres' envy for young Moses is very similar to Saul's for young David (1 Sam 18): in both cases, the king tries in vain to kill the young rival by sending him to a perilous military expedition.[21]

> When Chenephres saw the fame of Moses, he became jealous and sought to kill him on some reasonable pretext. Thus when the Ethiopian marched against Egypt, Chenephres, supposing that he had found the right moment, sent Moses against them as the commander of a force of troops. He conscripted a band of farmers for Moses, rashly supposing that Moses would be killed by the enemy because his troops were weak (*Praep. Ev.* 9.27.7).[22]

I think that this part of the story shows the strongest and deepest connection between Artapanus and the first book of *Bibliotheca Historica*: Chenephres' biggest fault is his lack of gratitude towards Moses, gratitude being the paradigmatic Egyptian value according to Diodorus.

> And they say the Egyptians in general are more gratefully disposed than the rest of the mankind for every benefit, holding that repayment of kindness to one's benefactors is of the greatest help in life: for it is evident that everyone will be eager to do kindness most especially to those whom they perceive will most graciously appreciate the favors. And it is apparently for this same reason that the Egyptians worship their kings and honor them as real gods (Diod. Sic. 1.90).

Moreover, Chenephres becomes king only because he married the daughter of the previous king[23]: he is presented, already from the onset, as less legitimate an heir than Moses who, as the adopted son of king Palmanothes' daughter,

---

Egyptian nomes. On the 'white chapel' of Karnak are listed all the provinces with their measurement, their main towns and their main deities." The religious ordinance of nomes was part of the instruction for temple's personnel and it's the subject of the so-called *Priestly Manual* and other texts from Tebtunis (Ryholt 2005: 149.161).

[19] Also according to Philo of Alexandria (*Vita Mosis* 1.27) and Josephus (*AgAp* 1.237) young Moses was deemed a god by the Egyptians. It's difficult to say if there is a relationship between this piece of information and the Jewish idea that Moses had a semidivine nature (see references in Moro 2010: 44 and Moro 2011: 136–139).

[20] This passage could remind Num 12:7 to a Jewish reader. See Yoyotte (1977: 148) for Thoth as vizier of Osiris.

[21] This is a very ancient tale type in the Ancient Near East: the first extant example of it is the so-called *Sumerian Sargon Legend*. Like David, Sargon is a young man living at court, chosen by the deity as future king and persecuted by Urzababa, king of Kish, who has received bad omens and feels that his power is at end because he lost the favor of the gods (translation in Cooper and Heimpel 1983). In Artapanus, Chenephres tries again to plot against his stepson, but the courtier appointed to assassinate Moses, Chanethothes, is killed by him in self-defense (*Praep. Ev.* 9.27.13–18; this episode is the author's version of Exod 2:11–15).

[22] See for comparison 1 Sam 18:21.25.

[23] In the Bible this seems to be the case with the second opponent of Moses: the discomfort of the Israelites at the death of the first Pharaoh (Exod 2:23) seems incomprehensible if we do not assume that Moses, the legitimate heir, has gone for good (cf. in *Sinuhe* the discomfort of the country at the death of Amenenhat, whose succession is similarly uncertain at this point of the narrative).

is a potential king.[24] This lack of proper legitimation makes Chenephres a villain also from an Egyptian point of view. About the identity of the second Pharaoh in Artapanus we know very little: in the beginning of the third fragment he says that the Egyptian princess can't conceive,[25] so the new king can't be a stepbrother of Moses.[26]

The fate of Egypt under the plagues can be considered, in Egyptian terms, as a "chaos from above," as Assmann calls it, i.e., a political and cosmic unrest caused by a kingship not conforming to divine law and will (Assmann 2002: 385). Artapanus never intends to negate the idea that the struggle to free Israel from Egypt is bound to be a violent one (God in the burning bush asks Moses to "wage war against Egypt"; *Praep. Ev.* 9.27.21), nor does he attempt to sugarcoat the tragedy of the plagues (the final earthquake is as cruel and indiscriminate a punishment as the death of the firstborn; *Praep. Ev.* 9.27.33), but he wants to emphasize the point that Moses did it all according to Egyptian rules: he is a divinely appointed king trying to rescue the country, as the Nubian Piye-Pi'ankhy did (see Gardiner 1961: 335–340), even though he will do it as a miracle worker[27] and not as a warrior—as his father-in-law would prefer, according to Artapanus.

> Reguel wanted to wage war against the Egyptian because he wanted to return Moses from exile and thereby establish the throne for his daughter and son-in-law (*Praep. Ev.* 9.27.19).[28]

The Historical Pharaoh

Various scholars[29] identify Chenephres with Khaneferra, the royal name (*nsw bity*) of Sobekhotep IV of the XIII dynasty,[30] a king unknown to Greek historiography (including Manetho)[31] but mentioned by contemporary documents and Egyptian king lists.[32] Artapanus might have used a source lost to us, perhaps a kind of fictionalized biography of this sovereign. Sobekhotep IV is generally considered the last great "king of Upper and Lower Egypt" of the Middle Kingdom before foreign dynasties, the so-called Hyksos, took possession of the whole country. Until the new findings of seal impressions of Khaneferra and Khayan (a *ḥḳ3 ḫ3swt* or *heqa khasut* of the XV dynasty) in the same archeological stratum at Tell Edfu, scholars considered his reign to be contemporary with the Delta's kinglets of foreign origins of the XIV dynasty.[33] During his career King Khayan abandoned the title of "prince of the foreign lands" for a full royal titulary (with praenomen and nomen); according to Ryholt, Khayan was the predecessor of King Apophis, defeated by Ahmose of Thebes (Ryholt 1997: 118–121).

Ryholt also writes that this change of names was "presumably connected with his conquests, eventually of the whole of Egypt, by which he became its sole ruler" (ibid.: 304). Other researches on the Second Intermediate Period pottery have, nevertheless, questioned the

[24] Hendel (2001: 618) (on biblical Moses). On king's daughters' right of succession in Egypt see Troy (1986: 107–114).

[25] See also Josephus *Ant* 2.232 and Philo of Alexandria *Vita Mosis* 1.13.

[26] See Holladay (1983: 218–219) for the quotation of Clement of Alexandria (*Stromateis* 1.23[154]2–3) that calls the second opponent Chenephres.

[27] See in this volume the contribution by Gary Rendsburg, Chap. 18. In Artapanus' account of the plagues the points of contact with the story of Setne Khaemwas are even stronger than in Exodus.

[28] This corresponds to Reguel's interpretation of Moses' prophetic dream in the *Exagoge* (vv. 84–85): Moses is going "to overthrow a big throne," that is, to supplant the Pharaoh as legitimate heir.

[29] Among them Waddell (1964: 72–73), Fraser (1972: 704–706), Grajetzki (2006: 72–73), Ryholt (1997: 352). Weill fancied a story of Khaneferra fighting the Hyksos (Weill 1918: 722–728).

[30] On the times and family of Sokehotep IV see Franke (1994: 69–77), Habachi (1981b), Grajetzki (2006: 68–75), Ryholt (1997: 69–93, 296–299).

[31] On Manetho see Moyer (2011: 84–141).

[32] The preserved fragments of Manetho give no names for the kings of the XIII dynasty ("sixty kings from Diospolis, who reigned for 453 years"; Waddell 1964: 72–75). The names are preserved in the *Turin King List*, columns VI and VII (see transcription and discussion in Ryholt 1997: 69–75).

[33] Moeller et al. (2011). This synchronism is discussed in this volume also by Michael Dee (Chap. 6).

likelihood of a real massive occupation of Memphis by West-Semitic rulers and population (Bader 2008). The loss of power of the XIII dynasty kings seems to have begun under Mennefer Iy, a successor of Sebekhotep IV, whose name is found only on objects from Southern Egypt (Grajetzki 2006: 73–75; see also Mynářová 2007: 70). The reign of Khaneferra, as the reigns of his two brothers before him, was a steady one, if compared with other kings of the same dynasty. It is evident that a new chronological and historical assessment of this period is desirable, and that Artapanus cannot be used at face value as a historical source to claim that Khaneferra reigned in a time of relative peace between two dynasties, a West-Semitic one in the Delta and an Egyptian one in Memphis. Artapanus has an ideological point to make: the king was powerful and prosperous before plotting to eliminate Moses, who deserved all the credit for the stability of his throne (*Praep. Ev.* 9.27.5).

The extant witnesses from Sobekhotep's times attest a relative prosperity of the country and a great commitment of the sovereign to the cult of Amun in Thebes. From the inscription on the statue of the vizier Iymeru (Neferkara) from Karnak (now at the Louvre) we learn that King Sobekhotep[34] erected in Karnak a "Temple of Million-of-years", a kind of sanctuary connected with coronation rituals or with the Sed Jubilee.[35] The king granted Iymeru that statue when the canal connecting the temple with the Nile was opened, maybe on the last inaugural ceremony:

> (front) The chief of the town and vizier, the overseer of the Six Great Mansions, Neferkare-Iymeru.
> (dorsal pillar) Given as a favor form the king to the prince and governor, the chief of the town and vizier, the overseer of the Six Great Mansions, Neferkara Iymeru, possessor of honour (*vacat*) (2) after the great opening of the canal, giving the house to his lord in the Temple-Million-Years (called) May-the-ka-of-Sobekhotep-be-Pacified (*vacat*).[36]

Sobekhotep's longest extant inscription is Cairo E 51911, a royal decree on the construction of two precious doors of cedar wood for the Amun temple in Karnak and the institution of a new *ꜥ3bt* offering. (Text in Helck 1975: 31–34; Miosi 1981: 4–11; text and transl. Helck 1969.) An interesting passage exalts the Theban roots of the king—despite the fact that, like XII dynasty kings (Hayes 1947: 10–11; Gardiner 1961: 154; Ryholt 1997: 79), he reigned from Itjet-Tawi—and his devotion to the local god.

> My majesty [came] to the Southern City since I wanted to see the august god; it's my city, it's in it that I was born, (lacuna), I saw the vigor of his majesty (sc. Amun) at every single feast when I was a child without discernment. (Commentary in Ryholt 1997: 226; Vernus 1989)

Even though Artapanus treats Chenephres as a villain, his Egyptian source probably depicted him as a hero. In fact, his name was remembered in New Kingdom's times as a revered predecessor of Tuthmosis (Thutmose) III in the list of Karnak (Redford 1986: 29–34). The list (now at the Louvre) was inscribed in the SW corner of the complex known as *3ḫ-mnw*. The royal ancestors, not in chronological order, are pictured in the form of seated statues. According to Wiedermann and Lacau, the kings included had reigned at Thebes, or at least left traces of building activity in the temple (as in the case of Sobekhotep IV) (ibid.: 31). The purpose of the chamber was stated in the dedication: "to inscribe the names of the fathers, to set down their offering portions, to fashion their imagines in all their likeness, and to offer to them great, divine oblations" (*Urk.* IV 607: 8–11; English transl. Redford 1986: 32). In the list of kings from the tomb of the priest Tjuloy in Saqqara, which dates to the reign of Ramesses II, the praenomen Khaneferra is recorded in the place of Neferefre of the V dynasty, who had an

[34] According to the inscriptions on the statue of Heidelberg, the Sobekhotep he served was the fourth of this name (Habachi 1981a: 261–263; see also Grajetzki 2012: 38–39).

[35] Gabolde (1998: 148). This kind of temple appears also in the dedicatory inscription of the temple of Tuthmosis III in Karnak.

[36] English translation by Habachi (1981a: 263–265). In another statue from Karnak, Iymeru calls himself "the one who fills the heart of Maat in restraining the patricians and in humiliating the rebels" (ibid.: 267).

ephemeral reign.[37] The praenomen and nomen of Sobekhotep IV are also inscribed on a great number of amulets, mostly in the form of scarab-seals, dating to the New Kingdom (Matouk 1971: 38, 207–208; Ryholt 1997: 352). A late scarab published by Petrie commemorated Tuthmosis IV and Khaneferra-Sobekhotep together.[38] Sobekhotep is remembered as a royal ancestor also in Tanis (XI century BC), where three ancient statues of this ruler were erected (Redford 1986: 56; Grajetzki 2006: 72).

Khaneferra was part of a family of usurpers: he and his two brothers and predecessors were sons of commoners (see Franke 1984: 260–261 on Khaneferra's father). They were apparently proud of it and stressed this lack of royal ancestry in the so-called genealogical scarabs, where their parents are called simply "father of the god" and "mother of the king" (Ryholt 1997: 284–286). Maybe they were trying to dissociate themselves from his predecessor, Seth, who had his name erased from all his monuments after his death (ibid.). If the biography of Khaneferra mentioned his nonroyal origins, that would result functional to a main point of Artapanus: Moses was a legitimate member of a royal family (being the son of the daughter of a king) more than Chenephres.

The Temple in Karnak as Example of Idealization and as Chronological Clue

The biography we have postulated as a source of Artapanus might have exalted Khaneferra Sobekhotep IV beyond his real merits, for example crediting to him the destruction of a temple in mud bricks in Karnak (*Diospolis*) and its reconstruction in stone.

> When the war was over, Chenephres welcomed him back in words but plotted against him in deeds. In fact, after taking away Moses' troops, Chenephres sent some of them to the borders of Ethiopia as a defense garrison and ordered others to destroy the temple in Diospolis. This temple was constructed with baked bricks, but he ordered them to build another one of stone quarried from the mountain nearby (*Praep. Ev.* 9.27.11).[39]

The real founder of the "great castle in beautiful limestone" of Karnak on the site of an earlier brick temple was Kakheperra Sesostris I, and his name was remembered for a long time: in the New Kingdom temple there was a copy of the relief of south wall, with his name (Gabolde 1998: 36) and the historical text on the foundation of the *ḥwt ꜥ3t*.[40] The memory of the founder was kept in Karnak at least until the times of Ramesses IX (XII century BC), when a high priest remembers Sesostris I as the builder of priestly house near the sacred lake (Gabolde 1998: 122). So, while I am tempted to date the composition of our pseudo-biography of Khaneferra to the New Kingdom, the blurring of names in this part of the narrative is too difficult to explain at a date as high as this, and the element is too important in a king biography to dismiss it as a later addition.[41] On the other hand, the first book of Diodorus shows us that in the Late Period there was a great confusion about the identity of (historical or mythical) builders of monuments: the foundation of the Amun temple in Karnak is attributed to Osiris (Diod. Sic. 1.15.3) or to mythical king Busiris (1.45.4).[42] Regarding Sesoosis (Senusret) Diodorus condenses his impressive building activity saying that he built in every town of Egypt a sanctuary to the most revered god (1.56.2).[43] It is likely that the biography of Khaneferra

[37] Ryholt (1997: 352). On this list, now in Cairo Museum, see Redford (1986: 21–24).

[38] Petrie (1976, no. 956) (this author calls them Tuthmosis III and Sobekhotep III).

[39] The limestone for the Sesostris' temple came from Tura, but here probably Artapanus is alluding to the leprous working in stone quarries in Manetho (*AgAp* 1.235–237).

[40] Gabolde (1998: 40–41, 163–164). The text, now fragmentary, must have been similar to literary works such as the *Leather Scroll* from Berlin, which narrates the foundation of a temple in Heliopolis by Sesostris I (De Buck 1938; Goedicke 1974).

[41] Among royal names in Late Period literature, many are those associated with construction or enlargement of temples in Thebes (Ryholt 2009: 232).

[42] Busiris in Diod. Sic. 1.17.3 is a companion of Osiris, in 1.67.11 and 1.88.5 the enemy of Heracles.

[43] On the impressing list of temples built or restored under the rule of Sesostris I, see Grajetzki (2006: 36–41).

originally narrated also other traditional "Sesostrian" deeds, attributed to Moses by Artapanus, such as political and religious reforms or victorious military campaigns in the South.[44]

The Historical Novel in I Millennium BC Egypt

Uncertain attributions and confusion of historical names were at home in Egyptian historical and pseudo-historical literature of Late Period,[45] now witnessed by an increasing number of documents, also in Aramaic, like AP 71 (maybe from Saqqara), with a story about Hor Bar Punesh and a king,[46] and the text painted in the tomb of Sheik-Fadl,[47] very fragmentary, that mentions Taharqa ("king of Nubians"), Necho, Psammeticus and Inaros, alongside Heliopolis, the god Atumnebu (Atum the Lord) and an Amun festival.[48] In the Tebtunis library, the only temple library of Ancient Egypt from which extensive remains have survived, all the literary-narrative material concerns historical people or events (Ryholt 2009: 231). Since now this kind of late Egyptian literature (with the exception of Setne Khaemwase) has received little or no attention by biblical scholars, but I think it might have inspired, in terms of borrowed material and genre, not only Artapanus but also biblical authors who wanted to give a historical veneer to their narratives.

Even if the most represented characters seem to be the kings of recent past, the name of Sesostris, the king who according to Manetho "was esteemed by the Egyptians as the next in rank to Osiris" (fragments 34–36: Waddell 1964: 66–73), appears in two Roman Period Demotic papyri, P. Carlsberg 411 and 412, in quite fragmentary state, in which is mentioned a "royal son" and several geographic names—most of the text seems to deal with Nubia and Ethiopia, but there are scattered allusion to other countries such as Syria and Arabia (Widmer 2002: 387–393). Even if the documents are poorly preserved, they prove the existence of a native tradition on this king and his military deeds.[49] Since I cannot find among texts published or announced any trace of "my" novelized biography of Chenephres, yet I can fancy for it a similar environment and an audience composed of both Egyptians and foreigners.

By the way, I always thought that Nacheros, the name of the functionary who presided to the reconstruction of the Amun temple in Artapanus, was a deformation of Neferkara (Iymeru), the vizier of Khaneferra mentioned above. Still this name is also reminiscent of ינחרו, the Aramaic spelling[50] for Inaros, the rebel against foreign rule of Late Egyptian tradition.[51]

Manetho's Enemies and Sacred Animals

> The Egyptian versions (of Exodus) are notable for their almost phobic concern with purity and their meticulous care for the cultic images and the sacred animals. (Assmann 2002: 401)

[44] On the control on Nubia by XIII dynasty (till Sobekhotep IV) see Ryholt (1997: 77, 91–93), Franke (1994: 75), Grajetzki (2006: 72).

[45] See the examples in Ryholt (2009: 234–235). Note Sesostris *(s-n-wsr.t)* who becomes *s-ws*, the Sesoosi of Diodorus. "Several of the names are so garbled that it would not have been obvious which historical king pertained to. (...) This situation may help explain the garbled chronology encountered in many historical narratives" (p. 235). See also Ryholt (2011b: 127) ("It may be noted that several of these stories confuse historical details and thus come to represent a *pot pourri* of historical information").

[46] Text and translation in Porten and Yardeni (1993: 54–57); see also Porten (2004).

[47] A village situated 186 km south of Cairo.

[48] According to Lemaire (1995: 110–111), the text of Cheick Fadl was composed before 591 BC, when Psammeticus II waged war against Ethiopians and the name of Taharqa was subject to *damnatio memoriae*.

[49] In 1982 Lloyd wrote that "the Sesostris Romance does not survive in any Egyptian source" (1982: 37).

[50] E.g., in the inscription of Sheik-Sadl (Ryholt 2004: 496; Holm 2007: 201–202).

[51] On the various historical figures bearing this name and their relationship with literature, see Holm (2007: 202–220), Ryholt (2009: 236–238). According to Diodorus Inaros built the pyramid of Mycerinus (1.64.13).

In deep contrast with the highly critical attitude against animal worship expressed by Jewish authors,[52] or even by pagan authors like Strabo,[53] Artapanus attributes to Moses the foundation of animal cult in Egypt:

> Thus, Moses and those with him, because of the size of army, founded a city in this place, and they consecrated the ibis in the city because of its reputation for killing those animals that were harmful to men. They named it "the city of Hermes" (*Praep. Ev.* 9.27.9).

> When he came with Moses to Memphis, Chenephres inquired from him whether there was anything else useful to mankind, and he suggested a breed of oxen because of their usefulness in tilling the land. Chenephres named a bull Apis and commanded the people to dedicate a temple to it. He also ordered that the animals which had been consecrated by Moses be brought there and buried, wishing thereby to conceal the ideas of Moses (*Praep. Ev.* 9.27.12).

The author represents this foundation as dictated by Moses' political cunning (*Praep. Ev.* IX 27, 5), or by desire to honor and show gratitude to useful creatures, and Egyptian animal cult as a misunderstanding of his intentions (*Praep. Ev.* 9.27.12).[54] Once again Artapanus' argument is very similar to what we can read in the large section of *Bibliotheca Historica* dedicated to animal worship (1.83–90).[55] At any rate, the presence and the importance of this theme, along with the previous considerations on the identity of Chenephres, seems to indicate that Artapanus knew a narrative on Exodus written by Manetho which was very similar to the version we have in Josephus' *Contra Apionem*, presenting *two* connected stories, one about the Hyksos conquest and the fight of the Egyptians for freedom, and another on the leprous Egyptians and their Hyksos allies, where we read about the order to sacrifice the sacred animals and about the devastation of temple and the killing of sacred animals during the following war.

> (*AgAp* 1.239)
> First of all, he [Osarsiph] made it a law that they should neither worship the gods nor refrain from any of the sacred animals prescribed as especially sacred in Egypt, but should sacrifice and consume all alike. (Stern 1974: 82)

> (*AgAp* 1.249)
> For not only did they [the Solymites] set towns and villages on fire, pillaging the temples and mutilating images of the gods without restraint, but they also made a practice of using the sanctuaries as kitchen to roast the sacred animals which the people worshipped; and they would compel the priests and prophets to sacrifice and butcher the beasts, afterwards casting the man forth naked.[56]

I suggest that Artapanus did not want to promote a form of syncretism, but an attitude of general respect for sacred animals, that in Egypt had replaced pharaoh in representing the divine presence in a living being (Assmann 2002: 407–408). A benevolent attitude towards animal worship was also part of the Ptolemies' effort to be considered as legitimate pharaohs by the Egyptians.[57] In Artapanus' times there might already have been exegetes who propounded aggressive interpretations of Jewish ritual and claimed, for example, that the sacrifices of the Hebrews, and especially the slaughter of the *pesach* lamb, were hostile acts towards the Egyptian gods and their sacred animals. We find this notion quoted in Jewish sources as an interpretation of Exodus 8:22, as in Philo *De fuga*, the Palestinian Targums,[58] and the Vulgate (v. 26).

---

[52] Smelik and Hemelrijk (1984: cols. 1910–1920). On Philo see Pearce (2007: 279–308).

[53] According to Strabo (*Geographica* 16.2.35), Moses was an Egyptian priest who abandoned his native religion since he disapproved of the representation of gods as animals (Stern 1974: 294).

[54] I see no contradiction with this attitude and *Praep. Ev.* 9.27.35: if sacred animals supposedly carried by Pharaoh's retinue drowned in the Red Sea (the context is corrupt; cf. *AgAp* 1.243), they would be victims of king's *hybris* as well as Egyptian population.

[55] See especially 1.87.2 (for the bull) and 1.87.6 (for the ibis). Burton (1972: 248–253), Smelik and Hemelrijk (1984: cols. 1898–1903), Pearce (2007: 250–253).

[56] Stern (1974: 83). In Artapanus, on the contrary, it is Pharaoh (the second, anonymous one) that has a blasphemous attitude, threatening Memphite priests with death and destruction of their temples unless they can duplicate the miracles of Moses (*Praep. Ev.* 9.27.30).

[57] Smelik and Hemelrijk (1984: cols. 1891–1895).

[58] Díez Macho (1980: 46:47) (*Pseudo-Jonathan*: "we are going to take and sacrifice before the Lord our God the lambs, that are the idols of Egyptians").

> *We will sacrifice the abominations of Egypt to the Lord our God* (Ex 8:22): for victims perfect and free from blemish are the virtues and virtuous conduct, and these the Egyptian body, in its devotion to the passions, abominates. For even as in this passage, understood in accordance with reality, things which Egyptians reckon profane are called sacred in the estimation of the keen-sighted [the Israelites], and are all offered in sacrifice (Philo *De fuga* 18–19).[59]

> Et ait Moses: Non potest ita fieri abominationes enim Aegyptiorum immolabimus Domino Deo nostro; quod si mactaverimus ea quae colunt Aegyptii coram eis lapidibus nos obruent (Ex 8:26).

The origin of this interpretation could have been an appropriation, on Jewish part, of the concept of "inverted sacrifice" (*AgAp* 1.239): here in Philo, being different from the Egyptians receives a positive instead of a negative connotation.[60] Still, I do not believe that Artapanus fails to mention the institution of Passover out of respect for animal worship: the sacrifice of the *pesach* lamb seems to have been also the object of internal polemics in Second Temple Judaism.[61]

Not all the scholars agree that Artapanus' work was a refutation of Manetho: I think that his use of Egyptian history proves his intention to create a story in which the Exodus from Egypt has nothing to do with the Hyksos conquest, an event still perceived as the first archetypical violation of Egyptian independence (and of order, kingship, religious ordinance), or with the Theban rescue of the land. Also the memory of famines and pestilence in the last part of the XIII dynasty (Ryholt 1997: 300–301), either narrated in his source or mentioned in a prophecy of doom, might have served well this shifting of chronology: to Artapanus it must have sounded like a memory of the plagues. With his synchronism, Artapanus tried to create a relative chronology for the Exodus different from the chronology of Manetho (and also higher than his, but this doesn't seem to have been the main point) or of Ptolemy of Mendes.[62] Probably he used the same tools as the Egyptian priest-historian: king lists, chronicles, and popular tales on the past. I think that Artapanus may have perceived the polemical use of Hyksos' story as a serious stumbling block for people who want to stay at peace probably not (or not only) with Egyptian population, but with their inner Egyptian, the part of themselves that wanted to be part of this cultural ideal, an Utopia that authors like Herodotus and Hecateus contributed to create.

[59] English transl. Colson (1949: 19). See comment in Pearce (2007: 123). An article on this argument (*Il sacrificio ebraico come "abominio degli Egiziani"*) will appear in the second issue of the review *Rationes Rerum*.

[60] See Niehoff (2001: 45–74). In Philo "the Egyptians emerge as the ultimate Other whose perversion implicitly defines the positive characteristics of the Jews" (ibid.: 46).

[61] While Philo of Alexandria thought that it could be performed at home by non-priests (see, e.g., *Spec.* 2.145–6, 148) other texts like *Jubilees* 49 and *Temple Scroll* XVII 6–9 forbade it outside the Jerusalem Temple. The *Liber Antiquitatum Biblicarum*, too, doesn't mention Passover in its chapter on Exodus from Egypt.

References

Assmann, Jan. 2002. *Mind of Egypt: History and Meaning in the Time of the Pharaohs*. Trans. Andrew Jenkins from German. New York: Metropolitan Books.

Bader, Bettina. 2008. Avaris and Memphis in the Second Intermediate Period in Egypt (*ca.* 1770–1550/40 B.C.). In *Proceedings of the 5th International Congress on the Archaeology of the Ancient Near East. Madrid, April 3–8 2006*, ed. Joaquín Mª. Córdoba et al., 207–223. Madrid: UA Ediciones.

Borgeaud, Philippe, Youri Volokhine, and Thomas Römer (eds.). 2010. *Interprétations de Moïse. Judée, Egypte, Grèce et Rome*. Leiden: Brill.

Burstein, Stanley. 1992. Hecataeus of Abdera's History of Egypt. In *Life in a Multi-Cultural Society: Egypt from Cambyses to Constantine and Beyond*, ed. Janet H. Johnson, 45–49. Chicago, IL: Oriental Institute.

Burton, Anne. 1972. *Diodorus Siculus Book I: A Commentary*, Études préliminaires aux religions orientales dans l'empire romain, vol. 29. Leiden: Brill.

Colson, Francis H. 1949. *Philo with an English Translation V*. Cambridge, MA: Harvard University Press.

[62] Stern (1974: 391). According to this author, the Exodus from Egypt took place under the reign of Amosis (Ahmose).

Cooper, Jerrold S., and Wolfgang Heimpel. 1983. The Sumerian Sargon Legend. *Journal of the American Oriental Society* 103: 67–82.

De Buck, Adriaan. 1938. The Building Inscription of the Berlin Leather Scroll. In *Studia Aegyptiaca I*, Analecta Orientalia, vol. 17, ed. A. De Buck, 47–57. Roma: Pontificium Institutum Biblicum.

Denis, Albert-Marie. 1987. Le portrait de Moïse par l'antisémite Manéthon (IIIe s. av. J.-C.) et la réfutation juive de l'historien Artapan. *Muséon* 100: 49–65.

Díez Macho, Alejandro. 1980. *Biblia Polyglotta Matritensia. Series IV. Targum Palaestinense in Pentatheucum. Additur Targum Pseudojonatan ejusque Hispanica versio L. 2 Exodus*. Matriti: Consejo Superior de Investigaciones Científicas.

Franke, Detlef. 1984. *Personendaten aus dem mittleren Reich (20.–16. Jahrhundert v. Chr.): Dossier 1–796*, Aegyptologische Abhandlungen, vol. 41. Wiesbaden: Harrassowitz.

———. 1994. *Das Heiligtum des Heqaib auf Elephantine*, Studien zur Archäology und Geschichte Altägyptens, vol. 9. Heidelberg: Heidelberg Orientverlag.

Fraser, Peter M. 1972. *Ptolemaic Alexandria*, 3 vols. Oxford: Clarendon Press.

Gabolde, Luc. 1998. *Le «Grand château d'Amon» de Sé sostris Ier à Karnak. La décoration du temple d'Amon-Rê au Moyen Empire*. Paris: Boccard.

Gardiner, Alan H. 1961. *Egypt of the Pharaohs: An introduction*. Oxford: Clarendon.

Goedicke, Hans. 1974. The Berlin Leather Roll (P Berlin 3029). In *Festschrift zum 150jährigen Bestehen des Berliner ägyptischen Museums*, ed. W. Müller, 87–104. Berlin: Akademie Verlag.

Grajetzki, Wolfram. 2006. *The Middle Kingdom of Ancient Egypt: History, Archaeology and Society*. London: Duckworth.

———. 2012. *Court Officials of the Egyptian Middle Kingdom*. London: Bristol Classical Press.

Gruen, Eric S. 1998. *Heritage and Hellenism. The Reinvention of Jewish Tradition*. Berkeley, CA: University of California Press.

Gruen, Erich S. 2002. *Diaspora: Jews Amidst the Greek and the Romans*. Cambridge, MA: Harvard University Press.

Habachi, Labib. 1981a. New Light on the Vizier Iymeru, Son of the Controller of the Hall, Iymeru, in Habachi 1987: 261–271. Reprint of *BIFAO* 81: 29–39.

———. 1981b. New Light on the Neferhotep I Family, as Revealed by Their Inscriptions in the Cataract Area, in Habachi 1987: 279–287. Reprint of William K. Simpson and Whitney M. Davis, 77–81, *Studies in ancient Egypt, the Aegean, and the Sudan: Essays in Honor of Dows Dunham on the Occasion of his 90th Birthday, June 1, 1980*. Boston, MA: Department of Egyptian and Ancient Near East Art, Museum of Fine Arts.

———. 1987. *Studies on the Middle Kingdom*, Studia Aegyptiaca, vol. 10. Budapest: Université Lorànd Eötvös, Chaire de l'Égyptologie.

Hayes, William C. 1947. Horemkha'uef of Nekhen and his Trip to Iṯ-Towe. *Journal of Egyptian Archeology* 33: 3–11.

Helck, Wolfgang. 1969. Eine Stele Sebekhoteps IV. Aus Karnak. Mitteilungen des deutschen archäologischen Instituts. *Abteilung Kairo* 24: 194–200.

———. 1975. *Historisch-biographische Texte der 2. Zwischenzeit und neue Texte der 18. Dynastie*. Wiesbaden: Otto Harrassowitz.

Hendel, Ronald S. 2001. The Exodus in Biblical Memory. *Journal of Biblical Literature* 120: 601–622.

Holladay, Carl R. 1983. *Fragments from Hellenistic Jewish Authors*, Historians, vol. 1. Atlanta, GA: Scholars Press.

———. 1996. *Fragments from Hellenistic Jewish Authors*, Orphic hymns, vol. 4. Atlanta, GA: Scholars Press.

Holm, Tawny L. 2007. The Sheikh Fadl Inscription in Its Literary and Historical Context. *Aramaic Studies* 5: 193–224.

Jacobson, Howard. 2006. Artapanus Iudaeus. *Journal of Jewish Studies* 42: 210–221.

Ladynin, Ivan L. 2010. Sesostris–Sesonchosis–Sesoosis: The Image of the Great King of the Past and Its Connotations of the Lybian Time in Egypt. *Cultural Heritage of Egypt and Christian Orient* 5: 122–142.

Lemaire, André. 1995. Les inscriptions araméennes de Cheick-Fadl (Égypte). In *Studia Aramaica: New Sources and New Approaches*, Journal of Semitic Studies Supplement, vol. 4, ed. Markham J. Geller et al., 77–132. Oxford: University Press.

Lewis, Brian. 1980. *The Sargon Legend. Study of the Akkadian Text and the Tale of the Hero Who Was Exposed at Birth* (American School of Oriental Research. Dissertation Series 4). Cambridge MA: American School of Oriental Research.

Lloyd, Alan B. 1982. Nationalist Propaganda in Ptolemaic Egypt. *Historia* 31: 33–55.

Mathews, Danny. 2012. *Royal Motifs in the Pentateuchal portrayal of Moses*, Library of Hebrew bible/Old testament studies, vol. 571. New York: Bloomsbury.

Matouk, Fouad S. 1971. *Corpus du scarabée égyptien*, Les scarabées royaux, vol. 1. Beirut: Matouk.

Miosi, Frank T. 1981. *A Reading Book of Second Intermediate Period Texts*, The Society for the Study of Egyptian Antiquities, vol. 9. Toronto: Benben Publications.

Moeller, Nadine, Gregory Marouard, and Natasha Ayers. 2011. Discussion of Late Middle Kingdom and Early Second Intermediate Period History and Chronology in Relation to the Khayan Sealings from tell Edfu. *Egypt and the Levant* 21: 87–121.

Moro, Caterina. 2010. L'historien Artapan et le passé multiethnique. In *Interpretations de Moise*, ed. P. Borgeaud et al., 43–56. Leiden: Brill.

———. 2011. *I sandali di Mosè. Storia di una tradizione ebraica*, Studi Biblici, vol. 167. Brescia: Paideia.

Moyer, Ian S. 2011. *Egypt and the Limits of Hellenism*. Cambridge: University Press.

Murphy, Edwin. 1985. *Diodorus* on Egypt. Trans. from the Ancient Greek by Edwin Murphy. Jefferson, NC: McFarland.

Mynářová, Jana. 2007. Thebes in the Second Intermediate Period. In *Thebes: City of Gods and Pharaohs*, ed. Pavel Onderka, 70–79. Praha: Národní Muzeum.

Niehoff, Maren R. 2001. *Philo on Jewish Identity and Culture*, Texts and Studies in Ancient Judaism, vol. 86. Tübingen: Mohr Siebeck.

Obsomer, Claude. 1989. *Les campagnes de Sésostris dans Hérodote: Essais d'interprétation du texte grec à la lumière des réalités égyptiennes*. Bruxelles: Connaissance de l'Egypte ancienne.

Pearce, Sarah J.K. 2007. *The Land of the Body: Studies in Philo's Representation of Egypt*, Wissenschaftliche Untersuchungen zum Neuen Testament, vol. 208. Tübingen: Mohr Siebeck.

Petrie, William M.F. 1976. *Historical Scarabs: A Series of Drawings from the Principal Collections*. Reprinted Chicago: Ares.

Porten, Bezalel. 2004. The Prophecy of Hor bar Punesh and the Demise of Righteousness. An Aramaic Papyrus in the British Library. In *Res Severa Verum Gaudium. Festschrift für Karl-Theodor Zauzich zum 65. Geburststag am 8. Juni 2004*, Studia Demotica VI, ed. F. Hoffmann and H.J. Thissen, 427–466, plates XXXV–XXVI. Leuven: Peeters.

Porten, Bezalel, and Ada Yardeni. 1993. *Textbook of Aramaic Documents from Ancient Egypt*, Literature, Accounts, Lists, vol. 3. Jerusalem: The Hebrew University.

Propp, William H.C. 1989. *Exodus. A New Translation with Introduction and Commentary*, The Anchor Bible, vol. 2. New York: Doubleday.

Redford, Donald B. 1986. *Pharaonic King-Lists, Annals and Day-books: A Contribution to the Study of the Egyptian Sense of History*. Mississauga, ON: Benben.

Rendsburg, Gary A. 2006. Moses as Equal to Pharaoh. In *Text, Artifact, and Image: Revealing Ancient Israelite Religion*, Brown Judaic Studies, vol. 346, ed. Gary M. Beckman and Theodore J. Lewis, 201–219. Providence: Brown Judaic Studies.

Ryholt, Kim. 1997. *The Political Situation in Egypt during the Second Intermediate Period (C. 1800–1550 B. C.)*, Carsten Niebuhr Institute Publications, vol. 20. Copenhagen: Museum Tusculanum Press.

———. 2004. The Assyrian Invasion of Egypt in Egyptian Literary Tradition: A Survey of the Narrative Source Material. In *Assyria and Beyond: Studies Presented to Mogens Trolle Larsen*, ed. Jan G. Dercksen, 484–511. Leiden: Nederlands Instituut voor het Nabije Oosten.

———. 2005. On the Contents and Nature of the Tebtunis Temple Library: A Status Report. In *Tebtynis und Soknopaiu Nesos: Leben im römerzeitlichen Fajum*, ed. S. Sandra Lippert and Maren Schentuleit. Wiesbanden: Harrassowitz.

———. 2009. Egyptian Historical Literature from Greco-Roman Period. In *Das Ereignis: Geschichtsschreibung zwischen Vorfall und Befund*, Internet-Beiträge zur Ägyptologie und Sudanarchäologie, vol. 10, ed. Martin Fitzenreiter, 231–238. London: Golden House.

———. 2011a. New Light on the Legendary King Nechepsos of Egypt. *Journal of Egyptian Archeology* 97: 61–72.

———. 2011b. King Necho I Son of King Tefnakhte II. In *Von Theben nach Giza. Festmiszellen für Stefan Grunert zum 65. Geburtstag*, Göttinger Miszellen, vol. 10, ed. Frank Feder et al., 123–127. Göttingen: Hubert.

Smelik, Klaas A.D., and Emily A. Hemelrijk. 1984. "Who Knows Not What Monsters Demented Egypt Worship?" Opinions on Egyptian Animal Worship in Antiquity as Part of the Ancient Conception of Egypt. In *Aufstieg und Niedergang der Römishen Welt II: Principat VII.4*, ed. Wolfgang Haase, cols. 1852–2000, 2337–2357. Berlin: De Gruyter.

Stern, Menahem. 1974. *Greek and Latin Authors on Jews and Judaism*, From Herodotus to Plutarch, vol. 1. Jerusalem: The Israel Academy of Sciences and Humanities.

Tiede, David L. 1972. *The Charismatic Figure as Miracle Worker*, Society of Biblical Literature Dissertations Series, vol. 1. Missoula, MT: Society of Biblical Literature.

Troy, Lana. 1986. *Patterns of Queenship in Ancient Egyptian Myth and History*, Acta Universitatis Upsaliensis. Boreas, vol. 14. Uppsala: Almqvist & Wiksell.

Vernus, Pierre. 1989. Sur deux inscriptions du Moyen Empire. *Bulletin de la Société d'Égyptologie de Genève* 13: 178–181.

Vogler, Christopher. 2007. *The Writer's Journey. Mythic Structure for Writers*[3]. Studio City: Michael Wise Productions.

Waddell, William G. 1964. *Manetho*, Loeb Classical Library. London: Heinemann.

Weill, Raymond. 1918. *La fin du moyen empire égyptien*, 2 vols. Paris: Pichard.

Widmer, Ghislaine. 2002. Pharaoh Maâ-Reê, Pharaoh Amenemhat and Sesostris: Three Figures form Egypt's Past as Seen in Sources of the Graeco-Roman Period. In *Acts of the Seventh International Conference of Demotic Studies*, ed. Kim Ryholt, 377–393. Copenhagen: The Carsten Niebuhr Institute of Near Eastern Studies.

Yoyotte, Jean. 1977. Une notice biographique du roi Osiris. *Bulletin de l'Institut Français d'Archéologie Orientale* 77: 145–149.

Zamagni, Claudio. 2010. La tradition sur Moïse d'"Hécatée d'Abdère", in Borgeaud. Volokhine and Römer 2010: 133–169.

Zellentin, Holger M. 2008. The End of Egypt. Artapanus and the Second Exodus. In *Antiquity in Antiquity: Jewish and Christian Past in the Greco-Roman World*, Texts and Studies in Ancient Judaism, vol. 123, ed. Gregg Gardner and Kevin L. Osterloh, 27–73. Tübingen: Mohr Siebeck.

Zlotnick-Sivan, Helena. 2004. Moses the Persian? Exodus 2, the "Others" and Biblical "Mnemohistory." *Zeitschrift für die Alttestamentliche Wissenschaft* 116: 198–205.

The Exodus in Islam: Citationality and Redemption

28

Babak Rahimi

Abstract

Classical Muslim exegetes, drawn from both Quranic and non-Quranic sources, have described the exodus as an illustration of divine punishment imposed on the Israelites for their transgression against God. This study, however, understands the Quranic accounts of the exodus in terms of a salvational drama. The revelation of Torah, central to the exodus story, is about the deliverance of God's will in the act of law giving. Moses as both a prophet and a legislator plays a key role in manifesting God as the word in the citation of an authentic divine intention through the Torah. Divine presence is also found through miracles when God orders Moses to return the sea to its original form, and so the Israelites would be saved from Pharaoh. For their lack of gratitude for God's help, the Israelites are punished for their transgression against his command. In 5:20–25, God commands the Israelites to enter the "holy land," but they refuse because of giants. In turn, God condemns the Israelites with 40 years of wandering (5:26). In 7:148–158 and 20:80–98 the Israelites are described to transgress God's command for worshiping the golden calf when Moses was absent for 40 nights. In turn, Moses orders the killing of those who worshiped the golden calf. However, while the Israelites are punished for their disobedience, they are also blessed with God's mercy and generosity. When Moses's anger subsides after throwing down the tables after finding the Israelites worshiping of the golden calf, he took up the tablets for "those who fearful of their Lord" (7:154). Throughout the Quran, the exodus narrative provides numerous instances when God would provide numerous blessings to the Israelites. Beyond punishment and blessing, however, the exodus identifies a metanarrative of spiritual liberation. In such account, the Israelites partake in a redemptive experience of a trial through adversity that ultimately reveals divine grace, a self-reflexive reference that unravels the

B. Rahimi (✉)
University of California, San Diego (UCSD),
La Jolla, CA, USA
e-mail: brahimi@ucsd.edu

T.E. Levy et al. (eds.), *Israel's Exodus in Transdisciplinary Perspective*,
Quantitative Methods in the Humanities and Social Sciences, DOI 10.1007/978-3-319-04768-3_28,

God it cites into existence, and hence a promise for salvation. The exodus story therefore becomes a chronicle about God's presence in the enactment of his will through the performance of delivering the laws, even as he appears to abandon his people, even as he appears to be invisible to all.

The argument proposed here offers an account of exodus in the context of Islamic traditions. While the opinion of overwhelming scholars rejects the Biblical exodus as historically valid, due to lack of archaeological evidence, this paper argues that the significance of exodus, in its Islamic context, is essentially one of a theological account of redemptive suffering rather than historiography. As an example of a non-Biblical account, and yet distinct from other sources such as Hecataeus of Abdera set in the fourth century BC, Islamic traditions bespeak of the exodus in terms of a model to serve a new community of spiritual orientation united by divine law. The story of exodus, I argue, is a performative formula, a self-reflexive reference that makes explicit a new possibility in the discourse of redemption through suffering.

The story of Israelites' wandering in wilderness in connection with Moses' apostolic mission is a significant part of the Islamic traditions. In the Quran Moses is considered a prominent prophet, along with Adam, Abraham, Noah, Jesus, and Muhammad, who is the last messenger of God. Moses, like Abraham, exemplifies an ideal Muslim whose submission to God at the Mount Sinai (*Tur*) testifies to divine guidance and ultimate act of mercy. Moses is also a lawgiver whose guidance and teachings to the Israelites about the oneness of God was characterized with the ability to have direct conversation with the divine. Though he was also able to see God (7:143), Moses was the one Prophet whose bond with the divine was marked through language, a bond that would be later modified by the gospel of Jesus.[1] Hence, in the Islamic literature Moses is known as the *kalim Allah*, the one who has directly conversed with God.

In the schema of prophetic traditions in Islam, the story of exodus plays an integral role in how Moses attained the ability to directly talk with God. Under his leadership, the ancient Hebrews were able to flee persecution and arrive at the promised land. In doing so, they participated in a grander sacred drama that, though the Quran does not provide a coherent chronology (Wheeler 2009: 252), involved plagues, splitting of the Red Sea, arrival at the Promised Land, followed by years of wilderness, and revelation of the Torah at Mount Sinai. In particular, the Israelite arrival at Canaan, the Promised Land, only proved to be part of the grander schema of a divine will. When the Israelites refused to trust God and fight on his behalf (5:24), God prohibited them to the land, making them wander for 40 years in the wilderness. But wandering in the wilderness served as only a stage in the teleos toward salvation for the few chosen followers of Moses who repented and accepted God's command.

In terms of discursive traditions, Muslim exegetes, drawn from both Quranic and non-Quranic sources, have primarily described the exodus story as a tale of divine punishment on the Israelites for their disobedience to Moses's authority and, by extension, transgression against God. As Brannon Wheeler has shown, the Muslim commentaries have primarily alluded to such sources in order to advance criticism against the Jews and also elevate him to a higher status than Moses, blamed for the sins of the Israelites (Wheeler 2002a: 124). Here I argue, however, that the Quranic exodus is ultimately an account of redemption and not punishment.[2] The account of the Israelites wandering in the desert and Moses leading them to the Mount Sinai is one

[1] In the Quran, Jesus explains, "I shall confirm the Torah that was before me, and will make lawful for you some of the things that were before unlawful for you." (3:50).

[2] Here, my argument is similar to Thomas Dozeman's account of exodus as a salvation history. See Dozeman 2010.

of divine expression through revealed law as a binding sacred commandment. The revelation of God's command, central to the exodus story, signifies an act of citationality or a reenactment that is brought to bear on performatively signifying the divine in the discursive act of textual law. The exodus, in this dramaturgical rather than historic sense, serves as a self-reflexive reference that makes explicit a new possibility for redemption through the experience of suffering. Moses as prophet (i.e., guide) and legislator (i.e., law-giver) plays the key role of a signatory and an enforcer in citing and giving witness to the presence of an authentic divine intention.

Also, Moses as both a prophet and a legislator plays a key role in manifesting God as the word in the citation of the presence of an authentic divine intention through divine law. God's presence is also found in miracles as divine signs when God orders Moses to return the sea to its original form, and so the Israelites would be saved from Pharaoh. Throughout the Quran, the exodus narrative articulates instances when God would provide numerous blessings to the Israelites, a feature that is also noted in Muslim exegesis. Beyond punishment and blessing, however, the exodus identifies a metanarrative of spiritual liberation. In such account, the Israelites overall partake in a redemptive experience of a trial through adversity that ultimately reveals divine grace, a self-reflexive reference that unravels the God it cites into existence, and hence a promise for salvation.

Equally important in the Quranic exodus is how Moses achieves the ability to converse with God through a complex set of divinely inspired events that involve encounters with temptation, transgression and, finally, reconciliation through the utterance of law as manifestation of God on earth. Moses is more than the messenger of God but also the prophet–conqueror. Unlike the biblical version, however, Quran describes Moses's mission not only in terms of seeking the emancipation of the Israelites but also delivering the message of God to Pharaoh in order so to submit to God. Also important to note here is that Moses's prophetic mission is similar to what Muhammad sought to achieve as he migrated from Mecca, where he and his followers were persecuted, and eventually led the formation of a new religious community in Medina as a legislator and the prophet of God. Muhammad like Moses is to whom the scripture is revealed and responsible for being a protector of the covenant, and also which believers "will have faith in him and assist him" (3:81) for fulfilling the divine will on earth.

Moses in the Quran

Moses appears more than 200 times in the Quran (Ayoub 2004: 36) and referenced more than that of any other prophets (Keeler 2005: 55; Tottoli 2002: 31–35; Wheeler 2009: 249). He is among the so-called prophets of power (46:35) whose ability to directly speak with God (4:164), assigned with the difficult task of delivering the revealed book, the "tablets" (7:150–154), or divine revelation to humanity. Moses is also the performer of miracles, especially while in Egypt where he began his prophetic mission as the leader of the children of Israel.

As Annemarie Schimmel has noted, most Quranic descriptions of Moses were revealed to Muhammad in the middle of his career, as he faced considerable persecution from the Meccans (Schmimel 1995: 15). Moses is a divine informer whose prophetic experience resembles Muhammad's prophetic mission as he travelled from Mecca to Medina to establish a new spiritual community with its distinct set of laws and rules. In the earlier surahs (chapters) of the Quran, during the Meccan revelations (610–622), Moses' mission is described to be as a warner to Pharaoh and deliverer of God's signs to humanity (79–15:25). He is, similar to Mohammad in Mecca, a prophet who informs about God and cautions others about disobeying him. But the Quranic descriptions of Moses as a legislative figure increasingly become apparent in the Medinan surahs when Muhammad began to lead a new community of monotheists after he migrated from Mecca to Medina in 622. It was in Medina where Mohammad led a growing community of faithful followers and where he, similar to Moses, upheld the legislative authority to implement divine law.

Such stories are marked with long plot-driven episodes, scattered throughout the Quran. Descriptions of Moses in Egypt from childhood to Prophethood revolve around miraculous events or turning points when the young Moses is able to survive death by the grace of divine will. For example, the story of the wife of Pharaoh rescuing Moses to be adopted into the house of Pharaoh and nursed by his own mother (20:38–41; 28:7–13) bespeaks of an unfolding miracle that sets the stage for an escape narrative, caused and guided by God, which essentially characterizes the story of exodus. Moses' flight to Midian after the death of the Egyptian man also underlines the theme of divine benevolence as Moses vows to do God's will.

After years of exile in Midian, Moses encounters God through various miraculous signs. When Moses leaves Midian he receives instruction from God (20:9–24; 27:7–12; 28:29–35), the most important sign being the fire on a tree (28:29:30), which marks the first instance of divine revelations.[3] Other signs appear when Moses holds a rod moving like a snake (20:17–22; 27:10; 28:32) and when he puts his hand into his clothes and see it unharmed and shine in whiteness (20:22; 27:12; 28:32). Such signs are of miraculous significance for they not only affirm divine agency but also interruptions in perceived order of things in terms of a stable, natural law of existence. It is through these miracles that the pre-exodus Moses gives witness to an alternative truth identified in supernatural interventions that reveal divine presence. Yet it is with the power of miracles, interruptions of supernatural disposition, which enable Moses to contest Pharaoh and undermine his claims to divine authority. Upon arriving at the court when Moses, along with his brother, Aaron, as his minister and representative, challenged Pharaoh, who claimed to be a deity (17:101–103; 20:49–63; 26:10–29), he mostly did so through the display of "great signs" from God (79:20).

In contrast to magic, as paranormal activities performed by Pharaoh's sorcerers to legitimize his authority, Moses won over the Egyptian magicians in performing acts of miracle that went beyond the mere *manipulation* but a *transformation* of the natural world as willed by God.[4] At the heart of miracles is the imprint of the sublime on the mundane in contrary to the laws of nature. In fact, the events leading up to the exodus of the Israelites from Egypt, and afterwards, can be recognized as a metanarrative of miracles, unfolding through episodic events (26:52–68; 44:17–33). All episodes leading up to the exodus, such as Pharaoh's ordering Haman to build the Tower of Babel so to arrogantly attempt to see God (28:36–42) or refusal to let the Israelites leave Egypt and its subsequent punishment with the plagues and other natural afflictions (7-127-36) serve as key stages of leading up to a significant endpoint that rests with the realization of divine power.

In the Quranic sense, Moses and Muhammad are both prophets of miracles. As two messengers who encounter the voice of God, directly or indirectly, in a mountain, they both act as revealers of God's intention to humanity. The miracle is the ability to deliver God's voice to humanity, which transcends human language. But such ability also entails the authority to perform prescribed practices that confirm God's command on earth. As depicted in the hadith traditions of the *Miraj* story in the Quran, for instance, Moses plays a key role in granting Muhammad the authority to designate daily prayers (limited to five) to his followers while ascending through the heavens in the night

[3] It maybe interesting to note that in contrast to the Biblical account of the bush the Quran describes a burning tree.

[4] The Quranic distinction between magic and miracles is articulated in 17:101–103: "To Moses we gave ten clear signs. Ask the Israelites. When he came to them, Pharaoh said to him: 'I consider you, Moses, to be affected by magic.' He [Moses] said: 'You know that these things have been sent down only the lord of the heavens and the earth as something to behold. I consider you, Pharaoh, to be doomed."

journey to the Masjid al-Aqsa in Jerusalem (Sells 1996: 48).

Meanwhile, Moses is also a liberator–prophet. As a prophet of "firm resolve" (46:35), Moses leads the Israelites to miraculously escape Egypt and find the Promised Land. He is seen as the ideal prophet with the judicial power to govern over a community in quest of spiritual salvation. He provides divine guidance and political legislation; he is both a spiritual and a worldly leader.

Exodus in the Quran

Exodus is ultimately a miracle story. It is a story of enslavement and an eventual emancipation from the bondage of oppression, led by a liberator and depicted in a journey toward a homeland, where God's covenant can be established and accordingly recognized. The Quranic version in many ways resembles the Biblical one found in the Exodus, Deuteronomy, Leviticus, and Numbers, mostly complied during and after the Babylonian exile (sixth century BCE). But there are some differences. The Quranic version, found in 26:52–68 and 44:17–33, also describes the story of a lost fish in 18:61 and 18:63, plus the mysterious figures in 18:60–65 and 18:65–82. Though Muslim exegetes have linked these verses to the Bible as a way to criticize Christian and Jewish narratives (Wheeler 2002a: 8), such descriptions, as argued by Arent Jan Wensinck, show possible references to non-Biblical sources and legendary tales.[5] The lost fish story in particular, with its emphasis on the missing fish and the flowing water, are symbolic of the quest for the water of life, with the quest for eternity playing a central role in the exodus narrative.

In the complex relationship between eternity and damnation, disobedience to God plays an important role. In several episodes, the Israelites transgress God's command. For instance, in 2:58 God orders the Israelites to enter the gates of a city by saying, "relieve us of our burdens," though 2:59 depicts some Israelites to change God's wording and accordingly are penalized by a plague. According to 2:51–56, Moses leaves the Israelites in the wilderness for 40 nights while he received the "book" and returns to find his people worship the golden calf (7:148–158 and 20:80–98). In turn Moses orders the killing of those who worshiped the golden calf and, in a way, as some Muslim commentators have argued, the laws signifies a curse for worshiping the golden calf (Wheeler 2009: 253). In 5:20–25, God commands the Israelites to enter the "holy land" but they refuse because of giants. In turn, God condemns the Israelites to 40 years of wandering (5:26). However, while the Israelites are punished for their disobedience, they are also blessed with God's mercy and generosity. When Moses's anger subsides after throwing down the tables after finding the Israelites worshiping of the golden calf, he took up the tablets for "those who fearful of their Lord" (7:154). After some are punished for changing the word of God (2:59) in 2:60 Moses strikes the rock and producing 12 flowing springs of water for the Israelites. But the Israelites continue to disobey. During the exodus the Israelites show ungratefulness for the luxuries God has granted them.

In this sense, the exodus story could be viewed in terms of the Israelite's failure to recognize God's mercy. But such understanding is not constant and undergoes various stages of trials, challenges and experiences of endurance toward salvation. Endurance and punishment play an integral role in this process. The plagues of Egypt in the form of floods, locust, and pestilence of lice, described in 7:133–136, for example, serve as natural calamities caused by spiritual wrath caused by Pharaoh's failure to recognize God's authority. They are a reminder of how misrecognition of God becomes manifest

[5] Wensinck argues that base of the fist verse is the Alexander romance story. According to the Alexander romance, the dried fish becomes alive when Alexander's cook washes it in the spring of life (Wensinck 1978: 902–903). Wheeler has correctly criticized Wensinck and Biblical scholars such as Israel Friedländer by arguing that the source of the fish episode based on the Alexander stories was alluded later in history by Muslim commentators of the Quran and that, historically speaking, it is incorrect to assume that the Quranic verses were based on this story (See Wheeler 2002a: 10–19 and also Wheeler 1998: 195–196).

in the natural world where the experience of endurance becomes evident.

But endurance can also about deliverance from hardship. Here, the story of Moses and the emancipation of the Israelites from Egypt parallels in what Muhammad and his followers underwent in the course of their emigration (*hijra*) from Mecca to Medina (Juan Eduardo Campo 2009: 483). The experience of liberation entails a *flight* from ignorance and ultimately triumph over the oppressors. Similar to the Pharaoh's defeat as his army was drowned in the sea, the enemies of early Muslims too encountered a similar fate as they faced defeat and eventual succeeded to triumph over God's enemies at the end.

But there are also ambiguities in the saga of flight and conquest. As Wheeler has noted, descriptions of the Quranic Moses, especially in relation to the Israelites, lack a single coherent narrative that articulates a common rational behind the story (Wheeler 2009: 252). Such incoherence lies in the fact that the Quran primarily provides an image of Moses who is both prophet–liberator and a legislator–conqueror, who seeks to unshackle his people from the bondage of oppression and also engage in the mission to fulfill God's command on earth while in exile. In this dual prophetic mission, God and his prophet are described to switch back and forth in punishing and rewarding the Israelites for wobbling on the spiritual path set forth for them.

From one perspective, the Quranic exodus is a story of emancipation. As explained earlier, the Israelites flee Pharaoh's oppression, wandering in the wilderness where they receive blessings and gifts from God (2:47–61). According to 2:57, the Israelites are blessed with the cloud, manna, and quails, though they fail to be thankful to God for his charity. In 2:60 Moses strikes a rock and 12 springs of water gush out for the Israelites. When Moses again strikes and this time divide the sea with his rod to help the Israelites escape while drowning Pharaoh and his army (26:63–66) he does so under the command of God, who offers his greatest miraculous intervention in the world of nature.

Yet the most significant episode is when Moses leaves the Israelites while he seeks the "book" (2:51–56). Moses spends 40 days and nights on the mountain, where he asks to see God, and God shows himself to the mountain that eventually crashes. According to 7:142–147, Moses's experience in the mountain comes to an end when finally he has the Torah, as God's revelation. On this crucial moment, the story of exodus becomes a tale of punishment just when God reveals himself through the written word.

The rise of tension with the triumph of Moses to deliver God's message and yet the failure of the Israelites to recognize the truth underscores the breakdown of a straightforward path toward salvation. The breakdown is so significant that not even Moses can repair it by interceding on the behalf of the Israelites (2:48). This is so because a significant act of transgression is committed which revolves around the act of denying God, especially when he shows his mercy. In 7:148 and 20:80–98, the Israelites are described to worship the golden calf while Moses is away, hence rejecting God who helped them to be liberated from Egypt. Upon arriving at the Canaan, the Promised Land (5:20–25), the Israelites refuse to enter it because of giants (5:24). As Moses and Aaron distanced themselves from the ungrateful Israelites (5:25), God then punishes them with 40 years of wandering for their sins in rejecting God's signs and failing to appreciate his generosity (5:26).

Most Israelites are punished for their disobedience to God. But Moses selects a few Israelites. Known as the "people of Moses" (7:159), the few chosen are those who repent and seek forgiveness from God, despite attempts to see him and transgress his will. The chosen Israelites are able to partake in a trial of larger magnitude that can only be realized through endurance.

Islamic Exegeses and Exodus

Drawn from both Quranic and non-Quranic sources, Muslim exegetes of Moses and lessons of exodus are essentially about how, as Wheeler explains, "Islam and Muhammad demand a simple obedience, not a questioning of the reasons for God's instructions." (Wheeler 2009: 264). Knowledge about God and his plans, however difficult they maybe to understand, lie in the prophets whose authority determines what is divinely sanctioned. In many ways, the intimate relationship between knowledge (i.e., light), law (i.e., tablet), and prophetic persona (i.e., Moses) plays an integral role in Muslim exegetical literature of the exodus. Revealed message is more than receiving knowledge about the mercy of God and focusing on his grace alone, but executing such knowledge under the sole authority of a prophet–legislator who can fulfill God's plan in the world.

While the Quran does not provide a clear account about the relationship between God and the Israelites in the context of exodus story, Muslim exegetes emphasize key themes related to receiving God's blessings and also the failures to appreciate such gifts. Abul al-Qasim al-Balkhi (d. 931), for example, describes the wilderness years for the Israelites in terms of gifts bestowed by God. With the blessing of God, the Israelites had indulgences amid the wasteland such as clouds, fire, food, and water coming out of rocks (Wheeler 2009: 253). Likewise, Al-Suddi (745) writes about how the Israelites had clothes on their bodies from birth without making the effort to make them. Verse 2:57 reference to honey and quail also identifies a paradisiac state of existence, which shows God's presence even amid wilderness.

Muslim exegetes on divine punishment also emphasize the motif of transgression understood in terms of the Israelites' rejection of God's commands, and its consequence. Here, the role of conquest is central to the story. The reference to God's command for the Israelites to enter the holy land, for example, underlines the authority of Moses as the conqueror–prophet who earlier confronted Pharaoh, parallel to Muhammad who also engaged in military campaigns against the Meccans. Some Muslim exegetes state that it was Moses who conquered the Promised Land, which was Jerusalem (e.g., Abi Karimah Suddi), Damascus or Palestine, along with parts of Jordan (e.g., Ziyad al-Farra).[6] In some accounts, Moses is depicted to be responsible for slaying the main giant Og b. 'Anaq left from Noah's flood (Wheeler 2009: 255).

The tale of Og giants in Muslim exegetes brings to light the significant role of myth in the Quranic exodus. In the account of Muhammad b. Jarrir Tabari (838–923), the giant Og is surprised to see the 12 spying Israelite, who were sent by Moses, and placed them in his belt and took them to his wife (Wheeler 2002b: 211). When the wife requested her giant husband to return them to their people, so they could report in what they saw, the 12 Israelites agreed only to inform Moses and Aaron about the giants. But ten of the informers broke the agreement and told their families, which led to the Israelites' refusal to enter the Promised Land because of the giants (5:24) and the subsequent punishment with 40 years of wandering throughout the wilderness (5:26). The story of giants and 12 spies bespeaks of the Israelites failure to uphold the authority of Moses as God's commander on earth. Their punishment is not just caused by disobedience to God, but the failure to recognize the authority of Moses as representative of God on earth.

A similar story can be found in reference to other exodus episodes. In the account of the golden calf by Abd Allah Ibn Abbas (619–687), for example, when angel Gabriel arrives on a horse to take Moses to God, Samiri, the man later responsible for convincing the Israelites to worship the golden calf when Moses was away (20:85; 20:87), throws dust left by the hoofprint of Gabriel's horse in fire, which he requested to make in order to burn Jewelry and ornaments of the Israelites so to purify them from the Egyptian experience of enslavement. According to Ibn Abbas, Aaron, who was in charge during the absence of Moses, gives permission to Samiri to throw the dust into the fire thinking that they are ornaments. Samiri then says, "Be an embodied

[6] According to Abu al-Hajjaj Mujahid (d. 722), a famous Quranic commentator under in the Ummayd period, the Promised Land, also known as "holy land" in Islamic literature, was Mount Sinai and its surrounding areas.

calf which bellows!"(Wheeler 2002b: 204 and in reference to 20:88) and God caused to come forth from the jewelry a golden calf as a way to test the Israelites. The Israelites then loved and worshiped the calf.

As punishment, those who did not worship the calf began to kill those who had with dagger in their hands, leading to 70,000 deaths (Wheeler 2002b: 205). Later, as Ibn Ishaq describes, the calf is burnt and dispersed into the sea. But at the end of the golden calf story, the sinful are forgiven when Moses is asked by angel Gabriel to stop the killing of the Israelites. God had earlier commanded Moses and Aaron that the sins of the Israelites will be forgiven only if "they [Israelites] kill themselves." (Wheeler 2002b: 205). At the end, according to Ibn Abbas, "all those who were killed were absolved, and all those who remained were absolve" (Wheeler 2002b: 205). Self-violence by the Israelites brings about a new prospect for redemption.

As for the story of the refusal to enter the Promised Land, after repeated acts of transgression, the Israelites are punished for a longer span of time. While only a few select are saved, as a sign of God's mercy for those who follow and accept Moses as God's authority on earth, the Israelites are cursed for their repeated subversion of the divine command. Following Christian commentaries on exodus, Muslim exegetes view the sins of the Israelites as so damning that not even Moses would be willing to intercede (Wheeler 2009: 255). For these accounts, the exodus represents a tale of divine punishment due to the consistency of disobedience, and serving as a reminder of the threat of transgression and its consequences to those who reject the will of God.

Exodus as a Tale of Redemption

But is punishment a mere retribution for violation of a norm ordained by God? In the Quranic sense, punishment is a manifestation of God's fury for disobedience and yet a way to show his mercy. Miracles, likewise, in the exodus story mirror God's generosity and also his wrath, against transgressors such as Pharaoh. The Quranic exodus, seen in this way, is less concerned about how divine law is implemented and more about how God and his commands are abided by his people.

Signs of God are key to the Quranic exodus. It is with the exodus that Moses encounters God's greatest sign, the "book" as signatory to divine manifestation in the mundane world. It is also in the failure to recognize such signs that the Israelites, at least the few chosen ones (7:155), are able to eventually understand God's mercy. The "people of Moses" (7:159) are selected precisely because they obey Moses who delivers the will of God. In the unconditional recognition of signs lies the compassion of God.

The motif of perpetual forgiveness, in close connection with punishment, resonates throughout Quranic depictions of Moses and his prophetic mission. In sura al-A'raf (The Heights), Moses seeks God's forgiveness for Aaron in failing to prevent the Israelites from worshiping the calf (7:151). God grants mercy, including those who repent and believe in him as the true God (7:153).[7] When Moses throws the tablet, he offers the sinful Israelites the promise of guidance and mercy (7:154). When Moses chooses from his people 70 men, he does so in order to instruct the few about divine mercy. God, the Quran reminds us, is the "best of forgivers" (7:155).[8]

The exodus is about how God can be merciful even when he appears to punish his people because of their acts of defiance. The law as sanctioned by God to provide correct belief and

[7] "And decree for us in this world [that which is] good and [also] in the Hereafter; indeed, we have turned back to you. [Allah] said, my punishment—I afflict with it whom I will, but my mercy encompasses all things." (7:156).

[8] "And Moses chose from his people 70 men for our appointment. And when the earthquake seized them, he said, "My Lord, if you had willed, you could have destroyed them before and me [as well]. Would you destroy us for what the foolish among us have done? This is not but your trial by which you send astray whom you will and guide whom you will. You are our protector, so forgive us and have mercy upon us; and you are the best of forgivers (7: 155).

conduct for the Israelites becomes an event of citation, performatively delivering the divine into the world through a reportive frame to establish a new order. The law in a way *enacts* the universality of the divine to purify the world from evil. By extension, it represents a formula, a self-reflexive reference that makes explicit a new possibility for salvation through the experience of suffering.

In this short study, I examined the Quranic accounts of exodus with the aim to rethink its significance in the Muslim traditions. I argued that the Quranic version could be viewed as a type of salvational history, a tale of redemption realized through trial and hardship. In terms of the thematic concepts of emancipation from slavery to encounters with disobedience, the exodus serves as a metanarrative of suffering for the ultimate recognition of God, the essential reality of which transcends human perception.

References

Ayoub, Mahmoud. 2004. *Islam: Faith and History*. Oxford: Oneworld.

Campo, Juan Eduardo. 2009. Moses. In *Encyclopedia of Islam*, 482–484. New York, NY: Infobase Publishing.

Dozeman, Thomas B. 2010. *Methods for Exodus*. Cambridge: Cambridge University Press.

Keeler, Annabel. 2005. Moses from a Muslim Perspective. In *Abraham's Children: Jews, Christians, and Muslims in Conversation*, ed. Solomon Norman, Harries Richard, and Winter Time, 55–66. London: T & T Clark.

Schmimel, Annemarie. 1995. *Und Muhammad ist sein prophet: Die verehrung der propheten in der islamischen frömmigkeit*. Munich: Eugen Diederichs Verlag.

Sells, Michael A. 1996. *Early Islamic Mysticism: Sufi, Qur'an, Mi'raj, Poetic and Theological Writings*. New York, NY: Paulist Press.

Tottoli, Roberto. 2002. *Biblical Prophets in the Qur'ān and Muslim Literature*. Richmond: Curzon.

Wensinck, Arent Jan. 1978. al-Khadir. In *Encyclopedia of Islam*, vol. IV, 2nd ed, ed. H.A.R. Gibb, 902–903. Leiden: E.J. Brill.

Wheeler, Brannon M. 2002a. *Moses in the Quran and Islamic Exegesis*. Richmond: Curzon.

Wheeler, Brannon M. 2002b. *Prophets in the Quran: An Introduction to the Quran and Muslim Exegesis*. London: Bloomsbury Academic.

Wheeler, Brannon M. 2009. Moses. In *The Blackwell Companion to the Quran*, ed. Rippin Andrew, 248–265. Malden, MA: Blackwell.

Wheeler, Brannon M. 1998. Moses Or Alexander? Early Islamic Exegesis of Qur'an 18:60–65. *Journal of Near Eastern Studies*, 57(3): 195–196

From Liberation to Expulsion: The Exodus in the Earliest Jewish–Pagan Polemics

29

Pieter W. van der Horst

Abstract

The exodus from Egypt played a pivotal role in Jewish–pagan polemics from the beginning of the Hellenistic period till far into the Imperial period. Pagan polemicists stood the biblical story of the liberation of the Israelites from bondage in Egypt upon its head and portrayed an extremely negative image of Israelite origins. They also pictured the Jewish people as misanthropes and atheists. Jewish–Hellenistic authors reacted to these attacks in a wide variety of ways (e.g., novels, drama, and philosophical treatise).

Whereas many of the other contributions deal with historical questions relating to the exodus, this paper focuses on the reception of the biblical exodus story in non-Jewish circles and on Jewish reactions to the ways non-Jewish authors abused the story in order to blacken the Jews.

From the very beginning of the Hellenistic period, the story of the exodus[1] played a pivotal role in Jewish–pagan polemics. This raises several questions. Why was it the exodus, why wasn't it the stories about Abraham and the other patriarchs[2] or another story that the pagan polemicists focused upon? Where did they get their information about the stories of the exodus? Did the Jews react to the pagan versions of the exodus story and, if so, how did they? Many other questions as well come to mind when one takes a look at the evidence. In the modest compass of this paper only a few of these questions can be dealt with, and only briefly at that.

Let us first present a quick survey of the evidence. We have 14 exodus stories from pagan Greek and Roman writers.[3] Obviously it is impossible to pay attention to all of them. For that reason I have decided to restrict my survey to the Greek authors with an Egyptian, more specifically an Alexandrian, background. These are Manetho, Lysimachus, Chaeremon, and Apion. They span more than three centuries and

[1] When I write "exodus" without a capital, I refer to the event; when I write "Exodus," I refer to the biblical book.

[2] See, e.g., van der Horst 2010.

[3] See the survey in Barclay 2007: 344.

P.W. van der Horst (✉)
Theology, University of Utrecht,
Utrecht, The Netherlands
e-mail: pwvdh@xs4all.nl

T.E. Levy et al. (eds.), *Israel's Exodus in Transdisciplinary Perspective*,
Quantitative Methods in the Humanities and Social Sciences, DOI 10.1007/978-3-319-04768-3_29,

demonstrate both the continuity and diversity in pagan anti-Jewish exodus traditions.[4]

Manetho,[5] an Egyptian priest who wrote a history of Egypt in Greek, tells the story of the exodus[6] in such a way that it became a counter-history that proved to be "one of the most powerful anti-Jewish statements in ancient history."[7] But in order to put his story into perspective, it should be noted that before presenting his version of the exodus, Manetho had informed his readers about the Hyksos (Shepherds), a western-Asiatic nation that had occupied and ruled over Egypt from about 1650 to 1540 BCE.[8] There he tells us[9] about the invasion of the country by the Hyksos and their brutal reign of terror. They burned the cities, demolished the temples of the gods, slaughtered many male Egyptians, and enslaved their wives and children. One of their kings founded the city of Avaris, a place having very strong religious associations with Set/Typhon, the god of confusion and chaos.[10] After a long time there was a revolt of the natives against the Hyksos, which resulted in the siege of Avaris. Finally they left Egypt and built a city they called Hierosolyma in the region of Judea. It is this mention of Jerusalem that induces Josephus (who quotes this passage) to happily claim that Manetho rightly regarded the Hyksos as the ancestors of the Jews, which proves—in his view—that the Jews could never have been originally Egyptians. And this is relevant for what follows. For when later on Josephus quotes Manetho's exodus story, the latter seems to imply exactly that, much to the anger of Josephus who asserts that here Manetho contradicts himself. In this second passage we read the following (*Ag.Ap.* 1.228–252 abbreviated):

King Amenophis felt a desire to see the gods and said so to his namesake the prophet Amenophis (Amenhotep). The latter told the king that he would be able to see the gods only if he cleansed the country of lepers and other unclean people. The king then rounded up all those in Egypt with deformed bodies, some 80,000 of them, and put them to work in the stone quarries to the east of the Nile so that they would be productive as well as isolated from the other Egyptians. There were even some learned priests among them who were afflicted with leprosy. After those in the quarries had suffered there for a long time, they asked the king to give them as a refuge the city of Avaris that had been abandoned by the Shepherds (i.e., Hyksos) and had been sacred to Typhon from the beginning. The king granted their request, and once they had settled there, they appointed Osarseph, a priest from Heliopolis, as their leader and took an oath of total obedience to him. He legislated—on the principle of normative inversion[11]—that they should not worship the gods or revere any of the sacred animals. Rather, they should consume them and attach themselves to no one except their fellow conspirators. Osarseph laid down other laws as well that were completely contrary to Egyptian customs and prepared his people for

[4] Some pagan authors cherished positive (not anti-Jewish) exodus traditions, such as Strabo, whose version is "an example of one possible direction the Exodus tradition could take, admittedly the most positive and friendly towards the Jews that we know;" see Schäfer 1997: 25. For positive assessments of Moses by pagan authors, see Feldman 1993: 233–242.

[5] On Manetho see Stern 1974: 62–65; Sterling 1992: 117–136; Schäfer 1997: 17–21 *et passim*; Verbrugghe and Wickersham 2001: 95–120.

[6] Whether Manetho knew the text of the biblical book of Exodus is a moot point. At least he had a Jewish informant whose knowledge was based upon the Bible. See Cook 2004: 6–8.

[7] Schäfer 1997: 15. As Amos Funkenstein (1993: 36) says, the polemical method of counter-history is "the systematic exploitation of the adversary's most trusted sources against their grain."

[8] See Redford and Weinstein 1992; Seidlmayer 1998 for further references.

[9] The fragment, which is partly paraphrase, is preserved in Josephus, *Ag. Ap.* 1.75–90.

[10] See van der Toorn 1999 and van Henten1999.

[11] Assmann 1997: 31: "The principle of normative inversion consists in inverting the abominations of the other culture into obligations and vice versa." The clearest example of "normative inversion" is what Tacitus says about the Jews (*Hist.* 5.4.1): *Profana illic omnia quae apud nos sacra, rursum concessa apud illos quae nobis incesta* (with the comments in Heubner and Fauth 1982: 43–44).

war against the king. He formed a council with other unclean priests; they sent ambassadors to the Shepherds in Jerusalem and asked them to unite in an expedition against Egypt. The latter were delighted and, some 200,000 men, they set off and arrived in Avaris. King Amenophis was frightened, so he assembled a great number of Egyptians and consulted with their leaders. He ordered the most sacred animals and the images of the gods to be brought into safety. He avoided battle but set out for Ethiopia with his whole army where he was welcomed by the king who provided them with anything they needed for their daily life. The mixed army of unclean Egyptians and Jerusalemites treated the population of Egypt in a very cruel and sacrilegious way. They burned the cities and villages, pillaged the temples, mutilated the images of the gods, used the sanctuaries as kitchens to prepare the sacred animals for consumption and forced the priests to slaughter these. The man who laid down their laws, Osarseph, when changing his allegiance, also changed his name and was henceforth called Moses. Later Amenophis advanced from Ethiopia, attacked the Shepherds and the polluted people and expelled them to Syria. This, according to Manetho, explains the origins of the Jewish people.

The following should be noted here: It is obvious that the whole story is patterned upon Manetho's own story of the Hyksos as summarized above. To put it another way, originally this was not a story about the Jews but one about the Hyksos that only at the very end was made into a story about the Jews by means of a gloss, most probably added by Manetho himself,[12] in which he identifies Osarseph with Moses. But there is more. As has been suggested, most forcefully by Jan Assmann, the traumatic experience of Akhenaten's fanatically monotheistic experiment in the Amarna period[13] is at the background here and shaped the Hyksos tradition because Manetho's story is about a religious confrontation and "there is only one episode in Egyptian history that corresponds to these characteristics: the Amarna period."[14] As Assmann argues,

> dislocated Amarna reminiscences began to be projected onto the Hyksos and their god Baal, who was equated with the Egyptian god Seth. (...) The Hyksos conflict was thus turned into a religious conflict. This process of distortion continued through the centuries as events occurred that fit into the story of religious otherness and its dangerous semantics of abomination and persecution.[15]

It is important to realize that cultural memories of both the Hyksos and the Amarna period shaped a template or mold into which later experiences with the Jews could be made to fit. The exodus story of the Jews and the conflict that arose between Jewish monotheism and pagan polytheism in the Hellenistic period are here made to fit into the template of the amalgamated memory of both the Amarna and the Hyksos period. Moreover, the mention of Avaris makes clear that the Jews are seen as adepts of the evil god Typho.[16]

In the four centuries after Manetho, we find in the writings of Graeco-Roman authors a whole series of more or less similar stories about the exodus in which the divergencies should be explained by assuming that they are not just copying one another but also using different sources, whether oral or written. As I said, I will only briefly mention those of Egyptian or Alexandrian provenance.

Lysimachus probably lived in the second century BCE, but we do not know his exact dates.[17] He was an Alexandrian, although not a native Egyptian but a Greek. He is important in that he shows that Manetho's version of the exodus story is modified in the sense that the Jews' supposed anti-Egyptian stance is broadened to an attitude

[12] See Assmann 1997: 34 and 224 n. 22.

[13] For a quick overview see Redford 1992.

[14] Assmann 1997: 30.

[15] Assmann 1997: 28-29.

[16] See van Henten and Abusch 1996.

[17] On Lysimachus see Stern 1974: 382–388; Gager 1972: 118–120; Goodman in Schürer 1986: 600–601; Schäfer 1997: 27–28. On Lysimachus' disputed dates see especially Fraser 1972: 1092–93 note 475, and Bar-Kochva 2010: 306–316, 333–337.

that is more generally opposed to the values of humankind at large. Lysimachus tells us[18] that in the time of king Bokchoris the Jews, who were afflicted with leprosy, scurvy, and other maladies, took refuge in the temples and lived a mendicant existence. The spread of their diseases caused a failure of the crops, whereupon the oracle of the god Ammon told the king to purge the temples of these impure people, drive them out into the desert, and to drown those inflicted with leprosy and scurvy. The latter were indeed drowned, while the others were exposed in the desert to die. But those in the desert assembled, kept a fast, and implored the gods to save them. On the next day a certain Moses advised them to take a straight course through the desert until they reached inhabited territory. He also instructed them to show goodwill to no one, never to give the best but always the worst advice, and to destroy sanctuaries and altars of the gods wherever they found them. The others agreed and after a difficult journey they arrived in an inhabited land. There they maltreated the people and plundered and burned the temples. Thereafter they came to Judaea where they founded a city they called Hierosyla [temple plunder or sacrilege] because of their propensities, but later they changed its name to Hierosolyma in order not to spoil their reputation. Later (*C.Ap.* 2.145 and 236) Josephus adds the information that Lysimachus had called Moses "a wizard and deceiver" and labeled the Jewish people as "the lowest of humankind" (a kind of *Untermenschen*).

In its full version, this account bristles with inconsistencies, but we leave that matter out of account because it does not affect our argument; moreover, the matter has been adequately dealt with at length by others.[19] Apart from other differences with Manetho (which may go back to different oral traditions), what is most striking about this report is that it is made clear right from the start that this is a story about the Jews and Moses and hence is a specimen of outspoken anti-Jewish propaganda. The drowning of the lepers is nothing but a reversal of the drowning of the Egyptian army in the sea according to the book of Exodus. Furthermore, Lysimachus makes clear that what he has to say about the Jews has much wider implications than just for Egypt and the Egyptians. "[Moses] also instructed them to show goodwill to no-one, never to give the best but always the worst advice, and to destroy sanctuaries and altars of the gods wherever they found them." It is this generalization that for the first time turns the Jews into enemies of humankind as a whole[20] and adversaries of religion and piety. It is from this writer onwards that the image of the Jews as misanthropes and atheists becomes a stock element in the anti-Jewish propaganda of pagan antiquity.[21] Lysimachus is the first author in whom we can clearly discern how the originally anti-Hyksos and anti-Amarna reminiscences have grown out into a fully fledged anti-Jewish version of the exodus, with wide-ranging implications.[22]

If we now jump to the first century CE, we again meet two Alexandrian scholars, both of Egyptian descent, whose exodus stories have developed into decidedly anti-Jewish propaganda, Chaeremon[23] and Apion.[24] Their stories follow the pattern familiar by now. Chaeremon has his own variants[25]: the goddess Isis appeared to king Amenophis blaming him for the fact that her temple had been destroyed. This frightened him but the sacred scribe Phritobautes said that

[18] Preserved in Josephus, *Ag.Ap.* 1.304–311; here abbreviated.

[19] Most recently by Bar-Kochva 2010: 320–325.

[20] Cf. what Tacitus (who echoes much of what Lysimachus reports) says about the Jews in *Hist.* 5.5.1: they cherish *adversus omnes alios hostile odium*.

[21] See Berthelot 2003.

[22] On the charge of Jewish misanthropy see Feldman 1993: 125–131, but esp. Berthelot 2003.

[23] On Chaeremon see Stern 1974: 417–421; Gager 1972: 120–122; van der Horst 1984a (2nd rev. ed. 1987); Goodman in Schürer 1986: 601–604. Aziza 1987: 60–61.

[24] On Apion see Stern 1974: 389–416; Gager 1972: 122–124; Goodman in Schürer 1986: 604–607; Aziza 1987: 61–63; van der Horst 2002: 207–221.

[25] Quoted in Josephus, *Ag. Ap.* 1.288–292.

he would find relief if he purged Egypt of those who had impurities. He then collected some 250,000 of these noxious people and expelled them. They were led by two scribes, Moses and Joseph, their Egyptian names being Tisithen and Peteseph.[26] They went to Pelusium and met 380,000 people there, whom Amenophis had not wanted to give access to Egypt. They made an alliance with these people and marched against Egypt. Amenophis fled to Ethiopia, but later his son Ramesses expelled "the Jews" (sic!) and drove them into Syria, so much for Chaeremon's account.

In spite of the differences, it is clear that Chaeremon followed the same basic structure as Manetho's exodus story. What is striking, however, is that the motifs of misanthropy and atheism do not play a role in his story, at least as far we can judge on the basis of Josephus's abridged version of it. We also do not find the motif of the reign of terror by the impure people but, again, that may be due to Josephus's abridgement. Be that as it may, here the Jews (explicitly so called by Chaeremon) are originally Egyptians who had to be expelled from Egypt at a divine command.

Apion, the inventor of the notorious blood libel,[27] "displayed hatred towards the Hebrews," as Clement of Alexandria says (*Strom.* 1.21.101.3), and he composed a book *Against the Jews*. In this work he also dealt with "the departure of our ancestors from Egypt," as Josephus states (*C.Ap.* 2.6), but unfortunately the Jewish historian does not give details of this version,[28] even though Apion was one of the main targets in his *Contra Apionem*. I mention Apion, however, because he was the last in the long range of Alexandrian writers, stretching from the first half of the third century BCE to the second half of the first century CE, to deal with the exodus story in a way hostile to the Jews. After him, we have only the famous historian Tacitus who sums up, as if in a repository, all the elements and features in the hostile exodus versions of the four previous centuries in the extremely hostile excursion in book 5 of his *Historiae*.[29] We should not forget that the pagan versions of the exodus most probably enjoyed considerable popularity and "were widely accepted as authentic accounts of Jewish origins."[30]

We now turn to the question: Did the Jews, in Alexandria or elsewhere, react to these versions of the exodus and, if so, how did they respond? The answer is: yes, they did respond and their responses took a variety of forms. Again, it is not my purpose to be exhaustive here; I will focus on a very limited number of Jewish authors, namely, Artapanus, Ezekiel the Tragedian, and the unknown author of the *Wisdom of Solomon*.[31]

Artapanus is a Jewish author who most probably lived in Egypt in the second century BCE.[32] In his work *On the Jews*, of which only three fragments have been preserved, he discusses *inter alia*[33] the early history of Moses.[34] There we read[35] that pharaoh Palmanothes, who dealt harshly with the Jews in Egypt, had a daughter named Merris whom he gave as a wife to a certain Chenephres. Because she remained childless, she adopted as her own child one of the Jews, a baby she named Moses. When he reached manhood, he was called Mousaios by the Greeks

[26] It is only Chaeremon who makes Moses and Joseph into contemporaries.

[27] See van der Horst 2008.

[28] Except for a minor detail in *Ag.Ap.* 2.20: Apion states that the expelled people numbered 110,000.

[29] Schäfer 1997: 32 calls this passage in Tacitus "the essence and climax of all the motifs which in antiquity were concerned with the Jews in general and with the Exodus tradition in particular." The literature on Tacitus' excursion on the Jews is vast; see, exempli gratia, Bloch 2002. On Tacitus' indebtedness to predecessors see Feldman 1996: 401–405.

[30] Gager 1972: 113.

[31] See van der Horst 1988.

[32] On Artapanus see Holladay 1983: 189–243; Goodman in Schürer 1986: 521–525; van der Horst 1985: 144–161; Barclay 1996: 127–132; Gruen 1998: 153–160 *et passim*; Gruen 2002: 201–212. Further bibliography in Denis 2000: 1135–1146; DiTommaso 2001: 1009–1015.

[33] In the first two fragments he deals with Abraham and Joseph.

[34] This fragment (3) has been preserved in Eusebius, *Praep. Ev.* 9.27.1–37, partly paralleled in Clement of Alexandria, *Strom.* 1.23.154.2–3.

[35] What follows is a much abbreviated version of Artapanus' text.

and became the teacher of Orpheus. He bestowed many useful discoveries on humankind: he invented ships, machines for lifting stones, Egyptian weapons, devices for drawing water, fighting equipment, and philosophy. He also divided Egypt into 36 nomes, and to each of the nomes he assigned the god to be worshipped—cats, dogs, and ibises—and he entrusted the sacred writings to priests. He did all these things in order to keep the monarchy stable for Chenephres, his stepfather. He was loved by the masses and deemed worthy of divine honor by the priests, who called him Hermes because of his skills in interpreting the sacred writings. But Chenephres became jealous and tried to kill Moses. He sent him with weak troops into a war with the Ethiopians, but Moses won every battle. Then he (Moses) and his troops built a city named after him, Hermoupolis. Chenephres made new attempts to assassinate Moses but his own Egyptians began to hate him and reported his plots to Moses. Moses' brother Aaron advised him to flee to Arabia, which he did. There he married Raguel's daughter. He prayed to God that he might liberate his people from oppression and a divine voice answered him from a mysterious fire that appeared out of the earth; it told him to rescue the Jews and return them to their ancient fatherland. Then Moses led a fighting force against Egypt and informed the king that he had come because the Lord of the universe had commanded him to liberate the Jews. The king imprisoned him but at night the doors of the prison opened of their own accord. When the king asked Moses the name of the God who had sent him, Moses whispered the name into the king's ear whereupon the king fell down speechless, but Moses revived him. What follows then is a free version of the stories of the Ten Plagues until finally the king released the Jews. Moses guided them to the Red Sea where, according to some, he watched for the ebb tide and then led them through the dry part of the sea but, according to others, he struck the water with his rod at divine command so that it divided. When the Egyptians arrived at the sea in hot pursuit, a fire blazed in front of them and the sea again flooded their path. All the Egyptians were consumed by the fire and the flood. This is Artapanus' version.

Now it may seem a bit hazardous to start with Artapanus because his Jewishness has become a matter of debate lately.[36] No less a scholar than Howard Jacobson has argued that Artapanus cannot have been a Jew because he not only adds details that are not found in the Bible (which is a common feature of Jewish ways of retelling the Bible), but he also adds elements that glaringly contradict the Bible, Moses' instituting animal worship in Egypt being the prime example of anti-biblical material. As Jacobson says, "No Jew would have used the tale of Moses as institutor of Egyptian animal worship" (219–220). He compares Artapanus to the second century CE philosopher Numenius who, being a pagan Platonist, shows his great admiration for Moses by saying: "What else is Plato than a Moses writing in Attic [Greek]?"[37] If Artapanus is a pagan author, it would seem I cannot use him in the present argument. But is he? I for one cannot imagine what could possibly have been the motive of a pagan writer for depicting the figure of Moses as not only the institutor of animal worship but also as the inventor of such major cultural assets as ships, machines for lifting stones, devices for drawing water, fighting equipment, and—last but not least—philosophy! Moses as the greatest cultural hero of human history can hardly be a product of pagan imagination. What *is* imaginable, and even probable, is that a Jew, confronted with Egyptian slander about Moses as the leader of a bunch of Egypt-hating rebels, tries to silence these opponents in an act of radical revisionism by arguing—most probably tongue in cheek[38]—that actually the Egyptians owe their own culture, including their religion, to this very Moses. The founder of the Jewish people is also the founder of Egyptian civilization! Even if, for the sake of the

[36] Jacobson 2006: 210–221.

[37] Numenius, fr. 8 (*ap*. Eusebius, *Praep. Ev*. 11.10.14); in des Places 1973: 51.

[38] On the humorous aspect of Artapanus' text see Gruen 1998: 159–160.

argument, we were to assume Artapanus to be non-Jewish, then it still makes most sense to take this depiction of Moses to be a response to the Manetho-style of exodus stories. True, the emphasis is more on the other accomplishments of Moses than on the exodus *stricto sensu*, but even so, it is a powerful response to the pervading motifs of misanthropy and atheism as characteristics of Moses and his followers in the Graeco-Egyptian texts we have seen so far. It is not completely unthinkable, indeed, that a "philosemitic" pagan author stood up for the Jews and defended them against attacks such as we have seen. But it seems more likely that we have to do here with a rather exceptional Jewish author.

Let us now turn to an author who is undoubtedly Jewish and whose main topic is the exodus itself, Ezekiel the Tragedian.[39] Ezekiel is a most remarkable man. His name is Hebrew but he wrote Greek tragedies in the language and style and with the dramatic techniques of Aeschylus, Sophocles, and Euripides. His subject, however, is fully biblical. Unfortunately, only one of his dramas has been preserved, and that only in part (269 iambic trimeters are extant), but fortunately, that is his work on the exodus, the *Exagôgê*. It is the only Jewish play known to us from antiquity, and it was probably intended to be staged, not just read.[40]

The drama follows the LXX version of Exodus 1–15 fairly closely but adds several interesting haggadic motifs, such as the appearance of the mythical phoenix, which heralds the inauguration of a new era in (salvation) history. The most striking and controversial scene is a dream of Moses in which he sees God upon his throne on the top of Mount Sinai; God beckons him to come to his throne, hands him his regalia and, leaving the throne himself, seats Moses upon it, whereupon all heavenly powers prostrate before Moses. It would seem that this indicates that according to Ezekiel all power in heaven and on earth has been handed over to Moses, who acts here as the viceregent or plenipotentiary of God.[41] The synthesis of biblical story, Greek literary form, post-biblical haggada, and theological speculation makes this play into one of the most typical products of Jewish Hellenism.

The extant fragments present us with a monologue by Moses recounting the events in Exodus 1–2; a brief dialogue between Moses and Zipporah; an even briefer dialogue between a certain Chum and Zipporah; then follows Moses' dream vision and Raguel's interpretation of the dream; next is a long dialogue between God and Moses at the burning bush including the precepts for Passover; further a long monologue by a messenger reporting the exodus and the Egyptians' destruction in the Red Sea; finally, we have a scout's report describing the oasis at Elim and the appearance of a spectacular bird.

As Howard Jacobson has remarked, we find here several elements of a polemic against anti-Jewish exodus traditions.[42] For instance, in line 130 the text says that, when God had ordered Moses at the burning bush to place his hand in his bosom and then remove it, Moses' hand "became like snow." The Hebrew Bible text has here that "it became leprous, white as snow" (Ex. 4:6). Now the omission of *leprous* is already found in the LXX, and after Ezechiel we also find it in Philo's *Life of Moses* 1.79 and Josephus' *Jewish Antiquities* 2.273. So the motif is not distinctive for our author, but it can hardly be doubted that all these sources omit the element of leprosy because it loomed so large in anti-Jewish propaganda: Manetho and Lysimachus spoke about the mob of leprous persons that had to be expelled from Egypt (*C. Ap.* 1.233 and 304). And Moses himself is said to have been leprous by

[39] On Ezekiel see Jacobson 1983 with the *addenda et corrigenda* by van der Horst 1984b 354–375; Lanfranchi 2006; further DiTommaso 2001: 1035–1042; Denis 2000: 1201–1216.

[40] Lanfranchi 2006: 39–56, presents the evidence for Jewish attendance of theaters.

[41] On this rather striking answer to the vilification of Moses by pagan authors and the parallel with postbiblical traditions about Enoch see van der Horst 1983.

[42] Jacobson 1983: 18.

several ancient Greek authors.[43] The omission of the motif of Moses' leprous hand seems to be directed at this slander.

Another well-known example is the passage in which God says to Moses: "When you are about to leave, I will make the Egyptians well-disposed to you and each of your women will receive from her neighbor vessels and raiment of all kinds. Gold, silver, and garments, so that the Egyptians will render payment to these mortals for all the work they have done" (160–166, transl. Jacobson). This is a matter that was of concern to the Jews because the biblical text, although it repeatedly speaks of the Israelites' "borrowing" things from the Egyptians at the exodus (Exod 11.2–3:12.35–36), was sometimes interpreted by non-Jews in the sense that the Jews had robbed the Egyptians of their legal possessions when they left Egypt. This negative interpretation was made possible because the very last words of Exod 12:36 are: "Thus they [the Israelites] stripped (or plundered)[44] the Egyptians." This element apparently was seized upon by anti-Jewish writers who accused the Jews of theft and robbery. For instance, Pompeius Trogus, a first-century BCE historian, writes that Moses secretly stole sacred objects from the Egyptians.[45] Elsewhere in Jewish literature, we find that this accusation very much bothered the Jews (e.g., b.*Sanh.* 91a; *Esther Rabba* 7; etc.) and for that reason they set forth the defense offered here by Ezekiel: What the Israelites took from the Egyptians was no more than their overdue wages!

A final example: in vv. 167–169 we read, immediately following the lines quoted above, that God says: "When you reach your own land, (. . .) you will have had a journey of 7 days from that morning on which you left Egypt" (transl. Jacobson). But the Bible says the journey took 40 years! The reason for this drastic change of the biblical story is probably as follows: "In a work one of whose purposes was to propagandize for the Jews among the Greeks it would not have been productive for the dramatist to include or even mention the wanderings of 40 years in the desert. In itself 40 years of difficult wanderings might have diminished the sense of God's concern and aid for the Jews which Ezekiel was eager to promote. To include the cause of these wanderings, that is the sinfulness of the Jews which impelled God to so punish them, would scarcely have increased the luster of these ancient Jews in the eyes of the pagan audience."[46] No less telling is the fact that towards the end of the play, the author has the Egyptians shout: "Let us run back home and flee the hands of the Most High, for He is helping them, but on us he is wreaking destruction" (239–241). Here Ezekiel has the Egyptians confess that Israel's God is not only the most powerful but also the Most High. In my opinion we can safely assume that one of Ezekiel's purposes was to react to anti-Jewish versions of the exodus story by means of a pro-Jewish play.

The *Wisdom of Solomon* is a semi-philosophical wisdom text from the early first century CE[47] that devotes no less than nine long chapters to the exodus.[48] In a series of seven antitheses, the author compares the Egyptians and the Israelites. For instance, the Egyptians were being slain by locusts and flies, while the Israelites survived a serpent attack through the agency of the bronze serpent; the Egyptians were unable to eat because of the hideousness of the

[43] For example, Nicarchus, Ptolemy Chennus, Helladius (see Stern 1974: no. 248; Stern 1980: nos. 331, 472).

[44] The verb *nitsel* (pi.) has the meaning "to strip, plunder, take spoil." The LXX has *eskyleusan*.

[45] *Ap.* Justinus, *Hist. Phil.* 36, epitome 2.13 (Stern 1974 no. 137).

[46] Jacobson 1983: 134–135. Lanfranchi 2006 neglects this anti-pagan aspect of the play and overemphasizes its role as "reenactment" of the exodus during celebrations of the Jewish Passover festival.

[47] See Winston 1979.

[48] On these chapters see esp. S. Cheon 1997. Cheon, too, sees these chapters as directed against anti-Jewish sentiments in Alexandria but he connects this (too closely) to the riots in Alexandria in 38 CE. In striking contrast to *Wisdom of Solomon*, the slightly later Pseudo-Philonic *Liber Antiquitatum Biblicarum* summarizes the whole exodus story in just one very short chapter (10). But it is of Palestinian not of diasporic provenance.

beasts sent against them, while Israel, after briefly suffering want, enjoyed exotic quail food; on the same night that the Egyptian first-born were destroyed, Israel was summoned to God and glorified. And, last but not least, in the final chapter the Egyptians are accused of *misoxenia*, "hatred of foreigners" or "hostility toward strangers" (19.13), exactly the same accusation that was leveled against the Jews by Alexandrian Jew-haters from the very beginning. This cannot be sheer coincidence. "In styling the conduct of the Egyptians as *misoxenia*, the author is reversing the very charge made against the Jews by pagan contemporaries."[49] And the reversal of the charge is here made in the context of the exodus story. Apparently, this story in its anti-Jewish form was still in the air in the time of this author, as writers such as Chaeremon and Apion prove.

It would be worthwhile to investigate whether Philo and Josephus yield evidence for polemics between Jews and pagans in their exodus stories, but time and space forbid.[50] What little has been presented here hopefully suffices to demonstrate that the foundational event and myth of the Jewish people, the exodus from Egypt, remembered by Jews all over the world every year at Passover up till the present day,[51] was a bone of contention right from the moment Jews and Greeks, especially Graeco-Egyptians, came into contact with each other in the early Hellenistic period. The two opposite versions of the exodus story easily lent themselves to give expression to the social tensions between these groups.[52]

[49] Winston 1979: 327–328. Winston points out that even the otherwise friendly Hecataeus of Abdera, one of the earliest Greek authors on Judaism (ca. 300 BCE), claimed that Moses introduced "a form of life encouraging a certain seclusion from humankind and hatred of aliens," *apanthrôpon tina kai misoxenon bion* (*ap*. Diodorus Siculus 40.3.4).

[50] One could think of Philo's remark that the Israelites took objects from the Egyptians not out of avarice or, *as someone might say to accuse them*, in covetousness of what belonged to others; it was the wage due for all their time of service (*Vit. Mos.* 1.141); or of Josephus' rebuttal of the rumor that Moses was a leper (*Ant. Jud.* 3.265–268).

[51] For the origins and early development of the Passover rite see esp. Bokser 1984 (p. 20 on Ezekiel).

[52] I owe thanks to my friend Dr. James N. Pankhurst for the correction of my English.

References

Assmann, J. 1997. *Moses the Egyptian: The Memory of Egypt in Western Monotheism*. Cambridge, MA: Harvard University Press.

Aziza, C. 1987. L'utilisation polémique du récit de l'Exode chez les écrivains alexandrins (IVème siècle av. J.-C.-Ier siècle ap. J.-C.). *Aufstieg und Niedergang der Römischen Werlt* II 20.1: 41–65.

Barclay, J.M.G. 1996. *Jews in the Mediterranean Diaspora from Alexander to Trajan (323 BCE–117 CE)*. Edinburgh: Clark.

Barclay, J.G.M. 2007. *Against Apion*, Flavius Josephus. Translation and Commentary, vol. 10. Leiden: Brill.

Bar-Kochva, B. 2010. *The Image of the Jews in Greek Literature*. Berkely, CA: University of California Press.

Berthelot, K. 2003. *Philanthrôpia judaica: Le débat autour de la 'misanthropie' des lois juives dans l'Antiquité*, Supplements to the Journal for the Study of Judaism, vol. 76. Leiden: Brill.

Bloch, R.S. 2002. *Antike Vorstellungen vom Judentum: Der Judenexkurs des Tacitus im Rahmen der griechisch-römischen Ethnographie*, Historia Einzelschriften, vol. 160. Stuttgart: Franz Steiner.

Bokser, B.M. 1984. *The Origins of the Seder: The Passover Rite and Early Rabbinic Judaism*. Berkeley, CA: University of California Press.

Cheon, S. 1997. *The Exodus Story in the Wisdom of Solomon*, Supplements to the Journal for the Study of the Pseudepigrapha, vol. 23. Sheffield: Sheffield Academic Press.

Cook, J.G. 2004. *The Interpretation of the Old Testament in Greco-Roman Paganism*, Studien und Texte zum Antiken Christentum, vol. 23. Tübingen: Mohr Siebeck.

Denis, A.-M. 2000. *Introduction à la littérature religieuse judéo-hellénistique*, vol. 2. Turnhout: Brepols.

des Places, E. 1973. *Numénius: Fragments*. Paris: Les Belles Lettres.

DiTommaso, L. 2001. *Bibliography of Pseudepigrapha Research 1850–1999*, Supplements to the Journal for the Study of the Pseudepigrapha, vol. 39. Sheffield: Sheffield Academic Press.

Feldman, L.H. 1993. *Jew and Gentile in the Ancient World: Attitudes and Interactions from Alexander to Justinian*. Princeton, NJ: Princeton University Press.

Feldman, L.H. 1996. *Studies in Hellenistic Judaism*, Arbeiten zur geschichte des Antiken Judentums und Urchristentums, vol. 30. Leiden: Brill.

Fraser, P.M. 1972. *Ptolemaic Alexandria*, vol. IIb. Oxford: Clarendon.

Funkenstein, A. 1993. *Perceptions of Jewish History*. Berkeley, CA: University of California Press.

Gager, J.G. 1972. *Moses in Greco-Roman Paganism*. New York, NY: Nashville.

Gruen, E.S. 1998. *Heritage and Hellenism: The Reinvention of Jewish Tradition*. Berkeley, CA: University of California Press.

Gruen, E.S. 2002. *Diaspora: Jews amidst Greeks and Romans*. Cambridge, MA: Harvard University Press.

Henten Van, J.W. 1999. Typhon. In *Dictionary of Deities and Demons in the Bible*, 2nd ed, ed. K. van der Toorn, B. Becking, and P.W. van der Horst. Ann Arbor, MI: Brill.

Heubner, H. and Fauth, W. 1982. *P. Cornelius Tacitus: Die Historien*, Band V, Heidelberg: Carl Winter.

Holladay, C.R. 1983. *Fragments from Hellenistic Jewish Authors, Volume 1: Historians*. Chico: Scholars Press.

Jacobson, H. 1983. *The Exagoge of Ezekiel*. Cambridge, MA: Cambridge University Press.

Jacobson, H. 2006. Artapanus Judaeus. *Journal of Jewish Studies* 57: 210–221.

Lanfranchi, P. 2006. *L'Exagoge d'Ezéchiel le Tragique*, Studia in Veteris Testamenti Pseudepigrapha, vol. 21. Leiden: Brill.

Redford, D.B. 1992. Akhenaten. *Anchor Bible Dictionary* 1: 135–137.

Redford, D.B., and J.M. Weinstein. 1992. Hyksos. *Anchor Bible Dictionary* 3: 341–348.

Schäfer, P. 1997. *Judeophobia: Attitudes towards the Jews in the Ancient World*. Cambridge, MA: Harvard University Press.

Schürer, E. 1986. The History of the Jewish People in the Age of Jesus Christ, rev. ed. by G. Vermes, F. Millar, M. Goodman, vol. III/1, Edinburgh: Clark

Seidlmayer, S.J. 1998. Hyksos. *Der Neue Pauly* 5: 780.

Sterling, G.E. 1992. *Historiography and Self-Definition: Josephos, Luke-Acts and Apologetic Historiography*, Supplements to Novum Testamentum, vol. 64. Leiden: Brill.

Stern, M. 1974–1980. *Greek and Latin Authors on Jews and Judaism*, vol. 1 and 2. Jerusalem: The Israel Academy of Sciences and Humanities.

van der Horst, P.W. 1983. Moses' Throne Vision in Ezekiel the Dramatist. *Journal of Jewish Studies* 34: 21–29.

van der Horst, P.W. 1984a. *Chaeremon, Egyptian Priest and Stoic Philosopher. The Fragments Collected and Translated with Explanatory Notes*, Etudes Préliminaires des Religions Orientales dans l'Empire Romain, vol. 101. Leiden: Brill. 2nd rev. ed. 1987.

van der Horst, P.W. 1984b. Some Notes on the Exagoge of Ezekiel. *Mnemosyne (ser. IV)* 37: 354–375.

van der Horst, P.W. 1985. Schriftgebruik bij drie vroege joods-hellenistische historici: Demetrius. *Artapanus, Eupolemus, Amsterdamse Cahiers voor exegese en bijbelse theologie* 6: 144–161.

van der Horst, P.W. 1988. The Interpretation of the Bible by the Minor Hellenistic Jewish Authors. In *Mikra. Text, Translation, Reading and Interpretation of the Hebrew Bible in Ancient Judaism and Early Christianity*, Compendia Rerum Iudaicarum ad Novum Testamentum, vol. II/1, ed. M.J. Mulder, 519–546. Assen: Van Gorcum.

van der Horst, P.W. 2002. Who Was Apion? In *Idem, Japheth in the Tents of Shem. Studies on Jewish Hellenism in Antiquity*, Contributions to Biblical Exegesis and Theology, vol. 32, 207–221. Leuven: Peeters.

van der Horst, P.W. 2008. *The Myth of Jewish Cannibalism: A Chapter in the History of Antisemitism*, Proceedings of the Israel Academy of Sciences and Humanities, vol. VIII no. 3. Jerusalem: The Israel Academy of Sciences and Humanities.

van der Horst, P.W. 2010. Did the Gentiles Know Who Abraham Was? In *Abraham, the Nations, and the Hagarites: Jewish, Christian, and Islamic Perspectives on Kinship with Abraham*, Themes in Biblical Narrative, vol. 13, ed. M. Goodman, G.H. van Kooten, and J.T.A.G.M. van Ruiten, 61–75. Leiden: Brill.

van der Toorn, K. 1999. Seth. In *Dictionary of Deities and Demons in the Bible*, 2nd ed, ed. K. van der Toorn, B. Becking, and P.W. van der Horst, 748–749. Leiden: Brill.

van Henten, J.W., and R. Abusch. 1996. The Jews as Typhonians and Josephus' Strategy of Refutation in Contra Apionem. In *Josephus' Contra Apionem: Studies in Its Character and Context with a Latin Concordance to the Portion Missing in Greek*, Arbeiten zur geschichte des Antiken Judentums und Urchristentums, vol. 34, ed. L.H. Feldman and J.R. Levison. Leiden: Brill.

Verbrugghe, G.P., and J.M. Wickersham. 2001. *Berossos and Manetho Introduced and Translated: Native Traditions in Ancient Mesopotamia and Egypt*. Ann Arbor, MI: University of Michigan Press.

Winston, D. 1979. *The Wisdom of Solomon (Anchor Bible 43)*. Garden City: Doubleday.

Part VII

The Exodus as Cultural Memory

The Exodus and the Bible: What Was Known; What Was Remembered; What Was Forgotten? 30

William G. Dever

Abstract

This chapter offers an archaeological critique of the current model of the Hebrew Bible as "cultural memory" with particular reference to the exodus–conquest narrative. Instead of asking how these texts functioned socially, religiously, and culturally, this chapter asks "What Really Happened?" This approach will facilitate a critique of the literary tradition based on external rather than internal evidence, attempting to isolate a "core history."

Introduction

The notion of "cultural memory" is now in vogue among some biblical scholars and historians.[1] As far as I am concerned, as an archaeologist, this simply designates the fundamental concept with which we deal: "culture" which is memory. Culture is formed by the patterned repetition of thoughts and actions in a social context that gives them meaning and reinforces that meaning, until the whole becomes "tradition," eventually enshrined in literary form, in this case in the Hebrew Bible.

In my judgment, the question who eventually wrote down the biblical story of the Exodus—when, where, or why—is of secondary importance for the historian, whose primary concern is the original events and their documentation. We must begin with the text that we happen to have, not an Urtext that we try to reconstruct, or the text we might wish to have. Reception history, now in vogue, is of little help.

Neither are many biblical scholars, who after the "literary turn," deal mostly with the transmission of the story, not any reality behind it.[2] This seems to me yet another myth and myth-making,

[1] For convenient orientation and bibliography, see Barstad (2010). The theme of this symposium on the Exodus reflects the current trend. One of the participants, Jan Assmann, is a major player; cf. Assmann (2007): cf. also the useful essay of Hendel (2010).

W.G. Dever (✉)
Near Eastern Studies, University of Arizona,
Tucson, AZ, USA

Lycoming College,
Williamsport, PA, USA
e-mail: gaber@lycoming.edu

[2] "Reception history" seems to be little more than an extension of the "reader response" approach popularized some time ago by New Literary Criticism. It simply brings the response up to the present moment, including popular media. To most archaeologists, antiquarians as they are, that is irrelevant. The impact of the "literary turn" on biblical studies, an offshoot of postmodernist

T.E. Levy et al. (eds.), *Israel's Exodus in Transdisciplinary Perspective*,
Quantitative Methods in the Humanities and Social Sciences, DOI 10.1007/978-3-319-04768-3_30,

ancient and modern—another legacy of postmodernism.[3] We archaeologists deal with the reality, "frozen in time," artifacts that have no subsequent "memory" to compromise then, since they are inanimate, and once buried they are invisible and have no observers. These artifacts are what Albright famously called "realia," a superior witness than later texts, since they are contemporary. The biblical text "refers" to the reality from a great distance; the artifact *is* the reality.

That is where history-writing begins. Theology, which is "historicized myth," comes later; and so does "cultural memory." Therefore, archaeologists are understandably not much concerned with "cultural memory," even though it is a theme of this symposium.[4]

Hans Barstad has recently observed of current cultural history studies among Biblicists that while history and memory are always intertwined, they are not identical, and they must always be carefully separated. And, as he puts it, "to the historian, everything one does will be governed by the quest to find out 'what really happened.'" He makes a plea for just such a positivist approach, however outdated that may seem to many today (Barstad 2010: 8).

Contrast that with Lemche, in the same volume, who says rightly that "cultural memory is not history," but is a way that people construct a history. He then goes on to declare that "whether or not this construction has much or little to do with what actually happened [is] something of little interest to students of cultural memory" (Lemche 2010: 12). That is why some of us think the revisionists nihilists, at least where history-writing is concerned (unless one means simply the history of ideology).

From the perspective of the historian—and that's what archaeologists are, "historians of things"—all the current emphasis on cultural memory is simply another way of talking about tradition.[5] There is really little new here, and nothing very promising for the history of a real Israel in the Iron Age. The fundamental issue is what it was at the beginning of critical biblical studies 150 years ago: *historicity*. What if anything, may lie behind the traditions? What actual events may have given rise to the stories?

The only thing that has changed is that archaeology is now being recognized as a "primary source"—indeed our only source of external data, which alone may (or may not) verify the events in question.[6]

If we are to talk about history and historical method, we must begin by defining what we intend, as well as separating the several tasks before us. I would suggest that we distinguish between several kinds of history-writing and the specific approaches that best characterize each where archaeology and the Hebrew Bible are concerned.[7]

Type of history	Method
1. A history of events	Archaeology
2. A history of traditions	Form and redaction criticism
3. A history of literature	Literary or source criticism
4. A history of ideology	History and philosophy of religion
5. A history of institutions	Political history
6. A history of later interpretations	Reception history; theology

Obviously all these approaches overlap, and all have their place, but we must specialize. As an archaeologist, I will address only the first category—the historicity of any "Exodus events"—while most other contributions here

epistemology, is too well known to need documentation, but cf. the balanced critique of Barr (2000).

[3] Barstad, one of Europe's best biblical scholars, has made the distinction clear, showing how "cultural memory" may mean the end of real history, as in Davies (2008). See Barstad (2010); and cf. Barstad (2007, 2008). For an extensive critique of biblical revisionism and its background in postmodernism, see Dever (2001).

[4] For the superiority of artifacts over texts as primary data, see Dever (2001: 81–95, 2010) (a review of Grabbe 2007, a leading biblical scholar who has advocated seeing the archaeological data as primary).

[5] Lemche (2010: 12) has made the same point, although with no misgivings.

[6] See footnote 4 above.

[7] On types of history, see further Dever (1997b, c). There are few if any other discussions by archaeologists.

will apparently deal with the supposed cultural meanings of the biblical story.

Let me turn now to the question in my title.

What Did the Biblical Writers Really Know?

I addressed this question in a 2001 book entitled *What Did the Biblical Writers Know, and When Did They Know It?* This was a deliberate challenge to the biblical revisionists, for whom no real history of Israel is possible, because it would be inconvenient for their minimalist ideology.

My answer was (1) that there was a "historical Israel" in the Iron Age; (2) that many of the biblical stories are firmly anchored in the context of the Iron Age, not in the revisionists' imaginary "Persian" or "Hellenistic" era; and (3) that this flesh-and-blood Israel is being dramatically brought back to life by archaeological discoveries, now our primary source.

Thus the biblical writers *could* be good historians by the standards of their day, when they chose to be, despite their obvious theocratic program. The Hebrew Bible overall may be *Heilsgeschichte*; but behind the literary construct that we now have, there undoubtedly lie older oral and written traditions. That means that historians, with the aid of archaeology as a control and corrective, may sift out from the written record some genuine historical information here and there (Dever 2001).

Can we do that, however, with the Exodus story, when our extant written sources are at least six or seven centuries later than the purported events? The dilemma is reflected in the theme of this symposium: the Hebrew Bible's "cultural memory" and history.

I begin to address the question by summarizing what I call "convergences," lines of evidence from both our sources—textual and archaeological data—that come together to create a portrait of past events that seems realistic, i.e., "true beyond a reasonable doubt." As an archaeological historian, I can attest confidently to several things that the Biblical writers did actually know, points at which their story, however late and tendentious, has gotten it right.

1. First, an "Israel" as a state and a people did exist in Canaan (or our southern Levant) in the Iron Age, and it had long historical roots there. The biblical writers knew that; and we now know it too, not only from the rich and detailed archaeological record but also from extrabiblical texts like the Merneptah inscription, the Mesha stele, the Tel Dan stele, and numerous Neo-Assyrian and Neo-Babylonian annals. Despite the biblical revisionists' discomfort, this Israel was *not* "invented": it has been discovered.[8]
2. Second, the biblical writers knew that this Israel had arisen partly out of conflict with the age old Canaanite culture of the region, in our Bronze Age. Despite acknowledging some continuities, the biblical writers, however, saw their Israel as revolutionary—a new and different "ethnicity," with a distinct sense of national identity and destiny. Today, despite some skeptics, we can specify Israelite ethnicity in detail on the basis of an independent analysis of the archaeological record. The ancient Israelites were demonstrably different from contemporary peoples such as the Philistines, Phoenicians, Arameans, Ammonites, Moabites, and Edomites. Israel was not unique, in the sense; but it was distinctive in material culture, and, therefore, by necessity distinctive in culture in general (including, of course, religious beliefs and practices).[9]
3. Third, the biblical writers knew that their origins were intertangled with the appearance of the "Sea Peoples" in Canaan, in particular the Philistines. They remembered the latter as their enemy for centuries; and they saw their early expansion as the event that triggered the rise of the Israelite monarchy after several

[8] "Invention" is a favorite term of the biblical revisionists; cf. Whitelam (1996), Thompson (1999), and especially Liverani (2005) (although not necessarily a member of the revisionist school).

[9] For positive views, see Killebrew (2005), Faust (2006, 2010), Dever (2007). Literature on negative views will also be found in these works.

generations of charismatic leadership in an agropastoral, village-based society.

Today, we can easily show that the biblical characterization of the Philistines, although minimal and late, is surprisingly accurate. Furthermore, their "prehistory" of both peoples, and their trajectory toward more complex or "state-level" formations, accords well with our current archaeological knowledge (Gitin 2010).

4. Finally, the cultural memory of the Hebrew Bible includes an indisputable sense that there had been "Egyptian connections." This memory is preserved in Egyptian-style names; in the stories of Joseph and Moses; in the fact that knowledge of the god "Yah" came perhaps from Midian and the Shasu people there; and, above all, in the notion that Yahweh, the God of Israel, was greater than the Pharaohs of Egypt and had the power to liberate and create a new, free, and sovereign people. The biblical writers knew considerable Egyptian lore and literature.[10]

All the above reflects *genuine* historical knowledge, the formation of a tradition that, however much refracted by later authors and editors, was rooted in reality. That reality is what we now know archaeologically as the "Iron Age of ancient Palestine"; and it confirms in broad outline what the biblical writers knew, at least thus far.

What Did the Biblical Writers "Remember"?

Memories do not necessarily correspond to reality; they are constructed out of some genuine recollections, but they are also embellished by later details, as well as enhanced by subsequent life experiences. For instance, our "memories" of our childhood are obviously compounded of things that we do actually recall, plus stories our parents have told us (*their* memories), and probably also things like mementoes and photographs that we still possess. But we all know that these memories, these stories of who we are, tend to grow with constant retelling. And they often become "larger-than-life" narratives—still true, but mostly metaphorically. The biblical "memories" are like that, as a recapitulation of some of the above memories and events will show.

But what about the *facts*? As historians, whether Biblicists or archaeologists, we must deal with such facts as we have, not mere speculations, which are the stuff of philosophy or theology. Thus we must ask: Are the biblical "memories" of the Exodus grounded in historical events for which we have actual evidence?

1. There may well have been a long period of slavery for some of Israel's ancestors in Egypt, since we know from pictorial representations and texts that Asiatics ("Canaanites") had been present in the Delta and sometimes enslaved there from at least the early second millennium onward. The cities of "Pi-Rameses" and "Pithom" are well attested in the time of Ramses II; and we have Egyptian depictions of slaves making mud bricks, even a portrait of Ramses himself beating a foreign slave.

Thus the biblical portrait of Hebrew servitude in Egypt for some 400 years (roughly the 18th and early 19th Dynasties) is not essentially fantastic. It may rest on some genuine historical memories of the long-term movement of Amorite and Canaanite peoples into the Delta, who were known to Egyptians as "Amu," or "sand-dwellers" from southwestern Asia. Nevertheless, as is well known, there is not a single reference in the whole of Egyptian literature to these "Hebrews" in the sixteenth to thirteenth century BCE. When we do meet them, they are Merneptah's "Israelites" ca. 1208 BCE somewhere in Canaan, described nonchalantly simply as a loosely organized people, who are said to have been exterminated (Dever 2007). Had Egyptian intelligence known these Israelite or Hebrew people as escaped slaves who had shortly before

[10] See this volume, Chaps. 8 (Moshier & Hoffmeier), 15 (Hoffmeier), and 34 (Redford).

humiliated the Egyptian Pharaoh, they surely would have further identified them and boasted specifically of the Pharaoh's revenge, as Egyptian literature typically does.

2. The biblical accounts of Yahweh's spectacular miracles, enacted through the mediator of Moses, are extraordinarily detailed and memorable (thus the tenacity of the central festival of Pesach). But no modern, sophisticated reader can give any credence to these stories of Yahweh's dramatic intervention in nature and history. Attempts have been made to rationalize these miracles, but that misses the point. They cannot *be* "explained": that's what makes them miraculous. You believe them, or you do not! The ancient storytellers "remembered" them and thus believed them; most of us do not. There is no way of getting around that.[11]
3. The biblical memory of wandering through the Sinai (that "great and terrible wilderness") for some 40 years is also fraught with difficulties. The biblical story as it now stands features "a mixed multitude" that, to have fielded a fighting force of some "600,000," would have numbered some three million, an enormous group that the Sinai could never have supported. Attempts to rationalize this inflated figure by playing on the Hebrew term *'alûf* are unpersuasive.[12]
4. Of the many sites named on the Delta-Sinai itinerary, only a few have been positively identified: (1) the fortress of "Migdol," plausibly located in Hoffmeier's excavations in the Tell el-Borg area; (2) "Pithom," probably at Tell Retabeh; (3) "Ramses," probably Qantir; and (4) "Kadesh-barnea," where Israeli excavations at 'Ain el-Qudeirat have discovered a tenth to ninth century BCE fort (probably a later pilgrimage site in "cultural memory"), but nothing other than a few earlier sherds of the twelfth century. There cannot have been a sojourn of "38 years" there in the late thirteenth century BCE; even a few Bedouin-like folk would have left some remains.[13]
5. As for the numerous places and peoples the Israelite tribes are said in Numbers to have encountered in Transjordan, few can be identified archaeologically, despite several determined and hopeful efforts by scholars who were highly motivated. The silence of excavations at securely identified sites such as Dibon (*Dhibân*) and Heshbon (*Hesbân*) speaks volumes. In Edom, the biblical sites probably did not come into visible existence as part of a state until the seventh century BCE.[14]

   Attempts of a few scholars like Redford, Na'aman, Rainey, and Faust (the latter tentatively) to connect the early Israelites with the pastoral-nomadic "Shasu," known from Egyptian 19th Dynasty texts are largely speculative. Such a theory flies in the face of almost everything that we know historically and ethnographically about the sedentarization of pastoral nomads. As Zvi Lederman, a staff member at 'Izbet Ṣarṭah, once wrote: "Nomads They Never Were."[15]
6. The follow-up of the exodus and the passage through the Sinai and Transjordan was, of course, the conquest of all of Canaan beyond the Jordan, as recounted in Joshua (less explicitly in Judges). Here we need not delay. Of the 31 sites said in the biblical narrative to have been taken (i.e., overrun or conquered) by invading Israelites, only two or three show any signs of destruction at the requisite LB/Iron I transition. Hazor was indeed violently destroyed; but the acting

[11] The papers here by natural scientists are welcome, but they do not provide an explanation of what really happened.

[12] The fact that the earlier sources allow for a smaller number does not resolve this problem—or the many others in the biblical narratives.

[13] See Hoffmeier (2005). On Kadesh-barnea, see now the final publication, Cohen and Bernick-Greenberg (2007).

[14] How Levy's metal-working installations in the Wadi Fidan, dated as early as the eleventh to tenth century BCE, will affect this date is not yet clear; see Levy (2010) and references there.

[15] See the full discussion in Dever (2001: 54–71, 1977).

field Director, Sharon Zuckerman, thinks that this was due to internecine warfare. Bethel shows a destruction at this time, followed by a presumably Israelite squatter occupation, but inadequate excavation and publication preclude any explanation.[16]

To make a long story short, today not a single mainstream biblical scholar or archaeologist any longer upholds "biblical archaeology's" conquest model. Various theories of indigenous origins prevail, in which case there is neither room nor need for an exodus of significant proportions. To put it succinctly, if there was no invasion of Canaan by an "Exodus group," then there was no Exodus. A small "exodus group" may have existed, perhaps a few hundred or thousand, and they could have come later to be identified with the biblical "House of Joseph." It was the view of these two southern tribes, who had a dominant influence in shaping the later literary tradition, that "all Israel" had come out of Egypt. In time that became quite understandably part of the foundation-myth of the Hebrew Bible, particularly in the Exile and hereafter. But the ancestors of the majority of ancient Israelites and Judeans had never been in Egypt. They were essentially Canaanites, displaced both geographically and ideologically.

As for the implications the biblical notion of a pan-military Israelite conquest of Canaan, a few scholars have sought to rationalize the story in several ways. Some evangelical scholars have argued that the biblical narrative of "destroyed Canaanite cities" specifies that only three are said actually to have been "burnt." Thus the absence of archaeologically attested destruction layers at many other sites means nothing. Yet this is disingenuous. Are we to suppose that the Israelite armed forces drew up at the gates of major Canaanite cities, upon which the civilian population surrendered and conveniently disappeared into the hinterland? The overall biblical narrative is clear; and it is about genocide, the extermination of the entire Canaanite population, men, women and children. And this is said to be Yahweh's will. I would reject that, historically and morally. Fortunately, it didn't really happen, as the authors of Judges acknowledge (and the Deuteronomic historians accepted, by putting Joshua and Judges back-to-back in the Canon).

7. Finally, the notion of a "twelve-tribe" league that had persisted throughout the enslavement in Egypt, characterized the early settlement in Canaan, and even continued into the monarchy, may be a late literary construct. This notion probably crystallized only in the exile, when there was a desperate attempt to create an identity for a people who were now without a state, a temple, or any other national institutions. But the rallying cry "To your tents, O Israel" is little more than nostalgia for a past that never was.[17]

The previous "events," while perhaps based on some cultural memory (and all memory *is* cultural), were invented, rather than actually being remembered. Of course, many biblical scholars are no longer interested in the issue of historicity. Their histories are only histories of the literature, or of the Hebrew language—not of the living community. They ask only "how the text is able to say what it says," how it functioned as "cultural meaning," not whether it is true. As a historian, however, I am asking the latter question—the historical question and what it means—with no apologies.

As for how the Exodus narrative might have functioned in the actual Israelite society of the Iron Age, the answer is that it probably did not function at all until perhaps near the end of the monarchy, when the Pentateuchal and Deuteronomist traditions were first reduced to writing (not earlier than the late eighth to seventh century BCE). Even then, however, the biblical narrative

[16] See Dever (2011) and the full discussion there; cf. also Dever (1997a). For the latter, see Lederman (1992).

[17] The notion of a "nomadic ideal" persists in the literature; but for independent refutations cf. Dever (1995), Hiebert (2009).

would have constituted cultural meaning only for the handful of elites who wrote the text in priestly and scribal circles, and the equally few literate people who could have or would have read such texts. Ordinary people would have had little recourse to these traditions.[18] There may have been, of course, an older oral tradition, one that could have stretched back even to the days of the settlement in the highlands. But speculating about how that oral tradition may have had a part in shaping Israel's ethnogenesis does not seem very useful to the historian. On the other hand, the revisionists' notion that virtually all of Israel's history was "invented" goes much too far; even though a good deal of her prehistory was invented. (I gladly leave wrestling with that fact to theologians and clerics.)

What Did the Biblical Writers Forget?

Here I can only speculate, in contrast to asking in my 2001 book, "What Did They Know"? We must look for a few clues that hint at remnants of a subconscious, largely lost (or suppressed) knowledge of a remote past, the memory of which was now fading with the textualization of tradition (and soon its scripturalization). Here the test would be whether this fleeting knowledge conforms to what we know with reasonable certainty, thanks to archaeological illumination of the facts on the ground.

Without mining the entire Hebrew Bible for such clues, often of necessity reading between the lines, I would suggest that in general what the biblical writers had forgotten by the late monarchy consisted of the original background of the themes outlined in the section "What Did the Biblical Writers Really Know?" above. That is, the writers knew the consequences of some of the events that they recalled, but they no longer recalled what may actually have happened. Here we have the advantage of hindsight, as well as detailed knowledge of both their past and their future, knowledge not available to them.

Here I will highlight only a few possible instances of "lost" knowledge, preserved perhaps only in intuition (or counter-intuition?).

1. It is well known that in the older strands of the Pentateuchal literature, it is El—"the God of the fathers"—who predominates, not Yahweh. This accords well with local Late Bronze Age Canaanite traditions, as illustrated particularly well in the Ugaritic texts. The basic sacrificial system of later Israel is likewise well attested at Ugarit. Here we have at least implicit knowledge of the biblical writers concerning elements of cultural continuity, which was the reality.
2. A rare acknowledgement of Canaanite backgrounds is found in Ezekiel 16:2, where the prophet complains: "You are of the land of the Canaanites; your father was an Amorite, and your mother was a Hittite." This is a tacit acknowledgement of indigenous, not foreign, origins of the Israelite peoples. One might even argue that the relentless polemics against Canaanite culture are the best evidence that this culture was indeed primeval and was still influential. Insisting constantly that "We are not like them" suggests that we actually are, or had been, something we would rather forget, but cannot.
3. Remnants of an Aramean connection are attested by the refrain "my father was a wandering Aramean." This sentiment is often projected back upon a presumed "patriarchal" epoch. But it fits much better in the Iron I period in the southern Levant, when we know that the Aramean peoples contemporary with early Israelites were becoming sedentary in Syria and they would soon experience a similar trajectory toward statehood (or city statehood in this instance). If our earliest Israelites had a history similar to that of the

[18] Few biblical scholars, elitists themselves, appreciate just how elitist the biblical texts are—limited not only by their late date but by a limited perspective. People could not have had any biblical texts before the seventh century BCE or so; and since at least 95% of them were illiterate, they could not have read these texts in any case. For a full exposition of the lives of ordinary people, see Dever (2012).

Arameans, this may indicate that both peoples had emerged out of the collapse of local Late Bronze Age culture.[19]

4. Finally, the overall continuity of Israelite and Canaanite culture, increasingly well documented and widely acknowledged, attests further to indigenous origins. By contrast, there is virtually no Iron I material culture in evidence at any of our early Israelites sites that betrays any Egyptian influence. Needless to say, our earliest Hebrew script and its language are essentially Canaanite (not "Moabite," as Rainey had argued).[20]

Conclusion

This inquiry began by asking several questions about the Hebrew Bible's "cultural memory" of supposed events in which the Israelite people began their existence as Hebrew slaves in Egypt, were miraculously liberated, then wandered through the Sinai for some 40 years before finally invading and occupying Canaan, the "Land of Promise" of the patriarchs. But the fact that there was no "Conquest" means that there was no "Exodus."

The archaeological and extra-biblical textual evidence adduced here shows that these larger-than-life stories cannot be read literally as history. They are "foundation myths," similar in function to those of other peoples we know in antiquity. Does that mean that these stories contain no truth? Not at all. Some memories may have been authentic. And even myths can be *profoundly* true, at least metaphorically, especially when they are couched in a form that has gripped the imagination of countless millions of people for more than 2,000 years. These are stories that still resonate with us—stories about liberation from tyranny; about the power of an apparently insignificant people to change the course of history; about a "New Israel," and a promised land here and now.

Some, reluctant to abandon a literal reading of the treasured biblical narrative, have resorted to desperate measures. They have insisted that a story of a nation's humble origins as slaves is not a story that anyone would simply have made up. However, such a story is the *perfect* foil against which to portray the Bible's *Magnalia dei*—Yahweh's "mighty acts," intervening in history to care for and glorify his people, *despite* all the odds, the "normal" and predictable course of events.

Others have argued that such a miraculous, detailed account as the exodus story could hardly have been "invented" out of whole cloth. On the contrary, its "fantastic" character means that the story cannot be read literally and given any credence, at least by modern critical thinkers.

In the end, "cultural memory" is about who we *think* we are. And that—not the bare facts—is what matters. We can be entirely wrong about what really happened to us, about the past; but what we *make* of the past as remembered is what may come to define us. The ancient Israelites thought that they were different, that they had a unique destiny. They were; and they did have. And that has become part of our "cultural memory."

Nevertheless, the Exodus–Conquest story overall is fiction—the stuff of legend. Whoever the early Israelites were, they were not invaders from Egypt, the Sinai, or Transjordan. They were indigenous peoples, displaced Canaanites, though possibly some had been slaves in Egypt, passing on genuine historical memories. The American and Israeli scholars who have written the most extensively on Israelite origins have virtually ignored the biblical "Exodus"; we have neither room nor need for it. These "events" are not remembered except perhaps in a few details. They are mostly invented. Like the Pilgrim Story for earlier Americans, it is a "foundation myth" and functioned as such. It is like other biblical stories, which Ernst Axel Knauf once aptly described as "pseudo-histories of non-events."

[19] Cf. Dever (2003), Faust (2006), Sader (2010).

[20] Cf. Rainey (2007), Dever (2011).

The real task of modern scholars may be to explain how and why such myths as the Exodus and Conquest stories ever developed in the first place—and, above all, why they later became so tenacious and influential. But I suspect that task is best left to folklorists; historians of comparative religion; students of the philosophy of religion; literary critics; and theologians.

I close with an observation of my colleague and friend, Ron Hendel:

> The collective memories of a culture recall and recreate a past that is relevant for the present. It is not the past of the historian, nor is it a wholly fictional past. It is a representation of the past that serves as a foundation and charter for collective values and identify, and as such is true existentially and morally, if not true historically. (Hendel 2010: 255).

References

Assmann, J. 2007. *Das kulturelle Gedächtnis, Schrift, Erinnerung und politische Identität in frühen Hochkulturen*, 6th ed. Munich: Beck.

Barr, J. 2000. *History and Ideology in the Old Testament: Biblical Studies at the End of a Millennium*. Oxford: Oxford University Press.

Barstad, H. 2007. The history of Ancient Israel: What directions should we take? In *Understanding the history of Ancient Israel*, ed. H.G.M. Williamson. Oxford: Oxford University Press.

Barstad, H.M. 2008. *History and the Hebrew Bible: Studies in Ancient Israelite and Ancient Near Eastern Historiography*, FAT, vol. 61. Tübingen: Mohr Sieback.

———. 2010. "History and Memory: Some Reflections on the 'Memory Debate'" in Relation to the Hebrew Bible. In *The Historian and the Bible: Essays in Honour of Lester L. Grabbe*, ed. P.R. Davies and D. V. Edelman, 1–10. New York: T & T Clark.

Cohen, R., and H. Bernick-Greenberg. 2007. *Excavations at Kadesh Barnea (Tell el-Qudeirat) 1976–1982*. Jerusalem: Israel Exploration Society.

Davies, P.R. 2008. *Memories of Ancient Israel: An Introduction to Biblical History—Ancient and Modern*. Louisville, KY: Westminster John Knox.

Dever, W.G. 1977. Palestine in the Second Millennium BCE: The Archaeological Picture. In *Israelite and Judean History*, ed. J.H. Hayes and J.M. Miller, 70–120. Philadelphia, PA: Westminster.

———. 1995. Will the Real Israel Please Stand Up: Part II: Archaeology and the religions of Ancient Israel. *Bulletin of the American Schools of Oriental Research* 298: 37–58.

———. 1997a. Is There Any Archaeological Evidence for the Exodus? In *Exodus: The Egyptian Evidence*, ed. E.S. Frerichs and L.H. Lesko, 67–86. Winona Lake, IN: Eisenbrauns.

———. 1997b. Philology, Theology, and History. What Kind of History Do We Want and What's Possible? In *The Archaeology of Israel: Constructing the Past, Interpreting the Present*, ed. N.A. Silberman and D. Small, 290–310. Sheffield: Sheffield Academic.

———. 1997c. On Listening to the Texts-and the Artifacts. In *The Echoes of Many Texts: Reflections on Jewish and Christian Traditions: Essays in Honor of Lou H. Silberman*, ed. W.G. Dever and J.E. Wright, 1–12. Atlanta, GA: Scholars Press.

———. 1998. Israelite Origins and the 'Nomadic Ideal': Can Archaeology Separate Fact from Fiction? In *Mediterranean Peoples in Transition: Thirteenth to Early Tenth Centuries B.C.E.* ed. S. Gitin, A. Mazar, and E. Stern, 230–237. Jerusalem: Israel Exploration Society.

———. 2001. *What Did the Biblical Writers Know and When Did They Know It? What Archaeology Can Tell Us About the Reality of Ancient Israel*. Grand Rapids, MI: Eerdmans.

———. 2003. *Who Were the Early Israelites and Where Did They Come From?* Grand Rapids, MI: Eerdmans.

———. 2007. Ethnicity and the Archaeological Record: The Case of Early Record. In *The Archaeology of Difference: Gender, Ethnicity, Class and the "Other" in Antiquity: Studies in Honor of Eric M. Meyers*, ed. D.R. Edwards and C.T. McCollough, 49–66. Atlanta, GA: American Schools of Oriental Research.

———. 2009. Merneptah's 'Israel', the Bible's and Ours. In *Exploring the Longue Durée: Essays in Honor of Lawrence E. Stager*, ed. J.D. Schloen, 89–96. Winona Lake, IN: Eisenbrauns.

———. 2010. Review of L. L. Grabbe, Ancient Israel: What Do We Know and How Do We Know It? London: T & T Clark. *Bulletin of the American Schools of Oriental Research* 357: 77–83.

———. 2011. Earliest Israel: God, Warriors, Revolting Peasants, or Nomadic Hordes? *Eretz-Israel* 30 (the Ben-Tor volume, 2011): 4*–12*.

———. 2012. *The Lives of Ordinary People: Where Archaeology and the Bible Intersect*. Grand Rapids, MI: Eerdmans.

Faust, A. 2006. *Israel's Ethnogenesis: Settlement, Expansion, and Resistance*. London: Equinox.

———. 2010. Future Directions in the Study of Ethnicity in Ancient Israel. In *Historical Biblical Archaeology and the Future: The New Pragmatism*, ed. T.E. Levy, 55–68. London: Equinox.

Gitin, S. 2010. Philistines in the Book of Kings. In *The Books of Kings: Sources, Composition, Historiography and Reception*, SVT, vol. 129, ed. A. Lemaire and B. Halpern, 301–364. Leiden: Brill.

Grabbe, L.L. 2007. *Ancient Israel: What Do We Know and How Do We Know It?* London: T & T Clark.

Hendel, R. 2010. Culture, Memory, and History: Reflections on Method in Biblical Studies. In *Historical Biblical Archaeology and the Future: The*

New Pragmatism, ed. T.E. Levy, 250–261. London: Equinox.

Hiebert, T. 2009. Israel's Ancestors Were Not Nomads. In *Exploring the Longue Durée: Essays in Honor of Lawrence E. Stager*, ed. J.D. Schloen, 199–205. Winona Lake, IN: Eisenbrauns.

Hoffmeier, J.K. 2005. *Ancient Israel in Sinai: The Evidence for the Authenticity of the Wilderness Tradition*. New York: Oxford University.

Killebrew, A.E. 2005. *Biblical Peoples and Ethnicity: An Archaeological Study of Egyptians, Canaanites, Philistines, and Early Israel, 1300–1100 B.C.E.* Atlanta, GA: Society of Biblical Literature.

Lederman, Z. 1992. Nomads They Never Were. In *Biblical Archaeology Today* [Proceedings of the Second International Congress on Biblical Archaeology, Jerusalem, June–July 1990], ed. A. Biran and J. Aviram, 483–484. Jerusalem: Israel Exploration Society.

Lemche, N.P. 2010. Did a Reform Like Josiah's Happen? In *The Historian and the Bible: Essays in Honor of Lester L. Grabbe*, ed. P.R. Davies and D.V. Edelman, 11–19. London: T & T Clark.

Levy, T.E. 2010. Integrating Anthropological, Digital, and Historical Biblical Archaeology. In *Historical Biblical Archaeology and the Future: The New Pragmatism*, ed. T.E. Levy, 3–42. London: Equinox.

Liverani, M. 2005. *Israel's History and the History of Israel*. London: Equinox.

Rainey, A. 2007. Whence Came the Israelites and Their Language? *Israel Exploration Journal* 57: 41–67.

Sader, H. 2010. The Arameans of Syria: Some Considerations on Their Origin and Material Culture. In *The Books of Kings: Sources, Composition, Historiography and Reception*, SVT, vol. 129, ed. A. Lemaire and B. Halpern, 273–300. Leiden: Brill.

Thompson, T.L. 1999. *The Mythic Past: Biblical Archaeology and the Myth of Israel*. New York: Basic Books.

Whitelam, K.W. 1996. *The Invention of Ancient Israel: The Silencing of Palestinian History*. London: Routledge.

31 Exodus as a *Mnemo-Narrative*: An Archaeological Perspective

Aren M. Maeir

Abstract

I discuss possible archaeological correlates from the second and first millennia BCE Levant and Egypt—spanning the Middle and Late Bronze Ages, the Iron Age I and II, and the Persian and Hellenistic periods—which may have served as background(s) for the formation, preservation, and transformation of the biblical and extra-biblical Exodus traditions. I will attempt to assess the character and relevance of strands of evidence from diverse periods and contexts and discuss the possible interface, and/or lack thereof, between these artifactually-based cultural events and the various Exodus narratives as reflected in the biblical texts and traditions.

The study of the Exodus, as can be seen from the chapters of this volume, is a complex and often highly-debated realm of research. In an attempt to suggest what might present a fresh approach to this broad topic, in this chapter, I try to combine two specific perspectives—an archaeological perspective on the one hand and the study of memory on the other.

I would like to start with what may seem a somewhat pessimistic archaeological view—I firmly believe that the study of the "Exodus" as an explicit event (or events)—is first and foremost not an archaeological exercise—but rather something that should be dealt with in the context of literary studies of the Bible, departments of theology, or perhaps even more productively—through the analytic perspectives of the various theoretical understandings on the formation of collective memories (for a recent collection of diverse perspectives on Cultural Memory, see, e.g., Erll and Nünning (2008)).

Only then can we turn to the archaeological/historical perspectives—since one can't deal with an "event" from a historical viewpoint—until one has defined the primary sources of this "event"—in the case of the Exodus—the biblical text.

To this can be added that I don't believe one can define a specific, single period or context in which the Exodus, as narrated in the biblical text, occurred per se. Rather, as will be expanded upon below, it should be understood as a literary matrix of *mnemo-narratives*—a narrative (or rather narratives) of memories.

As such, the Exodus does not represent a single set of temporally defined events that

A.M. Maeir (✉)
Tell es-Safi/Gath Archaeological Project, Institute of Archaeology, Bar-Ilan University,
Ramat-Gan 52900, Israel
e-mail: arenmaeir@gmail.com

T.E. Levy et al. (eds.), *Israel's Exodus in Transdisciplinary Perspective*,
Quantitative Methods in the Humanities and Social Sciences, DOI 10.1007/978-3-319-04768-3_31,

occurred even vaguely along the lines of the biblical *peshat* (the "simple" reading of the text—the so-called thin description). Rather, if one wishes to distill the historical core behind this story—one must look at numerous periods, contexts, and events, wide ranging in time, character, and levels of evidence—which were gathered together in this *mnemo-narrative* to eventually form the well-known Exodus story.

Over 150 years of intense study of the Hebrew Bible has resulted in perspectives that completely changed the understanding of this complex text, in a way very much different from traditional viewpoints. While many of the various facets of biblical research are highly debated (see for example the different perspectives on the Exodus narrative by various biblical scholars in this conference or volume—such as Friedman, Römer, Propp, and Schmid), all agree that the biblical text in general, and the Exodus narratives in particular, is the product of a long and complex process of formation, collation, editing, and reception. While biblical scholars today may disagree over the details—the complexity and multi-faceted nature of these texts is without doubt (see, e.g., Berner 2010, for a recent textual study of the Exodus narrative).

I believe that a particularly compelling way to look at the complexity of the biblical text is to compare it to a multi-period archaeological site, a "tell"—with all its layers, contexts, disturbances, and artifactual complexity (e.g. Brettler 1995: 113). And who if not archaeologists can connect to, understand, and utilize for their research such analogies? This is the approach that I suggest below.

This broad agreement among biblical scholars appears to have insufficiently filtered down, certainly to the general lay public (see regular notices in the media of finds that will "exonerate" the historicity of this or that biblical event and/or person), but more disturbing, to some of my archaeological colleagues. Time and again, one sees cases in which archaeologists who deal with sites and finds from periods in which portions of the biblical texts were formed (the Iron Age in particular) claim that a certain find, being it a specific site, building, or inscription, has the potential to "prove" the veracity of the straightforward biblical text and, overnight, cause 150 years of biblical studies to implode, collapse, and/or evaporate.

While all this may not be news to many of you in the present company—I believe that since this misconception is so prevalent—one must clearly and simply state: Sorry guys—it ain't going to happen—the modern research perspectives of the biblical text are here to stay!

There are various reasons for this yearning for a return to "simple," "Sunday-school" understanding of the biblical text. Most often it can be traced to ideological and/or religious stances—as complex, multilayered understandings of these texts drastically confound major portions of the core narratives—and that at times it deeply disturbs religious and ideological convictions.

Although I firmly believe that one can comfortably combine between many traditional religious/ideological tenets and the perspectives of modern research on the biblical text (and I say this as a committed, practicing, traditional Jew)—this is not the time and the place to delve into this.[1]

Rather, I would stress in general, and for the purpose of our discussion in particular (and I'm sure there is wall-to-wall agreement on this point in this volume), that when dealing with the historical study of even such an emotionally-laden topic as the story of the Exodus, we must shed, as much as we possibly can, our emotional, ideological, and religious baggage—and concentrate on distilling the evidence, and its meaning, that can be brought to the discussion, from our respective fields of scientific enquiry.

No less important is the need to counter something that I have time and again heard and read—the claim that those who look at the biblical text from a uniform, monolithic perspective relate to the biblical text "seriously"—while scholars who understand it as a complex and multi-faceted corpus of texts—don't take it seriously—or

[1] See, e.g., A. Maeir, "An Archaeological Perspective on Shavuot" (http://thetorah.com/archaeological-perspective-shavuot).

worse are ideologically driven in their "negative" attitude toward biblical evidence.

So let me be completely clear—I take the Bible VERY seriously. If anything, I have a suspicion—regarding those who profess a more simplistic view of the biblical text, it's not the Bible that they take more seriously—but rather their specific ideological/religious convictions.

Having said this, I would now like to return to the main issues at hand.

The study of memory, its effect on the formation of personal and collective traditions, and its interface with history are hardly a new topic. This has been broached often in the last century. Literary examples abound—perhaps the best known Proust's (1913–1927) *À la recherche du temps perdu* ("In Search of Lost Time").

The study of the complex interface between collective and cultural memory and history is also a well-worn track, whether "classical" treatments such as by Maurice Halbwachs (1941, 1950, 1952) and Aby Warburg (1932, 2008) or more recent ones—as by Schottroff (1967) Yerushalmi (1989), Funkenstein (1989), Nora (1989), Yael Zerubavel (1995) and Eviatar Zerubavel (2003), Confino (1997), Ricoeur (2004), and Jonker (2008)—to mention just a few and of course two central figures in this field—Jan Assmann (1992, 1995) and Aleida Assmann (1999, 2008). R. Hendel (2001, 2005), who has written on the Exodus from just this perspective, should be kept in mind as well. All these scholars have demonstrated the cardinally important perspectives which the study of the interface of memory/history in general and of collective memory in particular can provide. Many of these studies have stressed the importance of differentiating between history and memory—or as Megill (2007) has suggested "that memory is an Other that continually haunts history."

But we must go beyond this as well. The bourgeoning field of cognitive neuroscience, and its insights into human memory, has substantially altered the understanding of the biological foundations of memory—personal and collective—and how much more complex this is than nonscientists suppose.

J. Assmann (1995) pointed out that when Halbwachs and Warburg produced their seminal studies in the early 1920s, they attempted to diverge from then current biological/psychological understandings of memory (which were simplistic, and as we know today—simply wrong) and create alternative cultural frameworks for its study. Today though, this circle seems to have come around, and the cultural and biological perspectives on memory can now be integrated—and are mutually enhancing (see Jonker 2008).

Quoting from an overview by a cognitive neuroscientist—Charles Fernyhough—a leader in the field of memory study, one can get an idea of neuroscience's current understanding of memory:

> Memory is an essential part of who we are. But what are memories, and how are they created? A new consensus is emerging among cognitive scientists: rather than possessing a particular memory from our past, like a snapshot, we construct it anew each time we are called upon to remember. Remembering is an act of narrative as much as it is the product of a neurological process (Fernyhough 2012: backcover).

Or, as Daniel Schacter, another important figure in the contemporary study of memory, writes:

> We now know that we do not record our experiences the way a camera records them. Our memories work differently. We extract key elements from our experiences and store them. We then re-create or reconstruct our experiences rather than retrieve copies of them. Sometimes, in the process of reconstructing we add on feelings, beliefs, or even knowledge we obtained after the experience. In other words, we bias our memories of the past by attributing to them emotions or knowledge we acquired after the event (Schacter 2003: 9).

To this we can add that there is compelling and clear evidence that one can clearly identify several important types of "memories" that, from a simplistic, "nonprofessional" definition of memories, are not memories at all:

1. "False memories" of various kinds—"the misinformation effect"—in which due to various external and/or internal factors, people experience a memory—of an event that either never occurred at all—or at least not to that

person (e.g., Okado and Stark 2005; Loftus 2005).

2. Disputes over memory ownership and truth—where two or more people argue—at times even in court, about the veracity and "truthfulness" of their memories—and to whom these memories belong (e.g., Sheen et al. 2001, 2006).
3. The cardinal need to forget—in order to remember—is well known. Examples of unique cases of those who cannot forget and the largely tragic results of this have been documented by researchers (known as "hypermnesia," "hyperthymestic syndrome," or "highly superior autobiographical memory"). This is beautifully described in Jorge Luis Borges's (1967) story *Funes the Memorious*—where the protagonist suffers a riding accident that leaves him both crippled and unable to forget.
4. "Constructed group memories"—and this is perhaps the most directly relevant for the study of the Exodus—where groups can construct a memory based on various real and "fabricated" events—and weave them into a seemingly uniform collective memory.

I would like to refer to two personal examples from the time of my service in the Israeli Army:

1. The daring 1976 rescue by Israeli commandos of the hijacked Air France plane in Entebbe was an event that electrified the world. I recall being in Paris the day that this happened and the envious reactions of the Parisians—and as an Israeli, I felt immense pride—and of course direct credit—for the success of this mission. But more so, when I began my service in the IDF a few years later, you could see that just about every soldier in the IDF saw himself as either having been at Entebbe, or at the very least—he should have been there. An acquaintance of mine, who was serving at that time in a run-of-the-mill infantry unit—hardly the seasoned "commando-type" at all—told me the following story: he was once riding home on an intercity bus when a young female tourist sat down next to him. They began to converse, and at some point she asked him whether he had participated in the Entebbe raid. My friend, a very straightforward person, said that of course not—he was just a regular soldier and only elite soldiers had participated in the raid! To which she naively and incredulously replied: "You are the first Israeli soldier that I have met who was not at Entebbe!"
2. The second story, much less humorous, is nevertheless revealing. In 1977, right before the beginning of my army service, there was a horrendous helicopter crash in the Jordan Valley in which 54 Israeli paratroopers and airmen were killed. The helicopter had been overfilled and immediately after takeoff crashed. In fact—two soldiers were turned away from entering the craft since it was so full before takeoff—and thus their lives were saved. During my service, I personally met three soldiers who told me that they were one of the two who had been saved—and I've been told of at least another two from other people!

While both these examples could be argued away as being irrelevant for our purposes, since they appear to simply be the machismo and bravado of young, hot-blooded males—they serve as excellent examples of just how memory of events can be tailored and changed—and from there, the distance to the creation of completely new narratives—and memories—on a personal and a group level is not that far. Think of how many people would have been on the original *Mayflower*—if all those who claim to be descendants of these first colonists in North America were in fact on board—not to mention all the other transformations that the story of the arrival of these initial European settlers of the New World went through; or, from much more recently, how large *Woodstock* would have been—if all those who claim to have participated in this iconic rock concert actually had been on those grassy fields in upstate New York!

As Yosef Halevi Yerushalmi (1989: 5) stated: *Memory is always problematic, usually deceptive, sometimes treacherous.*

This though is not an attempt to claim that group memories of long-passed events are erased and cannot provide evidence with some historical value. Much has been written on how collective memories are "remembered," both in literate and nonliterate societies. There is no question of the ability of nonliterate societies to retain oral memories of historical events over extensive periods, with well-known examples from Scandinavia, the Balkans, and Australia (e.g., Lord 1960; Vansina 1965; Parry 1971; Hamacher and Norris 2009).

Nevertheless—one cannot, and should not—suppose that in the process of transmission of oral and written memories, even in highly structured frameworks (such as epic storytelling and sacred texts), these memories "froze" and did not change. Rather, with each "telling of the story" (orally or in writing), in the specific context(s) in which it is told, the story changes according to the needs of the people telling, and hearing, the story. One then has to look both at the original events behind memories and the history of their reception, and appropriation, over time (see, e.g., Confino 1997: 1395–96).

Memory, whether of an individual, a group, and of course that of very ancient times, is something that should be viewed as the product of ongoing, long-term bio-psychosocial processes.

To understand such group memories, one must approach them not as history per se.

To quote J. Assmann (1995: 9–10):

> Mnemohistory is not concerned with the past as such, but only with the past as it is remembered... It concentrates exclusively on those aspects of significance and relevance which are the product of memory... The present is 'haunted' by the past, and the past is modeled, invented, reinvented, and reconstructed by the present... Mnemohistory analyzes the importance which a present ascribes to the past.

While Assmann has suggested the term *mnemohistory*, I am somewhat hesitant to involve the term "history" in the description of these memories. I thus prefer to use the term *mnemo-narratives*[2]—as by its very definition it acknowledges the diverse and even competing narratives, the *rashumon* characteristics, of such group memories.

Let us now apply some of this to the topic at hand—the Exodus.

Over the years—and until this very day (in this volume), attempts have been made to reconstruct the historical background of the Exodus.

Many of these suggested that scenarios can perhaps be defined as "singular explanations"—in which an attempt is made to define the primary, specific historical, cultural, and/or geological context on which the Exodus story is based.

Without enumerating all of them, "Egypt-originating" views have been suggested for the late EB, terminal MB, mid-LB, terminal LB, early Iron I, and late Iron II, and to a certain extent—even the Persian and Hellenistic periods.

Other explanations have even removed the presence in Egypt as a central facet of the story but still attempt to define a "singular context," such as Knohl's (2012) recent suggestion that much of Exodus centers around events in north-eastern Canaan or Na'aman (2011), elaborating on Hendel's (2001) suggestion, who believes that it reflects a Canaanite perspective on the Egyptian oppression in Canaan during the LB/Iron Age transition.

The problem with all these perspectives is quite simple—the archaeological evidence does not corroborate them—and in fact often confounds them! While one can claim that this historical event, this geological process, or that destruction level may be reflected in the Exodus story—none of them add up to the overarching, supposedly coherent story as related to us in the Exodus narrative.

For the literal understandings—those that profess that the story occurred, more or less as described in the biblical text—the archaeological and inscriptional evidence does not corroborate most of the story. Why do the Egyptian texts not mention this explicitly? Why do the Levantine texts (when relevant, such as the el Amarna letters) not relate to this—or show evidence of the aftermath of these supposed events? The same goes for the archaeological remains—Where is the evidence of a large group (even if it was less than 600,000 men as suggested for

[2] I suppose that this would overlap somewhat with Assmann's (1992) *kulturelle gedächtnis* (cultural memory).

example by Kitchen (2003) traversing the desert (and archaeologists can identify the camps of small groups of ephemeral prehistoric hunter-gatherers)? Why is there no evidence of results of the Exodus—the conquest of Canaan? And why is there no material reflection of Egyptian influence in early Iron Age Israel?

For those who see a specific, and limited, time frame or context as the core element of the story—which is then further expanded—there are no fewer questions to ask: How does one explain the many layers of evidence emanating from the biblical text—seemingly connected to different periods, cultures, and backgrounds? For example, early and late Egyptian cultural and linguistic facets seen in the text and reflections of the *realia* of many time periods in the archaeological remains from Egypt, Sinai, and the Levant? And why did this particular event, and not others, become the core focus of this story?

I believe then that a more complex understanding should be recognized.

The ongoing connections between Egypt and the early peoples of the Southern Levant—the Israelites, their ancestors (real or imagined), as well as other groups in the region—undoubtedly left a long trail of memories, some of real events, over time. As it appears that significant parts of the populations that later came to be identified as Israelites in the later Iron Age originated from within the Southern Levant (which, by the way, is part and parcel of the biblical traditions as well), it is highly logical to assume that various traditions regarding these pre-Iron Age connections between the peoples of the Levant and Egypt were retained among the groups that formed the Israelite identities.

On the other hand, one could hardly say that all aspects of the later Israelite culture solely relate to internal Canaanite elements. There are items, whether artifactual or as remembered in biblical traditions, which seem to reflect external facets of the Israelite/Judahite identities.

Similarly, attempts to distill the core of the early Israelite identity as being related primarily to a wish to defy cultural alterity (such as in relationship to the Philistines [as for example A. Faust, in this volume, suggests]) are no less problematic—as they (a) minimize other factors that might have been just as important (while "otherness" is important, additional aspects can play a role in identity formation) and (b) assume that the biblical descriptions of conflicts between the Israelites and their neighbors, such as with the Philistines, did in fact occur. But are all these supposed confrontations reflected in the archaeological record?

The biblical traditions themselves point to a complex origin of the Israelites. While we concentrate in this volume on the Exodus from Egypt—other traditions are manifested as well—such as north-Syrian/Mesopotamian or internal Canaanite origins.

Since, despite repeated attempts, one cannot find a "perfect" archaeological/historical fit for the Exodus narrative, we must accept that it too reflects the diverse, complex, and multi-period origins of the Israelites. This tradition, or matrix of cultural memories, was woven together and altered over a long period (perhaps, periodically, unwoven as well—*Penelope's shroud* perhaps serving as an analogy), containing "snippets of yarns of memory" from many sources. This explains why this "amazing technicolor dreamcoat" does not dovetail with any specific, limited set of events; in fact, by definition it cannot fit into a restricted historical horizon! We are not looking for the "tree" that will provide the "ultimate" definition of the "forest"—but rather we must realize that this "forest" comprises many trees—each reflecting another "snippet" of collective memory.

Quoting Alon Confino:

> I would like to view memory as an outcome of the relationship between a distinct representation of the past and the full spectrum of symbolic representation available to a given culture. This view posits the study of memory as the relationship between the whole and its component parts, seeing society as a global entity – social, symbolic, political – where different memories interact (Confino 1997: 1391).
>
> ...memory as a whole that is bigger than the sum of its parts (Confino 1997: 1399).

Thus I believe that we must speak about a matrix of events, memories, and traditions, emanating from many periods and contexts, in which the Exodus *mnemo-narratives* were formed—a multi-faceted "narrative complex"—Daniele Hervieu-Leger's (2000) "Chain of memory."

But I should stress—this does not mean a completely ahistorical myth . . .

The ongoing connections that existed between Egypt and the Southern Levant and its diverse cultures are well known (for a general survey, see, e.g., Redford (1992)).

In the Early Bronze Age (and even in the late Chalcolithic) we have archaeological (and limited inscriptional and iconographic) evidence of ongoing connections—trade, warfare, and perhaps migration, and following Steiner (2011)—even direct evidence of Semitic influence on early Egyptian religion.

The extensive evidence of connections during the MB and LB need not be enumerated. Bidirectional influences, movements, and intermingling were common, as were various events which might have contributed to the formation of the Exodus narrative. While the Hyksos story per se is not the story of the Exodus, some of its components are reminiscent. Even if Freud's *Moses and Monotheism* (1955) is a simplistic (and largely incorrect) interpretation of the el Amarna origins of Israelite monotheism—various facets of this period may possibly be reflected in both the Exodus story and in later biblical materials (best known being Psalm 104; for different views of this, see recently, e.g., Krüger (2010) and Reichmann (2009)). Likewise, various biological and geological events that occurred during the LB are somewhat reminiscent of some of the biblical plagues (and see relevant discussions in this volume).

During the Iron Age, these connections continued. While throughout the early Iron Age there is little evidence of connections between the early Israelites in the highlands of Canaan and the Egyptians, this is not so for other regions of the Southern Levant. In addition to the evidence of the late NK Egyptian presence in parts of Canaan well into the second half of the twelfth century BCE, clear material evidence of connections, trade, and cultural influence can be seen in Philistia (Ben-Dor Evian 2011, 2012) and in Phoenicia (Gilboa and Sharon 2008: 159).

During later phases of the Iron Age, these connections expanded—and this is seen in many facets of southern Levantinian culture (e.g., Redford 1992; Schipper 1999; Wimmer 2008). As I have demonstrated in the past (Maeir 2003; see as well Holladay 2004) late Iron Age objects originating from the Southern Levant are found at sites in Egypt—indicating trade and perhaps even the presence of Judahites/Israelites in Egypt.

There is also extensive evidence for ongoing connections between the Levant and Egypt during the Persian period: whether connected to trade, to the Persian campaigns to, and conquests of, Egypt, and no less importantly the clear-cut inscriptional evidence of ongoing connections between the Jewish diaspora (in Elephantine) and the Jews in the province of YHD (e.g., Kratz 2006)—all of which argue strongly for substantive influences that very likely filtered into the Exodus traditions during this period—a crucial time as far as the formation of the biblical text is concerned (e.g., Greifenhagen 2002; Fantalkin and Tal 2012).

And finally, during the Hellenistic period, when the more or less finishing touches of the biblical corpus were woven together, the extensive connections between Egypt and the Levant—and Judea and the Jews in particular—were well known (e.g., Tcherikover 1963; Collins 2000).

The various events that occurred during these periods were remembered, and at times forgotten, by the various groups in the region. When some of these components, during various stages of the Iron Age and early Second Temple period, constructed a collective identity, time and again, these various strands of memory were combined—each serving the needs and fulfilling the requirements of the group constituents at each particular point in time. Some of these memories gained more prominent positions (such as the Egyptian components of the Exodus), while others were marginalized—or even forgotten. Additional events, whether cultural or environmental, were also added into this complex, woven matrix. And this eventually became the Exodus story as represented in the biblical text. Historical accuracy is not the issue here—but rather the processes of formation and ongoing reception and appropriation of these memories—and if we can, perhaps then we can identify some of the historical kernels.[3]

But make no mistake—the cessation of major textual revisions of the biblical text in the early Roman period did not mean that the collective memories relating to the Exodus and their reception stopped evolving. Throughout western history, as Jan Assmann discusses in Chap. 1, and for example Walzer (1985) and Pardes (1994, 2000) have demonstrated, the Exodus motif was used far and wide in very different and at times conflicting manners—to create collective memories for diverse groups.

Just as Halbwachs (1941) described the transformation of the collective memory of early Christian Palestine as an evolving "commemorative landscape"—so the Exodus story, and its meaning, was formed and transformed with time. Capitalizing on a term that Yerushalmi (1989) suggested, let us view the Exodus story as a "vehicle of memory"—perhaps even—as vehicles of memories; or conceivably we should view the entire Exodus "narrative complex"—or "chain of memory"—utilizing Aleida Assmann's (1999) term and see it as an *Erinnerungsraume* (a place of remembering).

I would like to end this chapter with a quote by the great cultural Zionist of the late nineteenth/early twentieth centuries CE, Achad ha-Am (1917: 3–4), who has a lovely essay on Moses, in which he explains his understanding of the historical figure of Moses and of the Exodus in general. His viewpoint, though written more than a century ago, in my mind resonates strongly today, from both a scholarly and personal perspective:

> [W]hen I read the Haggadah on the eve of Passover ... I care not whether this man Moses really existed ... We have another Moses of our own, whose image has been enshrined in the hearts of the Jewish people for generations and whose influence on our national life has never ceased from ancient times till the present days ... Even if you succeeded in demonstrating conclusively that the man Moses never existed, you would not thereby detract one jot from the historical reality of the ideal Moses – the Moses who has been our leader not only for forty years in the wilderness of Sinai, but for thousands of years in which we have wandered since the Exodus.

[3] See Lowenthal's (1998) in which he shows how "heritage"—as opposed to history—is something that is fabricated to meet the needs of the society whose "heritage" it describes. To a certain extent, "heritage" can be the modern equivalent of premodern collective memories.

Acknowledgement I would like to thank the organizers of the excellent conference and the volume, and in particular Tom Levy, for inviting to participate in and contribute to this fascinating venture. I had the honor of participating in the conference when I was a visiting professor at UCSD in the Spring of 2013.

References

Achad ha-Am. 1917. *Moses*. Trans. L. Simon. London: The Zionist.

Assmann, A. 1999. *Erinnerungsräume: Formen und Wandlungen das kulturellen Gedachtnisses*. Munchen: Beck.

———. 2008. Canon and Archive. In *Cultural Memory Studies: An International and Interdisciplinary Handbook*, Media and Cultural Memory, vol. 8, ed. A. Erll and A. Nünning, 97–107. Berlin: De Gruyter.

Assmann, J. 1992. *Das kulturelle Gedächtnis: Schrift, Erinnerung und politische Identität in frühen Hochkulturen*. Munich: Beck.

———. 1995. Collective Memory and Cultural Identity. *New German Critique* 65: 125–133.

Ben-Dor Evian, S. 2011. Egypt and the Levant in the Iron Age I-IIA: The Ceramic Evidence. *Tel Aviv* 38: 94–119.

———. 2012. Egypt and Philistia in the Iron Age I: The Case of the Philistine Lotus Flower. *Tel Aviv* 39: 20–37.

Berner, C. 2010. *Die Exoduserzählung*, Forschungen zum Altes Testament, vol. 73. Tübingen: Mohr Siebeck.

Borges, J.L. 1967. Funes the memorious. In *A personal anthology*, 35–43. Trans. A. Kerrigan. New York, NY: Grove.

Brettler, M.Z. 1995. *The Creation of History in Ancient Israel*. London: Routledge.

Collins, J.J. 2000. *Between Athens and Jerusalem: Jewish Identity in Hellenistic Diaspora*. Grand Rapids, MI: Eerdmans.

Confino, A. 1997. Collective Memory and Cultural History. *American Historical Review* 102: 1386–1403.

Erll, A., and A. Nünning. 2008. *Cultural Memory Studies: An International and Interdisciplinary Handbook*. Berlin: De Gruyter.

Fantalkin, A., and O. Tal. 2012. The Canonization of the Pentateuch: When and Why? (Continued, Part II). *Zeitschrift für die Alttestamentliche Wissenschaft* 124 (2): 201–212.

Fernyhough, C. 2012. *Pieces of Light: The New Science of Memory*. London: Profile Books.

Freud, S. 1955. *Moses and Monotheism*. Trans. K. Jones. New York, NY: Vintage.

Funkenstein, A. 1989. Collective Memory and Historical Consciousness. *History and Memory* 1(1): 5–26.

Gilboa, A., and I. Sharon. 2008. Between the Carmel and the Sea: Tel Dor's Iron Age Reconsidered. *Near Eastern Archaeology* 71(3): 146–170.

Greifenhagen, F.V. 2002. *Egypt on the Pentateuch's ideological map: Constructing biblical Israel's identity. JSOT Sup. 361*. Sheffield: Continuum.

Halbwachs, M. 1941. *La Topographie légendaire des évangiles en terre sainte: étude de mémoire collective*. Paris: Presses Universitaires de France.

———. 1950. *La mémoire collective*. Paris: Presses Universitaires de France.

———. 1952. *Les cadres sociaux de la mémoire*. Paris: Presses Universitaires de France.

Hamacher, D.W., and R.P. Norris. 2009. Australian Aboriginal Geomythology: Eyewitness Accounts of Cosmic Impacts? *Archaeoastronomy* 22: 60–93.

Hendel, R. 2001. The Exodus in Biblical Memory. *Journal of Biblical Literature* 120(4): 601–622.

———. 2005. *Remembering Abraham: Culture, Memory, and History in the Hebrew Bible*. Oxford: Oxford University Press.

Hervieu-Leger, D. 2000. *Religion as a Chain of Memory*. Piscataway, NJ: Rutgers University Press.

Holladay Jr., J. 2004. Judaeans (and Phoenicians) in Egypt in the Late Seventh to Sixth Centuries B.C. In *Egypt, Israel, and the Ancient Mediterranean World: Studies in Honor of Donald B. Redford*, Probleme der Ägyptologie, vol. 20, ed. G. Knoppers and A. Hirsch, 405–437. Leiden: Brill.

Jonker, G. 2008. *The Topography of Remembrance: The Dead, Tradition and Collective Memory in Mesopotamia*, Studies in the History of Religions, vol. 68. Leiden: Brill.

Kitchen, K. 2003. *The Reliability of the Old Testament*. Grand Rapids, MI: Eerdmans.

Knohl, I. 2012. *Ha-Shem: The Secret Numbers of the Hebrew Bible and the Mystery of the Exodus from Egypt* (In Hebrew). Tel Aviv: Dvir.

Kratz, R.G. 2006. The Second Temple of Jeb and of Jerusalem. In *Judah and the Judeans: In the Persian Period*, ed. O. Lipschits and M. Oeming, 247–264. Winona Lake, IN: Eisenbrauns.

Krüger, A. 2010. *Das Lob des Schöpfers: Studien zu Sprache, Motivik und Theologie von Psalm 104*. Neukirchen-Vluyn: Neukirchener.

Loftus, E.F. 2005. Planting Misinformation in the Human Mind: A 30-Year Investigation of the Malleability of Memory. *Learning and Memory* 12: 361–366.

Lord, A.B. 1960. *The Singer of Tales*. Cambridge: Cambridge University Press.

Lowenthal, D. 1998. Fabricating History. *History and Memory* 10(1): 5–24.

Maeir, A.M. 2003. The Relations between Egypt and the Southern Levant During the Late Iron Age: The Material Evidence from Egypt. *Ägypten und Levante* 12: 235–246.

Megill, A. 2007. *Historical Knowledge, Historical Error: A Contemporary Guide to Practice*. Chicago: University of Chicago.

Na'aman, N. 2011. The Exodus Story: Between Historical Memory and Historiographical Composition. *Journal of Ancient Near Eastern Religions* 11: 36–69.

Nora, P. 1989. Between History and Memory: Les Lieux de Mémoire. *Representations* 26: 7–24.

Okado, Y., and C.E.L. Stark. 2005. Neural Activity During Encoding Predicts False Memories Created by Misinformation. *Learning and Memory* 12: 3–11.

Pardes, I. 1994. Imagining the Promised Land: The Spies in the Land of the giants. *History and Memory* 6(2): 5–23.

———. 2000. *The Biography of Ancient Israel: National Narratives of the Bible*, Contraversions, vol. 14. Berkeley: University of California Press.

Parry, M. 1971. *The Making of Homeric Verse: The Collected Papers of Milman Parry*. Oxford: Clarendon.

Proust, M. 1913–1927. *À la recherche du temps perdu*. Paris: Bernard Grasset.

Redford, D.B. 1992. *Egypt, Canaan and Israel in Ancient Times*. Princeton, NJ: Princeton University Press.

Reichmann, S. 2009. Psalm 104 und der Große Sonnenhymnus des Echnaton: Erwägungen zu ihrem literarischen Verhältnis. In *Israel zwischen den Mächten*, Alter Orient und Alten Testament, vol. 364, ed. M. Pietsch and F. Hartenstein, 257–288. Münster: Ugarit-Verlag.

Ricoeur, P. 2004. *Memory, History, Forgetting*. Chicago: University of Chicago Press.

Schacter, D.L. 2003. *How the Mind Forgets and Remembers: The Seven Sins of Memory*. London: Souvenir Press.

Schipper, B. 1999. *Israel und Aegypten in der Konigszeit: Die Kulturellen Kontacte von Salomo Bis Zum Fall Jerusalems*, OBO, vol. 170. Freiburg: Universitatsverlag.

Schottroff, W. 1967. *'Gedenken' im alten Orient und im Alten Testament. Die Wurzel zakar im semitischen Sprachkreis*, Wissenschafliche Monographien zum Alten und Neuen Testament, vol. 15. Neukirchen-Vluyn: Neukirchener Verlag.

Sheen, M., S. Kemp, and D. Rubin. 2001. Twins Dispute Memory Ownership: A New False Memory Phenomenon. *Memory and Cognition* 29: 779–788.

———. 2006. Disputes Over Memory Ownership: What Memories Are Disputed. *Genes, Brain and Behavior* 5 (Supplement 1): 9–13.

Steiner, R.C. 2011. *Early Northwest Semitic Serpent Spells in the Pyramid Texts*, Harvard Semitic Studies, vol. 61. Winona Lake, IN: Eisenbrauns.

Tcherikover, A. 1963. *The Jews in Egypt in the Hellenistic-Roman Age in the Light of the Papyri*. Jerusalem: Hebrew University.

Vansina, J. 1965. *Oral Tradition: A Study in Historical Methodology*. Trans. H.H. Wright. Chicago, IL: Aldine Publishers.

Walzer, M. 1985. *Exodus and Revolution*. New York, NY: Basic.

Warburg, A. 1932. *Die Erneuerung der heidnischen Antike: Kulturwissenschafliche Beitrage zur Geschitchte der europaischen Renaissance*. Leipzig: Teubner.

———. 2008. *Der Bilderatlas Mnemosyne*. Berlin: Akademie Verlag.

Wimmer, S. 2008. *Palästinisches Hieratisch: Die Zahl- und Sonderzeichen in der altherbäischen Schrift*, Ägypten und Altes Testament, vol. 75. Wiesbaden: Harrassowitz.

Yerushalmi, Y.H. 1989. *Zakhor: Jewish History and Jewish Memory*. Seattle: University of Washington Press.

Zerubavel, Y. 1995. *Recovered Roots: Collective Memory and the Making of Israeli National Tradition*. Chicago: Chicago University Press.

Zerubavel, E. 2003. *Time Maps: Collective Memory and the Social Shape of the Past*. Chicago: Chicago University Press.

Remembering Egypt

32

Victor H. Matthews

Abstract

Within the range of studies that were offered at the conference and included in this volume, this chapter attempts to separate itself from efforts to determine the possible historicity of the exodus event. Instead, it concentrates on how and why collective memories are created, perpetuated, used, and reused. While questions are raised on when and where the exodus tradition originated, current data makes it impossible to provide a definitive answer. The conclusions that are drawn have more to do with why the exodus tradition is important to the Israelite community at various times and why it ultimately became a major facet in the identity and ethos of that people.

Memories resonate with individuals and communities. Simple, personal memories can be evoked by a sound, a smell, or a casual remark. More complex memories, those associated with group identity, survive because they are attached to various forms of ritual commemoration and specific sites or objects that are patterned to provide a constant reminder of an event, a hero, or a natal moment in a nation's history. Of course, it is possible for memories to fade or for a group to pass out of history taking their memories with them. For those memories that survive, it is the process and purpose of remembering that is worth exploring and analyzing. In this study, the goal is to gain some clarity on how and why ancient Israel chose to remember Egypt.

In fact, what I have discovered during the course of researching this topic is that there are a variety of ways of "remembering Egypt." It is possible to think of Egypt as a place, a concept, and a framework for memories. Within the biblical narrative, Egypt in various guises is preserved and used/manipulated by the biblical writers throughout the course of Israelite history. And, regardless of what may have actually happened in Egypt or in Egyptian-controlled Canaan during various periods,[1] the point is that a collective memory

V.H. Matthews (✉)
College of Humanities and Public Affairs,
Missouri State University,
Springfield, MO, USA
e-mail: VictorMatthews@MissouriState.edu

[1] See Nadav Na'aman (2011) for a careful examination of Egyptian sources that document to some extent the activities of the Egyptian pharaohs in Canaan during the Nineteenth and Twentieth Dynasties and suggests that the bondage and the delivery from slavery as related in the biblical story actually took place in Canaan and that

T.E. Levy et al. (eds.), *Israel's Exodus in Transdisciplinary Perspective*,
Quantitative Methods in the Humanities and Social Sciences, DOI 10.1007/978-3-319-04768-3_32,

evolved over time through the process of retelling that transformed the original events into attitudes that became the basis for emotional responses whenever the word/place Egypt is uttered. In that sense, the recounting process conforms to the sociologist Michael Schudson's (1989: 105) recognition that "the past is constantly being retold in order to legitimate present interests."

With that in mind then, it can be said that peoples, tribes, and nations are constantly telling stories about themselves, their ancestors and their heroes. In the process they create memories as cultural building blocks that eventually become the foundation or essence of who they are and how they portray themselves to others. But memories like stories are fragile things. They can fracture or dissolve into obscurity if there is no reliable mechanism or ritual to keep them intact and alive. Perhaps even more of a concern is that memories are fractals, capable of spinning off into many different directions and in the process so diluting the original story line that it becomes almost impossible to recapture it. As a result, in the biblical narrative Egypt became less of a place than a state of mind and the exodus story became a paradigm for oppression and liberation that could be applied and reapplied as needed.

Memory Studies

Among the conclusions drawn by scholars who have created a foundation for the study of memory and history is that it is not possible to separate individual and group memories since individual memories are "formed in relation to a society's various (group) articulations of collective memory" (Halbwachs 1992: 53).[2] That conclusion, however, should be tempered by noting that many individual memories are quick to fade since they involve only episodic interactions, such as a casual social encounter that have no lasting effect (Nienass and Poole 2012: 91). Collective memories tend to be more long-lasting since they are comprised of "narratives of past experience constituted by and on behalf of specific groups within which they find meaningful forms of identification that may empower" them to action or emotional response (Weeden and Jordan 2012: 143).[3] Cultural artifacts, such as school books, sacred texts, ritual performances assist these narratives to survive, if not entirely intact, in mental storage.

For a memory to be more than just a recitation (either immediately after an event or long after) of what a people say happened, the study of cultural memory also "recognises that cultures reproduce and reform themselves and that, in this process, understandings of the past are transformed" (Bal 1999: 57). As Michael Schudson (1989: 105) points out, "Memories are not credible unless they conform to an existing structure of assumptions about the past—an 'available past' that people accept as given and that possesses a self-sustaining inertia." Furthermore, I would suggest that Barry Schwartz, in his study of George Washington (1991: 222), is correct in saying that "a true community is a 'community of memory,' one whose past is retained by retelling the same 'constitutive narrative,' by recalling the people who have always embodied and exemplified its moral values."[4]

the memories were later transferred from Canaan to Egypt.

[2] See Barstad (2010) for a critical examination of the ongoing debate between those who advocate an "absolute dichotomy/discontinuity between memory and history" (p. 7) and those who continue to advocate for a relationship between history and memory. A wider summary by Nienass and Poole (2012) annotates the current views on memory studies within the social sciences.

[3] I agree with Jan Assmann (1995: 126) that we must differentiate between collective memory and "communicative memory," the latter comprised simply of everyday communications that are "characterized by high degree of non-specialization, reciprocity of roles, thematic instability, and disorganization."

[4] For a fuller exposition of this idea, see Robert N. Bellah et al. (1985: 152–55). See also Yael Zerubavel (1995: 6–7) and her use of the term "master commemorative narrative" that "focuses on the group's distinct social identity and highlights its historical development."

One way collective memory can be anchored within the community is through its attachment to ritual performance and to what Pierre Nora (1989: 7) calls "sites of memory" (*lieux de mé moire*) where he states memory is able to crystallize. These sites become particularly important over the course of time since the original environment of memory no longer exists or is accessible. Among the sites of memory he identifies (1989: 12) are "museums, archives, cemeteries, festivals, anniversaries, treaties, depositions, monuments, sanctuaries, fraternal orders." In the ancient Israelite context that would translate to significant places such as Jerusalem, Shechem, or Bethel that contain important sanctuaries and sacred objects such as the ark of the covenant.[5]

Of course, some cultural memories manage to remain sharper in the mind than others—usually because the events they contain are fresher or more recent or because they have been deemed too valuable to the community to let them fade away. Still, over time those memories that are older and thus hazy in the collective consciousness are subject to either being banished from commemoration or to reinterpretation and manipulation by those in authority who can make use of the power that they represent. Even so, there are some memories that are so important to the people's ethos that they must be memorialized or that people will have to give up a basic facet of their identity. That appears to be the case with the story of the exodus. It seems almost inconceivable that a people would choose to create and perpetuate a foundation myth in which they were slaves. Setting that in juxtaposition with the saving acts of a deity who has chosen to "bring them out of Egypt" (Exod 20:2) transforms a shameful beginning into an empowered future that allows for a multitude of subsidiary adaptations to fit Israel's situation in later time periods.

With that said, I would point to two facets of memory as guiding principles in this study. First, the role of memory is "to remind us—either individually or collectively—of those episodes in the past that we *ought* to take account of in the present" and secondly, "memory has *force*; that is, it moves us to act" (Nienass and Poole, 89). One way to approach the latter point is found in Marianne Hirsch's use of the term "postmemory." As part of her study of the Holocaust she describes how memory impacts "the relationship of children of survivors of cultural or collective trauma to the experience of their parents." While they did not experience these events themselves, they remember them "as stories and images with which they grew up" and as such they "are so powerful, so monumental, as to constitute memories in their own right" (Hirsch 1999: 8). Thus, like these "children" of the Holocaust, who experience only through story and the emotions transmitted to them by their elders, recipients of the exodus "postmemory" are allowed to feel its trauma through ritual performance and through a collectively created emotional labeling of the name/place/people of Egypt.

Of course, some memories are intentionally manufactured. They spring into being at the moment when they are voiced to serve the purposes of the author or those he serves. They may draw on fragments of reality that then blossom into a full-blown set of images and stories. For example, the story of the archetypal hero, whether he be Gilgamesh or Moses, draws on a set of type-scenes like miraculous birth narratives or the miraculous passage through a sea of death to germinate into a narrative that both draws the interest of the audience and functions as a metaphor for the emergence of a nation (see Kluger 1991: 207–208 and Assmann 1997: 4). Ultimately, the reality of what a hero may have actually accomplished is clouded or submerged into the overriding narrative that creates his persona and value to those who keep his memory alive. In that way the manufactured memory, which has been pieced together to form a sort of "grand narrative" or myth becomes normative since it embodies "ideals to which

[5] See Matthews (2009), for a discussion of the tendency in biblical narrative for major events to take place at particular sites or to use the memory of these sites to add authority to the actions of priests, kings, and prophets.

members of the group aspire" (Nienass and Poole 2012: 93).

Other memories are reshaped to fit a new political, theological or social context, such as Hosea's frustrated response to Israel's desire to "return to the land of Egypt" (Hos 11:5), which is both a veiled reference to the unrealistic defeatism voiced by the complaining Israelites in Num 14:1–4 and a warning not to trust the Egyptians in his own time, who are attempting to persuade Israel once again to revolt against their Assyrian overlords (Wolff 1974: 200).[6] Such an interpretation fits well with Halbwachs (1941: 7) assertion that "collective memory is essentially a reconstruction of the past [that] adapts the image of ancient facts to the beliefs and spiritual needs of the present."

Memories also become the basis for the creation of ritual performances or theological understandings that reinforce the authority of those who commission and use them for their own ends. Thus the Passover ritual detailed in Exod 12:1–28 and reviewed in Deut 16:1–8 provides a didactic vehicle for recollection and an interactive means of infusing this memory into the consciousness of each successive generation. Randall Collins refers to this process as an "interaction ritual chain," a performance that generates "powerful emotional energy" that makes it possible not only to internalize particular beliefs and create a sense of solidarity for the insider group, but to keep the memory of the event being ritualized vibrant and useful over time (2004: 41–44).

It really does not matter if the commemorative events that form the heart of the ritual activities have any historical reality or stretch the truth to the breaking point. What matters is that the memory lives on to provide a foundation myth or charter story for a people in need of a national and ethnic identity (van der Toorn 2001). For Israel to remember Egypt as the land of slavery is a way of providing a people, who may well have never been in Egypt, with a common memory to recite on ritual occasions and to teach their children. It also provides them with a continuous reminder of divine intervention in times of need. For instance, when they prepare to go into battle they are called on to remember that the mighty hand of "the Lord your God is with you, who brought you up from the land of Egypt" (Deut 20:1).

While some memories are kept close within the insider community as mysteries only to be shared with the initiated members of a group, others serve a broader purpose. It is therefore not enough to have the people perform rituals and swear to remember Yahweh's saving acts in bringing the nation out of Egypt and protecting them on their way to the Promised Land (Deut 29:16; Josh 24:17). These memories must also be placed in the mouths of non-Israelites so that the memory grows beyond the adherents to a particular ethos. That is why it is so important for Rahab, when speaking to Joshua's spies in Jericho, to tell them that "we have heard how the Lord dried up the water of the Red Sea before you when you came out of Egypt, and what you did to the two kings of the Amorites that were beyond the Jordan, to Sihon and Og, whom you utterly destroyed" (Josh 2:10). Her speech represents universal recognition of the power of Yahweh and by extension the distinctiveness of the chosen people of Israel who are about to begin their conquest of the land (Sherwood 2006: 60).

Remembering Egypt in the Biblical Narrative

Based on these general principles of memory studies, I now turn to the various ways in which Egypt is documented in a biblical narrative. What is particularly important here is how the collective memory is conjured up for the benefit of the ancient audience. For instance, from the perspective of spatiality theory Egypt can be and

[6] In this regard, Weeden and Jordan (2012: 144) note that "collective memory is shaped by specific interests and power relations, and the constitution of memory is above all a terrain of cultural politics." In the case of Hosea, it is religious interests that draw on political realities to create or call to mind a memory of another case of Israel's infidelity to Yahweh.

is described in physical/topographical terms (first space). It has a distinct geographical location that has a defined set of coordinates, but keep in mind that most Israelites never visited that region in their lives. As a result, while their memory may be based on a particular spatial reference, it has also been transfigured by the community's lack of familiarity with its physical dimensions into something that is more malleable and therefore useful for political or religious purposes (Truc 2011: 148).

That in turn allows Egypt, a place whose structures, monuments (see Jer 43:13) exist in one sphere to become imagined resources (second space), an arena in which many daily tasks are performed by its inhabitants (third space). However, most importantly, Egypt is remembered (fourth) space in which all its historical or mythological dimensions, political, economic, or culturally shaped facets, can and do function as part of collective memory (see Matthews 2003 and 2013).

Ultimately, then, what the Israelites know about Egypt the place, the people, and its history is based on a general sense of direction, word-of-mouth reports, anecdotal information, and the substance of those memories about Egypt which contribute to a social bonding of the Israelites (Assmann 2006: 5). Of course, all of this information is clouded entirely by cultural bias, past and current political climate between the two peoples, and any long-standing "memories" associated either with the exodus foundation myth or other "events" in which the two peoples interacted over time. What then results is a set of impressions, perhaps to be termed "common knowledge" about Egypt that has some association with reality but more to do with the sum total of collective memory as it is known in their own time.

One way to get a sense of how collective memories associated with Egypt were formed and then applied by the storytellers and editors is to map out what the biblical text contains:

- Egypt, the Nile River, or the Red Sea serves as a part of the formula statement defining the geographic dimensions of the Promised Land (or restored land), Solomon's kingdom, and the regions ruled by the King of Babylon (see Gen 15:18; Exod 23:31; Josh 15:4; 1 Kgs 4:21; 2 Kgs 24:7; 2 Chron 9:26; Ps 72:8; Isa 27:12; Micah 7:12).
- It is a rich land, well-watered by the Nile River (Jer 46:7–8; see comparison between Egypt and the less favored land of Canaan in Deut 11:10), and thus a resource for obtaining grain during times of famine in other parts of the ancient Near East (see Gen 12:10; 26:1–5; 41:57; 42:1–2; 45:18)—compare the reversal of this image and the resulting drought in Isaiah's oracle against Egypt in Isa 19, especially vv. 5–10.
- Egypt can serve as a temporary haven for the Israelites, but it is not to serve as a permanent place of settlement (Gen 46:1–4). Egypt also serves as a temporary haven for political dissidents: Hadad (1 Kgs 11:18–22); Jeroboam (1 Kgs 11:40; 2 Chron 10:2); Jer 24:9 (Zedekiah's officials); Johanan son of Kareah (Jer 42–43; 44:24, 26).
- While Egypt represents life because of its natural resources, it also has the potential for death when the Israelites must interact with its people and the pharaoh (Gen 12:10–13:1 (wife-sister); Gen 40–41 (interpretation of dreams)) and this interaction is the catalyst for the contest between gods motif in the exodus event (plague sequence in Exod 7–12; Neh 9:10; Ps 135:5).
- The contest between gods motif also serves as proof of Yahweh's power and becomes a short-hand for deliverance from Egypt: Josh 24:5; 1 Sam 4:8 (Philistines' cry); Judith 5:12; 2 Esdras 15:11.
- In later periods according to the narrative, reference to Yahweh's saving act of "bringing the people out of Egypt" becomes a theological scribal catch-phrase (Num 22:5; 23:22; Deut 4:34; Lev 19:36; 1 Kgs 8:9, 16; Ps 81:10; Jer 7:22; Amos 9:7).
- In political terms, the danger of alliance with Egypt is decried by the Assyrian ambassador, the Rabshakeh (2 Kgs 18:21; Isa 36:6), by Isaiah (30:1–5) and Ezekiel (29:6–7). Egypt

is labeled an unreliable ally, a "reed" that breaks and harms those who rely on it. Playing their own political game, Egypt entices Judah and other nations to rebel (Isaiah 20) and then switches sides leading to the death of Josiah (2 Kgs 23:28–29).

And, these facets of collective memory are coupled with the command to remember (*zākar*) and to perpetuate it through the generations:

- An authoritative injunction is made by Moses/Deuteronomist to instruct the children—Exod 13:14; Deut 6:20–21; 32:7—and that includes transmission of oral tradition by the elders (Judg 6:13).
- Ritual performance often includes a recitation of salvation history—Lev 23:43 (feast of booths); Deut 16:1–12 (Passover); 26:1–11 (first fruits sacrifice); Deut 29:2–29 (covenant renewal; Josh 24:4–7).
- Obedience to the law and perpetuation of ritual practice (Sabbath) is justified by reference to the period of slavery in Egypt—Exod 13:3 (festival of unleavened bread); Lev 25:42 (prohibit slavery); Deut 5:15 (Decalogue); 7:7–11; 16:1 (calendar); 24:17–18 (social justice to aliens, widows, and orphans); Num 15:41 (fringes); Jer 34:13–14 (sabbatical year).
- The label or title for the power and person of Yahweh includes the phrase "who brought you out of the land of Egypt and redeemed you from the house of slavery"—Exod 20:2; Deut 4:34–37; 5:6; 6:12; 8:14; 13:5, 10; Josh 24:17.
- The people's obligation to Yahweh includes the reminder that this is the god who "led you up from Egypt, and brought you out of the house of slavery"—Judg 2:1; 6:8; 1 Sam 10:18; Jer 34:13; Hos 13:4; Micah 6:4.
- The reminder of God's saving event is coupled with the need not only to remember but to engage in right behavior (covenant obedience)—Deut 6:22; 7:8, 18; 11:3; 29:2; 34:11 (Moses obit); 1 Sam 6:6; 2 Kgs 17:7.
- As in the covenant renewal in Moab (Deut 29:10–29), some prophets use the Egyptian episode as part of their enumeration of God's acts and their theodicy for destruction—Amos 2:6–10; 3:1–2; 9:7–8; Micah 6:3–5; 7:15 (restoration).
- Thus the theodicy for the fall of Israel to Assyria (Hos 11:1–5) and Judah to the Babylonians is tied to the failure of the people to obey the covenant even though Yahweh had "showed signs and wonders in the land of Egypt" and "brought your people Israel out of the land of Egypt with signs and wonders" (Jer 32:20–23).
- The act of remembering has a corresponding opposite: forgetting (Deut 8:14; 9:7; Ps 78:10–16; 106:21; Hos 13:5–6). In the biblical context these two polar opposites serve as fidelity to God and the covenant and the basis for God's lawsuit or theodicy against an unfaithful nation.

Reflections on Remembering Egypt

Among the chief questions raised by this mapping of the text is why, during certain periods of Israelite history and in certain sections of the country, there was a determined effort to remember Egypt and the exodus event. Ordinarily, a clearly defined, shared memory only lasts as long as the lifetimes of its witnesses or as long as it remains of cultural value. However, a transmitted/edited version of the memory can live on for generations if it is considered to be important to the group's identity. Of course, it would be expected that details would change depending on the storyteller's skills or political/theological agenda, but that in turn would depend on the strength of the memory and to an extent the ability to verify or believe in its basic substance. This process corresponds with Barry Schwartz (1991: 232) conclusion that "the earliest construction of an historical object limits the range of things subsequent generations can do with it." In other words, while later interpreters may make use of a memory, its basic fabric has an enduring quality that cannot be completely submerged to serve the needs of the present.

There are consequences when a shared memory of a foundational event like the exodus is revived long after the event in a new incarnation and generally for political purposes. To begin with, when Egypt, a land and a shared sense of relationship (for good or bad), is transformed from its first space (physical) realities into second space (imagined) possibilities and fourth (remembered) space, threads of tradition that represent new or rejuvenated purposes can be fulfilled. In a similar way, when the descriptive is added to Yahweh's name as the god "who brought them out of Egypt, out of the hand of slavery" (Exod 13:3, 14; Deut 5:6; 6:12; 13:5; Judg 6:8) that makes a case for the ultimate power of a particular deity over all other gods and all other peoples. Furthermore, that adds an authoritative dimension to the idea of being chosen by a deity, a collective memory which no other people can share.

But the question remains, why did the ancient Israelites choose this particular foundation myth? If it is a local, tenth-century creation, is it possible to simply consider it the creation of the northern kingdom during its efforts to separate itself from Jerusalem and the Davidides (Na'aman 2011: 56–59)? Or, was it part of the eighth century anti-Jeroboam faction within the prophetic community, who saw Israel's decline in terms of its idolatry and lack of adherence to the covenant (Hos 8:11–14, Amos 3:1–11)?

Is the impetus for emphasizing or renewing memories of the past, in this instance the exodus and Egypt, based on present or immanent circumstances? For example, is Hosea's use of the exodus as the typological event in Israelite history based on the destruction of the northern kingdom and the need to assure the survivors that there is hope for another exodus out of oppression (Hoffman 1989)? The problem is the nagging suggestion that there would have to be an awareness of the exodus tradition in Israel prior to Hosea so that his audience could both grasp his warning and appreciate the symbolism.

Is it possible that the lack of emphasis on the exodus tradition in Judah prior to the eighth century (other than in the stylized scenes associated with Solomon's dedication of the temple) is a reflection of their relatively safe political position prior to that time? Do the Assyrian invasions of 711 and 701 mark the turning point for Judah coupled with the appearance of refugees from Israel after 721? Did these refugees restore the exodus tradition to Judah's consciousness? Does that then help explain Micah's, Jeremiah's, and Ezekiel's use of the exodus tradition to frame their situation as well as the use of that liberation tradition by Second Isaiah?

Is the exodus story then a revived creation of the southern kingdom editor, the Deuteronomist? Could it be designed as a theological vehicle by the retrospective historian to place in juxtaposition the people's call for the creation of the monarchy with their rejection of the collective memory of God's ability to bring them out of Egypt (1 Sam 8:8; 10:18; 12:6–8)? Does it serve the purposes of the Deuteronomist's editing of events to tie Solomon's dedication of the new temple (1 Kgs 8:21, 53) and Yahweh's fidelity to the House of David (1 Kgs 8:16) to the memory of Yahweh's rescuing the people from Egypt, "from the midst of the iron-smelter" (1 Kgs 8:51; reiterated in Jer 11:4)? Finally, is there an explicit motive attached to the way in which the Deuteronomist pairs the theodicy of the fall of Samaria with Hezekiah's reform (2 Kgs 17:21–23; 18:1–8) and empowers Josiah's kingship and reform by cleansing the temple and restoring the celebration of the Passover (2 Kgs 23:21–23; Becking 1997)? After all, Josiah was facing Egyptian opposition to his control over an important border state—opposition that would eventually bring about his death at the battle of Megiddo (2 Kgs 23:28–30).

Alternatively, is the exodus story a convenient vehicle created during the exilic period? Was Second Isaiah (52:3–6) using Egyptian liberation as a metaphor for the eventual end to Babylonian captivity (Hamlin 1991)? Is the exodus story intended to revitalize the hopes of the exiles that they will indeed be freed to return to Jerusalem and once again be able to claim the Promised Land (Isa 11:16; Zech 10:10)? Or, is it a fifth century post-exilic creation intended to encourage a reluctant Golah community to return to Judah and rebuild the Jerusalem temple (Hag 2:5) or to provide encouragement to the Egyptian diaspora communities (Isa 19:16–17; Blenkinsopp 2000: 319)?

If not a late creation, is there any chance that the Egyptian oppression and the exodus are actual events that are perpetuated through memorial commemoration? The answer to that question is more the task of archaeologists, historians, and philologists, who have worked over the past century to determine or to debunk the reality of the exodus account and the role of Moses in the history of ancient Israel, and who are also represented in this volume.[7] While there is no way to know for sure given current data, it seems unlikely that it occurred in precisely the way that the story is ultimately crafted.

How much truth does a foundation myth require? Is it enough that the story resonates as true to its audience? It seems doubtful that a story that is too out of character, too fantastic (10 plagues and Red Sea crossing being demonstrations of divine power after all) to be believed would live for long. Therefore, what is it about the story that makes it live and makes it useful enough to be memorialized in ritual and story and psalm (see Ps 80:8; 81:10; 114:1)? If there is a grain of truth here perhaps it initially served a more local purpose until its possibilities as a national foundation story caused it to be adopted or revised to serve new purposes?

Realistically, all of what is contained in the biblical text about Egypt is in fact "remembered" data. While Egyptian records provide some hints at the relation between Israel and Egypt, they do not contain a truly coherent picture (see Dijkstra 2011). Even in those instances in later narratives in which Egypt played the role of erstwhile political ally (2 Kgs 18:21–24; Isa 30:1–2; 31:1), enemy nation (1 Kgs 14:25–26) or economic partner (1 Kgs 10:28) or supplier of goods (Prov 7:16), services are chosen from official reports and may well have been collectivized rather than be specific accounts of single occurrences.

Within this amalgamated world view, based on collective memory, the past is always immediately present. More than a simple recollection, collective memory functions as an essential aspect of lived reality. Should it require a means of preservation and recovery, then that conveniently occurs in the form of commemoration, recitation, and ritual. Thus, the substance of a cherished collective memory remains vibrant as long as it continues to be told "to your children" so that they will know that "We were Pharaoh's slaves in Egypt, but the Lord brought us out of Egypt with a mighty hand" (Deut 6:21). Furthermore, it can be elicited whenever the people face oppression from a stronger nation. In a practical sense, then, for the ancient Israelites Egypt becomes synonymous with whichever nation is the current foe (Assyria in Isa 10:24). Egypt also functions as the label for any location from which God will free and restore the captive people (Micah 7:11–15). In the end, the admonition to remember Egypt, whenever and for whatever original purpose it was created, becomes a theological survival mechanism for the Israelites. It provided a theodicy for God's punishment of the people and an ultimate reassurance and incentive that in time they would be liberated and restored to their covenant relationship with all its obligations and benefits.

[7] I would direct readers in particular to the contributions of Manfred Bietak on the question of the historicity of the exodus and Ron Hendel on the association of the exodus with memory and history (see also Hendel 2001). Jan Assmann's recently translated work *Religion and Cultural Memory* (2006) is another excellent starting point for this topic.

References

Assmann, Jan. 1995. Collective memory and cultural identity. *New German Critique* 65: 125–133.

———. 1997. *Moses the Egyptian: The Memory of Egypt in Western Monotheism*. Cambridge, MA: Harvard University Press.

———. 2006. *Religion and Cultural Memory*. Stanford, CA: Stanford University Press.

Bal, Mieke. 1999. *Acts of Memory: Cultural Recall in the Present*, Edited by Mieke Bal, et al. Hanover: University Press of New England.

Barstad, Hans. 2010. History and Memory: Some Reflections on the "Memory Debate" in Relation to the Hebrew Bible. In *The Historian and the Bible: Essays in Honour of Lester L. Grabbe*, ed. Philip Davies and Diana Edelman, 1–10. New York, NY: T & T Clark.

Becking, Bob. 1997. From Apostasy to Destruction: A Josianic View on the Fall of Samaria (2 Kings 17, 21–23). In *Deuteronomy and Deuteronomic Literature: Festschrift C.H.W. Brekelmans*, ed. Marc Vervenne and Johan Lust, 279–297. Leuven: Leuven University Press.

Bellah, Robert N., et al. 1985. *Habits of the Heart: Individualism and Commitment in American Life*. Berkeley: University of California Press.

Blenkinsopp, Joseph. 2000. *Isaiah 1–39*. New York, NY: Doubleday.

Collins, Randall. 2004. *Interaction Ritual Chains*. Princeton, NJ: Princeton University Press.

Dijkstra, Meindert. 2011. Origins of Israel between History and Ideology. In *Between Evidence and Ideology*, ed. B. Becking and Lester L. Grabbe, 41–82. Brill: Leiden.

Halbwachs, Maurice. 1941. *La Topographie Legendaire des Evangiles (The Legendary Topography of the Gospels)*. Paris: Presses Universitaires de France.

———. 1992. *On Collective Memory*. Chicago, IL: University of Chicago.

Hamlin, E. John. 1991. Deutero-Isaiah's Reinterpretation of the Exodus in the Babylonian Twilight. *Proceedings, Eastern Great Lakes and Midwest Biblical Societies* 11: 75–80.

Hendel, Ron. 2001. Exodus in Biblical Memory. *Journal of Biblical Literature* 120: 601–622.

Hirsch, Marianne. 1999. Projected Memory: Holocaust Photographs in Personal and Public Fantasy. In *Acts of memory: Cultural recall in the present*, ed. Mieke Bal et al., 3–23. Hanover: University Press of New England.

Hoffman, Yair. 1989. A North Israelite Typological Myth and a Judaean Historical Tradition: The Exodus in Hosea and Amos. *Vetus Testamentum* 39: 169–182.

Kluger, Rivkah S. 1991. *The Archetypal Significance of Gilgamesh: A Modern Ancient Hero*. Einsieden: Daimon Verlag.

Matthews, Victor H. 2003. Physical Space, Imagined Space, and "Lived Space" in Ancient Israel. *Biblical Theology Bulletin* 33: 12–20.

———. 2009. Back to Bethel: Geographical Reiteration in Biblical Narrative. *Journal of Biblical Literature* 128: 151–167.

———. 2013. Remembered Space in Biblical Narrative. In *Constructions of Space IV: Further Developments in Examining Social Space in Ancient Israel*, ed. Mark George, 61–75. T & T Clark International: New York, NY.

Na'aman, Nadav. 2011. The Exodus Story: Between Historical Memory and Historiographical Composition. *Journal of Ancient Near Eastern Religions* 11: 39–69.

Nienass, Benjamin, and Ross Poole. 2012. The Limits of Memory. *International Social Science Journal* 62: 89–102.

Nora, Pierre. 1989. Between Memory and History. *Representations* 26: 7–12.

Olick, Jeffrey K., and Joyce Robbins. 1998. Social Memory Studies: From Collective Memory to the Historical Sociology of Mnemonic Practices. *American Review of Sociology* 24: 105–140.

Schudson, Michael. 1989. The Present in the Past versus the Past in the Present. *Communication* 11: 105–113.

Schwartz, Barry. 1991. Social Change and Collective Memory: The Democratization of George Washington. *American Sociological Review* 56: 221–236.

Sherwood, Aaron. 2006. A Leader's Misleading and a Prostitute's Profession: A Re-examination of Joshua 2. *Journal for the Study of the Old Testament* 31: 43–61.

Truc, Gérôme. 2011. Memory of Places and Places of Memory: For a Halbwachsian Socio-Ethnography of Collective Memory. *International Social Science Journal* 62: 147–159.

van der Toorn, Karel. 2001. The Exodus as Charter Myth. In *Religious Identity and the Invention of Tradition*, ed. Jan W. van Henten and Anton Houtepen, 113–127, Koninklijke Van Gorcum: Assen.

Weeden, Chris, and Glenn Jordan. 2012. Collective Memory: Theory and Politics. *Social Semiotics* 22: 143–153.

Wolff, Hans W. 1974. *Hosea*. Philadelphia, PA: Fortress Press.

Zerubavel, Yael. 1995. *Recovered Roots: Collective memory and the Making of Israeli National Tradition*. Chicago, IL: University of Chicago Press.

The Exodus and History

33

William H.C. Propp

Abstract

The Exodus, however we define it, cannot be called "historical." So diffuse is the evidence, the biblical narrative cannot be adequately tested by the historical method. This paper compares and contrasts another mythic tale of improbable, miraculous salvation: the "Angel(s) of Mons" from World War I, where abundant information enables us precisely to sift truth from fiction and set both in historical context. For the Exodus, we must simply resign ourselves to ignorance.

Archaeologists and textual historians agree that the biblical narrative is not contemporary with purported events, has a complex literary prehistory, and does not fit comfortably with known ancient Near Eastern history. Over the millennia, it has become collective cultural memory—but memory of what?

In order to address the question, "Is the Exodus history?" we must define "Exodus" and "history." According to the Hebrew Bible, the Exodus is the departure, in a single night sometime in the fifteenth century BCE, of 600,000 adult Hebrew males and their families, embarking upon a trek from slavery in Egypt to freedom in Canaan. It is hard to imagine how that would work spatially, but that is what the text says.

As for History, Thucydides, the father of the historical method, begins his work as follows: "Thucydides, an Athenian, wrote the history of the war between the Peloponnesians and the Athenians."[1] It is easy to overlook the importance: an actual author is claiming personal credit and responsibility for what he is about to relate.[2] In place of an anonymous, omniscient narrator, such as one finds in Homer—who, admittedly, claims inspiration from a Muse[3]—Thucydides interposes himself and his new, critical method between the facts and the reader.

As Thucydides demonstrates, both time and space tend to distort the transmission of

W.H.C. Propp (✉)
Department of History, University of California, San Diego, CA, USA
e-mail: wpropp@ucsd.edu

[1] All quotations from Thucydides, *The History of the Pelopponesian War Book I*, trans. Richard Crawley. http://classics.mit.edu/Thucydides/pelopwar.1.first.html

[2] This *per se* is not new; Herodotus had begun his own History similarly.

[3] Unlike Homer, Hesiod does name himself in the opening of his *Theogony*, in addition to claiming the authority of the Muses.

T.E. Levy et al. (eds.), *Israel's Exodus in Transdisciplinary Perspective*, Quantitative Methods in the Humanities and Social Sciences, DOI 10.1007/978-3-319-04768-3_33,

information. And so does poetry. For this reason, although he begins his history in the Age of Heroes, Thucydides is skeptical of the supernaturalistic myths prevalent among the Greeks—even if he assumes (as we do, too) some historical basis behind the traditions of migration and the Trojan War.

For events of recent history, Thucydides proposes a better method: research.

> And with reference to the narrative of events, far from permitting myself to derive it from the first source that came to hand, I did not even trust my own impressions, but it rests partly on what I saw myself, partly on what others saw for me, the accuracy of the report being always tried by the most severe and detailed tests possible. My conclusions have cost me some labour from the want of coincidence between accounts of the same occurrences by different eye-witnesses, arising sometimes from imperfect memory, sometimes from undue partiality for one side or the other. The absence of romance in my history will, I fear, detract somewhat from its interest; but if it be judged useful by those inquirers who desire an exact knowledge of the past as an aid to the interpretation of the future... I shall be content.

In contrast with Homer, Hesiod, Herodotus, and the tragedians, and as part of his critical stance toward the mythic mindset, Thucydides avoids attributing divine causation to epochal events such as migrations, wars and plagues. What, then, would he have made of the Torah and its story of the Exodus, which probably reached its current form across the Mediterranean in the Athenian's own day, in the fifth century BCE? As he did with Homer, Thucydides would no doubt have assumed a historical core to the tradition, minus the fantastical elements: hungry Hebrews from an unstable climatic zone migrated to fertile Egypt, where they thrived, threatened to overrun the country and were enslaved. Eventually they rebelled and departed for their native country, which they conquered with the sword. But, as with Homer, Thucydides would have felt frustrated by the narrative's fanciful aspects, the prominent divine role, and the lack of authentication.

Unlike theologians, Thucydides would *not* be chagrined that the Torah contradicts itself. His method takes that into account. He would also expect the Torah to be biased and selective. He would perhaps suspect that, like Homer, the Torah just tells too good a story, and he might well reiterate his insight that the truth tends to be blander than fiction. He would surely assume that the dialogue, at least, had been fabricated, inasmuch as he himself admits to making up his own.

Thucydides would probably be more troubled by the Torah's anonymity. There is no historian claiming responsibility for what is recounted, nor does the mysterious author cite sources. Homer and Hesiod at least invoke Muses; the Torah just demands (and generally has obtained) readers' credulity. For two-and-a-half millennia Jews, joined later by Christians, have assumed that the narrator is omniscient because, as an article of faith, the Author is Omniscient.

Even though he would no doubt concede the possibility of real events underlying the Exodus tradition, Thucydides would feel much more at home in the biblical books of Samuel and Kings. During the account of David's reign, there is almost no supernatural causation, but much political chicanery. For First and Second Kings, there is a detailed chronological framework, albeit slightly inconsistent in details, with many events susceptible of investigation from multiple perspectives—both within the Bible itself and from what we now know from archaeology and ancient inscriptions. Even if there is no "I" taking responsibility, the author of Kings at least cites written sources, especially the Chronicles of the Kings of Israel and Judah. For other, less mundane material, such as the tales of the prophets, he hints at the oral sources: "Now the king was talking with Gehazi the servant of the man of God, saying, 'Tell me all the great things that Elisha has done.'" (2 Kings 8:4 RSV).

A naif might ask, "But didn't Moses himself write the Torah?" Since the Renaissance, however, especially the literary detective work of Lorenzo Valla, we no longer accept traditional claims of authorship on trust; we use the tools of philology. As Baruch Spinoza (*Tractatus Theologico-Politicus*) argued at great length in the seventeenth century, and certain medieval Jewish commentators may already have suspected, when we apply the "Valla test" to the Mosaic authorship of the Torah, the tradition fails miserably. After all, the Torah never claims to be written by a participant in the Exodus; its stance is consistently retrospective, looking backward from a time when there were no longer Canaanites, when the Edomites had kings, when the location of Moses' tomb had been forgotten, etc. Thus, vis-à-vis the supposed events of the Exodus, the Torah is at best a secondary source.

The fact that the Torah is a secondary source, indeed one that cites no documentary evidence of its own, makes it of limited value for the historian. Our best estimate for the final compilation of the Torah, around 400 BCE, is a millennium later than tradition dates the Exodus itself. To be sure, two centuries of philological research have revealed older sources hidden in the Torah, commonly called J, E, D, and P. This analysis does take us slightly closer to the events, but these reconstructed documents, too, are secondary sources: D and P are based upon J and E and thus possess little independent historical value, while J and E are themselves from the Iron Age. And many scholars today would argue that some or all of these supposed documents are chimerical and that the texts are even further removed from the end of the Late Bronze Age, when the Exodus supposedly occurred (Chaps. 20, 22 and 24).

However, inside of the Torah are a body of poems that some scholars think are older than their prose context. Among these is Exodus 15:1–18, the Song of the Sea (Russell 2009: 133–143; Chap. 5). Now, not everyone agrees that Exodus 15:1–18 is archaic; it resembles no form of Hebrew known from inscriptions and might rather be pseudo-archaic in the manner of the poetry of Job. But it is our best hope for a contemporary or near-contemporary documentary source. The poem describes Pharaoh's chariotry sinking into the Red Sea (or possibly a papyrus marsh [Chap. 8]). While Thucydides' reservations about poets would still apply, a few modern scholars have taken Exodus 15:1–18 to be an eyewitness response to a historical event. The problem is that we have no other data by which to interpret it, apart from a prose framework of questionable reliability and uncertain dating.

Another potentially archaic reference to the Exodus is found within the oracles of Balaam. Although the occasional archaism and obscurity of these texts have been cited as evidence of antiquity (Albright 1944; cf. Russell 2009: 81–103), these aspects are overbalanced by the reference to the Assyrian empire in Numbers 24:22. Without considerable apologies and gymnastics, we cannot date the oracles of Balaam earlier than the ninth century BCE.

If our sources do not elicit much confidence in terms of antiquity, might they still not contain real information, however passed down? To judge from his attitude toward Greek mythology, Thucydides would have given them the benefit of the doubt. There may after all be lost written sources; oral tradition is sometimes reliable—though, tellingly, we are always surprised when it is. In this case, how could we know? We could look for details in the text, such as techniques of brick making, place names, etc., that could have been known only in the second millennium BCE; or, alternatively, we could contemplate the overall sweep of the narrative, without binding ourselves to specifics of time, place, and numbers. I am not a fan of the first method, which has been used to both support (Halpern 1993; Hoffmeier 1996) and discredit (Chap. 34) the narrative in its details. I would look rather to the large events that might have left a trace in the historical-archaeological record.

While the Torah itself does not provide a chronological anchor—unless you think the world really is 6,000 years old—1 Kings 6:1 gives a date for the Exodus relative to the reign of Solomon, placing the departure from Egypt in the mid fifteenth century BCE, the richly documented heyday of the Egyptian Empire. Do we look for evidence of an Exodus there? Or do we suspend judgment over the date, and just seek more broadly?

Now, in fact there are numerous known historical events from the ancient Near East that rather resemble the Bible's account of the Exodus (Chaps. 39 and 4). In chronological order, here are the most prominent[4]:

Eighteenth–twelfth centuries BCE. Throughout these centuries, we find ample evidence of sheep- and goat-herding seminomadic pastoralists whose migrations recall the Israelite

[4] For more detail, see Propp 2006: 735–746.

ancestors as portrayed the Torah (Chap. 29). In addition, unruly groups called ʻAbiru or ʻApiru are attested throughout the ancient Near East. To judge from their names (often Hurrian, sometimes Semitic), they do not constitute an ethnic group but are rather a social class, their name meaning something like "freebooter." The word may survive in biblical *ʻibri*, "Hebrew," the term Israelites used to describe themselves to foreigners (Greenberg 1955).

Seventeenth century BCE. The huge volcano of Thera erupts, creating perturbations in Mediterranean weather and tides (Chaps. 7 and 9). Egypt is ruled by the foreign, Asiatic Hyksos (Chap. 2), whom later propaganda, prefiguring Roman critiques of Judaism and Christianity, will describe as near atheists, worshiping only Seth—here a cipher for the Canaanite storm god (de Moor 1997: 76, 102). The Bible still remembers the foundation of their capital, Tanis (Numbers 13:22). And one of their rulers, Yaʻqub-har, could be an inspiration for the Jacob–Joseph traditions of Genesis.

Sixteenth century BCE. Ca. 1550, the Hyksos are expelled from Egypt, with some sources mentioning accompanying meterological portents (Redford 1992: 420).

Sixteenth–fifteenth centuries BCE. Most of the large cities in Canaan are destroyed by the 18th dynasty Pharaohs as they create an empire in Asia.

Sixteenth–eleventh centuries BCE. Many Asiatics live in Egypt, whether as slaves, hostage princes or ordinary citizens. Some of these are probably Hyksos who stayed behind (Chap. 2), others would be new immigrants. Whether they lived in Canaan or Egypt, for much of this time all Canaanites were essentially slaves to Pharaoh (Chap. 5).

Fourteenth century BCE. East of the Arabah lurk various nomads (Shasu), including one group called the Nomads of *Yhw3* (Weinfeld 1987: 304–305; Redford 1992: 272–273; Chap. 41). In Egypt, Akhenaten claims there is no god but the sunlight; one may worship no graven images, Akhenaten himself being the sole divine form and embodiment (Propp 2006: 762–794). Simultaneously, some ʻApiru are making trouble in Canaan (on the "Amarna Period," see Chap. 36).

Thirteenth–twelfth centuries BCE. Ramesses II builds the cities of Pithom and Ramesses, mentioned in Exodus 1:11.

Thirteenth century BCE. The Late Bronze Age international order collapses. In Canaan a new, burgeoning population occupies the highlands; to judge from their artifacts and language, they are largely indigenous and foster an egalitarian ethos (cf. Chaps. 30, 37 and 41). Overoptimistically, Pharaoh Merneptah claims to have eradicated an ethnic group called "Israel" in this region.

Twelfth century BCE. The Philistines, mentioned in Exodus 13:17: 15:14, arrive and settle the Canaanite littoral. About this time, the Egyptians end their rule over Canaan; the natives are free at last (Chap. 5).

All this might sound like good news: we have so much evidence of the Exodus! In fact, it is bad news. An event cannot be spread across half a millennium! Therefore, many scholars nowadays argue that the Israelites and their traditions reflect the amalgamation of diverse groups and their experiences throughout the second millennium, and even into the first (Chaps. 36, 37, 5, 41 and 34). We cannot prove that either, but it is much easier to imagine than a historical Exodus event.

To complicate matters even further, we know that over the course of time the Israelites came to associate their departure from Egypt with an old Amorite myth, first attested at eighteenth-century BCE Mari, according to which in primordial times the storm god had established his kingship by defeating and drying up the sea and slaying the sea's minions, especially a seven-headed serpent the Hebrew Bible calls Rahab, Leviathan, or simply the Serpent (Propp 1999: 554–561). This myth, recovered over the past 150 years from Mesopotamian and Canaanite sources, is alluded to dozens of times in the Hebrew Bible and has survived into later Judaism (Cassuto 1975: 69–109). Several biblical passages make an explicit connection between the Combat Myth and the Exodus: Isaiah 51:9–10; Psalm 77:11–20, etc. Other texts equate Pharaoh with Leviathan (Ezekiel 29:3; 32:2). Thus, the tradition that

Yahweh created his people and established his kingship by fighting a battle against the sea, in the process drying or splitting it, was likely influenced by a pre-Israelite myth (cf. Chap. 14).

Some scholars, perhaps the majority, suppose that the similarity of the Exodus to the creation myth dawned only gradually upon the Israelites. For others, a historic salvation by the sea immediately assumed mythic resonances for its very participants (cf. Cross 1973: 79–90). For yet other scholars, the Exodus story is simply another version of the myth, recast as pseudo-history (e.g., Kloos 1986: 158–212).

I wish to leave the riddle of the Exodus for a moment in order to develop a contrast with a miraculous redemption much more amenable to historical analysis. Over the 5,000 years of known human history, there are hundreds if not thousands of records of supernatural salvations during military crises, of which the Exodus is but one example. Here is another, almost within living memory:

On August 22–23, 1914, the British Expeditionary Force first engaged German troops at Mons, Belgium, in order to cover a French retreat. Despite outnumbering the British by more than 2.5:1, the Germans were initially repulsed. They quickly regrouped, however, and forced the British into a long retreat, during which the Allies also reassembled and eventually defeated the Germans at the Battle of the Marne (September 5–12, 1914), thereby saving Paris. The Expeditionary Force's near brush with annihilation, at first concealed by censorship, was reported on August 30, 1914, and converted by propaganda into a qualified victory (which it was), as well as a call to enlistment.

The salvation was seen literally as miraculous. On April 24, 1915, the *Spiritualist* magazine published reports of visions of a supernatural force that intervened to help the British at the decisive moment of the battle. Other corroboratory accounts followed: Some described further wonders throughout the retreat of the British and the French. Some recalled a luminous cloud, and others described medieval longbow archers fighting alongside St. George, while most reported a host of angelic warriors. By May and thereafter, the miracle was bruited in publications and sermons as evidence of divine favor for the Allied cause, and soon the tale was circulating internationally (Clarke 2004: 37–63). Curiously, French soldiers had similar memories, except that instead of St. George they had seen St. Michael and St. Joan of Arc—the latter apparently having made her peace with the English (Clarke 2004: 135–145). Today this curious episode in the Great War is referred to as either the Angel or the Angels of Mons.

Right from the start there were cynics. Soldiers are notoriously superstitious, frightened, hungry, battle-fatigued men, high on adrenaline and intensely indoctrinated, and do not make the best witnesses. Stories of divine interventions on the battlefield, moreover, are as old as Homer, as old as the Bible, as old as the royal inscriptions of Egypt and Mesopotamia, doubtless as old as war itself. Further investigation proved that all early accounts were second-hand hearsay, some from soldiers who had not even been at Mons. Thus, the Society for Psychical Research concluded in 1915 that there was no credible eye-witness testimony or any evidence at all of the supernatural (Clarke 2004: 122–124).

In the early twentieth century, there was also the added factor of new methods of controlling information via censorship and propaganda. For the true believers in the miracle, any skepticism was simply evidence of a cover-up. On their part, skeptics assumed that the War Office was exploiting a supposed mass hallucination for its own propagandistic purposes. Besides the Angels of Mons, we know of numerous wild rumors circulating in wartime Great Britain, most of which had nothing supernatural about them but were nonetheless mistaken, even absurd. Some may indeed have been deliberately implanted to assist the war effort (Clarke 2004: 67–85).

And there is one more factor to consider:

On September 29, 1914—notably, the Feast of St. Michael and All Angels—1 month after the first press reports of Mons, Welsh author and journalist Arthur Machen, who wrote both factual accounts of the war and also fantasy and fiction, and was himself inclined toward neo-medieval, mystical forms of Christianity, published a very short story in the London *Evening News* (Clarke 2004: 85–99, 247–250). "The Bowmen," conceived while Machen was dozing at church, drew inspiration from recent published reports of the Battle of Mons, and also the antiquarian Machen's own familiarity with English traditions of angels and saints back through the Middle Ages. Machen probably also knew that, since the fifteenth century, St. George had been particularly venerated in Mons, the supposed site of the battle with the dragon (Clarke 2004: 17–36). And Machen was also a fan of the ghost stories of Rudyard Kipling.

The style of "The Bowmen" is first-person verisimilitude. The opening, "It was during the Retreat of the Eighty Thousand, and the authority of the censor is sufficient cause for not being more explicit," hints at a truth even more fabulous than what is to be revealed. The narrator, a soldier in a beleaguered British army unit, calls upon St. George, whereupon phantom bowmen materialize and destroy the German foe. The 1227-word story ends,

> In fact, there were ten thousand dead German soldiers left before that salient of the English army, and consequently there was no [German conquest of] Sedan. In Germany, a country ruled by scientific principles, the Great General Staff decided that the contemptible English must have employed shells containing an unknown gas of a poisonous nature, as no wounds were discernible on the bodies of the dead German soldiers. But the man who knew what nuts tasted like when they called themselves steak knew also that St. George had brought his Agincourt Bowmen to help the English.[5]

What happened next is best told in Machen's own words, published in a 1915 reprint of the story:

> [I]n a few days from its publication the editor of THE OCCULT REVIEW wrote to me. He wanted to know whether the story had any foundation in fact. I told him that it had no foundation in fact of any kind or sort.... Soon afterwards the editor of LIGHT wrote asking a like question, and I made him a like reply....
>
> A month or two later, I received several requests from editors of parish magazines to reprint the story. I–or, rather, my editor–readily gave permission; and then, after another month or two, the conductor of one of these magazines wrote to me, saying that the February issue containing the story had been sold out, while there was still a great demand for it. Would I allow them to reprint THE BOWMEN as a pamphlet, and would I write a short preface giving the exact authorities for the story? I replied that they might reprint in pamphlet form with all my heart, but that I could not give my authorities, since I had none, the tale being pure invention. The priest wrote again, suggesting–to my amazement–that I must be mistaken, that the main "facts" of THE BOWMEN must be true, that my share in the matter must surely have been confined to the elaboration and decoration of a veridical history. It seemed that my light fiction had been accepted by the congregation of this particular church as the solidest of facts; and it was then that it began to dawn on me that if I had failed in the art of letters, I had succeeded, unwittingly, in the art of deceit. This happened, I should think, sometime in April, and the snowball of rumour that was then set rolling has been rolling ever since, growing bigger and bigger, till it is now swollen to a monstrous size.... It was at about this period that variants of my tale began to be told as authentic histories....[6]

The more Machen protested, the more resistance he received. He was antipatriotic, he was anti-Christian; he must have at least subconsciously channeled soldiers' actual experiences!

Moreover, much later, in 1931 Brigadier-General John Charteris published his memoir *At G.H.Q.*, according to which the wondrous apparition was a popular rumor among the troops as of September 5, 1914, *nineteen days before the appearance of Machen's story*. Here we would seem to find proof that, despite his claims, Machen did not invent the miracle at Mons. In his preface, however, Charteris admits, "I had not kept a formal diary...where records were incomplete, *I have amplified them by my recollections*" (Clarke 2004: 215, emphasis added)—and the General's surviving correspondence from early September 1914 makes no reference to supernatural visions. Machen's claim stands.

[5] http://www.aftermathww1.com/bowmen.asp

[6] http://www.aftermathww1.com/bowmint2.asp

Incidentally, though I have not researched the matter thoroughly, I am not aware than anyone ever asked *Germans* present at Mons what they thought had happened. In 1930, however, a former member of the Imperial German Intelligence Service reportedly claimed that that the Germans had projected images onto white cloudbanks in a counterproductive effort to spook the British—even though no such technology existed, the battle occurred by day, and the German military could not confirm that the informant had actually been present or even existed (Clarke 2004: 206–209).

The Angels of Mons was not the only literary fiction that was converted into actual memories of World War I. In 1915, the Reverend W. H. Leathem published a first-person story "In the Trenches," in which a wounded soldier is nursed by a mysterious Comrade in White, who proves to be Christ. The whole pattern then recurred: numerous but nebulously attested visions of the Comrade in White, the celebration of the miracle in church sermons and religious and psychical publications, and the clerical author's chagrined acknowledgment that, while miracles may happen, he had in fact made this one up (Clarke 2004: 177–183).

Various morals might be drawn from the comparison between the Angels of Mons and the biblical Exodus. The optimist would admit, okay, Machen made up the supernatural bit, but there really was a battle of Mons, the British really did escape from the Germans, there really was a Great War. Likewise the Exodus.

The biblical ultra-skeptic might rejoin, however, that just as Machen converted immemorial Christian miracle stories into a purported report from the battle lines, which then entered British popular consciousness as stone-cold fact, despite (or rather because of) the supernatural content—so the ancient myth of the combat between the storm god and the sea was transmogrified into a pseudo-historical tale of Israel's miraculous salvation.

But here is the moral I would draw. We know that the Battle of Mons occurred. We know its precise dates, we know its exact location, we know the historical context. We can date Machen's story to the day. We can supply oral and written testimony from literally thousands, probably tens of thousands of diverse sources to gain a stereoscopic image of the times. In other words, we have precise reference points to support historical analysis.

From my perspective, the Exodus is *not historical* by definition, because it simply is not susceptible to the historical method. There is no paper trail of evidence, what literary sources survive are of uncertain date, and the story lacks a clear anchor in time. Our uncertainty factor is measured not in days, weeks or months, but in centuries. And *then*, there is also the problem of the supernatural—do you believe in miracles, or in the human disposition to believe in miracles (or both)? *Mundus vult decipi*, "The world wishes to be deceived."[7]

In such a situation, the only sane response—the philosopher's response, the Buddha's response—is to cultivate apathy. We should not care whether there was an Exodus, because wanting what we do not have, i.e., evidence, will make us unhappy.

For now, the cultural memory of the Exodus is alive and well, whereas the shining Angels of Mons are fading fast. They still have their devotees, but far fewer than the Old Testament enjoys. Why the difference? Obviously, the story of the Exodus is far older than the Angels of Mons and has had more time to diffuse. Because it arose in an age of technology and literacy, moreover, the twentieth century legend spread like wildfire—but then incurred immediate doubt for exactly the same reason, whereas the story of the Exodus, whatever its true origins, arose and was propagated in an age when credulity was even greater than the early twentieth century. And, crucially, the story of the Exodus has been embalmed in ritual—Bible reading among Christians, Bible reading plus the Passover Seder among Jews—whereas World War I commemorations do not to my knowledge emphasize or reenact the Mons miracle.

---

[7] On neurology and cultural memory, see Chap. 31.

To conclude: the historian must avoid the Exodus, put it into a black box, lock it up, and then hide but not discard the key, keeping it against the unlikely event that new evidence should emerge. Perhaps there was an actual event, or several events, that are commemorated in the Exodus tradition. But the data are just too diffuse, too sparse. In contrast, to doubt the historicity of the World War I Battle of Mons, solely because one rejected the attendant tale of supernatural rescue, would be to posit such a gargantuan hoax that it would be far easier to believe in the Angels.

References

Albright, William F. 1944. The Oracles of Balaam. *Journal of Biblical Literature* 63: 207–233.

Cassuto, Umberto. 1975. *Biblical and Oriental Studies*, vol. 2. Jerusalem: Magnes Press.

Clarke, David. 2004. *The Angel of Mons*. Chichester: John Wiley & Sons.

Cross, Frank M. 1973. *Canaanite Myth and Hebrew Epic*. Cambridge, MA: Harvard University Press.

de Moor, Johannes C. 1997. *The Rise of Yahwism*. Leuven: University Press.

Greenberg, Moshe. 1955. *The Hab/piru*, AOS, vol. 39. New Haven, CT: American Oriental Society.

Halpern, Baruch. 1993. The Exodus and the Israelite Historian. In *Abraham Malamat volume*, ed. S. Ahituv, and B.A. Levine. Eretz Israel 24. 89*–96*, Jerusalem: Israel Exploration Society.

Hoffmeier, James K. 1996. *Israel in Egypt*. New York, NY: Oxford University Press.

Kloos, Carola. 1986. *Yhwh's Combat with the Sea*. Amsterdam: van Oorschot.

Propp, William H.C. 1999. *Exodus 1–18*, Anchor Bible, vol. 2. New York, NY: Doubleday.

———. 2006. *Exodus 19–40*, Anchor Bible 2A. New York, NY: Doubleday.

Redford, Donald B. 1992. *Egypt, Canaan, and Israel in Ancient Times*. Princeton, NJ: Princeton University Press.

Russell, Stephen C. 2009. *Images of Egypt in early biblical literature: Cisjordan-Israelite, Transjordan-Israelite, and Judahite portrayals*. Berlin: Walter de Gruyter.

Weinfeld, Moshe. 1987. The Tribal League at Sinai. In *Ancient Israelite Religion*, Fs. F. M. Cross, ed. P.D. Miller Jr., P.D. Hanson, and S.D. McBride, 303–314. Philadelphia, PA: Fortress Press.

The Great Going Forth: The Expulsion of West Semitic Speakers from Egypt

34

Donald B. Redford

Abstract

The memory of a southward descent to northeast Africa, apparently sometime in the Bronze Age, by West Semitic-speaking peoples, coupled with a great going forth four generations later, is indelibly marked in the folk memory of the ancient Near East. In unlikely places, from Saite Egypt to Phoenicia and Greece, and in the Hebrew prophets, traces of the movement have crept into the mix of reminiscence and survived in myth and legend, if only in “one-liners.” The recollection, however vague, sometimes contradictory and distorted, is best known from the descent tradition of Gen. 37–50 and the tradition of Exodus 1–15; but neither text by any means has claim to a prime, authoritative position among the several versions. Far from being an account centered upon a believable and identifiable “Egypt,” Exodus 1–15 should better be labeled by the appellative “Moses: The Shepherd from the East.” Moses “The Egyptian” is nowhere in evidence. We shall explore the traditions as we have them and try to ascertain their roots in time and space. The “Osarseph” and “Bocchoris” tradition foci will be dissected as well as the role of the central Delta in complementing Biblical tradition of the sixth to fourth centuries BC. In terms of plot patterns, *dramatis personae*, and backdrop, it will soon appear that any attempts to interpret the Biblical material historically and to fix a date in the New Kingdom are illusory.

First of all, I must register disagreement with the title of this symposium. The *Bene-Yisrael*, if that is what is meant by “Israel,” experienced no great exodus from Egypt. The inhabitants of Israel or Judah in the Iron Age could no more claim that their fathers had participated in an expulsion from northeast Africa than a present-day immigrant to the USA from, say, southeast Asia could claim that his forefathers had taken part in the War of Independence. That there *was* a great exodus from Egypt by *somebody* is a demonstrable fact; but the question is by whom.

Second of all, while detailed accounts once existed in Phoenician and Greek sources (in particular on Io, see *inter alia* Edwards 1979; Griffiths

D.B. Redford (✉)
Pennsylvania State University,
University Park, PA, USA
e-mail: mendes34@comcast.net

T.E. Levy et al. (eds.), *Israel's Exodus in Transdisciplinary Perspective*,
Quantitative Methods in the Humanities and Social Sciences, DOI 10.1007/978-3-319-04768-3_34,

1986; Davison 1989; Gershenson 1996) and in the Hecataean and Manethonian sources, the only connected account today survives in the Joseph-Moses account, Genesis 37–Exodus 15 (Schmid 2012; for a counterargument see Baden (2012). Gen 1–36 anticipates in a disjointed fashion the *descent–sojourn–exodus* tradition, but the latter in no way is dependent upon, or incomplete without, these early "patriarchal" narratives. The compiler has fastened upon the personal names of early heroic figures of "Amorite" descent, memories of whom were current from Beersheva to the Negeb (Abram) (Epigraphic Survey 1954: pl. 3–4, no. 71–72), and in central Palestine and Transjordan (Jacob, Joseph) (*Urk*. IV: 785 [102], 784 [78]),[1] and made them the heroes of a series of aetiologies and narrative scripts.

There are extant in Middle Eastern traditions, both at the paraphrased and/or oral registers as well as writings *in extenso*, several traditions regarding a "coming-out" from Egypt (Russell 2009).[2] These include (1) the Egyptian king-list (Manethonian) account, (2) the Bocchoris "leper" tradition, (3) the Phoenician reminiscence, (4) the asides in the Hebrew prophets, and (5) the stories in Genesis 37 to Exodus 15. Issues regarding the first four are addressed by the author below and elsewhere (Redford 2011), but the last is by far the lengthiest and best preserved. While the Joseph Story shows a style, idiolect, and plot integrity second to none in the Hebrew Bible, the Exodus narrative is hopelessly inferior. "P" has done a poor job in integrating his sources,[3] whether oral or written, none of which in any case predate the seventh century by much.[4]

What has become increasingly certain over time is that all of this material reflects the unmistakable backdrop of the Hyksos occupation and expulsion. Thus we have the phenomenon of a West Semitic-speaking people with an "Amorite" onomasticon (note the presence of יוסף, יעקב, Khayan, Yansas-adon, etc.), occupying the fringes of the Delta, the Negeb, and the Sinai (see Appendix 1) for a period of 108 years (Gardiner 1959; Gen 15:16) and refusing to integrate with the autochthonous population in either culture or religion. Consider in addition that the denouement of Hyksos hegemony was attended by a famine (Brugsch 1891) and their expulsion by foul weather (Helck 1975: no. 124). Finally, the "deliverer" bore a name which arguably was compounded with *-msi*.

What we have here is the *Asiatic* folk memory of this great event, adjusted, distorted, and inverted, with motivation reversed or imputed. In contrast to the witness of such contemporaries, or near contemporaries, as Kamose, Ahmose, Ahmose-si-Abina, Hatshepsut, or Thutmose III, this memory suffers from distance in time and the co-opting of recent prejudices. It is, however, legitimate to ask what contribution, if any, the likes of Manetho's *Aegyptiaca*, the Piankhy legend, or the Exodus narrative make either to elucidating the original event or to its evolution through time.

The Traditions of the Exodus: Origins and Location

The "Manethonian" Tradition of the Hyksos

The format adopted in the king-list tradition which is represented by Manetho's *Aegyptiaca*, or at least the tradition this document represents, is that of a king-list with personal names and "exact" lengths of reign, interlarded with narrative (Helck 1956; von Beckerath 1980; Redford

[1] From the beginning of the seventh century BCE on, the scribal practice reflected in ס—č was done away with, and, under the influence of Aramaic, ס was used for any sibilant (See Appendix 2.).

[2] Russell argues that there was no unitary core Exodus story.

[3] It is overly charitable in analyzing sources to claim that the poor joins and lack of integrity are examples of intentional irony consciously introduced into the story (Wimmer 2011: 120).

[4] It is scarcely conceivable how one branch of scholarship continues to insist on widespread literacy and a flowering of belletristics in the time of Solomon (cf. Leuchter 2011). When we can assign the preservation of "E" to the flight of a refugee from the northern kingdom to the south with the "Book of E" in his hand, we have entered the realm of comic opera.

1986; Kitchen 2001). The latter, in the main, curiously eschewed formal hieroglyphic inscriptions (which the compilers including Manetho himself must have been able to read) in favor of tales, extant in cursive scripts, to be found in temple libraries.

The following form a consistent tradition reflected in Josephus' quotations and the glosses in Manetho, reflective both of "Demotica" and Judaic chronography: (1) The descent of the Hyksos, a Manethonian precis of a longer account, possibly colored by the details of the Assyrian invasion of 671 BCE (Redford 1997); (2) a note (Eusebius) regarding Joseph's *floruit* under Dynasty 17 ("the Shepherds"), and undoubtedly building upon the chronographic computation within the Judaeo-pagan polemic; and (3) an explicit assertion *apud* classical writers to the effect that it was under Amosis, founder of the 18th Dynasty, that Moses exited Egypt (Ptolemy of Mendes, Julius Africanus, etc.: Berkowitz and Squitier 1990: 337, no. 1647). By the outgoing fourth century BCE, therefore, there existed a strong belief that Moses and the Exodus were to be linked with the expulsion of the Hyksos.

Also underlying the section of the epitome of Manetho dealing with the Hyksos was a tale recounting the end of the 15th Dynasty. "Misphragmouthosis" besieged the Hyksos in Avaris, but could not carry the day. It was left to his son Tethmosis to resolve the blockade by signing a treaty with the besieged and allowing them to go wherever they chose. Subsequently they went to Judaea and founded Jerusalem. Somewhere early in the transmission of the king-list, confusion has been introduced by false identification. "Tethmosis" is clearly Thutmose III, but the historical figure has been reduplicated: "Misphragmouthosis" is the garbled prenomen-cum-nomen of the same Thutmose III (von Beckerath 1999: 137). Two famous sieges have been confused and treated as one: the siege of Avaris by Kamose and Ahmose and the siege of Megiddo by Thutmose III. The besieged have been wrongly identified in the latter case, but this is understandable: the Megiddo episode likewise involved *ḥḳꜣw-ḫꜣswt* (Hyksos: Barguet 1962: 161; *Urk.* IV: 555, no. 44–45; 593:10).

Amarna Rationalized: Osarseph

This hermeneutic on the end of the Amarna period is quoted *in extenso* by Josephus (*Contra Apionem* i.26–31). Mythomorphemically it is rooted in a clear memory of certain major, historic details of the period, which have survived, perhaps initially at an oral register, despite the anathematization of the anti-Amarna reaction (H. D. Schneider 2001; Gaballa 1977: 25; von Beckerath 1997: 128). The setting is the reign of Amenophis *alias* "Hor," that is, Amenophis III who frequently took "Horus" as an epithet (Redford 1986: 248). The king evinced a desire to "see the gods." Viewing the gods is a desirable act of piety, often induced by magic, which runs through the history of Egyptian religion (Johnson 1977; Betz 1986). Interestingly it is in the reign of Amenophis III that explicit reference is made to the king's intent to make the gods visible on earth by "giving birth" to their images in his Theban temples: Amun journeys "to see the gods" (*Urk.* IV: 1650:8); the king is "refurbisher of the [temples of all the gods], fashioning their bodies of gold" (*Urk.* IV: 1667:13); the monuments (scil. of the gods) "are seen as they occupy their places" (*Urk.* IV: 1672:18); "the divine images that emanated from you (Amun), My Majesty gave birth to all of them" (*Urk.* IV: 1673:3); "I had (them) rest in the august shrine, as they had been [. . ..] My Majesty has acted for you (Amun) to the utmost, knowing indeed that on earth they should exist" (*Urk.* IV: 1673:8–9). The gods themselves answer Amun: "Amenophis III it is your son that has done this; so praise him for it, for he has 'given us birth,' knowing that our lord would rejoice when he sees us existing on earth" (*Urk.* IV: 1676:9–13).

The king's confidant who advises him is the wise man "Amenophis son of Pa-apis." This of course is the historical Amenophis son of Hapu, a native of Athribis in the Delta, who served Amenophis III as "king's scribe, secretary of recruits." Even before he died—before his sovereign in the Osarseph story—he had become a legend for wisdom (Wildung 1977). Amenophis advised that seeing the gods would be possible only if the land were cleansed of lepers and the diseased. This allusion to an epidemic recalls the evidence of a plague which swept over the eastern Mediterranean during Akhenaten's reign and later (Helck 1971: 187–188; Redford 1984; Moran 1992 *passim*; Moran 2003: 304; Hoffner 2009: 187–188) and may have carried off many of the Amarna principals. The role played by quarries in the rounding up and forced labor of the lepers represents a striking aetiology on the visual prominence of commemorative texts and reliefs in quarries under Amenophis III and Akhenaten. Aswan, the Red Mountain, the eastern desert, and Gebel Silsileh all show large, unmistakable memorials to the extensive work done at this time period in the extraction of stone. The presence among the lepers of a prominent priest, Osarseph from Heliopolis (Redford 1983), sets the stage for the solar aspect of the cult which is about to be introduced. But it also finds a stunning historical parallel in the assignment to the quarries in the Wady Hammamat of the high priest of Amun, Maya, in Akhenaten's fourth year (Goyon 1957: no. 90).

The lepers are given an unoccupied territory in which to settle, a parallel to the identification of the new city of Akhetaten by the sun god as his new "horizon." The period during which the lepers were prophesied to occupy Egypt, viz., 13 years, is exactly the time span Akhenaten occupied Amarna. Akhenaten's iconoclastic reforms, with the resultant abandonment of the temples, underlie the account of the introduction of aniconic monotheism and the branding of animal worship as anathema.

The detailed knowledge of the Amarna fiasco, evinced in this account, and the manipulation of the mythemes militate in favor of a lively memory rendered into a *Märchen* perhaps in Ramesside times. The point would have been to stress the saving role of the new dynasty in solving the problems left by the 18th. But such is not quite the ending the present version enjoys. In fact the integrity and historicity of the story are seriously compromised at this point by a sort of "narrative afterthought." The "shepherds" are invited back in! They subject the country to *another* orgy of destruction, and sacred animals are slaughtered, roasted, and eaten. Help eventually comes from Ethiopia, and the shepherds are expelled from the land via Pelusium. Notwithstanding the eventual outcome, at one point the Egyptian king retreats to Memphis without offering battle. This addendum owes nothing to the New Kingdom. It is a product of the Saite and Persian period, a loose doublet to the original tale.

The Bocchoris/Leper Tradition

A version of the same motifs as that appeared in the Osarseph tale is to be found appended to the name of Bocchoris, second king of Dynasty 24 (717–711 BCE) by Lysimachus via Josephus (Kitchen 1973: 376–377; de Meulenaere 1973; Ray 2001). Why this king should have been singled out, in defiance of chronology, is difficult to say. There are, however, points of general similarity between the aetiology on Amarna and the period of Bocchoris. Both stories feature a wise man—Bocchoris was noted as a wise lawgiver (Redford 2001: 136–137); both narratives made mention of plagues or epidemics (Lloyd 1988: 102–104). The "sight" of the gods plays a pivotal role in the Osarseph story: if Bocchoris is equated with Ἀνύσις, blindness is a prominent factor (Herod. 2.137; Lloyd 1988: 91–92). Akhenaten laid waste temples and their cults, as did the Assyrians (Onasch 1994). In Osarseph the wise man prophesied dire events to come, as did the ram of Mendes in the Bocchoris account (Kákosy 1981; Thissen 2002).[5]

The general shape of folklore and oral and written narrative in the period from the late seventh through fourth century BCE owes a great deal to the Saite revival in the interest in the past as a source of models. Prescriptive writings too, of which the Third Intermediate Period had

[5] In the new Egyptian king-list (Popko and Rücker 2011) the prophecy of the ram receives a notice between a Psonsames(?) and Sabaco. The former name *may* incorporate the Egyptian *P3 S3w*, "the Saite," a believable epithet of Bocchoris. But equally *P3 šry n …*, "the child of …," could underlie the Greek.

spawned numerous variants, now underwent redaction in the interest of producing standard texts. This is the period when, sensitive to the new personae of recent history and contemporary events, tellers of tales fashioned unexpected plots and roles and adopted new moral judgments. Now appear the following newcomers: the tyrant king as the source of the ills of the story, as his opposite the ineffectual king, well-meaning or deceitful; the people caught in a crisis their king cannot resolve; the hero, the *magus*, who saves the day; the unclean or the diseased who morph into evil northerners. It is to this constellation of mythegmata that the Joseph and Exodus narratives belong.

The use of the reign of Bocchoris as the chronological venue of the "leper/expulsion" motif explains a good deal of "telescoping" in the folkloric sequence of events in ancient collective memory. At a certain register in oral transmission, and certainly in the time period of Bocchoris, archival facts disappear into oblivion and cede place to spurious equations. Exploits and meaningless associations take precedence over sober chronology, and folk memory, uncorrected by written record, is seen to be very short indeed. By the time the Table of Nations was composed Shabaka and Shabtaka of the 25th Dynasty had been wrongly identified as tribes (Gen 10:7; Astour 1965; Leclant 1983). Terms for Upper Egypt (פתרוס < *P3 t3 rsi*, "the Southland") and Lower Egypt (*נפתוח < *N3w-Ptḥ*, the Memphite environs) are misinterpreted as ethnic blocks (Gen 10:13-14), and Ramesses II transmogrified through his hypocoristicon Sesy-re into a Canaanite general (Judges 6). Into the Exodus genealogies personal names of late currency are insouciantly retrojected: Mushi (Exod 6:19), Puti-el (Exod 6:25), Pinehas, David, and Solomon are made contemporaries of Apries (Eupolemus: Charlesworth 1985: 866), and workers from Mendes and Sebennytos are sent to help build the temple in Jerusalem (Charlesworth 1985: 867). Pelusium, the venue of events of 674, 671, 525, and 375 BCE, figures prominently as the spot where the "shepherds" are expelled; and the retreat from Pelusium to Memphis, instead of a frontier defense (cf. the actions of Taharqa and Nektanebo II), is a known and familiar tactic. Like the Israelites, Tefnakhte and Bocchoris wander in the Sinai, have to subsist on peasant fare, and indulge in "murmuring" (Plutarch *De Iside* 8.354b; Diodorus 1.45; Athenaeus 10.418e).

It is into this confused chronological mêlée that a version was thrust of the great going forth of alien, socially unacceptable people to the north. The combined reigns of Tefnakhte and Bocchoris, founders of the Saite house, totaled 13 years (714–711 BCE), the same number as the occupation of Amarna. This constellation of stories is reflected in Artapanus, Eupolemus, and Josephus, all aware of Biblical tradition, but in receipt of earlier, independent material, dating from the Saite period. They are not indulging in *midrash* beholden solely to Hellenistic predilection but are writing under the influence of the strong, surviving memory of the great siege of Hermopolis (Grimal 1981; Redford 2011: 309–315). It may well be that the hero of the incident was commemorated by the title *mōsu*, "leader, chief" (cf. the Berber title, ms, "master, prince," in use in Egypt in the Late Period: Yoyotte 1958; Yoyotte 1961: 123 n. 3; Chevereau 1985: 41 n. b).[6] Ethiopians were thus introduced into the narrative glorifying a hero named *Mosou*[7] who marries *Θάρβις*, the daughter of the Ethiopian king (see Appendix 2). The same tradition incorporates the prediction of the birth of the Hebrew deliverer that occasions the slaughter of the male children (Josephus *Ant.* II.9.2), the trampling of the Egyptian crown Josephus *Ant.* II.9.7), the dream of Moses (Ezekiel the Tragedian: Charlesworth 1985: 803–819), the miracle of the jail, and the plot to kill Moses that provides the rationale for his flight.

It must be stressed that none of this material is necessarily to be construed as *midrash* on a pre-existent and primary scriptural narrative. The Biblical record (cf. Num 12:1) already knows of

[6] Interestingly, the bearer of the title investigated by Yoyotte has the personal name *P3 S3sw*, "the shepherd."

[7] For a Syrian settlement at Hermopolis in the fourth century BCE see Winnicki (2009: 166–167).

the Bocchoris-Leper-"Ethiopian" stories, the Oracle of the Potter (Exod 9:8), and the magic tricks with the rod. While Moses, as "Musaeus," introducer of civilized life to Egypt, is a Hellenic trope, the other turns of the plot derive from native Egyptian folklore and historical memory. Their roots descend into pre-Hellenistic times and reach as far back as the Kushite-Saite period, the mind-set of which they clearly reflect.

Appendix 1: The Hyksos and the Sinai

Text no. 357 at Mine L at Serabit el-Khadim (Albright 1969: 23–24, Fig. 4; Rainey 1975) has been read and interpreted in several ways; but Rainey's photograph and hand copy clear up some illusory readings. The vertical column reads: 'NTTPNDKML'BBML[8]K, for which two possible vocalizations and translations suggest themselves: I. "I am Teshupna (A), liegeman (B) to Apop (C) the king"; II. "O An! (D) Mayest thou protect me (E), liegeman to Apop the king."

(A) For Teshup, see Gordon (1965: no. 2623); for the *-na* ending, see Huffmon (1965: 236); for *-an* as a preformative used in hypocoristica, see Pagan (1998: 259).

(B) Rainey derived the word from *DKK, "to crush," a term he thought appropriate for mining operations. But it seems odd that self-identification would stoop to the lowliest of occupations. Egyptian officials under such circumstances would have identified themselves under the prestigious rubrics of *ỉmy-r mšʿ*, "expedition leader" (Seyfried 1981), or "prospector" (*Wb.* IV: 35:18–19). Alternatively it is tempting to derive from *dakāmu*, "to submit," or *daqāqu*, "to knock, break, crush" (*CAD* IV: 34, 107; Murtonen 1989: 152) with the meaning "surrogate, enforcer," the rough equivalent of Egyptian *šmsw*, "attendant, bodyguard." Cf. also *dêku*, "to summon, arouse, muster": *CAD* IV: 123–129.

(C) The writing is clearly *b*, but the grapheme is identical with Egyptian hieroglyphic *p*. A West Semitic /*b*/ is not often rendered by Egyptian /*p*/, but examples are sometimes found: Hoch (1994: 401–402), T. Schneider (1992: 124–145), and Lipinski (2001: 116). The root of the name is not certain, although several possibilities suggest themselves: *'B (B)*, "to shine, flourish" (Murtonen 1989: 79); *ebēbu*, "to become clean, cleansed" (*CAD* V: 4–7); and *'ubab* (< אבה), "much desired" (pu'al participial formation sans *m*: Gesenius § 52 S).

(D) On An(u) in the West Semitic pantheon, see del Olmo Lete (2008: 42–44, 148).

(E) Construing from *swp*, "excite, observe, protect": Murtonen (1989: 415).

Of the two translations offered above, the first (I) would appear preferable. The horizontal text beneath, undoubtedly retrograde, does not seem to be a direct continuation of the introductory column. Read ארבע אמר שמע [... F.], "Hear the words of 'Arba [... (?)]."

(F) For ארבע, derivatives, and personal names, see *K-B* I: 80–81; T. Schneider (1992: 34).

Appendix 2: The Figure and Name of Moses

Moses in Hebrew scripture is a composite figure (Van Seters 1994). In terms of overall "column space" devoted to him, the role of desert leader and lawgiver looms largest, a self-contained source (Schart 1990). Second comes the figure of "Moses the shepherd from the east" and a distant third "Moses the Egyptian" (Redford 2011: 324).

The "Shepherd from the East" motif is precisely the burden of Exodus 1–15, not "Moses the Egyptian." Chapters 1 and 2 may be a literary unity (Siebert-Hommes 1992), but the material is forced and the author/editors show no familiarity with Egypt. It is the account of shepherds descending on Egypt from the east to the assistance of fellow countrymen under threat or oppression. As such it finds its point of departure

[8] The photograph (Rainey 1975: pl. 11B) clearly shows an L, not N.

in the mistranslation "shepherds" in the Egyptian king-list tradition, qualifying the rulers of the 15th Dynasty,[9] and the reduplication of the descent upon Egypt which attends the cycle of stories centering upon Bocchoris.

A fortiori behind the figure of Moses there lurks a bifurcation in the tradition which is not always consistent. "Moses" in the Biblical tradition sometimes labors under a less than righteous appearance (Hepner 2011). There remains the distinct possibility, moreover, that the name has, along with such words as Goshen, Horeb, and perhaps Joseph, suffered dysphemism in its vocalization. A derivation long since favored by Egyptologists (Griffiths 1953; Vergote 1981), viz., a hypocoristicon derived from a stative of the root *msi*, "to give birth," (DN + *msw*), gains some support from New Kingdom transcriptions (Redford 1992: 20). But if the name entered the picture in the Later Period, the sibilant certainly would have been rendered by -ס-, as in the case of רעמסס (Vergote 1981: 92; Lipinski 2001: 136–137) under Aramaic influence (Garr 2004: 28–29).

It is tempting to see some significance in the fact that two individuals responsible for the expulsion of the Hyksos bore *-māsě* names (Hartmann 2004 with some reservations; Redford 2001), viz., Kamose and Ahmose. The tradition represented by the Egyptian king-list was aware of this identification and later connected the theophoric element with Moses. As this figure evolved in the person of the victor of the siege of Hermopolis and the great lawgiver of Egypt, additional associations came into play derived from false equations: *mosou*, "leader" (Yoyotte 1958), perhaps *masya* (and derivatives), "dark" (Hohenberger 1988: 135; cf. the "Ethiopian" connection). As the "shepherd" motif took root and underwent development *apud* the inhabitants of the coastal and southern Levant, the word was linked to the serpent *Mṯ* (*Mūsh*), the "snake" deity (Astour 1965: 229–231; Puech 2012).

[9] The meaning turns on the interpretation of the term *ḥḳ3w-ḫ3swt* and in particular in the confusion arising from a misunderstanding of the second element *ḫ3swt*. The latter, pronounced something like **šōse* in the Late Period, is here confused with *sōs* (< *šāsu*), "shepherd": Westendorf (1977: 201, 327).

References

Albright, W.F. 1969. *The Proto-Sinaitic Inscriptions and Their Decipherment*. Harvard Theological Studies, vol. 22. Cambridge: Harvard University.

Astour, M.C. 1965. *Hellenosemitica: An Ethnic and Cultural Study in West Semitic Impact on Mycenaean Greece*. Leiden: Brill.

Baden, J.S. 2012. The Continuity of Non-priestly Narrative from Genesis to Exodus. *Biblica* 93: 161–186.

Barguet, P. 1962. *Le temple d'Amon-Rê à Karnak: Essai d'exégèse*. Recherches d'archéologie, de philologie et d'histoire, vol. 21. Cairo: IFAO.

Berkowitz, L., and K.A. Squitier. 1990. *Thesaurus Linguae Graecae Canon of Greek Authors and Works*, 3rd ed. Oxford: Oxford University.

Betz, H.D. 1986. *The Greek Magical Papyri in Translation: Including the Demotic Spells*. Chicago, IL: University of Chicago.

Brugsch, Heinrich. 1891. *Thesaurus Inscriptionum Aegyptiacarum. Band 6*. Leipzig: Hinrichs.

CAD = University of Chicago, Oriental Institute. 1956–present. *The Assyrian Dictionary of the Oriental Institute of the University of Chicago*. Chicago: University of Chicago.

Charlesworth, J.H. 1985. *The Old Testament Pseudepigrapha and the New Testament: Prolegomena for the Study of Christian Origins*. Society for New Testament Studies Monograph Series, vol. 54. Cambridge: Cambridge University.

Chevereau, P.-M. 1985. *Prosopographie des cadres militaires égyptiens de la Basse Époque: Carrières militaires et carrières sacerdotales en Égypte du XIe au IIe siècle avant J.C.*. Paris: Antony.

Davison, J.M. 1989. Egyptian Influence on the Greek Legend of Io. In *The Archaeology, Geography and History of the Egyptian Delta in Pharaonic Times: Proceedings of Colloquium held in Waldham College 29–31 August, 1988, Oxford*, 61–79. Oxford: Cotswold.

De Meulenaere, H. 1973. Bakenrenef. *Lexikon der Ägyptologie* 1: 846.

Del Olmo Lete, G. 2008. *Mythologie et religion des sémites occidentaux II: Émar, Ougarit, Israël, Phénicie, Aram, Arabie*. Orientalia Lovaniensia Analecta, vol. 162. Louvain: Peeters.

Edwards, R.B. 1979. *Kadmos the Phoenician: A Study in Greek Legends and the Mycenaean Age*. Amsterdam: Hakkert.

Epigraphic Survey. 1954. *Reliefs and Inscriptions at Karnak III: The Bubastite Portal*. Oriental Institute Publications, vol. 74. Chicago, IL: University of Chicago.

Gaballa, G.A. 1977. *The Memphite Tomb-chapel of Mose*. Warminster: Aris & Phillips.

Gardiner, A.H. 1959. *The Royal Canon of Turin*. Oxford: Oxford University.

Garr, W.R. 2004. *Dialect Geography of Syria-Palestine, 1000–586 B.C.E.*. Winona Lake, IN: Eisenbrauns.

Gershenson, D.E. 1996. A Greek Myth in Jeremiah. *Zeitschrift für die Alttestamentliche Wissenschaft* 108: 192–200.

Gesenius, W. 1960. *Hebrew Grammar*, 2nd ed. Ed. E. Kautzsch. Trans. A.E. Cowley, from German. Oxford: Clarendon.

Gordon, C.H. 1965. *Ugaritic Handbook*. Analecta orientalia, vol. 38. Rome: Pontificium Institutum Biblicum.

Goyon, G. 1957. *Nouvelles inscriptions rupestres du Wadi Hammamat*. Paris: Adrien-Maisonneuve.

Griffiths, J.G. 1953. The Egyptian Derivation of the Name Moses. *Journal of Near Eastern Studies* 12: 225–231.

———. 1986. Lycophron on Io and Isis. *Classical Quarterly* 36: 472–477.

Grimal, N.-C. 1981. *La stèle triomphale de Pi('ankh)y au Musée du Caire, JE 48862 et 47086–47089*. Etudes sur la propagande royale égyptienne, vol. 1. Cairo: IFAO.

Hartmann, T.A.G. 2004. Mose und Maria—'Amuns Kind und Liebling'—auf den ägyptischen Spuren zweier biblischer Namen. *Zeitschrift für die Alttestamentliche Wissenschaft* 116: 616–622.

Helck, W. 1956. *Untersuchungen zu Manetho und den ägyptischen Königslisten*. Untersuchungen zur Geschichte und Altertumskunde Ägyptens, vol. 18. Berlin: Akademie-Verlag.

———. 1971. *Die Beziehungen Ägyptens zu Vorderasien im 3. und 2. Jahrtausend v. Chr*. Ägyptologische Abhandlungen, vol. 5. Wiesbaden: Harrassowitz.

———. 1975. *Historische-biographische Texte der 2. Zwischenzeit und neue Texte der 18. Dynastie*. Kleine ägyptische Texte, vol. 6. Wiesbaden: Harrassowitz.

Hepner, G. 2011. The Mockery of Kings and Prophets: The Balaam Narrative Contains an Implied Critique of Moses. *Revue Biblique* 118: 180–186.

Hoch, J.E. 1994. *Semitic Words in Egyptian Texts of the New Kingdom and Third Intermediate Period*. Princeton, NJ: Princeton University.

Hoffner Jr., H.A. 2009. *Letters from the Hittite Kingdom*, Writings from the Ancient World. vol. 15. Atlanta, GA: Society of Biblical Literature.

Hohenberger, J. 1988. *Semitische und Hamitische Wortstämme im Nilo-Hamitischen: Mit phonetischen Analysen*. Marburger Studien zur Afrika- und Asienkunde: Afrika, vol. 42. Berlin: Reimer.

Huffmon, H.B. 1965. *Amorite Personal Names in the Mari texts: A Structural and Lexical Study*. Baltimore, MD: Johns Hopkins University.

Johnson, J.H. 1977. Louvre E 3229: A Demotic Magical Text. *Enchoria* 7: 90–91.

K-B = Köhler, L., and W. Baumgartner. 1967. *Hebräisches und aramäisches Lexikon zum Alten Testament*, vol 1. 3rd ed., 6 vols (1967–1996). Leiden: Brill.

Kákosy, L. 1981. Prophecies of the Ram gods. *Studia Aegyptiaca* 7: 139–154.

Kitchen, K.A. 1973. *Third Intermediate Period in Egypt (1100–650 B.C.)*. Warminster: Aris & Phillips.

———. 2001. King Lists. In *The Oxford Encyclopedia of Ancient Egypt*, vol. 2, ed. D.B. Redford, 234–238. New York: Oxford University.

Leclant, J. 1983. Schabaka. *Lexikon der Ägyptologie* 5: 499–520.

Leuchter, M. 2011. Eisodus as Exodus: The Song of the Sea (Exod 15) reconsidered. *Biblica* 92: 321–346.

Lipinski, E. 2001. *Semitic Languages: Outline of a Comparative Grammar*. Orientalia Lovaniensia Analecta, vol. 80. Louvain: Peeters.

Lloyd, A.B. 1988. *Herodotus, Book II: Commentary 99–182*. Études préliminaires aux religions orientales dans l'empire romain, vol. 43. Leiden: Brill.

Moran, W.L. 1992. *The Amarna Letters*. Baltimore, MD: Johns Hopkins University.

———. 2003. *Amarna Studies: Collected Writings*. Winona Lake, IN: Eisenbrauns.

Murtonen, A. 1989. *Hebrew in its West Semitic Setting I: A Comparative Lexicon*. Studies in Semitic languages and linguistics, vol. 13. Leiden: Brill.

Onasch, H.-U. 1994. *Die assyrischen Eroberungen Ägyptens*, 2 vols. Ägypten und Altes Testament, vol. 27. Wiesbaden: Harrassowitz.

Pagan, J.M. 1998. *A Morphological and Lexical Study of Personal Names in the Ebla Texts*. Archivi reali di Ebla Studi, vol. 3. Rome: Missione Archeologica Italiana in Siria.

Popko, L., and M. Rücker. 2011. P.Lips. Inv. 1228 und 590: Eine neue ägyptische Königsliste in griechischer Sprache. *Zeitschrift für Ägyptische Sprache und Altertumskunde* 138: 43–62.

Puech, E. 2012. L'inscription canaanéenne en alphabet cunéiforme sur un ivoire de Tirynthe. *Revue Biblique* 119: 321–330.

Rainey, A.F. 1975. Notes on Some Proto-Sinaitic Inscriptions. *Israel Exploration Journal* 25: 106–116.

Ray, J.D. 2001. Bakenrenef. In *The Oxford Encyclopedia of Ancient Egypt*, vol. 1, ed. D.B. Redford, 162. New York: Oxford University.

Redford, D.B. 1983. Review of W. Helck, Die Beziehungen Ägyptens und Vorderasiens zur Aegäis. *Journal of the American Oriental Society* 103: 481–483.

———. 1984. *Akhenaten, the Heretic King*. Princeton, NJ: Princeton University.

———. 1986. *Pharaonic King-lists, Annals, and Daybooks: A Contribution to the Study of the Egyptian Sense of History*. Mississauga: Benben.

———. 1992. *Egypt, Canaan, and Israel in Ancient Times*. Princeton, NJ: Princeton University.

———. 1997. Textual Sources for the Hyksos Period. In *The Hyksos: New Historical and Archaeological*

Perspectives, ed. E.D. Oren, 1–44. Museum Monographs, vol. 96, Studies in Jungian Psychology, vol. 96, University Museum Symposium Series, vol. 8. Philadelphia, PA: University Museum.

———. 2001. The So-called 'Codification' of Egyptian Law under Darius I. In *Persia and Torah: The Theory of Imperial Authorization of the Pentateuch*. SBL Symposium Series, vol. 17, ed. J.W. Watts, 135–160. Atlanta, GA: Society of Biblical Literature.

———. 2011. Some Observations on the Traditions Surrounding 'Israel in Egypt'. In *Judah and the Judeans in the Achaemenid Period*, ed. M. Oeming, O. Lipschits, and G.N. Knoppers, 279–364. Winona Lake, IN: Eisenbrauns.

Russell, S.C. 2009. *Images of Egypt in Early Biblical Literature: Cisjordan-Israelite, Transjordan-Israelite, and Judahite Portrayals*. Berlin: de Gruyter.

Schart, A. 1990. *Mose und Israel im Konflikt: Eine redaktionsgeschichtliche Studie zu den Wüstenerzählungen*. Orbis Biblicus et Orientalis, vol. 98. Louisville, KY: Saint-Paul.

Schmid, K. 2012. Genesis and Exodus as Two Formerly Independent Traditions of Origins for Ancient Israel. *Biblica* 93: 187–203.

Schneider, H.D. 2001. Horemheb. In *The Oxford Encyclopedia of Ancient Egypt*, vol. 2, ed. D.B. Redford, 115. New York: Oxford University.

Schneider, T. 1992. *Asiatische Personennamen in ägyptischen Quellen des Neuen Reiches*. Orbis Biblicus et Orientalis, vol. 114. Freiburg, Göttingen: Universitätsverlag, Vandenhoeck & Ruprecht.

Seyfried, K.-J. 1981. *Beiträge zu den Expeditionen des Mittleren Reiches in die Ost-Wüste*. Hildesheimer Ägyptologische Beiträge, vol. 15. Hildesheim: Gerstenberg.

Siebert-Hommes, J. 1992. Die Geburtsgeschichte des [illegible] von Exodus I und II. *Vetus Testamentum* 42: 398–404.

Thissen, H.J. 2002. Das Lamm des Bokchoris. In *Apokalyptik und Ägypten: Eine kritische Analyse der relevanten Texte aus dem griechisch-römischen Ägypten*. Orientalia Lovaniensia Analecta, vol. 107, 113–138. Louvain: Peeters.

Urk. = Sethe, K. 1906. *Urkunden der 18. Dynastie*, vols. 1–4. Leipzig: J. C. Hinrichs.

Van Seters, J. 1994. *The Life of Moses: The Yahwist as Historian in Exodus-Numbers*. Contributions to Biblical Exegesis and Theology, vol. 10. Louvain: Peeters.

Vergote, J. 1981. A propos du nom de Moise. *Bulletin de la Société d'Égyptologie Genève* 4: 89–93.

von Beckerath, J. 1980. Königslisten. *Lexikon der Ägyptologie* 4: 534–535.

———. 1997. *Chronologie des pharaonischen Ägypten: Die Zeitbestimmung der ägyptischen Geschichte von der Vorzeit bis 332 v. Chr.* Münchner ägyptologische Studien, vol. 46. Mainz: von Zabern.

———. 1999. *Handbuch der ägyptischen Königsnamen*. Münchner ägyptologische Studien, vol. 49. Mainz: von Zabern.

Wb. = Erman, A., and H. Grapow. 1926–1961. *Wörterbuch der ägyptischen Sprache*, 6 vols. Berlin: Akademie-Verlag.

Westendorf, W. 1977. *Koptisches Handwörterbuch*. Heidelberg: Universitätsverlag Winter.

Wildung, D. 1977. *Imhotep und Amenhotep: Gottwerdung im alten Ägypten*. Münchner ägyptologische Studien, vol. 36. Munich: Deutscher Kunstverlag.

Wimmer, J.F. 2011. Das Buch Exodus. *Catholic Biblical Quarterly* 73: 119–120.

Winnicki, J.K. 2009. *Late Egypt and Her Neighbours: Foreign Population in Egypt in the First Millennium BC*. Journal of Juristic Papyri Supplement, vol. 12. Warsaw: Warsaw University.

Yoyotte, J. 1958. Le dénommé Mosou. *Le Bulletin de l'Institut Français d'Archéologie Orientale* 57: 81–89.

———. 1961. *Les principautés du Delta au temps de l'anarchie libyenne*. Mélanges Maspero, vol. 1, Mémoires publiés par les membres de l'Institut français d'archéologie orientale du Caire, vol. 66. Cairo: IFAO.

Part VIII

The Exodus and the Emergence of Israel: New Perspectives from Biblical Studies and Archaeology

Har Karkom: Archaeological Discoveries in a Holy Mountain in the Desert of Exodus

35

Emmanuel Anati

Abstract

The archaeological survey of Mount Karkom and surrounding valleys brought to the discovery of 1,300 archaeological sites where not a single site was known before. The findings define a mountain in the middle of nowhere, where altars, small sanctuaries, and numerous other worship places reveal its identity as Holy Mountain, located in the Desert of Exodus. In the Bronze Age Har Karkom was a paramount cult high place surrounded by living sites where multitudes were camping at its foot. If indeed a Mount Sinai existed what would one expect to find as archaeological testimony in the site of Mount Sinai? This chapter just describes the archaeological findings of this holy mountain. They indicate its unique and specific character. The considerations that led to its identification as the biblical Mount Sinai are defined in another text (Anati, Emmanuel. Is Har Karkom the Mount Sinai? Capo di Ponte: Atelier, 2013).

Location

Mount Karkom is in the southern Negev desert, at the northern edge of the Paran wilderness, about 7 km (4 mi.) south of the well of Beer Karkom. The mountain is visible from as far as the Edom Mountain, up to a distance of about 70 km (45 mi.). The mountain ridge is a plateau 4.5 km (3 mi.) long and between 1 and 2.5 km (0.6 and 1.5 mi.) wide (map reference 123–126.964–968). Because of its sheer cliffs which rise about 300 m above the surroundings, access to the mountain is limited to a few trails. The prominent mesa, 800–850 m above sea level, is reached from the Western Valley mainly by means of two well-marked ancient paths: one (map reference 123.966) includes a passage of steps partly hewn in antiquity, and the other (map reference 124.968) is snakelike. Concentrations of engraved small rocks and pillars along its sides were documented in the early years of the survey. Other secondary access trails are less marked and are likely to have been in use mainly in more recent times. From the eastern side a ceremonial trail includes pillars and other ceremonial structures along its way with Bronze Age

E. Anati (retired) (✉)
Centro Camuno di Studi Preistorici, Capo di Ponte, Italy

University of Lecce, Lecce, Italy
e-mail: emmanuel.anati@gmail.com

T.E. Levy et al. (eds.), *Israel's Exodus in Transdisciplinary Perspective*,
Quantitative Methods in the Humanities and Social Sciences, DOI 10.1007/978-3-319-04768-3_35,

material culture, climbing from the Paran valleys to the plateau.

Archaeologically the mountain is primarily characterized by the presence of Bronze Age altars and shrines on its plateau and by stone-built basements of Bronze Age habitation sites at its foot. The site appears to have been a cult high place, a holy mountain, in the late fourth and third millennium BC. The tradition of cult is likely to have started already in the Paleolithic period. The cultic role of the mountain seems to have been abandoned during a hiatus due to a period of drought in the second millennium BC. Being a rather unique paramount Bronze Age cult site, the mountain has been proposed to be identified with the biblical Mount Sinai, raising a broad debate which did not find as yet an agreed solution among scholars. The main considerations for this identification have been defined in "*The Riddle of Mount Sinai*" (2001) and summarized in "*Is Har Karkom Mount Sinai?*" (2013). Such important high place is unlikely to have been ignored by the Pentateuch. If not Mount Sinai, what would be its biblical name?

Exploration

Rock engravings on the mountain's plateau were first recorded by E. Anati in 1954 (E. Anati 1956). In 1980, a survey of the mountain was begun by the Italian Archaeological Expedition, in the frame of the Archaeological Survey of Israel, under Anati's direction. In over 30 years of fieldwork the survey has examined and recorded more than 1,300 sites, in a concession area of 200 km^2, with trial excavations carried out at some. In the course of the survey and the excavations, archaeological remains, mostly from the Paleolithic, Chalcolithic, Early Bronze Age, and the beginning of the Middle Bronze Age, were found. After a gap covering most of the second millennium BC, a few Iron Age sites precede a dense presence of remains from Roman and Byzantine times (Anati and Mailland 2009, 2010).

If the traditional concept that the biblical narration of Exodus may refer to the thirteenth century BC would be correct, the identification of Har Karkom with the biblical Mount Sinai would be out of question. Similarly, every other mountain of the Sinai Peninsula would have to be excluded as no evidence of Late Bronze Age tribal life is present in the Peninsula. The only exceptions are regime structures like military stations and minerary exploitations.

The Paleolithic Sites

More than 200 Paleolithic sites, mostly from the Middle Paleolithic and the early phases of the Late Paleolithic period, were recorded on and around Mount Karkom. An abundance of excellent-quality flints is widespread at the surface. Over 50 flint-tool workshops, containing numerous cores and flakes, as well as over 500 traces of hut floors from the same periods were recorded on the mountain alone. Because of desert conditions, the in situ sites and the flakes and tools scattered around cores are in an excellent state of preservation, some of them, apparently, in the same position they were left thousands of years ago.

Several sites were identified with large quantities of bifacial hand axes of the Lower Paleolithic. Many of the Middle and Upper Paleolithic sites have remains of basements of huts, of fireplaces, and of flint workshops.

An Early Upper Paleolithic site (HK 86B) includes an area where several large flint boulders have been erected. Early man selected and gathered to this spot 42 stones with natural vaguely anthropomorphic shapes. Some of them were enhanced by human hands adding engraved details like eyes and pattern incisions. Only a few of them were still in situ. Most of them were fallen. The original place of some of the fallen ones could be recorded thanks to holes in the fossil stepping soil recorded by Arch. L. Cottinelli (Cottinelli 1992; Anati, Cottinelli and Mailland 1996b). This impressive site appears to be some sort of social and cult place. Unfortunately, recently it was vandalized.

The lithic industry of this site, like that of 24 other sites on the plateau, has a typical flint industry characterized by both flake and blade tools with conspicuous presence of Levallois flaking technique and with points of Chatelperron type. It has been defined as the Karkomian industry and is considered to belong to an initial phase of the Upper Paleolithic.

One of the best preserved Upper Paleolithic sites (H.K.16) includes infrastructure remains of seven huts, six of them oval and one, larger than the others, having an almost rectangular shape. The plan of this outstanding site appears to reflect the social structure of a group of hunters of ca. 20,000 years ago, counting 30–40 peoples. The site includes traces of pits near the basements of the huts. Numerous cores and a large quantity of flakes, in its eastern part, together with a flat stone showing marks of cuttings, have defined a workshop for flint tools. Several thousand flint implements of Gravettian type were recorded from an area of 12 m^2 between the flint workshop and the dwelling site. The site includes also a large, double circle of postholes likely to have been a ceremonial structure.

There are many sites from the early phases of the Upper Paleolithic, while the later phases and the Epi-Paleolithic have revealed so far, besides site HK16, only a sporadic presence. These variations in the density of sites may be attributed to changing climatic conditions. This region of desert has today an average annual rainfall of ca. 50 mm. Even minor changes in climatic conditions may determine the presence or the absence of grazing areas and consequent resources for game and hunters alike to survive.

Neolithic to Bronze Age

The Neolithic period on Mount Karkom is represented by only a few sites. In one of them a typical flint axe was found next to a rock engraving, which may well belong to the same period. At the foot of Har Karkom a small village of the pre-pottery Neolithic was partly excavated by Dr. M.E. Peroschi and I. Mailland. Radiocarbon tests date it to ca. 10,000 B.P. Flint arrowheads indicate the use of bows. Oval basements of huts had a rich flint industry and grinding stones, illustrating a way of life of semi-sedentary mixed economy, including food collecting and hunting, likely to be accompanied by incipient agriculture.

The next most prevalent period represented has been defined as BAC or "Bronze Age complex." The same sites appear to have been reused during centuries. Similar traditions of the material culture persisted with only secondary changes. Altars, small shrines, and groups of standing pillars were built on the Har Karkom plateau. Two hundred and twelve BAC sites were recorded, documenting the presence of multitudes camping at the foot of the mountain. They include remains of living sites, stone constructions, stone circles, various types of structures defined as shrines, and massebot (standing pillars). A large variety of living sites and sanctuaries seem to indicate that different tribal groups have stationed at the foot of the mountain.

The BAC material culture is typical to the Sinai Peninsula and the Negev. It is basically a Chalcolithic tradition characterized by the abundance of fan scrapers, points, and borers. It had a long persistence of the same pastoral and some garden agriculture, half nomadic way of life, lasting until the beginning of the Middle Bronze Age. Pottery is rare and mostly undecorated. Ledge handles and flat bases relate to the Chalcolithic and the Early Bronze Age. "Metallic ware" similar to that found further north at the end of the Early Bronze and the beginning of the Middle Bronze appears in a few sites.

Some of the tumuli yielded remains of human bones in secondary burials; one of them included Early Bronze Age pottery (EBII) and a perforated shell pendant. Others of the excavated tumuli did not include burials but had altar-like stones and traces of fire which had been buried under a heap of stones after having been used for some ceremony.

Standing stones (massebot), stone circles, private shrines and other megalithic structures, and large concentrations of rock engravings, were recorded on the plateau and in the surrounding valleys.

Mention should be made of the interestingly shaped summit of the mountain. It consists of two tops; the higher one is narrow and long (2–3 m by c. 130 m). A pile of stones and seven concentrations of flint pebbles were found on it as well as a menhir or standing pillar. The lower hill (ca. 10 m lower than the first one) has three small summits. The largest one is circular, about 8–12 m in diameter, with a stone circle. On the top of the second summit there is a small cave, with remains of hearths at its opening. The third summit is just a protruding rock. The presence of a stone circle and the small cave on this top, faced by the standing pillar of the other top, is likely to give this couple of tops the determination of male and female. While the small cave is a natural feature, the stone circle on one top and the standing menhir on the other are intentional issues of human hands. It has to be noticed that among the many visited mountains in the entire Sinai Peninsula, this is the only one, known to the present writer, having a small cave on its top.

In the valleys west and north of the mountain there are numerous structure remains from the BAC period comprising extensive encampments, which include hundreds of foundations of huts, and other stone structures.

A structure with the base of a nearly rectangular platform (altar?) was found at the foot of one of the slopes. In front of it were a dozen standing stones arranged in two rows of six each. In several cases alignments count 12 standing stones which appear to have been a recurring number in Bronze Age ceremonial sites. In other sites 12 standing stones have been found in rows and in semicircles.

Private Shrines and Testimony Tumuli or "Gal-'ed"

Two peculiarities of Har Karkom are small shrines defined "private shrines" and "testimony tumuli" defined also "Gal-'ed." The private shrines (at least 18 of them on the plateau alone) consist of one or more standing stones with natural anthropomorphic shapes. Occasionally they have been completed by human hands adding the eyes or a mouth. They are surrounded by smaller stones which form a sort of circle around these so-called idols or ceremonial pillars. On the front side of the pillar, in two cases, blackened stones and other traces of fire have been found. In some cases they are related to BAC flint implements. They indicate the interesting presence of a worship of anthropomorphic stones, widespread in the Bronze Age but present also before, since the Paleolithic and persisting thereafter.

The testimony tumuli or Gal-'ed are heaps of stones which cover an altar or a standing stone. In a few cases excavations revealed the presence of burning around such stones located under the heap. In one case, after the heap was excavated by Dr. V. Manfredi, a heavy rectangular black stone was found to have been positioned on the base rock together with other smaller stones, forming a sort of platform. On top of it was laid a white stone intentionally cut by human hands in the shape of a half circle or half moon (length 60 cm, weight 44 kg). Near to it a flint fan scraper and other Early Bronze Age flint tools were found on the platform. At the foot of the platform the presence of blackened stones indicates that a fire had been lit (Manfredi 1995).

It was possible to reconstruct the actions of the Early Bronze Age executors. A heavy almost rectangular boulder with a flat top was positioned on the bedrock. The moonlike white stone was intentionally cut and shaped and was positioned on top of it. Fire was lit at the feet of the boulder. Then the platform with the "white moon" on top of it was covered by the heap of stone which was over 7 m in diameter and 2 m high. This Gal-'ed appears to have been made to commemorate or to dedicate the site to the moon or to the moon god.

Such testimony heaps, together with the private shrines, standing pillars, and clusters of pillars, reveal the role of the mountain plateau as an immense cult site. It is noteworthy that all this wealth of findings has a very modest appearance. There are no "monumental" monuments, nothing made to be impressive; apparently everything was just functional for its purpose. Most of these structures seem to have been done on the spot to be used at once. A peculiarity of the mountain plateau is the presence of traces of very large fires, some of them over 20 m in diameter. It must have been a sort of mythic site for ages. People were coming there to make agreements, to worship, and to have pilgrimages. It was a sort of Bronze Age Mecca where different tribes arrived for centuries.

Living hamlets and other remains of habitation and camping sites are located in the valleys surrounding the mountain, while the mountain plateau in the BAC period appears to have been reserved to cult and worship structures. In the center of the plateau there is a BAC structure with a platform or an altar facing east which may have been a small temple (site HK/24).

It has to be mentioned that so far, according to our limited ability of decoding, we found no evidence to say whether among these various groups of worshippers there may have been people related to the biblical account of Exodus. Besides the Hebrews, would it be possible to recognize in these remains traces of the various peoples mentioned in the biblical accounts like Midianites, Amalekites, or Edomites?

Later Sites

No site from the Late Bronze Age has been discovered so far in the survey. For the believers in the dating of Exodus according to the traditional exegetic school, this is an obvious handicap. One Iron Age site, two Hellenistic sites, and more than 50 sites from the Nabataean, Roman, and Byzantine cultures were recorded on and around the mount. Most of them contained remains of stone-built basements of huts. The largest habitation site belongs to the Hellenistic period and includes more than 100 stone-built basements of building units. It appears to be a well-planned site which had a short life.

The Roman and Byzantine periods are characterized by the presence of terraced agricultural fields, stone-built animal enclosures, and other structures revealing a widespread agrarian and pastoral use of the territory. Several of the Byzantine living sites are likely to reflect some sort of collective communities. It has been suggested by Arch. Cottinelli that they may have been monastic settlements. And, approaching our times, for the last 1,300 years the entire area has remains of nomadic campsites of the Islamic period.

Both climatic fluctuations and variations in the political situations may have influenced the demographic changes in the density of population from period to period.

Rock Engravings

The mount has a rich array of rock engravings, more than 150 groups counting thousands of engraved images. The periods represented range from hunter-gatherers considered to be pre-Neolithic to late engravings of Roman, Byzantine, and Islamic periods. Style I has large animal figures of Upper Paleolithic character. It is known already from the site of Kilwa in Arabia and a few other sites in Arabia and in the Sinai Peninsula. Style II, identified in earlier research in the Negev and in the Sinai Peninsula, is ascribed to the Neolithic period although no evidence is provided of agricultural activities. It is primarily characterized by elegantly stylized figures of wild animals, mainly ibexes. Human figures are rare. There are no scenes, just assemblages of figures of the animal preys.

Style III displays beautiful scenes of ibex hunting. Human figures are active protagonists. Hunters wear skins and use double-bent bows. They are helped by domesticated dogs. This style attests to the way of life of late hunters which may have lasted a long time in the desert regions. Some of these figures display bows similar to those depicted in Egypt in Protodynastic and Early Dynastic times.

The makers of this style may have coexisted with another kind of people, the makers of style IVA, in which domesticated animals are widespread for the first time. Style IVA is the only style in which domesticated oxen are present. This might well be an indication of better climatic conditions in the period of this style.

Several courtyard buildings, tumuli, and stone circles belonging to the BAC period are decorated with rock engravings. The enclosures of the Chalcolithic, Early Bronze, and the beginning of the Middle Bronze Ages have remains of material culture together with rock engravings. In several instances the rock engravings appear to be directly connected to man-built structures. The relation of a number of rock art sites to the cult of the moon god Sin has been evidenced by Dr. R. Bastoni Brioschi (Bastoni 1996, 1998).

Style IVB is unknown here but is common in the central Negev and well known from the "Chariot's Cave" of Timna in the Arava Valley. Weapons and tools and figures of wheeled chariots are associated to the Late Bronze Age. Such style is so far absent in the Har Karkom area. There seems to be a gap in rock art production during the second millennium BC.

Domestic camels and horses appear for the first time with style IVC. From figures of tools and weapons this style had a long duration, covering the Iron Age and persisting in Hellenistic and Roman times. It represents scenes of people devoted to hunting, animal raising, and caravan trade. The pictures of the later phases of style IVC are accompanied by inscriptions in various Semitic scripts.

Style IVC represents a way of life likely to have been introduced in the Iron Age which does not seem to have changed when the inscriptions started. The inscriptions are unlikely to mark the arrival of new people. Rather they appear to indicate the spread of literacy among local tribes.

Styles V and VI belong to the Roman–Byzantine and the beginning of the Early Islamic periods; most of the Nabataean and Thamudic inscriptions are related to images of these styles. In the course of these periods a gradual schematization of the figures is followed, reaching an extreme with *wassum* and tribal marks of style VII mostly belonging to the second millennium. The rock art appears to be a record of events produced by the protagonists. It is a promising new field of research.

Geoglyphs

On the mountain and in the surrounding valleys pebble drawings and geoglyphs have been discovered. They are large figures drawn on the ground with stones or with the clearing of the surface from stones. They are being studied by Dr. F. Mailland. A few of them can be identified as animal figures, some being over 30 m long. In several sites fragmentary traces of such geoglyphs have been found which appear to represent animal and human figures as well as geometric patterns. Such geoglyphs are better visible from the air than on the ground. They have been considered to be symbolic offerings to whoever could see them from above (sky gods?). An alternative concept makes of them marks of totemic ground for ceremonial events.

Some of the geoglyphs are related to flint implements belonging to the Chalcolithic and the Early Bronze Age. Others may be older as they seem to represent extinct animals such as rhino and elephant which belong to the Pleistocene fauna.

Conclusions

The material collected so far indicates that in the Paleolithic period the mountain was an excellent source of raw material for the production of flint tools and an important meeting place. The Upper Paleolithic "sanctuary" (HK 86B) and the ceremonial circle (HK 16) indicate that the mountain was a place of worship from very early times. In the Neolithic there are a few campsites and rock art sites but so far no clear evidence of ritual sites with the exception of one site to the north of the mountain where a stone circle had engraving of an anthropomorphic face.

In the BAC period the mountain became a paramount ceremonial and cult high place: numerous rock engravings of religious significance were carved, and standing stones "massebot" were set up. Private shrines, stone circles, and tumuli were also erected. A structure likely to be a small temple on the plateau and numerous structures having ritual functions both on the plateau and in the surrounding valleys evidence the intense cult activities. Several types of shrines have been defined.

So far there is no evidence of human activities on the mountain and around it from the beginning of the Middle Bronze Age II to Iron Age I. After the period of intense occupation, the plateau was abandoned for over 800 years. It is considered that this gap may be due to a climatic episode of extreme draught. Archaeological traces appear again in the course of Iron Age II. The human presence consistently grew in Hellenistic, Roman, and Byzantine times with large villages and terraced agricultural fields. It sharply decreased in Early Islamic times. The area was then occupied by desert nomads, who left behind the remains of numerous campsites.

The role of the mountain is indicated by its finds, mainly from the BAC. Shrines, tumuli, standing pillars, stone circles and other megalithic structures, geoglyphs, and rock engravings reveal that the mountain was a cult and religious center, a sort of prehistoric Mecca. As specified in previous works, in this writer's opinion, the site responds to the topographic and landscape descriptions that the Bible attributes to Mount Sinai. The archaeological finds confirm the presence of multitudes, at its feet, in the Bronze Age. The period of its archaeological remains awakens a debate on the age of the memories narrated by the biblical Exodus. The considerations at the base of such opinions have grown in the course of 30 years; they are described in several texts mentioned in the bibliography and summarized in the booklet "***Is Har Karkom the Biblical Mount Sinai?***" (Capodiponte, Atelier Editions, 2013).

There may be no doubt that the archaeological evidence indicates that Har Karkom, in the Bronze Age, was a paramount cult high place and that multitudes camped at its feet. How many other such mountains are known? The question is if it is the Mount Sinai that inspired the biblical narration or it is just a holy mountain which the Pentateuch may have mentioned with another name.

What would one expect to find as archaeological testimony in the site of Mount Sinai? Bronze Age remains of camping sites at its foot and evidence on the mountain of cult practices of the same age. Har Karkom provides such requirements and beyond.

Note

The present text does not enter into the exegetic analyses which are discussed elsewhere; it summarizes the archaeological results of the survey which finds its preliminary publication in two volumes: Anati and Mailland (2009, 2010).

References

Anati, Emmanuel. 1956. Rock Engravings from the Jebel Ideid (Southern Negev). *Palestine Exploration Quarterly* 83(1–2): 5–13.

———. 1984a. Har Karkom, dans le désert de Paran: un site sacré. *Le Monde de la Bible* 35: 53–55.

———. 1984b. I nomi del Monte Sinai e il problema dell'Horev alla luce dell'archeologia. *Bibbia e Oriente* 26/3(141): 151–158.

———. 1985a. Har Karkom: a Holy Mountain in the Paran Desert (Israel). *BCSP* 22: 129–131.

———. 1985b. Has Mt. Sinai been Found? *Biblical Archaeology Review* 11(4): 42–57.

———. 1985c. Elementi per una analisi paleodemografica di Har Karkom. *BCN* 2(4): 28–31.

———. (1986a). *La Montagna di Dio. Har Karkom.* Milan: Jaca Book; Eng. Ed.: *The Mountain of God.* New York: Rizzoli; Fr. Ed.: *La Montagne de Dieu Har Karkom*. Paris. Jaca Book, 338, 245 ill.

———. (1986b). Har Karkom: problemi cronologici ed esegetici. In *Estudios en Homenaje al Dr. Antonio Beltrán Martinez*. Zaragoza: Facultad de Filosofia y Letras, Univ. De Zaragoza. 439–451.

———. 1986b. Israël: Nouvelles decouvertes a Har Karkom. *Archeologia* 219: 30–43.

———. 1987a. Har Karkom e la cronologia dell'Esodo. *BCN* 4/2: 25–36.

———. 1987b. *I Siti a Plaza di Har Karkom*, 240. Capo di Ponte: Edizioni del Centro.

———. 1988. New discoveries at Har Karkom, the Negev Desert, Israel. *BCSP* 24: 113–117.

———. 1994. Il santuario paleolitico di Har Karkom: Arte o protoarte? XII *Valcamonica Symposium.*

———. (1996a)- I geoglifi di Har Karkom, *Valcamonica Symposium '96*, 6 pp.

———. 1996b. With L. Cottinelli & F. Mailland, Il santuario più antico del mondo. *Archeologia Viva* 15/56: 26–38.

———. 1996c. The Rock Art of Har Karkom. *BCSP* 29: 13–48.

———. 1997a. Il linguaggio delle pietre, geoglifi nel deserto israeliano del Negev. *Archeo* 13/2-144: 46–50.

———. 1997b. *Esodo tra mito e storia*, 300. Capo di Ponte: Edizioni del Centro.

———. (1999a)- La montagna del Dio della Luna, *Archeo*, vol. 15/1 (167), Gennaio 1999, 46–52.

———. 1999b. The Rock Art of the Negev Desert. *Near Eastern Archaeology* 62/1: 22–34, ill.

———. 2000a. L'art rupestre du mont Karkom. In *Des signes pictographiques à l'alphabet*, ed. Viers Rina, 43–72. Paris: Karthala, 21 ill.

———. (2000b). Messaggi del mito e contesto "Undici giorni da Horeb, per la via del monte Seir, a Kadesh Barnea" (*Deuteronomio* 1,2). In ed. E. Anati, *Arte rupestre e tribale: Conservazione e salvaguardia dei messaggi*, XVIII Valcamonica Symposium, 6 pp.

———. 2000c. *Les mystères du Mont Sinaï*, 225. Paris: Bayard Éditions.

———. 2001. *The Riddle of Mount Sinai. Archaeological discoveries at Har Karkom*, 189. Capo di Ponte: Edizioni del Centro, 205 ill.

Anati, E. and Mailland, F. 2009. *Archaeological Survey of Israel: Map of Har Karkom '(229)*. Geneva (Esprit de L'Homme) CISPE edit, 256+LI

———. 2010. *La riscoperta del Monte Sinai, Ritrovamenti archeologici alla luce del racconto dell'Esodo*, 247. Padova: Editrice Messaggero di Sant'Antonio. 205 ill.,dv.

———. 2013. *Is Har Karkom the Mount Sinai?* 92. Capo di Ponte: Atelier, 40 ill.

———. 2010. *Archaeological Survey of Israel: Map of Beer Karkom (226)*. Geneva (Esprit de L'Homme) CISPE edit, 208+XXVI

Barbiero Flavio. 2000. I cristiani ad Har Karkom. In E. Anati, *Arte rupestre e tribale: Conservazione e salvaguardia dei messaggi*, XVIII Valcamonica Symposium, 1 p.

Bastoni Brioschi Rosetta. 1996. Sinai e il dio Sin, *Valcamonica Symposium '96*.

———. 1998. Arte rupestre: Har Karkom e il dio Sin, in F. Mailland (ed.) (Settore Cultura e Musei, Civiche raccolte Archeologiche), *Har Karkom e Monte Sinai: Archeologia e Mito*, 25–34, 119. Milan: Comune di Milano.

Cottinelli Luigi. 1992. Santuario Paleolitico in Har Karkom, *Valcamonica Symposium '92*, 16 pp.

———. 1993. Tipologia strutturale dei siti BAC di Har Karkom, *Valcamonica Symposium '93*, 8 pp.

Diamond Larryn. 1984. The Geology of Har Karkom, Southern Negev Desert, Israel. In E. Anati, *The Mountain of God*, (It. ed.: *La Montagna di Dio. Har Karkom*; Fr. Ed.: *La Montagne de Dieu. Har Karkom*), pp. 333–341.

Fornoni Giorgio. 1998. La Grotta HK/244. In F. Mailland (ed.), *Har Karkom e Monte Sinai. Archeologia e Mito*, Milan (Comune di Milano, Settore Cultura e Musei, Civiche raccolte Archeologiche), 35–38.

Girolomoni Gino. 1998. Il Monte Sinai all'epoca dei Patriarchi. Racconto biblico e ritrovamenti. *Quaderni valtellinesi*, n. 76/3, 5–13.

Horowitz Aharon. 1984. The Climate of the Central Negev in the Third Millennium BC. In E. Anati, *The Mountain of God*, (It. ed.: *La Montagna di Dio. Har Karkom;* Fr. Ed.: *La Montagne de Dieu. Har Karkom*), p. 346.

Mailland Federico (ed.) Archeologica, 18 January 1997, Milan (Comune di Milano, Settore Cultura e Musei, Civiche Raccolte Archeologiche), 127 pp.

Mailland Federico (ed.). 1998. *Har Karkom e Monte Sinai: Archeologia e Mito*, Atti del Convegno di Studi, Associazione Lombarda.

Mailland Federico & Ida. 1993. The Har Karkom Plateau During the Paleolithic, *Valcamonica Symposium '93*, 13 pp.

Manfredi Valerio. 1995. Ricognizione e scavo del sito 221 bis nella zona di Har Karkom (Negev Israeliano), *BCSP*, vol. 28, 106–114.

———. 1992. Exodus et Sinai: Une réconsideration, *Dort ziehen Schiffe dahin... Collected Communications to the XIVth Congress for the Study of the Old Testament, Paris, 1992*, Frankfurt am Main (Peter Lang), 1996, pp. 11–25.

———. 1998. Il Monte Sinai nella cartografia antica. In F. Mailland (ed.), *Har Karkom e Monte Sinai: Archeologia e Mito*, Milan (Comune di Milano, Settore Cultura e Musei, Civiche raccolte Archeologiche), pp. 69-83, 125–127.

———. 2001. Una nuova via sacra di accesso all'altipiano di Har Karkom, *BCN*, vol. Marzo 2001, pp. 25–28.

Pirelli Paola & Giovanna Davini. 1984. The Vegetation at Har Karkom. In E. Anati, *The Mountain of God*, (It. ed.: *La Montagna di Dio. Har Karkom*; Fr. Ed.: *La Montagne de Dieu. Har Karkom*), 1986, pp. 347–349.

Pitkäaho Risto. 1999. Joseph and Imhotep. In "Valcamonica Symposium '97, Grafismo e semiotica", *BCSP*, vol. 30–31, pp. 197–206, ill.

Ravasi Gian Franco. 1997. Il dibattito su Har Karkom, *BCN*, vol. IV/3, 28–31.

Samorini Giorgio. 1984. Hallucinogenic Plants at Har Karkom. In E. Anati, *The Mountain of God*, (It. ed.: *La Montagna di Dio. Har Karkom*; Fr. Ed.: *La Montagne de Dieu. Har Karkom*), 1986, p. 350.

Tchernov Eitan. 1984. The Environmental History Biogeography and Fauna of Har Karkom, the Negev, Israel. In E. Anati, *The Mountain of God*, (It. ed.: *La Montagna di Dio. Har Karkom*; Fr. Ed.: *La Montagne de Dieu. Har Karkom*), 1986, pp. 342–345.

In Search of Israel's Insider Status: A Reevaluation of Israel's Origins

36

Brendon C. Benz

Abstract

In an effort to better understand Israel's origins, scholarship has proposed several models for its emergence in the land, many of which call into question the idea of a pan-Israelite exodus from Egypt. In spite of their differences, these models adhere to a common set of presuppositions and ultimately conclude that Israel consisted of a group of geographical, economic, and/or political outsiders. Drawing on several theoretical insights regarding the nature of social power and the makeup of ancient states in my analysis of the Amarna letters, I propose another alternative. The polities and populations of the Late Bronze Age Levant were more diverse and fluid than is generally recognized. Social power was widely distributed and often negotiated among a range of political players in an "egalitarian" manner. Because of this, political entities and alliances took a number of forms, including those that consisted of populations that were defined by settled centers and those that were not. This reconstruction highlights several points of continuity between the Levantine landscape of the Late Bronze Age and the constituents of early Israel as they are depicted in some of the core passages in the Bible, suggesting that early Israel included a contingent of geographical, economic, and political insiders.

By and large, the contributors to this volume reject the Bible's depiction of a mass exodus from Egypt followed by a settlement process in the Promised Land. Though many embrace the likelihood that a group of Israelites took flight from Egypt and that this historical event informed some of the traditions retained in the biblical narrative, there is a growing consensus that a majority of the populations that came to be identified with Israel stemmed from the southern Levant. In spite of their status as geographical insiders, the models supporting this reconstruction still contend that these "proto-Israelites" were economic and/or political outsiders. By proposing that at least a contingent of the early Israelites were participants in the economic and political systems of the so-called Canaanites, I offer a historical parallel to the literary

B.C. Benz (✉)
William Jewell College, Liberty,
MO 64068, USA
e-mail: benzb@william.jewell.edu; bcb269@nyu.edu

T.E. Levy et al. (eds.), *Israel's Exodus in Transdisciplinary Perspective*,
Quantitative Methods in the Humanities and Social Sciences, DOI 10.1007/978-3-319-04768-3_36,

dichotomy highlighted by Jan Assmann. Following Assmann's reading, just as the inclusivity of the patriarchal traditions balances the exclusivity of the exodus tradition, the historical data that point to the insider status of some of the early Israelites balances the data in support of Israel's outsider status.

Introduction

Ancient Israel is famously known for being set apart from the nations, representing a unique sociopolitical entity in the ancient world. There are numerous biblical traditions that promote this identity. First and foremost, the Israelites are regarded as a monotheistic community, called to worship the God who delivered them from Egypt and provided them with a code of social ethics that countered the oppression they faced there. United by a shared tribal identity, the history of a bucolic lifestyle, and a common experience of an exodus from Egypt, Israel was prepared to establish an egalitarian community in the southern Levant that stood in opposition to the hierarchical, urban-centered polities of their polytheistic, Canaanite counterparts who originally inhabited the land.

In spite of the number of biblical claims to the contrary, most (modern) scholars recognize the points of continuity between the so-called Canaanite religion and Israelite religion, concluding that these traditions emerged out of the same "cosmic pool." The question remains, however, as to whether or not there are points of continuity between the social and political structures of the Late Bronze Age Levantine populations and those of the early Israelites. If there are, it seems likely that at least part of Israel stemmed from the same "sociopolitical pool" as well.

While the parallel seems straightforward, pursuing a solution to this question is complicated. In addition to the influence that the exodus and other biblical traditions have had in shaping our perception of Israel's social and political character, there are several traditions deeply embedded in the history of thought that impact our understanding of the social and political character of the ancient Near East in general. For example, modern appraisals of ancient Near Eastern polities are influenced by Herodotus' evaluation of the Persian wars, which he cast as a conflict between Greek democracy and "oriental despotism."[1] At first glance, the earlier Sumerian and Akkadian literary traditions seem to draw a sharp contrast between the pastoralists of the steppe and the culturally advanced societies of urban-centered populations (see, however, Porter 2012: 252, 280–325). While Michael Rowton mitigated this distinction by suggesting that it reflects the biases of the urban populations from which these traditions stemmed, scholarship largely continues to regard these two categories of people as opposing sectors of society that remained politically distinct.

Though the ancient data provide a unique window into the social and political landscape of the Levant during the Late Bronze Age, these long-standing assumptions—both biblical and non-biblical—have directly influenced the way scholarship has interpreted the evidence. Most descriptions are characterized by several critical dichotomies, including sedentary vs. non-sedentary, state vs. tribe, and hierarchical vs. egalitarian. Ultimately, these dichotomies support and are supported by later biblical traditions that set Israel in opposition to Canaan and have resulted in the common belief that early Israel originally consisted of a group of geographical, economic, and/or political outsiders.

While I believe that these models contribute to our understanding of some of the groups that were later defined with Israel, the Amarna letters (*EA*) indicate that the standard expectations of a sharp cultural and political contrast between Late Bronze Age Canaan and early Israel are overrated. In what follows, I focus on three phenomena attested in these letters. They include (a) collective governance, (b) the prominence of a cooperative political structure that consisted of independent political units, and (c) the integration of populations that are not identified with a settled center into urban-centered polities. When compared with the biblical depictions of Israel's

[1] Cf. Marfoe 1979: 16; Porter 2012: 39–42, 202.

formative stages, these phenomena suggest that a portion of Israel consisted of geographical, economic, and political insiders. Ultimately, the historical and social implications of this conclusion will further our understanding of how the exodus tradition has functioned in the past and how it may continue to function effectively in the present.

While archaeology plays a very important role in this evaluation, I have chosen to focus on texts for this brief study because they contain direct accounts of the dynamics that occurred on the ground, identifying and depicting in relatively consistent terms the various entities, categories of people, and players involved. In this way, they provide an interpretive framework for the critical information derived from the archaeological record, which, in turn, can help disentangle some of the nuances of the biblical texts.

Collective Governance

Scholarship generally recognizes that two distinct political units dominated the landscape of the Late Bronze Age Levant. The first was the city (Akkadian *alu*), which is marked by the determinative URU. It consisted of a single urban center and its immediate hinterland. The second, which is marked by the determinative KUR, is what I refer to as a centralized "land" (Akkadian *mātu*). As with cities, centralized lands were administered from a single urban center from which they generally derived their names. However, the authority of a centralized land extended beyond its royal center and immediate hinterland to include other cities. In other words, they were politically integrated units that included multiple dependent populations.

In general, both of these polity types are regarded as having been hierarchically administered by a small cadre of the political elite over and against the large sedentary populations in a manner comparable to medieval feudalism or a Marxist Asiatic mode of production. This reconstruction has been used to support the three models of Israel's origins that dominate the field. As geographical outsiders, some of whom may have fled from Egypt, it was into this dominant "Canaanite" structure that the Israelites entered. As economic outsiders, it was out of this dominant structure through peasant revolutions and flight into the hill country that the Israelites emerged. As political outsiders, it was next to this dominant structure through a large-scale settlement process in the central highlands that Israel took shape. In each case, it was against this structure that Israel defined itself.

While the political will of cities and centralized lands was often identified with and articulated by their individual leaders in the Amarna letters, there are instances in which the citizens and/or representative decision-making bodies of a city or a centralized land took collective political action. Though they are often overlooked, these examples reflect the nature of social power articulated by several anthropologists, sociologists, and historians. As Edward Lehman (1996), Anthony Giddens (1984), and Michael Mann (1986) have pointed out, rather than being a purely hierarchical phenomenon, power is administered through multiple social spheres and resources, making it available to multiple social players. Because of this, the administration of power is best understood as a dialectical process, consisting of what Richard Blanton and his colleagues (1996) refer to as both exclusionary and corporate forms.

One of the most striking examples of this revolves around a category of people attested in the letters from Gubla referred to as the *ḫupšū*. The universal translation of this term as "peasant" or "serf"[2] indicates that scholars regard the *ḫupšū* as a politically disempowered group of economically depressed rural denizens. A nuanced reading of their status informed by the aforementioned understanding of power, however, demonstrates that they were an economically viable, urban-centered group who played a significant role in the political scene at Gubla

[2] See, for example, Mercer 1939; Albright 1926: 107; 1963: 25–6; 1975: 110; Campbell 1960: 15; Mendenhall 1962: 77–8; Astour 1964; Knudtzon 1964; Altman 1978; Liverani, 1979; Moran 1992; Zertal 1994: 67; Gottwald 1999: 212–13; Adamthwaite 2001: 247–9; Killebrew 2006: 571; Grabbe 2007: 66, 118–19.

(Benz 2013: 97–113). Their corporate activities are specifically demonstrated in their decision to reject their leader and banish him from their city.

Rib-Addu details the events that led to his exile in *EA* 138 (cf. *EA* 136 and 137). According to his testimony, the constant assault that his city faced from Aziru had depleted the resources of the Gublites, a category that included the *ḫupšū* (lines 36–38). Intent on improving their situation, the Gublites "moved against" (*ti-na-mu-šu*; line 39) Rib-Addu, with the result that several of them died at his hands ("and I killed them" *ù a-du-uk-šu-nu*; line 39). This act of force temporarily quelled their rebellion and bought Rib-Addu enough time to dispatch another request for military support to Egypt. However, the failure of the Egyptians to respond led the Gublites to proclaim, "Abandon him (the king of Egypt)! Let us align with Aziru!" (lines 44–45). Again, Rib-Addu rejected their demands and traveled to Beirut to forge a military alliance with Ammunira (lines 51–53). In his absence, Rib-Addu's brother capitalized on the frustration and fatigue of the Gublites. According to Rib-Addu, he spoke and swore to the city with the result that the lords of the city rejected Rib-Addu's authority and aligned themselves with the sons of 'Abdi-Aširta (lines 47–50).

A second example of corporate political activity in the Amarna letters responds to the popular claim that only the ruler of a vassal state could correspond with the Egyptian king (Na'aman 2005b: 148, 167; Finkelstein 1996b: 224; cf. Strange 2000: 74; Bunimovitz 1994: 3). While this activity required political authority and legitimate representation to occur, the data indicate that it was not limited to exclusionary forms of power. With the rejection and exile of Rib-Addu, *EA* 139 and140 introduce a new mode of political discourse at Gubla. Rather than being officially represented by a single individual, the political voice of this polity is articulated by a certain Ili-rapiḫ and the city Gubla (^{uru}Gubla).[3] According to Pinhas Artzi, "this joint leadership means to us a re-establishment at Gubla of the ancient city-organization" (1964: 162). However, this conclusion is unnecessary. The letters from Gubla indicate that corporate and exclusionary political strategies coexisted throughout the Amarna period and at times struggled against one another.

There are two other examples of the citizens of a city communicating with their Egyptian overlord. In the introductory greeting of *EA* 59, the correspondents address "the king of the land of Egypt, our lord" (*a-na* LUGAL KUR-*ti*$_4$ *mi-iṣ-ri be-lí-ni*; line 1), referring to themselves as "the sons of Tunip, your servant" (DUMU^{meš} ^{uru}*du-ni-ip*^{ki} ^{lú}ÌR-*ka-ma*; line 2). Similarly, the only extant letter from Irqata is identified as the "tablet of Irqata" (*ṭup-pí* ^{uru}*ir-qa-ta*; *EA* 100: 1–2) and contains the message of "Irqata and its elders" (*um-ma* ^{uru}*ir-qa-ta ù* $^{\text{lú.meš}}$*ši-b* <*u*> *-ti-ši*; lines 3–4).[4] As with the lords of Gubla and the sons of Tunip, Irqata's body of elders represented the political voice of the city in general.[5]

A final example of corporate political activity is reflected in the various ways in which the citizens of a city formed political alliances with other independent political entities. These alliances took several forms, including intercity alliances, alliances between the citizens of a city and individual leaders, and alliances between the citizens of a city and the *'apirū*. In addition to the alliance forged between the Gublites and Aziru highlighted above, one of the more compelling cases of this phenomenon is detailed in a letter

[3] See also *EA* 138: 9–10, 122, where Rib-Addu reports that the Gublites wrote to him during his stay in Beirut.

[4] For the reading *ši-b* <*u*> *-ti-ši*, see Albright 1946: 23.

[5] As a political entity, this body attributes a number of important political activities to themselves throughout their letter that can be added to the act of communicating with the Egyptian king. They "do obeisance at the feet of the king" (lines 5–6), they "guard Irqata for him" (lines 9–10, see also line 30) by going to war against its enemies (lines 27–28), and they "keep the city gate barred until the breath of the king reaches" them (lines 39–41). In fact, contrary to the claim that the Egyptian king only communicated with the "hereditary princes" of his vassal cities, the elders report an occasion in which a messenger of the king addressed this body directly: "When the [ki] ng, our lord, sent D[UMU]-Biḫa, he said to [u]s, 'Thus says the king: Guard Irqata.'" (lines 11–14).

from 'Abdi-Ḫeba, the ruler of Jerusalem. In *EA* 290, he informs his Egyptian overlord that the citizens of the city Bit-NIN.URTA seceded from his centralized land and forged a political alliance with the citizens of the city Qilṭu (lines 5–18).[6]

These examples illustrate that there was a wide spectrum of political representation in the Levant during the Amarna period. One end of the spectrum was occupied by entities that were represented by the political voice and actions of their respective leaders alone. At the other end were those that took corporate political action in the absence of a single leader. There are also examples of political entities falling somewhere in the middle, where exclusionary and corporate forms of power operated in tandem. In the end, the so-called Canaanites of the Amarna age did not solely reflect a hierarchical political organization that stood in opposition to the egalitarian structure of early Israel. Rather, the power dynamics involved reflect Blanton's definition of an "egalitarian" social structure, which "does not imply an absence of hierarchical control," but "any behavior that aims to establish and uphold restrictions on the exercise of exclusionary power, whatever its social setting in simpler or more complex societies" (1998: 151).[7]

Multi-polity Decentralized Lands

Political organization in the Late Bronze Age Levant is a complex matter that resists conceptualization, even by the ancient scribal tradition. This is illustrated in the different polity types to which the determinative KUR referred in the Amarna letters. In addition to centralized lands, there is a second polity type that takes this determinative, which I have dubbed a multi-polity decentralized land (Benz 2013: 158–180). As opposed to a centralized land, which was a discrete political unit under the authority of a single leader who governed his domain from a central administrative hub, this type of land consisted of a political coalition of cities and centralized lands that retained their local independence and identities under the authority of their respective leaders and/or collective representative bodies.

In the historical introduction of the treaty between Muršili II of Ḫatti and Tuppi-Teššup (c. 1312–1285 B.C.) of Amurru, Muršili recalls a time when Tuppi-Teššup's grandfather Aziru supported the Hittites when "the kings of the land of Nuḫašši" became hostile.[8] The conflict between the affiliate members of this multi-polity decentralized land and Aziru is reflected in several letters recovered from Amarna. Depicted as an alliance of independent kings who cooperated with one another against a common enemy, Aziru informed his Egyptian overlord on several occasions that the "kings of Nuḫašši" were at war with him (*EA* 160: 24–25; 161: 36–37; see also 169: 17–23).[9]

The corporate activities of the kings of the land of Nuḫašši are paralleled by those of the constituent leaders of the land of Gina, an entity that plays a more prominent role in the Amarna corpus. According to Nadav Na'aman (2005a: 238–239; 1986: 469), rather than representing a single polity or city (contra Moran 1992: 389; Morris 2005: 227), the land of Gina referred to the entire Jezreel Valley during the 18th Dynasty in Egypt. He goes on to suggest that Megiddo and Taḫnaka were the most important cities in the region (2005a: 238; contra Strange 2000: 72). In *EA* 245, Biridiya, the *ḫazannu* of Megiddo, and Yašdata, the *ḫazannu* of Taḫnaka, are listed as allied leaders of a coalition of cities that took up

[6] See also *EA* 73: 26–29; 74: 19–21; 76: 34–37; 81: 12–13; 84: 11–13; 87: 19–20; 104: 40–54; 105: 7–13; 114: 11–15; 116: 37–38; 118: 24–32; 138: 44–45, 80–93; 144: 22–30; 149: 54–63; 280: 9–24.

[7] See also Marfoe 1979: 33; Lemche 1985: 119–24; Giddens 1984; Wolf 1990; Blanton et al. 1996; Fleming 2004: 178; Porter 2012: 41–2.

[8] Cf. Fleming 2004: 105–6, 124–8, 132. For a translation of this treaty see Beckman 1999: 59.

[9] See also *EA* 51: 5–6, where Addu-nirari recalls the time when the king of Egypt made his forefather Taku "a king in Nuḫašše" (*i-na* ^{kur}*nu-ḫa-aš-še a-na* LUGAL-*ru-tú i-ip-p*[*u-š*]*a-aš-šu*; ll. 5–6). Contra Na'aman (1994: 412), who refers to Nuḫašše as a "kingdom," implying that it was centralized under the leadership of a single king.

arms against Lab'ayu. In *EA* 250, this coalition is specifically referred to as "the land of Gina," indicating that it was not simply a geographical reference, but a political body that could take collective action. The list of affiliate members is expanded as *EA* 250 unfolds to include the cities Šunama, Burquna, Ḫarabu, and Gittirimmunima.

Though collective action was a defining characteristic of multi-polity decentralized lands, affiliate members could choose whether or not they would participate in these corporate activities. In *EA* 365, Biridiya of Megiddo complains that he alone had furnished the required labor to cultivate the land around Šunama, accusing the other leaders in the land of Gina of not doing the same. This accusation indicates that as members of this land, they were expected to reciprocate. Their refusal to do so, however, illustrates that this type of political organization was not rigid. Independent constituents and their leaders retained the ability to make decisions for themselves, even if these decisions had negative repercussions.

Admittedly, the Amarna letters provide a relatively limited view of the Levantine political landscape during the Late Bronze Age. However, this has its benefits. The brevity of the period represented highlights the fact that political affiliation and identity were characterized by a high degree of variability that calls into question any attempt to explain group identity on the basis of ethnicity.[10] As we have already witnessed, though the citizens of Bit-NIN.URTA originally identified themselves as members of the centralized land administered from Jerusalem, in the face of increasing hostilities that threatened their own immediate interests, they appealed to their local identity as citizens of Bit-NIN.URTA at the expense of their identity as Jerusalemites. In addition, the members of the land of Gina had the freedom to choose whether or not they would participate in the corporate activities of their coalition.

The Integration of Populations Not Identified with a Settled Center with Urban-Centered Polities

The complexity of the Late Bronze Age political landscape increases when we consider the role of populations that were not identified with an urban center. Following Rowton's dimorphic model of ancient society (1973, 1974, 1976, 1977), scholarship often casts these populations as political outsiders. There are two groups attested in the Amarna letters that fall into this category—the *'apirū* and the Sutû. Recently, Daniel Fleming has called into question the tendency to regard the *'apirū* as disenfranchised urban dwellers who took to the hills as brigands and mercenaries in order to escape their former creditors and/or overlords.[11] Pointing to the association between the *'apirū* and 'Abdi-Aširta, Lab'ayu, and Biryawaza of Damascus (2012: 260–264), Fleming contends that they "have something in common with the Hana of Zimri-Lim and the Binu Sim'al, a population named by its mobile potential and yet carrying on the identity of a whole city-based kingdom" (2012: 260; cf. Porter 2012: 319–320). If one accepts this reconstruction, the *'apirū* consisted of populations involved in what Anne Porter calls broad-range herding (2000, 2009, 2012) that were fully integrated into polities administered from settled centers.

A similar line of thought can be used to explain the role of the Sutû (Benz 2013: 203–217). Most contend that they represented a "pastoralist, nomadic element in Canaanite society" (Rainey 2008; Finkelstein 2007: 81; Mazar 2007: 94; Fleming 2009: 230). Living on the fringes of settled society, the Sutû fell outside the hegemony of urban-centered polities (Redmount 1998: 86).

[10] For a carefully articulated argument in support of this approach, see Faust 2006 and Chap. 37. For a compelling argument against it, see Fleming 2012: 246–55; see also Finkelstein 1988: 27; 1996a; Dever 1991; 2003: 124; Whitelam 1994; Porter 2012: 80.

[11] See, for example, Mendenhall 1962; Redford 1990: 39; Pitard 1998: 47–9; Redmount 1998: 86; Stager 1998: 103–4; Gottwald 1999: 401–4; Dever 2003: 73–4, 179, 181; Lehmann 2003: 129; Faust 2006: 184.

According to the Amarna letters, however, they were affiliated with them. For example, in *EA* 195 Biryawaza of Damascus lists the Sutû, whom he identifies as "mine," among several groups that are prepared to fight with him on behalf of Egypt's interests (lines 24–30). In a context where Biryawaza also refers to "my troops" (^{lú}ÉRIN^{meš}-*ia*) and "my chariots" (GIŠ. GIGIR^{meš}–*ia*), he appears to be including the Sutû as part of his domain.

The Sutû are also named as an important element of Aziru's base of power in *EA* 169, a letter dispatched by Aziru's son while Aziru was in Egypt. In Aziru's absence, the kings of Nuḫašše began threatening hostilities against his son. According to the latter's account "all of the lands and all of the troops of the Sutû said to me, Aziru is not going to come forth out of Egypt" (lines 24–28). This position was likely informed by the belief that Aziru had been deemed "unfaithful" by his Egyptian overlord and, as a result, met his fate while there. Because of this, the Sutû threatened to depart (*i-pa-*[*ṭá*]-*ru-nim*) from the land of Amurru (lines 29–30). This notice indicates that the Sutû constituted an important element of Aziru's authority whose cooperation and participation had to be negotiated, a situation reminiscent of the political structure at Mari (Fleming 2004) and the events that took place between Rehoboam and the assembly of Israel at Shechem in 1 Kings 12. This evidence suggests that neither the *'apirū* nor the Sutû were simply "stateless" and "disruptive" elements of society. Instead, at least some of their numbers participated in the framework of urban-centered polities.

The Bible

It is my hope that the above evidence will give us fresh eyes for viewing some of the biblical depictions of early Israel. Many have questioned the validity of using the Bible for reconstructing the early history of Israel. This is largely based on the contention that it reflects the ideological and/or theological concerns of exilic and post-exilic Judean authors, editors, and redactors. As a number of the articles in this volume have pointed out, in many cases this assessment is correct. There are, however, several striking points of continuity between some of the biblical accounts of Israel's formative stages and the sociopolitical landscape of the Late Bronze Age that are so foreign to what the Bible as a whole promotes regarding the premonarchical period that they suggest an alternate political reality. These parallels call into question the notion that Judean scribes, who were removed temporally, ideologically, and geographically from Israel, fabricated them.[12]

When it comes to the Bible's vision of Israel in general, Judges 5 and Judges 9 are unusual. The Song of Deborah identifies several populations with Israel. This is not uncommon to the biblical depiction of Israel as a whole. However, the nature of Israel's constituents is unique. While the roster includes many well-known groups, it excludes Judah and Manasseh and introduces a unit referred to as Machir. In addition, it indicates that a city—Meroz—was also expected to participate in the coalition. Finally, the list is not limited to the constituents who participated, but also includes those who chose to remain aloof.

The depiction of Israel in Judges 5 closely corresponds to the nature of the multi-polity decentralized lands attested in the Amarna letters. As was the case with the members of the land of Qina, the constituents of Israel had the autonomy to choose whether or not they would participate in the collective. It is difficult to know why Meroz is the only member who is cursed by Yahweh for not responding to the call. Perhaps it is because she is not remembered in the Bible's later renditions of Israel's early history. Nevertheless, her decision to transfer her allegiance in a tense political situation calls to mind the decision of the citizens of Bit-NIN.URTA, who, in the face of an external threat, seceded from the land of Jerusalem and aligned themselves with the citizens of Qilṭu.

The idea that pre-monarchical Israel shared the characteristics of a multi-polity decentralized land sheds light on some of the oddities of Judges 9.

[12] For an example of this methodological approach, see Fleming 2012.

Another text whose compositional history is intensely debated, it depicts a political entity identified with an important urban center that played host to corporate and exclusionary forms of power that cooperated with and struggled against one another. The core of this narrative opens with the lords of Shechem agreeing to underwrite Abimelek's bid for power (vv. 3–4). After his successful campaign, the lords of Shechem, together with all Beth-millo, ratify Abimelek's authority as king (v. 6). Finally, in a manner reminiscent of the Gublite's rejection of Rib-Addu in *EA* 138, Ga'al, who appears to have operated at a distance from Shechem only to return to the city for a festival at the temple of their god (v. 27), convinces the Shechemites assembled with him to reject Abimelek's kingship.

These unique details have led many to conclude that Judges 9 is a Canaanite rather than an Israelite narrative.[13] As Baruch Halpern points out, when compared to the opposition posed by the religious establishment with the formation of the Israelite monarchy under Saul (1 Sam 8:4–22; 12:1–25), this "contrary tradition is ... somewhat jarring," particularly during "the supposedly pre-monarchical period of the Judges" (1978: 81). However, if the political structure of early Israel reflected that of the Late Bronze Age multi-polity decentralized lands, the idea of a monarchy within Israel before the establishment of the Israelite monarchy would not be out of place. In fact, several have suggested that Gideon[14] and even Jephthah[15] functioned in a similar capacity on a local level within the larger framework of the Israelite collective.

Conclusion

In spite of the weight that the Bible's final form has had on defining the identity of Israel, this brief survey of the evidence suggests that at least a proportion of the early Israelites were heirs to the social and political structures of the Late Bronze Age Levant. While some may have been geographical outsiders who participated in an exodus from Egypt, and others may have been economic and/or political outsiders, some were geographical, economic, and political insiders. In each case, as these groups were identified as "Israelite," so too were their historical memories and the traditions that developed around them. As this identity took shape and solidified over time, some of these traditions, including those revolving around an exodus from Egypt and life in the land before the monarchy, were retained, reworked, and applied to the people as a whole.

This conclusion offers a biblical counterbalance to an "outsider" identity that has been misappropriated and misused by the members of faith communities and those hostile toward them. While the Bible clearly promotes caution when engaging the world, overemphasizing the "outsider" status of a community can lead to isolationism and even violence. Recognizing the Bible's claims regarding Israel's "insider" status compels us to take part in meaningful discussions of what it looks like to operate in the world in just and fruitful ways and to recognize the contexts in which we are called to be outsiders and those in which we are called to be insiders. As is demonstrated in the competing visions of who is to be admitted into the Israelite assembly in Nehemiah 13 and Isaiah 60, this dynamic is reflected throughout the Bible.

[13] See, for example, Martin 1975: 10, 113–4; Moore 1976: 268; Mayes 1977: 316; Soggin 1981: 194; Würthwein 1994: 22; de Castelbajac 2001: 166–73.

[14] Burney 1918: 183; Martin 1975: 108–9; Olson 1998: 810; Heffelfinger 2009: 286; Benz 2013: 529–33.

[15] Benz 2013: 533–9.

References

Adamthwaite, M.R. 2001. *Late Hittite Emar: The Chronology, Synchronisms, and Socio-Political Aspects of a Late Bronze Age Fortress Town*, Ancient Near Eastern Studies Supplement, vol. 8. Louvain: Peeters.

Albright, W.F. 1926. Canaanite Ḫapši and Hebrew ofšî Again. *Journal of the Palestine Oriental Society* 6: 106–108.

———. 1940. *From the Stone Age to Christianity*. Baltimore, MD: The Johns Hopkins University Press.

———. 1946. Cuneiform Material For Egyptian Prosopography 1500–1200 B.C. *Journal of Near Eastern Studies* 5: 7–25.

———. 1963. *The Biblical Period from Abraham to Ezra*, 3rd ed. New York, NY: Harper & Row.

———. 1975. The Amarna Letters from Palestine. In *The Cambridge Ancient History*, vol. 2, Pt. 2, 3rd ed, ed. I. E.S. Edwards et al., 98–116. Cambridge: Cambridge University Press.

Altman, A. 1978. The Revolutions in Byblos and Amurru During the Amarna Period and their Social Background. In *Bar-Ilan Studies in History*, ed. P. Artzi, 3–24. Ramat-Gan: Bar-Ilan University Press.

Artzi, P. 1964. "Vox Populi" In the El-Amarna Tablets. *Revue d'assyriologie et d'archéologie orientale* 58: 159–166.

Astour, M.C. 1964. The Amarna Age Forerunners of Biblical Anti-Royalism. In *For Max Weinreich on his Seventieth Birthday*, ed. M. Weinreich and L.S. Dawidowicz, 6–17. The Hague: Mouton.

Beckman, G. 1999. *Hittite Diplomatic Texts*, SBL Writings from the Ancient World, 2nd ed. Atlanta, GA: Scholars Press.

Benz, B. C. (2013). *The Varieties of Sociopolitical Experience in the Late Bronze Age Levant and the Rise of Early Israel*. Ph.D. dissertation, New York University.

Blanton, R.E. 1998. Beyond Centralization: Steps Toward a Theory of Egalitarian Behavior in Archaic States. In *Archaic States*, ed. G.M. Feinman and J. Marcus, 135–172. Santa Fe, NM: School of American Research.

Blanton, R.E., G.M. Feinmanm, S.A. Kowalewski, and P. N. Peregrine. 1996. A Dual-Processual Theory for the Evolution of Mesoamerican Civilization. *Current Anthropology* 37: 1–14.

Bunimovitz, S. 1994. The Problem of Human Resources in Late Bronze and its Socio Economic Implications. *Ugarit-Forschungen* 26: 1–20.

Burney, C.F. 1918. *Book of Judges*. London: Rivingtons.

Campbell, E.F. 1960. The Amarna Letters and the Amarna Period. *Biblical Archaeologist* 23: 1–22.

De Castelbajac, I. 2001. Histoire de la redaction de Juges ix: une solution. *Vetus Testamentum* 51: 166–185.

Dever, W.G. 1991. Archaeological Data on the Israelite Settlement: A Review of Two Recent Works. *Bulletin of the American Schools or Oriental Research* 284: 77–90.

———. 2003. *Who Were the Early Israelites and Where Did They Come From?* Grand Rapids, MI: Eerdmans.

Faust, A. 2006. *Israel's Ethno-Genesis*. London: Equinox.

Finkelstein, I. 1988. *The Archaeology of the Israelite Settlement*. Jerusalem: Israel Exploration Society.

———. 1996a. Ethnicity and the Origin of the Iron I Settlers in the Highlands of Canaan: Can the Real Israel Stand Up? *Biblical Archaeologist* 59: 198–212.

———. 1996b. The Territorial-Political System of Canaan in the Late Bronze Age. *Ugarit-Forschungen* 28: 221–255.

———. 2007. When and How did the Israelites Emerge? In *The Quest for the Historical Israel: Debating Archaeology and the History of Early Israel*, ed. B. Schmidt, 73–83. Leiden: Brill.

Fleming, D.E. 2004. *Democracy's Ancient Ancestors: Mari and Early Collective Governance*. Cambridge: Cambridge University Press.

———. 2009. Kingship of City and Tribe Conjoined: Zimri-Lim at Mari. In *Nomads, Tribes, and the State in the Ancient Near East*, University of Chicago Oriental Institute Seminars, vol. 5, ed. J. Szuchman, 227–240. Chicago: The Oriental Institute of the University of Chicago.

Fleming, D.E. 2012. *The Legacy of Israel in Judah's Bible*. Cambridge: Cambridge University Press.

Giddens, A. 1984. *The Constitution of Society*. Berkeley, CA: University of California Press.

Gottwald, N. K. (1999). *The Tribes of Yahweh: A Sociology of the Religion of Liberated Israel, 1250-1050 B.C.E.* Biblical Seminar 66. Reprinted with an expanded introduction. Sheffield: Sheffield Academic Press.

Grabbe, L. 2007. *Ancient Israel: What do We Know and How do We Know It?* New York, NY: T&T Clark.

Halpern, B. 1978. The Rise of Abimelek ben-Jerubbaal. *Hebrew Annual Review* 2: 79–100.

Heffelfinger, K.M. 2009. 'My father is King': Chiefly Politics and the Rise and Fall of Abimelech. *Journal for the Study of the Old Testament* 33: 277–292.

Killebrew, A.E. 2006. The Emergence of Ancient Israel: The Social Boundaries of a 'Mixed Multitude' in Canaan. In *I Will Speak Riddles of Ancient Times': Archaeological and Historical Studies in Honor of Amihai Mazar on the Occasion of his Sixtieth Birthday*, ed. A.M. Maeir and P. de Miroschedji, 555–572. Winona Lake, IN: Eisenbrauns.

Knudtzon, J.A. 1964. *Die El-Amarna-Tafeln, Erster Teil. Die Texte*. Aalen: Otto Zeller Verlagsbuchhandlung.

Lehman, E.H. 1969. Toward a Microsociology of Power. *American Sociological Review* 34: 453–465.

Lehmann, G. 2003. The United Monarchy in the Countryside. In *Jerusalem in Bible and Archaeology: The First Temple Period*, Society of Biblical Literature Symposium Series, vol. 18, ed. A.G. Vaughn and A. E. Killebrew, 117–162. Leiden: Brill.

Lemche, N. P. (1985). Early Israel: Anthropological and Historical Studies on the Israelite Society Before the Monarchy. *Vetus Testamentum*, Supplements 37. Leiden: Brill.

Liverani, M. 1979. Social Implications in the Politics of Abdi-Aširta of Amurru. In *Three Amarna Essays*, Monographs on the Ancient Near East, vol. 1. Fasc. 5, 14–20. Malibu, CA: Undena.

Mann, M. 1986. *The Sources of Social Power*. Cambridge: Cambridge University Press.

Marfoe, L. 1979. The Integrative Transformation: Patterns of Socio-Political Organization in Southern Syria. *Bulletin of the American Schools or Oriental Research* 234: 1–42.

Martin, J. (1975). *The Book of Judges: Commentary*. Cambridge Bible Commentary, New English Bible. Cambridge: Cambridge University Press.

Mayes, A. 1977. The Period of the Judges and the Rise of the Monarchy. In *Israelite and Judean History*, The Old Testament Library, ed. J.H. Mayes and J.M. Miller, 285–331. Philadelphia, PA: Westminster Press.

Mazar, A. 2007. The Israelite Settlement. In *The Quest for the Historical Israel: Debating Archaeology and the History of Early Israel*, ed. B.B. Schmidt, 85–98. Leiden: Brill.

Mendenhall, G. 1962. The Hebrew Conquest of Palestine. *Biblical Archaeologist* 25: 66–87.

Mercer, S.A.B. 1939. *The Tell el-Amarna Tablets*. Toronto, ON: Macmillan.

Moore, G.F. 1976. *A Critical and Exegetical Commentary on Judges*, International Critical Commentary. Edinburgh: T. & T. Clark.

Moran, W. 1992. *Amarna Letters*. Baltimore, MD: The Johns Hopkins University Press.

Morris, E.F. 2005. *The Architecture of Imperialism: Military Bases and the Evolution of Foreign Policy in Egypt's New Kingdom*, Probleme der Ägyptologie, vol. 22. Leiden: Brill.

Na'aman, N. 1986. The Canaanite City-States in the Late Bronze Age and the Inheritances of the Israelite Tribes. *Tarbiz* 55: 463–488. Hebrew.

———. 1994. The Canaanites and Their Land, A Rejoinder. *Ugarit-Forschungen* 26: 397–418.

———. 2005a. Pharaonic Lands in the Jezreel Valley in the Late Bronze Age. In *Canaan in the Second Millennium B.C.E*, 232–251. Winona Lake, IN: Eisenbrauns.

———. 2005b. The Network of Canaanite Late Bronze Kingdoms and the City of Ashdod. In *Canaan in the Second Millennium B.C.E*, 145–172. Winona Lake, IN: Eisenbrauns.

Olson, D. 1998. The Book of Judges. In *The New Interpreter's Bible*, ed. L.E. Keck, 721–888. Nashville: Abingdon.

Pitard, W. 1998. Before Israel: Syria-Palestine in the Bronze Age. In *The Oxford History of the Biblical World*, ed. M.D. Coogan, 25–57. Oxford: Oxford University Press.

Porter, A. (2000). *Mortality, Monuments, and Mobility: Ancestor Traditions and the Transcendence of Space*. Ph.D. Dissertation: The University of Chicago.

———. 2009. Beyond Dimorphism: Ideologies and Materialities of Kinship as Time-Space Distanciation. In *Nomads, Tribes, and the State in the Ancient Near East*, University of Chicago Oriental Institute Seminars, vol. 5, ed. J. Szuchman, 201–226. Chicago: The Oriental Institute of the University of Chicago.

———. 2012. *Mobile Pastoralism and the Formation of Near Eastern Civilizations: Weaving Together Society*. Cambridge: Cambridge University Press.

Rainey, A.F. 2008. Shasu or Habiru? Who Were the Early Israelites? *Biblical Archaeology Review* 34: 51–55.

Redford, D.B. 1990. *Egypt and Canaan in the New Kingdom*, Be'er-Shev', kerekh, vol. 4. Beer-Sheva: Ben-Gurion University of the Negev Press.

Redmount, C.A. 1998. Bitter Lives: Israel in and out of Egypt. In *The Oxford History of the Biblical World*, ed. M.D. Coogan, 58–89. Oxford: Oxford University Press.

Rowton, M.B. 1973. Urban Autonomy in a Nomadic Environment. *Journal of Near Eastern Studies* 32: 210–215.

———. 1974. Enclosed Nomadism. *Journal of the Economic and Social History of the Orient* 17: 1–30.

———. 1976. Dimorphic Structure and Topology. *Oriens Antiquus* 15: 17–31.

———. 1977. Dimorphic Structure and the Parasocial Element. *Journal of Near Eastern Studies:* 36: 181–198.

Soggin, A. J. (1981). *Judges: A Commentary*. Trans. by John Bowden. Old Testament Library. Philadelphia: Westminster Press.

Stager, L.E. 1998. Forging an Identity: The Emergence of Ancient Israel. In *The Oxford History of the Biblical World*, ed. M.D. Coogan, 90–131. Oxford: Oxford University Press.

Strange, J. 2000. The Palestinian City-states of the Bronze age. In *A Comparative Study of Thirty City-State Cultures*, Historisk-filosofiske skrifter, vol. 21, ed. M.H. Hansen, 67–75. Copenhagen: Kongelige Danske Videnskabernes Selskab.

Whitelam, K.W. 1994. The Identity of Early Israel: The Realignment and Transformation of Late Bronze-Iron Age Palestine. *Journal for the Study of the Old Testament* 19: 57–87.

Wolf, E.R. 1990. Facing Power: Old Insights, New Questions. *American Anthropologist* 92: 586–596.

Würthwein, E. 1994. *Studien zum Deuteronomistischen Geschichtswerk*, Beihefte zur Zeitschrift für die alttestamentliche Wissenschaft, vol. 227. Berlin: de Gruyter.

Zertal, A. 1994. To the Land of the Perizzites and the Giants': On the Israelite Settlement in the Hill Country of Manasseh. In *From Nomadism to Monarchy*, ed. I. Finkelstein and N. Na'aman, 47–69. Jerusalem: Israel Exploration Society.

The Emergence of Iron Age Israel: On Origins and Habitus

37

Avraham Faust

Abstract

The question of Israel's origins is reexamined within the broader framework of Israel's emergence in the late second millennium BCE. Some methodological difficulties are outlined, and then the author's view of Israel's emergence as an ethnic group in the Iron Age is summarized. A more detailed discussion follows on the possible "origins" of the members of this group, and especially that of earliest Israel—the group that is mentioned in Merneptah's stele. It appears that while many individuals, families, and groups were involved in the process of Israel's ethnogenesis throughout the Iron Age, and that many of those who eventually became Israelites were of Canaanite origins, the first group was composed mainly of Shasu pastoralists. Other groups, probably including a small "Exodus" group that left Egypt, joined the process, and all were gradually assimilated into the growing Israel, accepting its history, practices and traditions, and contributing some of their own. Traditions and practices that were useful in the active process of Israel's boundary maintenance with other groups were gradually adopted by "all Israel." It appears that the story of the Exodus from Egypt was one such story.

The Exodus–Conquest narrative(s), which describes the escape of the Israelites from Egypt, their 40 years' wandering and their conquest and settlement in Canaan, has resulted in a plethora of studies that examine the story as whole, as well as many of its components, in great detail. The present study touches on this thorny issue by attempting to reconstruct the "origin" of the Iron Age Israelites in general and that of Merneptah's Israel in particular, and by reconstructing the development of Israel as an ethnic group. While such a study cannot yield definite answers about the Exodus event, it does allow us to evaluate the possible significance of an Exodus group, and perhaps also the possible mechanisms that enabled the Exodus story to be accepted by the Israelites and to achieve its "national" standing.

A. Faust (✉)
The Institute of Archaeology, The Martin (Szusz)
Department of Land of Israel Studies and Archaeology,
Bar-Ilan University, Ramat Gan, 52900, Israel
e-mail: avraham.faust@biu.ac.il

T.E. Levy et al. (eds.), *Israel's Exodus in Transdisciplinary Perspective*,
Quantitative Methods in the Humanities and Social Sciences, DOI 10.1007/978-3-319-04768-3_37,

Did the Israelites come from Egypt and conquer Canaan in a military campaign? Did they enter Canaan from Transjordan in a slow and peaceful process instead? Did some of the Israelites originally come from Mesopotamia or Syria, as is indicated by the Patriarchal narratives? Could the Israelites simply be Canaanite peasants who rebelled against their landlords? Were they descendants of Canaanite villagers who migrated from the lowlands and settled in the highlands? Or were they merely the local, seminomadic population of the central highland that (re)settled as part of a long and cyclic process? Were they local Canaanite outcasts, or newcomers from afar? All of the above scenarios, and various combinations of them, have been suggested for the origins of the Israelites. This is one of the most hotly debated questions in biblical and archaeological research of ancient Israel and hundreds of books and articles address it.[1]

Identity and Origin: A Methodological Note

The issue of Israel's origins is usually discussed together with that of Israel's identity and ethnicity. Their presumed interdependence can be seen in most studies of ancient Israelite ethnicity (e.g., Dever 1993, 1995, 2003; Finkelstein 1996; Ahituv and Oren 1998). The two are indeed related (see below), but it must be stressed that we are discussing two distinct issues.

It is clear today that ethnicity is subjective and in an endless process of change (e.g., Barth 1969; Banks 1996; Jones 1997; Emberling 1997; Faust 2006; Eriksen 2010; Wimmer 2013). What makes an ethnic group is "self ascription and ascription by others" (Barth 1969: 11, 13). Archaeologically, what is important is not the total, or sum, of traits shared by a group (the "archaeological culture"), but the boundaries it maintains with other groups—the elements that are chosen to transmit the message of difference. These are usually some traits or patterns of behavior that were deemed suitable to demarcate the differences between the groups and their neighbors in a specific context (McGuire 1982; Faust 2006, and references).

If one wishes to know when there was a group with the name "Israel," one is looking for the first time people said of themselves "we are Israel," and of whom others said they are "Israel." The question of their origin, or even what was their identity prior to this point, as important and interesting as it is, is not directly relevant to the question posed above. Every group develops a story about its origin, but the story does not have to be "real" and we must differentiate the question of a group's existence from the question of its "origin" (Stager 1985b: 86; Bloch 1953: 29–35).

Although many have fallen into the trap of interconnecting the two questions in a non-productive way (e.g., Skjeggestand 1992: 171; Coote and Whitelam 1987: 125–127), the problem of studying a people's identity through recourse to origins is clear to most scholars.[2] As far as the first question is concerned, what is important is to identify the point in time in which people first viewed themselves as "Israel." Once there were such people, there was an Israel, regardless of the question of their origin or previous identity. The other question deals with their "actual" origin, or in other words, their "prehistory." The questions posed at the beginning of this chapter deal with this second issue, and will be our main interest here. First, however, we need to say a few words on the related first question, i.e., since when was there a group by the name Israel, and how did it define itself?

[1] This chapter is based to a large extent on my previous works, and especially Faust (2006), and develops some of its themes.

[2] The study of the origins of artifacts, which has received much archaeological attention (and which is used in some cases to trace peoples' origins), actually pose an even greater problem. Cohen (1974: 3); Thomas (1991: 4); see also Goody (1982: 211); Dietler (2010: 190); Stager (1985b: 86).

On the Emergence of Israel

There is no place within the limits of this chapter to discuss the vast literature on Israel's emergence, and I will therefore present my views on Israel's emergence in a nutshell (Faust 2006). The Israelites first appear on the historical scene in the late thirteenth century, when a group by this name is mentioned in Merneptah stele. It is quite clear that this "Israel" corresponds with the beginning of the settlement process in the highlands, on both sides of the Jordan River (e.g., Bloch-Smith 2003; Killebrew 2003; Miller 2004; Faust 2006). This process probably started at some point in the second half of the thirteenth century BCE (Faust 2006: 160, 200; see also Finkelstein 1988: 320). It was accompanied by antagonistic relations between the highland settlers and the Egyptian rulers and administration in Canaan and the Canaanite city-states system that was subordinated to them. The new settlers were apparently pushed (or restricted) to the hilly and remote regions by the Egyptian administration that strengthened its hold over Canaan at the time (Bunimovitz 1994).[3] The highland settlers had an asymmetrical relationship with the powerful Egyptian overlords and the Canaanite cities.[4] Asymmetrical relations between groups, especially within the orbit of a state (e.g., Shennan 1989: 13–17; Emberling 1997: 304), typically result in the creation of groups with ethnic consciousness (Comaroff and Comaroff 1992; Faust 2006: 147–156), and it is therefore expected that the highland settlers would develop such an identity under those circumstances. This is probably the Israel that is mentioned in the Merneptah Stele.

This highland group defined itself as egalitarian in contrast to the highly stratified and diverse Canaanite society, which was comprised of many groups and classes and was highly divided both vertically (between various classes) and horizontally (between different social and political groups). Those are reflected in the texts (e.g., Aharoni 1979: 168–169; see also Rainey 2003: 172–176), and the archaeological finds, which exhibit, for example, various types of palaces and elite dwellings along with smaller houses and evidence for settlement hierarchy and more (e.g., Gonen 1992: 219–222; see also Bunimovitz 1990, 1995). The highland settlers avoided the use of imported or decorated pottery that was prevalent in Canaan at the time, and which was an important feature of nonverbal communication in Canaanite society during the Late Bronze Age. While differences in decoration could convey messages of difference within the Late Bronze Age Canaanite society (the significance of decoration was acknowledged even by processual archaeologists, which otherwise downplayed the significance of studying group identities, see, e.g., Binford 1962, 1965; Sacket 1977; Jones 1997: 111), complete avoidance of imported and decorated wares transmitted a much stronger message of difference. The egalitarian ethos was expressed by a very limited ceramic repertoire used by the settlers, their use of simple inhumations in the ground for the dead, and more.

During the twelfth century the Egyptian rulers withdrew from the Land of Israel. The Canaanite city-states system that characterized the Late Bronze Age was weakened, and lost whatever influence it had had in the highlands. At this point the highland settlers (Israel) had little interaction with the people of the lowland. In the absence of any significant external "other," the highland settlers maintained mainly a symmetrical relationship among themselves, i.e., each group of settlers interacted mainly with similar groups, and had no connection with a larger or stronger group from outside the highlands. Since ethnic consciousness is promoted by asymmetrical or hierarchical relations between groups, it is likely that the symmetrical relationship that characterized this time period put stress on the "simpler" or "local" forms of

[3] While any central government might try to limit the movement of nomadic groups and para-social elements, it appears that the Egyptians in particular had a negative view of such groups (Redford 1992: 271).

[4] The term Canaanites is used here to refer to the indigenous societies that existed in Canaan, although it is clear that it incorporates various distinct groups.

identity (sometimes labeled totemic identities; Comaroff and Comaroff 1992), which became more important at the expense of the broader, ethnic one, even if not replacing it altogether.

During the later part of the Iron Age I the highland population once again confronted a powerful external "other"—the Philistines. By that time the Philistines had an economic interest in various regions of Judah and probably also southern Samaria. This strong external pressure and the resulting asymmetrical relations led the highlanders to re-stress their ethnic identity, this time in relation to the Philistine "other." In the new ethnic negotiation that ensued many of the former relevant traits were renegotiated and were vetted with new meanings (i.e., undecorated pottery, avoidance of imported pottery, and the egalitarian ethos), along with new components that were deemed appropriate in the new context (e.g., the strict avoidance of pork and circumcision). All this left its mark on Israelite identity for centuries, often through a repetitive process of negotiation and renegotiation.

Israel's Origins

The Debate Over the Israelite Settlement and Israel's Origins

The question of Israel's origin is usually intertwined with the question of the settlement process. This was an intensively discussed issue for many years, with the two dominant schools espousing the conquest and the peaceful infiltration models (Finkelstein 1988: 295–306, and references). The conquest theory claimed, on the basis of the narratives of the Exodus and conquest, that the Israelites entered Canaan as a unified group and conquered it. The peaceful infiltration model, assumed that the Israelites were seminomads who entered Canaan from Transjordan as part of a long process of migration, and settled mainly in regions that were empty of Canaanite settlement.

Later, a third school was established, advocating an idea known by the name of "the social revolution" or "peasant revolt" (Mendenhall 1962; Gottwald 1979). The basic idea was that the Israelites were Canaanite peasants who rebelled against the exploiting Canaanite elite, left their houses and went to the highlands. There they met a small group who fled Egypt and brought with them a God of liberation. These rebels were, or became, the Israelites.

Finkelstein (1988, 1994) and Bunimovitz (1994), though differing on many details, have raised a new scenario. In light of the data from the new surveys conducted in the 1980s, and influenced by the *annales* school, they examined the Iron I settlement wave as part of long-term cyclic processes in the highland. The new settlers were seen as the descendants of sedentary Canaanites who became pastoralists following the destruction of the urban system of Canaan in the sixteenth century BCE, and resettled after some 400 years.

A final approach to be mentioned is one that views Israel's emergence as an evolutionary process in which local, mainly sedentary Canaanites moved to the highlands and settled there. There are many variations among the supporters of this view (called "evolutionary," "symbiotic," etc.), and it should be noted that it is held by various scholars who disagree on many if not most of the details (e.g., Lemche 1985; Dever 1995; and many others).

While, broadly speaking, the unified conquest and the peasant revolt theories seem to have been disproved (Finkelstein 1988; Weinstein 1997; Faust 2006), the consensus today is that all previous suggestions have some truth regarding the origins of the ancient Israelites (Dever 1995: 210–211; 2003; Finkelstein 1991: 57; Finkelstein and Na'aman 1994: 13; Kempinski 1995; Miller and Hayes 1986: 85; Gottwald 1983: 6; 1992: 72; see also Rainey 2001: 75; Knohl 2008). The debate today is over "the ratios of the various groups in the Iron I population…" (Finkelstein 1991: 57; see also Dever 1992: 54; Killebrew 2003). Many scholars agree that among the various groups that constituted Israel was also a group that came from Egypt (more below), and recently the suggestion that some came from Syria–Mesopotamia was also revived (Knohl 2008).

It is clear today that the settlement process was a long one, and in its course many groups, families and even individuals joined in and became part of Israel, adopting the various traditions, and sometimes contributing some of their own. This gradual process, which recreated Israel over and over again, was intertwined with Israel's interaction with other groups. While the former, in a long process of inner negotiations added potential *habitus* (see definition below) and "traditions" to Israel, the latter was responsible for the traditions that became important and meaningful and were consequently accorded more significance.

The Beginning: Merneptah's Israel and Onward

But when did "Israel" begin? We know that there was an ethnic (or "identity") group by the name "Israel" already at the time of Merneptah. It is likely that during the long processes through which many groups joined in to become Israel (vis-à-vis other groups) and were briefly described (above), the "Israel" group (the one mentioned in Merneptah's stele) was very dominant, even if in the long-run its demographic significance was limited. While the importance of the identity of the early group will be further explained below, we must stress that it probably gave its name to the new ethnic group that was formed, and the other groups, families, and individuals that were incorporated into it throughout the Iron Age I became part of this "Israel." The growing Israel therefore received some (or much) of its habitus, traditions and "history" from this "core" group, though it is likely that the history and myths of other groups were included as well (see, e.g., Dever 1995: 210; see also Na'aman 1994: 231–247; Tubb 1998: 168–169; Knohl 2008).

We do not know the size of Merneptah's Israel,[5] but it absorbed new members throughout its existence. Many likely joined in a slow process throughout the Iron Age I, while some joined only under the monarchy (Iron Age II). Israel's demographic growth was probably a result of both newcomers and natural growth (more below).

Before the Beginning: The Origins of Merneptah's Israel

But what was the origin of Merneptah's Israel itself? Where did the original group that settled in the highland in the late thirteenth century come from? What was their identity ***before*** they settled down in the highlands?

We have noted above that both the military conquest and the social revolution theories seem to have been discredited by the vast majority of scholars (see, e.g., Finkelstein 1988; Weinstein 1997; Faust 2006, and references) and the main archaeological debate regarding Israel's origins can therefore be divided into two questions: (1) whether the first Israelites were pastoralists/seminomads or a sedentary group, and (2) whether or not they came from outside Cisjordan (i.e., that they came from outside the area between the Jordan River and the Mediterranean). Obviously, those who believe that the Israelites were sedentary also claim that they were mainly indigenous to Cisjordan (at least these who settled in the central highlands). Those who claim that they were seminomads are divided into those who believe they came from outside Cisjordan or were local to the area.

Evaluating the Local Nomads School

The local nomads school views the settlers as seminomads who lived in Cisjordan for several

[5] And since an ethnic group is not necessarily a kinship group (despite what ethnic groups usually claim to be) and can absorb new members rather quickly (cf., the growth of the Zulu; Thompson 1969: 342–345), even knowing the number of Israelites at a later phase (cf., Finkelstein 1988: 330–335, although I think such attempts are not really feasible) would not allow us to estimate the number of Israelites at the time of Merneptah.

hundred years, until—as a result of certain circumstances on which Finkelstein (1990, 1994) and Bunimovitz (1994) differ—they were forced to settle. This school views it as part of a cyclic process of sedentarization and nomadization in the highlands. The idea that all the nomads were local, however, is unlikely. The end of the Late Bronze Age was a period of decisive population movements (e.g., Kuhrt 1995: 375–377; Na'aman 1994: 239), which seem to have impacted the entire region. Na'aman writes accordingly: "[T]he claim that there was a limited reservoir of manpower in the peripheral areas of Canaan and that the settlement [of Israel] was necessarily an inter-Palestinian process, ignores the historical moment in which the settlement was taking place" (1994: 239). It is, therefore, extremely unlikely that only the highland west of the Jordan was left untouched by the social upheavals and migration of the time.

Moreover, the settlements in Transjordan need to be addressed. The Iron Age I settlement process, after all, is not unique to Cisjordan (see also Rainey 2001: 67). Transjordan is part of the same geographic region, so there should be no reason to limit the settlement process of nomads to one side of the Jordan River (cf. Rainey 2001; Van der Steen 1995). Thus, even those who do not accept "foreign intrusion" should not exclude Transjordan. This means that a "local nomads" theory, which limits the potential origin of the settlers to Cisjordan, is extremely unlikely.

Evaluating the Canaanite Origins School

The Canaanite origins school, while clearly not homogeneous, rejects all evidence for nomadic (or seminomadic) origins for the highland settlers. As an alterative, they propose that the settlers came from within Canaanite society. This is a very wide school of thought (or better, an approach), and includes scholars such as Lemche (1985) and Dever (e.g., 1993), who would disagree on many, if at all, of the details. Its view can be exemplified by the following from Dever (1992: 52–53):

> [T]he early Israelites are best seen as homesteaders—pioneer farmers settling the hill-country frontier of central Palestine, which had been sparsely inhabited before Iron I. They were not pastoral nomads who had originally migrated all the way from Mesopotamia... Nor were the early Israelites like the modern Bedouin... For the most part, the early Israelites were agriculturalists from the fringes of Canaanite society.

It should be noted that Dever agrees that the group perhaps also included some pastoral nomads, and maybe even a small group from Egypt, but in the end asserts that they were mostly "indigenous Canaanites" (ibid.: 54; see also 1993b: 31*; Dever, Chap. 30).

The first "positive" line of argument of the Canaanite origins school is based on material evidence of continuity with the Late Bronze Age Canaanite society. If this continuity in material traits would be complete and uninterrupted, then there would be no real doubt that we are discussing the same peoples. But it is clear that such is not the case. There are some marked differences between the Late Bronze Age material culture and that of the Iron I highlands. The differences are expressed in almost every aspect—settlement form and patterns, burials, ceramic repertoire, etc. (more below)—and it is clear that one cannot speak of straightforward and complete continuity.

Thus, even Dever's (1993b) examination of continuity between the two eras resulted in 23 points on the side of continuity, and 47 for discontinuity (Dever 1993: 23*). Dever examined continuity in settlement type and pattern, subsistence, technology (architecture, ceramics, metallurgy, and "other," i.e., terraces and cisterns), demography, social structure, political organization, art, ideology (language, burials and religion) and international relations. This seems to indicate marked discontinuity, and even more so when we take into account that many of the elements of continuity need a reassessment, and that many elements of discontinuity were not examined (see detailed discussion in Faust 2006: 179–181). Consequently, the ***first*** Israelites may have been "local" in a loose meaning of the term, but they were most likely *not* settled Canaanites, and it is clear, at least, that the evidence does not suggest this.

Some scholars suggest that the Iron I settlement seems to reflect a developed agriculture, and hence cannot be attributed to pastoralists (e.g., Ahlstrom 1986: 36; Lederman 1992, 1993; Dever 2003). This line of argument is very problematic, even if one accepts the unlikelihood that seminomads had a good knowledge of agriculture (and seminomads do practice some agriculture: e.g., Khazanov 1994). In most sites we cannot differentiate between the initial phase of Iron Age I settlement and the last phase, as exposed in the excavations. It is likely that all of the so-called advanced agricultural techniques belong to the last phase of Iron Age I settlement, or, at least, do not belong to the first. Therefore the evidence of terraces and other indicators, even if dated to the Iron Age I (cf. for other opinions see de Groot 2000; Gibson 2001; Faust 2005, and references), cannot prove the origin of the settlers, since we do not and cannot know from which phase of the settlement they came, thus rendering their existence meaningless to the present debate. Moreover, one cannot find such evidence in the preceding Late Bronze Age Canaanite society either, as Dever himself has demonstrated (1993: 24*; also Faust 2005).

Additional Problems with the Canaanite Origins School

What is missing from the various suggestions raised by this school, is an explanation of *how* the Israelites became a different ethnic group. In what manner did these people come to view themselves as Israelites, and separate from the Canaanite society of which they were part? What was the process of ethnogenesis? I am not familiar with any explanation suggested by supporters of this school.

The problem becomes even clearer when one examines the issue of burials. If the first Israelites were Canaanites, why did they not bury their dead like their ancestors? After all, burials seem to have been an important facet of life in Middle and Late Bronze Age Canaan, *including* in the highlands (e.g., Gonen 1992; Faust and Safrai 2008), but the Iron Age settlers used simple inhumations (Kletter 2002; Faust 2004, and references). If the Israelites were Canaanites they must have consciously chosen to cease this habit; otherwise they would have continued to bury their dead exactly as in preceding centuries (more below).

In addition, from where did the settlers in Transjordan arrive? And why did they settle in these remote areas? The currently available explanations of this school are not sufficient.

While I agree that many Canaanites became Israelites in the course of the Iron Age (and that in the bottom line they might have even been the majority, see below), and that as far as the "later" Israel is concerned the above reconstruction (which views the Canaanites as the main population source from whom the Israelites evolved) might be correct, it is clear that they did not constitute the original "core" of this group—Merneptah's Israel.

Evaluating the Seminomadic Origin

This school views Israel as having originated from seminomads who came from outside Cisjordan. This is basically the original "Peaceful Infiltration school," although in a more sophisticated form. Following the Egyptian records, these seminomads are now identified mainly with the Shasu (e.g., Weippert 1979: 32–34; Rainey 1991, 2001; Redford 1992; Van der Steen 1999; Levy and Holl 2002, and others). It should be noted, however, that these scholars do not claim that all Shasu were Israelites, but that Israel was one Shasu group (e.g., Rainey 2001; Levy and Holl 2002).

It has been suggested that this reconstruction is based on nothing but a romantic perception of Bedouin life (Dever 1992: 30). While such "romantic" views no doubt influence modern scholarship, and had immense impact on past studies as stressed by Dever, there is a wealth of modern ethnographic and ethnoarchaeological evidence that can support this notion (e.g., Van der Steen 1999; Levy and Holl 2002). The existence of modern analogies by no means proves seminomadic origins for the settlers, but they show that it is possible.

Furthermore, there are several additional lines of evidence that were raised in the past and might suggest a seminomadic origin for Merneptah's Israel, though they are admittedly not conclusive. The first is the form of the earliest settlement sites, like 'Izbet Sartah III and Giloh, which might indicate that herding was an important aspect of the economy (e.g., Finkelstein 1988; Levy and Holl 2002: 91–93). I agree that evidence of sheep/goat husbandry as reflected in the faunal assemblage does not prove nomadic origins, but it clearly supports it. The Levy and Holl (2002) analysis, particularly of the archaeological finds in regards to the oval courtyards, adds ethnographic data that seem to support the view that a seminomadic beginning for Israel is likely, or at least possible.[6] One should also note the traditions regarding Reuben's seniority (e.g., Cross 1988). This tribe is also described as pastoralist,[7] and as being located in Transjordan.

Although the above is not conclusive evidence for the pastoral origins of Merneptah's Israel, it supports it. And as the above discussion shows that Merneptah's Israel is unlikely to have developed from local nomadic or sedentary Canaanites, we are left with a modified version of the traditional peaceful infiltration model as the only likely candidate. It appears that at least many of those who constituted what the Merneptah stela called Israel were of pastoralist background and had their origins with the Shasu. The Shasu were quite widespread (Levy et al. 2004: 65–67; Redford 1992: 272–274) and we are probably discussing only one such Shasu group. Hence, the majority of the settlers in the hill country in the thirteenth century on both sides of the Jordan River were most likely of pastoralist background, though it is more than likely that already at this stage a few para-social elements (like the 'Apiru) were integrated into it.

The seminomadic origin of Merneptah's Israel is even strengthened by a more direct line of evidence, which examines the process by which Israel's ethnic traits were chosen.

Israel's Origins and Israel's *Habitus*

When groups interact with one another, they choose traits that are then used to demarcate the boundaries between them. Groups therefore usually choose a trait that is very different from those of their "other," i.e., the traits that are used by the group in relation/contrast to whom it defines itself. But how are the traits selected? Groups usually do not begin to do something new only because it is different from what their opponents do. Meaningful traits are chosen or developed from the *habitus* (Jones 1997: 275; Shennan 1989: 20; Faust 2006: 152–155). According to Bourdieu (1977: 72):

> The structures constitutive of a particular type of environment (e.g., the material conditions of existence characteristic of a class condition) produce *habitus*, systems of durable, transposable *dispositions*, structured structures predisposed to function as structuring structures, that is, as principles of the generation and structuring of practices and representations which can be objectively "regulated" and "regular" without in any way being the product of obedience to rules.

According to Shenan, the habitus is, unconsciously, what individuals learn to do and think from birth onwards, merely by virtue of their having been brought up in one place rather than another (1989: 20; see also Jary and Jary 1995: 275; Jones 1997: 88). The habitus, while clearly not synonymous with ethnicity, is, in a sense, the tool-kit from which ethnicity chooses its traits (also Jones 1997: 120–121, and references).[8] The habitus provides the source of the traits, which are then vested with new meanings (Faust 2006: 152–155).

[6] Although the "nomadic" origin of the four-room house seems unlikely.

[7] Some connect the seniority of Reuben with the first settlements in Transjordan, some of which are even a continuation of Late Bronze Age sites (Cross 1988; Herr 2000; but see Finkelstein 2011). It appears that those settlements reflect the more sedentary component of a pastoralist group (cf., Khazanov 1994). The nomadic groups had some connection with the Late Bronze Age sites in the nearby region, and might have been a different segment of the same group.

[8] Significant traits might be imposed on a group (Faust 2006: 154), but the habitus is quite probably the *major* channel for the development of ethnic traits and behaviors.

If the observation regarding the importance of the habitus is correct, then the above discussion of Israelites ethnic traits might allow us to learn about earliest Israel's habitus, and hence its setting. As noted above, Israel's ethnic traits or behaviors include the use of simple, undecorated pottery, avoidance of imported pottery, use of limited ceramic repertoire, burials in simple inhumations, and more—all associated with an ethos of simplicity and egalitarianism (see Faust 2006: 92–107; 2013, for many references). All this seems to suggest that the earliest Israelite settlers did not originate directly from Canaanite, or at least, mainstream, settled Canaanite society. They more likely came from groups who did not usually use imported and decorated pottery, had a limited ceramic repertoire, used simple inhumations, and embraced a relatively egalitarian ethos.

Burial practices can serve as an important example. Why did the highland settlers use simple inhumations? After all, even when considering Israel's egalitarian ethos and the need to contrast with the Late Bronze Age ideological system, the setters could still have used multiple burials in natural caves, as was common throughout the second millennium BCE. Their avoidance of even this practice—which could suit the egalitarian ideology—and their preference of simple inhumations imply that this custom came from a different "source." While we do not know much about the Shasu, it is likely that they did not bury their dead in a monumental way, but used simple inhumations (as might be indicated by the lack of archaeological indications for this group in the past; Finkelstein 1995).

Levy et al. (2004: 83) published a cemetery in southern Transjordan they attribute to the Shasu population. It is therefore worth noting that the burials were relatively simple, and the authors explicitly wrote that "we assume some kind of 'egalitarian' principle was at work in the burial tradition." They believe that this was the typical burial by seminomads in this region throughout history (ibid.: 71).[9]

[9] Indeed, in many cases (though not always of course) simple societies do not have elaborated death or burial ideology, and use simple inhumations (e.g., Woodburn 1982; Bloch 1982: 230).

It follows then that the first settlers were, to a very large extent at least, seminomads, as ***all*** of the above qualities are expected to be found among them and were not present, as far as we know, among any known Late Bronze Age sedentary Canaanite group. These seminomads came, most probably, from among the Shasu groups (perhaps including small groups of "local" ʿApiru, or outcast Canaanites). This is, most likely, the core of Merneptah's Israel.

Other Groups and the Formation of Israel

As already noted, over the years many groups, families, and individuals joined Merneptah's Israel. Those groups, each in its turn, assimilated into the growing body of Israel. This process enlarged the group on the one hand, but continuously changed it on the other hand. The core of Israel's self-assertion was probably the same, and the changes in this regard were slow, but more and more traditions, customs and stories were added. Some of those were shared only by parts of the growing Israel, while others were gradually absorbed and eventually became part of the history and tradition of all Israel. A central role in determining the fate of the various customs and traditions was probably the process of Israel's boundary maintenance. A custom or trait that fitted at some context into the process of self-definition vis-à-vis other groups was more readily adopted, invested with additional meaning, and became a marker of all Israel. Thus, for example, the interaction with the Philistines vested new meaning into the customs of circumcision and avoiding pork (Faust 2006: 35–40, 85–91, 147–156, and references), although it is likely that both were practiced earlier.

This process continued throughout the Iron I, and deep into the Iron II when Canaanite groups assimilated and became Israelites (e.g., Faust and Katz 2011). By the end of the process, it is possible that most of those who joined were of sedentary Canaanite background. But as we have seen, the core group—Merneptah's Israel—was probably of pastoralist background, and it determined much of Israel's "core" values and traditions. The late-comers of whatever origins continuously assimilated into this "core." The changes—and they were continual—were nevertheless slow, and their reference point was a "core" created by the world of the seminomads.

The importance of a "core" group can be illustrated by the situation in the USA. While the vast majority of the ancestors of present day Americans did not come from English-speaking countries, the USA is still an English-speaking country, by virtue of its original core. This core group spoke English, all the other newcomers joined in, and within one generation they spoke English. They changed America in the process, but they themselves changed faster. This is how, despite the fact that many of those who gradually became Israelites were probably of settled Canaanite origin, their world was so different from the world of their forefathers.

The Exodus Group

While there is a consensus among scholars that the Exodus did not take place in the manner described in the Bible, surprisingly most scholars agree that the narrative has a historical core, and that some of the highland settlers came, one way or another, from Egypt (cf. Bietak 2003; Gottwald 1979; Herrmann 1985: 48; Mazar 2001: 76; Na'aman 1994: 245; Stiebing 1989: 197–9; Friedman 1997: 82–83; Halpern 1992: 104, 107; Halpern 2003; Dever 1993: 31*; 1995: 211; Tubb 1998: 169; Williamson 1998: 149–150; Hoffmeier 1997; Weisman 1984: 15–16; Malamat 1997; Yurco 1997: 44–51; Machinist 1991: 210; 1994; Hendel 2001, 2002; Knohl 2008; see also Levy and Holl 2002; and see many contributions to this volume).

In this, I am not referring to the various traditions of Israel's interaction with Egypt resulting from the era of Egyptian control in Canaan or from some relations with the Hyksos, which found their way into the Bible (Russell 2009; see also Hendel 2001; Knohl 2008; Na'aman 2011; more below), but to the possibility that there was a group which fled Egypt, and brought this story of Exodus with it. Though the size of this group is debated, most of the above scholars agree that it was in the range of a few thousands, or even hundreds (some give it more weight, e.g., Hoffmeier 1997). Still, despite the limited size of this group, it appears that during the process of Israel's ethnogenesis its story became part of the common history of all the Israelites.

Most of those who accept some historical core for the story of the Exodus from Egypt, date it to the thirteenth century (e.g., Hoffmeier, Chap. 15), at the time of Ramses II, while others date it to the twelfth century, during the time of Ramses III (e.g., Halpern 1992; Rendsburg 1992; cf. Bietak, Chap. 2).[10] Archaeology does not really contribute to the debate over the historicity or even historical background of the Exodus itself, but if there was indeed such a group, it contributed the Exodus story to that of all Israel. While I agree that it is most likely that there was such a group, I must stress that this is based on an overall understanding of the development of collective memory and of the authorship of the texts (and their editorial process). Archaeology, unfortunately, cannot directly contribute (yet?) to the study of this specific group of Israel's ancestors.[11] So was this Exodus group

[10] Some scholars date the Exodus to the fifteenth century (e.g., Bimson 1991; cf. Geraty, Chap. 4). Since, however, there is no evidence for Israel in Canaan before the thirteenth century, there is no need to address this approach here. This is also true for other early dates for the Exodus (e.g., Anati 2013 and this volume). If the Exodus is somehow related to such episodes, its incorporation in Israel's history was through other channels (more below).

[11] The lack of finds in Sinai, for example, led some scholars to doubt the historicity of the event altogether (e.g., Finkelstein and Silberman 2001: 61–63). Others, e.g., Cross (1988), suggested that the Sinai of the Exodus was in Midian (see also Stager 1998: 142–149; Knohl 2008). This was a brilliant idea, which was based on the evidence for Late Bronze Age activity in some desert oases in Midian, but recent archaeological work suggests that there was continuous human activity in this region throughout the Late Bronze—Iron Age II time frame (e.g., Hausleiter 2012: 229–230). Hence, even if this is the Sinai of the biblical text, the finds cannot help in dating the Exodus. There is also a significant group of scholars who doubt the existence of an Exodus group.

also Merneptah's Israel, or at least part of it? Clearly, if there was an Exodus in the thirteenth century this group of people could have been part of Merneptah's Israel. However, despite the assumed significance of this group (the Exodus as a "national" epic, more below), it is likely that this group was incorporated at a later stage, only after Merneptah's time, or at least that it was distinct from Merneptah's Israel. After all, although this group clearly brought with it some of what became the history of Israel, it wasn't Merneptah's Israel, or any "Israel" for that matter. While many scholars agree that the Exodus group brought with it YHWH as a new deity (Cross 1988; Knohl 2008; cf., Römer, Chap. 22), the name Israel has the component "El," rather than "Ya" or "Yahu."[12] Thus, Israel could preceded the arrival of the Exodus group, and it is likely that the latter was not Israel's "core" group.

The Exodus Story and All Israel

The process by which all these groups came to share this history is illustrated by Dever (1995: 211):

> The "Exodus-Conquest" story is perhaps really about a small group... A simple analogy may help us to understand this phenomenon. In mainstream American tradition, we all celebrate Thanksgiving as though we ourselves had come to these shores on the Mayflower. That is the myth; yet in fact, most of us got here some other way...

[See also Halpern 1992: 107; Dever 2003; for another reconstruction, see Knohl 2008.]

While Dever's words seem to give a general idea on how an important story of one group can become that of a much larger one, we still need to ask ***why*** was the story of a late-coming group accepted by the earlier components of Israel, and turned into a "national epic." We have seen (also Faust 2006) that Merneptah's Israel defined itself against the Egyptian empire (which forced them to settle) and its subordinated Canaanite city states, and this in itself made the Exodus story easy to accept by the original "core" group. As suggested by Hendel (Chap. 5), Egyptian oppression of Canaan, which lasted until about 1150 BCE, made the Canaanite groups also prone to accept the story. Memories of oppression in Egypt, or by Egypt, were therefore shared by all components of emerging Israel (and to this one can also add the memory of the Hyksos expulsion; Redford 1992, and Chap. 34), so this made the reception of the Exodus story quite easy, regardless what constituted the story at the early stages.

Conclusion

Although the question of Israel's origins is related to the question of its identity, the two issues should be dealt with separately, as they are different questions. The origin of either a people or a trait will not necessarily tell us much about the people's identity or the trait's

---

They connect the development of the story to a later background, e.g., the seventh century BCE (Finkelstein and Silberman 2001: 48–71), perhaps relying on some vague memories of the expulsion of the Hyksos (Finkelstein and Silberman 2001: 69; Redford 1992: 419–422; Na'aman 2013; see Finkelstein, Na'aman, Redford, Chaps. 3, 42, and 34, respectively), but not as related to the flight of a group of slaves. Some have also suggested that the story is connected with the Egyptian control in Canaan during the Late Bronze Age (Na'aman 2011 and this volume). This story was gradually reworked until it received its final shape (Knohl 2008 also connects the emergence of Israel to the Hyksos expulsion, but for him there was also an Exodus group from Egypt). However, the idea that the memories of an Exodus are based *only* on Egyptian presence in Late Bronze Age Canaan is unlikely, and so is the idea that the memories are based *only* on the expulsion of the Hyksos.

[12] While there were many Shasu groups throughout the region (Levy et al. 2004: 65–67; Redford 1992: 272–274), it appears that some of them (in the south) were affiliated with YHWH (e.g., Redford 1992: 272), and it is possible that this is where YHWH came from (also in line with many biblical verses; Knohl 2008: 73–76; Römer, Chap. 22). While it is clear that those who brought YHWH were not Merneptah's Israel, it is possible that the small Exodus group from Egypt met the YHWH Shasu along their travels, and that this interaction influenced the Exodus group. It is also possible that some of these Shasu joined the Exodus group, or that the latter were composed of many of the former. This background might help us understand how the new Exodus group merged so smoothly into Merneptah's Israel.

meaning at any given moment. Separating the discussion of the two issues might later allow us to integrate them! I therefore first briefly presented my view on the development of the Israelite ethnic identity, from the second half of the thirteenth century (Merneptah's Israel) and onward. It is agreed that ancient Israel was composed of peoples who came from various backgrounds: a seminomadic population who lived on the fringe of settlement, settled Canaanites who for various reasons changed their identity, outcast Canaanites, tribes from Transjordan and maybe even from Syria, and probably even a group who fled Egypt. In the end it is likely that many, if not most, Israelites had Canaanite origins. This was clearly the case in the period of the monarchy, in which many Canaanites in the lowland gradually became Israelites. The intake of people of various backgrounds was at times the main source of Israel's population increase, in addition to natural growth. They were all integrated and assimilated into the main group of Israel (which was subsequently in a constant process of change).

But can we say something on the "origins" of the Israel of the thirteenth century—about Merneptah's Israel—the group that "shaped" the growing Israel? Various lines of evidence, and especially an attempt to examine the groups whose *habitus* could produce Israelite-sensitive ethnic traits seem to indicate that this original group—those who came on the "Mayflower" to use Dever's metaphor—likely included mainly pastoralists, most likely Shasu who were perhaps accompanied by others, para-social groups. It is possible that the small Exodus group, which most scholars agree existed, was also part of this early Israel (Merneptah's Israel), but it is more likely that it joined in only later. This Exodus group along with other groups, families, and individuals assimilated into Israel, adopted its customs and traditions and most importantly its self-perception and world-view. Those newcomers also contributed some habitus, stories and traditions to Israel. Most of the latter probably remained local and were not adopted by "all" Israel, while a few seemed relevant to all Israel in some historical contexts, and were adopted and became part of the story of Israel. The Exodus story was apparently one of the latter.

Excursus: Israel in Merneptah's Stela and Reliefs

All of the above is closely dependent on our understanding of Israel in Merneptah's stela and Karnak reliefs. While there is a consensus that the term Israel on the stela refers to a people, or an ethnic group (Stager 1985a; see also Faust 2006: 163, and references), there is less agreement regarding the nature of this group. Some claim that this Israel refers to a sedentary population, as was deduced from the use of "*prt*" (Hasel 1994; see also 2003). This claim was refuted in a very detailed discussion by Rainey (2001: 57–66), who concluded that Israel "is defined by the [Egyptian hieoglyphic] determinative for a socio-ethnic group. The group thus designated might be living on the level of village culture, or could be pastoralist still in the nomadic stage" (Rainey 2001: 65–66). However, Hasel (2003) also showed that, in other places, the determinative used for Israel was not used in reference to the Shasu. While it is likely that this is not the final word on the debate, it should be stressed that even if the stela does refer to a sedentary group, such would be meaningless for the debate of origins. Nobody would argue that the settlers in Giloh were not sedentary; the question is not what they were in the late thirteenth century, when the stela was inscribed, but what they were several dozens of years earlier. The stela, therefore, does not provide information on Israel's origins.

The Karnak wall reliefs, previously attributed to Ramses II, were identified by Yurco more than 20 years ago as dating to the time of Merneptah. Yurco also showed that they describe the same events mentioned in the Israel stela. His view is generally accepted (Yurco 1990, 1991; Stager 1985a; Rainey 1991, 2001; although not by all: see, e.g., Redford 1986, 1992: 275, n. 85), and led to a heated debate on the manner and the place where the Israelites are depicted. Yurco (e.g., 1990, 1991), followed by others, claimed that the Israelites are depicted as Canaanites (cf. Stager 1985a). Rainey (2001), however, strongly opposes this interpretation and believes the Israelites were depicted as Shasu captives. He claims that the Israelites did not have chariots, while the Canaanites with which Yurco identifies Israel are depicted with them. Yurco basically agreed (1991), but suggested several scenarios by which the Israelites could have acquired them, which Rainey did not accept.[13] But again, the reliefs depict the Israel of the late thirteenth century, not that of a generation earlier or their ancestors, and hence limiting the significance of the discussion, regardless of its outcome.

References

Ahituv, S., and E. Oren. 1998. *The Origin of Early Israel—Current Debate: Biblical, Historical and Archaeological Perspectives*. Beer-Sheva: Beer-Sheva University.

Ahlstrom, G.W. 1986. *Who Were the Israelites?* Winona Lake, IN: Eisenbrauns.

Aharoni, Y. 1979. *The Land of the Bible, a Historical Geography*. Philadelphia, PA: Westminster Press.

Anati, E. 2013. *Is Har Karkom the Biblical Mount Sinai?* Brescia: Atelier.

Banks, M. 1996. *Ethnicity: Anthropological Constructions*. London: Routledge.

Barth, F. 1969. Introduction. In *Ethnic Groups and Boundaries*, ed. F. Barth, 9–38. Boston, MA: Little, Brown and Co.

Bietak, M. 2003. Israelites Found in Egypt: Four-Room House Identified in Medinat Habu. *Biblical Archaeology Review* 29(5): 41–49. 82–83.

Bimson, J.J. 1991. Merneptah's Israel and Recent Theories of Israelite Origins. *Journal for the Study of the Old Testament* 49: 3–29.

Binford, L.R. 1962. Archaeology as Anthropology. *American Antiquity* 28: 217–225.

———. 1965. Archaeological Systematics and the Study of Culture Process. *American Antiquity* 31: 203–210.

Bloch, M. 1953. *The Historian's Craft*. New York: Albert A. Knopf.

[13] While Yurco claimed that the Israelites are not to be identified with the Shasu, he agreed that some of the Israelites came from a pastoralist background (which, in turn, could be Shasu) (Yurco 1997: 41–42).

———. 1982. Death, Women and Power. In *Death and the Regeneration of Life*, ed. M. Bloch and J. Parry, 211–230. Cambridge: Cambridge University Press.

Bloch-Smith, E. 2003. Israelite Ethnicity in Iron I: Archaeology Preserves What is Remembered and What is Forgotten in Israel's History. *Journal of Biblical Literature* 122: 401–425.

Bourdieu, P. 1977. *Outline of a Theory of Practice*. Cambridge: Cambridge University Press.

Bunimovitz, S. 1994. Socio-Political Transformations in the Central Hill Country in the Late Bronze-Iron I Transition. In *From Nomadism to Monarchy*, ed. I. Finkelstein and N. Na'aman. Jerusalem: Yad Ben Zvi.

Bunimovitz, S. 1990. The Land of Israel in the Late Bronze Age: A case study of socio-cultural change in a complex society. Ph.D. dissertation, Tel Aviv Univeristy [Hebrew].

Bunimovitz, S. 1995. On the Edge of Empires—The Late Bronze Age (1500–1200 BCE). In *The Archaeology of Society in the Holy Land*, ed. T.E. Levy, 320–331. London: Leicester University Press.

Cohen, A. 1974. *Two Dimensional Man*. London: Routledge and Kegan Paul.

Comaroff, J., and J. Comaroff. 1992. *Ethnography and Historical Imagination*. Boulder, CO: Westview Press.

Coote, R., and K.W. Whitelam. 1987. *The Emergence of Ancient Israel in Historical Perspective*. Sheffield: Almond Press.

Cross, F.M. 1988. Reuben, First-Born of Jacob. *Zeitschrift für die Alttestamentliche Wissenschaft* 100(Suppl): 46–65.

De Groot, A. 2000. On the Dating of Terraces in the Central Hill Country. In *26th Archaeological Congress in Israel*, Abstracts (Hebrew), 20–21.

Dever, W.G. 1992. How to Tell a Canaanite from an Israelite? In *The Rise of Ancient Israel*, ed. H. Shanks, 26–56. Washington, DC: Biblical Archaeology Society.

———. 1993. Cultural Continuity, Ethnicity in the Archaeological Record, and the Question of Israelite Origins. *Eretz-Israel* 24: 22*–33*.

———. 1995. Ceramics, Ethnicity, and the Questions of Israel's Origins. *Biblical Archaeologist* 58(4): 200–213.

———. 2003. *Who Were the Israelites and Where Did They Come From?* Grand Rapids, MI: Eerdmans.

Dietler, M. 2010. *Archaeologies of Colonialism: Consumption, Entanglement and Violence in Ancient Mediterranean France*. Berkeley, CA: University of California.

Emberling, G. 1997. Ethnicity in Complex Societies, Archaeological Perspectives. *Journal of Archaeological Research* 5(4): 295–344.

Eriksen, T.H. 2010. *Ethnicity and Nationalism*, 3rd ed. London: Pluto Press.

Faust, A. 2004. Mortuary Practices, Society and Ideology: The Lack of Highlands Iron Age I Burials in Context. *Israel Exploration Journal* 54: 174–190.

———. 2005. The Israelite Village: Cultural Conservatism and Technological Innovation. *Tel Aviv* 32: 204–219.

———. 2006. *Israel's Ethnogenesis: Settlement, Interaction, Expansion and Resistance*. London: Equinox.

———. 2013. Decoration Versus Simplicity: Pottery and Ethnic Negotiations in Early Israel. *Ars Judaica* 9: 7–18.

Faust, A., and H. Katz. 2011. Philistines, Israelites and Canaanites in the Southern Trough Valley during the Iron Age I. *Egypt and the Levant* 21: 231–247.

Faust, A., and Z. Safrai. 2008. Changes in Burial Patterns in the Land of Israel through Time in Light of the Findings from Salvage Excavations. In *The Hill-Country, and in the Shephelah, and in the Arabah (Joshua 12, 8): Studies and Researches Presented to Adam Zertal in the Thirtieth Anniversary of the Mennasseh Hill-Country Survey*, ed. S. Bar, 105–121. Ariel (Hebrew): Jerusalem.

Finkelstein, I. 1988. *The Archaeology of the Period of Settlement and Judges*. Jerusalem: Israel Exploration Society.

———. 1990. The Iron I in the Land of Ephraim—A Second Thought. In *From Nomadism to Monarchy*, ed. I. Finkelstein and N. Na'aman, 101–130. Jerusalem: Yad Ben-Zvi. Hebrew.

———. 1991. The Emergence of Israel in Canaan: Consensus, Mainstream and Dispute. *Scandinavian Journal of the Old Testament* 2: 47–59.

———. 1994. The Emergence of Israel: A Phase in the Cyclic History of Canaan in the Third and Second Millennium BCE. In *From Nomadism to Monarchy*, ed. I. Finkelstein and N. Na'aman, 152–178. Jerusalem: Yad Ben-Zvi.

———. 1995. *Living on the Fringe, the Archaeology and History of the Negev, Sinai and Neighbouring Regions in the Bronze and Iron Ages*. Sheffield: Sheffield Academic Press.

———. 1996. Ethnicity and the Origin of the Iron I Settlers in the Highlands of Canaan: Can the Real Israel Stand Up? *Biblical Archaeologist* 59(4): 198–212.

———. 2011. Tall al-Umayri in the Iron Age I: Facts and Fiction with an Appendix on the History of the Collared Rim Pithoi. In *The Fire Signals of Lachish: Studies in the History and Archaeology of Israel in the Late Bronze Age, Iron Age and the Persian Period in Honor of David Ussishkin*, ed. I. Finkelstein and N. Na'aman, 113–128. Winona Lake, IN: Eisenbrauns.

Finkelstein, I., and N. Na'aman. 1994. *From Nomadism to Monarchy*. Jerusalem: Yad Ben Zvi.

Finkelstein, I., and N.A. Silberman. 2001. *The Bible Unearthed: Archaeology's New Vision of Ancient Israel and the Origin of its Sacred Text*. New York: Free Press.

Friedman, R.E. 1997. *Who Wrote the Bible*. New York: Harper San Francisco.

Gibson, S. 2001. Agricultural Terraces and Settlement Expansion in the Highlands of Early Iron Age Palestine: Is There a Correlation Between the Two. In *Studies in the Iron Age of Israel and Jordan*, ed. A. Mazar, 113–146. Sheffield: Sheffield Academic Press.

Gonen, R. 1992. *Burial Patterns and Cultural Diversity in Late Bronze Age Canaan*. Winona Lake, IN: Eisenbrauns.

Goody, J. 1982. *Cooking, Cuisine and Class: A Study in Comparative Sociology*. Cambridge: Cambridge University Press.

Gottwald, N.K. 1979. *The Tribes of Yahweh*. New York: Orbis Books.

———. 1983. Two Models for the Origins of Ancient Israel: Social Revolution or Frontier Development. In *The Quest for the Kingdom of God: Studies in Honor of George E. Mendenhall*, ed. H.B. Huffmon, F.A. Spina, and A.R.W. Green, 5–24. Winona Lake, IN: Eisenbrauns.

———. 1992. Response to William G. Dever. In *The Rise of Ancient Israel*, ed. H. Shanks, 70–75. Washington, DC: Biblical Archaeology Society.

Halpern, B. 1992. The Exodus from Egypt: Myth or Reality? In *The Rise of Ancient Israel*, ed. H. Shanks et al., 86–113. Washington, DC: Biblical Archaeology Society.

———. 2003. Eyewitness Testimony: Parts of Exodus Written Within Living Memory of the Event. *Biblical Archeology Review* 29(5): 50–57.

Hasel, M.G. 2003. Merenptah's Inscription and Reliefs and the Origins of Israel. In *The Near East in the SouthWest: Essays in Honor of William G. Dever*, ed. B. Alpert Nakhai, 19–44. Boston, MA: ASOR.

Hasel, M.G. 1994. Israel in the Merneptah Stela. *BASOR* 296: 45–62.

Hausleiter, A. 2012. The Oasis of Tayma. In *Roads of Arabia: Archaeology and History in the Kingdom of Saudi Arabia*, ed. A.I. Al-Ghabban, B. Andre-Salvini, F. Demange, C. Juvin, and M. Cotty, 218–261. Washington, DC: Smithsonian.

Hendel, R.S. 2001. The Exodus in Biblical Memory. *Journal of Biblical Literature* 120: 601–622.

———. 2002. Exodus, A Book of Memories. *Bible Review* 18(4): 38–45.

Herr, L.G. 2000. The Settlement and Fortification of Tell al-'Umayri in Jordan during the LB/Iron I Transition. In *The Archaeology of Jordan and Beyond, Essays in Honor of James A Sauer*, ed. L.E. Stager, J.A. Greene, and M.D. Coogan, 167–179. Winona Lake, IN: Eisenbrauns.

Herrmann, S. 1985. Basic Factors of Israelite Settlement in Canaan. In *Biblical Archaeology Today—1984: Proceedings of the International Congress on Biblical Archaeology, Jerusalem, April 1984*, ed. J. Amitai, 47–53. Jerusalem: Israel Exploration Society.

Hoffmeier, J.K. 1997. *Israel in Egypt, The Evidence for the Authenticity of the Exodus Tradition*. New York: Oxford University Press.

Jary, D., and J. Jary. 1995. *Collins Dictionary of Sociology*. Glasgow: Collins.

Jones, S. 1997. *The Archaeology of Ethnicity: Constructing Identities in the Past and Present*. London: Routledge.

Kempinski, A. 1995. To What Extent were the Israelites Canaanites? *Archeologya, Bulletin of the Israel Association of Archaeologists* 4: 58–64. Hebrew.

Khazanov, A.M. 1994. *Nomads and the Outside World*. Madison, WI: University of Wisconsin Press.

Killebrew, A.E. 2003. *Biblical Peoples and Ethnicity: An Archaeological Study of Egyptians, Canaanites, Philistines and Early Israel 1300–1100 BCE*. Atlanta, GA: SBL.

Kletter, R. 2002. People Without Burials? The Lack of Iron Age Burials in the Central Highlands of Palestine. *Israel Exploration Journal* 52: 28–48.

Knohl, I. 2008. *The Bible Genetic Code*. Zmora-Bitan, Dvir (Hebrew): Or Yehuda.

Kuhrt, A. 1995. *The Ancient Near East (3000–330 BCE)*. London: Routledge.

Lederman, Z. 1992. Nomads They Never Were. Paper presented at the 18th Archaeological Congress in Israel, Lectures' Abstracts, Tel Aviv (Hebrew).

———. 1993. Response. In *Biblical Archaeology Today—1990: Proceedings of the Second International Congress on Biblical Archaeology, Jerusalem, June–July 1990*, ed. Z. Lederman, 483–484. Jerusalem: Israel Exploration Society.

Lemche, N.P. 1985. *Ancient Israel*. Leiden: Brill.

Levy, T.E., R.B. Adams, and A. Muniz. 2004. Archaeology and the Shasu Nomads: Recent Excavations in the Jabal Himdat Firdan, Jordan. In *Le-David Maskil: A Birthday Tribute to David Noel Freedman*, ed. R.E. Friedman and W.H. Propp, 63–89. Winona Lake, IN: Eisenbrauns.

Levy, T.E., and A.F.C. Holl. 2002. Migrations, Ethnogenesis, and Settlement Dynamics: Israelites in Iron Age Canaan and Shuwa Arabs in the Chad Basin. *Journal of Anthropological Archaeology* 21: 83–118.

Machinist, P. 1991. The Question of Distinctiveness in Ancient Israel: An Essay. In *Ah, Assyria. Studies in Assyrian History and Ancient Near Eastern Historiography Presented to Hayim Tadmor*, ed. M. Cogan and I. Ephal, 196–212. Jerusalem: The Magnes Press.

———. 1994. Outsiders or Insiders: The Biblical View of Emergent Israel and its Context. In *The Other in Jewish Thought and History: Constructions of Jewish Culture and Identity*, ed. L.J. Silberstein and R.L. Cohn, 35–60. New York: New York University Press.

Malamat, A. 1997. The Exodus: Egyptian Analogies. In *Exodus: The Egyptian Evidence*, ed. E.S. Frerichs and L.H. Lesko, 15–26. Winona Lake, IN: Eisenbrauns.

Mazar, A. 2001. On the Relation Between Archaeological Research and the Study of the History of Israel in the Biblical Period. *Cathedra* 100: 65–88. Hebrew.

McGuire, R.H. 1982. The Study of Ethnicity in Historical Archaeology. *Journal of Anthropological Archaeology* 1: 159–178.

Mendenhall, G. 1962. The Hebrew Conquest of Palestine. *Biblical Archaeologist* 25: 66–87.

Miller, R.D. 2004. Identifying Earliest Israel. *Bulletin of the American Schools of Oriental Research* 333: 55–68.

Miller, J.M., and J.H. Hayes. 1986. *A History of Ancient Israel and Judah*. Philadelphia, PA: The Westminster Press.

Na'aman, N. 1994. The 'Conquest of Canaan' in the Book of Joshua and in History. In *From Nomadism to Monarchy*, ed. N. Na'aman and I. Finkelstein, 218–281. Jerusalem: Yad Ben-Zvi.

———. 2011. The Exodus Story: Between Historical Memory and Historiographical Composition. *Journal of Ancient Near Eastern Religions* 11: 39–69.

———. 2013. The Inscriptions from Kuntillet Ajrud. *Zmanim* 121: 4–15. Hebrew.

Rainey, A.F. 1991. Can You Name the Panel with the Israelites? Rainey's Challenge. *Biblical Archaeology Review* 17(6): 56–60. 93.

———. 2001. Israel in Merenptah Inscription and Reliefs. *Israel Exploration Journal* 51: 57–75.

———. 2003. Amarna and Later: Aspects of Social History. In *Symbiosis, symbolism and the power of the past: Canaan, Ancient Israel and their Neighbors from the Late Bronze Age through Roman Palestine*, ed. W. G. Dever and S. Gitin, 169–187. Winona Lake, IN: Eisenbrauns.

Redford, D.B. 1986. The Ashkelon Relief at Karnak and the Israel Stela. *Israel Exploration Journal* 36: 188–200.

———. 1992. *Egypt, Canaan and Israel in Ancient Times*. Princeton: Princeton University Press.

Rendsburg, G. 1992. The Date of the Exodus and the Conquest/Settlement: The Case for the 1100's. *Vetus Testamentum* XLII: 510–527.

Russell, S.C. 2009. *Images of Egypt in Early Biblical Literature: Cisjordan-Israelite, Transjordan-Ammonite, and Judahite Portrayals*. Berlin: Walter de Gruyter.

Sacket, J.R. 1977. The Meaning of Style in Archaeology: A General Model. *American Antiquity* 42: 369–380.

Shennan, J.S. 1989. Introduction: Archaeological Approaches to Cultural Identity. In *Archaeological Approaches to Cultural Identity*, ed. J.S. Shennan, 1–32. London: Unwin Hyman.

Skjeggestand, M. 1992. Ethnic Groups in Early Iron Age Palestine, Some Remarks on the Use of the Term "Israelite" in Recent Literature. *Scandinavian Journal of the Old Testament* 6: 159–186.

Stager, L.E. 1985a. Mernephtah, Israel and the Sea Peoples: New Light on an Old Relief. *Eretz-Israel* 18: 56*–64*.

———. 1985b. Response. In *Biblical Archaeology Today—1984: Proceedings of the International Congress on Biblical Archaeology, Jerusalem, April 1984*, ed. J. Amitai, 83–87. Jerusalem: Israel Exploration Society.

———. 1998. Forging an Identity: the Emergence of Ancient Israel. In *The Oxford History of the Biblical World*, ed. M.D. Coogan, 123–175. New York: Oxford University Press.

Stiebing, W.H. 1989. *Out of the Desert? Archaeology and the Exodus/Conquest Narratives*. Buffalo: Prometheus.

Thomas, N. 1991. *Entangled Objects, Exchange, Material Culture, and Colonialism in the Pacific*. Cambridge, MA: Harvard University Press.

Thompson, L. 1969. Co-operation and Conflict: The Zulu Kingdom and Natal. In *The Oxford History of South Africa*, ed. M. Wilson and L. Thompson, 334–390. Oxford: Oxford University Press.

Tubb, J.N. 1998. Response to Israel Finkelstein. In *The Origins of Early Israel—Current Debate: Biblical, Historical and Archaeological Perspectives*, ed. S. Ahiyuv and E.D. Oren. Beer-Sheva: Ben-Gurion University of the Negev.

van der Steen, E.J. 1995. Aspects of Nomadism and Settlements in Central Jordan Valley. *Palestine Exploration Quarterly* 127: 141–158.

Van Der Steen, E. 1999. Survival and Adaptation: Life East of the Jordan in the Transition from the Late Bronze Age to the Early Iron Age. *Palestine Exploration Quarterly* 131: 176–191.

Weinstein, J. 1997. Exodus and Archaeological Reality. In *Exodus: The Egyptian Evidence*, ed. E.S. Frerichs and L.H. Lesko, 87–103. Winona Lake, IN: Eisenbrauns.

Weippert, M. 1979. The Israelite Evidence from Transjordan. In *Symposium Celebrating the Seventy-Fifth Anniversary of the American Schools of Oriental Research*, ed. F.M. Cross Jr., 15–34. Cambridge, MA: American Schools of Oriental Research.

Weisman, Z. 1984. The Period of the Judges in the Biblical Historiography. In *The History of Eretz Israel: Israel and Judah in the Biblical Period*, ed. I. Ephal, 85–98. Keter Hebrew: Jerusalem.

Williamson, H.G.M. 1998. The Origins of Israel: Can We Safely Ignore the Bible? In *The Origins of Early Israel—Current Debate: Biblical, Historical and Archaeological Perspectives*, ed. S. Ahituv and E.D. Oren, 141–151. Beer-Sheva: Ben-Gurion University of the Negev.

Wimmer, A. 2013. *Ethnic Boundary Making: Institutions, Power, Networks*. Oxford: Oxford University Press.

Woodburn, J. 1982. Social Dimensions of Death in Four African Hunting and Gathering Societies. In *Death and the Regeneration of Life*, ed. M. Bloch and J. Parry, 187–210. Cambridge: Cambridge University Press.

Yurco, F.J. 1990. 3,200-Year-Old Picture of the Israelites Found in Egypt. *Biblical Archaeology Review* 16(5): 20–38.

———. 1991. Can You Name the Panel with the Israelites? Yurco's Response. *Biblical Archaeology Review* 17(6): 61.

———. 1997. Merenptah's Canaanite Campaign and Israeli's Origins. In *Exodus: The Egyptian Evidence*, ed. E.S. Frerichs and L.H. Lesko, 27–55. Winona Lake, IN: Eisenbrauns.

Living by Livestock in Israel's Exodus: Explaining Origins over Distance

38

Daniel E. Fleming

Abstract

The exodus is first of all a biblical story about how a people called Israel abandoned their residence in Egypt in spite of stubborn opposition from the pharaoh. Little in the primary biblical account lends itself to direct archaeological or historical investigation as an ancient "event." Nevertheless, the story is historically interesting for its function as one of the core biblical explanations for how the people came into existence. It assumes a background outside their current land and conflict with the dominant power of an age when Canaan belonged to Egypt. I propose that the identity of the people as mobile herdsmen or pastoralists is both integral to the original narrative and crucial as a vehicle for explaining how a population could have moved to Canaan across substantial distance. As confirmed by references to Egypt in Hosea, this story was preserved in Israel before its eventual arrival in the separate kingdom of Judah, and it is Israel that maintained a memory that the people lived as mobile pastoralists before settling in their own land—whether based in Syria with Jacob or in Egypt with Moses.

This article shares the notion of other biblical scholars that evaluation of the "exodus" must proceed from the Bible's story. My argument that pastoralism is central to the core exodus narrative stands with older analyses and in tension with current European trends, even as I find European interpretation to pursue important and sometimes compelling approaches to how the exodus story developed. The perspectives of Römer, Schmid, and Berner are always in view, along with the proposal here by Russell. Among the archaeologists, my approach addresses the group that has worked on Israelite origins, especially Dever, Finkelstein, and Faust. Pastoralism may have played some role in this process, but my interest is rather in how the idea of background as mobile herdsmen was embedded in the origins traditions of Israel (not Judah) and served its formulation of identity.

For all the pleasure it is to be invited to take part in such discussion of an iconic biblical event

D.E. Fleming (✉)
New York University, 58 Quintard Avenue,
Norwalk, CT 6854, USA
e-mail: daniel.fleming@nyu.edu

T.E. Levy et al. (eds.), *Israel's Exodus in Transdisciplinary Perspective*,
Quantitative Methods in the Humanities and Social Sciences, DOI 10.1007/978-3-319-04768-3_38,

with such stimulating company, I have approached the meeting with an undercurrent of unease. Even the definition of the stimulating company, which is marvelously inclusive, revolves around treatment of Israel's exodus from Egypt as an event in history, to test against the context of Egypt's New Kingdom and the Levant's transition from the Late Bronze to early Iron Ages. This testing may be cast in terms of the story's historicity, a word and a procedure that I find intrinsically problematic for a historical study. Alternatively, it may be treated as a literary product with a long reception history in religious communities across centuries, after an interval of composition and shape-shifting in earlier times. Here, the history in question may become detached from the ideas at stake in the biblical account, that Israel could not be explained as permanent inhabitants of its land, whether under their own name or by identification with Canaan, and that their origins involved contacts with distant places, whether Syria to the north or Egypt to the west.

With my contribution to this discussion, I set aside the historicity of the exodus as an investigative cul-de-sac, because I am convinced that available evidence does not allow productive answers, even where much may be learned about Egypt, Canaan, and Israel (cf. Redford 1992). Instead, I have revisited the Bible's primary account of Israel's escape from Egypt with an eye toward isolating it from its narrative surroundings and with interest in what it says about the people who treasured it and made it central to defining their identity. Having occupied myself recently with separating biblical material transmitted among Israelites and Judahites (Fleming 2012), the peoples of two distinct kingdoms, I propose to take seriously the apparently Israelite origin of the exodus story. For historical study, this choice alone is extremely important and any further conclusions take shape with reference to it. Within what I perceive as an Israelite context, I focus on a feature of the exodus story that is more prominent in the ancestor lore of Genesis: the notion that these people lived by their livestock as economic pastoralists with capacity for long-range movement. This picture contrasts self-consciously with the agricultural and sedentary existence that typified life in monarchic Israel and Judah, as well as later. How would such ideas have reflected populations and politics long after such a past was imagined to have played out? Investigation of such a question is historically accessible and interesting, wherever it may lead.

My location of the exodus from Egypt among Israelites rather than Judahites is based in part on repeated references in the book of Hosea to one-time residence in Egypt (Russell 2009: 53–63). The same prophetic book preserves a separate tradition that shares this heritage of movement across long distances, with Jacob and his stay in Aram (Hos 12:13; de Pury 2001; Macchi 2001; Blum 2009). This juxtaposition suggests that the two background stories could coexist in the same scribal circle without being combined into a single text or narrative.[1] While its composition and reformulation must have crossed from Israel into Judah, the book of Hosea is preoccupied with the Assyrian crisis that brought down the kingdom in 722–720 and it distances the people from the regime. Just after reference to Jacob and Aram, the writer characterizes the leader of the escape from Egypt in terms remote from monarchy: "By a prophet Yahweh brought Israel up from Egypt, and by a prophet they were tended" (12:14). Israel is explained not as the glorious achievement of rulers in Samaria, legitimized by their permanence; rather, Israel must somehow have immigrated into its land after escaping the ancient power to its southwest. The oddity of this origin at distance has provoked much discussion, and it deserves to be central to historical inquiry without the filter of historicity. What does it mean that a people with immediate experience of kingship, a primarily settled population, and an agriculturally based economy could resort to explaining their background as foreign to their own land? This question is where story and history meet, and this intersection defines my own interest in the exodus.

[1] I take the form of this juxtaposition in Hos 12:13–14 as separate references to narratives that show no hint of combination into one text or origins story.

The Exodus Story in the Book of Exodus

The reference in Hosea to Yahweh bringing Israel up from Egypt by a prophet cites what I will call the exodus story itself, without reference to the larger collection that has come to be associated with the account of escape. There is no mention of a confrontation at the Reed Sea or an encounter with Yahweh at a divine mountain, named Sinai or otherwise. There is thought of neither disobedience and enforced wandering nor ultimate arrival in Canaan and its conquest. At most, the reference to Israel's "tending" or guarding (*šmr*) by a prophet may envision some period of oversight as a flock in movement, like those of Jacob in the preceding verse, though the point is ambiguous, especially given the book's own function.

In the book of Exodus, it is likewise worth distinguishing the exodus proper from the other episodes now joined to it. Here also, if the combined prose and poetry of the Reed Sea drama derive from the Song of the Sea, this poem celebrates Yahweh's victory over Egypt in terms that do not clearly link the occasion with an escape by Israel or even with location in Egypt's own neighborhood (Smith 1997: 205–26; Russell 2007).[2] The whole tradition of meeting Yahweh at his holy mountain may not of itself require a starting-point in Egypt, though if it does make such an assumption, the exodus account does not for its part look to such a destination. Yahweh demands that his people be allowed to go celebrate a festival for him in the wilderness (Exod 5:3; etc.). No further agenda is provided, save perhaps the reality understood by all involved that Israel will not be back. The biblical idea thus stands on its own that Israel once lived in Egypt and had to escape. For purposes of consideration as Israelite tradition, it is best not to conflate it with other celebrated features of the extended narrative.

In its finished form, the narrative that brings Israel from Egypt to Mount Sinai in the book of Exodus is layered with generations of scribal adjustments, so that earlier forms of individual blocks and combinations are lost to us, surviving only in partial state. The essential exodus story in this sequence is concentrated in the account of Moses' serial debate with Egypt's Pharaoh, as found mainly in Chaps. 7–11. Chapter 5 elaborates Israel's servitude as makers of mud bricks, where both slavery in general and this particular task are entirely absent from the plague cycle that follows.[3] Chapter 12 deviates from treatment of the final plague for detailed instruction regarding the feast of Passover, inspired by

[2] One important line of conversation that developed out of this conference involved the relationship between the Song of the Sea in Exodus 15 and the preceding narrative of Chap. 14. All three European biblical scholars (Römer, Schmid, and Berner) affirmed in private discussion that they consider the Song to be composed under inspiration of the prose and thus to be a later text (see Gertz 2000 and Berner 2010). Ever since the work of W. F. Albright, American scholars have generally regarded the Song as one of the earliest pieces of biblical writing (Cross 1973; Smith 1997), and in matters relating to the exodus, this means a significant divergence that will not easily be reconciled. For me, one reason to doubt the dependence of the Song on Chap. 14 is the degree to which it does not depend on the geographical and narrative reference points from the account of escape and crossing. Nothing in the poem envisions Israel or the people of Yahweh escaping from Egypt, and they do not pass through the waters of the Yam Suph (15:4). We are only told that pharaoh's chariots and troops were drowned at this site in an event that is given no context and that has no immediate audience. Egypt's defeat permits Yahweh to lead his people in view of the surrounding peoples of Philistia, Edom, Moab, and Canaan to his own sacred place (vv 14–17). Nothing suggests that they begin outside the land.

[3] This conclusion represents a major surprise for me in my preparation for the conference, and for me individually (if not for American scholars more generally), it offers a significant point of alignment with my European colleagues at the event. Christoph Berner (2010) in particular has argued that Exodus 5 is a later interpretation of the exodus narrative, part of what he and others would call "post-priestly" revision. While I would be inclined to assign less material than some of my European colleagues to such expansion after combination of priestly and non-priestly writing in Exodus and elsewhere, in this case, the proposal is both plausible and important. If the theme of slavery in Egypt, with the particular assignment of brick-making as part of large-scale construction projects, has its narrative home only in Exod 1:11 and Chap. 5, then the original idea of escape from Egypt did not depend on the notion of servitude, or at least not on service as construction workers. The theme of return from Egypt in the book of Hosea does not depend on bondage.

the notion of special protection from death. So far as exodus itself is escape from Egypt, it takes its form from Pharaoh's dialogue with Moses as the mouthpiece of Yahweh.

The one other story element that is occupied only with escape from Egypt is Moses' flight to Midian in Chap. 2. Conceptually, this tale of Moses as murderer on the lam imitates Israel's larger flight, which must be primary. It is possible that the anticipation of the exodus in Moses' escape could have been composed with reference to Israel's departure in a form not available to us, unaware of the plague narrative in the book of Exodus. Moses' family connections vanish once he is launched on his confrontation with Pharaoh, and nothing about the planned festival in the wilderness looks to Midian and his old home. The sojourn as a shepherd among Midianites then becomes a vehicle for reorganizing the exodus narrative around a mountain destination: Moses sees the burning bush while traveling with his flocks at the mountain of God (3:1), where his father-in-law will join him after escape from Egypt (Chap. 18). In this separate thread of a combined exodus-mountain narrative, the two episodes are linked by the name of Jethro as priest of Midian.[4]

Taken on their own, the two accounts of flight from Egypt have the same destination: the wilderness where shepherds move their flocks. In the first case, the narrator takes for granted that Moses will take refuge among herdsmen, seeking out a well in the grazing range of Midianites, in land accessible to Canaan, somewhere east of Egypt (cf. Monroe 2012: 228–30). The exodus of the plague cycle names only "the wilderness" (*midbār*) as Israel's destination, with no further definition.[5] Such generic identification recalls the book of Hosea, once again, where in first-person speech, Yahweh names the *midbār* as the place that recalls Israel's youth (2:16), the place where he "found" Israel as an unexpected treat, "like grapes" (9:10), the place where he "made the acquaintance" (*yd'*) of the one he had brought up out of Egypt (13:4–5). Intriguingly, the plague narrative of Exodus makes no effort to treat this as a return to Canaan, the land of Israel's ancestors, a move that would require intent to connect the tale with material now found in Genesis. Further, there is no Promised Land or its equivalent, no anticipation of Israel's eventual situation in its own land.

Both versions of flight to the back country envision some kind of encounter there with Yahweh. Moses' stated objective for Israel is worship in the wilderness, "the festival of Yahweh" in Exod 10:9. It appears that Yahweh is to be found in the wilderness; Israel cannot worship in Egypt as a collective act in the way that their god requires. Pharaoh never challenges this assertion. The various traditions of a wilderness mountain in the south share a rough geography with the exodus story, and perhaps some of its religious sensibility, though neither the plague cycle nor the book of Hosea shows awareness of such a mountain. Moses' time among the Midianites brings a very different form of contact with Yahweh. This short narrative shows a notable lack of interest in separating Israelites from their neighbors and in claiming an exclusive bond between Yahweh and this people. Moses' father-in-law is "a priest of Midian" (2:16), then identified as Reuel (verse 18), and Moses marries the daughter of this religious leader without hesitation. Nothing is made of Zipporah as daughter of a wilderness priest in this episode, which leaves Moses among the Midianites, needing some mechanism to restore him to his people in time for their own escape. Such is provided in Chap. 3 by the bush on fire at the mountain, which has displaced whatever may have brought Moses back to Egypt in the Midian narrative. A remnant of a tale that brought Moses back to Egypt without the mountain interlude may survive in the famously difficult reference to the "bridegroom of blood" in Exod 4:24–26 (cf. Propp 1993; Embry 2010; Luciani 2012). En route without stated destination, Moses is assaulted by Yahweh with deadly intent as if

[4] For 3:1–4:18 and Chap. 18 in one strand see Carr 2011: 118.

[5] The requirement of three days' journey comes up in the context of Pharaoh's offer to let Israel have their festival inside the land of Egypt (Exod 8:21–23) and is picked up in 5:3 as part of the plan from the beginning.

the two had never met, which in fact they have not, if we set aside the affair of the bush. Moses is saved by Zipporah, whose role offers a bridge to the account of their marriage both by her name and by her familiarity with Yahweh's needs, as daughter of Reuel. The one son named in Chap. 2 provides the object for the requisite circumcision. Somehow this rite identifies Moses with Yahweh, when he had up to that time maintained no commitment, somewhat like Jacob before his dream at Bethel.

Along with their dual destinations in the wilderness, the two escapes of Moses and Israel from Egypt both involve close identification with the people who inhabit this domain: those who live by their livestock and move across distance in order to provide for them. On meeting the daughters of Reuel, Moses presents as "an Egyptian man" (verse 19), though the background to his flight proves his real identification with the "Hebrews" (verses 11, 13).[6] He only becomes a shepherd by marriage, becoming an honorary Midianite. Behind this development, however, stands the question why Moses headed for Midianite country in the first place. The text offers this as simple fact, the basis for establishing a new life in the wilderness, yet the move into Yahweh's home territory is inevitable and retraces lines that already define the people of Midian.

In the account of Israel's own exodus, the people are separated from Egypt not as slaves but as shepherds. Setting aside Chap. 5 as an effort to highlight the urgency of Israel's suffering and the drama of Yahweh's extended dispute with Pharaoh, the people spend the period of plagues in silence, sheltered from the worst of the disasters by special dispensation.[7] After Pharaoh shakes off the initial plagues with their removal, the arrival of new calamities yields more serious negotiation with potential compromises. First, he offers a festival to Yahweh in Egypt itself (8:21), a position then modified to allow an event to take place close by (8:24). Then in a further round of talks, negotiation turns to the matter of who may take part. When Moses insists that the whole people will head into the wilderness, young and old, male and female, along with all their livestock, Pharaoh attempts to limit participation to men (10:8–11). Moses repeats the demand for full participation after the plague of darkness, emphasizing that every last animal will join them so as to guarantee selection of the best for sacrifice (10:26). Combined with the wilderness festival itself, Pharaoh's compromises underscore Israel's way of life as herdsmen, people whose livelihood depends essentially on their stock. Pharaoh is shown to understand what it means to let Israel leave Egypt with all their families and flocks: they have nothing left to hold them. Israel's mobility is the logical extension of their identity as a herding people.[8] However the Joseph narrative relates to this exodus tradition, it makes this identity explicit in what Jacob is instructed to tell Egypt's king when given an audience: that the family prefers to live in the land of Goshen because they are herdsmen (Gen 46:32–34; 47:1–6).

In its finished form, the exodus narrative combines materials from two distinct plague

[6] It may be impossible to distinguish which story is intrinsically older, though Moses' flight must have been composed to anticipate that of Israel as a whole. In its Exodus setting, Moses' escape caps a series of intertwined events that begin at least with his birth if not before. Moses' birth and adoption by Pharaoh's daughter (2:1–10) supply the necessary backdrop to his crime and flight. Whatever the process of the text's formation, the Moses cluster in Chap. 2 probably reflects a rendition of the exodus tradition that was once quite distinct from the plague cycle, and likewise the odd circumcision/Zipporah unit in 4:24–26.

[7] The silence of the Israelites holds true in both the priestly and the non-priestly components of Exodus 7–11, though the theme of special protection for Israel appears only in the latter. For this, see 9:4–7 (livestock); 9:26 (no hail in Goshen); 10:23 (darkness, versus light for the Israelites); 11:7 (death of firstborn).

[8] This analysis is bound up with the older conclusion that the priestly elements of the text are later than the other content and that this other content as a whole constitutes a coherent narrative with a smaller number of plagues. Berner (2010) argues that the non-priestly material in Chaps. 10 and 11 represents a post-priestly revision, so that the herding motif is secondary (or tertiary) to the exodus account, apparently inspired by this way of life in Genesis.

collections, one of which has apparent affinities with priestly writing and the other of which does not and may at least be designated "non-P," leaving aside the possibility of tracing more specific sources.[9] The driving pattern of negotiation between Moses and Pharaoh, along with the story's ultimate destination in the wilderness, follows the lines of this non-P material, which I still regard as the older element of the finished text. Hosea's treatment of the exodus confirms the impression that this tradition predates the priestly narrative by its identification of a prophet leader. Moses serves as Yahweh's mouthpiece, confronting Pharaoh when he leaves the palace (e.g., 8:16) in a way that recalls Isaiah meeting king Ahaz by the road when he has left his private space (Isa 7:3). Even his threats to bring down Yahweh's judgment on Egypt by physical calamity are the stuff of the prophetic repertoire. However widely such stories of escape from Egypt may have circulated, the references in Hosea and perhaps the centrality of Joseph to an affiliated Egypt tradition indicate a location in the central highlands of Israel.[10] These would have found their way into Judah, the eventual setting for all biblical content, but the exodus story appears to find its earliest situation in the framework of the larger kingdom to its north.

Living by Livestock and Explaining Origins

The biblical tradition that the people once lived in Egypt and escaped into the wilderness by divine intervention belongs to a cluster of stories that explain Israel's origins as foreigners to the land they finally inhabited. Both the Egypt and the Jacob lore understand Israel to have occupied their land only after a history of long-range migrations that took their forebears in and out of the southern Levant. Even if the Jacob material begins and ends in what will become the land of Israel, his flight from Esau into Syria assumes longstanding family ties with the herdsman Laban. After he succeeds in building a family of his own and substantial wealth, Jacob is able to move back to his original homeland with this wealth intact because it is mobile, embodied in his flocks. With their interest in northern and eastern sites, outside the territory of Judah, and the genealogical framework for interpreting tribal affiliation, along with the references in Hosea 12, the Jacob narratives are likewise rooted in Israel and its separate kingdom.

The exodus from Egypt equally builds its account of Israel's links to distant lands according to a framework that assumes herding as necessary to support this kind of movement, though it does so without reference to Jacob or genealogical ancestors for Israel or its peoples. Neither the flight of Moses to Midian nor the escape of Israel after the plagues is recounted in terms that evoke Jacob or the Genesis lore. Joseph and his descendants play no special role.[11] It seems rather that the herding of flocks over distance serves as a broad social matrix in Israelite tradition for explaining a background outside the land. This notion of ancestral pastoralism does not reflect the economy and social structures of monarchic Israel, and the isolation of the type to ancient times of origins suggests a conscious separation between then and now. Further, while we lack texts in direct service of royal ideology, the Jacob and Egypt images of distant contacts suggest nothing that serves kingship in itself.[12] These stories belong to a different social

[9] Both Carr (1996, 2011) and Gertz (2000) use this terminology, in spite of the contrasts between their analyses.

[10] Joseph is linked particularly to Bethel in Judg 1:22–26, and the central units of Amos 5–6 include two rare references to Joseph's fate (5:15 and 6:6).

[11] This constitutes my immediate objection to the notion that the herding motif belongs to a post-priestly revision of the exodus story that derived the identification of Israel as shepherds from the Genesis texts (against Berner, above).

[12] Consider the Mesha inscription as a comparison for royal claims in a context that includes tribal populations with likely backgrounds in inland pastoralism. Although there are hints of tribal groups, nothing invokes origin outside the land, and the king makes a case for his legitimacy through the standard achievements of military victory and construction projects, all under the blessing of a single god Kemosh (see Routledge 2004).

constituency, not necessarily pre-monarchic or post-monarchic but rather coexisting with kingship and not defined by it.[13]

At this point, I have allowed historical concerns to creep into my discussion of herding as an explanation for background outside the land. Where preoccupation with historicity and the question whether some portion of Israel really escaped from Egypt can lead to investigative dead ends, it may be useful to open our historical inquiry to matters of Israelite self-perception and the stories that informed it. If the idea of herding across long distances did not derive from the circumstances of first-millennium Israel, then where did it come from, and whom did it serve? It is possible that the model of pastoralism over distance was fleshed out with knowledge of contemporary practice among other groups. If so, the connections would have pointed inland, as indeed do both the Arameans with Jacob and the Midianites with Moses. Neither of these explanations is entirely satisfying, though they warrant further exploration. By the ninth century, Aram would have been encountered first of all through the kingdom based at Damascus, no more a setting for long-range pastoralism than was Israel. The wider Syrian landscape could have offered examples of long-distance mobility with flocks, though we know too little of the details. Other biblical tradition preserves a multiform memory of Midian as a longstanding enemy of Israel once the latter was established in the land, though this people could still have provided a picture of herding as a way of life. It is certain that if we go back in time to the second millennium and beyond, such patterns were essential to the fabric of Near Eastern society on a large scale, as discussed most recently by Anne Porter (2012).

In general, even the older stories in Genesis and Exodus only pick up bits of archaic data that are now embedded in later matrices.[14] Syria is viewed through the lens of Aram and the Arameans, categories that reflect people who only gave their name to the region in the first millennium. While I have proposed elsewhere details that appear to reach back to a second-millennium landscape, Genesis shows no awareness of the actual political powers and social structures of the Bronze Age (Fleming 1998 and 2004). Likewise, while some may argue that tidbits of Egyptian geography reflect New Kingdom realities, the book of Exodus includes no hint that Egypt ruled Canaan, arguably the fact most relevant to a story about how Israel began.

If the notion of herding over distance represents some such survival of earlier social strategies from Israel's background, it has likewise been forgotten as much as remembered. Israel in the exodus story is settled in Egypt as much as the Egyptians, for all that they live by their livestock and can thus bring their property with them. There is no suggestion that Israelite herdsmen have moved in and out of Egypt freely, as would be the habit of the type, and Pharaoh's negotiation over celebration of a festival in the wilderness treats this as something exceptional. It would seem that for the exodus writer, all pastoralism was local. Only certain elements of the deeper narrative structure follow the logic of long-range movement. Both Moses and Israel head into a wilderness that is perfectly suited for the pastoralism that has been their people's way of life, and they go specifically into a country that connects the Egyptian circle with herding bases in Canaan and the southern Levant. By the logic of long-range pastoralism, the exodus from Egypt amounts to a change of operating base, whereby a shepherd people abandons a situation that has become hostile in search of alternatives, following the lines of regular herding movement.[15]

[13] This is one logical implication of Macchi's hypothesis (1999) that the core tribal sayings of Jacob in Gen 49:13–21 derive from the Israelite kingdom of the ninth century (cf. Fleming 2012: 86–90).

[14] During this conference, much discussion of any possible memory of detail specific to the New Kingdom revolved around the Egyptian sites called Pithom and Ramses in Exod 1:11 and fragments of itineraries for the eastern Nile delta in 13:20 and 14:2.

[15] Something like this must explain the survival of Zimri-Lim after the expulsion of his family from Mari by Samsi-

What then is the origin of these traditions about Egypt and Aram? Because they do not support the right of kings and do not appear of post-monarchic date, I am drawn toward Israel's political tradition of balance between what Richard Blanton and Gary Feinman would call "exclusionary" and "corporate" strategies. In the work of Blanton and Feinman, all ancient political systems involved negotiation between forces that drew power toward a center and those that benefited from coordinated decision-making, with varying results.[16] Israel's "corporate" or coordinated aspect could take diverse forms in biblical portrayal, whether in collective action as Israel or as acknowledged in separate constituent peoples or "tribes." The traditions of deep connections with distant lands, made possible by migration with livestock, may somehow be attached to social circles in which such migration could be celebrated.[17] Whatever the answer, this line of inquiry can be tested against a range of existing evidence, textual and archaeological, and it offers something more than a methodological dead end. The exodus from Egypt was a strange way of defining an identity and a backstory, and it would be well worth figuring out whom it served.

Addu of upper Mesopotamia and Ekallatum. After his return to Mari and accession to the throne, we encounter references to the king of Aleppo taking credit for his success. On the general historical situation for this period in the Mari archives, see Durand (1997, 1998, 2000); Charpin and Ziegler (2003). Durand has addressed Zimri-Lim's Sim'alite tribe in detail (2004) and elsewhere (1997) treats ritual aspects of tribal mobility. Porter (2009) proposes a framework for multiple settled centers connected to individual tribal peoples. For the larger political landscape of the Mari archives, see Fleming 2004.

[16] See especially Blanton et al. (1996), Feinman 2000.

[17] My focus throughout this contribution has been on the herding interest of tales about Israel's life outside and before its current existence in the land. It is not my intent to demonstrate any specific relationship of these stories to actual Israelite origins. Among those present at this conference, William Dever has represented an analysis that emphasizes the realignment of settled Canaanite populations, while Avraham Faust and Thomas Levy are pursuing a key role for pastoralists. At one remarkable moment, Dever, Faust, and Israel Finkelstein acknowledged that in spite of their different emphases, their interpretations are not as far apart as might seem to be the case. While I would be happy to see this biblical tradition receive consideration as part of the aftermath of Israel's emergence, I do not perceive it to serve any single approach to the historical problem.

References

Berner, Christoph. 2010. *Die Exoduserzählung: das literarische Werden einer Ursprungslegende Israels*, Forschung zum Alten Testament 73. Tübingen: Mohr Siebeck.

Blanton, Richard E., Gary M. Feinman, Stephen A. Kowalewski, and Peter S. Perigrine. 1996. A Dual-Processual Theory for the Evolution of Mesoamerican Civilization. *Current Anthropology* 37: 1–14.

Blum, Erhard. 2009. Hosea 12 und die Pentateuchüberlieferungen. In *Der Erzväter in der biblischen Tradition: Festschrift für Matthias Köchert*, ed. Anselm C. Hagedorn and Pfeiffer Henrik, 291–321. Berlin: de Gruyter.

Carr, David M. 1996. *Reading the Fractures of Genesis: Historical and Literary Approaches*. Louisville, KY: Westminster John Knox.

———. 2011. *The Formation of the Hebrew Bible: A New Reconstruction*. Oxford: Oxford University Press.

Charpin, Dominique, and Nele Ziegler. 2003. *Mari et le Proche-Orient à l'époque amorrite: Essai d'histoire politique*, Florilegium Marianum [FM] V. Paris: Société pour l'Étude du Proche-Orient Ancien.

Cross, Frank M. 1973. *Canaanite Myth and Hebrew Epic*. Cambridge, MA: Harvard University Press.

de Pury, Albert. 2001. Situer le cycle de Jacob: Quelques réflexions vingt-cinq ans plus tard. In *Studies in the book of Genesis*, ed. André Wénin, 213–241. Peeters: Leuven.

Durand, Jean-Marie. 1997. *Documents épistolaires du palais de Mari, Tomes I*, Littératures du Proche-Orient Ancien 16, 17, and 18. Paris: du Cerf.

———. 1998. *Documents épistolaires du palais de Mari, Tomes II*, Littératures du Proche-Orient Ancien 16, 17, and 18. Paris: du Cerf.

———. 2000. *Documents épistolaires du palais de Mari, Tomes III*, Littératures du Proche-Orient Ancien 16, 17, and 18. Paris: du Cerf.

———. 2004. Peuplement et société à l'époque amorrite (I): Les clans Bensim'alites. In *Amurru 3: Nomades et sédentaires dans le Proche-Orient ancien*, Comptes Rendus de la Rencontre Assyriologique Internationale 46, ed. Nicolle Christophe, 111–197. Paris: Éditions Recherche sur les Civilisations.

Embry, Brad. 2010. The Endangerment of Moses: Towards a New Reading of Exodus 4:24–26. *Vetus Testamentum [VT]* 60: 177–196.

Feinman, Gary M. 2000. Dual-Processual Theory and Social Formations in the Southwest. In *Alternative Leadership Strategies in the Prehispanic Southwest*, ed. Barbara J. Mills, 207–224. Tuscon: University of Arizona Press.

Fleming, Daniel E. 1998. Mari and the Possibilities of Biblical Memory. *Revue d'Assyriologie et d'Archéologie Orientale [RA]* 92: 41–78.

———. 2004. Genesis in History and Tradition: The Syrian Background of Israel's Ancestors, Reprise. In *The Future of Biblical Archaeology: Reassessing Methodologies and Assumptions*, ed. James K. Hoffmeier and Alan Millard, 193–232. Grand Rapids, MI: Eerdmans.

———. 2012. *The Legacy of Israel in Judah's Bible: History, Politics, and the Reinscribing of Tradition*. Cambridge: Cambridge University Press.

Gertz, J.C. 2000. *Tradition und Redaktion in der Exoduserzählung: Untersuchungen zur Endredaktion des Pentateuch*, Forschungen zur Religion und Literatur des Alten und Neuen Testaments 186. Göttingen: Vandenhoeck & Ruprecht.

Luciani, Didier. 2012. Didier Luciani, Çippora (Ex 4,24–26): le petit oiseau va sortir... et il échappe encore aux interprètes. *Estudios Bíblicos* 70: 159–184.

Macchi, Jean-Daniel. 1999. *Israël et ses tribus selon Genèse 49*, Orbis Biblicus et Orientalis 171. Göttingen: Vandenhoeck & Ruprecht.

———. 2001. Genèse 31,24-42. La dernière rencontre de Jacob et de Laban. In *Jacob: Commentaire à plusieurs voix de Gen 25–36*, ed. Macchi Jean-Daniel and Römer Thomas, 144–162. Geneva: Labor et Fides.

Monroe, Lauren A.S. 2012. Phinehas' Zeal and the Death of Cozbi: Unearthing a Human Scapegoat Tradition in Numbers 25:1–18. *Vetus Testamentum [VT]* 62: 211–231.

Porter, Anne. 2009. Beyond Dimorphism: Ideologies and Materialities of Kingship in Time-Space Distanciation. In *Nomads, Tribes, and the State in the Ancient Near East: Cross-Disciplinary Perspectives*, ed. Jeffrey Szuchman, 201–225. Chicago: Oriental Institute.

———. 2012. *Mobile Pastoralism and the Formation of Near Eastern Civilizations*. Cambridge: Cambridge University Press.

Propp, William H.C. 1993. That Bloody Bridegroom (Exodus 4:24–6). *Vetus Testamentum [VT]* 43: 495–518.

Redford, Donald B. 1992. *Egypt, Canaan, and Israel in Ancient Times*. Princeton, N.J.: Princeton University Press.

Routledge, Bruce. 2004. *Moab in the Iron Age: Hegemony, Polity, Archaeology*. Philadelphia: University of Pennsylvania Press.

Russell, Brian D. 2007. *The Song of the Sea: The Date of Composition and Influence of Exodus 15:1–21*. New York: Peter Lang.

Russell, Stephen C. 2009. *Images of Egypt in Early Biblical Literature: Cisjordan-Israelite, Transjordan-Israelite, and Judahite Portrayals*. Berlin: de Gruyter.

Smith, Mark S. 1997. *The Pilgrimage Pattern in Exodus*. Sheffield: Sheffield Academic.

Exodus and Exodus Traditions After the Linguistic Turn in History

39

Garrett Galvin

Abstract

Since German higher scholarship started critically examining the texts of the Book of Exodus, there have been many who questioned the story as has been traditionally understood by various faith communities. This chapter seeks to examine the historicity of the Exodus event. I will start with the evidence of an Israelite presence in Egypt. Then, I will proceed to examine the possibility that the Exodus is solely a literary device consciously created from whole cloth. Much archaeological work concerns the Late Bronze Age, but there is no clear evidence of the Exodus event from a purely archaeological perspective. This has led certain scholars to reject modern Biblical scholarship, others to reach back to the Hyksos period, and still others to look at the Persian Period. Mainstream biblical scholarship acknowledges the archaeological limitation, but also takes the view of the prophets such as Hosea seriously. The linguistic turn in history has caused historians to view their sources more cautiously and led others in the direction of cultural memory. Some will question whether a historical kernel exists because of these sources, but scholars can only approach the question of a historical kernel from a biblical perspective.

This chapter focuses on the thirteenth century BCE and later biblical periods, as the Hyksos period was more than adequately covered by Manfred Bietak. It shares a similar concern with Jan Assmann: Exodus concerns not what really happened, but who remembers what happened. The chapter is very much informed by Ronald Hendel's contention that one cannot date folklore or cultural memory to one date, in contrast with Donald Redford's interest in a specific seventh century BCE Exodus context in Egyptian Saite geography. Bernard Batto's emphasis on history being in sync with myth helps guide this chapter as well as Thomas Römer's understanding of Exodus as *Urgeschichte*. Israel Finkelstein's discussion of a Northern Kingdom charter myth also informs this chapter.

G. Galvin (✉)
Franciscan School of Theology, 4050 Mission Avenue,
Oceanside, CA 92057, USA
e-mail: garrettmgalvin@gmail.com

T.E. Levy et al. (eds.), *Israel's Exodus in Transdisciplinary Perspective*,
Quantitative Methods in the Humanities and Social Sciences, DOI 10.1007/978-3-319-04768-3_39,

Aren Maeir contends that Middle Bronze Age people migrated without their material culture, like proto-Israelites. I would argue that like the proto-Israelites, they brought their material culture with them in the form of the Syrian style embankment city construction, especially at Hazor, Dan, Yavneh-Yam, Acco, Kabri, and Tel Haror. Secondly, they brought a fast pottery wheel, not known in the Early Bronze Age. I would seek to share René Bloch's understanding of the Exodus event as open to many interpretations as in the Book of Wisdom vis-à-vis the Book of Exodus.

Introduction

The historicity of an exodus event has been under debate for well over a century now. Since German higher scholarship started critically examining the texts of the Book of Exodus, there have been many who questioned the story as has been traditionally understood by various faith communities. This chapter seeks to examine the historicity of an exodus event. I will start with three examples of putative Israelite presence in Egypt. I will skip the fifteenth century BCE as it has been thoroughly dealt with in this UCSD conference. I will then proceed to an examination of the possibilities of an exodus in the thirteenth century and to the possibility that an exodus narrative is solely a literary device consciously created from whole cloth.

Hayden White argues "historical narratives are verbal fictions, the contents of which are as much *invented* as *found* and the forms of which have more in common with their counterparts in literature than they have with those in the sciences" (Iggers 1997: 119). Language theory argues that the text has no reference to an external reality, but is contained within itself. I am interested in looking at the impact of this linguistic turn on our understanding of the Book of Exodus. Many current histories of Israel and Judah reflect the increasing awareness of the limits of objectivity. On the other hand, the Bible does refer to people and events found in the archaeological record. This forces the historian to look more cautiously at biblical literature without completely discounting it.

With all due respect to the present scholars, I do not see White's ideas as a nihilistic venture, but as an important exercise in understanding what is invented and what is found. Being a Franciscan, I am a great devotee of the fourteenth century Franciscan William of Occam and his razor. I see great need for its use based on the field of Irish history. In 1988, the Oxford historian R. F. Foster wrote the book *Modern Ireland*. It fundamentally changed the controversial field of Irish history by reexamining what was invented and found. It moved away from Irish and British narratives. Many were saying at the end of the century that the former Yugoslavia needed an R. F. Foster as they fought over Serbian and Croatian narratives.

Israelite Presence in Egypt

Anastasi Papyri

While many different scholars highlight various possibilities for an Israelite presence in Egypt, I will focus on four possibilities. Abraham Malamat emphasizes the Anastasi Papyri. The exact meaning of these papyri remains open to interpretation, but they are from the Eastern Delta and the thirteenth cent. BCE. The various Anastasi Papyri are generally related to border crossings between the Eastern Delta and

Palestine around the time of Merenptah. For the purposes of the historicity of Exodus, Papyrus Anastasi V is the most interesting document. Malamat tells us that "it reports the escape of two slaves or servants from the royal residence at Pi-Rameses, on the western edge of Wadi Tumilat" (Malamat 1997: 21). Malamat claims four parallel features exist between Anastasi V and Exodus: (1) the escape of slaves from the city of Rameses, (2) the pursuit of Egyptian military forces, (3) the escape route in the Sinai is "roughly identical" with the biblical report, and (4) the flight took place at night (Malamat 1997: 21–22). Malamat describes this papyrus as the most "exciting" of our evidence (1997: 21).

The difficulty with this material appears obvious at some levels and more complicated at other levels. This "flight" only involves two runaway slaves (Malamat 1997: 21). We have no description of their ethnic origin. At this time, there were many slaves in Egypt from Nubia. We know that in later epochs, Israelites worked at the southern border of Egypt. The question of geography is a complicated one. The town of Pi-Rameses is well-known in the New Kingdom, but this is not what is described in Exodus 1:11. Redford tells us "If the Exodus account had intended the famous Ramesside foundation, I am at a loss to explain why the form did not emerge as פי־רעמסס" (Redford 1987: 139). Accordingly, while Malamat easily switched between Pi-Rameses and Rameses in his discussion, other scholars feel this cannot be done. I think we see an invented narrative here. Redford proposes that the town in Exodus actually references a Late Period town more commonly known as Bubastis (ibid.). My main problem with this document involves only the first of the four parallel features described above. Although Malamat describes an escape from Rameses, the document could only be referring to Pi-Rameses. The number of slaves involved is only two and we do not know their ethnic identity. This seems almost irrelevant when we compare it to the idea of Exodus as a foundational event in the Old Testament. We see a recurring motif here. Biblicists like Malamat are sanguine about the Egyptological evidence, but they are often corrected by Egyptologists.

Karnak Reliefs

Although a number of scholars have focused on the importance of the Karnak Reliefs as a source, I will use a more recent article by Frank Yurco, an Egyptologist who held similar views. Yurco maintained that the fourth battle relief of Merenptah "does retain clear evidence of battle in a hilly country, against a foe depicted as Canaanites that had no fortified city" (Yurco 1997: 29). He reasoned that these Canaanites have to be Israelites because of the Merenptah Stele. He states: "This combination, three battle scenes against city-states, one named Ashkelon, and a battle in open, hilly country, matches precisely the Merenptah 'Israel' Stela's retrospective account, and it is a reasonable analysis that the stela and the Karnak battle reliefs record one and the same military campaign" (ibid.). The actual relief never identifies them as Israelites. There were a number of other peoples mentioned on the stele. I think it is very debatable how reasonable his analysis is.

The other important element of this is the nature of Egyptian reliefs. Many would question whether these are straightforward analyses of history. There are highly symbolic aspects to the reliefs. Yurco's discussion of the reliefs does not bring any of these matters into play. I would also argue for the religious significance of Karnak. This is not a library, but a highly ideological center. I wonder if this is not like trying to reconstruct history from stained glass. In similar discussion of the reliefs concerning the Sea Peoples at Medinet Habu, there is a lively academic debate about their veracity. All parties seem to acknowledge the importance of the highly symbolic nature of these reliefs. The period of the Sea Peoples is only about 50 years after our material, so it would seem to have bearing on this material. Yurco's arguments have not won widespread acceptance in the field.

The 400-Year Stele

This stele seems to have been moved from Pi-Rameses to Tanis during Dynasty XXI (1075–948 BCE) (Halpern 1992: 97). It appears to have been written in the time of Ramses II and it "records the inauguration of the cult of the Hyksos god Seth 400 years earlier" (ibid.). Seth is a very old Egyptian god that came to be equated with some Semitic gods in the New Kingdom. Thus, Ramses tries to equate him with the Hyksos. Many compare him with an upstart god like Baal as opposed to the older El-like gods in the Canaanite pantheon. Professor Halpern posits that:

> . . . if Ramesses II's 400-year stele has inspired the biblical tradition of a 400-year sojourn in Egypt, a conviction that Ramesses II was the pharaoh of the oppression has actually shaped the development of Israel's traditions: The Semites subjected to forced labor in the Delta identified themselves with the traditions of their Hyksos predecessors, against the Egyptian pharaoh; thus Israel came to identify itself with the Hyksos. (Halpern 1992: 101)

Professor Halpern is very careful to put "if" here, but I think the historical narrative often gets the reification rather than the qualification. All we truly know about this stele is that it was written in the time of Ramses II and has domestic concerns in mind.

An examination of three possibilities for an Israelite presence offers clear evidence that Egypt dealt with foreign adversaries. In only one case do those adversaries appear to be Israelites. This is the Merenptah Stele. Yet, this in no way recognizes an Israelite presence in Egypt. The other pieces of evidence can only be linked to Israel in a very indirect way. Nahum Sarna puts it best when he says, "there is a complete absence of any external written documents testifying to Israel's presence and subjugation in Egypt, to her migration from that land, or to her conquest of Canaan and her settlement there" (1992: 2:696). Now what would Hayden White say about our field? We see historical narratives being invented that are not there after applying Occam's Razor to the artifacts.

Thirteenth Century BCE

For the Position

The idea of a thirteenth century BCE exodus is much more popular than a fifteenth century exodus. Fifteen years ago, Carol Redmount could say that most biblical scholars still support the basic historicity of the biblical narrative (Redmount 1998: 63). More recently, Megan Bishop Moore and Brad Kelle stated: "Most historians of ancient Israel no longer consider information about the Egyptian sojourn, the exodus, and the wilderness wanderings recoverable or even relevant to Israel's emergence" (Moore and Kelle 2011: 81). I shall now survey the extent of the historicity that many biblical scholars accept. There is a wide divergence. What changed? The field has become more tuned to the linguistic turn not just in peer-reviewed articles but also in general works.

Kenneth Kitchen and Frank Yurco cite each other's material. While Yurco appeared to agree with Kitchen on the thirteenth century dating, he opted for a much smaller number of people. Kitchen never gives a definitive number, but he seems to like the idea of about 30,000 to 50,000 people (1992: 2:705). Yurco stated: "As for the number of Israelites said to have left Egypt in the Exodus, 600,000 is clearly inflated highly. If *Bene Yisrael* initially included only the descendents of Jacob, then 6,000, or even 600, would be a more reasonable figure. Such a group might easily have escaped through a swampy or marshy papyrus lake. . . ." (Yurco 1997: 49). A number of other scholars echo Yurco's call for a smaller number.

Carol Redmount is an Egyptologist with a much stronger connection to mainstream biblical scholarship. While she supports the idea of "an actual historical origin for the Exodus events" (1998: 63), she maintains that "recent research indicates that even more of the extant Exodus account than previously thought comes from periods during or after the Israelite monarchy or even the exile" (ibid.). She focuses on how different ancient notions of history are to our own. Thus, what seems anachronistic to us is perfectly legitimate to them. When contemporary scholars focus on anachronistic elements within these stories, they are coming at it with a different sensibility that assumes a reference to earlier events. This could be part of a genre rather than an actual event. In many respects, she is making the opposite point than Kitchen made. Kitchen attacked non-literal readings of Scripture as the corrupt fruit of nineteenth century German scholarship. Redmount has a conception of Scripture that takes into account its theological nature as well as different conceptions of history.

Even scholars who accept a thirteenth century BCE date for the Exodus want to put limits around the event like Redmount does. Coogan makes the point that "the Bible itself is virtually devoid of concrete detail that would enable the Exodus to be securely dated" (Coogan 1993: 209). Like Redmount, he points out the generic nature of the Exodus story. No pharaoh is ever named from Joseph to Moses. Pharaohs are named in later books of the Old Testament that have a more easily confirmed history such as Shishak in 1 Kings 11:40, So in 2 Kings 17:4, Necho in 2 Kings 23:29, Tirhakah in Isaiah 37:9 and Hophra in Jeremiah 44:30. Ultimately, Coogan argues "it is impossible to determine what may actually have occurred to Hebrews in Egypt, probably during the thirteenth century" (1993: 211). Coogan's unique understanding of Exodus is that "literary analysis of the narratives suggests that what may in fact have been several movements out of Egypt by Semitic peoples have been collapsed into one" (ibid.). He seems to be suggesting a conflation of the Hyksos, an event in the thirteenth century perhaps mimicking Chap. 15 of Exodus, and "later writers who saw in the events of their own times a kind of reenactment of the original Exodus" (ibid.). Coogan's approach argues strenuously for recognizing the limits of what can be said about the material.

Against the Position

Another important insight concerns how far evidence of the Exodus now lags behind evidence of the settlement of Israel. James Weinstein suggests a "cynic might claim that nothing new has been learned about the departure of the Children of Israel from Egypt since the last redaction of the Old Testament. While many people might consider such a statement an unduly harsh assessment of the situation, the archaeological record has revealed little to dispute its validity" (1997: 87). Weinstein then makes the very important point that "the opposite situation is of course true in regard to the Israelite settlement" (ibid.). The important evidence from this settlement is that "the hill country sites lacked a Nilotic background and that their culture developed quite independently of that of Egypt" (ibid.: 90). Weinstein claims that the archaeological record is equally dismissive of a Hyksos influx, that there is simply no "substantial influx of Egyptian material culture features in the late-16th century B.C. Palestine" (ibid.: 95).

This allows Weinstein to categorically state: "There is no archaeological evidence for an exodus such as is described in the Bible in any period within the second millennium B.C." (ibid.: 97). He does not discount that there may have been a migration, but it involved such a small number that it currently "cannot be identified in the archaeological record" (ibid.: 98). Given the strides made in understanding the settlement of Israel, it is very significant that no Egyptian material culture can be traced in the hill country. Weinstein concludes that "the Exodus story cannot be considered a topic for productive archaeological research" (ibid.).

Philistine Analogy

The Philistines offer a powerful analogy for immigrant groups. They arrived in the Levant in early Iron Age I as Israel also emerged. They derive from the Sea Peoples, and there is considerable evidence of their conflicts with Egypt. "These Sea Peoples make their initial appearance in the fourteenth century BCE. The Lukka, Sherden, and Danuna were first mentioned during the reigns of Amenophis III and Amenophis IV (Akhenaten), often in the role of mercenaries" (Killebrew and Lehman 2013: 7). The powerful difference is the evidence on the ground. For Israelites no evidence exists of Egyptian cooking pots in the central hill country. The exact opposite is the case with the Philistines. They brought their Aegean-style pots and produced similar ones using local raw material. They also brought their own spool-shaped loom weights. We see a gradual spread of Aegean-inspired pottery that "provides evidence for the progressive expansion of the Philistines' influence throughout the country" (Finkelstein and Silberman 2001: 134). It would seem that we should expect similar evidence of Egyptian cooking pots in the central hill country.

The important counterargument to this analogy is that the Israelites never submitted to the use of Egyptian pottery while in Egypt, in order to maintain group identity and strengthen solidarity (Barako 2013: 43). Although evidence exists of this phenomenon in the archaeology of Syria-Palestine, these cultures were standing side-by-side rather than completely subjugated. We do not seem to have any evidence for Egyptian cooking pots or weaving techniques. They would never abandon these basic elements—they are absolutely essential to survival of an immigrant people.

Van Seters

Another important critic of the historicity of Exodus is John Van Seters. Van Seters' minimalist claims must be explored in more detail. He believes "the geography of J's exodus account will reflect his familiarity with the Egypt of his own day rather than preserve hoary traditions of place-names from the second millennium" (Van Seters 2001: 260). Van Seters claims that "Per Atum [Pithom] is the name used for a number of different temples to the god Atum in various locations, but none receives the town determinative in Egyptian until the inscriptions of the late period" (2001: 260). While this certainly seems to be the case, the Egyptologist Redmount proposes a simpler solution. She states:

> In Egyptian usage *Pr-'Itm* would not stand alone but would be followed by a specific location designator, identifying the Per Atum of a specific place. In effect, the biblical rendering of Pithom strips the reference of its specificity and thus its identifiability, and transforms it into a collective allusion equivalent to the generic references to 'Pharaoh'. (Redmount 1998: 65)

Redmount's simpler solution helps us see the generic nature of certain arguments for specificity, a special pleading that defines these arguments as having specificity.

Van Seters's scholarship also seems to be overly aggressive in his discussion of *yam suf.* He only discusses it in terms of it being the Red Sea. He will not allow it to mean Reed Sea and he does not even consider other options. Once again, Redmount can see grounds for both translations (1998: 65). More importantly, we must also consider *yam suf* as a mythological term rather than a geographic term (Dozeman 1996: 408). This viewpoint would coalesce nicely with Redmount's previous point that the author wanted people and places to be generic rather than specific.

Origin Stories

Recent historians take considerable interest in origins. Jill Lepore argues: "All nations are places, but they are also acts of imagination. Who has a part in a nation's story, like who can become a citizen and who has a right to vote, isn't foreordained, or even stable" (Lepore 2012: 3). Investigations into Israel's ethnogenesis confirm this view. Avraham Faust makes a similar point: "People can change their identity and, if necessary, 'reinvent' their origins" (2006: 170). Faust makes a argument similar to Redmount that the narrative has a historical core, but the Bible does not accurately describe how it took place (2006: 174). I would argue that an examination of the Egyptian material suggests a historical core may exist, but it is impossible to discern. The very idea of an origin story tells us more than an unrecoverable historical core can tell us.

In a 1987 article, Redford argues that the little Egyptian coloring in Exodus is "almost wholly toponymic in nature" (1987: 138). In a more recent work, Redford claims "the Exodus was part and parcel of an array of 'origin' stories to which the Hebrews fell heir upon their settlement of the land, and which, lacking traditions of their own, they appropriated from the earlier culture they were copying" (1992: 442). I would like to close this chapter with a short reflection on what origin stories can contribute to us.

Linguistic Turn

Hayden White encourages us to look at the counterparts of historical narratives in literature. Ann Killebrew has implicitly done this by following Herwig Wolfram's (1990: 30–31) definition of ethnogenesis (Killebrew 2005: 149). I would like to focus on only one aspect of this definition: "a story or stories of a primordial deed, which can include the crossing of a sea or river, an impressive victory against all odds over an enemy, or combinations of similar 'miraculous' stories (e.g., the exodus)" (Killebrew 2005). Rainer Albertz claims: "there is good evidence for the thesis that Jeroboam's revolt was found under the religious banner of the Exodus tradition and that the Exodus story found its oldest shape in this revolt" (2001: 141–142). John Collins' conclusion that "there seems to me to be no reliable evidence of pan-Israelite celebration of the Exodus prior to the time of Jeroboam" (2001: 150) explains why the literary echoes of Moses resound so strongly here. Commentators have referred to this as Israel's charter myth. Similar phenomenon can be found throughout Greek, Roman, and German history.

Numerous examples exist of the importance of river crossings. Potentially hostile river gods often needed to be placated before the crossing of a river. We see Alexander do this before both the Danube (Campbell 2012: 131) and the Indus (Campbell 2012: 371). Riverine legends are also recorded in Roman literature. Rivers were reported to flow backward on Nero's death (Campbell 2012: 149). When river levels fell at auspicious times, river gods were seen as intervening in Rome's favor (Campbell 2012: 444, n. 72). All these examples connect the crossing of a river with stories of how a people understand themselves and their important leaders.

Further evidence emanates from the origin stories of Germanic tribes like the Allemanni. Rivers represent geographical and political borders. When they successfully crossed them, expansion and greater unity are found in the tribe (Castritius 1990: 77).[1] They are named the Allemanni after these successful events. The Exodus material seems to play a similar role for Israel. Strong leadership emerged for the German tribes at times of river crossing, and we find the same thing for Israel. Moses fully emerges at the Red Sea, and Joshua emerges at the Jordan River. These Exodus stories may have been written much later when the strong leadership of figures like Hezekiah and Josiah was desperately needed. As Judah faced the Assyrian and Babylonian crises, historical writings that confirmed the efficacy of strong leadership in the past were appropriate. Strong leadership was lost for the Allemanni after the river events, and the Romans eventually subdued them. The Exodus Story also points to the importance of strong leadership.

The way Exodus discusses rivers or seas also must be taken into consideration. We can compare it with *Papyrus Westcar* from Egypt. Semitic literature has more troubling depictions of rivers than Egyptian literature, with the possible exception of *The Shipwrecked Sailor*. Beyond Moses, difficulties at rivers or bodies of

[1] "Die durch militärischen Druck und Erfolg erzwungene Aufhebung der Reichsgrenze im rechtsrheinischen Germanien, die erst relativ kurze Zeit zuvor unter ungeheurem Aufwand erneuert und verstärkt worden war, mag von den angrandenden Elbgermanen viel eher als die ephemeren Übertritte über den Oberrhein als primordiale Tat gefeiert worden sein und ein neues Selbstbewußtsein erzeugt haben, das direkt in die Ethnogenese eines neuen Großstamms mit demalten Namen der Alemannen einmündete."

water continue with Elijah, Elisha, and even Jesus in the New Testament. Egyptian literature seems to have no literary parallels with the Exodus. The image of the Nile as treated in Egyptian literature is generally quite positive and different from the role of rivers in Semitic sources.

Conclusion

Although evidence exists of an Asiatic presence in Egypt, definitive evidence does not exist for an Israelite exodus on the scale of the Book of Exodus. The Philistines offer a powerful analogy where we can trace the introduction of new pottery and new loom weights into the Holy Land. This lacuna with the Israelites puts the stress on their origin story. The Book of Exodus offers an origin story with many similarities to Greeks, Romans, and German tribes. Rather than try to mine the Book of Exodus for references within the archaeological record, we should appreciate it as a theological document operating within the conventions of historical narratives.

References

Albertz, Rainer. 2001. Exodus: Liberation History against Charter Myth. In *Religious Identity and the Invention of Tradition: Papers Read at a Noster Conference in Soesterberg, January 4–6, 1999*, ed. Jan W. van Henten and Anton Houtepen. Assen: Royal Van Gorcum.

Barako, Tristan J. 2013. Philistines and Egyptians in Southern Coastal Canaan during the Early Iron Age. In *The Philistines and Other "Sea People" In Text and Archaeology*, SBL archaeology and biblical studies 15, 37–52. Atlanta: Society of Biblical Literature.

Campbell, J.B. 2012. *Rivers and the Power of Ancient Rome: Studies in the History of Greece and Rome*. Chapel Hill: University of North Carolina Press.

Castritius, Helmut. 1990. Von politischer Vielfalt zur Einheit. Zu den Ethnogenesen der Alemannen. In *Typen Der Ethnogenese: Unter Besonderer Berücksichtigung Der Bayern: Berichte Des Symposions Der Kommission Für Frühmittelalterforschung, 27. Bis 30. Oktober 1986, Stift Zwettl, Niederösterreich*, eds. Herwig Wolfram and Walter Pohl, 71–84. Wien: Verlag der österreichischen Akademie der Wissenschaften.

Collins, John J. 2001. The Development of the Exodus Tradition. In *Religious Identity and the Invention of Tradition: Papers Read at a Noster Conference in Soesterberg, January 4–6, 1999*, ed. Jan W. van Henten and Anton Houtepen, 144–155. Royal Van Gorcum: Assen.

Coogan, Michael. 1993. The Exodus. In *The Oxford Companion to the Bible*, ed. Bruce Metzger and Michael Coogan, 609–612. New York: Oxford University Press.

Dozeman, Thomas. 1996. The *yam-sup* in the Exodus and the Crossing of the Jordan River. *Catholic Biblical Quarterly* 58: 407–416.

Drews, Robert. 2000. Medinet Habu: Oxcarts, Ships, and Migration Theories. *Journal of Near Eastern Studies* 59: 161–190.

Faust, Avraham. 2006. *Israel's Ethnogenesis: Settlement, Interaction, Expansion and Resistance*. London: Equinox.

Finkelstein, Israel, and Neil Asher Silberman. 2001. *The Bible Unearthed : Archaeology's New Vision of Ancient Israel and the Origin of Its Sacred Texts*. New York/London: Free Press.

Foster, Roy F. 1988. *Modern Ireland 1600–1972*. London: Allen Lane.

Halpern, Baruch. 1992. The Exodus from Egypt: Myth or Reality. In *The Rise of Ancient Egypt: Symposium at the Smithsonian Institution, October 26, 1991*, 86–117. Washington, DC: Biblical Archaeological Society.

Iggers, Georg G. 1997. *Historiography in the Twentieth Century : From Scientific Objectivity to the Postmodern Challenge*. Hanover, NH: Wesleyan University Press.

Killebrew, Ann E. 2005. *Biblical Peoples and Ethnicity: An Archaeological Study of Egyptians, Canaanites, Philistines, and Early Israel, 1300–1100 B.C.E*, SBL Archaeology and Biblical Studies 9. Leiden/Boston: Brill.

Killebrew, Ann E., and Gunnar Lehman. 2013. The World of the Philistines and Other "Sea Peoples". In *The Philistines and Other "Sea People" In Text and Archaeology*, SBL Archaeology and Biblical Studies 15, 1–17. Atlanta: Society of Biblical Literature.

Kitchen, Kenneth. 1992. The Exodus. In *Anchor Bible Dictionary*, 2:700–708. New York: Doubleday.

Lepore, Jill. 2012. *The Story of America: Essays on Origins*. Princeton, NJ: Princeton University Press.

Malamat, Abraham. 1997. The Exodus: Egyptian Analogies. In *Exodus: The Egyptian Evidence*, ed. Ernst S. Frerichs and Leonard H. Lesko, 15–26. Winona Lake, IN: Eisenbrauns.

Moore, Megan Bishop, and Brad E. Kelle. 2011. *Biblical History and Israel's Past: The Changing Study of the Bible and History*, 2011. Grand Rapids, MI: Eerdmans.

Redford, Donald. 1987. An Egyptological Perspective on the Exodus Narrative. In *Egypt, Israel, Sinai: Archaeological and Historical Relationships in the Biblical Period*, ed. Anson Rainey, 137–161. Tel Aviv: Tel Aviv University.

———. 1992. *Egypt, Canaan, and Israel in Ancient Times*. Princeton, NJ: Princeton University Press.

Redmount, Carol. 1998. Bitter Lives: Israel in and out of Egypt. In *The Oxford History of the Biblical World*, ed. Michael Coogan, 58–89. New York: Oxford University Press.

Sarna, Nahum. (1992). The Book of Exodus. In *Anchor Bible Dictionary*, ed. D.N. Freedman, 689–700. New York: Doubleday.

Van Seters, John. 2001. The Geography of the Exodus. In *The Land That I Will Show You: Essays on the History and Archaeology of the Ancient Near East in Honour of J. Maxwell Miller*, JSOTSS 333, ed. J.A. Dearman and M.P. Graham, 255–276. Sheffield: Sheffield Academic Press.

Weinstein, James. 1997. Exodus and Archaeological Reality. In *Exodus: The Egyptian Evidence*, ed. Ernst S. Frerichs and Leonard H. Lesko, 87–103. Winona Lake, IN: Eisenbrauns.

Wolfram, Herwig. 1990. Einleitung oder Uberlegungen zur Origo Gentis. In *Typen Der Ethnogenese: Unter Besonderer Berücksichtigung Der Bayern: Berichte Des Symposions Der Kommission Für Frühmittelalterforschung, 27. Bis 30. Oktober 1986, Stift Zwettl, Niederösterreich*, eds. Herwig Wolfram and Walter Pohl, 19–34. Wien: Verlag der österreichischen Akademie der Wissenschaften.

Yurco, Frank. 1997. Merenptah's Canaanite Campaign and Israel's Origins. In *Exodus: The Egyptian Evidence*, ed. Ernst S. Frerichs and Leonard H. Lesko, 27–55. Eisenbrauns: Winona Lake, IN.

Biblical Claims About Solomon's Kingdom in Light of Egyptian "Three-Zone" Ideology of Territory

40

Christopher B. Hays

Abstract

Biblical rhetoric about Solomon's empire shares some significant features with Egyptian royal ideology of territory. Both the Egyptians and the Israelites seem to have thought of their national boundaries in three zones: a well-defended "internal zone" of primary settlements, an "outer zone" of economic interests; and finally an "ideological zone" that was generally not controlled militarily, but rather an idealized expression (indeed an exaggeration) of royal power.

One reason for the ongoing quest to understand "the Exodus" is the recognition of the numerous cultural contacts between ancient Israel and Egypt. There is little doubt that some portion of the Iron Age proto-Israelite state handed down memories of Egypt, which later became literary traditions in the Hebrew Bible.[1] Furthermore, textual and material data related to the earliest Israelite monarchy indicate ongoing relationship with Egypt (Schipper 1999; Galvin 2011). This paper adds to the corpus of examples of Egyptian influence on early Israel by describing an apparent similarity between the ways the two nations conceptualized the extent of their kingdoms.

It has long been recognized that certain biblical descriptions of the extent of ancient Israel correspond with those of the LBA Egyptian empire in the Levant (Mazar 1986: 115; Weinfeld 1993: 64; Havrelock 2007), so the purpose of this essay is to add detail to the picture, and to explain how diverse "zones" coexisted.

As the emphasis on "rhetoric" and "ideology" should indicate, the argument of this essay is largely about how territories were perceived, not what they actually were. It does not make claims directly about the (hotly contested) extent and nature of Solomon's kingdom, not least because the outskirts of any ancient kingdom were fluid and poorly defined—at least, by any modern standard. Rather, the goal here is to correct faulty presuppositions about how ancient royal claims about territory developed.

As a starting point, I take a well-known passage describing the extent of Solomon's

[1] This view has been common in biblical scholarship since at least the mid-twentieth century. For a recent summary, see Russell 2009.

C.B. Hays (✉)
Ancient Near Eastern Studies, Fuller Theological Seminary, Pasadena CA, USA
e-mail: hays@fuller.edu

T.E. Levy et al. (eds.), *Israel's Exodus in Transdisciplinary Perspective*,
Quantitative Methods in the Humanities and Social Sciences, DOI 10.1007/978-3-319-04768-3_40,

kingdom, 1 Kgs 5:1–5 (ET 4:21–25). It is translated in this way in the NRSV:

> Solomon was sovereign (*hāyâ môšēl*) over all the kingdoms from the Euphrates to the land of the Philistines, even to the border of Egypt; they brought tribute and served Solomon all the days of his life.
>
> Solomon's provision for one day was thirty cors of choice flour, and sixty cors of meal, ten fat oxen, and twenty pasture-fed cattle, one hundred sheep, besides deer, gazelles, roebucks, and fatted fowl.
>
> For he had dominion (*hû' rōdê*) over all the region west of the Euphrates from Tiphsah to Gaza, over all the kings west of the Euphrates; and he had peace on all sides.
>
> During Solomon's lifetime Judah and Israel lived in safety, from Dan even to Beer-sheba, all of them under their vines and fig trees.

One finds in this passage at least three apparently different claims about the extent of Solomon's rule: in v. 1, he is said to be "sovereign over all the kingdoms from the Euphrates to the land of the Philistines"; in v. 4 he is said to have "dominion over all the region west of the Euphrates from Tiphsah to Gaza"; while v. 5 states that Judah and Israel "lived…from Dan to Beer-sheba."

Scholars have tended to either dismiss these claims as late fabrications with little connection to reality or attribute their diversity to extensive redactional activity. In *A History of Ancient Israel and Judah*, Max Miller comments, "Typical of the sweeping and generalized editorial statements of this Solomonic section of Genesis-2 Kings, the terminology is vague and the description based on later fantasies about Jerusalem's 'golden age'" (Miller and Hayes 2006: 210).[2] A more moderate view comes from John Gray, who hypothesized that "[i]n this case, 'and to the border of Egypt' may be a later accretion by a glossator who took the reference to 'the land of the Philistines' as inclusive" (1963: 141)[3]—in other words, an initially realistic claim about Solomon's kingdom was later expanded when its real historical extent had been forgotten.

These are only examples of what might be called the "Glory Days" theory of royal rhetoric: Kings themselves kept realistic records, whereas much later scribes tended to inflate their achievements in rose-tinted hindsight. This was Martin Noth's opinion on this passage in his seminal work on the Deuteronomistic History (Joshua-2 Kings), and it has been influential ever since.[4] Walter Dietrich, in his volume on the early monarchy in the Biblische Enzyklopädie, concludes that the general idea expressed by the ration list in 5:2 is the only part of 1 Kgs 5:1–5 that has any likelihood of being drawn from source material (Dietrich 1997: 147). It is almost certainly true that these verses have undergone redaction in their history of transmission, but it is probably incorrect to assume that later redactors were by nature responsible for the text's extravagant territorial claims. A new analysis of the text itself and comparison with imperial ideology from New Kingdom Egypt helps to put the interpretation of this text on a firmer empirical basis.

The Vocabulary of Domination

The interpretation of the passage in question will be hampered if one only reads it in translation. Most readers would take the English translations "was sovereign" and "had dominion" to indicate that Solomon was king of the whole region, but there are strong clues in the passage's language that these verses are not actually claiming an enormous empire for him. In the first place, the passage does not employ the very common verb √*m-l-k*, "to be king, to rule." (Indeed, *nowhere* in the biblical accounts of Solomon is he said to be king [√*m-l-k*] of his region.) Instead, 1 Kgs 5:1–5 uses √*m-š-l* and √*r-d-h*. That is because the passage does not mean to say that Solomon ruled the

[2] Sometimes, of course, this critique is extended to the entire idea of a united monarchy: "The bible's stories about a united monarchy are a late fictional reflection on a 'golden age,' much like the legends of Arthur" (Thompson 1995: 60).

[3] In fairness, some interpreters have recognized the polyvalence of these borders. For example: "[The description 'Dan to Bethel'] does not contradict the map of the full extent of Solomon's rule in v. [21], since only the home territories are meant here" (Cogan 2001: 214).

[4] In *The Deuteronomistic History*, Noth deemed the provision records of vv. 22–23 to be the original layer, with the land claims inserted later (Noth 1991: 93 n. 5), though he would later modify this view in his commentary (Noth 1968: 62, 75–78). There, he commented that the huge quantities of food would indicate an exaggeratedly large court to feed.

entire middle portion of the Near East, only that he was a dominant figure among the many kings of small, regional kingdoms like his own.

The Hebrew root √*r-d-h* does not exclusively or even generally indicate direct political rule. In Gen 1:26–28 it is used for the dominion of humankind over animals; in Lev 24:43–46, it is used to describe the control of slaves by their owners; and in Jer 5:31 for the rule by priests over their own people. The verb "dominate" would better suit the range of occurrences. And as one sees in texts such as Lev 26:17 or Isa 14:2, what is in view is not direct rule, such that the subjugated party ceases to exist as an independent entity, but rather the domination of one independent state by another.[5]

The case with √*m-š-l* is similar; it means "to exercise authority," not "to rule as king." The root has a wide ranges of applications,[6] but when used in political contexts, it commonly refers to the dominance of one independent party over another. For example, the Philistines were said to have dominance over Israel (Jdg 14:4; 15:11; 1 Sam 13:3), but Israel was not incorporated into a "Philistine Empire." The same is true of the root in Old Aramaic. In the Sefire Treaty, the dominant king (Bar-Ga'yah of KTK) instructs the less powerful king (Mati'ilu of Arpad) that he must give free passage to messengers to and from Bar-Ga'yah's court. "You will not be superior to me in this respect (*l tmšl by bz'*)," he instructs.[7] Bar-Ga'yah exercises authority over Mati'ilu, but does not rule his land. This sort of inequality of status among kings is well known in the ancient Near East from the title "King of Kish" in third-millennium Mesopotamia to the early Anatolian title "Great Prince" (*rubā'um rabi'um*). There are of course cases where in treaties and correspondence among Bronze Age kings, one denies the title "king" to another on the basis of their unequal power, but it is only with the rise of the larger empires in the first millennium that the idea of independent rule was completely effaced.

Since it is unnecessary to infer direct rule from √*r-d-h* and √*m-š-l*, the conclusion that this passage is making outlandish claims about the extent of Solomon's kingdom becomes very hard to substantiate. The passage itself indicates numerous "kings west of the Euphrates" continued to rule (5:4); in other portions of the biblical Solomon story, some of them are named: King Hiram of Tyre (1 Kgs 5:1, etc.), Hadad the Edomite (1 Kgs 11:14), King Hadadezer of Zobah (1 Kgs 11:23), the kings of the Hittites and Aram (1 Kgs 10:29), and the kings of Arabia (1 Kgs 10:15). The question of where Solomon's Israel ranked among these small Levantine nations is a harder question to answer. Solomon is portrayed as at best an equal of Hiram; as for the others, it would be difficult to determine their relative power without much more extensive archaeological and/or textual data than what presently exists.

If 1 Kgs 5:1–5 does not make claims about Solomon's rule over his region, then what does it say? And how do its seemingly contradictory claims about the extent of Solomon's Israel coexist? Recent scholarship in Egyptian ideology of territory may supply an answer.

Ideology of the Thutmosids' Territory

During at least the New Kingdom, Egypt appears to have conceived of its empire in terms of various zones. Typically, scholars have perceived two zones: an internal zone over which Egypt exercised firm and consistent military control, and a more flexible outer zone that was largely dedicated to ensuring the nation's economic interests.

The *internal zone* was marked by *ḫtm* fortresses that limited movement in and out of

[5] √*r-d-h* is a biform of √*r-d-d*, which can mean to hammer, or to beat gold. One would not want to essentialize the "root meaning" too much, but this image of battering down surely sheds light.

[6] It is used in a variety of contexts, in celestial bodies exercising authority over the day and the night (Gen 1:18), a husband exercising authority over his wife (Gen 3:16), men exercising authority over sin (Gen 4:7), a slave exercising authority over the property of his master (Gen 24:2), and Joseph exercising authority over his brothers (Gen 37:8), and over Egypt (Gen 45:8, 26).

[7] KAI (1971) 224:9. The translation cited, by Brent Strawn, accurately captures the nuance. See Chavalas 2006: 303.

the Nile Valley. (Interestingly, the Egyptian term *ḫtm* has the common meaning "lock, seal," and has a cognate in the Hebrew word *ḥtm*; such fortresses might have been thought to seal off the Egyptian homeland from external threats.[8]) The northern and southern boundaries of the internal zone were the fortress cities of Tjaru and Elephantine, respectively. Ellen F. Morris notes that men stationed at each of those fortresses carried out analogous duties,[9] and she adduces textual evidence that "the two fortresses could be invoked together to call to mind the entirety of Egypt" (Morris 2005: 196).

The boundaries of the *external zone* were dotted with fortress-towns, called *mnnw* in the South. Almost all New Kingdom *mnnw*-fortresses "served as population centers of some magnitude" (Morris 2005: 811)—not because their regions were densely populated with Egyptians, but just the opposite: to form a strong, defensible outpost in an otherwise borderland region (Morris 2005: 44). "Napata [probably] the farthest-flung Egyptian base in Upper Nubia…was also located at the head of the primary trade route that connected Upper Nubia to the Butana region" (Morris 2005: 76).

The administrative features in the North are somewhat less well understood, partly because Lebanon and southern Syria have not undergone archaeological exploration as intense as southern areas have (Morris 2005: 827),[10] but it is clear that bases were placed in strategic locations for defense of Egyptian interests, and various terms were used for them (Morris 2005: 809).[11] Remains of NK Egyptian bases have been discovered in locations such as Megiddo and Byblos. "North of Canaan proper," Morris notes, "only the coast appears to have been fortified by the Egyptians" (Morris 2005: 137). Indeed, beyond Byblos, even the most powerful pharaohs seem to have had to work cooperatively with local powers of significant stature. For example, the Egyptians seem to have had garrisons at Ugarit during certain periods, probably with the king's permission. Egypt also struggled with intermittent resistance from the city-state of Tunip, in the Orontes River valley in southern Syria.[12] This means that they did not come close to controlling the whole region that they sometimes claimed.

Those claims demarcated a third zone—an *"ideological zone"* that corresponded with the pharaoh's most ambitious expeditions, but covered territory that was not ruled (or even patrolled) by the Egyptians in any active way. Unlike the two previous zones, the ideological zone was marked not by fortresses, but simply by stelae (King 2010; Morris 2005: 181). The best example of such stelae is the one erected at Hagr el-Merwa (Kurgus) by Thutmose I (and revisited by Thutmose III). The hieroglyphic inscription on this massive rock formation between the fourth and fifth cataracts of the Nile calls down curses upon "any Nubian who shall violate this stela" (Davies 2001: 50). W. Vivian Davies describes it as "threatening royal and divine retribution upon a 'real' native population, but of a type—foreign, Nubian, desert-dwelling—traditionally counted among the 'mythic' forces of chaos" (Davies 2001: 57).

Despite the Thutmosids' desire to stake their claim at Kurgus, "permanent Egyptian presence does not seem to have reached beyond the region of Napata" (Török 2008: 17). There has been some speculation that further archaeological excavation might yet uncover a NK *mnnw* fortress at Kurgus (Morris 2005: 76, 82), but to date the only fort found there is significantly post-pharaonic (Sjöström 1998: 30–34). Thus, the looming inscription was a symbol and reminder

[8] HALOT (2001), 364. Cf. Morris 2005: 808.

[9] These duties were largely economic. Once the fortress cities had been established, the empire expanded further so that they ceased to have much martial function. "In almost all attested cases, *ḫtm*-fortresses were installed at precisely those locations at which entrance to the Nile Valley could be effectively monitored and controlled." (Morris 2005: 804, 824).

[10] She deems it "doubtful" that "a high density of Egyptian-style administrative headquarters will also be discovered in these areas."

[11] Morris finds "only two textual references to *mnnw* in Syria-Palestine."

[12] The exact location of Tunip is still contested; some identify it with biblical Hamath, others with a nearby tell.

of the pharaoh's secular and mythological power in an area where he would not actually be.[13]

Thutmose I and III are supposed to have erected other stelae at the bank of the Euphrates, and although they have never been found, it is widely accepted that they did so (Redford 2003: 74, 150). As noted above, the NK pharaohs never even fully controlled Lebanon, southern Syria, or the Orontes Valley, so the Euphrates was well beyond real Egyptian control. With the monument at Hagr el-Merwa, these would have created a bookending effect and marked the largest extent that a pharaoh could ever claim to control.

King poses the rhetorical question, *Why would the Egyptian state construct boundary markers that did not correspond to the actual area of control?* The answer appears to have been the pressure on rulers to expand their boundaries (King 2010: 42). The reasons for this pressure would have included immediate gain through plunder, lasting economic benefits through control of trade routes, as well as the straightforward prestige of victory and conquest. Expansionist ideology was characteristic of the Thutmosids, and stelae appear to have sufficed to allow these rulers to claim expansions, even where continuing rule was not viable.

Ideology of Solomon's Territory

Monarchic Israel, based on biblical and archaeological evidence, can be envisioned in three zones directly comparable to those of New Kingdom Egypt. The internal zone, as ancient inhabitants perceived it, is easily identified. The formula "Dan to Beer Sheba" is used repeatedly in the Bible to express (or specify) "all Israel" (Judg 20:1; 1 Sam 3:20; 2 Sam 3:10, 17:11; perhaps Amos 8:14). It is widely recognized that this is "the conventional description of the Israelite homeland" (Gray 1970: 143). Particularly instructive is the example of David's census in 2 Sam 24: "[David] said to Joab and the commanders of the army, who were with him, 'Go through all the tribes of Israel, from Dan to Beer-sheba, and take a census of the people, so that I may know how many there are.'" From the perspective of the author, "Dan to Beer-sheba" marked the extent of the populace of Israel. Furthermore, the cities of Dan and Beer Sheba have both been discovered and excavated, and they represent the northernmost and southernmost large cities in monarchic Israel and Judah.

The formula "Dan to Beer-sheba" was certainly a pre-exilic invention. The biblical histories of David and Solomon were of course subject to later redaction, but the formula "Dan to Beer-sheba" could be a diagnostic marker for earlier Hebrew, since the equivalent phrase in late phraseology is reversed: "From Beer-sheba to Dan" (1 Chr 21:2; 30:5).[14]

As with New Kingdom Egypt, the kings of Israel and Judah also had another, larger zone, that was intended to protect its economic interests, and also less populated. The extent of Israel's (and later Judah's) southern economic zone is marked by forts in the Negev desert (Faust 2006: 135–160; Meshel 1994: 39–67; Dever 1982: 281–286). Then as now, this would not have been a very desirable place to live, but it was a crossing for lucrative desert trade routes (Holladay 2006: 309–331). The recent discovery of Iron Age copper mining at Khirbet en-Naḥas has also suggested to some that Solomon and his successors would have had a need to control the Negev routes (Levy and Najjar 2006: 107–122). There has been a long-running dispute about who built and used these forts (as well as who controlled the mines), and it remains uncertain that Solomon's Israel controlled the Negev, but it stands to reason that whichever kings controlled Beer-sheba also extended their economic reach into the desert to its south. Beer-sheba was not a border fortress, but a large

[13] Thutmose I had only one known campaign to Nubia, and it remains to be determined whether Egypt ever had any significant presence as far south as Kurgus.

[14] In the same way, Late Biblical Hebrew uses "Judah and Jerusalem" instead of the Standard Biblical Hebrew phrase "Jerusalem and Judah."

city that would have drawn resources from the whole surrounding region.

Also as in Egypt, this economic zone was probably flexible, and partly for that reason it would not have been enshrined in formulaic expressions in surviving texts. There are various references to southern wilderness fortresses in the Bible (1 Sam 23:14; 1 Chr 12:1–9; 2 Chr 11:5–11; 17:12; 26:10), but nothing that clearly defines the farthest southern border.

Still less can be said about the northern economic zone, though trade with northern kingdoms of Hatti and Aram is mentioned in 1 Kgs 10:29. Israelite economic control in the north would have been limited and contested by the Aramean kingdoms based in Damascus (cf. 1 Kgs 11:24–25). A recent analysis by Jeffrey Blakeley suggests that the major fortresses in the north were no farther north than Dan, but were placed along major roads in order to enforce and collect tariffs—at for example Tel Hadar and Tell Abu Hawam (Blakely 2002: 49–54). This meshes well with Baruch Halpern's observation that Israelites seem to have complained "that Solomon was sacrificing the far north [near Damascus, and the tribal territory of Asher, south of Tyre] in order to protect his state" (Halpern 2001: 424).

When one widens the scope further, however, another formulaic pairing emerges: the river (or wadi) of Egypt and the river Euphrates. In Gen 15:18, the land between those borders is supposed to have been given to Abram by YHWH. Other passages indicate a special relationship between YHWH and this territory, such as Isa 27:12 ("On that day the LORD will thresh from the channel of the Euphrates to the Wadi of Egypt, and you will be gathered one by one, O people of Israel") and Jer 2:18, which condemns those who look elsewhere for blessing: "What then do you gain by going to Egypt, to drink the waters of the Nile? Or what do you gain by going to Assyria, to drink the waters of the Euphrates?"[15]

The Wadi of Egypt[16] and the Euphrates were commonly used ancient markers for delimiting the Levant. In addition to the Egyptian references to the Euphrates already noted, one finds references to campaigns to the *naḫal musur* from Assyrian rulers such as Tiglath-Pileser III, Sargon II, and Esarhaddon (Görg 1992: 321). A few biblical examples further substantiate the use of these borders by foreign rulers: Second Kings 23:29 reports that in 609 BCE, "Pharaoh Neco king of Egypt went up to the king of Assyria *to the river Euphrates*." This was seen as a crucial border. The campaign did not go well for Neco, however, so that afterward "the king of Egypt did not come again out of his land, for the king of Babylon had taken over all that belonged to the king of Egypt *from the Wadi of Egypt to the River Euphrates*" (2 Kgs 24:7). Still later, one reads that "[Antiochus IV] left Lysias, a distinguished man of royal lineage, in charge of the king's affairs *from the river Euphrates to the borders of Egypt*" (1 Macc 3:32).

Thus, the boast on Solomon's behalf in 1 Kgs 5:1 is a common sort of claim that ancient Near Eastern kings made: it claims that among the kings in the Levant at that time, Solomon was the dominant one. Whether this was factually true depends on conclusions about a number of other questions: *Did he indeed control rich copper mines? Did he embark on a shipping venture? Did he fortify numerous cities? Was his kingdom the equal of the Phoenicians'?* Those questions move well beyond the ambitions of this essay; but there are possible reconstructions in which these literary claims are merely exaggerated instead of completely fantastical. As in the case of the Thutmosids' placement of stelae at the outermost edges of their exploration, an Israelite claim to control to the Euphrates would have "generated tension" (Havrelock 2007: 656)—certainly tension with northern neighbors, but also the potentially productive tension of an unfulfilled goal.

[15] Of course, the political significance of the borders is that they are bounded by Mesopotamia in the north and Egypt in the south. In the Bible the literary pairing of Assyria and Egypt is often used to express "the far ends of the earth." Only in their most extravagantly universalizing rhetoric did biblical authors ever imagine an Israel or Judah that transgressed the boundaries of Egypt and Mesopotamia.

[16] The biblical *nḥl mṣrym* has been identified with the Wadi el-ʿArish and the Wadi Bezor. The resolution of this question is immaterial for the purposes of this essay.

Chart

	Zone 1: Inner/homeland	Zone 2: Outer/economic	Zone 3: Ideological
Solomonic Israel	"Dan to Beer-sheba"	Tel Hadar (?) to N. Negev	Wadi of Egypt to Euphrates
Thutmosid Egypt	Tjaru to Elephantine	Byblos (?) to Napata	Kurgus to Euphrates

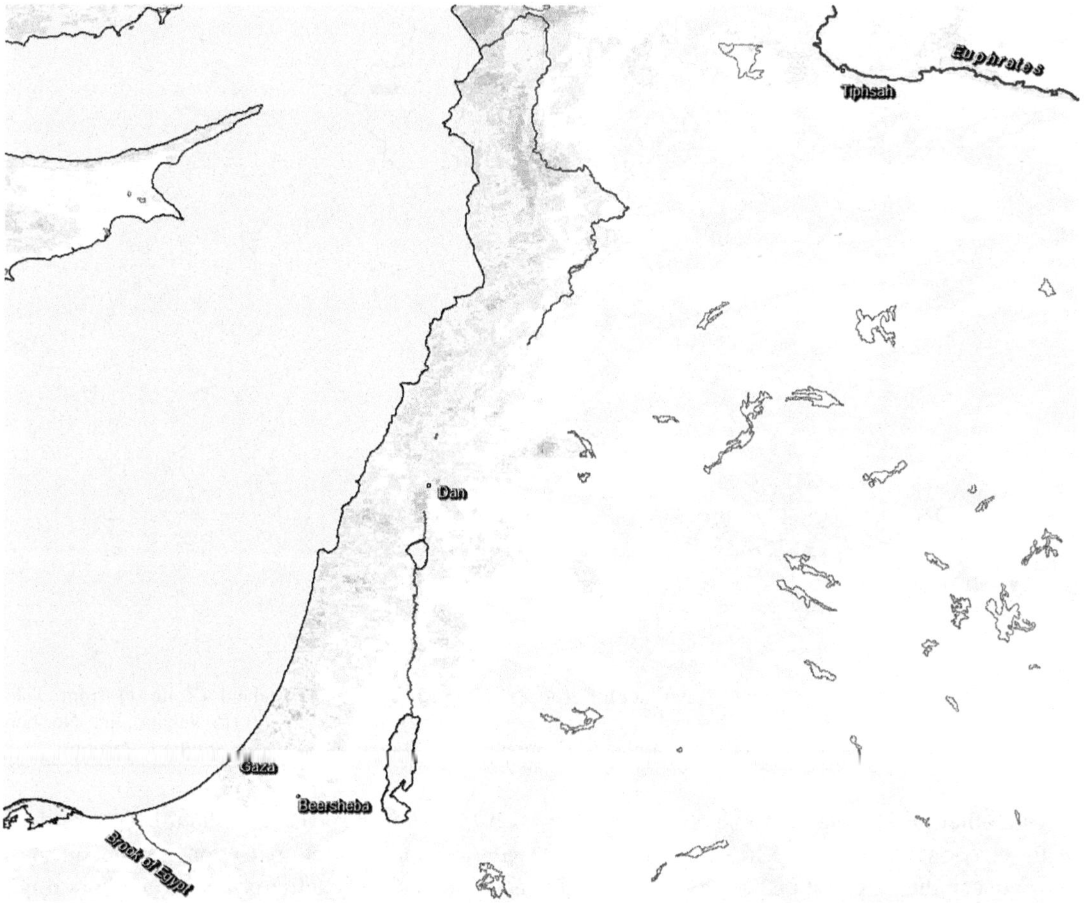

Fig. 40.1 Map of Solomomic Israel's Internal Zone / Homeland (Zone 1) from Dan to Beer-sheba and Ideological Zone (Zone 3) from Wadi of Egypt to Euphrates

In sum, one sees in the case of Solomon an ideology of territory very much like that of the Thutmosids (see Chart, above, and Figs. 40.1 to 40.3): (1) an inner zone of primary habitation identifiable both in archaeology and texts; (2) a flexible economic zone without much textual support or ideological significance; and (3) an ideological zone that maximized the ruler's claim on his region, to the point of exaggeration.

Concluding Reflection: Mechanisms and Periods of Influence

Although the argument that the biblical/Solomonic and Egyptian ideologies of territory are closely comparable draws on archaeological and historical data, it is essentially a literary claim; it is most clearly observable in surviving texts.

Fig. 40.2 Map of Thutmosid Egypt's Northern Portion of Internal Zone / Homeland (Zone 1) from Tjaru (to Elephantine), Northern Portion of External / Economic Zone (Zone 2) from Byblos (?) (to Napata), and Northern Portion of Ideological Zone (Zone 3) (from Kurgus) to Euphrates. See Map in Fig. 40.3 for Southern Portions.

As with other intertextual relationships, one may go further and ask about the historical relationship between the "texts." In other words, how did it come to be that Israelite and/or Judean ideology looked like that of Egyptian rulers?[17] There is no doubt in this case that the three-zone ideology is attested first in Egypt, so the question becomes when Israelite or Judean authors might have become aware of it. The essential question is whether the territorial ideology expressed by 1 Kgs 5:1, 4–5 was first expressed in the time of Solomon, or later, perhaps under one of the kings responsible for an edition of the Deuteronomistic History.

Although Egypt never again enjoyed a controlling presence in the Levant equivalent to what they had in the Late Bronze Age, Egyptian rulers certainly continued to stake their claim on it in ways that could have transmitted imperial ideology to Levantine nations. The most famous

[17] I set aside the possibility that there *was* no historical transmission, the idea that this shared ideology is the result of psychological processes common to humankind or derived strictly from natural features of the land. The close and regular contact between Egypt and the Levant throughout the Bronze and Iron ages makes such theories unnecessary and implausible. Nevertheless, Havrelock has made some helpful observations about the mythopoeic way in which ancient Near Eastern ideologies of territory make land claims correspond to physical geography. "The designation of seas and rivers as boundaries," she wrote, "conveys a sense that the order of the land reflects the structure of the cosmos" (Havrelock 2007: 657). Thus the selection of rivers (and seas) as borders would have been seen as "natural" by ancient rulers; it is important, however, to remember that in many cases what seems natural is *also* enculturated.

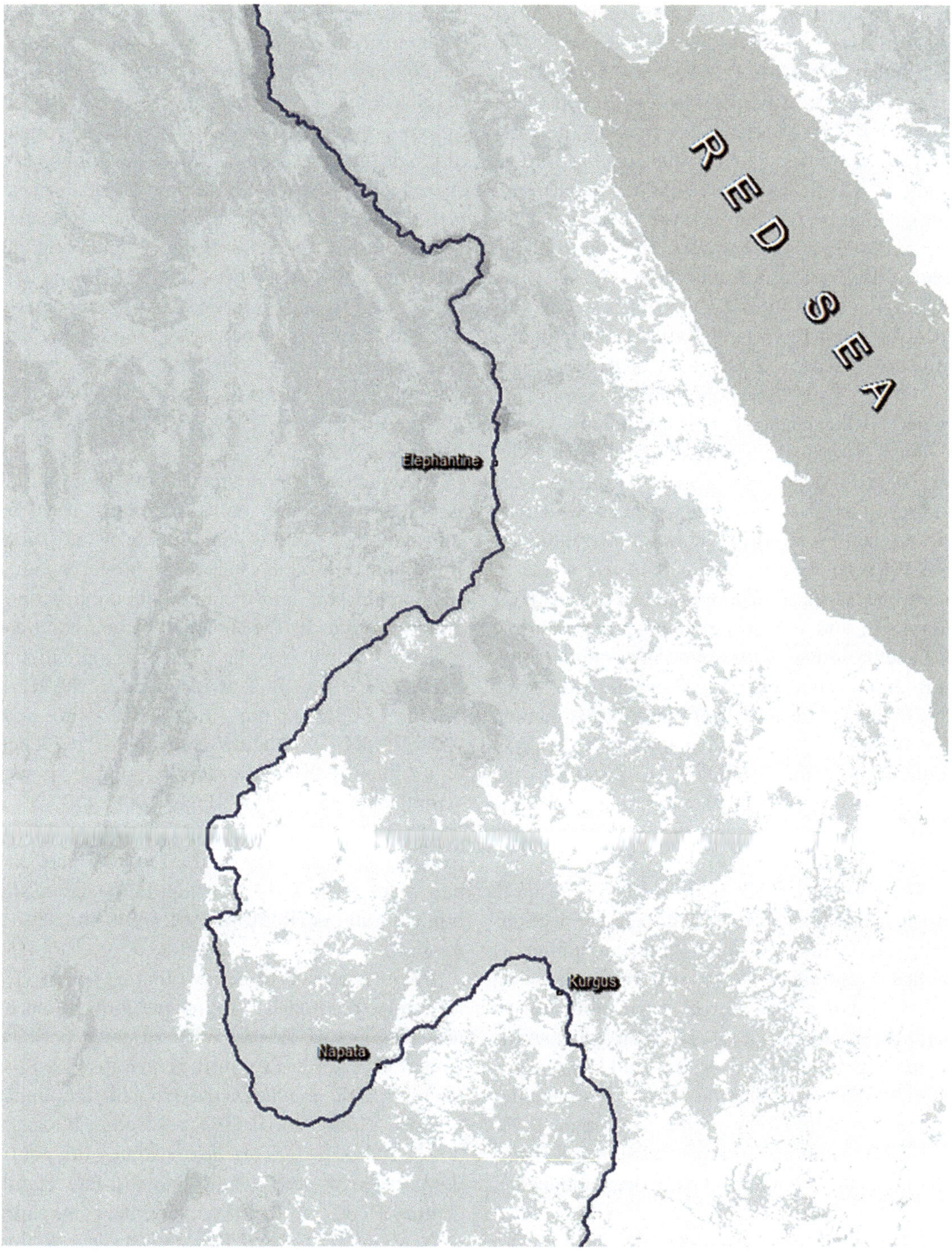

Fig. 40.3 Map of Thutmosid Egypt's Southern Portion of Internal Zone / Homeland (Zone 1) (Tjaru) to Elephantine, Southern Portion of External / Economic Zone (Zone 2) (from Byblos (?)) to Napata, and Southern Portion of Ideological Zone (Zone 3) from Kurgus (to Euphrates). See Map in Fig. 40.2 for Northern Portions.

Iron Age example of this is the campaign of Sheshonq (r. 945–924). In the rhetoric of his own monumental inscriptions, Sheshonq was a great conqueror of the Levant. On the Bubastite Portal relief at Karnak, he portrays himself "smiting the chiefs…of all inaccessible foreign lands, of all the lands of the Phoenicians, and foreign lands of the Asiatic back-country" (Ritner 2009: 201). However, records from the Levant itself tell a different story. An inscribed statue recovered from Byblos reflects "an alliance [between Egypt and Byblos] rather than a conquest" (Ritner 2009: 220; Kitchen 1996: 292–293). Thus Sheshonq probably *did not* smite "all the lands of the Phoenicians," as he claimed. Nevertheless, Sheshonq did place a victory stela at Megiddo, a heavily fortified city at an important pass in northern Israel which he sacked before returning home. This indicates that Egyptian rulers continued to use stelae to mark the farthest aspirational extent of their dominion, and it provides a potential example of Israelite contact with Egyptian royal ideology. Furthermore, that fact that Sheshonq was never able to rule the Levant,[18] but symbolically marked it as his territory with a stela anyway, indicates a continuation of the ideological use of stelae by the Thutmosids. The Egyptian aspirations were now smaller—the Euphrates was beyond their reach—but they continued to reach.

The biblical narratives of Solomon are quite insistent that his kingdom had deep connections to Egypt. At the level of plot, there are numerous biblical references to a marriage between Solomon and a daughter of a pharaoh (1 Kgs 3:1; 7:8; 9:16; 9:24; 11:1–2; 2 Chr 8:11). As is often pointed out, however, it would have been a unique event for a pharaoh to send his daughter to be married to a foreign king (Schulman 1979: 177–193). A number of other examples of Egyptian influence on Solomon have been adduced.[19] The most significant of these may be the system of tax districts attributed to Solomon in 1 Kgs 4, which has been compared to the system of Sheshonq I (Redford 1972: 141–156). If Solomon organized tax districts as the Egyptians did, this gives reason to think he might have shared other aspects of territorial ideology as well. Other examples of Egyptian influence include titles of administrators and officials (Mettinger 1971; de Vaux 1939: 394–405); numbers, weights, and measures that were clearly adopted from Egypt into Israel; and artistic and architectural styles, including such central symbols as the throne (1 Kgs 10:18–20; Currid 1997: 169; Williams 1975: 231–252) and temple (Strange 1985: 35–40).

It is true that not one of the above examples of Egyptian influence on Solomon can be substantiated by any data outside the Bible (Keel and Uehlinger 1998: 173); however, other scholarship points in the same direction. Iconographic studies by the so-called Fribourg School indicate that whereas the signs of Egyptianizing tendencies in Israelite art declined in the tenth century, during the Iron IIB era (ca. 925–725 BCE) Judahite crafts give "evidence of an intense fascination with Egyptian power symbols" (Keel and Uehlinger 1998: 173, 266). Redford has written that Dynasties 24–26 (roughly between 725 and 525 BCE) are the best place to look for Egyptian influence on the biblical authors (Redford 1992: 399–400; 1972: *passim*).

There is reason to believe that Egyptian victory stelae continued to follow the conventions of the Thutmosids' stelae even in later periods. Robert Gozzoli has recently shown that the Piye Victory Stele emulates the Nubian Triumphal Stela of Thutmose III at Gebel Barkal (Gozzoli 2003: 204–212).[20] This demonstrates the continuity of royal ideals well into the late eighth century BCE, and the aforementioned campaign of Neco in 609 BCE (2 Kgs 23:29) suggests that the Egyptians held on to the ideological notion of

[18] As Herbert Donner has remarked, the campaign was "…kaum mehr als eine Machtdemonstration, die zeigen sollte, dass es mit Ägypten nach langer Pause wieder aufwarts ging" (Donner 1984: 246).

[19] For a convenient summary and further citations, see Currid 1997: 159–171.

[20] The emulation is at the level of general thematic parallels rather than precise linguistic influence.

their imperial reach to the Euphrates even longer. Thus it is quite likely that Judean court continued to be privy to the same Egyptian rhetoric concerning territory in the Levant. In short, the enduring character of Egyptian territorial ideology makes it hard to determine when it was transmitted to Judah.

The internal evidence of 1 Kgs 5:1, 4–5 itself provides some clues, but the data are mixed. The formulation of 5:5, "*Judah and Israel* lived in safety," probably points to the perspective of a later redactor. Although Judah and Israel form a common merism throughout the narratives of the united monarchy, the ordering "Israel and Judah" is far more common in the Deuteronomistic History (1 Sam 17:52; 18:16; 2 Sam 5:5; 11:11: 21:2; 24:1; 2 Kgs 17:13; also Jer 30:3–4; 36:2; 51:5). Apart from 1 Kgs 4:20 and 5:5, "Judah and Israel" appears entirely in postexilic texts (2 Chr 16:11; 25:26; 28:26; 32:32; 35:18; Zech 11:14), which is in keeping with the Chronicler's emphasis on Judah as opposed to Israel. In the same verse (5:5), however, one finds the formula "Dan to Beer-sheba," which as noted above is typologically earlier than "Beer-sheba to Dan." The conflicting data within this one sentence, assuming that it is not random, would support the idea that this passage had a complex redactional history. In any case, the fact that an apparent marker of late redaction is connected to the most modest and historically grounded claim about territory[21] further problematizes the notion that exaggerated claims are late.

The goal of this study is not to settle the complex question of the redaction of 1 Kgs 5:1–5, but rather to point out that the descriptions of Solomon's kingdom and power, which originally seem contradictory, are not. They certainly may be inflated compared to Solomon's actual sphere of power and influence, but the foregoing analysis shows that such exaggerated claims were typical in the ancient Near East. For additional examples, one might compare slightly later examples of imperial rhetoric such as the Neo-Assyrian royal claim to be "king of the universe" and "king of the four corners of the earth," or the Neo-Babylonian boast that they controlled the whole land "from the Upper Sea to the Lower Sea." The Assyrians' claims were extravagant at face value, and even the Babylonian formulation is shown to be a "fiction of world domination" masking a "more limited imperial dominion" by the Neo-Babylonians' own detailed lists of provinces and territories (Vanderhooft 1999: 39).[22] The latter situation is very much like that of 1 Kgs 4–5.

Volkmar Fritz said of Solomon's portrayal in 1 Kings: "This picture of Solomon's magnificence leaves no room for doubt about the greatness of his rule. The king has become, as it were, an idolized replica of himself" (Fritz 1996: 187–188). The historian of the ancient Near East can only reflect that such self-idolization was precisely the goal of much ancient royal propaganda. Therefore it is methodologically faulty to assume that "the source materials taken up into this chapter...are sparse, practical, and based on historical fact; [while the late expansion materials] are high-blown, lyrical, and imaginatively laudatory" (Vries 1985: 74). One already knows that there is much in the redaction of the stories of the Jerusalem monarchy that is highly critical of kings, and of kingship in general. The Deuteronomistic History ends with the failure of the monarchy, and Solomon himself is elsewhere portrayed as a great sinner, and by no means a modest ruler (e.g., 1 Kgs 11). It appears to be a mere force of scholarly habit to attribute the aggrandizement of kings to later periods. In reality, such claims to

[21] The formulation *ʿēber hannāhār* in 1 Kgs 5:4 will suggest to some a Persian Period provenance, since this was the Hebraized version of the name for the region under Achaemenid rule (Ezra 8:38; Neh 2:7, 9; 3:7). However, "*Ebir Nāri*, 'Across-the-River,' was an Akkadian geographical proper name for part or all of Syria and Palestine at least as early as the eighth century B.C." (Stolper 1989: 288–289).

[22] It is possible that further research could show the three-zone ideology was shared by the Mesopotamians as well, although since they did not share the relatively narrow north-south orientation that Egypt and Israel did, this would be more difficult to show.

large influence are as likely to have been contemporary fantasies as late fantasies.

In light of the whole formation of the Deuteronomistic History, the text as we have it is not likely to have taken shape before the time of Hezekiah.[23] But one must reckon with more than three centuries of unbroken Davidic rule in Jerusalem. It is unlikely that the ideologies and inscriptions of Solomon were lost to Hezekiah or Josiah's scribes. If these later scribes preserved anything, such as lists of administrative districts and daily records, it is not unlikely that they also preserved the high-flown rhetoric of previous kings' ambitions. Whenever these stories were written down, nothing in their claims should have been unthinkable to Solomon himself, if even a portion of the biblical account of his connections with Egypt has an historical basis. As Halpern wrote, "[Solomon] was an emperor, if of a backwater. He gave himself the airs of an international player" (Halpern 2001: 424).

Acknowledgments I would like to thank Jeremy Hutton and Megan Bishop Moore for reading and commenting on drafts of this essay, and Nathan Yearian for creating the maps that accompany it (map data is under copyright: BibleWorks LLC, 2006–2011).

References

Blakely, Jeffrey A. 2002. Reconciling Two Maps: Archaeological Evidence for the Kingdoms of David and Solomon.*Bulletin of the American Schools of Oriental Research [BASOR]* 327: 49–54.

Chavalas, Mark W. (ed.). 2006. *The Ancient Near East: Historical Sources in Translation*. Oxford: Blackwell.

Cogan, Mordechai. 2001. *1 Kings: New Translation with Introduction and Commentary. Anchor Bible [AB]* 10. New York: Doubleday.

Currid, John D. 1997. *Ancient Egypt and the Old Testament*. Grand Rapids: Baker.

Davies, W. Vivian. 2001. Kurgus 2000: The Egyptian Inscriptions. *Sudan and Nubia* 5: 46–58.

De Vaux, R. 1939. Titres et fonctionnaires égyptiens à la cour de David et de Salomon. *Révue Biblique* 48: 394–405.

[23] On the likelihood of a Hezekian edition of the history, see Halpern and Vanderhooft 1991: 179–244; Weippert 1972: 301–339; Provan 1988.

De Vries, Simon J. 1985. *1 Kings Word Biblical Commentary 12*. Waco, TX: Word Books.

Dever, William G. 1982. Monumental Architecture in Ancient Israel in the Period of the United Monarchy. In *Studies in the Period of David and Solomon and Others Essays*, ed. Ishida Tomoo, 269–306. Yamakawa-Shuppansha: Tokyo.

Dietrich, Walter. 1997. *Die frühe Königszeit in Israel: 10. Jahrhundert v. Chr. Biblische Enzyklopädie 3*. Stuttgart: W Kohlhammer.

Donner, Herbert. 1984. *Geschichte des Volkes Israel und seiner Nachbarn in Grundzügen. Grundrisse zum Alten Testament 4*. Göttingen: Vandenhoeck & Ruprecht.

Faust, Avraham. 2006. The Negev 'Fortresses' in Context: Reexamining the 'Fortress' Phenomenon in Light of General Settlement Processes of the Eleventh-Tenth Centuries B.C.E. *Journal of the American Oriental Society [JAOS]* 126: 135–160.

Fritz, Volkmar. 1996. Monarchy and Re-Urbanization: A New look at Solomon's Kingdom. In *Origins of the Ancient Israelite States*, ed. V. Fritz and P.R. Davies, 187–195. Sheffield: Sheffield Academic.

Galvin, Garrett. 2011. *Egypt as a Place of Refuge Forschungen zum Alten Testament II [FAT/2] 51*. Tübingen: Mohr Siebeck.

Görg, M. 1992. Egypt, Brook of. *Anchor Bible Dictionary* 2: 321.

Gozzoli, Robert. 2003. Piye Imitates Thutmose III: Trends in a Nubian Historiographical Text of the Early Phase. In *Egyptology at the Dawn of the Twenty-First Century, volume 3: Language, Conservation, Museology*, ed. Zahi Hawass, 204–212. Cairo: American University in Cairo.

Gray, John. 1963. *I & II Kings: A Commentary*. Philadelphia, PA: Old Testament Library, Westminster.

John, G. 1970. *I & II Kings: A commentary*, 2nd ed. Philadelphia, PA: Old Testament Library; Westminster.

HALOT = Koehler, Ludwig and Baumgartner, Walter, 2001. *The Hebrew and Aramaic Lexicon of the Old Testament*, vol. 2. Leiden: Brill. Revised. Trans. M.E. J. Richardson.

Halpern, Baruch. 2001. *David's Secret Demons: Messiah, Murderer, Traitor, King*. Grand Rapids: Eerdmans.

Halpern, Baruch, and David S. Vanderhooft. 1991. The Editions of Kings in the 7th–6th Centuries BCE. *Hebrew Union College Annual [HUCA]* 62: 179–244.

Havrelock, Rachel. 2007. The Two Maps of Israel's Land. *Journal of Biblical Literature [JBL]* 126: 649–667.

Holladay, John S. 2006. Hezekiah's Tribute, Long-Distance Trade, and the Wealth of Nations ca. 1000–600 BCE: A New Perspective. In *Confronting the Past: Archaeological and Historical Essays on Ancient Israel in Honor of William G. Dever*, ed. Seymour Gitin et al., 309–331. Winona Lake, IN: Eisenbrauns.

KAI = Donner, H., and Rollig, W. 1971. *Kanaanaische und aramaische Inschriften (3 vols)*, 3rd ed. Wiesbaden: Harrassowitz.

Keel, Othmar, and Christoph Uehlinger. 1998. *Gods, Goddesses, and Images of God in Ancient Israel.* Minneapolis, MN: Fortress Press. Trans. T.H. Trapp, from German.

King, Meryl. 2010. *Walking the Line: Defining, Marking and Experiencing the Cardinal Boundaries of Ancient Egypt in the New Kingdom.* Unpublished M.Phil. dissertation, Cambridge.

Kitchen, K.A. 1996. *The Third Intermediate Period in Egypt, 1100–650 B.C*, 2nd ed. Warminster: Aris & Phillips.

Levy, Thomas E., and Mohammad Najjar. 2006. Some Thoughts on Khirbat En-Naḥas, Edom, Biblical History and Anthropology—A Response to Israel Finkelstein. *Tel Aviv* 33: 107–122.

Mazar, Benjamin. 1986. *The Early Biblical Period: Historical studies.* Edited by Shmuel Ahituv and Baruch A. Levine. Jerusalem: Israel Exploration Society.

Meshel, Ze'ev. 1994. The 'Aharoni Fortress' Near Quseima and the 'Israelite Fortresses' in the Negev. *Bulletin of the American Schools of Oriental Research [BASOR]* 294: 39–67.

Mettinger, Tryggve N. D. 1971. *Solomonic State Officials: A Study of the Civil Government Officials of the Israelite Monarchy Coniectanea Biblica: Old Testament [ConBOT] 5.* Lund: Gleerup.

Miller, J. Maxwell, and John H. Hayes. 2006. *A History of Ancient Israel and Judah*, 2nd ed. Louisville, KY: Westminster John Knox.

Morris, Ellen F. 2005. *The Architecture of Imperialism: Military Bases and the Evolution of Foreign Policy in Egypt's New Kingdom Probleme der Ägyptologie 22.* Leiden: Brill.

Noth, Martin. 1968. *Könige, 1. Teilband I Könige 1–16. BKAT 9/1.* Neukirchen-Vluyn: Neukirchener Verlag.

———. 1991. *The Deuteronomistic History*, 2nd ed. Sheffield: Sheffield Academic. JSOT Supp 15.

Provan, Iain W. 1988. *Hezekiah and the Books of Kings: A Contribution to the Debate About the Composition of the Deuteronomistic History. Beihefte zur Zeitschrift für die alttestamentliche Wissenschaft [BZAW] 172.* Berlin: W. de Gruyter.

Redford, Donald B. 1972. Studies in Relations Between Palestine and Egypt During the First Millennium B.C.: I. The Taxation System of Solomon. In *Studies on the Ancient Palestinian World*, ed. J.W. Wevers and D.B. Redford, 141–156. Toronto, ON: University of Toronto.

———. 1992. *Egypt, Canaan, and Israel in Ancient Times.* Princeton, NJ: Princeton University Press.

———. 2003. *The Wars in Syria and Palestine of Thutmose III.* Leiden: Brill.

Ritner, Robert K. 2009. *The Libyan Anarchy: Inscriptions from Egypt's Third Intermediate Period.* Society of Biblical Literature Writings from the Ancient World [SBLWAW] 21. Atlanta: Society of Biblical Literature.

Russell, Stephen C. 2009. *Images of Egypt in Early Biblical Literature: Cisjordan-Israelite, Transjordan-Israelite, and Judahite Portrayals.* BZAW 403; Berlin: Walter de Gruyter.

Schipper, Bernd Ulrich. 1999. *Israel und Egypten in der Königszeit: Die Kulturellen Kontakte von Salomo bis zum Fall Jerusalems.* OBO 170; Freiburg, Schweiz: Universitätsverlag. Göttingen: Vandenhoeck & Ruprecht.

Schulman, Alan R. 1979. Diplomatic Marriage in the Egyptian New Kingdom. *Journal of Near Eastern Studies [JNES]* 38: 177–193.

Sjöström, Isabella Welsby. 1998. New Fieldwork at Kurgus: The Cemetery and the Fort. *Sudan and Nubia* 2: 30–34.

Stolper, Matthew W. 1989. The Governor of Babylon and Across-the-River in 486 B.C. *Journal of Near Eastern Studies [JNES]* 48: 283–305.

Strange, John. 1985. The Idea of the Afterlife in Ancient Israel: Some Remarks on the Iconography of Solomon's Temple. *Palestine Exploration Quarterly [PEQ]* 117: 35–40.

Thompson, T.L. 1995. 'House of David': An Eponymic Referent to Yahweh as Godfather. *SJOT* 9: 60.

Török, László. 2008. *Between Two Worlds: The Frontier Region Between Ancient Nubia and Egypt.* Probleme der Ägyptologie 29. Leiden: Brill.

Vanderhooft, David S. 1999. *The Neo-Babylonian Empire and Babylon in the Latter Prophets.* Harvard Semitic Museum Monographs [HSM] 59; Atlanta: Scholars Press.

Weinfeld, Moshe. 1993. *The Promise of the Land: The Inheritance of the Land of Canaan by the Israelites.* Berkeley, CA: University of California Press.

Weippert, Helga. 1972. Die 'deuteronomistischen' Beurteilungen der Könige von Israel und Juda und das Problem der Redaktion der Königsbücher. *Biblica* 53: 301–339.

Williams, R.J. 1975. A People Come Out of Egypt. *Supplements to Vetus Testamentum [SVT]* 28: 231–252.

The Emergence of Israel in Retrospect 41

Robert A. Mullins

Abstract

Three models have been put forth to explain the appearance of Israel in the western highlands of the southern Levant: conquest, pastoral sedentarization, and social revolt. All of them attempt to reconcile archaeological data with the textual evidence; but a fourth model, which takes into account the dissolution of the Egyptian empire at the end of the Late Bronze Age, provides a more satisfying explanation for what must have been a widespread, complex, and lengthy process. Even so, the various events and forces that gave rise to formative Israel are far from the straightforward history that many assume in the Bible. What we find instead is a constructed history whereby later Israel both remembered and forgot essential dimensions of itself. In doing so, Israel was able to enshrine and reshape its past to create official memories of a culture and formulate a new vision for the future.

It is generally acknowledged that the evidence for an historical Exodus as described in the Hebrew Bible is meager. While this does not exclude the possibility of genuine memories of Asiatic slaves escaping from Ramesside Egypt,[1] in its present form, the Exodus account is part of an extensive foundation narrative whose purpose is to explain the origins of the Israelite nation extending all the way back to creation. Given the theological shape of this narrative, it is important to look more broadly if we want to better understand the historical circumstances that gave rise to Israel in the central hill country of Palestine.

Introduction

On a stele dated to the last decade of the thirteenth century BC, Merneptah, the son and successor of Ramesses II, recorded the details of a military campaign to Libya and Canaan. On line 27, Merneptah claimed to have subjugated a people called "Israel". Who were they? Where did they come from? What is the relationship between "Israel" of the Merneptah Stele and "Israel" of

[1] This argument was skillfully advocated in this conference by Manfred Bietak and James Hoffmeier. My own inclination has been to assume an historical core to this event, even if the present account in Exodus has been augmented and elaborated upon by later writers.

R.A. Mullins (✉)
Biblical Studies, Azusa Pacific University,
Azusa, CA 91702, USA
e-mail: ramullins@apu.edu

T.E. Levy et al. (eds.), *Israel's Exodus in Transdisciplinary Perspective*,
Quantitative Methods in the Humanities and Social Sciences, DOI 10.1007/978-3-319-04768-3_41,

the Hebrew Bible? These are not easy questions to answer, and over the years, various models have been set forth to explain Israel's appearance on the stage of history.

The three best-known models of Israelite origins are conquest, pastoral sedentarization, and social revolt.[2] All of them attempt to reconcile known archaeological data with the textual evidence; but how well do they explain the process of Israel's emergence historically? In the pages that follow, we will briefly examine these models and explore a fourth possibility, which views the rise of Israel as a response to the varying political and economic circumstances of Egyptian rule during the Late Bronze Age. This model has the advantage of not excluding valid observations made by the older theories, yet providing a more adequate explanation for what must have been a very complex and lengthy process.

A model that highlights the effects of Egyptian hegemony in Canaan, particularly during the Ramesside period, and its subsequent dissolution during the twelfth century BC, might also provide a basis for understanding the rise of other regional polities like the Phoenicians and Arameans.[3] Traditionally, investigations into Israel's appearance have been limited to those areas specifically mentioned in the Bible; however, as archaeological work continues to develop in neighboring regions, it is becoming apparent that we must also interpret Israel's emergence against the larger backdrop of political, economic, and environmental changes in the Levant during the Late Bronze to Iron Age transition.[4]

The Conquest Model

Given its congruence with the biblical narrative, the conquest model is the oldest and most familiar of the views on Israelite origins. At one time defended by such influential scholars as W.F. Albright, G.E. Wright, Y. Yadin, and J. Bright, the model of a unified military conquest was based largely on an uncritical reading of Joshua 1–12 (see especially 10:40), combined with archaeological evidence for the destruction of some Late Bronze Age sites towards the end of the thirteenth century BC.[5] Even though this model has been thoroughly discredited, it remains the dominant view in many religious circles. Indeed, the challenge faced by many conservative interpreters of the Bible is the sheer lack of evidence for such a military campaign.[6]

In the case of Jericho, archaeologists have yet to find any compelling evidence for the city's conquest as described in Joshua 6. The fortifications that John Garstang claimed were destroyed by the Israelites proved to be from the Early Bronze Age (Garstang 1931: 144–147; Kenyon 1970: 210–212).[7] Though scant Late Bronze Age

[2] For a discussion of these models, including their varying strengths and weaknesses, see Frick (2003: 247–263) and Gottwald (1979: 191–219).

[3] In his survey of Upper Galilee, Rafi Frankel noted the rise of many small villages following the demise of Hazor in the thirteenth century, a trend also observed in parts of Lebanon (Marfoe 1979). Thus, in the wake of the collapsed Late Bronze city-states and weakened Egyptian power, we witness the gradual appearance of smaller polities to fill the vacuum. These peoples include the Israelites, Phoenicians, Arameans, and Ammonites.

[4] For a discussion of the decline of the major powers in the eastern Mediterranean region during the thirteenth to twelfth centuries BC, see Ward and Sharp-Joukowsky (1992). For an analysis of the patterns of settlement during the Late Bronze to Iron Age transition in Lebanon, see Marfoe (1979). For a look at the role of climate in the collapse of Late Bronze Age civilization, see Kaniewsky et al. (2013) and Langgut et al. (2013).

[5] The majority of archaeologists and historians who hold to an historical conquest of Canaan under Joshua have argued for a late thirteenth century date referred to as the "Late Date Conquest" (Harrison 2004). A late fifteenth century date for the "Early Date Conquest" has always been problematic, despite sustained arguments by Waltke (1972), Merrill (1996: 119–121), and others.

[6] The conflicting traditions in the books of Joshua and Judges hint to their own complexity (Frick 2003: 240–248). At the very least the story of formative Israel was a prolonged process. Joshua probably contains an amalgamation of separate battle traditions woven into a unified and telescoped narrative centered on the promise-fulfillment motif. Its aim was to show that God gave Israel the land as promised to the Ancestors (Gottwald 1985: 230–260). Judges contains much of the same theological outlook as Joshua (a result of its Deuteronomistic shape), but its portrayal of village life tends to be more consistent with what archaeologists find on the ground.

[7] There is little question that Jericho was damaged over the centuries by local farmers who hauled away nitrogen-

remains have been found on the tell and in the cemetery, they cannot be attributed to either the fifteenth or the thirteenth centuries, when scholars of the early and late date views have generally argued for the city's demise. The conquest of Ai as recounted in Joshua 8 is equally problematic. Nothing from the period has been found at Khirbet et-Tell or other sites in the region. Recent attempts to relocate Ai at Khirbet el-Maqatir near Bethel are not convincing (Wood 2008).[8]

The foregoing problems of Jericho and Ai illustrate a major problem with the conquest model—there are no Late Bronze Age remains at several sites central to the biblical account of the conquest.[9] The same applies to Transjordan. Several sites mentioned in Numbers 21:21–32 do not have Late Bronze Age settlements (Dever 2003: 35). Tall al-ᶜUmayri at the northern edge of the Madaba plateau was destroyed at this time (Herr 2012: 209–213), but at present, there do not seem to be any certain Late Bronze IIB–Iron Age I remains south of Wadi Mujib. Most appear to be late Iron Age I or even Iron Age II (Herr 2012: 212).

The one notable exception is Hazor, described in Joshua 11:10–11 as "captured," its inhabitants "utterly destroyed by the sword," and the city itself "burnt with fire". Yadin (1972) argued for a late thirteenth century date for the Israelite destruction of Hazor, but the pottery suggests a time earlier in the thirteenth century than most assume for a "Late Date" conquest. Moreover, the discovery of a cartouche of Ramesses III in the gateway at Lachish, another city said to have been subjugated by Joshua, indicates that it was destroyed after 1170 BC. A span of nearly one century between the destructions of these two fabled cities makes a lightning conquest by Joshua and the Israelites highly unlikely. That said, it does appear that the biblical writers had access to genuine historical memories about Hazor, including that it had once been a major city-state, "the head of all these kingdoms" (Josh 11:10). So it is possible that some group(s) that became part of Israel was responsible for its destruction, explaining how the tradition found its way into the biblical account.

The Pastoral Sedentarization Model

The pastoral sedentarization model exists in two main variants. The first one, also known as the immigration or infiltration model, is identified with A. Alt (1989), M. Noth (1960), M. Weippert (1971), and others. According to this theory, Israel originated from pastoralist bands that migrated into the western highlands from Transjordan, where they eventually settled down and established villages. For the most part this was a prolonged and peaceful process. Military conflicts arose as the highland population grew and expanded its territory.

As with the conquest model, this theory also shares a correspondence with the biblical narrative, which alludes to a nomadic past for Israel's progenitors. While such stories sound suspiciously like the foundation myths of other peoples (Dever 2003: 51), it is hard to ignore the claims of a nonindigenous past (Amos 9:7), or to dismiss the traditions that associate Israel's earliest ancestors with the Arameans (Gen 24:10; 25:20; 31:20–21; Deut 26:5; Josh 24:2–4), especially since Hebrew shares more linguistic features in common with

rich earth from the mound to use as fertilizer. This explains why most of the site was denuded down to the Early Bronze Age levels. Even so, enough survives from the later phases of occupation to get a general sense of the city's occupational history. The Late Bronze Age pottery from the tell and cemetery belongs to the fourteenth century BC (Late Bronze IIA). Moreover, the Italian excavations have demonstrated that the Middle Bronze Age city is clearly MB IIC. There is nothing in the assemblage to suggest that the city survived into the fifteenth century BC (Late Bronze Age I), contrary to the claims of Wood (1990).

[8] The pottery that Wood has published from his "LB fortress" is actually MB IIC, making his settlement typical of Middle Bronze Age sites found elsewhere in the hill country at this time.

[9] For a list of cities claimed to have been taken by Israel, see Frick (2003: 253–255, Tables 8.1–8.3). For a list of destroyed sites in Late Bronze/Iron I Canaan, see Gottwald (1985: 263, Table 16). For a thorough discussion of these sites and their relevance to the conquest narrative, see Dever (2003: 54–71, especially Table 4.1 on pp. 56–57, which lists all of the cities recorded in Joshua 12:9–24).

Aramaic and Moabite than it does with Phoenician (Rainey 2008).

A second variant of the sedentarization model advocated by Israel Finkelstein (1996) is that native pastoralists already present in the western highlands settled down, established villages, and became the nucleus of biblical Israel. The strength of this argument lies in its *longue durée* perspective, which stresses how the region has been long characterized by recurring cycles of nomadism and sedentarization. Thus, what took place during the Late Bronze to Iron Age I transition was simply another phase in this ongoing cycle.

A more recent development of the sedentarization hypothesis equates Israel with the Shasu, a pastoral group that inhabited southern Transjordan and parts of Cisjordan as well. (Giveon 1971; Weippert 1979; Redford 1992; Rainey 2001).

The one element that has drawn the most attention to this group has been the occurrence of the toponym *t3 Š3sw yhw* in lists dating to the time of Amenhotep III and Ramesses II (fourteenth to thirteenth centuries BC), though it remains uncertain whether one can make a genuine linguistic connection between Egyptian *yhw* and Hebrew *yhwh* (Hasel 2008: 56–57). Even so, it is worth noting that a few archaic Hebrew texts locate Yahweh in this same region, where the toponyms Sinai/Seir/Edom/Teman/Mt. Paran all appear (Deut 33:2; Judg 5:4–5; Hab 3:3; Ps 68:7–8). As has been suggested, perhaps Yahweh came to be venerated by the hill country population through contact with the Shasu (Albertz 1994: 51; Redford 1992: 275–280).

Such a scenario is not unreasonable. The Shasu are no longer mentioned in texts after the end of the Late Bronze Age, suggesting that they assimilated into the general population during Iron Age I, whether in Transjordan, Cisjordan, or both.[10]

[10]Texts of Seti I and *Papyrus Anastasi I* indicate that those parts of Cisjordan inhabited by Shasu during the thirteenth century BC are the same areas known to be Patriarchal places of tradition: Shechem, Bethel, Gerar, and Beersheba (Rainey, pers. comm.).

It is also notable in this regard that out of 502 place names in the Hebrew Bible, none of them are Yahwistic. Toponyms are quite helpful because they help us identify the various deities venerated in Canaan. El was certainly the chief god of the pantheon and the divine element in "Israel".

It may be that Yahweh was not indigenous to the land of Canaan, but was a deity from the outside (Hess 2007: 274). It is also possible that Yahweh was a minor deity in the local Canaanite pantheon. Like Yahweh, Chemosh of Moab and Qaus of Edom are also previously unattested deities. Deuteronomy 32:8–9 is an archaic text that describes how each of the sons of El received its own realm. Yahweh was given Israel as an inheritance (Smith 2002: 32). Whether foreign or indigenous to Canaan, the real story of Yahweh seems to have been a centuries-long struggle for primacy over the traditional gods of Canaan. Except for a few brief moments in Israel's history, this preeminence was never fully achieved until the Babylonian exile.

The Social Revolt Model

The idea that Israel emerged as the result of a social revolt began with G. Mendenhall (1962), and was further developed by N.K. Gottwald (1979), and others. This model draws on elements from the conquest and sedentarization models, rearranging and nuancing them to form a new conception of Israel's rise (Gottwald 1985: 272). According to this theory, the existing city-state system created a sort of feudalism that was oppressive. Those who refused to participate in it removed themselves entirely by forming a separate society whom the city-dwellers regarded as "outlaws" (Habiru). The sociopolitical entity called "Israel" was composed in large part by these native Canaanites, who joined forces with a nuclear group of infiltrators from the desert, the "exodus Israelites," guided by a belief in its liberating god Yahweh (Gottwald 1985: 272–276).

For a long time scholars were skeptical of this model's reductionist Marxist undertones, though recent formulations have sought to temper the earlier overstatements (Gottwald 1985: 273). But for those who remain cautious, it is important to remember that one cannot easily dismiss the impact that a system of taxation, military conscription, and forced labor would have on a population, especially at a time when the Egyptian overlords were engaged in similar practices (Na'aman 1981, 1988; Redford 1992: 209). Indeed, the main argument of this paper is that Egypt's system of taxation, forced labor, and deportation prompted lowland Canaanites to withdraw into the central hill country, particularly during the nineteenth dynasty, when Egypt's grip on southern Canaan increased due to a growing concern over Hittite activity in the north.[11]

The advocates of social revolt were also among the first to seriously consider the *indigenous* character of the emerging highland population, an ancient memory expressly stated in Ezekiel 16:3, "Your origins and your birth are of the Canaanites..." While this declaration appears to contradict the traditions about Israel coming from outside of the Land of Canaan, such counterclaims may point to a more complex weaving of early traditions. They might also betray a desire on the part of later Israelite theologians to forget Israel's Canaanite past in favor of the pastoralist ideal, particularly in light of the Deuteronomistic polemic against other gods.

Though scholars may interpret the same evidence differently, the data seems to show that the highland population was mainly composed of farmers using an established agricultural tradition (Dever 2003: 107), though villages closer to the arid zones had a larger pastoral component (Coote and Whitelam 1987: 123). There is also a remarkable consistency in the type of house; all of which belong to the pillared variety with roots to some extent in the Late Bronze Age.[12] The technology for terraced agriculture and lime plastered cisterns go back to the Early Bronze Age and Middle Bronze Age, respectively. The Iron Age I pottery is distinctive but clearly in the Late Bronze tradition. The collared-rim jar, once regarded as a type fossil of newly arriving Israelites, also had roots in the Late Bronze Age. The limited range of vessel types in the highland pottery repertoire and their lack of decoration contrasts sharply with the painted assemblages of the lowlands. This simplicity indicates a subsistence lifestyle and egalitarian ethos (Faust 2006: 45–46).

The Dissolution Model

A fourth model put forth in various forms by L.E. Stager (2001), N.P. Lemche (1985), M. Liverani (2003) and others has been labeled the "digression" model by Albertz (1994: 72), though "dissolution" may be a better term to describe this phenomenon.

This theory stresses how a convergence of factors, including the political and economic weakening of the Canaanite cities, a marked decline in trade at the end of the Late Bronze Age, the chaos and devastation brought on by the Sea Peoples, and an erosion of Egyptian power and prestige in the Levant, prompted many peasants, pastoralists, and outlaws (*habiru*) to disassociate themselves from the urban sphere of influence and establish a new economic basis of existence by cultivating the hill country and marginal zones.[13]

[11] Redford (1990: 38) has suggested that this policy of uprooting and transporting whole communities to Egypt may lay behind some of the depopulation of Palestine during the Late Bronze Age.

[12] Mazar (1997: 52–69) excavated LB I-IIA pillared houses at Tel Batash, which has led some scholars to think that the Israelite four-room house must have Canaanite antecedents. However, Faust (2006: 75–78) has pointed to the prevalence of four-room houses in territories associated with Israel in the Bible. Future excavations will likely clarify whether this type of house is unique to Israel as a sociopolitical entity, or if it had a broader distribution as Bloch-Smith (2003: 407–408) and others have posited. What is clearly a four-room house has been excavated at Tall al-ᶜUmayri on the northern edge of the Madaba plateau (Herr 2012: 211–213; Figs. 2 and 4).

[13] This recalls the pattern at the end of the Early Bronze Age when the urban centers collapsed and there was a shift to pastoral and village life.

It was these farmers and shepherds who formed the tribal alliance known as "Israel" (Albertz 1994: 72; Coote and Whitelam 1987: 133–134; Lemche 1985: 431).

Egypt had two main concerns at the beginning of the thirteenth century: confront the threat of Hittite activity to the north (Faulkner 1975: 217–221; Weinstein 1981: 14–18), and deal with the problem of incursions into the Nile Delta from Libya. The latter contributed significantly to the demise of Egypt over time.

Seti I and Ramesses II both clashed with the Libyans. Later, Merneptah fought off a substantial Libyan invasion accompanied by their Sea People allies (the Sherden, Teresh, and Sheklesh). Although Merneptah successfully repelled them, the Libyans returned several decades later in 1180 BC with more allies and a large fleet of ships (Chadwick 2005: 205). Ramesses III managed to deter their attempt to take control of the Delta, bringing about Philistine settlement on the southern coastal plain. Egypt eventually withdrew from the Levant by the mid-twelfth century or slightly later.

The Libyans and Sea Peoples must have had a compelling reason to travel some 1,200 km (744 mi) to the Egyptian delta. Merneptah mentions drought as a factor, so it seems likely that they were spurred on by food shortages (Chadwick 2005: 206). New data-rich studies demonstrate that climate helped set the stage for this large-scale movement of peoples in search of food and suitable areas to colonize, contributing to the weakening of the Egyptian empire and to the demise of the Late Bronze Age (Kaniewsky et al. 2013; Langgut et al. 2013).[14]

The issue of drought is an important one. Like the theories of migration from years past, archaeologists have tended to dismiss climate as a decisive factor in historical change and have usually placed greater emphasis on internal changes and socioeconomic processes. But in this instance, Hittite and Ugaritic texts mention famines and the importation of grain from Syria to Anatolia, so climate *should* be a factor that one takes into account (Liverani 2003: 34).

Who and Where Was "Israel" of the Merneptah Stele?

Of all the theories presented above, the dissolution model provides a more satisfying explanation for the complex historical processes that lay at the heart of Israel's ethnogenesis. The identity of Israel as a socioethnic group was not only shaped by what took place during the twentieth dynasty as Canaan transitioned into the Iron Age, but is rooted in events extending back three centuries earlier (ca. 1500–1200 BC). This is where Lemche's term of "evolutionary Israel" is helpful, since the events described below may have represented the first stages of Israel's emergence.

Palestine had been a structure of city-states under Egyptian domination since the time of Thutmose III in the mid-fifteenth century BC. This was a time of great hardship on the local population. Written records mention the deportation of Canaanites to Egypt as slaves and other types of adversity in the form of tribute, taxation, military service, and forced labor. Although the nomadic population had a slight advantage in their mobility and seasonal movement which made them more difficult to control (Gottwald 1985: 273), even they were not immune to the varieties of hardships brought on by Egyptian policies. With a shift to more direct control at the beginning of the nineteenth dynasty, pressure on the lowland populace presumably increased, setting the stage for the appearance of Israel as mentioned in the Merneptah Stele.

This "Israel" is usually located in a geographic zone that can be narrowly defined as the Samaria

[14] Recent studies by Kaniewsky et al. (2013) for Cyprus, and Langgut et al. (2013) for Israel, emphasize the role of climate in the collapse of Late Bronze Age civilization. The results of the latter study were brought up during the conference by Israel Finkelstein. It seems that severe cold in the north destroyed crops and led to reduced precipitation in the steppe-lands to the east, damaging agricultural output and leading to droughts and famine. This climatic crisis, which occurred between 1250–1100 BC, prompted large groups of people to start moving further south in search of food.

arch, or more broadly as the central hill country.[15] Based on the survey material, Finkelstein proposed that the earliest settlements were located in the more arid eastern zone of Manasseh, and progressed through the highlands over time in an east to west and north to south direction. Given the dearth of *excavated* Iron I sites in the highlands, these conclusions have depended largely on the survey material, which doesn't always give an accurate reading of when these settlements were first established.

Even so, the data as we have it suggests that the earliest villages were in the Samaria arch, which is the same area biblical tradition locates the territories of Manasseh and Ephraim, the two sons of Joseph, favored son of Jacob/Israel.[16] Whatever the historical basis for these traditions, it is hard to ignore the correspondence between a "people" called "Israel" in the Merneptah Stele and the northern part of the hill country as the earlier, more populated zone.

The answer to the question of Israel's origins may lie in the Jezreel Valley and adjoining lowland regions. As Na'aman (1988: 237–238) has pointed out, the Jezreel Valley was regarded as crown property since the time of Thutmose III. These royal estates may have created conditions that prompted people to relocate elsewhere. If the earliest settlements were in Manasseh and Ephraim, perhaps it was people from the Jezreel Valley who first retreated into the hills to escape the hardships created by Egypt's policies; perhaps augmented by the climate change taking shape between 1250 and 1100 BC (Langgut et al. 2013). This does not exclude movement into the hills from regions further south, as the thirteenth century foundation dates for Izbet Sartah and Giloh indicate. However, the principal early settlement seems to be coming from further north.

The desire to escape Egyptian domination may be reflected in the name "Israel". If the name is derived from a root meaning "to fight," "to rule," or "to heal", then Israel might mean something like "El rules" (Albertz 1994: 76) or "Let El rule". Such a designation might have been used by the first highland settlers to both unify and express dissatisfaction with New Kingdom rule. We might assume that other disenfranchised groups fled to the hills for refuge as well.

The name "Israel" is clearly non-Yahwistic and presupposes a time when the population worshipped God as El, the supreme Canaanite deity (Smith 2002: 32). This was the group the Egyptians knew as "Israel". In the stele, they appear to have participated in a revolt against Egypt, together with the cities of Ashkelon, Gezer and Yeno'am. While difficult to prove, the dozen or so sites destroyed at the end of the thirteenth century may have been related to this resistance and its aftermath.

As the Egyptian empire began to disintegrate in the twelfth century, aggravated by the effects of climate change, an ever growing shift of Canaanite populations to the hills may have been accompanied by the migration of people from Transjordan. These immigrants may have been joined by the Shasu and/or the Moses group who brought the faith of Yahweh with them.[17] Perhaps the need to fight against the surrounding enemies contributed to a bond and united them.

[15] As noted by Hasel (2008: 53), the determinative signs only indicate Israel as a socio-ethnic entity. They say nothing about whether this group was settled or seminomadic. The lack of any reference to the Philistines, even though Ashkelon is well known as a Philistine city, indicates that Merneptah's campaign took place before they arrived on the southern coastal plain. Thus, the Philistines had no impact on Israel's earliest formation; however, later interaction and competition with the Philistines compelled those Israelites to redefine themselves through a number of distinct practices that helped them maintain clear ethnic boundaries (Faust 2006: 147).

[16] The prehistory of the Rachel and Leah tribes as representing two stages in Israelite settlement is a difficult one and beyond the scope of this paper. See also Lemche (1985: 430–431).

[17] John Bright (1981: 114) raised such a possibility when he noted that Israel of the Merneptah Stele may not have been the same as the exodus group who left Egypt, and while there is no proof for it, a tribal group named Israel may have already been present in the land. Moreover, texts from the time of Seti I and *Papyrus Anastasi I* indicate that some Shasu were already present in the hill country by this time (Rainey, pers. comm.).

At the very least, it would seem that under the pressure of Philistine expansion during the eleventh century, the various highland groups began to forge a national identity.

Conclusion

There is a complex interaction of events taking place during the Late Bronze to Iron I transition. Perhaps the Hebrew Bible retained some of this memory; not only by those who relocated from the lowlands into the highlands, but also by outsiders from Transjordan who came into conflict with the indigenous population.

The latter may be memorialized in the book of Joshua and in the competing religious visions of the god Yahweh, who will take over the land from the indigenous population worshipping El, Baal and Astarte. The former may be echoed in the book of Judges, in a person like Gideon from the tribe of Manasseh, who is cast as a Yahwist, but had no problem with an altar to Baal alongside an Asherah pole (Judg 6:25). When Gideon destroys the cultic furniture in response to Yahweh's command, the villagers are so angry that they are ready to kill him (Judg 6:28–30). Perhaps these beliefs were held so deeply because the religion of Canaan was the only religion they knew. Later on, during the Davidic monarchy, when Yahwistic names (rather than those of El or Baal) appear more commonly as a theophoric element, Yahwism gains ascendency. Still, many Israelites never abandoned their traditional beliefs or had any difficulty integrating them with a belief in Yahweh.

At some point it became important for Yahwists to distinguish themselves from the traditional religion of Canaan. Perhaps this is why Israel eventually forgot the Canaanite part of its identity. This process began in the pre-exilic period, but was not fully realized until post-exilic times when Jewish theologians reworked the earlier traditions. As a result, they were able to create official memories of the past and to articulate a new vision for the future.

References

Albertz, Rainer. 1994. *A History of Israelite Religion in the Old Testament Period. Vol. 1: From the Beginnings to the End of the Monarchy*. Louisville: Westminster/John Knox. Trans. J. Bowden, from German.

Alt, Albrecht. 1989. The Settlement of the Israelites in Palestine. *In: Essays on Old Testament History and Religion* (trans. Wilson, R.A., from German 1925). Sheffield: Sheffield Academic Press.

Bloch-Smith, Elizabeth. 2003. Israelite Ethnicity in Iron I: Archaeology Preserves What Is Remembered and What Is Forgotten in Israel's History. *Journal of Biblical Literature [JBL]* 122(3): 401–425.

Bright, John. 1981. *A History of Israel*, 3rd ed. Louisville, KY: Westminster John Knox.

Chadwick, Robert. 2005. *First Civilizations: Ancient Mesopotamia and Ancient Egypt*, 2nd ed. London: Equinox.

Coote, R.B., and K.W. Whitelam. 1987. *The Emergence of Early Israel in Historical Perspective*. Sheffield: Almond.

Dever, William G. 2003. *Who Were the Early Israelites and Where Did They Come From?* Grand Rapids, MI: Eerdmans.

Faulkner, Robert O. 1975. Egypt: From the Inception of the Nineteenth Dynasty to the Death of Ramesses III. In *Cambridge Ancient History, 3rd edn, vol 2, Pt 2*, ed. I.E.S. Edwards, C.J. Gadd, N.G.L. Hammond, and E. Sollberger, 217–251. Cambridge: Cambridge University.

Faust, Avraham. 2006. *Israel's Ethnogenesis: Settlement, Interaction, Expansion and Resistance*. Oakville, CT: Equinox.

Finkelstein, Israel. 1996. Ethnicity and the Origin of the Iron I Settlers in the Highlands of Canaan: Can the Real Israel Stand Up? *Biblical Archaeologist [BA]* 59: 198–212.

Frick, Frank S. 2003. *A Journey through the Hebrew Scriptures*, 2nd ed. Belmont, CA: Thomson/Wadsworth.

Garstang, John. 1931. *The Foundations of Bible history: Joshua, Judges*. London: Constable & Co.

Giveon, R. 1971. *Les Bédouins Shosou des documents Égyptiens*. Documenta at Monumenta Orientis Antiqui 18. Leiden: Brill.

Gottwald, N.K. 1979. *The Tribes of Yahweh: A Sociology of the Religion of Liberated Israel, 1250–1050 B.C.E.* Maryknoll, NY: Orbis Books.

———. 1985. *The Hebrew Bible: A Socio-Literary Introduction*. Minneapolis, MN: Fortress.

Harrison, R.K. 2004. *Introduction to the Old Testament*. Peabody, MA: Hendrickson.

Hasel, Michael G. 2008. Merenptah's Reference to Israel: Critical Issues for the Origin of Israel. In *Critical Issues in Early Israelite History*, ed. Richard S. Hess, Gerald A. Klingbeil, and Paul J. Ray Jr. Winona Lake, IN: Eisenbrauns.

Herr, Larry G. 2012. Jordan in the Iron I and IIA Periods. In *The Ancient Near East in the 12th–10th centuries BCE: Culture and History*, ed. Gershon Galil, Ayelet Gilboa, Aren M. Maeir, and Dan'el Kahn, 207–220. Proceedings of the International Conference held at the University of Haifa, 2–5 May 2010. Münster: Ugarit-Verlag.

Hess, Richard S. 2007. *Israelite Religions: An Archaeological and Biblical Survey*. Grand Rapids, MI: Baker Academic.

Kaniewsky, D., E. Van Campo, J. Guiot, S. Le Burel, T. Otto, et al. 2013. Environmental Roots of the Late Bronze Age Crisis. *PLoS ONE* 8(8): e71004. doi:10.1371/journal.pone.0071004.

Kenyon, Kathleen M. 1970. *Archaeology in the Holy Land*, 3rd ed. New York: Praeger.

Langgut, Dafna, Israel Finkelstein, and Thomas Litt. 2013. Climate and the Late Bronze Collapse: New Evidence from the Southern Levant. *Tel Aviv* 40(2): 149–175.

Lemche, Niels Peter. 1985. *Early Israel: Anthropological and Historical Studies on the Israelite Society Before the Monarchy*. Leiden: E.J. Brill.

Liverani, Mario. 2003. *Israel's History and the History of Israel*. London: Equinox.

Marfoe, Leon. 1979. The Integrative Transformation: Patterns of Sociopolitical Organization in Southern Syria. *Bulletin of the American Schools of Oriental Research [BASOR]* 234: 1–42.

Mazar, Amihai. 1997. *Timnah (Tel Batash) I: Text. Qedem 37*. Monographs of the Institute of Archaeology. Jerusalem: The Hebrew University of Jerusalem.

Mendenhall, George E. 1962. The Hebrew Conquest of Palestine. *Biblical Archaeologist [BA]* 25: 66–87.

Merrill, Eugene H. 1996. *Kingdom of Priests: A History of Old Testament Israel*. Grand Rapids, MI: Baker

Na'aman, Nadav. 1981. Economic Aspects of the Egyptian Occupation of Canaan. *Israel Exploration Journal [IEJ]* 31: 172–185.

———. 1988. Pharaonic Lands in the Jezreel Valley in the Late Bronze Age. In *Society and Economy in the Eastern Mediterranean (c. 1500–1000 B.C.)*, ed. M. Heltzer et al., 177–185. Uitgeverij Peters: Leuven.

Noth, Martin. 1960. *The History of Israel*, 2nd ed. London: SCM Press. Trans. A.C Black, from German.

Rainey, Anson F. 2001. Israel in Merneptah's Inscription and Reliefs. *Israel Exploration Journal [IEJ]* 51: 57–75.

———. 2008. Inside, Outside: Where Did the Early Israelites Come From? *Biblical Archaeology Review [BAR]* 34(6): 45–50.

Redford, Donald B. 1990. *Egypt and Canaan in the New Kingdom. Studies by the Department of Bible and Ancient Near East, Vol IV*. Beer-sheva: Ben Gurion University of the Negev.

———. 1992. *Egypt, Canaan and Israel in Ancient Times*. Princeton: Princeton University.

Smith, Mark S. 2002. *The Early History of God: Yahweh and the Other Deities in Ancient Israel*, 2nd ed. Grand Rapids, MI: Eerdmans and Dove.

Stager, Lawrence E. 2001. Forging an Identity: The Emergence of Ancient Israel. In *The Oxford History of the Biblical World*, ed. Michael Coogan, 90–129. London: Oxford University.

Waltke, Bruce K. 1972. Palestinian Artifactual Evidence Supporting the Early Date of the Exodus. *Bibliotheca Sacra* 129: 33–47.

Ward, William A., and Martha Sharp-Joukowsky (eds.). 1992. *The Crisis Years: The 12th Century B.C.: From Beyond the Danube to the Tigris*. Dubuque, IA: Kendall Hunt.

Weinstein, James M. 1981. The Egyptian Empire in Palestine: A Reassessment. *Bulletin of the American Schools of Oriental Research [BASOR]* 241: 1–28.

Weippert, M. 1971. *The Settlement of the Israelite Tribes in Palestine*. Naperville, IL: Alec R. Allenson.

———. 1979. The Israelite "Conquest" and the Evidence from Transjordan. In *Symposia Celebrating the Seventy-Fifth Anniversary of the Founding of the American Schools of Oriental Research (1900–1975)*, ed. F.M. Cross, 15–34. Cambridge, MA: American Schools of Oriental Research.

Wood, Bryant G. 1990. Did the Israelites conquer Jericho? *Biblical Archaeology Review [BAR]* 16(2): 44–58.

———. The search for Joshua's Ai. In *Critical Issues in Early Israelite History*, ed. Richard S. Hess, Gerald A. Klingbeil, and Paul J. Ray Jr, 205–240. Winona Lake, IN: Eisenbrauns.

Yadin, Yigael. 1972. *Hazor. Schweich Lectures on Biblical Archaeology*. Oxford: Oxford University.

Out of Egypt or Out of Canaan? The Exodus Story Between Memory and Historical Reality

42

Nadav Naʿaman

Abstract

The factual background of the Exodus story is the most perplexing issue in biblical historical studies. On the one hand, the Exodus tradition is very old, and its status as the central Israelite foundation story finds remarkable expression in every genre of biblical literature. On the other hand, most scholars doubt the historicity of the story, and generally consider it to be the vague memory of a small group, which was gradually adopted by all other Israelite tribal groups. The contrast between the central place of the Exodus in Israelite memory and its questionable historical status requires explanation. The chapter suggests that the bondage, the suffering, and the miraculous delivery from slavery actually took place in Canaan and that the locus of these memories was later transferred from Canaan to Egypt. The bondage and liberation were experienced by the pastoral groups that later settled in the highlands of the Northern Kingdom. Hence, its central place in the cultural memory of Israel's inhabitants. Since the process of settlement in the Judean highlands took place later and on a limited scale, the memory of the Exodus played only a minor role among Judah's inhabitants.

Scholars dispute the historicity of the Exodus narrative. The range of opinions stretches from those who suggest that the nucleus of the story is basically authentic and the episode reflects an important event in the early history of Israel to those who entirely dismiss the historicity of the story, emphasizing that it was written at a later time and suggesting that it mainly reflects the time of its composition. According to the latter view, the Exodus story is essentially a myth that was formulated at a later time and does not reflect the reality of the Israelites' early history (for recent discussion see: Redford 2011). Between the two extremes lie scholars who accept the historicity of a few details in the story and posit that the story includes a small nucleus of historical events that took place on Egyptian soil and on the way from Egypt to Canaan (for the history of research see recently: Davies 2004; Propp 2006: 735–762; Russell 2009: 1–23, with earlier literature; Schmid 2010: 117–139 and n. 556).

N. Naʿaman (✉)
Tel Aviv University, Tel Aviv-Yafo, Israel
e-mail: nnaaman@post.tau.ac.il

T.E. Levy et al. (eds.), *Israel's Exodus in Transdisciplinary Perspective*,
Quantitative Methods in the Humanities and Social Sciences, DOI 10.1007/978-3-319-04768-3_42,

Scholars do agree that the Exodus tradition was mainly accepted in the Northern Kingdom and finds clear expression in the prophecies of Amos and Hosea and in some early psalms of possible northern origin (Ps 77; 80; 81) (Hoffman 1989: 169–182; Blum 2012: 42–49, with earlier literature). Although the tradition is not mentioned in the original prophecies of Isaiah and Micah, it is dominant in various Judahite texts written in the late First Temple period (Davies 2004: 26–27). These references include the celebration of the Passover in Josiah's 18th year (2 Kings 23:21–23), the Song of the Sea (Exod 15:1–17), the Book of the Covenant (Exod 22:20; 23:9), the early layer of the Book of Deuteronomy (5:15; 15:15; 16:12; 24:18, 22; 26:5–8), and some psalms of possible Judahite origin (Ps 78; 114). It is thus inconceivable that the tradition was unknown in the Kingdom of Judah before the seventh century BCE and suddenly played such an important role in the consciousness of the late First Temple period's ruler and elite. We should better assume that the Exodus tradition was known in both kingdoms but occupied a more important role in the historical memory of the Northern Kingdom.

The prophecies of Amos and Hosea are the earliest available sources reflecting the memory of the Exodus in Israel. The coming out of Egypt is explicitly mentioned three times in the Book of Amos (2:10–11; 3:1; 9:7), five times in the Book of Hosea (2:15 [Heb. 17]; 11:1; 12:9 [Heb. 10], 13 [Heb. 14]; 13:4), and, though indirectly, in three additional Hoseanic references (8:13; 9:3; 11:5). The prophecies unambiguously indicate that the Exodus was considered to be a foundational event in the Northern Kingdom and had the status of a constitutive tradition.

The few eighth century texts available to us are separated by four centuries from the historical event they relate. Analyzing them indicates how the vague historical event was memorialized by the Israelites at the time of writing. The texts reflect the manner in which the eighth century Israelite society conceived of its past in a quest to make it relevant for the present (rather than to maintain historical accuracy of the "event").

What might have been the original event behind the memory of the Exodus? Since I already discussed the question in detail, I will present only the main contours of the solution I suggested (Na'aman 2011).

Analysis of the Exodus story *vis-à-vis* the historical data led scholars to conclude that the story contains a number of characteristics that reflect the late exilic or post-exilic date in which it was written, whereas it is difficult to find distinct characteristics that reflect the period to which the story is attributed. No evidence that might be connected to the Exodus story either directly or indirectly was ever detected in the Egyptian texts. Examination of the archaeological data from Iron Age I sites did not produce any evidence that sheds light on the Exodus story. Were this story not mentioned in the Bible, no scholar would have guessed that even a small group from among those who settled in the highlands on both sides of the Jordan arrived from Egypt (Dever 1997: 67–86; Weinstein 1997: 87–103).

In light of the absence of evidence, some scholars hypothesized that a small group of people of West Semitic origin was initially in Egypt and, upon migrating to Canaan, joined the pastoral groups that settled in the highlands and the adjacent peripheral areas. This group transferred the tradition of bondage to Egypt and the miraculous escape by divine help to the other groups that later settled in the highlands (e.g., Noth 1958: 110–138; Halpern 1992: 87–113, 1993: 89*–96*; Malamat 1997: 15–26; Davies 2004; Blum 2012: 49–63). According to this hypothesis, a certain elite group whose identity cannot be established migrated from Egypt, and in light of its political, social, or religious–cultural influence, was able to transform its exclusive historical memory into an all-encompassing Israelite historical consciousness.

However, the central place of the Exodus tradition in early biblical historiography, prophecy and psalms does not suit this assumption, which fails to explain how an event that was originally connected to a small group became the basis for the central claim of origin and establishment of the people of Israel. In my opinion, the supposition that such an early and deeply entrenched common Israelite tradition could have grown from the experience of a small group does not make sense. Another explanation for this central perception must be sought; and the place to look for it lies in the concept of "mnemohistory," namely, the past as memorialized by the people (Assmann 1998: 8–17). Differently stated, the discussion of the biblical story should not be directed at the past as it really happened, but rather at the manner in which Israelite society shaped its identity by creating a certain image of its own past (Hendel 2001: 601–604, 2005: 57–59). We must concentrate on the way in which the historical memory of the Exodus has emerged and developed. In this context we may inquire regarding the background of the perception that Israel emerged as the people of YHWH at a time when the Israelite peoples were subjugated to Egypt. How did the tradition of liberation from Egyptian bondage became a foundational story of the People of Israel?

In my opinion, the major event underlying the Exodus tradition is the dramatic Egyptian withdrawal from Canaan after the Egyptian bondage reached its peak during the Twentieth Dynasty (Na'aman 2011: 60–69). Over the course of the thirteenth to twelfth centuries, the Egyptians greatly expanded their grasp of Canaan, annexed large territories, and increased their pressure on the city-state rulers. They conducted campaigns, destroyed settlements and deported many local inhabitants to Egypt. They operated against the pastoral nomadic groups—among them the group called "Israel," which lived in the peripheral areas of Canaan. Elements of the Exodus story, such as bondage, suffering and arbitrariness of the government, well reflect the experience of all the inhabitants of Canaan in their contacts with the Egyptian government. Evidently, the Egyptian withdrawal brought relief to all those who lived in Canaan. This event would explain the strong feeling of freedom from the bondage of foreign power that engrained into the memory of the Exodus.

Moreover, the sense of a miracle that happened to the Israelites, so prominent in the Exodus tradition, well reflects the reality of mid-twelfth century Canaan. For hundreds of years, Egypt occupied Canaan and none of its inhabitants could have remembered a different reality than that of Egyptian governance of the land. Suddenly, Egypt retreated from Canaan and its inhabitants became free of foreign rule. No wonder that the withdrawal was conceived as a kind of miracle that the local inhabitants attributed to their God. My suggestion that the historical memory of the release from the bondage of Egypt was originally connected to Canaan fully explains the all-encompassing dimension of the memory involving the entire Israelite people.

It is well known that Egypt ruled Canaan uninterruptedly for about 350 years, and during that long period its involvement was gradually intensified and reached its zenith shortly before the Egyptian withdrawal from Canaan. Notwithstanding, Egypt is absent from all biblical texts that describe Canaan before the conquest and destruction of its major cities. How can we explain the black hole that was opened wide in the biblical memory, in a place where we would expect the preservation of memory? In my opinion, the riddle is resolved by the assumption that the vivid memory of the Egyptian presence in Canaan was absorbed within the Exodus tradition and thus disappeared from the collective Israelite memory. Memorization and forgetfulness complement and nourish each other so that the effect of one dictates in many ways the other. I suggest, with due caution, that the transfer of the memory of bondage and liberation from Canaan to Egypt, and with it the forgetfulness of the memory of the long Egyptian occupation of Canaan, is the result of the shaping of a new identity of the young Israelite society when it settled in the highlands of Canaan. Thus, a segment of the new settlers' past was removed from its local context in Canaan and transferred to the land of the subjugator.

An important element that might be connected to the shifting of memory from Canaan to Egypt is tradition of the YHWH's origin from the southern periphery of Canaan. This tradition is evident in the Song of Deborah (Judg 5:4–5), the Blessing of Moses (Deut 33:2), and the prophecy of Habakkuk (3:3a). Within the Exodus story as well, YHWH reveals himself to Moses in Midian, somewhere in the desert area south of Canaan. Several inscriptions dated to the mid-eighth century discovered at Kuntillet 'Ajrud mention "YHWH of Teman," namely, the God of Israel who is directly connected to the desert areas south of Palestine. According to all these sources, YHWH—the divine leader of the Exodus story—originated from the southern periphery of the settled country (Seir, Sinai, Paran, Teman, Midian). The God and his followers moved from south northward, to the Land of Canaan, and the south-northward movement might have influenced the memory of the Exodus as a movement that proceeded from Egypt to Canaan (Hutton 2010; Leuenberger 2010, with earlier literature; Na'aman 2011: 66–67; Blum 2012: 52–63).

No source illuminates the way in which the Exodus was remembered before the eighth century. Hence, the mechanism by which the memory was shifted from Canaan to Egypt and the identity of the agents of this change are unknown. We can only hypothesize that through a long process, the painful memory of the bondage to Egypt and the miraculous withdrawal of the Egyptians was severed from Canaan and attached to Egypt. The two rival sides of the historical memory—the Israelites and the Egyptians—remained at the center of the plot, but the arena was reversed, so that the retreat from Canaan to Egypt was replaced by a migration from Egypt to Canaan. According to this scenario, all elements of memory that originally were linked to Canaan were gradually absorbed within the memory that connected the Exodus with Egypt, and in this form reached the eighth century BCE writers and prophets.

How should we explain the prominent role of the Exodus memory in the Northern Kingdom and its lesser role in the Kingdom of Judah? Some scholars suggested that the Exodus story was a "charter myth" of the Northern Kingdom, written after the Kingdom's liberation from the Judahite yoke in the time of Jeroboam I and since then holding central place in the ideology of the kingdom (Crüsemann 1978: 111–127; van der Toorn 1996: 287–302, 2001; Albertz 1994: 140–145, 2001; Carr 2011: 477–479). According to this hypothesis, some elements of the biblical narrative of Solomon's oppression and Jeroboam's rebellion reflect the historical reality of the late tenth century BCE. The Exodus story was modeled after the main contours of the story of Israel's subjugation to Solomon and its liberation under the leadership of Jeroboam and served as an early paradigm of the establishment of the Israelite kingdom by divine support (for discussions of the relations of the Exodus and the rise of Jeroboam stories see: Dietrich 1986; Zakovitch 1991: 87–97; Särkio 1998: 165–173, 2000; Oblath 2000; Frisch 2000; Blanco Wissmann 2001; Russell 2009: 27–47). As a model of the establishment of the kingdom, the Exodus story held a prominent place in the Kingdom's ideology. However, the history of Solomon and the story of the foundation of the Northern Kingdom were written long after the period they refer to and were guided by literary and ideological considerations of the author's time. The historical chain of events that led to the foundation of the Northern Kingdom must have been entirely different from the one related in the Book of Kings. Moreover, we know next to nothing about the ideology of the early Israelite monarchy and about the outlines of the Exodus story as related in the early monarchical period. The Exodus as a "charter myth" of the Northern Kingdom was probably shaped only in the eighth century BCE, in the time of Jeroboam II, and the story of Jeroboam's rebellion and his rise to power was probably composed at about the same time.[1] Hence, the "charter myth" hypothesis does not really explain the original place of the Exodus tradition in the Northern Kingdom.

To account for the prominence of the Exodus memory in the Kingdom of Israel, we must emphasize that Israel and Judah were two distinct territorial, sociopolitical, and cultural phenomena. Israel was much larger, richer and more densely populated compared to its southern neighbor. The surveys revealed that the main concentration of Iron Age I–IIA sites was found in the northern part of the highlands, between the Jezreel Valley and Jerusalem and that most of the highland

[1] Some scholars dismissed the suggestion that the Exodus story reflects an authentic memory of the early history of Israel. See recently Berner (2011). Other scholars dismissed the idea that the story of Jeroboam's rise to power is based on a pre-Deuteronomistic text and suggest that it is a fictive description without any basis in a chronistic or novelistic source. See Hoffmann (1980: 59–73), Berlejung (2009: 16–24).

population in the eleventh to ninth centuries was located in Israel's territory. In contrast, the highlands of Judah were sparsely inhabited until the Iron Age IIB (Finkelstein 1988; Zertal 1994; Ofer 1994; Lehmann 2003). Moreover, Israel emerged as a full-blown state already in the early ninth century BCE, whereas Judah emerged as a full-blown state about a century later (Finkelstein 1995). It is evident that the tradition of the subjugation, suffering and liberation from Egypt was part of the experience of the pastoral groups that wandered in the peripheral areas of Canaan in the thirteenth to twelfth centuries BCE, among them the group called "Israel." When these tribal groups settled the highlands areas on both sides of the Jordan in Iron Age I, they carried with them the living memory of the release from the Egyptian bond as well as the belief in YHWH as their divine savior. About two centuries later, when the institution of monarchy was established in the Northern Kingdom, the experience of the early settlers became an integral part of the inhabitants of the kingdom's cultural tradition. Memory of the Exodus was transferred verbally for centuries until it was finally put in writing, either in the eighth or in the seventh century BCE.

In contrast to the early, large-scale settlement in the highland regions of the Northern Kingdom, the process of settlement and consolidation in the Judean highlands took place later and on a more limited scale, so the experience of subjugation and liberation from a foreign yoke might not have formed part of the early settlers' living memory. Moreover, the city of Jerusalem rose to power in rivalry with its northern neighbor. As the Exodus tradition played such prominent role in the Northern Kingdom, Jerusalem's rulers and elite must have shaped a different memory of the kingdom's past—one that emphasized its separate origin and distinct identity *vis-à-vis* its strong northern neighbor. The Exodus gained prominence in Judah only after the Assyrian conquest and annexation of the Northern Kingdom, and since then gradually integrated into Judah's historical memory and ideology. King Josiah, in his efforts to take over the ancient Israelite identity, was probably the first king who officially celebrated the Passover in the temple of Jerusalem and thereby sealed the Exodus official place in the historical memory of the Kingdom of Judah.

I suggest that both the centrality of the Exodus tradition in the Northern Kingdom and its marginality in the Kingdom of Judah demonstrate the tradition's great antiquity. Its acceptance in the north is the result of the early settlement in the highlands of many population groups that experienced the suffering, subjugation and liberation from the Egyptian yoke. In contrast, those who settled in the Judean highlands did not directly experience the Egyptian oppression. Hence, the memory of the Egyptian withdrawal from Canaan and the sudden release from the Egyptian bondage did not play significant role in their cultural memory.

In sum, the understanding that the story of the Exodus reflects the historical memory of the early Israelite society in Canaan opens a small window for recognition of the consciousness of the early settlers in the highlands. Assuming that this is indeed the case, the Exodus story might be considered the earliest source available for research into the cultural–religious worldview of early Israelite society in the twelfth to eleventh centuries BCE.

References

Albertz, R. 1994. *A History of Israelite Religion in the Old Testament Period*, From the Beginning to the End of the Monarchy, vol. 1. Trans. J. Bowden, from German. Westminster: John Knox.

———. 2001. Exodus: Liberation History against Charter Myth. In *Religious Identity and the Invention of Tradition*, Papers Read at a Noster Conference in Soesterberg, January 4–6, 1999, eds. J.W. van Henten, and A. Houtepen, 128–143. Studies in Theology and Religion, vol. 3. Assen: Royal van Gorcum.

Assmann, J. 1998. *Moses the Egyptian. The Memory of Egypt in Western Monotheism*. Cambridge, MA: Harvard University Press.

Berlejung, A. 2009. Twisting Traditions: Programmatic Absence-Theology for the Northern Kingdom in 1 Kgs 12:26–33* (the "Sin of Jeroboam"). *Journal of Northwest Semitic Languages* 35(2): 1–42.

Berner, C. 2011. The Egyptian Bondage and Solomon's Forced Labor: Literary Connections between Exodus 1–15 and 1 Kings 1–12? In *Pentateuch, Hexateuch, or Enneateuch? Identifying Literary Works in Genesis through Kings*, Society of Biblical Literature: Ancient Israel and Its Literature, vol. 8, ed. T.B. Dozeman, T. Römer, and K. Schmid, 211–240. Atlanta, GA: Society of Biblical Literature.

Blanco Wissmann, F. 2001. Sargon, Mose und die Gegner Salomos. Zur Frage vor-neuassyrischer Ursprünge der Mose-Erzählung. *Biblische Notizen* 110: 42–54.

Blum, E. 2012. Der historische Mose und die Frühgeschichte Israels. *Hebrew Bible and Ancient Israel* 1: 37–63.

Carr, D.M. 2011. *The Formation of the Hebrew Bible—A New Reconstruction*. Oxford: University Press.

Crüsemann, F. 1978. *Der Widerstand gegen das Königtum. Die antiköniglischen Texte des Alten Testament und der Kampf um der frühen israelitischen Staat*, Wissenschaftliche Monographien zum Altes und Neuen Testament, vol. 49. Neukirchen-Vluyn: Neukirchener Verlag.

Davies, G. 2004. Was There an Exodus? In *In Search of Pre-Exilic Israel*, Journal for the Study of the Old Testament Supplement, vol. 406, ed. J. Day, 23–40. London: T & T Clark.

Dever, W.G. 1997. Is There Any Archaeological Evidence of the Exodus? In *Exodus: The Egyptian Evidence*, ed. E.S. Frerichs and L.H. Lesko, 67–86. Winona Lake, IN: Eisenbrauns.

Dietrich, W. 1986. Das harte Joch (1 Kön 12,4). Fronarbeit in der Salomon Überlieferung. *Biblische Notizen* 34: 7–16.

Finkelstein, I. 1988. *The Archaeology of the Israelite Settlement*. Jerusalem: Israel Exploration Society.

———. 1995. The Great Transformation: The 'Conquest' of the Highlands Frontiers and the Rise of the Territorial States. In *The Archaeology of Society in the Holy Land*, ed. T.E. Levy, 349–365. London: Leicester University Press.

Frisch, A. 2000. The Exodus Motif in 1 Kings 1–14. *Journal for the Study of the Old Testament* 87: 3–21.

Halpern, B. 1992. The Exodus from Egypt, Myth or Reality? In *The Rise of Ancient Israel*, Symposium at the Smithsonian Institution, October 26, 1991, ed. H. Shanks, 87–113. Washington, DC: Biblical Archaeological Society.

———. 1993. The Exodus and the Israelite Historian. In *Avraham Malamat volume*, eds. S. Ahituv and B.A. Levine, 89*–96*. Eretz-Israel, vol. 24. Jerusalem: Israel Exploration Society.

Hendel, R. 2001. The Exodus in Biblical Memory. *Journal of Biblical Literature* 120: 601–622.

———. 2005. *Remembering Abraham. Culture, Memory, and History in the Hebrew Bible*. Oxford: University Press.

Hoffman, Y. 1989. A North Israelite Typological Myth and a Judaean Historical Tradition: The Exodus in Hosea and Amos. *Vetus Testamentum* 39: 169–182.

Hoffmann, H.-D. 1980. *Reform und Reformen: Untersuchungen zu einem Grundthema der deuteronomistischen Geschichtsschreibung*, Abhandlungen zur Theologie des Alten und Neuen Testament, vol. 66. Zürich: TVZ, Theologische Verlag.

Hutton, J.M. 2010. Local Manifestations of Yahweh and Worship in the Interstices: A Note on Kuntillet 'Ajrud. *Journal of Ancient Near Eastern Religions* 10: 177–210.

Lehmann, G. 2003. The United Monarchy in the Countryside: Jerusalem, Judah, and the Shephelah during the Tenth Century B.C.E. In *Jerusalem in Bible and Archaeology: The First Temple Period*, ed. A.G. Vaughn and A.E. Killebrew, 117–162. Atlanta, GA: Society of Biblical Literature.

Leuenberger, M. 2010. Jahws Herkunft aus dem Süden. Archäologische Befunde—biblische Überlieferungen—historische Korrelationen. *Zeitschrift für die Alttestamentliche Wissenschaft* 122: 1–19.

Malamat, A. 1997. The Exodus: Egyptian Analogies. In *Exodus: The Egyptian Evidence*, ed. E.S. Frerichs and L.H. Lesko, 15–26. Winona Lake, IN: Eisenbrauns.

Na'aman, N. 2011. The Exodus Story: Between Historical Memory and Historiographical Composition. *Journal of Ancient Near Eastern Religions* 11: 39–69.

Noth, M. 1958. *The History of Israel*. Trans. A and C. Black, from German. London: Adam & Charles Black.

Oblath, M.D. 2000. On Pharaohs and Kings—Whence the Exodus? *Journal for the Study of the Old Testament* 87: 23–42.

Ofer, A. 1994. 'All the hill country of Judah': From a Settlement Fringe to a Prosperous Monarchy. In *From Nomadism to Monarchy: Archaeological and Historical Aspects of Early Israel*, ed. I. Finkelstein and N. Na'aman, 92–121. Jerusalem: Israel Exploration Society.

Propp, W.H.C. 2006. *Exodus 19–40. A New Translation with Introduction and Commentary*, The anchor Bible. New York: Doubleday.

Redford, D.B. 2011. Some Observations on the Traditions Surrounding "Israel in Egypt". In *Judah and the Judeans in the Achaemenid Period. Negotiating Identity in an International Context*, ed. O. Lipschits, G.N. Knoppers, and M. Oeming, 279–364. Winona Lake, IN: Eisenbrauns.

Russell, S.C. 2009. *Images of Egypt in Early Biblical Literature. Cisjordanian-Israelite, Transjordanian-Israelite and Judahite Portrayals*, Beihefte zur Zeitschrift für die alttestamentliche Wissenschaft, vol. 403. Berlin: Walter de Gruyter.

Särkio, P. 1998. *Exodus und Salomo: Erwägungen zur verdeckten Salomokritik anhand von Ex 1–2; 5; 14 und 32*, Schriften der Finnischen Exegetischen Gesellschaft, vol. 71. Helsinki: Finnische Exegetische Gesellschaft.

———. 2000. Concealed Criticism of King Solomon in Exodus. *Biblische Notizen* 102: 74–83.

Schmid, K. 2010. *Genesis and the Moses Story. Israel's Dual Origin in the Hebrew Bible*. Trans. J.D. Nogalski, from German. Siphrut: Literature and theology of the Hebrew Scriptures, vol. 3. Winona Lake, IN: Eisenbrauns.

van der Toorn, K. 1996. *Family Religion in Babylonia, Syria, and Israel: Continuity and Change in the Forms of Religious Life*, Studies in the History and Culture of the Ancient Near East, vol. 7. Leiden: E.J. Brill.

———. 2001. The Exodus as Charter Myth. In *Religious Identity and the Invention of Tradition*, Papers Read at a Noster Conference in Soesterberg, January 4–6, eds. J.W. van Henten, and A. Houtepen, 113–127. Studies in Theology and Religion, vol. 3. Assen: Royal van Gorcum.

Weinstein, J.M. 1997. Exodus and Archaeological Reality. In *Exodus: The Egyptian Evidence*, ed. E.S. Frerichs and L.H. Lesko, 87–103. Winona Lake, IN: Eisenbrauns.

Zakovitch, Y. 1991. *"And You Shall Tell Your Son…": The concept of the Exodus in the Bible*. Jerusalem: Magnes Press.

Zertal, A. 1994. 'To the Land of the Perizzites and the Giants': On the Israelite Settlement in the Hill Country of Manasseh. In *From Nomadism to Monarchy: Archaeological and Historical Aspects of Early Israel*, ed. I. Finkelstein and N. Na'aman, 47–69. Jerusalem: Israel Exploration Society.

Part IX

Conclusion

43 Modern Scholarship Versus the Demon of Passover: An Outlook on Exodus Research and Egyptology through the Lens of Exodus 12

Thomas Schneider

Abstract

This study presents an example of the problems of Exodus research. Any attempt to trace and contextualize motifs of the narrative is obstructed by the complexity of the text's history. Exegetical certainties of the twentieth century have vanished in the crisis of Pentateuchal research and given way to multiple scenarios of text composition and redaction, the interrelationship of major themes, and the provenance and historical context of phenomena mentioned in it. The received text of Exodus 12 describes the last plague brought onto Egypt by Yahweh—the killing of Pharaoh's firstborn son and the firstlings of the country's livestock—by Yahweh or alternatively, his "destroyer" who strikes the Egyptians but spares the homes of the Israelites. Several aspects of the Passover protection ritual have not yet been explained in a satisfactory way. After giving an overview of the intricate exegetical situation, this study proposes a new approach to the text by drawing on parallels from Egyptian rituals which would have been appropriated by the text's authors for the Israelite cause. Particular attention will be given to Pap. Cairo 58027, a ritual for the protection of Pharaoh at night, and rituals aimed at the "Plague of the Year."

The grand narrative of the Exodus story has been of momentous significance for ancient Israel, for Jewish identity, for Western thought, and for the formation of the academic disciplines studying the ancient Near East. The foundation myth of a people and a religion, the exodus became a written and cultural artifact in the first millennium BCE, a canvas of continued ideological imagination. The proceedings gathered in this volume are testament to the diversity of attempts dealing with this grand narrative: from its textual genesis to its reception history, from the origin of its literary motifs to the narrative's purpose as a carrier of cultural memory, and beyond the horizon of the text, the question of historicity. There has been a certain consensus that, by virtue of the role this narrative acquired—retold, rethought and reworked over many centuries—the multifacetedness of later engagement with the received idea of a departure from Egypt is such that it eclipses that starting point of mnemohistory. While textual and literary

T. Schneider (✉)
University of British Columbia,
Vancouver, BC, Canada

T.E. Levy et al. (eds.), *Israel's Exodus in Transdisciplinary Perspective*,
Quantitative Methods in the Humanities and Social Sciences, DOI 10.1007/978-3-319-04768-3_43,

criticism are far from agreeing on the layers and dates of the narrative, there is today unanimity as to the fact that it grew significantly over time, and that some of its most iconic and distinctive parts are late additions and not part of the oldest tradition. Equally opaque remain the mechanisms that may have created the memory of a departure from Egypt in the first place: Was it the memory of a small group later transferred to Israelite society as a whole? Was the Exodus memory a matrix for repeated emigrations from Egypt to Palestine that became conflated? Was it more generally a space of memory where critical components of Israel's identity could be etiologically anchored? Did the memory of salvation from Egyptian oppression originally pertain not to Egypt but to the period of Egyptian dominance in LBA Palestine? What groups carried that memory and how can we explain its attachment to the Northern Kingdom? And how do those assumptions tie in with our evidence on the emergence of Israel in early Iron Age Palestine, and its later political and social history? Here again, the abundance of modern hypotheses multiplies the historical variables and potential scenarios.

Modern historiography has seen an intense debate about the historian's engagement with historical sources, the nature and existence of historical "facts," and the historian's (in)ability to reach the past. Carlo Ginzburg, in his *Menahem Stern Jerusalem lectures* held in 1999 and published as "History, Rhetoric and Proof," makes the following claim (Ginzburg 1999: 25):

> Sources are neither open windows, as the positivists believe, nor fences obstructing vision, as the skeptics hold: if anything, we could compare them to distorting mirrors. The analysis of the specific distortion of every specific source already implies a constructive element. But construction (...) is not incompatible with proof; the projection of desire, without which there is no research, is not incompatible with the restrictions inflicted by the principle of reality. Knowledge (even historical knowledge) is possible.

Few of us would probably share Ginzburg's positivism. Not only are sources such as the Biblical text much more complex and refracted as to comply with the simple metaphor of a distorted mirror. To rectify the distortion, we would need to be aware of the historical reality in the first place. And as individuals deeply embedded ourselves in the stream of tradition that had its beginnings in the emergence of Israel, it may be more difficult than imagined by Ginzburg to reach an objective understanding of the different realities of the exodus, its remembered and imagined histories. Many historians today would subscribe to the view that history is not knowledge of the past, but a responsible dialogue with what has been preserved from it (Goertz 1995: 88). In this perspective, Exodus research is not research about the factual exodus—which is beyond reach—but a dialogue with the exodus narrative and memory, and the evidence from Iron Age Israel that converted that memory into a feature of its identity (Hendel 2001).

The following case study about the Passover episode in Exodus 12, and in particular the killing of the firstborn by a nocturnal demon, as well as the ritual put in effect to protect the Israelites, is here presented as an example of this dialogue. It shows, on the basis of one central chapter of the Exodus narrative, the complex and often inconclusive debate about the layers of the received text and their date, about the origin and context of the rituals and regulations they contain, and about the people and interests responsible for their genesis and perpetuation. While the multiplicity of options for every single of these hermeneutic steps has so far prevented scholars from reaching a unanimous conclusion, I will adduce new evidence from the field of ancient Egyptian rituals that is apt to provide a new angle of view and to reorientate the dialogue. The struggle of modern scholarship to perceive, behind the multiplicity of exegetical approaches, the ancient reality of the demon, has inspired the title of this contribution: *Modern Scholarship versus the Demon of Passover*.

Exodus 12 in Recent Scholarship

In the received text of the Hebrew Bible, the institution of Passover in Exodus 12 is linked to the tenth and last plague imposed on Egypt,[1] the

[1] On the exegesis of the plague narrative, see Schmidt (1990); Lemmelijn (2009).

killing of the firstborn children and the firstlings of Egypt's livestock, either by YHWH or by his *mašḥit*, the "destroyer". EXODUS 12, 12f.23.29f. reads as follows in the NEW INTERNATIONAL VERSION (I have substituted the original 'Yahweh' for 'the Lord')

> 12 "On that same night I will pass through Egypt and strike down every firstborn of both people and animals, and I will bring judgment on all the gods of Egypt. I am Yahweh. 13 The blood will be a sign for you on the houses where you are, and when I see the blood, I will pass over you. No destructive plague will touch you when I strike Egypt (. . .). 21 Then Moses summoned all the elders of Israel and said to them, "Go at once and select the animals for your families and slaughter the Passover lamb. 22 Take a bunch of hyssop, dip it into the blood in the basin and put some of the blood on the top and on both sides of the doorframe. None of you shall go out of the door of your house until morning. 23 When Yahweh goes through the land to strike down the Egyptians, he will see the blood on the top and sides of the doorframe and will pass over that doorway, and he will not permit the destroyer to enter your houses and strike you down. (. . .) 29 At midnight Yahweh struck down all the firstborn in Egypt, from the firstborn of Pharaoh, who sat on the throne, to the firstborn of the prisoner, who was in the dungeon, and the firstborn of all the livestock as well. 30 Pharaoh and all his officials and all the Egyptians got up during the night, and there was loud wailing in Egypt, for there was not a house without someone dead.

This most momentous infanticide described in the Hebrew Bible is the narrative trigger for Israel's exodus to the promised land, "a defining moment, perhaps the defining moment in ancient Israelite tradition" (Gruen 1998: 93). I will first provide an overview of the debates on the exegesis of Exod 12 (2.1), on the relationship between the Passover and the Plague/Exodus narratives (2.2), and on the proposed origins of the protection ritual contained in the Passover narrative (2.3), before presenting ancient Egyptian rituals that can elucidate important elements of the Passover narrative.

The Textual, Literary, and Redaction History of Exod 12

Scholarship on Ex 12, 1–20 commonly distinguishes a P section in vv. 1–13. For the following verses 14–20, different interpretations have been proposed. I mention three possibilities: Propp identifies as P as well vv. 14–17a.18–20, and as a redactor's note, 17b (Propp 1999: 374).[2] Albertz views vv. 1–13 on the Passover law as the original priestly source (PB^1) with the *Maṣṣot* prescriptions (vv. 14–17: PB^2),[3] vv. 18–20 (PB^4) and v. 12b (PB^5) as later priestly additions (Albertz 2012: 21–23.25f.199f.). Utzschneider and Oswald regard vv. 1–13.18–20 as priestly and vv. 14 und 15–17 as late editorial additions at the time of the redaction of the Tora ("Tora-Komposition") (Utzschneider and Oswald 2013, 280–289). The following verses 21–27 are regularly set apart by scholars as a passage of different origin. Of all texts analyzed in Pentateuchal literary criticism, this section is among those with the most controversial assessments (Bar-On 1995: 18; Utzschneider and Oswald 2013: 281). The recent demise of former certainties in the historical-critical analysis of the Pentateuch has increased the difficulties, to the extent that Fischer and Markl, in their 2009 Exodus commentary, have decided to focus on the transmitted text, regarding it as impossible to reconstruct earlier stages of text formation (Fischer and Markl 2009: 24). Similarly, Diana Edelman decided "not to include the initial Exodus story in Exod 12–13 in my review because the date when this aetiological legend that commemorates the Exodus was first developed and then subsequently committed to its present form is unknown" (Edelman 2012). Traditional exegesis identified this text as J or partially J, such as Gerhard von Rad who recognized J in vv. 21–23.27b, with a dtr addition in 24–27a (von Rad 1961: 72). Propp regards the presence of "D-like material" as undeniable (1999: 377). He identifies the non-P section of Ex 12 as entirely Elohistic, "with the author quoting a D-like document or more likely

[2] He considers that alternatively, v. 14 might be R, as might be 18–20.

[3] According to Albertz, PB^2 is close to the redactor of the Holiness Law in Lev 17–26.

using D-like language himself."[4] The crisis of Pentateuchal criticism has now removed the Yahwist and the Elohist as undisputed interpretive models.[5] Instead, scholarship has adopted the umbrella term "Non-P" for the non-priestly stages of the Pentateuch. Dozeman notes the affinity of vss. 21–27 with Deuteronomy and the Deuteronomistic History and assigns them to the "Non-P History" which he places in the exile or more likely, in the post-exilic period, but before P (Dozeman 2009: 39f.).[6] At the same time, he assumes vv. 21–23 to include an earlier "ancient rite." David Carr sees vv. 25–27 as a secondary D-like instruction to the early non-P Passover law in Ex 12:21–24 (Carr 2011: 267).[7] According to Schmidt (2005), the older, non-P layer consists of 12, 21–23.27b.

Some scholars, unlike the majority of exegetes, consider vv. 21–27 to be younger than the priestly text in vv. 1–13(-20). Bar-On (1995) suggests that 21–27 is a commentary-like priestly continuation of 1–20. Gertz (2000: 38–50) recognizes both priestly and Deuteronomistic elements in vv. 21–27 and assigns them to the final redaction of the Pentateuch. Albertz (2012: 220f.212) ascribes the verses to the late-Deuteronomistic redactor D whom he places chronologically later than his priestly layers PB^1 and PB^2. Berner (2010: 331–339) sees in vv. 21–27 a post-priestly supplement in which three layers of halakhic regulations (12:21–23; 12:25; 12:26.27a) and a narrative addition (12:27b) appear amalgamated.

In turn, arguments for a date of these verses prior to the priestly passage of vv. 1–13.18–20 have again been proposed in Utzschneider and Oswald's latest (2013) evaluation of the literary history of Exodus 12 which identifies 12, 21–27 as a uniform piece belonging to the Deuteronomistic history.[8] While Dozeman (2009: 39f.) concedes that "a firm date for the Non-P History is not necessary for interpreting Exodus," this is certainly not true for a historical contextualization of the texts in their development during the first millennium BCE.[9] Given the common acceptance of Deuteronomistic material in Ex 12, 20–27, any dating depends on assumptions about the Deuteronomistic tradition where the state of

[4] "We might impute the D-like diction to a literary *topos* rooted in didactic tradition. It need not be specific to a single source" (Propp 1999: 377).

[5] For the history of scholarship, see Carr (2012: 11f.) Cf. Schmid (2006: 30 n. 4) (the observations that led Graupner to reemphasize the existence of the Elohist are true but they lead to the distinction between P and non-P texts, not between J and E).

[6] According to Dahm (2003: 137–141), both Ex 12,21–7 and Ex 12,1–14 originated in the same cercles ("gleicher Verfasserkreis"). She assigns 12,1–14 to an "Aharon layer" from the second half of the fourth century BCE, and 12,21–27 to her so-called "Zadokite Moses layer" from the end of the fifth century BCE.

[7] For his literary assessment of the Exodus narrative, see Carr (2012).

[8] See also the diagram in Utzschneider and Oswald (2013), p. 280.

[9] Cf. here the critical comments by Morrow (2012): (http://www.jhsonline.org/reviews/reviews_new/review584.htm):

> For the most part, Dozeman declines to reconstruct the literary history of the non-P and P Histories. One consequence of Dozeman's methodology, therefore, is a certain flattening in historical perspective when accounting for the origins of the book. For some readers this will stand out as a weakness in the commentary. Dozeman does an excellent job of describing the status of source critical debates on various segments of the text, but these concerns are subordinate to the author's stated aims, which are focused on mapping the dimensions of the non-P and P Histories and explaining their viewpoints and relationships. A good example of his approach is found in the discussion about the altar law in Exod 20:24–26. While he recognizes its associations with centralization ideas in Deuteronomy, ultimately he leaves the question open as to whether the law is an older expression of a theological perspective appropriated by Deuteronomy or derivative of Deuteronomic thinking. What the reader is presented with is a description of the function of the altar law within the final form of the non-P History. A particular bias emerges with discussions of proposals regarding post-Priestly redaction(s) of Exodus. While he acknowledges this hypothesis, Dozeman does not appeal to it in his interpretative comments. By implication, this means that post-P insertions or authorial activity are negligible and, for all practical purposes, either non-existent or non-identifiable.

the debate is similarly opaque (cf. Römer 2005; Person 2009).

The Relationship Between Passover and the Plague/Exodus Narrative

The literary and historical relationship between the celebrations commemorating the Exodus in the received text of Exodus 12 (*Pesaḥ* and *Maṣṣot*) and the exodus and plague narratives has been the subject of an intense debate. A view widely held in scholarship today sees the association of the two festivals with the exodus as secondary, the result of a gradual transfer of institutions of different origins to commemorative events of Israel's departure from Egypt (Propp 1999: 429). By way of example, I mention here Willi-Plein (1993: 119), Bar-On (1995),[10] Levinson (1997: 59),[11] Propp (1999: 429),[12] Dahm (2003: 136),[13] Russell (2009: 77),[14] and Albertz (2012: 197). Dozeman has called this process the "historization" of the festivals (2009: 262—"The Passover is clearly historized into the exodus in the P History, where it is merged with the Feast of Unleavened Bread"[15]) although it might be more appropriate to speak of the *mythologization of historical festivals*, festivals that had a historical origin and a history of practice.[16]

Proposals of how the exodus narrative and the festival prescriptions were interwoven over time are very complex. Here, I present the latest three scenarios by Albertz (2012), Utzschneider and Oswald (2013), and Dozeman (2009):

1. Albertz (2012: 200f.212) who assigns vv. 20–27 to the late-Deuteronomistic redactor D posits that this passage presupposes at least the first priestly version (*erste priesterliche Bearbeitung*, PB^1), and that the regulations set forth there make good sense in post-priestly times (which he later defines as post-PB^{1-2}). In his view, Passover was celebrated decentrally in Israelite families in pre-dtn times, probably reflecting a situation of decampment in a nomadic or seminomadic society, such as transhumance, in this case the movement of groups from winter to summer pastures. The blood rite's function would have been apotropaic, to prevent demons of illness to enter the houses (Albertz 2012: 212f.). At the end of the seventh century BCE, the Deuteronomic legislators would have linked Passover to the Festival of Unleavened Bread which according to Albertz had been connected with the exodus since the end of the eighth century BCE. This correlation would have tied Passover to the Exodus and to the temple cult (Albertz 2012: 213). In his view, the killing of the firstborn was genuinely part of the plague narrative and completely unrelated to Passover and *Maṣṣot* (Albertz 2012: 197)—it stands in marked contrast to the Passover ritual which provides protection for all the Israelites (Albertz 2012: 208). Albertz dates the genesis of the plague/exodus *narrative* to the first half of the seventh century, t.i., between the exodus/*Maṣṣot* correlation and the *Pesaḥ*/*Maṣṣot* one, and believes it was formed in response to the Assyrian menace (Albertz 2012: 216).

[10] "Das Passa als außerheiligtümliches Hausopfer mit seinem apotropäischen Hauskult wird zum historischen Passa der Auszugsgeschichte ("Pessach Mizraim") degradiert."

[11] He speaks of the apotropaic origin of the Passover blood ritual "with no inherent connection to the plagues or to the events of the exodus," and sees a sympathetic effect at work (p. 59).

[12] Stating that *Pesaḥ* and *Maṣṣot* only gradually evolved into commemorations of the Exodus.

[13] *Pesaḥ* and *Maṣṣot* were unconnected originally, and practiced independently from the exodus tradition.

[14] "The biblical descriptions of the Passover festival allow for the possibility that an original pastoralist festival may have been unconnected with the exodus, as has been argued. If so, it may have been through the Passover that pastoralist imagery came to be associated with the exodus."

[15] Cf. also p. 261: "The Passover was not originally associated with the Exodus."

[16] Cf. here the statement made by Albertz (2001: 135) according to which, in adopting the Exodus as a charter myth of the Northern kingdom, "a historical event was mythologized in some way."

At the end of the sixth century BCE, the priestly authors (Albertz' "PB[1]") would have restored the domestic celebration of *Pesaḥ* and established a different relation to the Exodus via the blood rite in vv. 12–13. Only when the temple was rebuilt did PB[2] insist on renewing the connection between Passover and *Maṣṣot*, at the beginning of the fifth century BCE (vv. 14–17). A later addition (PB[5]) would have expanded the Passover event to a judgement on all Egyptian gods in 12, 12b (Albertz 2012: 206). As this put the blood rite in danger of receding into the background, the late-Deuteronomistic redactor would have re-emphasized the ritual in modified form in vv. 21–27, for the sake of Israel's cultural memory.[17]

2. Utzschneider and Oswald's scenario (2013: 280–289) takes its point of departure from the pre-dtr "older Exodus account" which reported, after the failed negotiations with the pharaoh, the killing of the firstborn and the release of Israel (12, 29–33). Legal regulations would have been added in the Deuteronomistic *Fortschreibungsphase* for three reasons:
   (a) DtrH's definition of *Pesaḥ* as a pilgrimage festival. Within the context of the narrative, Ex 12, 21–23 pertain to the exodus occurence, whereas vv. 24–27a point beyond the exodus to a recurrent practice, the temple festival. Unlike other scholars, Utzschneider and Oswald regard the prohibition to leave the house as a motif confined to the Egypt event and thus not a contradiction to the Passover regulations in Dtn 16, 1–8. In the dtr version, the blood rite lacks any explanation. Since vv. 21–27 do not mention the firstborn but speak exclusively about Yahweh striking Egypt, it follows that the dtr Passover was linked to the exodus but not in any way to the killing of the Egyptian offspring and the salvation of the Israelite one (as in Dtn 16, 1–8 in contrast with P). While the theme of the firstborn was known to DtrH from the older Exodus narrative and/or the tradition, DtrH links it with the consecration of the firstborn in Ex 13, 11–16 and not with the Passover.
   (b) the establishment of *Maṣṣot* as a reenactment of the Exodus by the DtrH, and
   (c) the correlation of the consecration of the firstborn and the *Maṣṣot* festival. DtrH provides the history of Israelite faith as the justification for the reenactment of exodus (*geschichtstheologische Begründung*).

The priestly author discontinued this correlation by depriving the *Maṣṣot* festival of any *Vergegenwärtigungscharakter* and focussing all historical remembrance on *Pesaḥ* which was understood as a domestic celebration. In P's arrangement, it is no longer the consecration of the firstborn that reenacts the killing of the Egyptian firstborn and the salvation of the Israelite one, but the blood rite (12, 7.13) which was left without any explanation in vv. 20–27. Different layers of P can be distinguished in additions to the text.[18] At the stage of the Torah composition, the final redactor would have harmonized the dtr and P scenarios.

The main discrepancies regarding the earlier history of the rituals addressed in Ex 12 are these: In Utzschneider and Oswald's view, dtr Passover was linked to the Exodus but not to the killing of the Egyptian offspring and the salvation of the Israelite one. The blood rite is part of the dtr Passover account and thus the Exodus nexus. It was only DtrH who established *Maṣṣot* as a reenactment of the Exodus and correlated the consecration of the firstborn and the *Maṣṣot* festival. P's use of the blood rite is a reinterpretation of the blood rite transmitted by DtrH. By contrast, in

[17] Albertz (2012: 206), mentioning that both forms of Passover (home, temple) are attested in post-exilic times, e.g., in Elephantine.

[18] E.g., regarding the year reckoning, the minimization of meat consumption, the explanation of Passover, and additions to the exodus account. No theological explanation is given of the Feast of Unleavened Bread.

Albertz' view, there was a genuine connection between Exodus and *Maṣṣot* dating back to the eighth century. Passover was genuinely a domestic rite unrelated to them but containing an apotropaic blood rite. Deuteronomic legislators would have connected Passover to *Maṣṣot* (and thus implicitly, to the exodus). The blood rite was a mechanism re-employed by P to link Passover and Exodus when they redefined Passover as a domestic celebration.

3. Different again is a third option proposed by Dozeman (2009). He argues (unlike both Utzschneider/Oswald and Albertz) that while Passover is not originally linked to the exodus event, it yet commemorates an occurrence that happened in Egypt: the killing of the firstborn, and Israel's escape from the plague of death. He comments on the Non-P History version of Passover in Ex 12, 21–27 (which he assumes to include an earlier "ancient rite" in vv. 21–23) that if Passover was originally independent, then v. 23 must be seen as an addition to vv. 21–22 "to accommodate the ritual to the narrative context of the exodus" (Dozeman 2009: 273). In *God at War* he elaborates:

> The passover instruction in Exod 12:21-27 presents strong points of similarity to both Exod 13:1-16 and Deut 6:4-25. (. . .) Some of the differences can be accounted for by the place of passover within the deuteronomistic interpretation of the exodus. Passover occurs in Egypt at Ramses and is separated from the promulgation of *Maṣṣot*/firstlings at Succoth. This geographical separation indicates that passover is not a festival about the exodus. Rather it commemorates Israel's escape from the plague of death in Egypt. The absence of the exodus motif reinforces this conclusion. The exodus motif is defined narrowly in deuteronomistic tradition in relationship to Israels march out of Egypt—so narrowly, in fact, that its introduction into the larger story of the exodus has even been separated geographically form the passover legislation. (. . .) The placement of passover at Ramses in Egypt raises the suspicion that this festival is meant to function in the larger context of the plagues, of which the death of the Egyptian firstborn is the culmination. The instruction in 12:27 reinforces such a suspicion, since it states that passover commemorates Israel's protection from the plague of death. Once the close relationship between passover and the larger plague cycle is noted, then the deuteronomistic insertion of instruction at the outset of the plague of locusts (10:1b-2a) also takes on a larger function in relationship with passover (12:21-27) (. . .)" (Dozeman 1996: 57)

This view is also shared by Gertz who holds:

> Der überkommene nicht-priesterliche Textbestand zum Passa *[12,21b-23, as explained in the previous sentence, TS]* gehört vermutlich mitsamt seines szenischen Rahmens in 12,21a.27b—ursprünglich oder redaktionell—zu einer nichtpriesterlichen Plagenerzählung (Gertz 2000: 73).

The killing of the firstborn is part of the non-P plague narrative (Gertz 2000: 185f.).[19] This is precisely the assessment given by William Propp for the narrative thread of E (= non-P according to Dozeman and Gertz) and P: from the announcement of the imminent death of the Egyptian firstborn and that Israel should prepare for departure (10:24–11:8) to the instruction about how to escape Yahweh's destroyer by the *Pesaḥ* ritual in 12:20–27) and the hasty departure. Propp sees the priestly text as a commentary following this outline and supplying details absent from the older legislation, ending abruptly with the consecration of the firstborn in 13: 1–2 (Propp 1999: 379).

The Origins of the Protection Ritual of the Passover Narrative

The blood ritual employed to avert the destroyer (משחית) has seen an intense debate and often been described as a pastoralist and/or domestic ritual.[20] Such assumptions have sometimes acquired the status of firm declarations ("Passover, in contrast, was originally an apotropaic rite from Israel's nomadic period": McConville 2000: 49[21]) although research about Israel's early history has shown that nomadism might

[19] Although Albertz advocates a different literary-critical analysis of Ex 12 than Gertz and Dozeman, he also thinks that the killing of the firstborn was genuinely part of the plague narrative (and completely unrelated to Passover and *Maṣṣot*).

[20] See Albertz (2012: 197.200f.212f); Bar-On (1995); Dahm (2003: 168–172); Propp (1999: 440ff.); Willi-Plein (1993: 119).

[21] Cf. also Hess (2004: 249f).

be at most the form of subsistence of one of many elements that amalgamated in Israel (and, as has been remarked, contradicts the mention of doorposts). The term "domestic," in turn, is suggested by the application of the blood to the doorposts; it does not help in contextualizing the ritual in terms of its actual origin. In their recent Exodus commentary, Utzschneider and Oswald sum up the debate as follows (2013: 255):

> Das Wort מַשְׁחִית ist ein substantivisches Partizip Hifil der Wurzel שׁחת. Es kann als unpersönliches Nomen in der Bedeutung „Vernichtung" oder „Verderben" gebraucht werden oder personal als „Verderber" verstanden werden. Als Nomen wird es häufig „mit le konstruiert, um zum Ausdruck zu bringen, dass das jeweilige Geschehen auf die definitive Vernichtung der Betroffenen abzielt. . ." Diese Bedeutungsvariante liegt auch in Ex 12,13 vor. In Ex 12,23 hingegen ist das Wort mit Bezug auf eine personale Größe gebraucht, auf eben den „Verderber", der an JHWHs Stelle die Erstgeburt der Ägypter tödlich schlägt, von Gott aber daran gehindert wird, dies auch den Israeliten zuzufügen. Die religionsgeschichtliche Forschung ist sich darin einig, dass diese Gestalt als Dämon zu verstehen ist. Die ältere Forschung hat dazu die Hypothese eines Päsachfestes aus der nomadischen Frühzeit der Israeliten aufgestellt, das der Abwehr gefährlicher Dämonen im Wanderleben der Hirten diente. Der nächste „Verwandte" des personifizierten „Verderbers" ist aber eher der „Verderberengel" (מלאך המשחית) in 2Sam 24,16, der wegen einer als sündhaft empfundenen Volkszählung Davids das Volk mit der Pest schlagen sollte. In der Version der Plagenerzählung, die sich in Ps 78,48–51 findet, ist von „Unheilsengeln" (V. 49 מלאכי רעים) die Rede, in deren Gefolge Tod und Pest zunächst über das Vieh und dann auch über die Erstgeborenen der Ägypter hereinbrechen. So spricht viel dafür, dass das Modell für die Handlung, die Mose den Ältesten gebietet, „ein apotropäischer Ritus gegen den Pestdämon" gewesen ist. In Ps 78,48 (vgl. auch Dtn 32,24) erscheinen weitere Unheilsbringer, die רשפים (*rešāpîm*) genannt werden. Religionsgeschichtlich ist die Vermutung naheliegend, dass sich in diesen Gestalten die zum Dämon degradierte kanaanäische Pest- und Unheilsgottheit Rescheph spiegelt.

While many parallels have been adduced to explain the ritual and the destroyer from the modern Middle East (from the Arabic *fidya*—Propp 199: 434–439[22]—to the Moroccan *tfaska* ritual—Dahm 2003: 156–160) and several ancient Near Eastern cultures (e.g., Neo-Assyrian purification rituals; Dahm 2003: 151), it is surprising to notice that for a ritual linked in the plot to one of the plagues imposed on Egypt and purportedly employed there for the first time, none of the many commentators has ever signaled comparable ancient Egyptian rituals.[23] In the next paragraph, I will examine such rituals and explore to what extent they are conducive to the interpretation of Exod 12.

Egyptian Rituals as Comparanda

The formation of the grand exodus narrative in the first millennium BCE owes many of its themes and motifs to Egyptian texts and ideas. The editors of the narrative appropriated such texts and ideas for the Israelite cause, to mention only studies by Rendsburg 1988, 2006; and Schipper 1999, 2001, 2009a, b, c, 2012.[24] While Egyptian texts have often been used for comparisons with Biblical motifs, such texts were mostly taken from the well-published literature of the second millennium, rather than the millennium after the Egyptian New Kingdom (1,100–100 BCE), the time of Israel's historical existence, the composition of the Biblical texts and their later redactions. I also contend that we need to compare Egyptian magical and ritual texts as the most likely source of inspiration because major parts of the Exodus narrative expose magical and ritual activities (the contest between Moses and pharaoh, the plagues, the Passover ritual, the parting of the sea, the Golden Calf). Establishing this level of comparison can shed significant new light on the interpretation and origin of elements of the Exodus narrative,

[22] Like *fidya*, *Pesaḥ* and *Maṣṣot* are for Propp rites of passage (435.443).

[23] Propp (1999: 442) mentions bloodsmearings in fourth century CE Egypt in conjunction with the vernal equinox. This is one millennium after the time of late Israel and Judah and does not seem to have precursors in contemporary Late Period Egypt.

[24] For a comprehensive study of the significance of Egypt in Rabbinic Judaism, see Ulmer (2009).

and on the ritual of the Passover night that interests us here.[25]

A General View: Rituals and the Aversion of Plagues and Threats

In terms of how the ritual action is classified by the Egyptian titles of protection rituals,[26] the focus is either on the type of threat from which protection is sought ("To pacify Sakhmet"), or on the individual person or object that needs protection ("Protection of the flesh of pharaoh"). The most menacing threats were deities (such as Sakhmet, the goddess of plague and pestilence[27]) and demons[28] inflicting illnesses and death, particularly during dangerous periods of the year (e.g., after the annual Nile inundation when the "plague of the year" occurred; at night; during the five epagomenal days [361–365] of the Egyptian year). Such demons often acted in groups and were the executioners of the divine will of major deities, but they also acted individually and were described with their particular demonic features and skills. The most significant targets needing protection were the ordered existence of this world, and the Egyptian king as its earthly guarantor. I give here a number of representative quotes pertaining to the "plague of the year" caused by Sakhmet, from texts ranging from the New Kingdom to the Ptolemaic period:

> The might of the Flame appears (. . .)
> When she has smashed the Two Lands with the fear of her,
> When she has burned everything to ashes,
> As life and death belong to her at her discretion.
> When she has given disaster throughout the two lands
> And assigned all slaughterers to their massacres. (Goyon 2006: 27)[29]

> *(The goddess is addressed)*
> One performs for you the sacrifices in the form of what you love.
> When you turn your face to the south, north, west and east,
> The fear of you is in them,
> Wherever you emerge on the (valley-)roads and the mountains.
> You instill the fear of you in the gods.
> Your plague is throughout the country's populace.
> You devour blood,
> You equip the gods who are over the fetters
> On the days of your "wanderers".[30] (Goyon 2006: 28)

> You have split the mountains with your plagues,
> You have killed all the livestock through fear of you,
> Lady of fear, who causes great trembling,
> Who captures the gods in her fulgor. (Goyon 2006: 41)

> *(The goddess speaks)*
> I am the mighty one, I have power over my enemies,
> I make the demons go out on their tasks.
> Millions tremble before me, being docile. (Goyon 2006: 42)

> *A Spell for Purifying Anything in/from the Plague*
> May your emissaries be burned, Sakhmet!
> Let your slaughterers retreat, Bastet.
> No year(-demon) passes along to rage against my face!
> Your breeze will not reach me!
> I am Horus, (set) over the wanderers, oh Sakhmet,
> I am your Horus, Sakhmet, I am your Unique One, Uto!
> I will not die on account of you—I am the Rejoiced One.
> I am the Jubilated one, oh son of Bastet!
> Do not fall upon me, oh Devourer!
> Tousled ones, do not fall upon me, do not approach me—
> I am the King inside his shrine!
> *[an instruction follows of how to speak this spell over an amulet]*
> A means to scare away the plague,
> to ward off the passing of slaughterers along anything edible,
> as well as along a bedroom. (from Pap. Edwin Smith; Borghouts 1978: 17)

[25] For the debate, see Propp (1999: 427–461).

[26] For general treatments of Egyptian magic and rituals see Schneider (2000); Etienne (2000); Koenig (2002); for rituals in Graeco-Roman Egypt, Quack (2013).

[27] For a recent treatment of some aspects of Sakhmet see von Lieven (2003). For spells against Seth, Fiedler (2011).

[28] Lucarelli (2010, 2011).

[29] I follow Goyon's translation but have made some changes on the basis of the Ptolemaic text here and in the following passages.

[30] A type of demons, see the discussion below.

A correlation between the protection of the king and the protection of Egypt that is visible here occurs in many other texts, such as a Late Period naos from Bubastis (Rondot 1989) or a ritual papyrus in Vienna with the title "Protection of the flesh of pharaoh" (Flessa 2006). In the latter, Sakhmet's plague is additionally averted from cattle, birds, and the fish in the Nile:

Withdraw your Impure one, Sakhmet,
may you loosen your arrow, Bastet.
Your rage to the ground!
This fury of yours which is the plague,
Retreat! (...)
Her rage is in the cattle,
The plague is in the birds,
Her fury is in the fish in the Nile. (Flessa 2006: 54[31])

In Exod 12, the Passover demon slays human firstborns as well as the firstlings of the *b^ehema*, a term used for cattle, domestic livestock, and animals in general. Instead of all the firstborn and firstlings of Exodus 12, a note from the Egyptian *Calendar of Lucky and Unlucky Days* that will be quoted below more specifically says that all newborn children will necessarily die from the plague. In the context of the larger plague narrative of Exod 7:14–11:10 (continued in the killing of the firstborn in Exod 12), it is worth noticing that many motifs of the Biblical plagues have their precise equivalents in ritual texts describing the effects of the annual plague in Egypt, or the actions taken by particular demons.[32]

[31] My translation differs from Flessa's in several instances.

[32] By way of example, I mention the "seven arrows" of the goddess Bastet (a possible form of appearance of Sakhmet), a group of demons that both menaced Egyptians but could also be instrumentalized for the protection of Egypt against enemies (Rondot 1989). The fifth of these demons whose names are attested since the reign of Osorkon I (924–889 BCE) was called "The one who is in the Nile flood who makes blood" (Osing 1998: 253). This could be understood as a demon who creates carnage in the Nile, and thus turns the Nile into blood (Exod 7:17–20).

A Specific View: The Ritual for the Protection of Pharaoh at Night

For the more specific context of Exod 12, I propose to discuss an Egyptian ritual for the protection of pharaoh at night, called "Bedroom of the palace" (Pries 2009; for a general description of the text and its structure, 1–19). It is preserved in Theban Papyrus Cairo 58027 of the late Ptolemaic period and the mammisi (birth sanctuaries of divine children) of Dendera and Edfu where the ritual is transferred to the protection of the child gods of Graeco–Roman temples. When the original text was composed is unclear although the division of the text follows the New Kingdom (and later) Books of the Underworld; the language is Middle Egyptian, although with some interference of later Egyptian grammatical forms (Pries 2009: 2 and passim for the grammatical forms; cf. Flessa 2006: 14). Elements of the ritual are attested in a variety of other ritual texts that are more specifically conceived to offer protection from the annual plague after the Nile inundation, and for mothers and their newborn children, and which date back to the Middle and New Kingdoms. Comparable in structure and purpose is the late apotropaic ritual "protection of the house" which was equally carried out before nightfall, originally for the king and secondarily for the temple house of gods (Pries 2009: 4; for the text Jankuhn 1972).

The ritual of Pap. Cairo 58027 offers protection of the pharaoh during the twelve hours of the night. It contains a specific recitation for every subsequent hour of the night when a protective deity is invoked who has to ward off malevolent demons that would harm pharaoh.[33] Pharaoh

[33] Other texts invoke the divine protection of the king for both the day and the night such as in a ritual comprising four clay spheres ritually equated with four deities and deposited in the places that needed protection (preserved in Edfu and Pap. Vienna Aeg 8426): *One from among you is in the palace of the king of Upper and Lower Egypt/ Another one from among you is behind him./Another one from among you is the protection of his throne./Another one from among you is the protection of his bed chamber/ In order to make him safe day and night, and vice versa.* (Goyon 2006: 125).

himself is equated with a specific god in every hour so as to assume divine powers himself. The standard wish of protection is as follows:

> *May you be vigilant in your hour, may you watch over pharaoh, may you spread terror among (or: make tremble) those who are in the deep night, may you keep the wanderers away from his bedroom. (Pries 2009 passim)*

The malevolent demons are called *wš3.w* "Those of the night" and *šm3y.w* "wanderers," in related texts also *ḫ3ty.w* "slaughterers" (and other terms; see the texts cited above), and are seen as being sent by Sakhmet or Bastet, variant appearances of the goddess of plague and pestilence. The designation *wš3.w* used for the demons ("those who are in the night") uses an Egyptian term (*wš3*) denoting the hours around midnight (the deep night), when according to other texts Osiris was believed to have been killed and Re was most remote on his nocturnal journey. For the sixth hour leading to midnight, the Theban papyrus is not preserved. In the version from Dendara, the pharaoh is indeed identified with the underworldly (Osirian) form of Re and is protected by Isis and Min, whereas as the attacking demon the Seth animal is mentioned. In the seventh hour (the hour after midnight), the king is given the protection of the sun god in the form of a winged scarab amulet and is equated with the sun god himself in an attempt to dispel the darkness of the night: "*The appearance of pharaoh is that of a living lion whose eyes are fire, and whose face is daylight*" (Pries 2009: 44; cf. 46f. for parallels of this motif). The endangerment by demonic forces at night is also a recurrent theme in other texts: In spells for the "Lighting of the Torch," the torch is identified with Re and said to dispel "Seth's might which is in the darkness" (Luft 2009: 43), and Pyramid Text §1334 already invokes protective gods to "*slay Seth and protect the Osiris NN (the deceased king) from his hand until it dawns*" (Luft 2009: 37).

The text of the hourly recitations is followed by a detailed description of the ritual procedure: images of the protective deities invoked for the successive hours of the night were drawn in ochre around the king's bed, and additionally an *udjat* eye in front of the bed in whose pupil the king had to be seated at the beginning of the ritual (more likely, a figurine of the king). In addition, the recipient of the ritual was anointed with a special unguent whose ingredients are explicated in a prescription (Pries 2009: 87–90), and the same ointment was to be placed on "every window [or: opening, *wsj*] of his house" (Pries 2009: 89). The procedure of applying such unguents to windows and doors of the house reoccurs in a supplement to the papyrus aimed at the specific protection of the king in the night of New Year (Pries 2009: 95.98)[34] and in other ritual texts (Pries 2009: 89). In the case of bedrooms that were already contaminated with the plague of the year, brushing them with specific medicinal plants would cleanse them again (Pap. Edwin Smith 19, 18–20, 8: Pries 2009: 89 n. 511), a motif noteworthy with regard to the use of a bunch of marjoram (Syrian hyssop, Propp 1999: 407) to distribute the blood in Exod 12: 22.

The ointment contains 5 (or 6, in the second prescription) medicinal plants, anointing oil, goose fat, honey, and another liquid (Pries 2009: 88.94f.). In more general terms, anointing is an activity providing power and legitimacy (Martin-Pardey 1984). It is interessing to notice that ointments played a significant role in the protection of Osiris at night in the *Stundenwachen* performed for him (Junker 1910: 2). The fragrance of unguents also had a smoothing effect, maybe desirable in the appeasement of Bastet, a form of the plague goddess, who seems to have had a close genuine association with ointments (Capel 1996: 209[35]).

In contrast to the Egyptian ritual texts, in Exod 12 blood from sacrificed sheep or goat is applied on the doorposts. On the significance of this rite, opinions differ widely, ranging from proposals to see it as a repellent, to the use of blood in purification rites (see Propp 1999:

[34] This is interesting with regard to the proclamation of the Passover event as the beginning of a new system of time-reckoning in Exod 12,2, a reference either to a different calendrical system or a symbolic new era, a question extensively discussed in the literature.

[35] Her name means "The one from the city of ointment."

435–439; Eberhart 2002, 277f.; Dahm 2003: 141–168; Weidemann 2004: 427–429).[36] From an Egyptological viewpoint, the blood of sacrificial animals can signify the blood of killed enemies of the world order and signal the triumph over inimical forces.[37]

The Passover Demon in an Egyptian Perspective

On the demonic nature of the *mašḥit* who strikes blindly and does not tell the Israelite houses from the Egyptian ones, cf. the comments made by W.H.C. Propp and J.D. Levenson:

> The paschal blood rite seems out of character with biblical theology. Throughout Exodus 7-11, Yahweh easily distinguished between Hebrew and Egyptian households, without the help of blood. But now, even though the Destroyer is an aspect of God himself , Yahweh instructs Israel to treat it as an amoral being that slays blindly unless checked—i.e., a demon; compare Yahweh's quasi-demonic behavior in 4:24-26. The very name *Pesaḥ* 'protection' suggests inherent apotropaic powers. Not too far beneath the surface, then, we glimpse a primitive Israelite or pre-Israelite belief that, in some fashion, the paschal blood averts a supernatural threat. (Propp 1999: 436)

> The Destroyer is YHWH in his aspect of slayer of the first-born son. This is not an aspect of the Deity that the biblical tradition is inclined to celebrate, and for obvious reasons. It is, after all, an aspect that recalls Molech and the monster on the Pozo Moro Tower more than the gracious and delivering god of the Exodus. (Levenson 1993: 46)

Rather than to ascribe this feature as a "primitive Israelite belief" and "out of character with Biblical theology," it seems a genuine example of *divine* and *demonic ambiguity*, well represented in deities and demons engineering disaster in ancient Egypt, as we will see below. In a response to an article by Rita Lucarelli (Lucarelli 2011), David Frankfurter has recently commented on such *demonic ambiguity*:

> Lucarelli's paper demonstrates how the identification *of* the demonic and protection *from* the demonic in ancient cultures involved, first and foremost, the description of *liminality*: what lies *between* here and there, *beyond* the village or *before* sacred zones; how to evaluate the ambiguous theophany, the *wrath* of a temple god, and the vicious weaponry of a gate-protector. By describing, naming, and listing liminal zones and beings you can control their powers; you can *re-place* those beings as either minions of demon-masters, or as verbally vanquished by a protector god, or as fulfilling a *locative* function as a gate-keeper that the incomer can pacify with the right spell. As hard as it may be to understand from a modern (American) vantage, steeped as we are in apocalyptic Protestantism, demons in the ancient and late antique Mediterranean world were rarely "evil" but rather hovered in a zone of uncertainty. (Frankfurter 2011: 129)

As 2 Sam 24,16 and Ps 78,48–51 indicate (see Utzschneider and Oswald's comments cited above), the *mašḥit* of Exod 12 was most likely the demon of plague. William Propp also quotes *Jub 49:15* from the second century BCE as articulating *Pesaḥ*'s primal significance: "The plague will not come to kill or to smite during that year when they have observed the Passover in its (appointed) time" (Propp 1999: 437). This evidence correlates the *mašḥit* closely with the demons sent by the Egyptian goddess of plague, Sakhmet/Bastet, both of whom display merciless and all-encompassing rage. In the *Book of the Heavenly Cow*, Sakhmet has to be averted from killing the humans by a trick: a field is flooded with beer made to appear like blood by mixing it with red ochre, and the blood-thirsty Sakhmet thus becomes drunk and placated (Hornung 1997). The hemerologies of the New Kingdom indicate for the third month of the inundation season, day 20:

> *Coming forth of Bastet, the lady of Ankhtawy, in front of Re,*
> *so furious that the god could not withstand in her proximity.*
> *Whoever is born on this day dies of the plague of the year* (Leitz 1994: 134)

[36] According to C. Berner, it would be no more than a narrative invention by P to provide an etiology for Passover (Berner 2010: 84f.).

[37] Sacrifical animals are equated with evil forces; in a mythological episode preserved in the Ritual of "Breaking open the soil," Seth and his followers who had taken the form of goats to attack Osiris, were slaughtered; their blood soaked the soil (Guglielmi 1975).

As the most powerful divine source of destruction, the goddess was at the same time the most powerful potential defender of human life (cf. the hymns in Goyon 2006: 34, 92) if it was possible to placate her and to obtain her mercy. Sakhmet offers protection to whoever was not struck by the death inflicted by her "wanderers" (Goyon 2006: 76); Bastet protects the king from the slaughterers of Atum and their arrows; she is a bastion against any malignant fever, and against any evil wind of the year (Goyon 2006: 93). Their avatar Hathor/Ernutet is called "the shutter of a window that cannot be opened" (Goyon 2006: 75), as opposed to the window as a dangerous access way for demons that needed to be protected (as aimed at in the ritual). A supplication to Sakhmet-Bastet-Ernutet asks for the king to be spared:

> *O goddess, you have spoken with your own mouth,*
> *Spare your beloved son, the son of Re,*
> *From the wanderers who are in your retinue.*
> *Do not cause the destruction of the king of Upper and Lower Egypt*
> *In the uproar since he is a member of your crew.*
> *He knows your name and he knows the name of the wanderers and the slaughters in your following.*
> *May you smite, in his place, one man from the million in the plague of the year.* (Goyon 2006: 116)

This twofold nature was a preponderant theme in Egyptian theology and the need of appeasement a major cultic concern. The ritual texts describe in detail placation offerings for these goddesses that were carried out at dawn, mainly consisting of meat and bread.[38] In the first millennium BCE, this cultic worship for the purpose of appeasement was also extended to demons who were thus turned into protective deities; an interesting case for this *demonic ambiguity* comprises 12 gods as in the ritual "Bedroom of the Palace" adduced above (Raven 1997[39]).

Conclusion and Historical Context

Commentators on Exod 12 perceive it as an amalgamation of rituals from different original contexts that were secondarily reinterpreted and combined with the plague and Exodus narrative. In his encyclopedic Exodus commentary, William Propp identifies three major questions about *Pesaḥ* (Propp 1999: 440): "*Why is it limited to the evening, the doorway and the springtime?*" He concludes that these aspects of the festival can be answered by seeing it as a specialized form of the *fidya* ritual known from the Islamic Middle East—if the latter was indeed pre-Islamic and pre-Israelite. Both *Pesaḥ* and *Maṣṣot* (the latter maybe a rite of annual purification) would later have been attached to the Moses tradition and reinterpreted as commemorative events: "There arose the etiological legend of the first *Pesaḥ* of Egypt, when the rite redeemed Israel from two predicaments: a plague (*via* vicarious sacrifice) and servitude" (Propp 1999: 457). A similar attachment would have occured with the previously independent and pre-Israelite idea of God's special relationship with firstborn sons: "Again, there was an effort to attach the institution to Moses. The story arose that, on the paschal night, Yahweh killed Egypt's firstborn, while Israel, God's firstborn, was ransomed" (Propp 1999: 457).

As there is neither scholarly consensus on the historical context of these rituals and reinterpretations nor a less disparate situation of exegetical scholarship to allow for any clarity about the stages and the historical contexts of the narrative itself, the approach adopted here is more modest: It introduces to the scholarly dialogue Egyptian

[38] E.g., Goyon (2006: 63; 76: offerings of antelopes, cranes, ducks and other meat, bread and white bread, cakes, beer, and frankincense; 92: offerings of grilled meat, fat, and ducks; 104: offerings of different kinds of bread).

[39] Leiden I 346 starts with a copy of the *Book of the Last Days of the Year*, invoking 12 gods identified as "slaughterers who stand in waiting upon Sakhmet, who have come forth from the Eye of Re, messengers everywhere present in the districts, who bring slaughtering about, who create uproar, who hurry through the land, who shoot their arrows from their mouths, who see from afar," followed by a protection ritual "to save a man from the plague of the year (. . .) to placate the gods in the retinue of Sakhmet and Thoth." In the following "Book of the Five Epagomenal Days," a vignette depicts the 12 invoked demons/gods who are drawn on linen amulets and placed "at a man's throat" for his protection.

ritual texts that have been entirely absent from the debate in the past. The suggestion is that the rituals and motifs that the narrative places within an ancient Egyptian context could be indeed borrowed from the Egyptian ritual repertoire, rather than seeing in them a secondary reinterpretation of (pre-)Israelite institutions when there arose a narrative about Israel's salvation from Egypt. This repertoire provides Egyptian parallels for the figure and the ritual context of the demon of Passover sent by Yahweh in the form of the demons sent by the Egyptian goddess of plague Sakhmet during the "plague of the year" and other perilous times such as the epagomenal days and the night. The protection ritual "Bedroom of the Palace" protects the house of pharaoh and his life during the night; it is equally celebrated at nightfall and comprises a ritual not identical but similar to the Passover one in which the openings of the house were anointed in order to avert the attack of demons. The Egyptian rituals also presuppose a similar nexus between a more general plague and, more specifically, the death of pharaoh('s son) as it is visible in Exod 12's continuation of the plague cycle in Exod 7–11.

In this vein, the ritual to prevent the demon from entering the house of Israelites and instead to have the demon kill the king's son (and the other firstborn Egyptians), could be an *appropriation of Egyptian rituals aimed at that very protection*. By removing the divine protection from the Egyptians, assigning it to Israel, and dispatching his slaughtering demon to pharaoh's palace, Yahweh indeed "brings judgment on all the gods of Egypt" (Exod 12:12)—the gods who had failed as the protective deities of the twelve hours of the night.[40] The idea of an adoption of these elements of Egyptian rituals by Israelite authors suits particularly well certain of the exegetical models presented above, such as those suggesting a genuine connection of the plague cycle, the killing of the firstborn, and the blood rite,[41] but it does in no way preclude other scenarios. It is outside the scope of this article to suggest a precise historical scenario in which such an appropriation[42] would have taken place,[43] and probably outside of what can currently be inferred on the basis of the opaque situation of scholarship. If it occurred, it gave ancient Egyptian rituals, embedded and refracted within memories of the exodus story, a continuous reception until the present.[44]

[40] Admittedly, Exod 12:12 is commonly seen as a late addition to the text. As a side note, it is intriguing to notice the invocation of the protective god of the eleventh hour, Horus of Dawn, which states: "You are the perfect golden calf belonging to the breast of Hathor, the appearance of the lord of the sky and of the two lands, the lord of the land of turquoise (= the Sinai)" (Pries 2009: 72.74). It may be interesting to further explore this tradition of Horus as the "golden calf and lord of the Sinai" in the context of the discussion of the golden calf episode in Exodus 32. For the debate on the golden calf, see Schmitt (2000).

[41] E.g., Dozeman's according to which the killing of the firstborn was part of the non-P plague narrative, as was the older passage on the *Pesaḥ* ritual in 12: 20–23. These later verses comprise a protection ritual but lack elements (most importantly, the Passover meal) extant in the P version. It seems feasible that such elements are a secondary addition, and not part of the original ritual practice.

[42] Cf. the demonstration, on the later traditions of the "lepers" episode, by Gruen (1998: 113f.): "Jewish inventiveness expropriated Egyptian myth in order to insert their own heroes, their religious superiority, and even their military triumphs. (...) The Jews freely adapted the Exodus legend and infiltrated native fables in order to elevate their own part in the history of their adopted land." In a similar vein, David Frankfurter (1993: 203f.) has commented on the use of the plague motifs in the Exodus narrative: "But the Jewish use of the motif may be a case of counter-propaganda (attributing to YHWH the power to effect this most traditional horror of Egyptians)."

[43] A general familiarity with the Egyptian "plague of the year" can probably be inferred from the statement Amos 4,10, "I send a plague among you in the way of Egypt." New consideration also deserves the hypothesis that the term *Pesaḥ* is Egyptian ("the striking"; cf. Exod 12: 12, 23, 29; in modification of a hypothesis first suggested by Görg (1988) [festival of "the smiting god"]).

[44] William Propp has proposed that the plague demon itself lives on in a transfigured form: "Though demons may be repelled or ignored, they are harder to kill. Our paschal demon survived, so to speak, by donning various disguises. Most obviously, he was absorbed into Yahweh's persona as the Destroyer. I also suspect that the characterization of the Pharaoh of the oppression as a would-be baby-killer (1:16, 22) owes an unconscious debt to the paschal demon. And the antique sprite is still with us, but defanged, as it were. He has become a kindly being, in fact a Jew. The Hebrews of Egypt bloodied their door frames in order to

Acknowledgments I am indebted to several individuals for help in the preparation of this article: Sara J. Milstein read and commented on my original conference paper. Stephen C. Russell commented on the section on scholarship on Exodus 12. Helmut Utzschneider, Dominik Markl, and Christopher Berner sent me chapters of their publications on Exodus 12 that were unavailable to me.

References

Albertz, R. 2001. Exodus: Liberation History Against Charter Myth. In *Religious Identity and the Invention of Tradition: Papers Read at a NOSTER conference in Soesterberg, January 4–6, 1999*, ed. J.W. van Henten and A. Houtepen. Assen: Van Gorcum.

———. 2012. *Exodus 1-18*, Zürcher Bibelkommentare. Zürich: Theologischer Verlag.

Bar-On, Sh. 1995. Zur literarkritischen Analyse von Ex 12,21-27. *Zeitschrift für die Alttestamentliche Wissenschaft* 107: 18–30.

Berner, C. 2010. *Die Exoduserzählung. Das literarische Werden einer Ursprungslegende Israels*, FAT, vol. 73. Mohr Siebeck: Tübingen.

Borghouts, J.F. 1978. *Ancient Egyptian Magical Texts*. Leiden: Brill.

Capel, A.K., and G.E. Markoe (eds.). 1996. *Mistress of the House, Mistress of Heaven: Women in Ancient Egypt*. Cincinnati, OH: Cincinnati Art Museum.

Carr, D.M. 2012. The Formation of the Moses Story: Literary-Historical Reflections. *Hebrew Bible and Ancient Israel* 1: 7–36.

———. 2011. *The Formation of the Hebrew Bible: A New Reconstruction*. Oxford: Oxford University Press.

Dahm, U. 2003. *Opferkult und Priestertum in Alt-Israel. Ein kultur- und religionswissenschaftlicher Beitrag*, BZAW, vol. 327. Berlin: de Gruyter.

Dozeman, T.B. 1996. *God at War: Power in the Exodus Tradition*. New York, NY: Oxford University Press.

———. 2009. *Commentary on Exodus*, Eerdmans Critical Commentary. Grand Rapids, MI: Eerdmans.

Eberhart, C. 2002. *Studien zur Bedeutung der Opfer im Alten Testament: die Signifikanz von Blut- und Verbrennungsriten im kultischen Rahmen*. Neukirchener Verlag: Neukirchen-Vluyn.

Edelman, D. 2012. Exodus and Pesach/Massot as Evolving Social Memory. In *Remembering (and Forgetting) in Judah's Early Second Temple Period*, FAT, vol. 85, ed. C. Levin and E. Ben Zvi, 161–193. Tübingen: Mohr Siebeck.

Etienne, M. 2000. *Heka. Magie et envoûtement dans l'Egypte ancienne*. Paris: Éditions de la réunion des musées nationaux.

Fiedler, N. 2011. *Sprüche gegen Seth. Bemerkungen zu drei späten Tempelritualen*. Inauguraldissertation zur Erlangung der Doktorwürde. Universität Heidelberg 2011. www.ub.uni-heidelberg.de/archiv/13643.

Fischer, G., and D. Markl. 2009. *Das Buch Exodus*, Neuer Stuttgarter Kommentar. Altes Testament, vol. 2. Stuttgart: Katholisches Bibelwerk.

Flessa, N. 2006. *"(Gott) schütze das Fleisch des Pharao". Untersuchungen zum magischen Handbuch pWien Aeg 8426*, Corpus papyrorum Raineri archeducis Austriae, Bd. 27. München: K. G. Saur.

Frankfurter, D. 1993. *Elijah in Upper Egypt: The Apocalypse of Elijah and Early Egyptian Christianity*. Minneapolis, MN: Fortress Press.

———. 2011. Master-Demons, Local Spirits, and Demonology in the Roman Mediterranean World: An Afterword to Rita Lucarelli. *Journal of Ancient Near Eastern Religions* 11: 126–130.

Gertz, J.C. 2000. *Tradition und Redaktion in der Exodus-Erzählung. Untersuchungen zur Endredaktion des Pentateuch*, FRLANT, vol. 186. Vandenhoeck & Ruprecht: Göttingen.

Ginzburg, C. 1999. *History, Rhetoric and Proof. The Menahem Stern Jerusalem Lectures*. Hanover, NH: University of New England Press.

Goertz, H.-J. 1995. *Umgang mit Geschichte: eine Einführung in die Geschichtstheorie*. Rowohlt: Reinbek b. Hamburg.

Görg, M. 1988. Paesah (Pascha): Fest des "schlagenden" Gottes? In *Biblische Notizen*, vol. 43, 7–11. Reprinted in M. Görg (ed.). 1991. Aegyptiaca—Biblica. Wiesbaden: Harrassowitz, 168–172.

Goyon, J.-C.L. 2006. *Le rituel du sḥtp Sḫmt au changement de cycle annuel. D'après les architraves du temple d'Efou et textes parallèles, du Nouvel Empire à l'époque ptolémaique et romaine*, Bibliothèque d'étude, vol. 141. Le Caire: IFAO.

Gruen, E. 1998. The Use and Abuse of the Exodus Story. *Jewish History* 12(1): 93–122.

Guglielmi, Waltraud. 1975. Art. "Erdaufhacken". *Lexikon der Ägyptologie* 1: 1261–1263.

Hendel, R. 2001. The Exodus in Biblical Memory. *Journal of Biblical Literature* 120: 601–622.

Hess, R.S. 2004. Multiple-Month Ritual Calendars in the West Semitic World: Emar 446 and Leviticus 23. In *The Future of Biblical Archaeology: Reassessing Methodologies and Assumptions*, ed. J.K. Hoffmeier and A.B. Millard, 233–253. Grand Rapids, MI: Eerdmans.

Hornung, E. 1997. *Der Mythos von der Himmelskuh. Eine Ätiologie des Unvollkommenen*, OBO, vol. 46. 3rd edition. Fribourg: Academic Press.

Jankuhn, D. 1972. *Das Buch "Schutz des Hauses" (s3-pr)*. Bonn: R. Habelt.

shunt the Destroyer onto their enemies. Nowadays the traditional Passover Seder ends with the opening of a door and a call for the destruction of infidels. But then a genial spirit is symbolically invited into each house to share the "blood of the grape": none other than the prophet Elijah, harbinger of the Messiah." (Propp 1999: 439)

Junker Hermann. 1910. *Die Stundenwachen in den Osirismysterien nach den Inschriften von Dendera, Edfu und Philae*. Wien: A. Hölder.

Koenig, Y. 2002. *La magie en Egypte: à la recherche d'une définition. Actes du colloque organisé par le Musée du Louvre, les 29 et 30 septembre 2000*. Paris: Documentation française.

Leitz, C. 1994. *Tagewählerei: das Buch ḥ3t nḥḥ pḥ.wy ḏt und verwandte Texte*. 2 vols. Ägyptologische Abhandlungen, vol. 55. Wiesbaden: Harrassowitz.

Lemmelijn, B. 2009. *A plague of texts? A text-critical study of the so-called 'plagues narrative' in Exodus 7:14-11:10*. Brill: Leiden.

Levenson, J.D. 1993. *The Death and Resurrection of the Beloved Son: The Transformation of Child Sacrifice in Judaism and Christianity*. New Haven, CT: YUP.

Levinson, B.M. 1997. *Deuteronomy and the Hermeneutics of Legal Innovation*. Oxford: OUP.

von Lieven, A. 2003. Wein, Weib und Gesang: Rituale für die Gefährliche Göttin. In *Rituale in der Vorgeschichte, Antike und Gegenwart. Neue Forschungen und Perspektiven von Archäologie, Ägyptologie, Altorientalistik, Ethnologie und vergleichender Religionsgeschichte*, Interdisziplinäre Tagung vom 1.-2. Februar 2002 in Berlin, ed. C. Metzner-Nebelsick, 47–55. Rahden (Westf.): Leidorf.

Lucarelli, R. 2010. Demons (benevolent and malevolent). In *UCLA Encyclopedia of Egyptology*, eds. Jacco Dieleman and Willeke Wendrich. Los Angeles. http://escholarship.org/uc/item/1r72q9vv

———. 2011. Demonology during the late Pharaonic and Greco-Roman Periods in Egypt. *Journal of Ancient Near Eastern Religions* 11: 109–125.

Luft, D.C. 2009. *Das Anzünden der Fackel. Untersuchungen zu Spruch 137 des Totenbuches*, Studien zum altägyptischen Totenbuch, vol. 15. Harrassowitz: Wiesbaden.

Martin-Pardey, E. 1984. Salbung. *Lexikon der Ägyptologie* 5: 367–369.

McConville, J.G. 2000. Deuteronomy's Unification of Passover and Massot: A Response to Bernard M. Levinson. *Journal of Biblical Literature* 119: 47–58.

Morrow, W. 2012. Review of Dozeman, T.B.: *Commentary on Exodus* (Erdmans Critical Commentary. Grand Rapids, MI: Erdmans, 2009). *JHS* 12. http://www.jhsonline.org/reviews/reviews_new/review584.ht

Osing, J. 1998. *Hieratische Papyri aus Tebtunis I*. 2 Bde. Kopenhagen: Carsten Niebuhr Institute of Ancient Eastern Studies.

Person Jr., R.F. (ed.) 2009. In Conversation with Thomas Römer: The So-called Deuteronomistic History. *A Sociological, Historical and Literary Introduction* (London: T. & T. Clark, 2005). *JHS* 9.

Pries, A.H. 2009. *Das nächtliche Stundenritual zum Schutz des Königs und verwandte Kompositionen. Der Papyrus Kairo 58027 und die Textvarianten in den Geburtshäusern von Dendara und Edfu*, Studien zur Archäologie und Geschichte Altägyptens, vol. 27. Heidelberg: Heidelberger Orientverlag.

Propp, W.H.C. 1999. *Exodus 1-18. A New Translation with Introduction and Commentary*, The Anchor Bible. New Haven, CT: YUP.

Quack, J.F. (ed.). 2013. *Ägyptische Rituale der griechisch-römischen Zeit*, Orientalische Religionen in der Antike. Tübingen: Mohr Siebeck.

von Rad, G. 1961. *Das zweite Buch Mose: Exodus*, Das Alte Testament Deutsch, Teilband 5. 2nd edition. Göttingen: Vandenhoeck & Ruprecht.

Raven, M.J. 1997. Charms for Protection during the Epagomenal Days. In *Essays on ancient Egypt in honour of Herman te Velde*, ed. J. van Dijk, 275–292. Leiden: Brill.

Rendsburg, G.A. 1988. The Egyptian Sun-God Ra in the Pentateuch. *Henoch* 10: 3–15.

———. 2006. Moses as equal to Pharaoh. In *Text, Artifact, and Image: Revealing Ancient Israelite Religion*, Brown Judaic Studies, vol. 346, ed. G.M. Beckman and T.J. Lewis, 201–219. Providence: Brown Judaic Studies.

Römer, T. 2005. *The So-Called Deuteronomistic History. A Sociological, Historical and Literary Introduction*. London: T. & T. Clark.

Rondot, V. 1989. Une monographie bubastite. *Bulletin de l'Institut Français d'Archéologie Orientale* 89: 249–270.

Russell, S.C. 2009. *Images of Egypt in Early Biblical Literature: Cisjordan-Israelite, Transjordan-Israelite, and Judahite Portrayals*, BZAW, vol. 403. Berlin: de Gruyter.

Schipper, B. 1999. *Israel und Ägypten in der Königszeit. Die kulturellen Kontakte von Salomo bis zum Fall Jerusalems*, OBO, vol. 170. Academic Press: Fribourg.

———. 2001. Kultur und Kontext: Zum Kulturtransfer zwischen Ägypten und Israel/Juda in der 25. und 26. Dynastie. *Studien zur Altägyptischen Kultur* 29: 307–319.

———. 2009a. Ägypten und Israel. Erkenntnisse und Perspektiven. *Theologische Literaturzeitung* 134: 1153–1164.

———. 2009b. Die »eherne Schlange«. Zur Religionsgeschichte und Theologie von Num 21,4-9. *Zeitschrift für die Alttestamentliche Wissenschaft* 121: 369–387.

———. 2009c. Schlangenbeschwörung in Ägypten und Israel. In *Israel zwischen den Mächten—Studien zum Alten Testament in seiner altorientalischen Umwelt (FS S. Timm)*, AOAT, vol. 364, ed. F. Hartenstein and M. Pietsch, 265–282. Münster: Ugarit Verlag.

———. 2012. Egypt and Israel. The ways of cultural contact in the late Bronze- and Iron-Age. *Journal of Ancient Egyptian Interconnections* 4(3): 30–47.

Schmid, K. 2006. The So-Called Yahwist and the Literary Gap Between Genesis and Exodus. In *A Farewell to the Yahwist? The Composition of the Pentateuch in Recent European Interpretation*, SBL Symposium Series, vol. 34, ed. T.B. Dozeman and K. Schmid, 29–50. Atlanta, GA: SBL.

Schmidt, L. 1990. *Beobachtungen zu der Plagenerzählung in Exodus VII 14-XI 10*, Studia Biblica, vol. IV. Leiden: Brill.

———. 2005. Die vorpriesterliche Darstellung in Ex 11,1–13,16*. *Zeitschrift für die Alttestamentliche Wissenschaft* 117: 171–188.

Schmitt, H.-C. 2000. Die Erzählung vom Goldenen Kalb Ex. 32* und das Deuteronomistische Geschichtswerk. In *Rethinking the Foundations. Historiography in the Ancient World and in the Bible*, eds. S.L. McKenzie, Th. Römer, and H.H. Schmid. FS J. van Seters. BZAW 294, 235–250. Berlin: de Gruyter; reprinted in Schmitt H.-C. 2001. *Theologie in Prophetie und Pentateuch. Gesammelte Schriften*. BZAW 310, 311–325. Berlin: de Gruyter.

Schneider, T. 2000. Die Waffe der Analogie. Altägyptische Magie als System. In *Das Analogiedenken. Vorstöße in ein neues Gebiet der Rationalitätstheorie*, ed. K. Gloy and M. Bachmann, 31–85. Freiburg: K. Alber.

Ulmer, R. 2009. *Egyptian Cultural Icons in Midrash*. Berlin: De Gruyter.

Utzschneider, H., and W. Oswald. 2013. *Exodus 1-15*, Internationaler Exegetischer Kommentar zum Alten Testament. Stuttgart: Kohlhammer.

Weidemann, H.-U. 2004. *Der Tod Jesu im Johannesevangelium: Die erste Abschiedsrede als Schlüsseltext für den Passions- und Osterbericht*, BZNW, vol. 122. Berlin: de Gruyter.

Willi-Plein, I. 1993. *Opfer und Kult im alttestamentlichen Israel*. Stuttgart: Katholisches Bibelwerk.

About the Editors

Thomas Evan Levy is Distinguished Professor and holds the Norma Kershaw Chair in the Archaeology of Ancient Israel and Neighboring Lands at the University of California, San Diego. He is a member of the Department of Anthropology and Judaic Studies Program, and leads the Cyber-archaeology research group at the Qualcomm Institute, California Center of Telecommunications and Information Technology (Calit2). Elected to the American Academy of Arts and Sciences, Levy is a Levantine field archaeologist with interests in the role of technology, especially early mining and metallurgy, on social evolution from the beginnings of sedentism and the domestication of plants and animals in the Pre-Pottery Neolithic period (ca. 7500 BCE) to the rise of the first historic Levantine state-level societies in the Iron Age (ca. 1200–500 BCE). A Fellow of the Explorers Club, Levy won the 2011 Lowell Thomas Award for "Exploring the World's Greatest Mysteries." Levy has been the principal investigator of many interdisciplinary archaeological field projects in Israel and Jordan that have been funded by the National Geographic Society, the National Endowment for the Humanities, National Science Foundation, and other organizations. Tom also conducts ethnoarchaeological research in India. Levy, his wife Alina Levy, and the Sthapathy traditional craftsmen from the village of Swamimalai coauthored the book *Masters of Fire—Hereditary Bronze Casters of South India* (Bochum: German Mining Museum, 2008). Tom has published 10 books and several hundred scholarly articles. Levy's most recent book is entitled *Historical Biblical Archaeology—The New Pragmatism* (London: Equinox Publishers, 2010 that in 2011 won the "best scholarly book" from Biblical Archaeology Society (Washington, DC). Levy and his colleague Mohammad Najjar recently won *Biblical Archaeology Review's* "Best BAR Article" for "Condemned to the Mines: Copper Production & Christian Persecution."

Thomas Schneider is Professor of Egyptology and Near Eastern Studies at the University of British Columbia, Vancouver, a position he assumed in 2007. He studied at Zurich, Basel, and Paris, earning a Master's degree (Lizentiat), a doctorate, and a habilitation in Egyptology at the University of Basel. He was a Visiting Professor at the University of Vienna in 1999, and at the University of Heidelberg in 2003–2004. From 2001 to 2005, he was a Research Professor of the Swiss National Science Foundation at the University of Basel, and from 2005 to 2007, Professor and holder of the Chair in Egyptology at the University of Wales, Swansea. He was a visiting scholar at New York University in 2006 and at the University of California, Berkeley, in 2012, as well as an Associate Professor (Status only) in the Department of Near and Middle Eastern Civilizations, University of Toronto (2010–2012), and an Affiliate Professor in the Department of Near Eastern Languages and Civilization, University of Washington (2009–2012). He has published five self-authored books and more than 100 journal articles, book chapters, and book reviews, and edited or coedited another five books in his main areas of research: Egyptian

T.E. Levy et al. (eds.), *Israel's Exodus in Transdisciplinary Perspective*,
Quantitative Methods in the Humanities and Social Sciences, DOI 10.1007/978-3-319-04768-3,

interconnections with the Levant and the Near East, Egyptian history and chronology, and Egyptian historical phonology. His current research project is on the history of Egyptology in Nazi Germany. He is founder and editor-in-chief of the *Journal of Egyptian History*, as well as editor of *Near Eastern Archaeology*. He was the editor-in-chief of "Culture and History of the Ancient Near East" (2006–2013), and area editor on history for the *UCLA Encyclopedia of Egyptology*. He also serves on the editorial board of the *Journal of Ancient Egyptian Interconnections*, as well as the advisory boards of "*Egypt and the Levant*" and *Lingua Aegyptia*.

A native New Yorker (born at Mount Sinai hospital), **William H. C. Propp** is the Harriet and Louis Bookheim Professor of Biblical Hebrew and Related Languages at the University of California, San Diego, where he has taught since 1983 for the Departments of Literature and History, and for the Judaic Studies Program; he has also been a visiting professor of Religious Studies at Dartmouth College. His wide-ranging approach to the Hebrew Bible synthesizes philology, cultural anthropology, folklore studies, psychology, history, linguistics, and literary criticism. Having published on a multitude of topics in both scholarly and popular venues, he is best known for his two-volume, 1,600-page reference commentary on the Book of Exodus in the Anchor Bible series (Doubleday 1999, 2006).

Abbreviations

(Selective List)

(See also the Society of Biblical Literature standard abbreviations of Bible books, Classical sources, etc., in the *SBL Handbook of Style*, available online, not repeated here.)

^{12}C	carbon 12 (stable carbon isotope)
^{14}C	carbon 14 (radiocarbon)
2D	Two-dimensional (2 dimensions)
3D	Three-dimensional (3 dimensions)
4D	Four-dimensional (4 dimensions)
3ICAANE	Third International Congress on the Archæology of the Ancient Near East
4K	4,000 pixel resolution (resolution per edge of display)
ÄA	*Ägyptologische Abhandelungen* (Wiesbaden)
ÄAT	Ägypten und Altes Testament (monograph series)
AB	*Anchor Bible*
ABD	*Anchor Bible Dictionary* (later renamed *Anchor Yale Bible Dictionary*)
AD	Anno Domini (Year of the Lord) = CE (chronological era following BC/BCE)
Ag.Ap.	*Against Apion* (Josephus) (*Contra Apionem, C.Ap.*)
AMS	Accelerator Mass Spectrometry
ANE	Ancient Near East (or Ancient Near Eastern)
ANET	*Ancient Near Eastern Texts Relating to the Old Testament*, 3rd rev., ed. James B. Pritchard. Princeton, NJ: Princeton Univ. Press.
Ant.Jud.	*Antiquitates Judaicae* (*Antiquities of the Jews*) (Josephus) (also, *Ant.*)

AOAT	Alter Orient und Altes Testament
AOS	American Oriental Society
AP	Aramaic Papyri
API	Application program instruction(s)
ARCE	American Research Center in Egypt
ARCHAVE	Archaeological Virtual Environment
ASCII	American Standard Code for Information Interchange
ASOR	American Schools of Oriental Research
AT	Altes Testament
ATD	*Das Alte Testament Deutsch*
BA	*Biblical Archaeologist* (ASOR) (renamed *Near Eastern Archaeology, NEA*)
BAC	Bronze Age Complex (Sinai-Negev)
BAR	*Biblical Archaeology Review*
BAR	British Archaeological Reports (monograph series)
BC	Before Christ (years) = BCE (chronological era followed by AD or CE)
BC	British Columbia (Canada)
BCE	Before Christian (or Common) Era (years) = BC (chronological era followed by AD or CE)
BCN	*B.C. Notizie* (notiziario del Centro Camuno di Studi Preistorici)
BCSP	*Bollettino del Centro Camuno di Studi Preistorici*
BdE	*Bibliothèque d'Étude* (IFAO, Cairo) (also *BdÉ*)
BE	*Biblische Enzyklopädie*
BETL	Bibliotheca Ephemeridum Theologicarum Lovaniensium

T.E. Levy et al. (eds.), *Israel's Exodus in Transdisciplinary Perspective*, Quantitative Methods in the Humanities and Social Sciences, DOI 10.1007/978-3-319-04768-3,

BHS	Biblia Hebraica Stuttgartensia (edition of Masoretic Text of Hebrew Bible)
BIFAO	*Bulletin de l'Institut Français d'Archéologie Orientale* (IFAO)
BM	British Museum
BN	Bibliothèque nationale de France
BN	*Biblische Notizen*
BP	Before Present (years) [Note: review technical definition for radiocarbon dating: Year 0 BP is 1950 AD/CE]
b.Sanh.	Babylonian Talmud (=b.) tractate "Sanhedrin"
BZAR	Beihefte zur Zeitschrift für altorientalische und biblische Rechtsgeschichte
BZAW	Beiheft zur Zeitschrift für die alttestamentliche Wissenschaft (monograph supplement series)
CA	California
ca.	circa (approximately) (also, c.)
CAA	Computer Applications for Archaeology
CAD	*Chicago Assyrian Dictionary*
Calit2	California Institute for Telecommunications and Information Technology
C.Ap.	*Contra Apionem* (Josephus) (*Against Apion, Ag.Ap.*)
CAVE	Cave Automatic Virtual Environment
CBH	Classical Biblical Hebrew
CBQ	*Catholic Biblical Quarterly*
CdÉ	*Chronique d'Égypte*
CE	Christian (or Common) Era (years) = AD (chronological era following BC/BCE)
CG	Catalogue General du Musee du Caire (series, Cairo Museum)
CGLX	Cross Platform Cluster Graphics Library (UC-San Diego)
chap.	chapter
CISA3	Center of Interdisciplinary Science for Art, Architecture and Archaeology, UC-San Diego
CISPE	Centro Internazionale di Studi Preistorici ed Etnologici
CMS	Corpus der minoischen und mykenischen Siegel (Corpus of Minoan and Mycenaean Seals) (monograph series)
CMT	Complete Momentum Transfer
cm	centimeter (unit of linear measurement) (1 cm ≈ 0.4 in.)
CO_2	carbon dioxide (also CO2)
col.	column (plural: cols.)
ConBOT	*Coniectanea biblica, Old Testament*
COS	*Context of Scripture*
CPJ	Corpus Papyrorum Judaicarum
CPU	Computer Processing Unit
CT	Coffin Texts (numbering: ed. A. de Buck, *The Egyptian Coffin Texts*, 7 vols., 1935–1961, Chicago)
CT	Connecticut
D	Deuteronomist writer, tradition (JEDP Documentary Hypothesis)
DA	Drag Along
DAR	Daughters of the American Revolution
DEM	Digital Elevation Model
Diod.Sic.	Diodorus Siculus
DN	Divine Name; or Deity Name
DNA	deoxyribonucleic acid (genetic code material)
d'Orb	d'Orbiney (papyrus)
Dtr	Deuteronomy, Deuteronomic
DtrH	Deuteronomistic History
DVD	Digital Video Disc
DZA	Digitalisiertes Zettelarchiv (online *Wörterbuch*)
E	Elohist writer, tradition (JEDP Documentary Hypothesis)
EA	El-Amarna tablets (also EAT) (numbering: ed. J.A. Knudtzon, *Die El-Amarna-Tafeln,* 2 vols, 1915, Leipzig)
EB	Early Bronze Age
EB IV	Early Bronze age IV (also EB IV) (Intermediate Bronze Age, IBA, or sometimes Middle Bronze I, MB I)
ECC	*Eerdmans Critical Commentary*
EdF	Erträge von Forschung
EEF	Egypt Exploration Fund (later renamed Egypt Exploration Society, EES)

ELRAP	Edom Lowlands Regional Archaeology Project, UC-San Diego
EMODnet	European Marine Observation and Data Network
Eng.	English (also, Engl.)
EROS	Earth Resources Observation Systems (EROS) Data Center (EDC), U.S. Geological Survey
ESRI	Environmental Systems Research Institute
ET	English translation
Ex	Exodus, Book of (Bible) (also, Exod)
EX3	"Exodus, Cyber-archaeology and the Future" Exhibition, Qualcomm Institute, UC-San Diego, May-June 2013
Exod	Exodus, Book of (Bible) (also, Ex)
FAT	Forschungen zum Alten Testament
fig.	figure
FRE	Freshwater reservoir effect (radiocarbon dating)
FRLANT	Forschungen zur Religion und Literatur des Alten und Neuen Testaments
Fs.	Festschrift (also, fs.); plural Festschriften (Fss.)
ft	foot; feet (unit of linear measurement; 1 ft ≈ 0,30 m)
g	acceleration of gravity at earth's surface (~9.8 $m/s^2 \approx 32\ ft/s^2$)
GA	Georgia (U.S.)
GB	gigabyte (billion or 10^9 bytes of data)
GDAL	Geospatial Data Abstraction Library
Ger.	German (also, Germ.)
GIS	Geographic Information System
GKC	Gesenius' Hebrew Grammar, eds E. Kautzsch, A.E. Cowley (Oxford 1910 2nd ed.)
GMTED	Global Multi-resolution Terrain Elevation Data
GPS	Global Positioning System
GSI	Geological Survey of Israel
GUI	Graphical User Interface
H-	Hornung numbering of Amduat lines (1963)
h	hour (unit of time measurement) (also, hr)
ha	hectare (unit of area measurement, 1 ha ≈ 2.47 acres)
HALOT	*Hebrew and Aramaic Lexicon of the Old Testament* (Koehler, Baumgartner and Stamm)
HB	Hebrew Bible (Tanakh, Old Testament)
HBAI	*Hebrew Bible and Ancient Israel* (also HeBAI)
HCOT	Historical Commentary on the Old Testament
HD	hard drive (computer)
HD	High Definition
HDTV	High Definition TV
Heb.	Hebrew
HeBAI	*Hebrew Bible and Ancient Israel* (also HBAI)
H.I.H.	His Imperial Highness
HK	Har Karkom
HSM	Harvard Semitic Museum Monographs
HUCA	Hebrew Union College Annual
IAA	Israel Antiquities Authority
IAE	International Association of Egyptologists
IBA	Intermediate Bronze Age (also Early Bronze age IV EB IV, or sometimes Middle Bronze I, MB I)
ICAANE	International Congress on the Archæology of the Ancient Near East
ID	Identification, identity
IEEE	Institute of Electrical and Electronics Engineers
IEJ	*Israel Exploration Journal*
IEKAT	Internationaler Exegetischer Kommentar zum Alten Testament
IES	Israel Exploration Society
IFAO	Institut Français d'Archéologie Orientale (Cairo)
IGERT	Integrative Graduate Education and Research Traineeship, NSF
IL	Illinois
IN	Indiana
INTCAL	International Calibration Group (radiocarbon calculation program) (also IntCal, IntCal98, 1998, IntCal09, 2009, IntCal13, 2013)
I/O	Input/Output

ISBN	International Standard Book Number
ISPRS	International Society for Photogrammetry and Remote Sensing
ISSN	International Standard Serial Number
IS&T	Society for Imaging Science and Technology
ISVC	International Symposium on Visual Computing
IVP	Inter-Varsity Press
J	Jahwist (Yahwist) writer, tradition (JEDP Documentary Hypothesis)
JANER	*Journal of Ancient Near Eastern Religions*
JAOS	*Journal of the American Oriental Society*
JBL	*Journal of Biblical Literature*
JdE	Journal d'Entrée (Cairo Museum) (also, JE)
JE	Jahwist-Elohist tradition (JEDP Documentary Hypothesis) (also Non-P tradition)
JE	Journal d'Entrée (Cairo Museum) (also, JdE)
JEA	*Journal of Egyptian Archaeology*
JEDP	Jahwist-Elohist-Deuteronomist-Priestly writers (also JEPD) (Documentary Hypothesis)
JNES	*Journal of Near Eastern Studies*
JPS	Jewish Publication Society
JSON	JavaScript Object Notation
JSOT	*Journal for the Study of the Old Testament*
JSOTSS	JSOT Supplement Series
KAI	Kanaanäische und Aramäische Inschriften (Canaanite and Aramaic inscriptions)
KJV	King James Version (= Authorized Version, AV) (Bible translation)
kg	kilogram (unit of mass) (1 kg ≈ 2.2 pounds)
km	kilometer (unit of linear measurement, 1 km ≈ 0.62 statute mile)
km^2	square kilometer (unit of area) (1 km^2 = 100 hectares)
km^3	cubic kilometer (unit of volume) (also cu.km)
KRI	Kenneth Kitchen, *Ramesside Inscriptions* (Oxford, 1979)
KTU	*Die keilalphabetischen Texte aus Ugarit* (AOAT 24) (1976)
KV	King's Valley (Valley of the Kings, Thebes, Egypt)
KY	Kentucky
LB	Late Bronze age (also LBA)
LBH	Late Biblical Hebrew
LCD	Liquid crystal display
LH	Late Helladic age
LHIIA	Late Helladic II-A
Lit.	literally
LM	Late Minoan age
LMIA	Late Minoan I-A
LMIB	Late Minoan I-B
LNCS	Lecture Notes in Computer Science
LSJ	Liddell-Scott-Jones Greek-English Lexicon
LXX	Septuagint (70) (early Greek translation of the Bible)
m	meter (unit of linear measurement, 1 m ≈ 3.28 ft ≈ 39.4 in.)
M	Magnitude (earthquake)
m^3	cubic meter (unit of volume) (also cu.m)
MA	Massachusetts
MB	megabyte (million or 10^6 bytes of data)
MB	Middle Bronze age
MC	MediaCommons
MCF	MediaCommons Framework
MD	Maryland
MEDIBA	Mediterranean Basin
mg	milligram (unit of mass)
MK	Middle Kingdom (Egypt)
MI	Michigan
mi.	mile (unit of linear measurement) (1 statute mile = 5,280 ft ≈ 1.6 km)
MN	Minnesota
MO	Missouri
MS	manuscript (also, ms); plural MSS (mss)

m/s	meters per second (unit of velocity) (1 m/s ≈ 2 mph, statute miles per hour)
MT	Massoretic (Masoretic) Text, Hebrew Bible
N	North
N.B.	*nota bene* (note well, note especially)
NC	North Carolina
Neth	Netherlands
NGA	National Geospatial Intelligence Agency (U.S.)
NK	New Kingdom (Egypt)
NJ	New Jersey
NMT	No Momentum Transfer
NN	*nomen natus* (name of the deceased; usually tomb owner in ancient Egypt)
Non-P	Non-Priestly tradition (comparable or equivalent to JE in JEDP Documentary Hypothesis)
NOSTER	Nederlandse Onderzoekschool voor Theologie en Religiewetenschap (Netherlands School for Advanced Studies in Theology and Religion)
NRSV	New Revised Standard Version (Bible translation)
NSF	National Science Foundation (U.S.)
NSKAT	Neuer Stuttgarter Kommentar Altes Testament
NW	North West, northwest
NY	New York
NYU	New York University
ÖAW	Österreichische Akademie der Wissenschaften (Austrian Academy of Sciences)
OBO	Orbis Biblicus et Orientalis (monograph series)
OH	Ohio
OIP	Oriental Institute Publications (University of Chicago) (monograph series)
OK	Oklahoma
OK	Old Kingdom (Egypt)
OLA	Orientalia Lovaniensia Analecta (monograph series)
OREA	Institut für Orientalische und Europäische Archäologie (Institute of Oriental and European Archaeology, Austrian Academy of Sciences)
OSG	OpenSceneGraph
OT	Old Testament (Hebrew Bible, Tanakh)
OUP	Oxford University Press
OxCal	Oxford Calibration online radiocarbon calculation program (Oxford Radiocarbon Accelerator Unit)
p.	page (page number)
p	papyrus (also, P., Pap.)
P	Priestly writer, tradition
P.	Papyrus (also, Pap.)
PA	Pennsylvania
PB	Priesterliche Bearbeitung (Priestly Editor) (JEDP Documentary Hypothesis)
pBN	Papyrus Bibliothèque nationale de France (Paris)
PC	Personal computer
PE	Priestly-Elohist tradition (JEPD/JEDP Documentary Hypothesis)
PEQ	*Palestine Exploration Quarterly*
pers. comm.	Personal communication
PG	*Patrologia Graeca*, ed. Migne
PGM	Papyri Graecae Magicae (Greek Magical Papyri, Papyrus)
pl.	plate
PLoS	*Public Library of Science*
PNAS	*Proceedings of the National Academy of Sciences* (U.S.)
Praep.Ev.	*Praeparatio Evangelica* (Eusebius of Caesarea)
QI	Qualcomm Institute, UC-San Diego
R	Redactor (JEDP Documentary Hypothesis)
RA	*Revue d'Assyriologie et d'arché ologie orientale*
R&D	Research & Development
RGB	red-green-blue
RGBA	Red Green Blue Alpha
RSV	Revised Standard Version (Bible translation)

s	second (unit of time measurement) (sec)
S	South
SAGA	Studien zur Archäologie und Geschichte Altägyptens (monograph series) (Heidelberg)
SAGE	Scalable Adaptive Graphics Environment
SAOC	Studies in Ancient Oriental Civilizations (University of Chicago Oriental Institute) (monograph series)
SBA	Society of Biblical Archaeology (London)
SBL	Society of Biblical Literature
SBLMS	Society of Biblical Literature Monograph Series
SBLWAW	Society of Biblical Literature Writings from the Ancient World
SBS	*Stuttgarter Bibelstudien*
sc.	scilicet (=namely) (also, scil.)
SC	Sources Chrétiennes
SCA	Supreme Council of Antiquities, Egypt (superseded by Ministry of State for Antiquities, MSA)
SCIEM	Synchronisation of Civilisations in the Eastern Mediterranean in the Second Millennium BC
SIP	Second Intermediate Period (Egypt)
STDJ	Studies on the Texts from the Desert of Judah (monograph series)
SUNY	State University of New York
SVT	Supplements to *Vetus Testamentum*
SW	South West; southwest
TCP/IP	Transmission Control Protocol/ Internet Protocol
TEECH	Training, Research and Education in Engineering for Cultural Heritage Diagnostics, UC-San Diego
TIFF	Tagged Image File Format
TN	Tennessee
trans.	translator; translated by; translation (also, transl.)
TX	Texas
UC	University of California
UCIAMS	University of California, Irvine, Accelerator Mass Spectrometry (radiocarbon dating)
UCL	University College, London
UCLA	University of California, Los Angeles
UCSD	University of California, San Diego
UDP	User Datagram Protocol
UMI	University of Michigan
UML	Unified Modeling Language
uncal	uncalibrated (radiocarbon date)
Urk.	*Urkunden des ägyptischen Altertums*, 8 vols. ed. K.Sethe, et al.
USC	University of Southern California
USGS	U.S. Geological Survey
UZK	Untersuchungen der Zweigstelle Kairo (monograph series)
v	verse (also, vs.)
VA	Virginia
VAST	Virtual Reality, Archaeology, and Cultural Heritage
VEI	Volcanic Explosivity Index
Vit.Mos.	*De Vita Mosis* (The Life of Moses) (Philo)
VR	Virtual Reality
vs.	verse (also, v)
VT	*Vetus Testamentum*
vv	verses
WA	Washington
WACE	World Conference on Cooperative & Work-Integrated Education
WÄS	*Wörterbuch der ägyptische Sprache* (also, Wb, WB)
WAVE	Wide Angle Virtual Environment (UC-San Diego)
Wb	*Wörterbuch der ägyptische Sprache* (also, WB, *WÄS*)
WGB	Wissenschaftliche Buchgesellschaft (Darmstadt) (publisher)
WGS84	World Geodetic System 1984
W-K	Wiebach-Köpke numbering of Amduat lines (2003)
WMANT	Wissenschaftliche Monographien zum Alten und Neuen Testament

WMS	Web map service
WS	White Slip pottery
XML	Extensible Markup Language
y	year, years
YBP	Years BP (Before Present) (see BP)
YUP	Yale University Press
ZAW	*Zeitschrift für die Alttestamentliche Wissenschaft*
ZBKAT	Zürcher Bibelkommentare Altes Testament

Ancient Sources and Authors Index

(Selective List)

Note : "*See*" and "*See also*" cross-references are to the main Index for further cross-references and/or page citations.

T.E. Levy et al. (eds.), *Israel's Exodus in Transdisciplinary Perspective*, Quantitative Methods in the Humanities and Social Sciences, DOI 10.1007/978-3-319-04768-3,

Index

Quantitative Methods in the Humanities and Social Sciences, DOI 10.1007/978-3-319-04768-3,

CPSIA information can be obtained at www.ICGtesting.com
Printed in the USA
LVOW05*2331260415

436157LV00002B/67/P